Current Topics in Membranes, Volume 50

Gastrointestinal Transport
Molecular Physiology

Current Topics in Membranes, Volume 50

Series Editors

Douglas M. Fambrough
Department of Biology
The Johns Hopkins University
Baltimore, Maryland

Dale J. Benos
Department of Physiology and Biophysics
University of Alabama
Birmingham, Alabama

Contents

Contents

CHAPTER 12   Molecular Physiology of Mammalian Epithelial Na$^+$/H$^+$ Exchangers NHE2 and NHE3
Mark Donowitz and Ming Tse

CHAPTER 13   Molecular Aspects of Intestinal Brush-Border Na$^+$/Glucose Transport
Ernest M. Wright

Contributors

Numbers in parentheses indicate the pages on which the authors' contributions begin.

James Melvin Anderson (163), Section of Digestive Diseases, Department of Internal Medicine, Yale School of Medicine, New Haven, Connecticut 06520

Stephen A. Baldwin (329), School of Biochemistry and Molecular Biology, University of Leeds, Leeds, LS2 9JT United Kingdom

Kim E. Barrett (249), Division of Gastroenterology, Department of Medicine, School of Medicine, UCSD Medical Center, San Diego, California 92103

Henry J. Binder (77), Department of Internal Medicine, Yale University, New Haven, Connecticut 06520

Carol E. Cass (329), Department of Oncology, University of Alberta, Edmonton, Alberta T6G 2H7, Canada

Christopher I. Cheeseman (329), Department of Physiology, University of Alberta, Edmonton, Alberta T6G 2H7, Canada

Mark Donowitz (437), Department of Medicine and Physiology, GI Division, Johns Hopkins University School of Medicine, Baltimore, Maryland 21205

Malliga E. Ganapathy (379), Department of Medicine, Medical College of Georgia, Augusta, Georgia 30912

Vadivel Ganapathy (379), Department of Biochemistry and Molecular Biology, Medical College of Georgia, Augusta, Georgia 30912

Jean-Daniel Horisberger (413), Institut de Pharmacologie et de Toxicologie, Universite de Lausanne, CH-1005 Lausanne, Switzerland

Stephen J. Keely (249), Division of Gastroenterology, Department of Medicine, School of Medicine, UCSD Medical Center, San Diego, California 92103

Juha Kere (301), Finnish Genome Center, University of Helsinki, F-00014 Helsinki, Finland

Frederick H. Leibach (379), Department of Biochemistry and Molecular Biology, Medical College of Georgia, Augusta, Georgia 30912

John R. Mackey (329), Department of Oncology, University of Alberta, Edmonton, Alberta T6G 2H7, Canada

R. John MacLeod (1), Department of Medicine, Endocrine Hypertension Division, Brigham and Women's Hospital, Boston, Massachusetts 02115

Marshall H. Montrose (301), Department of Physiology and Biophysics, Indiana University, Indianapolis, Indiana 46202

Richard Rose (35), Chicago Medical School, North Chicago, Illinois 60064

Richard Rozmahel (187), Department of Genetics and Genomic Biology, The Hospital for Sick Children, and the Department of Pharmacology, University of Toronto, Toronto, Canada

Hamid M. Said (35), Department of Medicine/Physiology and Biophysics, UCI Long Beach Medical Program, Long Beach, California 90822

Erik M. Schwiebert (187), Department of Physiology and Biophysics, University of Alabama at Birmingham, Birmingham, Alabama 35295

Bellur Seetharam (35), Medical College of Wisconsin, Milwaukee, Wisconsin 53226

Satish K. Singh (77), Departments of Internal Medicine and Molecular Physiology, Yale University, New Haven, Connecticut 06520

Ming Tse (437), Department of Medicine and Physiology, GI Division, Johns Hopkins University, Baltimore, Maryland 21205

Christina M. Van Itallie (163), Section of Digestive Diseases, Department of Internal Medicine, Yale School of Medicine, New Haven, Connecticut 06520

Lon J. Van Winkle (113), Department of Biochemistry, Midwestern University, Downer's Grove, Illinois 60515

Ernest M. Wright (499), Department of Physiology, University of California, Los Angeles, School of Medicine, Los Angeles, California 90095

James D. Young (329), Department of Physiology, University of Alberta, Edmonton, Alberta T6G 2H7, Canada

Preface

Epithelial biologists, the primary audience for this volume, share a common passion for understanding the molecular mechanisms that underlie the specific transport of solutes across the barrier presented by the epithelial monolayer. In this regard, it is impossible to imagine a more fertile ground in which to study transport events than the gastrointestinal tract, where the wealth of transport mechanisms is truly remarkable. By virtue of its function as an organ in which diverse nutrients are assimilated, the intestine has evolved pathways for the uptake of a host of molecules with widely differing structural features. Moreover, electrolyte transport mechanisms, similar to those seen in other epithelial tissues, are also highly expressed in the intestine. Derangements in these various transport mechanisms form the basis for a number of important human diseases, such as cholera and nutrient malabsorption. Moreover, the ability of the intestine to transport a variety of solutes both into and out of the body is the target of intricate and complex regulatory mechanisms. Such regulation occurs both on a minute-to-minute basis, as is needed to subserve physiological function, and in the longer term, in response to changes in nutrient availability, for example.

The rate of advances in scientific knowledge and the level of sophistication of scientific studies continue to accelerate exponentially, and the area of gastrointestinal transport is no exception. This prompted us to gather the group of experts who have contributed chapters to this volume, in an attempt to synthesize and make sense of this burgeoning field at a pivotal time in its development. Just before this preface was written, Craig Venter announced, at a major lecture at the Johns Hopkins University School of Medicine, not only that his company, Celera, had completed sequencing the human genome, but that before the end of 2000 they would have ordered the sequences and the human genome sequence would be made publicly available, after a group of experts was invited to attempt to classify the genes as occurred so successfully for the *Drosophila* genome. In fact, Celera's sequencing army had already turned to the next project, sequencing the mouse genome. Thus, we stand at a time at which all human genes have been or shortly will have been cloned, adding this information to the previously completed sequencing of the genome of the fruit fly, *Drosophila*, the worm, *C. elegans*, and the yeast, *S. cerevisiae*, in addition to those of multiple bacteria. That is, we are now firmly at the transition to the postgenomic era. What does this mean, especially for the type of gastrointestinal research described in this volume?

From the perspective of membrane transport proteins, a major topic of this book, this transition implies that all transport proteins will have been cloned and most identified by sequence as transporters. In completed genomes, 5% of the cloned proteins embody transport functions. If this can be extrapolated to the human genome, with the final number of human genes predicted to lie between 50,000 and 140,000, this means that there will be 2,500–7000 transport proteins, clearly many more than we currently know about. What are the implications of these numbers, and what does the postgenomic era mean for future studies of epithelial transporters and their regulation and diseases? The clearest fact, as Venter suggested, is that scientists who have obtained recent NIH funding to clone human genes should consider other ways to use their funds. However, more to the point, there will be plenty left to do, seemingly more than ever before. There will be far, far more cloned transport proteins than there will be scientists to study them. New ways of studying groups of transporters via bioinformatics already are emerging. These methods are based on applying advances in one genome to others and applying general principles to classes of transporters rather than to individual proteins. Of course, experimental confirmation of the validity of such approaches will require experimental input. The ways in which bioinformatics can be applied to study transport proteins can only be guessed at by these editors, although major advances will certainly come from this sphere and will likely be tailored to the Venter philosophy that faster is better.

A second area of rapid advances that is just beginning is the identification of disease genes, especially relating to transport proteins. With the genomic structure known, linkage analyses of precise loci will undoubtedly reveal previously undiscovered links of transport proteins to diseases. However, the majority of studies of transporters will continue to focus on understanding how these fascinating proteins work and are regulated. That is, the same types of questions that have been the motivating force for studies until now will be asked at increasingly detailed and sophisticated levels, using combinations of approaches that certainly include study of the cloned proteins. What is different is the level of precision that can be brought to bear and application of the information gained to other similar or even unrelated transporters. Additionally, with microarray/cDNA chip technologies starting to be widely applied, the signaling molecules that impinge on transport proteins will be characterized. Moreover, although microarray technology is certain to reveal patterns of response at the mRNA/cDNA level, true interest will eventually lie in the protein response. Proteomic approaches are being developed and certainly will advance past the two-dimensional gel electrophoresis/mass spectrometry approach. However, such techniques are already revealing relationships of proteins to processes that were not recognized previously. In addition, protein–protein interactions are being identified by wide-scale yeast two-hybrid screens. The application of such approaches to transport proteins is in its infancy.

Most likely to thrive in the postgenomic period are detailed studies of recognized transport proteins. Knockouts, conditional knockouts, and knockins in mouse models will be studied using existing sophisticated physiological approaches as well as new ones. It is ironic that the sophisticated methodologies that need to be applied to mouse models are being delayed not only because miniaturization is needed to allow studies in mice, but also because the number of scientists who use these non-molecular approaches has been greatly diminished by previous emphasis on molecular and cell biological approaches. However, it is the importance of these modern physiological mouse models that caused Venter to state that the mouse genome is likely, over the short term, to be much more important than the human genome for understanding human disease. In addition, ultimate understanding of the structure and function of the cloned transporters awaits determination of structure—advances in the crystallization of membrane proteins continue to be made, but at a limited rate, which can only accelerate in the future. Thus, the challenges of the postgenomic era are similar to those that predated detailed knowledge of the genome, but with the added exactness of limiting studies to precisely defined and manipulable proteins. The only thing missing will be that scientists will no longer be able to bond with their proteins by being the first to clone them, but in return the rate of advances promises to be remarkably faster than in the past because of the many new approaches outlined here. The contributors to this volume have been at the forefront in applying such approaches to the study of the myriad transport processes that underpin both normal and abnormal function of the mammalian intestine. We are indebted to these individuals for their insights, which will undoubtedly inspire future work and current students in the postgenomic era at a time when our knowledge of the links of intestinal transport to human health and disease is poised to explode.

We thank all the authors for their outstanding contributions. We are also indebted to Ms. Glenda Wheeler-Loessel, who provided diligent, unfailingly upbeat, and well-organized administrative support for the project in San Diego. The book would also not have become a reality without the initial vision of Dr. Emelyn Eldredge, former Acquisitions Editor at Academic Press, and more recently the patience and forbearing of Ms. Jenny Wrenn, whose optimism and enthusiasm for the project moved things forward when the editors' own spirits were flagging. Finally, we thank Dr. Dale Benos for suggesting the book in the first place, and the members of our laboratories, whose hard work and dedication provided the impetus for much of the science that is discussed here.

Kim E. Barrett
Mark Donowitz

Previous Volumes in Series

Current Topics in Membranes and Transport

Volume 23   Genes and Membranes: Transport Proteins and Receptors*
(1985)
Edited by Edward A. Adelberg and Carolyn W. Slayman

Volume 24   Membrane Protein Biosynthesis and Turnover (1985)
Edited by Philip A. Knauf and John S. Cook

Volume 25   Regulation of Calcium Transport across Muscle
Membranes (1985)
Edited by Adil E. Shamoo

Volume 26 Na^+–H^+ Exchange, Intracellular pH, and Cell Function*
(1986)
Edited by Peter S. Aronson and Walter F. Boron

Volume 27   The Role of Membranes in Cell Growth and
Differentiation (1986)
Edited by Lazaro J. Mandel and Dale J. Benos

Volume 28   Potassium Transport: Physiology and Pathophysiology*
(1987)
Edited by Gerhard Giebisch

Volume 29   Membrane Structure and Function (1987)
Edited by Richard D. Klausner, Christoph Kempf, and Jos van
Renswoude

Volume 30   Cell Volume Control: Fundamental and Comparative
Aspects in Animal Cells (1987)
Edited by R. Gilles, Arnost Kleinzeller, and L. Bolis

Volume 31   Molecular Neurobiology: Endocrine Approaches (1987)
Edited by Jerome F. Strauss, III, and Donald W. Pfaff

Volume 32   Membrane Fusion in Fertilization, Cellular Transport, and
Viral Infection (1988)
Edited by Nejat Düzgünes and Felix Bronner

Part of the series from the Yale Department of Cellular and Molecular Physiology

Volume 33   Molecular Biology of Ionic Channels* (1988)
Edited by William S. Agnew, Toni Claudio, and Frederick J. Sigworth

Volume 34   Cellular and Molecular Biology of Sodium Transport*
(1989)
Edited by Stanley G. Schultz

Volume 35   Mechanisms of Leukocyte Activation (1990)
Edited by Sergio Grinstein and Ori D. Rotstein

Volume 36   Protein–Membrane Interactions* (1990)
Edited by Toni Claudio

Volume 37   Channels and Noise in Epithelial Tissues (1990)
Edited by Sandy I. Helman and Willy Van Driessche

Current Topics in Membranes

Volume 38   Ordering the Membranes Cytoskeleton Tri-layer* (1991)
Edited by Mark S. Mooseker and Jon S. Morrow

Volume 39   Developmental Biology of Membrane Transport Systems
(1991)
Edited by Dale J. Benos

Volume 40   Cell Lipids (1994)
Edited by Dick Hoekstra

Volume 41   Cell Biology and Membrane Transport Processes* (1994)
Edited by Michael Caplan

Volume 42   Chloride Channels (1994)
Edited by William B. Guggino

Volume 43   Membrane Protein–Cytoskeleton Interactions (1996)
Edited by W. James Nelson

Volume 44   Lipid Polymorphism and Membrane Properties (1997)
Edited by Richard Epand

Volume 45   The Eye's Aqueous Humor: From Secretion to Glaucoma
(1998)
Edited by Mortimer M. Civan

Volume 46   Potassium Ion Channels: Molecular Structure, Function,
and Diseases (1999)
Edited by Yoshihisa Kurachi, Lily Yeh Jan, and Michel Lazdunski

CHAPTER 1

The Role of Volume Regulation in Intestinal Transport: Insights from Villus Cells in Suspension

R. John MacLeod

Endocrine-Hypertension Division, Brigham and Women's Hospital, Department of Medicine, Harvard Medical School, Boston, Massachusetts 02115

I. INTRODUCTION

Absorptive villus cells of the small intestine swell when they absorb Na^+ cotransported solutes such as D-glucose and L-alanine. This villus cell swelling is followed by a volume recovery or regulatory volume decrease (RVD), which is due to the activation of K^+ and Cl^- channels. Salt loss, together with osmotically obliged water, returns the cell volume to normal, as well as maintains the electrical driving force for continued transport (Schultz et al., 1985; MacLeod and Hamilton, 1991a). These physiological, Na^+-solute–induced volume changes are small relative to those induced in most experimental models. While 30 years separate the demonstration that villus enterocytes cotransport Na^+ and glucose in osmotically active forms (Czaky and Esposito, 1969) from the demonstration that the Na^+–glucose cotransporter (SGLT1) when expressed in Xenopus oocytes behaves as a water cotransporter (Meinild et al., 1998), our understanding of the molecular determinants of Na^+-solute–induced volume regulation in villus cells is in its infancy.

How does a villus epithelial cell sense its changes in volume? To address this question, we note that three components account for the volume response. The first component is a volume sensor that detects the increase in volume, the second is the mechanism that activates the effector responsible for volume regulation, and the third is the effector itself. Activation of effectors by tyrosine phosphorylation, changes in intracellular calcium, membrane stretch and cytoskeletal changes, cytosolic ionic strength, or protein concentration, as well as a catalogue of effectors of volume regulation, has been recently reviewed (Lang et al., 1998; Summers et al., 1997; Li et al., 1998; Strange et al., 1996, 1998; Hoffmann and Dunham, 1995; Minton, 1994).

Jejunal villus epithelial cells are a unique model that helps us to understand how cells sense volume. This is because the signal transduction for activating the K^+ and Cl^- channels responsible for RVD depends on how these cells swell (Na^+-solute uptake vs hypotonic dilution). Indeed, different signal transduction pathways for K^+ channel activation are stimulated depending on the extent of the volume increase (modest hypotonic dilutions of 5 to 7% vs substantial, but conventional, hypotonic dilutions of 20 to 50%). Like the proverbial canoeist who

> paddled seven miles
> along a lake near here
> at night, with the trees like a pelt of dark
> hackles, and the waves hardly moving.
> In the moonlight the way ahead was clear
> and obscure both. I was twenty
> and impatient to get there, thinking
> such a thing existed.[1]

---

[1]The Ottawa River by Night from *Morning in the Burned House* by Margaret Atwood © 1995; published in Canada by McClelland & Stewart and in the U.S. by Houghton Mifflin. Reprint granted with permission of author.

"To get there," in this chapter, we shall first discuss how the signal transduction of RVD after Na^+-solute–induced swelling differs from hypotonic swelling. Then we will focus on how changes in pH_i are required to activate the charybdotoxin-sensitive K^+ channels required for volume regulation after modest volume increases. Finally, we will conclude with new evidence that demonstrates a physiological role for H^+ conductance during RVD, after Na^+-solute–induced swelling.

II. TWO MODELS OF VILLUS CELL SWELLING

A. Na^+-Coupled Solute Uptake or Hypotonic Medium Results in RVD

We first characterized the volume response of guinea pig jejunal villus cells exposed to hypotonic dilutions of 30 or 50% using electronic cell-sizing techniques (MacLeod and Hamilton, 1991b). The pharmacological sensitivity of the subsequent RVD, the altered rate and direction of RVD to extracellular potassium, and the generation of secondary volume changes were consistent with volume regulation being mediated by separate K^+ and Cl^- conductances. We then demonstrated that the addition of D-glucose or L-alanine to villus cells suspended in isotonic medium resulted in a transient volume increase of 5 to 7% of the cell's isotonic volume, which was followed by RVD (MacLeod and Hamilton, 1991a). Cell swelling did not occur in the absence of extracellular Na^+, nor after the addition of a non-Na^+-cotransported isomer (D-alanine or L-glucose). Villus cell swelling was also stimulated by α-methyl-D-glucoside, a preferred substrate of SGLT1, and this swelling was prevented by phlorizin, a well-characterized inhibitor of SGLT1. Together, these data were consistent with the transient increase in villus cell volume caused by the influx of Na^+ cotransported with D-glucose or L-alanine, along with water. Different classes of K^+ or Cl^- channel blockers prevented RVD after villus cell swelling that was caused by the addition of a Na^+ solute. Furthermore, when cells were permeabilized with the cation ionophore gramicidin, they continued to swell in Cl^--containing, but not Cl^--free, medium. These data suggested to us that the RVD, after cell swelling caused by Na^+-nutrient absorption, was also due to the activation of K^+ and Cl^- channels. Thus two models of villus cell swelling—hypotonic dilution and Na^+-solute addition—result in RVD mediated by loss of the same osmolytes. But these two models of cell swelling are not comparable.

B. Effects of Protein Kinase C Inhibitors on RVD

The first evidence that different signaling mechanisms were responsible for the activation of the K^+ and Cl^- channels for RVD in these two models of cell swelling came from volume and ^{36}Cl efflux measurements using protein kinase C (PKC) inhibitors (MacLeod *et al.*, 1992a). In the presence of 1-(5-isoquino-linylsulfonyl)-2-methylpiperazine (H-7), the RVD following cell swelling stim-ulated by L-alanine was blocked (Fig. 1B). However, the same concentration of H-7 had no effect on RVD after the villus cells were swollen by hypotonic dilutions of 5 or 10%, to duplicate the size reached by the cells because of L-alanine absorption (Fig. 1C) or a more substantial hypotonic dilution of 50% (Fig. 1A). The more selective and potent inhibitor staurosporine had the same effect. Increases in the rate of ^{36}Cl efflux that were stimulated by hypotonic di-lution were inhibited by a chloride channel blocker but were not affected by PKC inhibitors, while L-alanine–stimulated increases in this rate were inhibited an equivalent amount by either the PKC inhibitor or the Cl$^-$ channel blocker. Notably, gramicidin permeabilization of these cells in isotonic medium stimu-lated volume changes that were not influenced by the PKC inhibitors, suggest-ing Na$^+$ entry (and depolarization) alone was not activating the PKC-sensitive Cl$^-$ channel. These experiments were consistent with the concept that PKC was required to activate the Cl$^-$ channel required for RVD after cell swelling stim-ulated by L-alanine. In contrast, the Cl$^-$-channel–activated cell swelling after hy-potonic dilution did not require PKC for activation.

Glycine added to Ehrlich ascites cells caused a slow volume increase that was prevented in Na$^+$-free medium but was not followed by RVD (Hudson and Schultz, 1988). The Cl$^-$ channel activated because of glycine addition had the same characteristics as the Cl$^-$ channel activated by a substantial (50%) hypo-tonic dilution. The signaling pathways responsible for this activation have not been determined. The effect of PKC on RVD or volume-sensitive Cl$^-$ channels

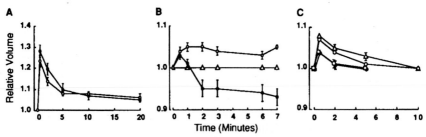

FIGURE 1   Effect of 1-(5-isoquinolinylsulfonyl)-2-methylpiperazine (H-7) on regulatory vol-ume decrease (RVD). (A) Hypotonic medium: O, 0.5× isotonic; ●, 0.5× isotonic with 100 μM H-7, *n* = 5. (B) Volume changes after addition of 25 mM L-Ala: ●, control; O, 100 μM H-7 and 25 mM L-Ala; △, 100 μM H-7 (*n* = 7). (C) Hypotonic medium: △, 0.9× isotonic; O, 0.9× isotonic with 100 μM H-7; □, 0.95× isotonic; ▼, 0.95× isotonic with 100 μM H-7 (*n* = 5). Volume mea-sured electronically, expressed relative to isotonic control. (From MacLeod *et al.*, 1992a, Fig. 1, p. C951, with permission.)

varies with the cell type. PKC is not required for RVD or the activation of volume-sensitive Cl^- channels in bovine (Szücs *et al.*, 1996), human endothelial (Nilius *et al.*, 1994), or HeLa cells (Hardy *et al.*, 1995) or rat hepatoma and osteoblasts (Schliess *et al.*, 1995; Goslin *et al.*, 1995). Yet, PKC regulation of RVD and volume-sensitive Cl^- channels has been observed in skate erythrocytes (Musch and Goldstein, 1990), guinea pig hepatocytes (Konmi *et al.*, 1995), proximal tubule cells of *Rana temporaria* (Robson and Hunter, 1994), NIH3T3 cells (Hardy *et al.*, 1995), ciliary epithelium (Coca-Prados *et al.*, 1995), and pancreatic duct cells (Verdon *et al.*, 1995). Villus epithelial cells are a distinctive model because depending on how the villus cell swells, PKC will be required for Cl^- channel activation and volume regulation.

C. The Ca^{2+} Requirement for RVD in Hypotonic and Na^+-Nutrient RVD

Studies of the Ca^{2+} dependence of RVD in these two models of cell swelling suggested that the signaling of the K^+ channels required for volume regulation depended on the mode of swelling (MacLeod *et al.*, 1992b). The evidence was as follows: Enterocytes subjected to a substantial hypotonic dilution (Fig. 2A) or a more modest dilution to duplicate the size reached during Na^+-solute–induced swelling (Fig. 2C) showed no RVD in Ca^{2+}-free medium. In contrast, volume changes stimulated by L-alanine addition were not influenced by the same Ca^{2+}-free conditions (Fig. 2B). When putative increases in $[Ca^{2+}]_i$ were buffered by loading the cells with a calcium chelator, BAPTA, RVD was prevented in both models of cell swelling. These experiments and others (MacLeod and Hamilton, 1999a,b) suggested that with hypotonic swelling, the Ca^{2+} requirement is extracellular. These findings are in accord with studies in other epithelia and various symmetrical cell types (Foskett, 1994; Lang *et al.*, 1998). However, the volume response of cells swollen by L-alanine cotransport suggested that intracellular Ca^{2+} was mobilized. In experiments described in Section III.C, we found that L-alanine addition caused a rapid increase in $[Ca^{2+}]_i$ compared with addition of nontransported D-alanine. In earlier experiments, the addition of thapsigargin, an inhibitor of microsomal Ca^{2+}-ATPase, had no effect on the swelling resulting from L-alanine addition. This drug did prevent the subsequent RVD, however. Treatment with thapsigargin had no effect on RVD of cells swollen by either a modest or a substantial hypotonic dilution (MacLeod, 1994). Consistent with this interpretation, both organic and inorganic Ca^{2+} channel blockers prevented RVD after hypotonic dilution, but the same inhibitors had no effect on cell swelling or RVD after L-alanine addition. These data suggested to us that the Ca^{2+} required for RVD after Na^+-solute–induced swelling was derived from a thapsigargin-sensitive pool, while the Ca^{2+} required for RVD after any extent of hypotonic swelling was predominantly from an influx pathway.

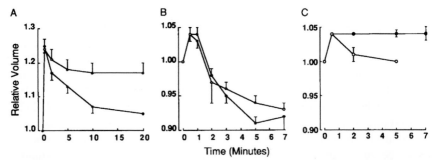

FIGURE 2   Effect of extracellular Ca^{2+} chelation on RVD. (A) Hypotonic medium: ●, $0.5\times$ isotonic (1 mM Ca^{2+}); ○, 0.1 mM EGTA in Ca^{2+}-free medium; $n = 8$. (B) Volume changes after addition of 25 mM L-Ala: ●, control (1 mM Ca^{2+}); ○, 0.1 mM EGTA in Ca^{2+}-free medium; $n = 20$. (C) ○, $0.95\times$ isotonic (1 mM Ca^{2+}); ●, $0.95\times$ isotonic in 0.1 mM EGTA, Ca^{2+}-free medium; $n = 5$. Volume was measured electronically, expressed relative to isotonic control. (From MacLeod *et al.*, 1992b, Fig. 1, p. 25.)

The Ca^{2+} dependency of RVD in these two models of swelling differed in two additional ways. Calmodulin antagonists and inhibitors of Ca^{2+}/calmodulin kinase II prevented RVD after substantial ($>12\%$) volume increases caused by hypotonic dilution (MacLeod and Hamilton, 1999b; MacLeod *et al.*, 1992b), but these inhibitors had no effect on RVD after Na^+-solute–induced swelling. Inhibition of RVD by these antagonists/inhibitors could be bypassed by gramicidin addition, suggesting that it was the K^+ channel that required Ca^{2+}/calmodulin kinase for activation. Charybdotoxin, a selective and potent inhibitor of large conductance, (maxi K^+) Ca^{2+}-activated K^+ channels, had no effect on RVD after substantial volume increases. Charybdotoxin did, however, block RVD after cell swelling caused by D-glucose addition or after a modest 5 to 7% volume increase caused by hypotonic dilution to duplicate the size reached during Na^+-solute–induced swelling (MacLeod and Hamilton, 1999a; Section II.C.). Clearly, the signaling of K^+ and Cl^- channels required for RVD when cell swelling is generated by hypotonic dilution is substantially different from that when the villus cell swells because of Na^+-solute cotransport.

III. INTRACELLULAR pH SIGNALING DURING "MODEST" VOLUME INCREASES

A. Activation of Na^+/H^+ Exchange Is Required for RVD after Physiological Volume Increases

To understand how intracellular pH (pH_i) changes relate to the signaling of RVD after villus cell swelling due to Na^+-solute cotransport, we assumed, because of the studies described above, that a slight (5 or 7%) hypotonic dilution would be an appropriate control for cells that increase their volume 5 to 7% by L-alanine absorption (MacLeod and Hamilton, 1996). When villus cells were slightly diluted 5 or 7% and their pH_i was assessed by fluorescent spectroscopy, this stimulated a cytosolic alkalinization which was prevented by a low concentration of N-methyl isobutyl amiloride (MIA, Fig. 3). The "standard" hypotonic dilution of 30% resulted in a slight acidification, which the amiloride derivative increased. However, the real surprise occurred when we examined the consequences of this alkalinization on RVD. As expected, the amiloride derivative or Na^+-free medium had no effect on RVD of cells hypotonically diluted 20 or 30%. But in villus cells diluted 5 or 7%, the amiloride derivative blocked the subsequent RVD (Fig. 4). Consistent with the idea that Na^+/H^+ exchange (NHE) was required for this volume regulation, Na^+-free medium also prevented RVD after this modest 5 to 7% volume increase (Fig. 4). These results contrasted with other reports of the activation of NHE during volume changes. First, "osmotic activation" of NHE generally refers to the concept that when cells are shrunken by increases in osmolarity or tonicity, NHE is activated (together with, Cl^-/HCO_3^- exchange), which results in Na^+ and Cl^- influx that drives a regulatory volume increase (reviewed in Orlowski and Grinstein, 1997; Lang *et al.*, 1998). Further, earlier studies (Livne and Hoffmann, 1990; Livne *et al.*, 1987) using ascites cells or platelets had observed that standard hypotonic dilution resulted in cytosolic acidification, which then activated amiloride-sensitive alkalinization. In these studies, amiloride accelerated the subsequent RVD. This acceleration was consistent with the idea that activated osmolyte (Na^+) influx was offsetting the osmolyte (K^+ and Cl^-) efflux required for volume regulation. Yet in the villus cells a modest volume increase, which duplicated the size these cells swell during Na^+-solute cotransport, resulted in activation of NHE. Furthermore, the subsequent RVD required activated NHE.

In additional experiments, modest swelling of the villus cells increased the rate of ^{22}Na influx that was sensitive to the low concentration of MIA. We then asked whether cytosolic acidification was responsible for activating NHE and, if so, which isoform of NHE was responsible for these volume changes. At least five isoforms of NHE have been cloned (reviewed in Orlowski and Grinstein, 1997). In villus cells, NHE-3 and NHE-2 are located on the apical membrane (Wormmeester *et al.*, 1998; Hoogerwerf *et al.*, 1996) while the ubiquitous isoform, NHE-1, is found on the basolateral membrane (Bookstein *et al.*, 1994). Studies using transfectants of NHE-1, 2 or 3 and nonamiloride derivatives found that clonidine was more potent than cimetidine in inhibiting NHE-2 and NHE-3. It is only with NHE-1 that this order of potency reversed (Orlowski, 1993; Yu *et al.*, 1993). We found that RVD was prevented by cimetidine and clonidine, and that cimetidine was six times more potent than clonidine. This was strong evidence that NHE-1 was the isoform activated by the 5 to 7% volume increase.

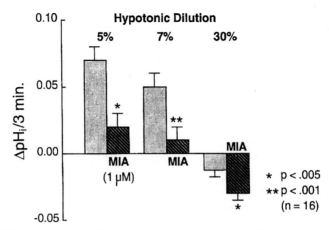

FIGURE 3   Summary of pH$_i$ changes in absence or presence of MIA after 0.95× isotonic, 0.93× isotonic, or 0.70× isotonic dilution. (From MacLeod, R.J., and Hamilton, J.R., 1996, Fig. 2, p. 23140, with permission.)

In separate experiments, we measured the rate of pH$_i$ recovery from an acid load (generated using ammonium prepulse) in cells that were resuspended in isotonic medium and compared it with 5% hypotonic medium. We found faster rates of MIA-sensitive pH$_i$ recovery in the 5% hypotonic medium. We also observed that amiloride-insensitive sources contributed to pH$_i$ recovery from an acid load in villus cells (Fig. 5). When bafilomycin A, a potent and selective inhibitor of type V H$^+$-ATPases (Dröse *et al.*, 1988), and Zn^{2+}, an inhibitor of H$^+$ conductance (Demaurex *et al.*, 1993), were present together, they inhibited about 75% of the pH$_i$ recovery. We therefore measured the MIA-sensitive rate of pH$_i$ recovery in the presence of these inhibitors. In 5% hypotonic medium, MIA-sensitive pH$_i$ recovery was faster than that in isotonic medium, while in 30% hypotonic medium, the rate was substantially diminished. Thus, the activation of NHE-1 by 5% hypotonic swelling was not due to cytosolic acidification. These experiments also alerted us to the possibility that H$^+$ conductance might play a role in pH$_i$ homeostasis or, alternatively, RVD after cell swelling by Na$^+$-solute absorption.

B. NH$_4$Cl-Induced Alkalinization Allowed RVD When NHE Was Inhibited

It was counterintuitive that increasing Na$^+$ influx was essential for RVD when the villus cells were signaling K$^+$ and Cl$^-$ loss. We therefore tested the hypothesis that it was the alkalinization occurring because of the activated

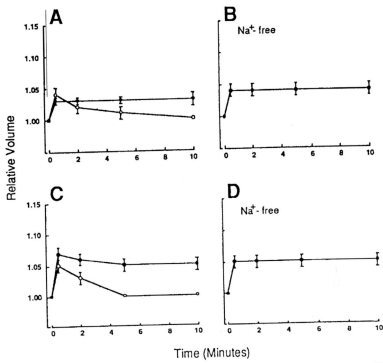

FIGURE 4 Effect of MIA and Na^+-free medium on RVD after 5 or 7% hypotonic dilution. (A) ○, 0.95× isotonic dilution; ●, 0.95× isotonic + MIA (1 μM). (B) ●, 0.95× isotonic dilution in Na^+-free medium. (C) ○, 0.97× isotonic dilution; ●, 0.93× isotonic + MIA (1 μM). (D) ●, 0.93× isotonic dilution in Na^+-free medium. $P < 0.001$ in all cases where RVD inhibited ($n = 6$). Volume measured electronically and is expressed relative to the isotonic control. (From MacLeod, R.J., and Hamilton, J.R., 1996, Fig. 3, p. 23140, with permission.)

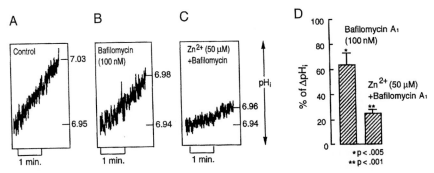

FIGURE 5 Effect of bafilomycin and Zn^{2+} on pH_i recovery from acidified cells. (A) Control, (B) bafilomycin (100 nM), (C) Zn^{2+} (100 μM) + bafilomycin, (D) cumulative effects of bafilomycin and of bafilomycin and Zn^{2+} on pH_i recovery from an acid load. Results are means ± SE of five experiments performed in duplicate. (From MacLeod, R.J., and Hamilton, J.R., 1996, Fig. 6, p. 23141, with permission.)

NHE-1 that was a determinant of RVD. We hypotonically diluted the cells 7% in the presence of MIA so that volume regulation was prevented, then added ammonium chloride to the cells (Fig. 6). Addition of 1 mM NH4Cl to the swollen cells caused RVD in the presence of MIA (Fig. 6A). The amount of alkalinization caused by ammonium addition (0.086 ± 0.010 pH unit, $n = 15$) was no different from the amount measured in these cells following a 5% hypotonic dilution (Fig. 3). As a converse of this experiment, we acidified the pH_i an equivalent amount in comparably treated cells (Figs. 6C and 6D) and found that the RVD remained inhibited. Thus, the Na^+ influx resulting from activated NHE-1 was osmotically neutral. It was the alkalinization of pH_i that occurred due to the activated NHE-1 that determined the osmolyte loss required for volume regulation.

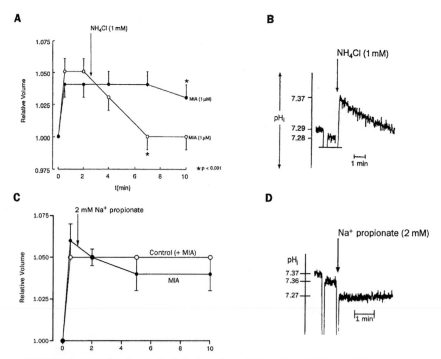

FIGURE 6 NH_4Cl-induced alkalinization allows RVD when NHE-1 is inhibited. (A) RVD blocked by MIA is bypassed with NH_4Cl addition. ●, MIA (1 μM) + 7% hypotonic dilution; ○, NH_4Cl (1 mM) added after 2 min volume measurement, MIA, + 7% hypotonic dilution. (B) pH_i tracing of NH_4Cl addition. Cells were diluted 7%; 1 min later NH_4Cl was added. Tracing is from a single representative experiment. (C) Na^+ propionate added to MIA-treated cells. (D) pH_i tracing of Na^+ propionate addition. (From MacLeod, R.J., and Hamilton, J.R., 1996, Fig. 9, p. 23144, with permission.)

C. Charybdotoxin-Sensitive K^+ Channels Require Cytosolic Alkalinization

Previously, we had found that RVD following swelling caused by uptake of D-glucose was sensitive to the high-conductance (maxi-K) K^+ channel blocker charybdotoxin (CTX), whereas RVD following substantial volume increases (>12%) was insensitive to the toxin (MacLeod et al., 1992b). Since the Ca^{2+} gating of the charybdotoxin-sensitive K^+ conductance is exquisitely sensitive to alkaline pH_i (Garcia et al., 1995; Chang et al., 1991), we thought that the activation of NHE-1 during the modest 5 to 7% volume increase might serve as the source of the required alkalinization for Ca^{2+} gating of the charybdotoxin-sensitive K^+ loss. Indeed, when villus cells were hypotonically diluted 5 to 7%, the subsequent RVD was charybdotoxin sensitive (MacLeod and Hamilton, 1999a,b). The modest hypotonic dilution increased the rate of ^{86}Rb efflux which was charybdotoxin sensitive. This rate was diminished an equal amount by MIA or Na^+-free medium, which was consistent with the NHE-1 activation being required for the charybdotoxin-sensitive K^+ channel activation. Consequently, when the cells were swollen to substantial volumes (>15%), alkalinizing these cells then made the subsequent volume responses charybdotoxin sensitive. We found this alkalinization also increased $[Ca^{2+}]_i$, and that the source of the increase was extracellular calcium. Experiments examining the role of Ca^{2+} and cytosolic alkalinization on RVD after the modest, "physiological" increase are summarized in Figure 7. A modest 7% hypotonic dilution caused a rapid increase in $[Ca^{2+}]_i$ above basal levels which then declined by 3 min to a level that was higher than the resting $[Ca^{2+}]_i$. These changes were in accord with observations of others (Foskett, 1994; McCarty and O'Neil, 1992). However, in the presence of MIA, or in Na^+-free medium after a 7% hypotonic dilution, the initial rapid increase of $[Ca^{2+}]_i$ was reduced by an equivalent amount. Both MIA and Na^+-free medium also block the NHE-1–induced alkalinization and, as described above, prevent RVD. In Ca^{2+}-free medium, the initial rapid increase of $[Ca^{2+}]_i$ was reduced to a level comparable to the presence of MIA or Na^+-free medium containing Ca^{2+}. Yet under Ca^{2+}-free conditions, the alkalinization caused by NHE-1 still occurred, but RVD was blocked. Together, these data suggested to us that the alkalinization caused by activated NHE-1 was increasing $[Ca^{2+}]_i$ by activating a Ca^{2+} influx pathway.

In nonepithelial cell types, cytosolic alkalinization has been shown to increase Ca^{2+} current or open channel probability of "L" type Ca^{2+} channels whereas acidification had the opposite effect (Yamakage et al., 1995; Schmigol et al., 1995; Kaibara et al., 1994). To date, however, the electrophysiology, molecular biology, and regulation of calcium channels and Ca^{2+} influx pathways in the absorptive villus cell have received scant attention (MacLeod et al., 1998). Nevertheless, in villus enterocytes, our studies suggested that activation of the charybdotoxin-sensitive K^+ channel for RVD required both an increase of

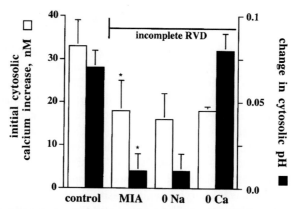

FIGURE 7   Initial changes in $[Ca^{2+}]_i$ and pH_i after 7% hypotonic dilution. MIA (1 μM); 0 Na$^+$ is Na$^+$-free medium and 0 Ca^{2+} is Ca^{2+}-free medium containing 150 μM EGTA. *$P < 0.05$ compared with control. Under all three conditions the Ca^{2+} transient was reduced and RVD was prevented but only in Ca^{2+}-free medium was the alkalinization preserved.

$[Ca^{2+}]_i \geq 27$ nM and a cytosolic alkalinization of ≥ 0.07 pH unit above basal levels after a modest but "physiological" volume increase. Because inhibiting either the Ca^{2+} increase or the pH_i change prevented CTX-sensitive K$^+$ channel activation (which then blocked the volume regulation), these findings suggest a direct role of $[Ca^{2+}]_i$ increases in RVD rather than parallel, unrelated consequences of volume perturbation (Jørgensen *et al.*, 1997; Altamirano *et al.*, 1998). Nevertheless, when these cells swell due to Na$^+$-solute absorption, an additional mechanism must be activated to generate sufficient alkalinization for CTX-sensitive K$^+$ channel activation.

D. A Physiological Size of Volume Increase

Beginning with the distinctive differences between the two models of villus cell swelling, we then found that the pH_i changes occurring during hypotonic swelling were also distinctive. If the villus cells were swollen to a size that duplicated the size they reach during Na$^+$-solute absorption, the NHE-1 isoform was activated and the subsequent volume regulation required cytosolic alkalinization. In contrast, swelling the cells amounts that are commonly used in most (if not all) volume regulation protocols caused inhibition of the NHE-1 isoform, and volume regulation did not require alkalinization. Because pH_i is a determinant of the signaling of the K$^+$ channels required for RVD, and because the size the cell swells determines the nature of the change in pH_i, these data suggested to us that the size these epithelial cells swell is a determinant of RVD.

Furthermore, because the size that results in NHE-1 activation is the same size these cells reach during Na^+-solute absorption, we believe there is a physiologically relevant size that is distinctive to Na^+-absorbing epithelial cells. Thus, both how an epithelial cell swells and the size that it swells are determinants of RVD signaling.

E. How Is NHE-1 Activated by a Modest Volume Increase?

We found that a G protein and intact microfilaments were required for the activation of NHE-1 after a 5 to 7% modest volume increase (MacLeod, 1996). Using villus cells loaded by electropermeabilization with an antagonist of G-protein activation (GDPβS), we found that after a modest volume increase, activation of NHE-1 was prevented, as was the RVD. In isotonic medium after photolysis, villus cells loaded with a caged G-protein agonist, DMNPE-GTPγS, exhibited cytosolic alkalinizations blocked by low concentrations of either MIA or GDPβS. Furthermore, the MIA-sensitive rate of pH_i recovery from an acid load in 5% hypotonic medium was equal to the rate in isotonic medium of cells loaded with GTPγS. When cells were preincubated in dihydrocytochalasin B to depolymerize the actin-based cytoskeleton, there was no activation of NHE-1 or RVD after a modest volume increase. This cytochalasin also inhibited the increased rate of pH_i recovery from an acid load in hypotonic medium. Together, these data suggested that an intact cytoskeleton and a G protein were required to activate NHE-1 after the modest volume increase. However, the mechanisms of activation of NHE-1 in other cell types are legion (Orlowski and Grinstein, 1997). Others have shown that NHE activity is regulated by both heterotrimeric and small GTP-binding proteins (Dhanasekaran et al., 1994; Lin et al., 1996; Hooley et al., 1996). The effect of $G_{\alpha 13}$ is mediated by RhoA and/or Cdc42, which then activate MEKK-1. Additional evidence that small GTP-binding proteins are involved was demonstrated by showing that oncogenic forms of Ras enhanced the $[H^+]$ sensitivity of NHE (Kaplan and Boron, 1994; Bianchini et al., 1997). Substantial hypotonic swelling of different cell types will activate the p38 MAP kinase signaling cascade (Tilly et al., 1996) and, in some cases, the nitrogen-activated protein (MAP) kinases Erk-1 and -2 (Schliess et al., 1995). The mechanism of action of these kinases is unknown. Tyrosine kinase $p56^{lck}$ appears to mediate activation of swelling-activated Cl^- channels in lymphocytes (Lepple-Wienhues et al., 1998), whereas activation of swelling-induced Cl^- channels in endothelial cells is dependent on tyrosine phosphorylation (Voets et al., 1998). An elegant study of the induction of tyrosine phosphorylation and NHE-1 activation in human neutrophils (Krump et al., 1997) will most likely lead to elucidating the mechanism of how NHE-1 is activated during cell shrinkage. These authors found that neutrophil shrinkage activated kinases fgr and hck of the src family and NHE-1 activation was prevented by inhibitors of tyrosine phosphorylation. They found that tyrosine phosphorylation was induced by shrinkage at constant osmolarity and ionic strength, but that increasing osmolarity and ionic strength at constant volume had a minimal effect on phosphotyrosine formation. This is consistent with an alteration in cell size being a determinant of tyrosine phosphorylation. Others have shown that osmotic shrinkage of HeLa cells induced clustering and activation of several tyrosine kinase receptors in the absence of their ligands (Rosette and Karin, 1996). In neutrophils, the cross-linking of Fc receptors or integrins activated the same tyrosine kinases as osmotic shrinkage. Consequently, their results suggested to them that stimulation of tyrosine kinases preceded activation of NHE-1. They speculated that osmotic activation of NHE-1 involved stimulation of RhoA through src-related tyrosine kinases. Others (Tomimaga et al., 1998) have found that p160ROCK, a Rho-associated kinase, stimulated NHE-1 activity. Experiments with a dominant negative mutant of p160ROCK demonstrated that p160ROCK acted downstream of RhoA and upstream of NHE-1 to mediate RhoA activation of the exchanger. Progress in identifying Rho-binding proteins and other intermediates should be rapid, given the rate at which auxiliary proteins responsible for the cAMP-mediated inhibition of NHE-3 have been identified (Lamprecht et al., 1998; Hall et al., 1998). In a recent study, we found that genistein, a potent tyrosine kinase inhibitor, had no effect on RVD following substantial ($\geq 12\%$) volume increases (MacLeod and Hamilton, 1999b) but this drug did block RVD following the modest 5 to 7% increases. This suggests tyrosine phosphorylation might be required for this distinctive volume regulation. A model to describe activation of NHE-1 for RVD, apart from identifying which G protein is required (heterotrimeric or small GTP-binding), must explain why an intact cytoskeleton is necessary for NHE-1 activation. For example, further experiments might focus on integrins, which provide a physical linkage between cytoskeletal structures and the extracellular matrix (Miyamoto et al., 1995; Howe et al., 1998). Integrins are heterodimeric transmembrane receptors that bind to proteins in the extracellular matrix. The engagement of integrins with their extracellular matrix and lateral clustering of integrins in the plane of the membrane result in the formation of organized complexes between integrins and cytoskeletal proteins. These complexes can activate cytoplasmic tyrosine kinases such as the focal adhesion kinase, pp125[FAK], as well as those in the mitogen-activated protein kinase (MAPK) cascade (Hemler, 1998; Chicural et al., 1998). It must be emphasized that an extracellular matrix is required to occupy integrins and that lateral diffusion and integrin clustering is always postoccupancy. Invoking integrin signaling of cells in suspension is putting the cart before the horse. Fortunately, two emerging paradigms might put the cart where it belongs. The first paradigm is from studies showing that the release of cytoskeletal constraints leads to increased integrin lateral mobility and clustering independent of ligand binding (Kucik et al., 1996;

Yauch *et al.*, 1997). The second model is that a subset of integrins associates with proteins of the transmembrane-4 superfamily (TMS4F). Evidence is accumulating to suggest that TMS4F proteins recruit signaling enzymes like PKC and PI4-kinase into complexes with integrins (Hemler, 1998). The involvement of TMS4F proteins and integrins may be more widespread than is currently appreciated. For example, CD98 has been shown to be required for the functional expression of an Na^+-independent amino acid cotransporter, LAT1 (Kanai *et al.*, 1998). It may be possible that the physical distortion occurring because of a 5 to 7% volume increase triggers sufficient release of cytoskeletal constraints to allow certain integrins to cluster in the absence of engagement with the extracellular matrix. Alternatively, could the physical distortion of the modest volume increase allow activation of a signaling cascade mediated by a TMS4F protein with a subset of integrins? Villus enterocytes may provide an ideal model to ask whether a volume sensor comprised of TMS4F proteins and integrins exists.

1. Glutamine Transport in Skeletal Muscle

Studies of the regulation of glutamine transport in rat skeletal muscle are consistent with the idea that the engagement of integrins during volume changes will influence glutamine transport (Low and Taylor, 1998), but the emerging picture is quite different from studies using enterocytes. Addition of glutamine to cultured rat myotubes increased the volume of myotubes, and the increase in volume was prevented in Na^+-free medium. This suggests that Na^+-dependent glutamine transport was the effector of the volume increase. A 50% hypotonic challenge also increased the rate of glutamine transport, whereas hypertonic shrinkage diminished this transport (Low *et al.*, 1996). Focusing on how hypotonic swelling alone influenced glutamine transport, it was then observed that wortmannin, a selective and well-characterized inhibitor of phosphatidylinositol 3-kinase (PI3-kinase), prevented the hypotonic activation of glutamine transport (Low *et al.*, 1997). This suggested that PI3-kinase was required for signal transduction. Subsequently, it was found that inactivating β1 integrin–extracellular matrix interactions using an integrin-binding peptide, GRGDTP, prevented the hypotonic activation of glutamine transport (Low and Taylor, 1998). Two cytoskeletal inhibitors, colchicine and cytochalaisin D, also prevented the hypotonic activation and, in isotonic medium, increased basal glutamine uptake that was wortmannin-sensitive. Together, these results suggested a pathway of extracellular matrix–integrin engagement during volume increases activated PI3-kinase, which then altered glutamine transport. While this system may offer important clues in identifying a putative volume sensor, it would be important to know if the activity of the glutamine cotransporter alone caused integrin engagement or activated PI3-kinase. We do not know if any extent of hypotonic swelling activates Na^+-solute cotransport or whether PI3-kinase is required to transduce RVD after Na^+-solute–induced swelling or modest volume increases

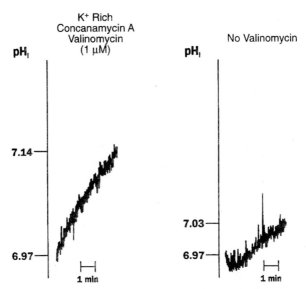

FIGURE 8 pH_i recovery from acid load is conductive. Left tracing: pH_i recovery (after NH_4Cl prepulse) in K^+-rich medium with concanamycin A (350 nM), tetraethylammonium (20 mM), and valinomycin (1 μM). Right tracing: Same conditions, no valinomycin. $\Delta pH_i/5$ min: 0.170 ± 0.020 vs 0.060 ± 0.010 pH unit, $P < 0.001$, $n = 6$. Tracings are from single representative experiment.

in villus cells. The muscle cell studies are interesting because hypotonic swelling rapidly activates solute transport, whereas swelling occurs in villus cells because of Na^+-solute cotransport.

IV. INTRACELLULAR pH AND Na^+-SOLUTE INDUCED VOLUME INCREASES

A. Evidence That Villus Cells Possess a Voltage-Dependent H^+ Channel

As described earlier, we found that Zn^{2+} substantially reduced pH_i recovery from an acid load in villus cells. This suggested that H^+ conductance contributed to pH_i homeostasis in these cells. H^+ conductance in different cell types has been reviewed (De Coursey and Cherny, 1994; Lukacs et al., 1993). Electrophysiological evidence of H^+ conductance is found in neurons (Byerly and Suen, 1989), oocytes (Barish and Baud, 1984), alveolar cells (De Coursey, 1991), phagocytic leukocytes (neutrophils and macrophages) (Demaurex et al., 1993; Kapus et al., 1993, 1994; Hendersen et al., 1987), and lymphocytes (Káldi et al., 1994). A conductive H^+ permeability has been shown in brush border membrane vesicles of kidney cortex and proximal tubule (Ives, 1985; Reenstra et al., 1987). The properties of H^+ conductance have been established by studies using either fluorometric determinations of pH_i changes or measurements of pH_i using H^+-selective microelectrodes (Meech and Thomas, 1987) and whole-cell voltage clamp techniques. Strong evidence that H^+ is the charge-carrying species for this conductive pathway came from studies measuring both the current and pH_i during application of depolarizing voltage steps. An alkalinization accompanied the appearance of the outward current. These two measurements had similar time courses and voltage dependence; only when the assumption was made that H^+ were the charge carriers could the magnitude of the current account for the pH_i change (Demaurex *et al.*, 1993; Kapus *et al.*, 1993). H^+ conductance is exquisitely selective for H^+ over other ions ($>10^6$ to 1). It is also voltage dependent, activated at depolarizing potentials, and outwardly rectifying. Intracellular acidification activates H^+ conductance, while extracellular acidification substantially reduces it. A hallmark of H^+ conductance is its inhibition by submillimolar concentrations of Zn^{2+} or Ca^{2+}. Inhibition by these heavy metals is both rapid and reversed on washing (Barish and Baud, 1984; Demaurex *et al.*, 1993; Kapus *et al.*, 1993; Mahant-Smith, 1989).

To demonstrate that villus cells possessed a Zn^{2+}-sensitive H^+ conductance, we used a fluorometric determination and an experimental protocol previously validated using neutrophils (Demaurex *et al.*, 1993). Villus cells were first acid loaded using an NH_4Cl prepulse. This would serve to generate an outwardly directed

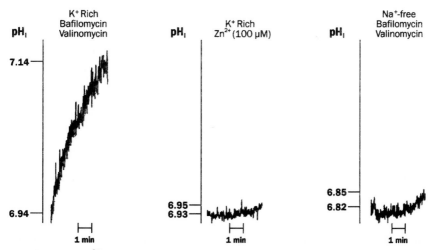

FIGURE 9 Zn^{2+}-sensitive pH_i recovery is responsive to membrane potential. Left-hand tracing: pH_i recovery after NH_4Cl (5 mM) prepulse in K^+-rich medium with bafilomycin A (350 nM), valinomycin (1 μM). Middle tracing: same as left-hand tracing but containing Zn^{2+} (100 μM). Right-hand tracing: pH_i recovery in Na^+-free medium containing valinomycin and bafilomycin A. Zn^{2+}-sensitive alkalinization is accentuated in depolarizing medium and inhibited in hyperpolarizing medium. Tracings from single representative experiment.

chemical gradient for H^+. These cells were then suspended in depolarizing K^+-rich medium that contained both concanamycin or bafilomycin A to inhibit vascular-type H^+ pumps and tetraethylammonium to diminish constitutive K^+ conductance. The effect of valinomycin, a conductive K^+ ionophone that would collapse the membrane potential of the cells, on the rate of pH_i recovery was determined. As illustrated in Figure 8, a gradual net H^+ (equivalent) efflux was observed; the presence of valinomycin increased the rate of pH_i recovery about threefold. The absence of valinomycin decreased the rate of alkalinization, which suggested that the endogenous counterion permeability was limiting net H^+ efflux. This finding was consistent with H^+ extrusion being conductive. The conductive nature of this alkalinization was further demonstrated by manipulating the membrane potential of the villus cells during pH_i recovery from an acid load (Fig. 9). The cells rapidly alkalinized when suspended in depolarizing K^+-rich medium containing concanamycin or bafilomycin A and valinomycin. The pH_i remained constant when cells were suspended in Na^+-free medium containing the V-type H^+ ATPase inhibitor and valinomycin, conditions that one would predict to hyperpolarize the membrane. These data suggest the alkalinization observed in K^+-rich medium is potential sensitive, consistent with a conductive H^+ flux. Furthermore, in the K^+-rich medium, this potential-sensitive alkalinization was abolished by 100 μM Zn^{2+}. These data strongly suggest that mammalian villus cells possess a Zn^{2+}-sensitive H^+ conductance that is responsive to changes in membrane potential. Others (Peral and Ilundain, 1995) have shown Na^+-independent alkalinization in chicken enterocytes during pH_i recovery from an acid load that was sensitive to Zn^{2+} but used valinomycin to bypass inhibition of this pH_i recovery by DCCD or NBD-Cl. Electrogenicity of H^+ conductance was inferred from the finding that in low K^+, Na^+-independent medium valinomycin inhibited alkalinization.[2]

B. RVD after L-Alanine Absorption Requires Cytoplasmic Alkalinization from Two Sources

After villus cell swelling caused by L-alanine absorption, we determined the effect of MIA on pH_i and volume regulation by measuring volume changes by electronic cell sizing and pH_i by fluorescent spectroscopy (Fig. 10). Addition of L-alanine to villus cells in suspension caused a transient volume increase that was followed by RVD (Fig. 10A). When MIA was added together with L-alanine, the cells swelled but the subsequent RVD was prevented. MIA alone had no effect

[2] H^+ channels were majestically cloned using a Caco-2 cDNA library (Baufi, B., Maturana, A., Arnaudeau, S., Laforge, T., Sinha, B., Ligetti, E., and Demaurex, N. (2000). A mammalian H^+ channel generated through alternative splicing of the NADPH oxidase homolog NOH-1. *Science* **287,** 138–142).

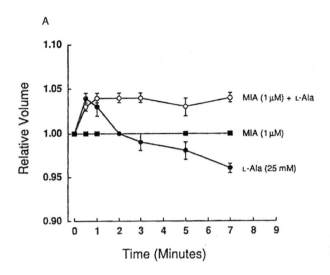

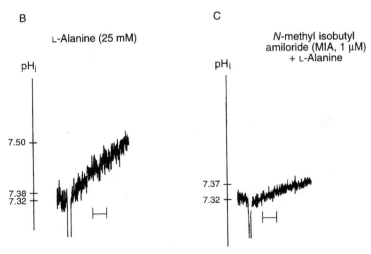

FIGURE 10 RVD after L-alanine absorption is MIA sensitive. (A) Volume response to L-alanine (25 mM) and MIA (1 μM). RVD prevented by MIA. (B) pH_i response after L-alanine addition. (C) Effect of MIA (1 μM) on pH_i response (ΔpH_i/5 min: 0.04 ± 0.01 vs 0.120 ± 0.010 pH unit, $P < 0.001$, $n = 9$). Tracings are from a single representative experiment and volume was measured electronically, expressed relative to isotonic control.

on the volume of the cells throughout the experiment (Fig. 10A). In this nominally HCO_3-free medium, the resting pH_i of these cells was 7.38 ± 0.04, $n = 40$. Addition of L-alanine caused the cells to alkalinize (Fig. 10B), and MIA partially inhibited the alkalinization (Fig. 10C). Thus, while the volume reduction was completely blocked, only a fraction of the pH_i response was inhibited by the amiloride derivative. When the relative pharmacological sensitivities of RVD for L-alanine–induced cell swelling to both amiloride and nonamiloride derivatives were determined, we found the EC_{50} values of cimetidine (20 μM) and clonidine (110 μM) were greater than that of MIA (230 nM), suggesting an order of potency as MIA > cimetidine > clonidine. This order strongly suggests that the basolateral NHE-1 isoform was responsible for a proportion of these pH_i changes.

We then determined whether Zn^{2+}-sensitive H^+ conductance was responsible for the MIA-insensitive alkalinization after L-alanine addition (Fig. 11). Zn^{2+} (100 μM) blocked RVD after cell swelling caused by L-alanine addition (Fig. 11A). Notably, when the villus cells were hypotonically diluted 5% to duplicate the size reached during L-alanine absorption, Zn^{2+} had no effect on the subsequent RVD (Fig. 11B). The alkalinization of pH_i observed after L-alanine addition was partially inhibited by Zn^{2+}. In the presence of both MIA and Zn^{2+}, the L-alanine–induced alkalinization was obliterated (Fig. 11C). This suggested that both an activated H^+ conductance and NHE-1 constituted the alkalinization of pH_i by L-alanine absorption. Furthermore, these findings strongly implied that H^+ conductance was not required for volume regulation after cell swelling induced by hypotonicity, but that RVD required alkalinization from both H^+ conductance and NHE-1 after L-alanine absorption.

After Na^+-solute–induced swelling, the RVD is charybdotoxin-sensitive (MacLeod et al., 1992b) and the activation of this maxi-K^+ or Ca^{2+}-activated K^+ channel requires alkalinization of pH_i (MacLeod and Hamilton, 1996, 1999a). We speculated that, if activation of CTX-sensitive K^+ channels after Na^+-solute–induced swelling required alkalinization, then inhibition of RVD by Zn^{2+} or MIA could be bypassed by increasing cytosolic alkalinization. Using the experimental protocol illustrated in Figure 6, we added NH_4Cl (2 mM) to villus cells swollen by L-alanine absorption but with RVD blocked by the presence of Zn^{2+}. We found that NH_4Cl addition caused the cells to shrink in the presence of Zn^{2+}. A comparable bypass of inhibition was observed when RVD had been blocked by MIA. Addition of NH_4Cl caused an alkalinization (0.120 ± 0.010 pH unit, $n = 5$) that was no different from the pH_i change observed after L-alanine addition in the absence of any inhibitors (see Fig. 14). These data suggested to us that, when villus cells swell with Na^+-solute absorption, two sources of cytosolic alkalinization were required to allow activation of the CTX-sensitive K^+ channel for the subsequent volume regulation. This contrasts with the hypotonic model of cell swelling whereby activation of basolateral NHE-1 alone is sufficient to activate the CTX-sensitive K^+ channel (MacLeod and Hamilton, 1999a).

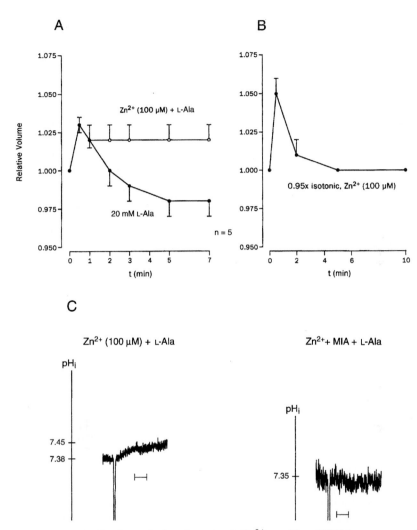

FIGURE 11 RVD after L-alanine absorption is Zn^{2+} sensitive. (A) Volume response to L-alanine and Zn^{2+} (100 μM). RVD prevented by Zn^{2+}. (B) Zn^{2+} had no effect on RVD after 5% hypotonic dilution. (C) pH_i response to Zn^{2+} and L-alanine ($\Delta pH_i/5$ min: 0.05 ± 0.01 pH unit, $P < 0.001$, $n = 9$). pH_i response of Zn^{2+} and MIA with L-alanine ($\Delta pH_i/5$ min: 0.005 ± 0.010, $P < 0.001$, $n = 9$). Volume measured electronically, expressed relative to isotonic control; pH_i tracings from single representative experiment. Zn^{2+} diminishes the pH_i response to L-alanine as well as blocks RVD but Zn^{2+} and MIA together obliterate the alkalinization.

C. How Does Na$^+$-Solute Swelling Activate H$^+$ Conductance?

As discussed above, both depolarization and cellular acidification will activate H$^+$ conductance. Elegant studies have also shown that exogenous arachidonic acid is a potent activator of H$^+$ conductance in macrophages and neutrophils (Kapus et al., 1994; De Coursey and Cherny, 1993; Susztak et al., 1997). Arachidonic acid release may be catalyzed by the 85-kDa cytosolic phospholipase A$_2$ (cPLA$_2$) (Exton, 1994). However, this precursor may also be generated by phospholipase C or phospholipase D when coupled with the appropriate diacyl- and monoecylglycerol lipases, respectively (Allen et al., 1992; Burgoyne and Morgan, 1990). The regulation of cPLA$_2$ has recently been summarized (Clark et al., 1995; Leslie, 1997). cPLA$_2$ is rapidly activated by increased [Ca^{2+}]$_i$ and phosphorylation mediated by the mitogen-activated kinase erk-1. The increased Ca^{2+} causes translocation to the membrane from the cytosol and it is currently believed that partial activation of the cPLA$_2$ occurs in the absence of phosphorylation (Lin et al., 1993). Evidence for both PKC-dependent and -independent pathways has been reported (Leslie, 1997). A cell-permeable trifluoromethyl ketone derivative of arachidonic acid (AACOCF$_3$) has been shown to be a potent and extremely selective inhibitor of cPLA$_2$ (Street et al., 1993; Trimblee et al., 1993). This inhibitor has been reported to decrease RVD and [^{3}H]arachidonic acid release from hypotonically swollen Ehrlich ascites cells (Thoroed et al., 1997). It has no effect on N-formylmethionyl-leucylphenylalanine (fLMP)-stimulated [Ca^{2+}]$_i$ transients or protein kinase C activation in neutrophils, yet will block fMLP-stimulated H$^+$ conductance in these cells (Susztak et al., 1997). Because L-alanine addition to villus cells does not cause cytosolic acidification, we focused our experiments on whether cPLA$_2$ was required to activate H$^+$ conductance in the enterocytes.

We first determined whether [Ca^{2+}]$_i$ changed with L-alanine addition using fluorescent spectroscopy of cells loaded with Indo-1 (Fig. 12). The basal level of [Ca^{2+}]$_i$ in the villus cells in suspension was 144 ± 5 nM, n = 20. Addition of L-alanine caused a rapid increase in [Ca^{2+}]$_i$ (Fig. 12A). Addition of D-alanine, which has been shown not to cause a volume increase (MacLeod and Hamilton, 1991b), had no effect on [Ca^{2+}]$_i$ (Fig. 12A). The L-alanine–stimulated change in [Ca^{2+}]$_i$/2 min was substantially greater than changes seen after D-alanine addition (29 ± 2 nM vs 2 ± 1 nM, P < 0.001, n = 10) (Fig. 12B). Thus, L-alanine–stimulated cell swelling resulted in an increase in [Ca^{2+}]$_i$.

To assess a potential role for cPLA$_2$ in regulating the Zn^{2+}-sensitive H$^+$ conductance required for RVD after L-alanine–stimulated cell swelling, we measured changes in pH$_i$ and volume in the presence of the cPLA$_2$ inhibitor, AACOCF$_3$ (Fig. 13). After cell swelling caused by L-alanine, the RVD was prevented by AACOCF$_3$ (Fig. 13A). In contrast, AACOCF$_3$ had no effect on volume regulation of villus cells that received a 5% hypotonic dilution to duplicate

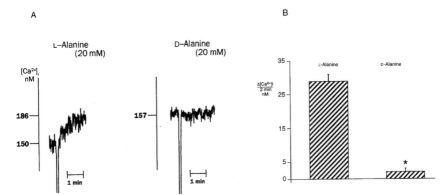

FIGURE 12 L-alanine swelling increases $[Ca^{2+}]_i$. Response of cells loaded with Indo-1 (described in MacLeod, R.J., and Hamilton, J.R., 1999a) to L-alanine or D-alanine (which does not cause swelling). Basal level was 144 ± 5 nM, $n = 20$. (B) Change in $[Ca^{2+}]_i/2$ min in response to L-alanine (29 ± 2 nM) or D-alanine (2 ± 1 nM, $P < 0.001$, $n = 9$). Tracings from single experiment.

the size these cells reached during Na^+-solute absorption (Fig. 13B). The pH_i increase stimulated by L-alanine addition was inhibited by the $cPLA_2$ inhibitor (Fig. 13C). We then added arachidonic acid (0.5 μM) to cells treated with $AACOCF_3$ and L-alanine (Figs. 13C and 13D). Addition of arachidonic acid to these cells increased the extent of cytosolic alkalinization compared with cells treated with L-alanine and $AACOCF_3$ (Fig. 13C). Arachidonic acid addition to cells swollen because of $AACOCF_3$ and L-alanine reversed inhibition of RVD caused by the $cPLA_2$ inhibitor (Fig. 13D). When arachidonic acid was added to isotonic villus cells, we observed a Zn^{2+}-sensitive alkalinization ($\Delta pH_i/5$ min: 0.040 ± 0.010 vs 0.005 ± 0.010, $P < 0.05$, $n = 8$), but cell volume did not change. In further experiments, we found that neither indomethacin, an inhibitor of cyclooxygenases, nor eicosatetraenoic acid, an inhibitor of cyclooxygenases and lipoxygenases, nor MK-866, an inhibitor of the 5-lipoxygenase activating protein, had an effect on the extent of the alkalinization stimulated by L-alanine addition or the RVD that followed swelling. Together, these data suggested to us that villus cell swelling caused by L-alanine absorption mobilized $cPLA_2$ and, subsequently, that arachidonic acid was activating the H^+ conductance allowing RVD to proceed. Furthermore, the putative increases in arachidonic acid occurring after mobilization of $cPLA_2$ during Na^+-solute absorption are not metabolized to prostaglandins or leukotrienes. The striking absence of an effect of Zn^{2+} or $AACOCF_3$ on RVD after modest hypotonic swelling strongly suggests that activation of H^+ conductance is not a requirement for RVD after villus cell swelling is caused by hypotonic dilution.

The findings of these experiments are summarized in Figure 14. The cytosolic alkalinization caused by L-alanine–stimulated swelling was partially inhibited

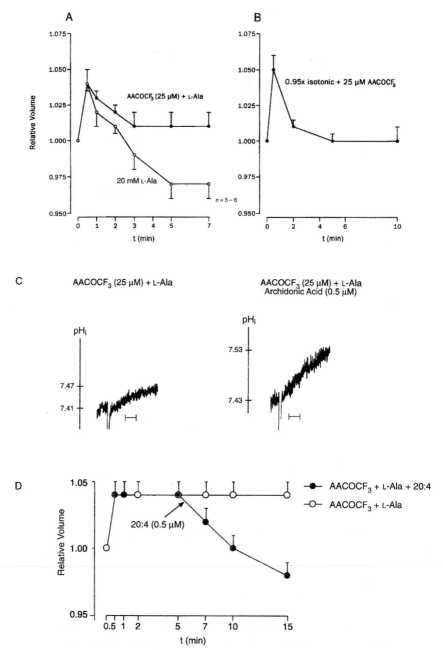

FIGURE 13 Effect of cPLA$_2$ inhibitor (AACOCF$_3$) on RVD and Δ pH$_i$. (A) Volume response to L-alanine and AACOCF$_3$ (25 μM). RVD blocked by AACOCF$_3$. (B) AACOCF$_3$ had no effect on RVD after 5% hypotonic dilution. (C) pH$_i$ responses: left tracing, L-alanine + AACOCF$_3$; right tracing, L-alanine + AACOCF$_3$ + arachidonic acid (0.5 μM). (D) Volume response to arachidonic acid addition of cells treated with AACOCF$_3$ + L-alanine. ○, L-alanine + AACOCF$_3$ (25 μM). ●, AACOCF$_3$ + L-alanine + arachidonic acid at 5 min: Final rel. vol.: 0.98 ± 0.01 vs 1.04 ± 0.01, $P < 0.001$, $n = 6$). Volume measured electronically, expressed relative to isotonic control. Tracings from single representative experiment.

by an amiloride derivative, MIA, or Zn^{2+}, an inhibitor of H^+ conductance. One of either of these inhibitors also prevented the subsequent RVD, but the presence of both inhibitors obliterated the Na^+-solute–stimulated alkalinization. A $cPLA_2$ inhibitor partially inhibited the L-alanine–induced alkalinization and blocked RVD, whereas arachidonic acid, the product of $cPLA_2$, restored the alkalinization and allowed complete volume regulation in the presence of the $cPLA_2$ inhibitor. These data suggest to us that one physiological role of H^+ conductance in absorptive villus cells is as an essential requirement for volume regulation after Na^+-solute cotransport. Because depolarization activates H^+ conductance in other cells, it could be argued that the Na^+ influx during solute cotransport might activate H^+ conductance. Indeed, in Na^+-rich medium (when all NHE isoforms had been inhibited), cell permeabilization caused by addition of the cation ionophore gramicidin caused a substantial alkalinization that was inhibited $>50\%$ by Zn^{2+}. The $cPLA_2$ inhibitor had no effect on this cytosolic alkalinization, which suggested that a $cPLA_2$–arachidonate cascade was not activated. Furthermore, if depolarization during cotransport alone were responsible for activating H^+ conductance, the $cPLA_2$ inhibitor would not have influenced volume regulation or the change in pH_i after L-alanine addition. These inhibited responses suggested second-messenger involvement. Since others have shown that PKC will activate H^+ conductance in neutrophils (Nanda and Grinstein, 1993) and chicken enterocytes (Calonge and Ilundáin, 1996), we added staurosporine together with L-alanine and observed that the remaining alkalinization could only be blocked with Zn^{2+}, and not amiloride. This would be consistent with PKC not being responsible for activating the H^+ conductance during L-alanine–induced cell swelling.

IV. REGULATION OF TIGHT JUNCTIONS DURING Na^+-SOLUTE COTRANSPORT

Evidence is accumulating that transjunctional solute movement in the intestine may be coupled to Na^+-solute cotransport (Turner and Madara, 1995; Turner et al., 1997; Madara, 1998). Using Caco-2 intestinal epithelial cells in monolayers that had been transfected with the intestinal Na^+–glucose cotransporter (SGLT1), it was found that activation of SGLT1 resulted in a fall in transepithelial resistance. This increase in tight junctional permeability was size selective, with increased permeability to nutrient-sized molecules such as mannitol, but not larger molecules such as inulin. Notably, the SGLT1-stimulated increases in permeability were prevented by inhibitors of myosin light chain kinase (MLCK). Further, activation of SGLT1 increased myosin light chain phosphorylation. These elegant studies suggested tight junction regulation was subsequent to SGLT1 activation. A central focus of further studies will be understanding how MLCK is activated. Previous work (Shrode

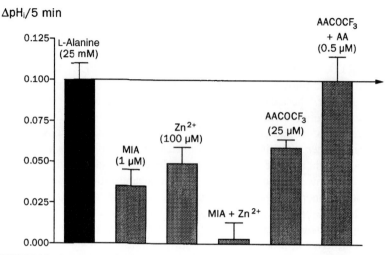

FIGURE 14   Summary of ΔpH_i in response to L-alanine. Amiloride derivative (MIA) which inhibits NHE-1, or Zn^{2+}, an inhibitor of H^+ conductance, partially inhibits alkalinization caused by L-alanine–stimulated swelling. The presence of both inhibitors completely blocks the response. AACOCF$_3$, a cPLA$_2$ inhibitor, partially inhibits alkalinization while exogenous arachidonic acid restores alkalinization in the presence of the cPLA$_2$ inhibitor. Results are means ± SE of nine experiments performed in duplicate.

et al., 1995) had found that MLCK inhibitors blocked osmotic activation (after cell shrinkage) of NHE-1, but the effect of MLCK inhibitors on NHE-1 activation after villus cell swelling is unknown. It would be of interest to know if hypotonically swelling Caco cells a modest 5 to 7% resulted in myosin light chain phosphorylation. Because both MAPK and Rho-kinase will phosphorylate and activate MLCK (Howe *et al.,* 1998; Mackay *et al.,* 1997), it has been suggested that Rho- and MAPK-mediated signals cooperate to stimulate actinomyosin contractility. Consequently we ask: Is it the volume change occurring because of SGLT1 activity (as discussed earlier, integrin engagement or clustering caused by modest swelling, then activating a MAPK cascade that leads to MLCK activation), or is the cotransporter itself coupled to an effector pathway? Regardless, it is clear that the tight junction-barrier function of the intestine must be considered an activity that is coupled to the volume changes occurring because of Na^+-solute absorption.

V. CONCLUDING COMMENTS

Table I summarizes the differences in signaling of RVD in villus cells after swelling by three different experimental maneuvers. Two concepts emerge from

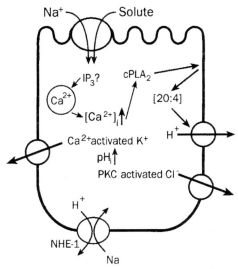

FIGURE 15 A working model of the signaling of RVD after Na^+-solute absorption. The volume increase is a modest 5 to 7%. Activation of Cl^- channels requires PKC. Activation of charybdotoxin-sensitive K^+ channel requires an increase in cytosolic Ca^{2+} as well as cytosolic alkalinization. The change in pH_i is because of activated basolateral NHE-1 and H^+ conductance. Cytosolic phospholipase A_2 must be mobilized to generate arachidonate to activate the H^+ conductance. In contrast, modest hypotonic swelling of 5 to 7% shows no requirement for PKC, $cPLA_2$, or H^+ conductance for RVD.

TABLE I

Differences between Hypotonic and Na^+-Solute–Induced RVD in Villus Cells

	L-Alanine (20 mM)	Modest (5–7%) hypotonic dilution	Substantial (20–50%) hypotonic dilution
Extent of volume increase	5–7%	5–7%	$\geq 15\%$
PKC-sensitive Cl^- conductance	Yes	No	No
Ca^{2+}	Yes, intracellular	Yes, extracellular	Yes, extracellular
Charybdotoxin-sensitive K^+ channels	Yes	Yes	No
CaMKII-sensitive K^+ channels	No	No	Yes
$cPLA_2$	Yes	No	No
H^+ conductance	Yes	No	No
pH_i	Alkalinization	Alkalinization	Acidification

these data. The first concept is that different signal transduction pathways are activated depending on how the villus cell swells and on the extent of the volume increase.

The second concept is that the activity of the cotransporter is coupled with the generation of second messengers. It is possible to activate some, but not all, of the signaling for volume regulation observed after Na^+-solute cotransport by duplicating the volume increase that occurs with L-alanine absorption with a modest hypotonic dilution. This implies that a putative volume sensor is activated by the extent of the volume increase. The RVD occurring after swelling caused by Na^+-solute absorption, however, requires several factors. It requires PKC to activate Cl^- conductance, intracellular calcium and $cPLA_2$ to be mobilized, and H^+ conductance to be activated, in addition to the pH_i changes that modest hypotonic swelling stimulates.

The signaling for RVD that follows Na^+-solute swelling is illustrated in Figure 15. A working model for how Na^+-solute cotransport activates CTX-sensitive K^+ channels, Cl^- channels, and H^+ conductance necessary for RVD must account for (1) the PKC requirement of the Cl^- conductance, (2) the mobilization of intracellular Ca^{2+} from a thapsigargin-sensitive source, and (3) the activation of $cPLA_2$. The simplest model would require the Na^+-solute cotransporter to be coupled with an effector pathway that would stimulate inositol lipid hydrolysis and diacylglycerol production. Such a model requires the Na^+-solute cotransporter to be coupled with phospholipase $\beta 1$. Accordingly, we would like to know: Does Na^+-solute addition to suspended villus cells stimulate PLCase? If diacylglycerol is released, which isozyme of PKC is required to activate the Cl^- channel? Is a MAPK cascade activated to stimulate $cPLA_2$? Are Na^+-solute cotransporters coupled to heterotrimeric or small GTP-binding proteins? Does activating an Na^+-solute cotransporter result in integrin clustering and engagement? Perhaps thinking about an Na^+-solute cotransporter as if it worked like a receptor is unusual, but the manifest differences in the signal transduction of volume regulation after Na^+-solute–induced swelling strongly suggest that second messengers are generated when the Na^+-solute cotransporter works in an intact villus cell. Further understanding of homocellular regulation (Schultz, 1981, 1989) in the intestine will require a more precise molecular understanding of the volume changes induced by changes in transcellular transport.

In conclusion, the absorption of Na^+-coupled solutes, D-glucose or L-alanine, by jejunal villus cells in suspension results in a volume increase of 5 to 7%. This increase is followed by regulatory volume decrease due to the activation of K^+ and Cl^- channels. These physiological volume increases are small relative to those induced by hypotonic dilution in other experimental models. With hypotonic dilution, villus cells also swell and exhibit RVD by activating K^+ and Cl^- channels. The signal transduction responsible for channel activation differs in these two models of cell swelling. After Na^+-solute–induced swelling, RVD requires protein kinase C to activate Cl^- channels, mobilization of Ca^{2+} from an intracellular source, and an increase in cytosolic alkalinization. The increase in intracellular pH is due to activation of basolateral NHE-1 and H^+ conductance, which together allow activation of K^+ channels for RVD. Only some of these signaling pathways are activated when cells are swollen by a modest hypotonic dilution to duplicate the size these cells reach because of Na^+-solute absorption. This difference in RVD in these two models of cell swelling suggests that additional second messengers are generated to regulate volume decrease when Na^+-solute transport occurs in an intact villus cell. Therefore, the mechanism of villus cell swelling and the extent of the volume increase are both determinants of signaling enterocyte volume regulation.

Acknowledgments

I thank Dr. Richard J. Hamilton, Department of Pediatrics, McGill University, who made it possible for me to study the volume determinants of villus and crypt epithelial cells, and Dr. Bill Powell, Department of Medicine, McGill University, for the generous gift of arachidonic acid. I also thank M.K. Badger for valuable editorial comments and Dr. Ed Brown, Brigham and Women's Hospital, Harvard Medical School, for allowing me to think about volume regulation while learning about the calcium-sensing receptor. This chapter was written while I was the recipient of a Postdoctoral Fellowship from the Medical Research Council of Canada.

References

Allen, A. C., Gammon, C. M., Ousley, A. H., McCarthy, K. D., and Morell, P. (1992). Bradykinin stimulates arachidonic acid release through the sequential actions of an *sn*-1 diacylglycerol lipase and monoacylglycerol lipase. *J. Neurochem.* **58**, 1130–1139.

Altamirano, J., Brodwick, M. S., and Alvarez-Leefmans, F. J. (1998). Regulatory volume decrease and intracellular Ca^{2+} in murine neuroblastoma cells studied with fluorescent probes. *J. Gen. Physiol.* **112**, 145–160.

Barish, M. E., and Baud, C. (1984). A voltage-gated hydrogen ion current in the oocyte membrane of the axolotol, *Ambystoma. J. Physiol. (Lond.)* **352**, 243–263.

Bianchini, L., L'Allemain, G., and Pouyssègur, J. (1997). The p42/p44 nitrogen-activated protein kinase cascade is determinant in mediating activation of the Na^+/H^+ exchanger (NHE-1 isoform) in response to growth factors. *J. Biol. Chem.* **272**, 271–279.

Bookstein, C., De Paoli, A. M., Yue, X., Niu, P., Musch, M., Rao, M. C., and Chang, E. B. (1994). Na^+/H^+ exchangers, NHE-1 and NHE-2 of rat intestine: Expression and localization. *J. Clin. Invest.* **93**, 106–115.

Burgoyne, R. D., and Morgan, A. (1990). The control of free arachidonic acid levels. *Trends Biochem. Sci.* **15**, 365–366.

Byerly, L., and Suen, Y. (1989). Characterization of proton currents in neurones of the snail, *Lymnae stagnalis. J. Physiol. (Lond.)* **413**, 75–89.

Calonge, M. L., and Ilundáin, A. A. (1996). PKC activators stimulate H^+ conductance in chicken enterocytes. *Pflügers Arch.* **431**, 594–598.

Chang, D., Kushman, N. L., and Dawson, D. C. (1991). Intracellular pH regulates K^+ and Cl^- conductances in colonic epithelial cells by modulating Ca^{2+} activation. *J. Gen. Physiol.* **98**, 183–196.

Chicurel, M. E., Chen, C. S., and Ingber, D. E. (1998). Cellular control lies in the balance of forces. *Curr. Opin. Cell Biol.* **10**, 232–239.

Clark, J. D., Schievella, A. R., Nalefski, E. A., and Lin, L.-L. (1995). Cytosolic phospholipase A₂. *J. Lipid Mediators Cell Signalling* **12,** 83–117.

Coca-Prados, M., Anguita, J., Chalfant, M. L., and Civan, M. M. (1995). PKC-sensitive Cl⁻ channels associated with ciliary epithelial homologue of pICln. *Am. J. Physiol.* **268,** C572–C579.

Czaky, T. Z., and Esposito, G. (1969). Osmotic swelling of intestinal epithelial cells during active sugar transport. *Am. J. Physiol.* **217,** 753–755.

De Coursey, T. E. (1991). Hydrogen ion currents in rat alveolar epithelial cells. *Biophys. J.* **60,** 1243–1253.

De Coursey, T. E., and Cherny, V. V. (1993). Potential pH and arachidonate gate hydrogen ion currents in human neutrophils. *Biophys. J.* **65,** 1590–1598.

De Coursey, T. E., and Cherny, V. V. (1994). Voltage-activated hydrogen ion currents. *J. Membr. Biol.* **141,** 203–223.

Demaurex, N., Grinstein, S., Jaconi, M., Schlegel, W., Lew, D., and Krause, K.-H. (1993). Proton currents in HL-60 granulocytes: Regulation by membrane potential and intracellular pH. *J. Physiol. (Lond.)* **466,** 329–344.

Dhanasekaran, N., Prassad, M. V. V. S., Wadsworth, S. J., Dermott, J. M., and van Rossum, G. (1994). Protein kinase C-dependent and independent activation of Na⁺/H⁺ exchanger Gα12 class of G proteins. *J. Biol. Chem.* **269,** 11802–11806.

Dröse, S., Bindseil, K., Bowman, E. J., Siebers, A., Zeeck, A., and Altendorf, K. (1993). Inhibitory effect of modified bafilomycins and concanamycins on P and V-type adenosinetriphosphatases. *Biochemistry* **32,** 3902–3906.

Exton, J. H. (1994). Phosphatidylcholine breakdown and signal transduction. *Biochim. Biophys. Acta* **1212,** 26–42.

Foskett, J. K. (1994). The role of calcium in the control of volume regulatory transport pathways. *In* "Cellular and Molecular Physiology of Cell Volume Regulation" (K. Strange, ed.), pp. 259–278. CRC, Boca Raton, Fla.

Garcia, M. L., Knaus, H.-G., Munujos, P., Slaughter, R. S., and Kaczorowski, G. J. (1995). Charybdotoxin and its effect on potassium channels. *Am. J. Physiol.* **269,** Cl–C10.

Goslin, M., Smith, J. W., and Poymer, D. R. (1995). Characterization of a volume-sensitive Cl⁻ current in rat osteoblast-like (ROS 17/2.8) cells. *J. Physiol. (Lond.)* **485,** 671–682.

Hall, R. A., Premont, R. T., Chow, C.-W., Blitzer, J. T., Pitcher, J. A., Claing, A., Stoffel, R. H., Barak, L. S., Shenolikar, S., Weinman, E. J., Grinstein, S., and Lefkowitz, R. J. (1998). The beta2-adrenergic receptor interacts with the Na⁺/H⁺ exchanger regulatory factor to control Na⁺/H⁺ exchange. *Nature* **392,** 626–630.

Hardy, S. P., Goodfellow, R. H., Valverde, M. A., Gill, D. R., Sepulveda, F. V., and Higgins, C. F. (1995). The multidrug resistance P-glycoprotein modulates cell regulatory volume decrease. *EMBO J.* **14,** 68–75.

Hemler, M. E. (1998). Integrin associated proteins. *Curr. Opin. Cell Biol.* **10,** 578–585.

Henderson, L. M., Chappell, J. B., and Jones, O. T. (1987). The superoxide-generating NADPH oxidase of human neutrophils is electrogenic and associated with an H⁺ channel. *Biochem. J.* **246,** 325–329.

Hoffmann, E. K., and Dunham, P. B. (1995). Membrane mechanisms of intracellular signalling in cell volume regulation. *Int. Rev. Cytol.* **161,** 173–262.

Hoogerwerf, W. A., Tsao, S. C., Deruyst, O., Levine, S. A., Yun, C. H. C., Yip, J. W., Cohen, M. E., Wilson, P. T., Lazenby, A. J., Montgomery, J., Tse, C.- M., and Donowitz, M. (1996). NHE2 and NHE3 are human and rabbit ileal brush border proteins. *Am. J. Physiol.* **270,** G29–G41.

Hooley, R., Yu, C. Y., Symons, M., and Barber, D. L. (1996) G alpha 13 stimulates N⁺-H⁺ exchange through distinct Cdc42-dependent and RhoA-dependent pathways. *J. Biol. Chem.* **271,** 6152–6158.

Howe, A., Aplin, A. E., Alahari, S. K., and Juliano, R. L. (1998). Integrin signalling and cell growth control. *Curr. Opin. Cell Biol.* **10**, 220–231.

Hudson, R. L., and Schultz, S. G. (1988). Sodium-coupled glycine uptake by Ehrlich ascites tumor cells results in an increase in cell volume and plasma membrane channel activities. *Proc. Natl. Acad. Sci. U.S.A.* **85**, 279–283.

Ives, H. E. (1985). Proton/hydroxyl permeability of proximal tubule brush border vesicles. *Am. J. Physiol.* **248**, F78–F86.

Jørgensen, N. K., Christensen, S., Harbak, H., Brown, A. M., Lambert, I. H., Hoffmann, E. K., and Simonsen, L. O. (1997). On the role of calcium in the regulatory volume decrease (RVD) response in Ehrlich mouse ascites tumor cells. *J. Membr. Biol.* **157**, 281–299.

Kaibara, M., Mitarai, S., Yano, K., and Kameyama, M. (1994). Involvement of Na^+-H^+ antiporter in regulation of L-type Ca^{2+} channel current by angiotensin II in rabbit ventricular myocytes. *Circ. Res.* **75**, 1121–1125.

Káldi, K., Szászi, K., Susták, K., Kapus, A., and Ligeti, E. (1994). Lymphocytes possess an electrogenic H^+-transporting pathway in their plasma membrane. *Biochem. J.* **301**, 329–334.

Kanai, Y., Segawa, H., Miyamoto, K., Uchino, H., Takeda, E., and Endon, H. (1998). Expression cloning and characterization of a transporter for large neutral amino acids activated by the heavy chain of 4F2 antigen (CD98). *J. Biol. Chem.* **273**, 23629–23632.

Kaplan, D. L., and Boron, W. F. (1994). Long-term expression of c-H-ras stimulates Na-H and Na(H)-dependent Cl^- HCO_3 exchange in NIH-3T3 fibroblasts. *J. Biol. Chem.* **269**, 4116–4124.

Kapus, A., Romanek, R., and Grinstein, S. (1994). Arachidonic acid stimulates the plasma membrane H^+ conductance of macrophages. *J. Biol. Chem.* **269**, 4736–4745.

Kapus, A., Romanek, R., Qu, Y., Rotstein, O. D., and Grinstein, S. (1993). A pH-sensitive and voltage-dependent proton conductance in the plasma membrane of macrophages. *J. Gen. Physiol.* **102**, 729–760.

Konmi, S., Sato, R., and Aramaki, T. (1995). Activation of the plasma membrane chloride channel by protein kinase C in isolated guinea pig hepatocytes. *J. Physiol. (Lond.)* **487**, 379–394.

Krump, E., Nikitas, K., and Grinstein, S. (1997). Induction of tyrosine phosphorylation and Na^+/H^+ exchanger activation during shrinkage of human neutrophils. *J. Biol. Chem.* **272**, 17303–17311.

Kucik, D. F., Dustin, M. L., Miller, J. M., and Brown, E. J. (1996). Adhesion-activating phorbol ester increases the motility of leukocyte integrin LFA-1 in cultured lymphocytes. *J. Clin. Invest.* **97**, 2139–2144.

Lamprecht, G., Weinman, E. J., and Yun, C.-H. C. (1998). The role of NHERF and E3KARP in the cAMP-mediated inhibition of NHE3. *J. Biol. Chem.* **273**, 29972–29978.

Lang, F., Busch, G. L., Ritter, M., Völkl, H., Waldegger, S., Gulbins, E., and Häussinger, D. (1998). Functional significance of cell volume regulatory mechanisms. *Physiol. Rev.* **78**, 247–306.

Lepple-Wienhues, A., Szabo, I., Laun, T., Kaba, N. K., Gulbins, E., and Lang, F. (1998). The tyrosine kinase p56[lyk] mediates activation of swelling-induced chloride channels in lymphocytes. *J. Cell Biol.* **141**, 281–286.

Leslie, C. C. (1997). Properties and regulation of cytosolic phospholipase A_2. *J. Biol. Chem.* **272**, 16709–16712.

Li, C., Breton, S., Morrison, R., Cannon, C. L., Emma, F., Sanchez-Olea, R., Bear, C., and Strange, K. (1998). Recombinant pI_{cln} forms highly cation-selective channels when reconstituted into artificial and biological membranes. *J. Gen. Physiol.* **112**, 727–736.

Lin, L.-L., Wartmann, M., Lin, A. Y., Knopf, J. L., Seth, A., and Davis, R. J. (1993). cPLA₂ is phosphorylated and activated by MAP kinase. *Cell* **72**, 269–278.

Lin, X., and Barber, D. L. (1996). A calcineurin homologous protein inhibits GTPase-stimulated Na-H exchange. *Proc. Natl. Acad. Sci. USA* **93**, 12631–12636.

Livne, A., Grinstein, S., and Rothstein, A. (1987). Volume-regulating behavior of human platelets. *J. Cell. Physiol.* **131**, 354–363.

Livne, A., and Hoffmann, E. K. (1990). Cytoplasmic acidification and activation of Na^+/H^+ exchange during regulatory volume decrease in Ehrlich ascites tumor cells. *J. Membr. Biol.* **114**, 153–157.

Low, S. Y., Rennie, M. J., and Taylor, P. M. (1997). Signalling elements involved in amino acid transport responses to altered muscle cell volume. *FASEB J.* **11**, 1111–1117.

Low, S. Y., and Taylor, P. M. (1998). Integrin and cytoskeletal involvement in signalling cell volume changes to glutamine transport in rat skeletal muscle. *J. Physiol. (Lond.)* **512.2**, 481–485.

Low, S. Y., Taylor, P. M., and Rennie, M. J. (1996). Responses to glutamine transport in cultured rat skeletal muscle to osmotically induced changes in cell volume. *J. Physiol. (Lond.)* **492.3**, 877–885.

Lukacs, G. L., Kapus, A., Nanda, A., Romanek, R., and Grinstein, S. (1993). Proton conductance of the plasma membrane: Properties, regulation, and functional role. *Am. J. Physiol.* **265**, C3–C14.

Mackay, D. J. G., Esch, F., Furthmayr, H., and Hall, A. (1997). Rho- and Rac-dependent assembly of focal adhesion complexes and actin filaments in permeabilized fibroblasts: An essential role for ezrin/radixin/moesin proteins. *J. Cell Biol.* **138**, 927–938.

MacLeod, R. J. (1994). How an epithelial cell swells is a determinant of the signalling pathways that activate RVD. *In* "Cellular and Molecular Physiology of Cell Volume Regulation" (K. Strange, ed.), pp.191–200, CRC, Boca Raton, FL.

MacLeod, R. J. (1996). Signal transduction of volume regulation in villus epithelial cells. Ph.D. thesis. McGill University, Montreal, Canada.

MacLeod, R. J. and Hamilton, J. R. (1991a). Volume regulation initiated by Na^+-nutrient cotransport in isolated mammalian villus enterocytes. *Am. J. Physiol.* **260**, G26–G33.

MacLeod, R. J., and Hamilton, J. R. (1991b). Separate K^+ and Cl^- transport pathways are activated for regulatory volume decrease in jejunal villus cells. *Am. J. Physiol.* **260**, G405–G415.

MacLeod, R. J., and Hamilton, J. R. (1996). Activation of Na^+/H^+ exchange is required for regulatory volume decrease after modest physiological volume increases in jejunal villus epithelial cells. *J. Biol. Chem.* **271**, 23138–23145.

MacLeod, R. J., and Hamilton, J. R. (1999a). Increases in intracellular pH and Ca^{2+} are essential for K^+ channel activation after modest 'physiological' swelling in villus epithelial cells. *J. Membr. Biol.* **172**, 47–58.

MacLeod, R. J., and Hamilton, J. R. (1999b). Ca^{2+}/calmodulin kinase II and decreases in intracellular pH are required to activate K^+ channels after substantial swelling in villus epithelial cells. *J. Membr. Biol.* **172**, 59–66.

MacLeod, R. J., Lembessis, P., and Hamilton, J. R. (1992a). Effect of protein kinase C inhibitors on Cl^- conductance required for volume regulation after L-alanine cotransport. *Am. J. Physiol.* **262**, C950–C955.

MacLeod, R. J., Lembessis, P., and Hamilton, J. R. (1992b). Differences in Ca^{2+}-mediation of hypotonic and Na^+-nutrient regulatory volume decrease in suspensions of jejunal enterocytes. *J. Membr. Biol.* **130**, 23–31.

MacLeod, R. J., Lembessis, P., James, S., and Bennett, H. P. J. (1998). Isolation of a member of the neurotoxin/cytotoxin peptide family from *Xenopus laevis* skin which activates dihydropyridine-sensitive Ca^{2+} channels in mammalian epithelial cells. *J. Biol. Chem.* **273**, 20046–20051.

Madara, J. L. (1998). Regulation of the movement of solutes across tight junctions. *Ann. Rev. Physiol.* **60**, 143–159.

Mahant-Smith, M. P. (1989). The effect of zinc on calcium and hydrogen ion currents in intact snail neurones. *J. Exp. Biol.* **145**, 455–464.

McCarty, N. A., and O'Neil, R. G. (1992). Calcium signalling in cell volume regulation. *Physiol. Rev.* **72**, 1037–1061.

Meech, R. W., and Thomas, R. C. (1987). Voltage-dependent intracellular pH in the *Helix aspersa* neurones. *J. Physiol. (Lond.)* **390**, 433–452.

Meinild, A.-K., Klaerke, D. A., Loo, D. D. F., Wright, E. M., and Zeuthen, T. (1998). The human Na$^+$-glucose cotransporter is a molecular water pump. *J. Physiol. (Lond.)* **508.1**, 15–21.

Minton, A. P. (1994). Influence of macromolecular crowding on intracellular association reactions: Possible role in volume regulation. *In* "Cellular and Molecular Physiology of Cell Volume Regulation" (K. Strange, ed.), pp. 181–190. CRC, Boca Raton, FL.

Miyamoto, S., Teramoto, H., Coso, O. A., Gutkind, J. S., Burbelo, P. D., Akiyama, S. K., and Yamada, K. M. (1995). Integrin function: Molecular hierarchies of cytoskeletal and signalling molecules. *J. Cell Biol.* **131**, 795–805.

Musch, M. W., and Goldstein, L. (1990). Hypotonicity stimulates phosphatidyl choline hydrolysis and generates diacylglycerol in erythrocytes. *J. Biol. Chem.* **365**, 13055–13059.

Nanda, A., and Grinstein, S. (1991). Protein kinase C activates an H+ (equivalent) conductance in the plasma membrane of human neutrophils. *Proc. Natl. Acad. Sci. USA* **83**, 10816–10820.

Nilius, B., Oike, M., Zahradnik, I., and Droogmans, G. (1994). Activation of a Cl$^-$ current by hypotonic volume increase in human endothelial cells. *J. Gen. Physiol.* **103**, 787–805.

Orlowski, J. (1993). Heterologous expression and functional properties of amiloride high affinity (NHE-1) and low affinity (NHE-3) isoforms of rat Na$^+$/H$^+$ exchanger. *J. Biol. Chem.* **268**, 16369–16377.

Orlowski, J., and Grinstein, S. (1997). Na$^+$/H$^+$ exchangers of mammalian cells. *J. Biol. Chem.* **272**, 22373–22376.

Peral, M., and Ilundáin, A. A. (1995). Proton conductance and intracellular pH recovery from an acid load in chicken enterocytes. *J. Physiol. (Lond.)* **484**, 165–172.

Reenstra, W. W., Warnock, D. G., Yee, V. J., and Forte, J. G. (1981). Proton gradients in renal cortex brush-border membrane vesicles. Demonstration of a rheogenic proton flux with acridine orange. *J. Biol. Chem.* **256**, 11663–11666.

Robson, L., and Hunter, M. (1994). Role of cell volume and protein kinase C in regulation of a Cl$^-$ conductance in single proximal tubule cells of *Rana temporaria*. *J. Physiol. (Lond.)* **480**, 1–7.

Rosette, C., and Karin, M. (1996). Ultraviolet light and osmotic stress: Activation of the JNK cascade through multiple growth factor and cytokine receptors. *Science* **274**, 1194–1197.

Schliess, F., Schreiber, R., and Haussinger, D. (1995). Activation of extracellular signal-regulated kinases Erk-1 and Erk-2 by cell swelling in H4IIE hepatoma cells. *Biochem. J.* **309**, 13–17.

Schmigol, A., Smith, R. D., Taggart, M. J., Wray, S., and Eisner, D. A. (1995). Changes of pH affect calcium currents but not outward potassium currents in rat myometrial cells. *Pflügers Arch.* **431**, 135–137.

Schultz, S. G. (1981). Homocellular regulatory mechanisms in sodium-transporting epithelia: Avoidance of extinction by "flush-through." *Am. J. Physiol.* **241**, F579–F590.

Schultz, S. G. (1989). Volume preservation: Then and now. *News Physiol. Sci.* **4**, 169–172.

Schultz, S. G., Hudson, R. L., and Lapointe, J.-Y. (1985). Electrophysiological studies of sodium cotransport in epithelia: Towards a cellular model. *Ann. N.Y. Acad. Sci.* **456**, 127–135.

Shrode, L. D., Klein, J. D., O'Neil, W. C., and Putnam, R. W. (1995). Shrinkage-induced activation of Na$^+$/H$^+$ exchange: Role of cell density and myosin light chain phosphorylation. *Am. J. Physiol.* **269**, C257–C266.

Strange, K. (1998). Molecular identity of the outwardly rectifying, swelling-activated anion channel: Time to re-evaluate pI$_{cln}$. *J. Gen. Physiol.* **111**, 617–622.

Strange, K., Emma, F., and Jackson, P. S. (1996). Cellular and molecular physiology of volume-sensitive anion channels. *Am. J. Physiol.* **270**, C711–C730.

Street, I. P., Lin, H. K., Laliberté, F., Ghomashi, F., Wang, Z., Perrier, H., Tremblay, N. M., Huang, Z., Weech, P. K., and Gelb, M. H. (1993). Slow- and tight-binding inhibitors of the 850-kDa human phospholipase A$_2$. *Biochemistry* **32**, 5935–5940.

Summers, J. C., Trais, L., Lajvardi, R., Horgan, D., Buechler, R., Chang, H., Peña-Rasgado, C., and Rasgado-Flores, H. (1997). Role of concentration and size of intracellular macromolecules in cell volume regulation. *Am. J. Physiol.* **273**, C360–C370.

Susztak, K., Mócsai, A., Ligeti, E., and Kapus, A. (1997). Electrogenic H+ pathway contributes to stimulus-induced changes of internal pH and membrane potential in intact neutrophils: Role of cytoplasmic phospholipase A_2. *Biochem. J.* **325,** 501–510.

Szücs, G., Heinke, S., Droogmans, G., and Nilius, B. (1996). Activation of the volume-sensitive Cl^- current in vascular endothelial cells requires a permissive intracellular Ca^{2+} concentration. *Pflügers Arch.* **431,** 467–469.

Thoroed, S. M., Lauritzen, L., Lambert, I. H., Hausen, H. S., and Hoffmann, E. K. (1997). Cell swelling activates phospholipase A_2 in Ehrlich ascites tumor cells. *J. Membr. Biol.* **160,** 47–58.

Tilly, B., Gaestel, M., Engel, K., Edixhoven, M. J., and deJonge, H. R. (1996). Hypotonic swelling activates the p38 MAP kinase signalling cascade. *FEBS Lett.* **395,** 133–136.

Tomimaga, T., Ishizaki, T., Narumiya, S., and Barber, D. L. (1998). p160ROCK mediates RhoA activation of Na-H exchange. *EMBO J.* **17,** 4712–4722.

Trimblee, L. A., Street, I. P., Perrier, H., Tremblay, N. M., Weech, P. K., and Bornstein, A. (1993). NMR structural studies of the tight complex between a trifluromethyl ketone inhibitor and 85-kDa human phospholipase A_2. *Biochemistry* **32,** 12560–12565.

Turner, J. R., and Madara, J. L. (1995). Physiological regulation of intestinal epithelial tight junctions as a consequence of Na^+-coupled nutrient transport. *Gastroenterology* **109,** 1391–1396.

Turner, J. R., Rill, B. K., Carlson, S. L., Carnes, D., Kerner, R., Misny, R. J., and Madara J. L. (1997). Physiological regulation of epithelial tight junctions is associated with myosin light-chain phosphorylation. *Am. J. Physiol.* **273,** C1378–C1385.

Verdon, B., Winpenny, J. P., Whitefield, B. E., Argent, B. B., and Gray, M. A. (1995). Volume-activated chloride currents in pancreatic duct cells. *J. Membr. Biol.* **147,** 173–183.

Voets, T., Manolopoulos, V., Eggermont, J., Ellory, C., Droogmans, G., and Nilius, B. (1998). Regulation of swelling-activated chloride current in bovine endothelium by protein tyrosine phosphorylation and G proteins. *J. Physiol.* **506.2,** 341–352.

Wormmeester, L., Sanchez de Medina, F., Kokke, F., Tse, C.-M., Khurana, S., Bowser, J., Cohen, M. E., and Donowitz, M. (1998). Quantitative contribution of NHE2 and NHE3 to rabbit ileal brush-border Na^+/H^+ exchange. *Am. J. Physiol.* **274,** C1261–C1272.

Yamakage, M., Lindeman, K., Hirshman, C., and Croxton, T. (1995). Intracellular pH regulates voltage-dependent Ca^{2+} channels in porcine tracheal smooth muscle cells. *Am. J. Physiol.* **268,** L642–L646.

Yauch, R. L., Felsenfeld, D. P., Kraeft, S.-K., Chen, L. B., Sheetz, M. P., and Hemler, M. E. (1997). Mutational evidence for control of cell adhesion through integrin diffusion/clustering, independent of ligand binding. *J. Exp. Med.* **186,** 1347–1355.

Yu, F. H., Shull, G. E., and Orlowski, J. (1993). Functional properties of the rat Na^+/H^+ exchanger NHE-2 isoform expressed in Na/H exchanger-deficient Chinese hamster ovary cells. *J. Biol. Chem.* **268,** 25536–25541.

CHAPTER 2

Intestinal Absorption of Water-Soluble Vitamins: Cellular and Molecular Aspects

Hamid M. Said,* Richard Rose,† and Bellur Seetharam**

*University of California–School of Medicine, Irvine, California, and VA Medical Center, Long Beach, California 90822; †Chicago Medical School, North Chicago, Illinois 60064; **Medical College of Wisconsin, Milwaukee, Wisconsin 53226

I. Introduction
II. Ascorbic Acid
    A. Mechanisms of Intestinal Ascorbic Acid Absorption
    B. Mechanism of Intestinal Dehydro-Ascorbic Acid (DHAA) Transport
III. Biotin
    A. Digestion of Dietary Biotin
    B. Mechanisms of Biotin Absorption in the Small Intestine
    C. Regulation of the Intestinal Absorption Process of Biotin
    D. Absorption of Bacterially Synthesized Biotin in the Large Intestine
IV. Cobalamin
    A. Hydrophobic Protein Ligands of Cbl (IF and TC II)
    B. Structure of IF and TCII
    C. IF and TCII Receptors
    D. Receptor-Mediated Endocytosis of Cbl
    E. Inherited Disorders of Cbl Transport
V. Folate
    A. Digestion of Dietary Folates
    B. Mechanisms of Intestinal Absorption of Folate Monoglutamates
    C. Regulation of Intestinal Folate Absorption
    D. Absorption of Bacterially Synthesized Folate in the Large Intestine
VI. Pantothenic Acid
VII. Riboflavin
VIII. Thiamine
IX. Vitamin B_6
X. Concluding Remarks
    References

I. INTRODUCTION

The water-soluble vitamins represent a structurally and functionally diverse set of organic molecules that are essential for human health because they play a crucial role in maintaining the metabolic, energy, differentiation, and proliferation status of cells. Humans and other higher mammals have lost their ability to synthesize these compounds and thus must obtain them from the diet and other exogenous sources. The cellular assimilation and function of these compounds are, to a large extent, dependent on their normal absorption via the gastrointestinal tract. A deficiency of any of these vitamins can occur from a variety of causes that affect intestinal uptake and trans-epithelial transport into the circulation. These include both inherited disorders of absorption and transport and secondary causes such as intestinal disease, drug interactions, excessive alcohol intake, and surgery. The focus of this chapter is to update our recent understanding of the molecular and cellular aspects of absorption and regulation of water-soluble vitamins. The focus is also to discuss the available evidence for the existence of carrier systems involved in the transport of vitamins synthesized locally by normal microflora in the large intestine for the nutriture of colonocytes. It is hoped that the chapter will stimulate and foster further research on the regulation, evolutionary relationship, structure, and cellular function of many of these fascinating vitamin transporters.

II. ASCORBIC ACID

Ascorbic acid (vitamin C) is required by all living organisms; it acts as an antioxidant and serves in a variety of hydroxylation reactions. Among mammals, only primates and guinea pigs lack the ability to synthesize ascorbic acid from glucose, and therefore have a dietary requirement for it. Without a dietary source of vitamin C, a clinical state of scurvy is reached within a few weeks or months. Most studies on intestinal absorption of this nutrient have been carried out using these animal species that have a specific dietary requirement.

A. Mechanisms of Intestinal Ascorbic Acid Absorption

Ascorbic acid, a water-soluble compound, exists as an anion ($pK_a = 4.2$) in the aqueous environment of intestinal chyme. Compounds with these characteristics generally do not cross the intestinal epithelium rapidly by simple diffusion. Intestinal absorption of ascorbic acid has been examined by Stevenson and Bush (1969) using rings of guinea pig ileum. Ascorbic acid was found to be taken up rapidly when the rings were incubated at 37°C in a physiological buffer. Ninety-four percent of the transported ascorbic acid was found to be in the original reduced state and was free to diffuse from the tissue back to the media. Because the tissue ascorbate concentration was higher than the media level, a form of active transport was suggested. In contrast to the ileum, proximal portions of the small intestine had much less accumulation, whereas no accumulation was seen in the stomach or the cecum. Absorption of ascorbic acid also was evaluated by stripping the epithelial layer from the underlying serosa and musculature and incubating the mucosa in media that contained a radioactive marker of the vitamin. After determining the water content of the tissue (wet weight minus dry weight) and the vitamin content, a comparison was made of the vitamin concentration on the inside and the outside of enterocytes. In considering the mechanism of absorption of a charged molecule, the negative intracellular electrical potential (-40 mV) of the absorptive cell must be taken into account (Mellors et al., 1977). From these data, a tentative conclusion can be made as to whether the substrate is actively transported into the tissue. In a study on muscle-free strips of guinea pig ileal mucosa, it was established that cellular accumulation reached fivefold the buffer level of the vitamin (Patterson et al., 1982).

In another useful preparation, the intestine was cut open along the mesenteric border and mounted as a flat sheet in chambers that expose known amounts of mucosal surface to a physiological buffer. The kinetics of unidirectional substrate uptake across the brush border into absorptive cells was conveniently determined. The kinetics of ascorbic acid uptake were evaluated in sections of ileum (Mellors et al., 1977). In Na^+ free buffer, the uptake was found to be slow and linear as a function of ascorbic acid concentration in the incubation medium. In physiological buffer, however, influx was saturable as a function of substrate concentration.

The mechanism of ascorbic acid transport was further evaluated with the use of brush border and basolateral membrane vesicles (BBMV and BLMV, respectively). When incubated in media representing the extracellular environment of the cell, these vesicles were found to be a good model to study the kinetics and other features of enterocyte transport independent of cellular metabolic events. These studies have confirmed that ascorbic acid crosses the brush border membrane in conjunction with Na^+ via a carrier-mediated process (Bianchi et al., 1986), probably in a one-to-one cotransport process that is electroneutral (Siliprandi et al., 1979). Transfer across the basolateral cell membrane also was found to be via a carrier-mediated system, but the process was Na^+-independent.

Our understanding of the molecular mechanism of ascorbic acid transport has been significantly advanced since the recent cloning of a family of Na^+-dependent L-ascorbic acid transporters, the so-called sodium-dependent vitamin C transporter 1 and 2 (SVCT1 and SVCT2, respectively) from rat tissues (Tsukaguchi et al., 1999). SVCT1 was found to encode a protein of 604 amino acids, whereas SVCT2 encodes a protein of 592 amino acids. Both transporter proteins were found to have a similar predicted hydropathy profile, with 12 putative membrane spanning domains and multiple potential sites for N-linked glycosylation, protein kinase C (PKC), and protein kinase A (PKA) phosphorylation. SVCT1 was found to share 65% identity with SMVT2 at the amino acid level; however, no homology was found between these two transporters and any other known mammalian membrane transporter cloned thus far. Both SVCT1 and SVCT2 were found to have higher selectivity for L-ascorbic acid compared with D-isoascorbic acid and dehydroascorbic acid, and both transport L-ascorbic acid via an electrogenic Na^+-dependent process (the stoichiometric ratio of Na^+ to ascorbic acid was reported to be 2:1). An interesting observation was the ability of these membrane carriers to also act as Na^+ uniporters in the absence of their substrate (ascorbate), allowing Na^+ leakage to take place across the cell membrane. With respect to the distribution of SVCT1 and SVCT2 in different rat tissues, striking differences were observed. In the small intestine, both SVCT1 and SVCT2 were found to be expressed, with the former being the predominantly expressed form in enterocytes, as shown by Northern blot analysis and hybrid depletion (with antisense oligonucleotide against SVCT1) of poly $(A)^+$ RNA isolated from rat small intestine following micro-injection into *Xenopus* oocytes.

B. Mechanism of Intestinal Dehydro-Ascorbic Acid (DHAA) Transport

The oxidized form of vitamin C, DHAA, is of interest in part because it is speculated that the ratio of the reduced/oxidized levels of various antioxidants in the body might be an indicator or a determinant of a person's state of health (Rose and Bode, 1993). This is based on the recognition that DHAA is chemically very different from ascorbic acid, and has toxicity similar to alloxan, which has long been used by endocrine investigators to induce diabetes in laboratory animals. The amount of DHAA in the human diet is not fully documented, but it is likely to increase during storage of food. Thus, a more complete study of intestinal vitamin C absorption must incorporate information on metabolism of the nutrient. The enterocyte is characteristic of several cell types that take up DHAA and metabolize it to the reduced form (Bianchi *et al.*, 1986; Choi and Rose, 1989). An evaluation was made in the small and large intestinal mucosa of guinea pigs and rats. Enterocytes were homogenized, and the supernatant was fractionated by ammonium sulfate precipitation. The dialyzed 55–70% saturated fraction was processed through Sephadex G-100 and evaluated for DHAA-reductase activity, as reported in detail for other tissues (Rose *et al.*, 1988; Schell and Bode, 1993). A high-molecular-weight component that uses NADPH, glutathione, or other reducing equivalents to bring about regeneration of reduced ascorbic acid was present. It was concluded that a soluble intestinal enzyme aids to maintain the intracellular concentration of DHAA at a low level.

Studies on the cellular uptake of DHAA have shown that the enterocyte takes up DHAA across the brush border membrane by a Na^+-independent transport process (Bianchi *et al.*, 1986). Substantial uptake of DHAA was also shown to occur across the serosal surface of the intestinal tissue (Rose and Choi, 1990). Serosal uptake might occur in exchange for the reduced form of vitamin C as it leaves the cell. The combination of DHAA uptake followed by intracellular enzymatic reduction possibly serves to clear DHAA from both dietary sources and the blood, while also maintaining a low intracellular level. In more recent studies, Vera *et al.* (1993) have shown the mammalian facilitative hexose transporters to be mediators of DHAA cellular uptake. Using the *Xenopus laevis* oocyte expression system, these investigators showed that the mammalian hexose transporters expressed in these cells were efficient transporters of the oxidized form of vitamin C, and that they represented a physiologically significant pathway for the uptake and accumulation of vitamin C by cells.

Regulation of whole-body vitamin C content is worthy of brief mention in view of wide public interest in micronutrient supplementation and interest of scientists in homeostatic control mechanisms. Ascorbic acid supplementation in guinea pigs was shown by Rose and Nahrwold (1978) to result in the downregulation of intestinal absorption capacity of vitamin C. This finding was subsequently confirmed, and found to be applicable also to vitamin administration by intramuscular injection (Karasov *et al.*, 1991). This concept helps to substantiate the very incompletely evaluated concept that "vitamin homeostasis" might apply on a whole-body basis in mammals.

III. BIOTIN

In mammalian cells, biotin acts as a coenzyme for four carboxylases essential for the metabolism of several branched-chain amino acids, gluconeogenesis, and fatty acid synthesis. These cells cannot synthesize biotin; rather, the vitamin is obtained from exogenous sources via absorption in the intestine. The intestine is exposed to two sources of biotin: (1) dietary and (2) bacterial, wherein the vitamin is synthesized by the normal microflora of the large intestine. Following is a discussion of our current understanding of absorption of biotin from these two sources.

A. Digestion of Dietary Biotin

Dietary biotin exists in free and protein-bound forms (Lampen *et al.*, 1942). The latter is digested by gastrointestinal proteases and peptidases not to generate free biotin but to generate biocytin (*N*-biotinyl-L-lysine) and biotin-containing short peptides (Wolf *et al.*, 1984). These biotin derivatives are then converted to free biotin by the enzymatic action of biotinidase prior to absorption (Wolf *et al.*, 1984). This hydrolysis step to form free biotin appears to be essential for efficient absorption and optimal bioavailability of dietary biotin (Said *et al.,* 1993c). The source of intestinal biotinidase is believed to be the pancreatic juice (Wolf *et al.,* 1984). This enzyme has been recently cloned by Cole *et al.* (1994). The cloned cDNA was shown to have an open reading frame of 1629 bases, which encodes a protein of 543 AA residues. Results of Southern blot analysis suggested that biotinidase is a product of a single-copy gene. Furthermore, mRNA transcripts corresponding to the cloned biotinidase were identified by Northern blot analysis in human pancreas, liver, heart, brain, kidney, and skeletal muscles (Cole *et al.,* 1994), suggesting the necessity to form free biotin in all tissues/cells.

B. Mechanisms of Biotin Absorption in the Small Intestine

The mechanism of absorption of free biotin in the small intestine has been studied using a variety of intact intestinal preparations (Dyer and Said, 1997; Said, 1999, and references therein). These studies have demonstrated the involvement of a specialized, carrier-mediated, Na^+-dependent system in the absorption process. Subsequent studies using purified intestinal BBMV and BLMV preparations have shown that the Na^+-dependent, carrier-mediated process is localized in the BBM domain of the absorption cells (Said *et al.,* 1987). Furthermore, it was shown that the Na^+ gradient (out > in), and not the presence of Na^+ alone, is needed to drive the transport of biotin across the BBM against a concentration gradient. The process was found to be electroneutral in nature and occurred with a $biotin^-/Na^+$ stoichiometric coupling ratio of 1:1 (Said *et al.,* 1987). Using group-specific reagents, other studies have shown the involvement of histidine residues and sulfhydryl groups in the functioning of the intestinal BBM biotin transporter (Said and Mohammedkhani, 1992). The histidine residues were suggested to be localized at (or near) the substrate binding site, whereas the sulfhydryl groups were suggested to be localized at a site(s) other than the substrate binding region. As to the exit of biotin from the enterocyte, that is, transport across the BLM domain, this process also was found to involve a carrier-mediated system. This system, however, was found to be Na^+ independent (Said *et al.,* 1988).

Regional differences in the ability of the small intestine to absorb biotin has been observed and found to vary according to the developmental stage of the intestine. In adult humans and rats, biotin uptake was found to be significantly greater in the duodenum and the jejunum than in the ileum (Said and Redha, 1988). The regional variation of biotin uptake was found to be due to differences in the V_{max}, and not the apparent K_m of the transport carrier, suggesting that the number of biotin carriers is higher in the proximal part of the small intestine than in the distal region.

Recent studies have shown that the intestinal transport system for biotin is shared by another water-soluble vitamin, pantothenic acid (Said, 1999) (see

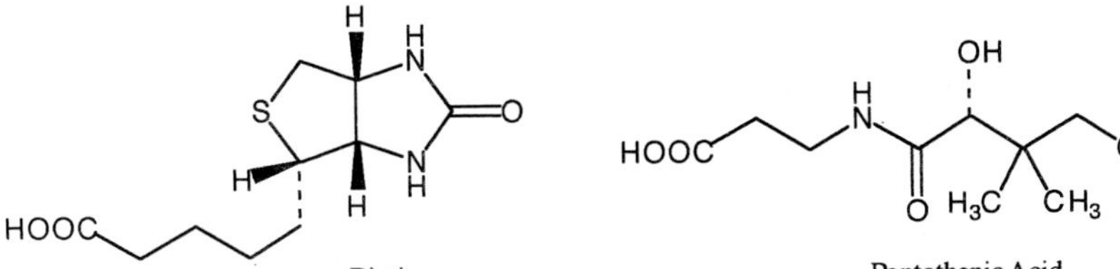

FIGURE 1   Chemical structure of biotin and pantothenic acid.

Fig. 1 for structure of these two vitamins). Studies with intestinal epithelial Caco-2 cells have shown that biotin and pantothenic acid act as competitive inhibitors of each other's transport. Furthermore, both compounds also induce trans-stimulation in the efflux of the other from preloaded Caco-2 cells (Said, 1999; unpublished observations). The physiological and nutritional significance of this interaction deserves further investigation. A similar type of interaction between these two water-soluble vitamins also has been reported for other tissues such as the colon (see later), the blood–brain barrier (Spector and Mock, 1987), the heart (Beinlich *et al.,* 1990), and the placenta (Grassl, 1992).

The molecular characteristics of the intestinal biotin absorption process of rats, humans, and rabbits have been recently delineated (Chatterjee *et al.,* 1999; Prasad *et al.,* 1999). The open reading frame of the cDNA cloned from the intestine was found to be identical to that identified for the vitamin transporter in placental tissue (the so-called sodium-dependent multivitamin transporter [SMVT]) of the corresponding species and encodes for a protein with 12 predicted transmembrane domains, multiple potential glycosylation sites, and multiple consensus sequences for phosphorylation by PKC and PKA. Studies with the rat intestinal clone (Chatterjee *et al.,* 1999) have shown the existence of significant heterogeneity in the 5′ untranslated region of the cloned cDNA, with four distinct variants (I, II, III, IV) being identified. Variant II was found to be the predominant form expressed in rat small and large intestine. The existence of multiple variants suggests possible involvement of multiple promoters in driving the transcription of the cloned biotin transporter. Functional identity of the cloned intestinal cDNAs was established by expression in COS-7 and HRPE (human retinal pigment epithelial) cell lines, which showed a significant increase in the uptake of biotin and pantothenic acid in transfected cells compared with controls. The induced biotin uptake was found to be Na^+-dependent, saturable as a function of biotin concentration with an apparent K_m similar to that of the native intestine, and inhibited by unlabeled biotin and pantothenic acid and their structural analogs. The distribution of mRNA transcripts complementary to the ORF of the cloned intestinal cDNA along the vertical (villus vs crypt) and longitudinal (duodenum, jejunum, ileum, proximal colon, and distal colon) axes of the intestinal tract also has been delineated. Expresson was 2.6-fold higher in villus cells than in crypt cells, which corresponded with the higher carrier-mediated biotin uptake found in villus cells (Chatterjee *et al.,* 1999). As to the longitudinal distribution of mRNA complementary to the ORF of the cloned intestinal cDNA, a discrepancy in expression, compared to functional biotin transport activity, was found. Although biotin transport activity is known to be higher in the proximal small intestine than in the ileum or the colon (Said, 1999, and references therein), a similar level of expression of mRNA complementary to the cloned intestinal cDNA was found along the length of the small intestine and colon (Chatterjee *et al.,* 1999; Prasad *et al.,* 1999). This paradox may suggest involvement of a cell-specific posttranslational event(s) that regulates the expression of the functional protein in the different areas of the intestinal tract. The identification of mRNA transcript complementary to the cloned intestinal cDNA in the colon corroborates the recent finding of Said *et al.,* (1998) on the existence of a functional Na^+-dependent carrier-mediated biotin uptake system in the colon that may be involved in the absorption of bacterially synthesized pantothenic acid by the normal microflora of the large intestine (see later).

C. Regulation of the Intestinal Absorption Process of Biotin

The small intestinal absorption process of biotin was found to be regulated by both the extracellular substrate level and specific intracellular protein kinase–mediated pathways (Said, 1999; Said *et al.,* 1989). Biotin deficiency in rats was found to lead to significant up-regulation in biotin intestinal uptake (compared with pair-fed controls). The up-regulation occurred in both the jejunum and the ileum, and was specific for biotin. On the other hand, oversupplementation of rats with pharmacological amounts of biotin was found to lead to a specific and significant down-regulation of biotin uptake. These adaptive changes in the intestinal uptake of biotin by substrate level were found to be mediated through changes in the V_{max} of the uptake process, with minimal changes in the apparent K_m. This suggests that up-regulation is mediated via changes in the number/activity of the transporter, but not its affinity. Further studies are needed to determine the molecular basis of these effects, that is, whether they involve transcription and/or posttranscription mechanisms.

The intestinal uptake of biotin has been found to be under the regulation of intracellular protein kinase C and Ca^{2+}/calmodulin-mediated pathways (Said, 1999). Using specific modulators of these pathways, it has been shown that activation of PKC leads to significant inhibition in biotin uptake by the human-derived intestinal epithelial Caco-2 cells, while inhibition of PKC leads to a stimulation in vitamin uptake. This regulatory effect was found to be mediated via a decrease in the V_{max}, but not in the apparent K_m, of the biotin uptake process, suggesting a decrease in the activity/number of biotin carriers with no changes in the system affinity. It is interesting to note that the cloned intestinal biotin transporter contains two consensus sequences for PKC phosphorylation (Chatterjee *et al.*, 1999; Prasad *et al.*, 1999). Similarly, inhibiting the Ca^{2+}/calmodulin-mediated pathway was found to lead to significant inhibition in biotin uptake; this effect was again found to be mediated via an inhibition in the V_{max} with slight changes in the apparent K_m. Although both PKC and Ca^{2+}/calmodulin-mediated pathways appear to exert their effects on the V_{max} of biotin uptake process, they appear to utilize different mechanisms. This is suggested by the observation that simultaneous activation of PKC and inhibition of Ca^{2+}/calmodulin-mediated pathways leads to an additive inhibition in biotin uptake (Said, 1999).

D. Absorption of Bacterially Synthesized Biotin in the Large Intestine

It has been recognized for some time that the normal microflora of the large intestine synthesizes a considerable amount of biotin and that a substantial portion of this biotin exists in the free unbound form, that is, is available for absorption (Burkholder and McVeigh, 1942; Wrong *et al.*, 1981). Furthermore, *in vivo* studies in humans, rats, and minipigs have shown that the colon is capable of absorbing significant amounts of luminally introduced biotin (Barth *et al.*, 1986; Brown and Rosenberg, 1987; Sorrell *et al.*, 1987). The mechanism of biotin transport in the colon has been recently characterized with use of the human-derived, nontransformed colonic epithelial cell line NCM460 (Said *et al.*, 1998). The results showed the involvement of a specialized, carrier-mediated, Na^+-dependent uptake mechanism that again is shared with the vitamin pantothenic acid; this system also was found to be under the regulation of a PKC-mediated pathway. These findings are similar to those in the small intestine and suggest that a common mechanism may be involved in the transport of biotin in the small and large intestinal epithelia.

IV. COBALAMIN

All mammalian cells require cobalamin (Cbl; vitamin B_{12}), which they use in its coenzyme forms (methyl-Cbl and 5′-deoxyadenosyl Cbl) for the enzymatic conversion of homocysteine to methionine by the enzyme methionine synthase, and conversion of methyl malonylCoA to succinyl CoA by the enzyme methyl malonyl CoA mutase. However, Cbl is a highly polar, water-soluble molecule, and thus is practically impervious to mammalian plasma membranes. Its transport across cellular plasma membranes occurs bound to two hydrophobic protein ligands, gastric intrinsic factor (IF), and plasma transcobalamin II (TC II). Cbl bound to these two proteins is endocytosed via distinct cell surface receptors, the intrinsic factor–cobalamin receptor (IFCR), and the transcobalamin II–receptor (TC II-R).

Intestinal epithelial cells are able to mediate transepithelial transport of Cbl when internalized bound to IF from the apical domain, but retain and utilize it as Cbl coenzymes when internalized from the basolateral side bound to TC II. This property of cultured intestinal epithelial cells is reminiscent of the *in vivo* situation in which the enterocytes transport exogenous Cbl (diet) presented to the lumen bound to IF, to provide Cbl to other tissues of the body and to derive Cbl for their own use from endogenous sources (circulation) bound to TC II. In this section, we will discuss recent studies on the molecular and cellular aspects of Cbl transport proteins, and the Cbl-sorting pathways in polarized epithelial cells.

A. Hydrophobic Protein Ligands of Cbl (IF and TC II)

IF is a secretory glycoprotein secreted in a regulated manner. In humans, IF is localized mainly to the parietal cells, but it also can be detected at the margins of the anatomic regions in clusters of chief cells, enteroendocrine cells, and endothelial cells (Howard *et al.*, 1996). In contrast, in rats, IF is mainly localized to the chief cells, but a small percentage of parietal cells also stain for IF (Shao *et al.*, 1998). These ultrastructural observations have suggested that IF synthesis could occur in more than one gastric cell type in any given species. This hypothesis was tested using a fusion gene consisting of a nontranscribed mouse intrinsic factor gene and human growth hormone gene (IF-1029 to +55/hGH+3). In mice, the transgene was expressed in the parietal but not in the chief cells (Lorenz and Gordon, 1993), the site of IF expression in mice stomach. In contrast to humans, rats, and mice, the IF gene in dogs is transcribed by the pancreatic duct cells (Simpson *et al.*, 1993), and IF activity has been reported in the opossum pancreas (Ramanujam *et al.*, 1993) indicating that the IF gene can be transcribed in more than one tissue across species. These studies indicate that the synthesis of IF is controlled by a complex set of *cis-trans* interactions, and further studies are needed to understand how these interactions regulate cell- and tissue-specific expression of the IF gene. Such studies may help to further understand the molecular mechanism of the increases in the steady-state IF mRNA levels noted in rat stomach by cortisone (Diekgraefe *et al.*, 1988) and growth hormone (Lobie *et al.*, 1992).

In contrast to IF, TC II is secreted from many cell types in a constitutive manner (Said *et al.*, 1993c, and references therein). TC II mRNA is expressed in many human and rat tissues, but in a tissue-specific manner (Li *et al.*, 1994c). In humans, TC II mRNA levels are highest in the kidney and could arise from a mixed population of cells, mainly endothelial and epithelial cells. Human TC II gene promoter is positively regulated by a distal GC box and negatively regulated by a proximal GC/GT box (Li *et al.*, 1998) when transfected in human epithelial and leukemic cells. Both these elements were bound by transcription factors Sp1 and Sp3, and cotransfection studies using Sp1 and Sp3 expression plasmids have shown that although Sp1 activated the weak promoter activity of the TATA-less TC II gene, Sp3 suppressed Sp1-mediated transactivation of TC II transcription. These studies have suggested that TC II gene expression in human tissues/cells may be controlled by the relative ratios of transcription factors Sp1 and Sp3 that bind to the GC/GT box and that the weak promoter activity is due to transcriptional suppression caused by the binding of Sp3 to the proximal GC/GT box.

B. Structure of IF and TCII

Both IF and TC II are single polypeptides with a molecular mass of approximately 43 kDa and contain one Cbl binding site/mole. Although mature processed IF is a glycoprotein of molecular mass of 50 kDa, TC II is not a glycoprotein. The carbohydrate on IF has no role in the binding of Cbl or its receptor, but it may affect its folding to protect it from proteolytic degradation (Gordon *et al.*, 1991). This is not surprising considering the ability of IF to remain structurally and functionally intact even after its exposure to proteases of the stomach and intestinal lumen.

Studies using monoclonal antibodies to IF and TC II and patients' plasma have established that both IF and TC II have two separate binding sites for binding Cbl and their respective cell surface receptors (Seetharam, 1994, and references therein). However, these sites are not characterized completely. The receptor binding site of rat and human IF has been localized to residues 25–62 at the NH_2 terminus (Tang *et al.*, 1992). Within this short stretch, the minimum length of the receptor binding is limited to consecutive residues 25–44. Although this site in IF makes the initial contact with IFCR, evidence exists (Wen *et al.*, 1997) that additional sites on IF may interact with IFCR. It is not known whether the receptor binding of TC II also is localized to the amino terminus.

Similar high-affinity Cbl binding, a property common to both IF and TC II, suggests that Cbl binding by these two proteins must share some common structural features. Sequence alignment (Li *et al.*, 1993) of mammalian Cbl binding proteins has shown some interesting features of their structure that may help explain how Cbl binding to these proteins can occur. Essential features of these proteins are that they all contain six high homology (60–80%) hydrophobic regions, which are flanked on both sides by nonconserved hydrophilic regions (Fig. 2, regions I–VI). These conserved hydrophobic regions contain 29 of the 44 residues (Fig. 2, asterisks) that are identical in these proteins, including four cysteines, and many of these regions, especially regions VI and V, are rich in residues with small side chain (Ala, Gly) and OH-bearing residues (Ser, Thr, Tyr). Based on these sequence analyses, a new model for Cbl binding by either IF or TC II can be proposed. Essential features of this modified model (Fig. 2), which is based on a model proposed earlier by Grasbeck (1969), are as follows: (1) some or all of the conserved hydrophobic regions of IF and TC II form the interior of a pit, and the nonconserved hydrophilic regions form the exterior (surface); (2) the hydrophobic pit may be conformationally stabilized by one or two S—S bonds, and the presence of four Cys residues in identical position of these proteins support this suggestion; (3) Cbl enters the pit with its nucleotide portion facing in and its axial ligand facing out;

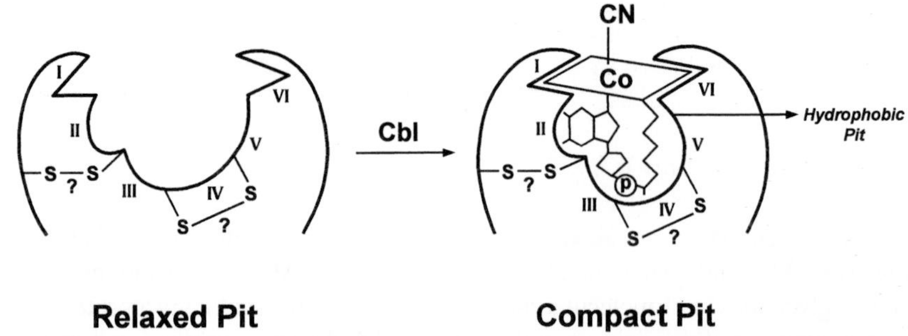

Relaxed Pit   **Compact Pit**

FIGURE 2 A model for the binding of Cbl by mammalian Cbl-binding proteins. Illustration of the Cbl-binding pit before and after Cbl binding. Dimensions of the pit may be defined by one or two disulfide bonds formed by four cysteine residues that are conserved in all Cbl-binding proteins.

and (4) the occurrence of small (Ala/Gly) and OH-bearing (Ser, Thr, Tyr) residues in some of the highly conserved hydrophobic regions lining the interior of the pit may aid in the proper positioning of the nucleotide moiety and in developing hydrogen bonding with the amide groups of the corrin ring. As the nucleotide moiety enters the pit, the nonconserved hydrophilic regions, which may be important in receptor recognition, might be exposed later after Cbl binding. This model proposed for Cbl binding by IF and TC II explains several of the hydrodynamic properties noted earlier for Cbl binding by these proteins (Seetharam, 1994). However, further *in vitro* mutagenesis, peptide antibody studies, and x-ray analysis of these proteins are needed to confirm the validity of this model.

Both IF and TC II share similarities at the genomic level despite their different chromosomal location, that is, at 11 (Hewitt *et al.,* 1991) and 22 (Li *et al.,* 1995), respectively. A comparison of the exon size and the boundaries of IF and TC II gene has shown similar-sized genes (about 20 kb) containing an identical number of exons of about the same size in their coding region. Of the eight exon boundaries, four are conserved. Interestingly, five of the six conserved hydrophobic regions (Fig. 3) implicated in Cbl binding are localized to different exons, suggesting that the Cbl binding property of these proteins evolved earlier through gene duplication from a common ancestral gene and that their different receptor binding regions evolved later.

C. IF and TCII Receptors

Consistent with its physiological role in the absorption of dietary and biliary Cbl transport, IFCR is expressed in the distal intestine of all mammals. In addition to the intestinal mucosa, high levels of IFCR also have been detected in the

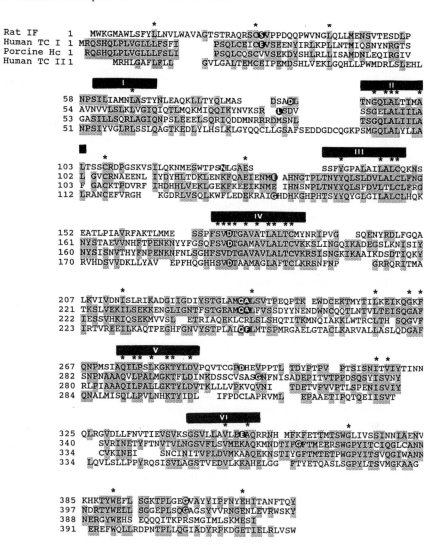

FIGURE 3   Amino acid alignment of mammalian Cbl binding proteins. The amino acid sequences of rat IF, human TC II, and haptocorrins (human TC I and porcine Hc) were aligned using the pileUp program of the GCG system. Bars (I–VI) indicate the higher homology hydrophobic regions, asterisks indicate the identical residues.

mammalian kidney, including that of humans (Schron *et al.,* 1988), and its synthesis has been studied in the rat kidney (Seetharam *et al.,* 1992) and yolk sac (Ramanujam *et al.,* 1993a). The physiological role of renal IFCR in Cbl transport, or of rat yolk sac IFCR in Cbl delivery to the fetus, if any, is not clear since IF is present only in very minuscule amounts in the circulation. Evidence that renal and yolk sac IFCR may have a role in the endocytosis of proteins has come from the recent demonstration that IFCR expressed in these tissues is identical to a protein known as gp280 (Seetharam *et al.*, 1997). gp280 identified in the rat kidney and yolk sac is a teratogenic antigen localized mainly in the clathrin-coated areas of the proximal tubule and the yolk sac epithelia (Sahali *et al.*, 1998).

At present there is no direct evidence that the teratogenic activity of the IFCR/gp280 in pregnant rats is caused by an antibody-mediated inhibition of Cbl delivery to the fetus, or that renal IFCR plays a major role in Cbl transport to the kidney. However, there is evidence to suggest that IFCR/gp280 plays a role as a scavenger receptor in the apical endocytic pathway (Christensen *et al.*, 1998, and references therein). In the intestine, a similar role for IFCR also may exist because ileum is a site for high endocytic activity.

Evidence that intestinal, renal, and yolk sac IFCR/gp280 are products of the same gene has come from two sets of studies. One involved simultaneous depletion of IFCR activity and protein levels in both renal and intestinal apical brush border membranes isolated using harvested tissues from a family of dogs with selective intestinal malabsorption of Cbl (Fyfe *et al.*, 1991b). In the second, a cDNA clone isolated from rat yolk sac carcinoma BN/MSV cells identified a single transcript of 11.6 kb in the intestine, kidney, and yolk sac—a transcript size sufficient to encode a protein of 460 kDa (Moestrup *et al.*, 1998). By immunoblotting of these tissue membranes, IFCR/gp280 antisera identified a single band of 460 kDa (Moestrup *et al.*, 1998), a size similar to that determined for pure rat renal IFCR (457 kDa) by quantitative amino acid analysis (Moestrup *et al.*, 1998). The earlier estimates of the size of IFCR/gp280 based on mobilities on SDS-PAGE of 230 kDa and 280 kDa for IFCR and gp280, respectively, were only an approximation.

The structure of IFCR/gp280 as revealed by its predicted cDNA sequence (Moestrup *et al.*, 1998) has shown some interesting features of this protein. It encodes a protein of 396,953 Da containing three distinct domains, but lacks a transmembrane domain. The name "cubulin" is given to this protein because 88% of its polypeptide mass of 397 kDa is due to 27 repeated domains of 110 residues each. These domains, known as CUB domains, are widespread in developmentally regulated proteins and have a characteristic hydrophobicity pattern predicted to form antiparallel B-barrels (Bork and Beckmann, 1993). It encodes a protein with a plasma membrane topology of a peripheral protein (Moestrup *et al.*, 1998) confirming earlier similar predictions (Seetharam *et al.*, 1981) for intestinal IFCR, based on the ability of various reagents to release it following its reconstitution in artificial lipid bilayers.

The site of IF-Cbl binding to IFCR, limited to a 70 to 80–kDa region (Seetharam, 1994, and references therein), is not known. It is unlikely that it is localized in the repeated CUB domains, since IFCR contains only one IF-Cbl binding site/mole. However, it cannot be ruled out since one of the CUB domains (domain 13) was different in that it contained only two cysteines instead of the usual four present in the other CUB domains (Moestrup *et al.*, 1998). Alternatively, IF-Cbl binding may occur with the EGF type-B repeats localized upstream of the CUB domains or to the unique 110 residue N terminus. Additional studies are required to confirm that cubulin actually encodes functional IFCR and to identify which of the three regions of IFCR bind to IF-Cbl.

In contrast to IFCR, TC II-R, because of its role in promoting plasma transport of Cbl to all tissues/cells, is expressed in many tissues. Immunoblot studies have shown that TC II-R protein is expressed in many human (Bose *et al.*, 1995a) and rat tissues (Bose *et al.*, 1995b) in a tissue-specific manner, being highest in the human kidney and the rat intestine and kidney. In the rat, TC II-R activity and protein levels, and tissue transport of plasma Cbl, are regulated by cortisone (Bose *et al.*, 1995b). In addition, unlike IFCR, TC II-R is not regulated during postnatal development of the intestine and kidney, and in the intestine its expression is uniform throughout the entire length of the gut (Bose *et al.*, 1995b). These studies have emphasized the importance of TC II-R in mediating plasma transport of Cbl to all tissues and cells throughout postnatal development.

IFCR/gp280 is synthesized in a number of epithelial cells (Christensen, 1998, and references therein). Domain-specific biotinylation studies using OK cells have shown that IFCR expression occurs only in the apical brush borders and that expression at this site requires an intact microtubular network (Ramanujam *et al.*, 1991a). Its posttranslational processing includes maturation of its several N-linked oligosaccharides and palmitoylation (Ramanujam *et al.*, 1994). IFCR turnover in OK cells is slow with a $T_{1/2}$ of 48 h (Ramanujam *et al.*, 1994), and its $T_{1/2}$ is reduced to 24 h when cells are treated with tunicamycin, an inhibitor of N-glycosylation during pulse-chase labeling experiments. These results have indicated that core N-glycosylation of IFCR is important for its intracellular stability.

Studies (Bose *et al.*, 1998; Bose *et al.*, 1997; Bose and Seetharam, 1997) using polarized intestinal epithelial Caco-2 cells have provided important information, particularly on the synthesis, turnover, and polarized distribution of TC II-R. At steady state, TC II-R is distributed between the basolateral and apical membranes of Caco-2 cells in the ratio of 6:1, a ratio similar to that observed using basolateral and apical membranes isolated from human intestinal mucosa (Bose *et al.*, 1997). Metabolic labeling and domain-specific biotinylation studies have shown that the targeting of TC II-R to the basolateral membranes is a rapid process (Bose and Seetharam, 1997) and is regulated by post–trans-Golgi trafficking. Disruption of the Golgi by the macrocyclic antibiotic brefeldin A (Bose *et al.*, 1998) or of disulfide bond formation of TC II-R by treating Caco-2 cells with low concentrations of N-ethylmaleamide (Bose and Seetharam, 1997) inhibited its post–trans-Golgi trafficking but not its transport from the endoplasmic reticulum to the trans-Golgi network. During its trafficking to the basolateral surface of Caco-2 cells, TC II-R is heavily glycosylated, and one of its *N*-linked oligosaccharide chains is converted to the complex type.

One interesting aspect of TC II-R trafficking in Caco-2 cells is its ability to change its physical state from that of a monomer (62 kDa) to that of a dimer (124 kDa) during trafficking from the endoplasmic reticulum to the basolateral plasma membranes (Bose *et al.*, 1998). TC II dimerization occurs only in the basolateral membranes, not in the Golgi or the trans-Golgi network, and is due to its selective interaction with cholesterol-rich (>10 mole %), highly ordered membrane domains of the basolateral membranes. Lack of TC II-R dimerization in the endoplasmic reticulum and the trans-Golgi network results from low (<10 mole %) cholesterol in these cholesterol-poor intracellular organelle membranes. *In vitro,* the physical state of TC II-R can be converted *in situ* in isolated tissue membranes, from one form to the other, by manipulating cholesterol content (Bose *et al.*, 1996a). The functional significance of dimerization is not known, but the dimers, like the monomers, are functional in ligand binding and internalization.

D. Receptor-Mediated Endocytosis of Cbl

IF-mediated Cbl endocytosis has been studied using *in vivo* animal models and a number of polarized epithelial cells (Seetharam, 1994, and references therein). Common findings from these studies include (1) apical endocytosis of IF-Cbl, (2) degradation of IF, (3) intracellular formation of TC II-Cbl, and (4) delay in Cbl transcytosis. Studies using Caco-2 cells (Dan and Cutler, 1994; Ramanujam *et al.*, 1991b) have shown that the $T_{1/2}$ of TC II-Cbl formation (Ramanujam *et al.*, 1991b) and IF degradation (Dan and Cutler, 1994) is about 4 h indicating, that formation of TC II may be a rate-limiting factor in Cbl transcytosis. Both *in vivo* and *in vitro* studies (Seetharam, 1994, and references therein) have shown that the acid pH and the proteolysis of IF (Gordon *et al.*, 1995; Ramanujam *et al.*, 1992) are essential for Cbl release, degradation of IF, and Cbl transfer to TC II, suggesting that IF-Cbl is processed by the classical endosomal-lysosomal pathway. This has been confirmed by light and electron microscopic autoradiography studies that have shown the movement of Cbl from the endosomal invagination to the lysosomes following microinjection of IF-57[Co]Cbl into a single proximal tubule (Birn *et al.*, 1997) in about 20 min.

One unresolved issue during IF-mediated Cbl transcytosis is the site of formation of TC II-Cbl complex. Intracellular localization of TC II in polarized epithelial cells is difficult because of the low levels of its synthesis and rapid constitutive secretion. Its location in the lysosomes is unlikely because apo or halo TC II would be degraded. Thus, the most likely cellular location of TC II is a vesicle that is distinct from lysosomes. It is known that after synthesis, TC II is targeted for rapid secretion via the basolateral membranes in polarized epithelial

Caco-2 cells (Ramanujam *et al.*, 1991b), and it is likely that the site of TC II-Cbl formation during IF-mediated Cbl transcytosis occurs in these vesicles. Free Cbl that egresses from the either pre-lysosomes (late endosomes) or lysosomes may enter such a vesicle. However, what is not clear in this scenario is how Cbl gains access to TC II. It could occur by simple diffusion of free Cbl or by fluid-phase transfer during a potential fusion event involving late endosomes or lysosomes with the TC II–containing secretory vesicle.

As to the TC II–mediated endocytosis of Cbl, polarized epithelial cells (Caco-2) express TC II-R in both basolateral and apical membranes (ratio 6:1), and TC II is able to mediate Cbl endocytosis in proportionate amounts from both directions, but with one difference. Following basolateral endocytosis of TC II-Cbl, Cbl is retained and utilized as Cbl enzymes (Bose *et al.*, 1997; Bose *et al.*, 1996b), while apically internalized TC II-Cbl is transcytosed intact bypassing the lysosomes (Bose *et al.*, 1997). How these cells are able to sort Cbl bound to the same ligand by two distinct pathways depending upon the side of entry is not known, but what is interesting is that after apical entry of Cbl bound to either IF or TC II, its fate is transcytosis bound to TC II. This observation suggests and argues for the role of a secretory vesicle (apical endosome?) that is targeted for basolateral secretion as a common carrier of intracellular halo TC II. It is likely that even the endogenous apo TC II that is secreted constitutively in these cells across the basolateral surface of these cells is carried by the same vesicle that carries halo TC II.

The apical pathway for TC II-Cbl endocytosis noted in Caco-2 cells also appears to be functional in the intact rat intestine, because orally administered labeled TC II appeared intact in the circulation (Bose *et al.*, 1997). However, the relevance of this pathway *in vivo* is not evident because the presence of TC II in the intestinal lumen has not been demonstrated, and patients with either IF or IFCR defects do develop Cbl deficiency (Fenton and Rosenberg, 1995; Seetharam, 1994). On the other hand, there is enough TC II-R activity in the apical brush border membranes (Bose *et al.*, 1995a; Bose *et al.*, 1995b) that can be exploited to transport Cbl in patients with malabsorption of Cbl due to a variety of inherited disorders or to secondary causes (Seetharam, 1994, and references therein). Further studies are required to assess the potential use of apical TC II-Cbl as a possible Cbl delivery system. The apical and basolateral sorting pathways of Cbl are shown in Fig. 4.

E. Inherited Disorders of Cbl Transport

Human Cbl deficiency can result from defects in the transport proteins (IF, TC II, IFCR) or defective passage of Cbl through the ileal cell (Cbl F). Disorders of IF and TC II are inherited as autosomal-recessive traits based on the genetic cri-

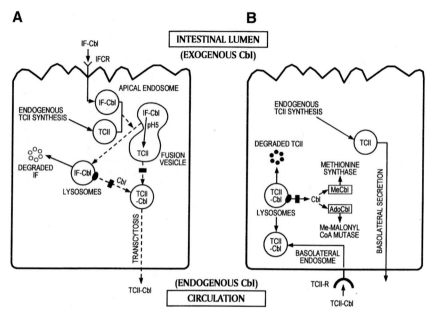

FIGURE 4 Proposed cellular sorting pathways of Cbl in a polarized epithelial cell. (A) The transcytosis of Cbl following the apical internalization of Cbl bound to IF. Broken lines represent incompletely defined pathways. (B) Retention and utilization of Cbl internalized from the basolateral side bound to plasma TC II. The solid oval shown in both panels represents the lysosomal transporter of Cbl; the solid rectangle indicates the block in Cbl sorting in Cbl F patients. For details, see the text.

teria, and occur in three forms: (1) lack of synthesis, (2) lack of Cbl binding, and (3) inability to promote cellular transport (Fenton and Rosenberg, 1995; Seetharam, 1994). The molecular basis for the different types of IF defect is not known because of the rarity of these disorders and the difficulty in obtaining gastric biopsies from otherwise normal patients. However, Southern blotting of genomic DNA obtained from a patient with pernicious anemia showed normal restriction fragments, suggesting that in congenital pernicious anemia, major deletions of the IF gene do not occur (Hewitt *et al.,* 1991). Since the presence of IF mRNA was not tested, it is not possible to ascertain whether the lack of IF in patients with pernicious anemia is due to a transcriptional defect.

The molecular basis for the most common form of TC II deficiency, lack of immunoprecipitable TC II in plasma, has been delineated using the RT-PCR technique and RNA isolated from cultured fibroblasts of TC II–deficient patients. In these patients, TC II is not synthesized because of a great reduction in its TC II mRNA levels that cannot be detected even by the very sensitive ribonuclease protection assays (Li *et al.,* 1994c). Sequencing of eight RT-PCR amplified mutant alleles from four patients has shown that the mRNA deficiency is caused by heterogeneous forms of mutations and that these mutations were private and occurred in individual families. The mutations included small (4-nucleotide) and big deletions (most of the defective allele) of the gene (Li *et al.,* 1994b), nonsense mutations, and a possible transcription defect (Li *et al,* 1994a).

In patients with Cbl F, Cbl deficiency occurs because of poor absorption of Cbl (Laframboise *et al.,* 1992) and a failure to utilize Cbl intracellularly, due to its retention in the lysosomes (Rosenblatt *et al.,* 1985; Vassiliadis *et al.,* 1991). Transport of Cbl out of the lysosomes, destined either for transcytosis bound to TC II (intestinal cells) or for utilization as Cbl coenzymes, is mediated by a Cbl-specific membrane transporter (Idriss and Jonas, 1992). In patients with Cbl F, it is not known whether the Cbl transporter is defective.

In patients with selective intestinal malabsorption of Cbl, Cbl deficiency develops because of decreased IFCR activity at the cell surface, perhaps caused by synthesis of an unstable receptor (Burman *et al.,* 1985; Eaton *et al.,* 1998; Gueant *et al.,* 1995) which, like in the canine model of the disease (Fyfe *et al.,* 1991b), may be retained (Fyfe *et al.,* 1991a) inside the cell and degraded. The molecular basis of IFCR deficiency in the human selective Cbl malabsorption syndrome is not known. With the availability of the human IFCR gene (Kozyraki *et al.,* 1988), which is localized to the same gene locus as the autosomal recessive megaloblastic anemia gene (Aminoff *et al.,* 1995), further studies are possible not only to delineate the molecular basis for the development of Cbl deficiency due to its malabsorption, but also to further understand its role in causing proteinuria in these patients.

It is important to recognize Cbl deficiency in children early, so they can be treated with Cbl to prevent complications leading to the development of megaloblastic anemia or neuronal damage. When treated with large doses (mg), frequently a very small percentage of Cbl diffuses across the cell membranes to maintain normal cellular function. Table I summarizes the pathophysiology of Cbl malabsorption owing to various inherited causes.

V. FOLATE

The coenzyme derivatives of folic acid are necessary for the synthesis of purine and pyrimidine precursors of nucleic acids, for the metabolism of certain amino acids, and for the initiation of protein synthesis in mitochondria. As with other water-soluble vitamins, humans and other mammals have lost their ability to synthesize folate, and thus must obtain the vitamin from exogenous sources via intestinal absorption. The intestine is exposed to two sources of folate: (1) dietary source, which is absorbed in the small intestine, and (2) bacterial source, wherein the vitamin is synthesized by the normal microflora of the large intestine.

TABLE I

Inherited Causes of Cbl Malabsorption

Disorders	Pathophysiology
1. Defective IF	(a) Lack of IF synthesis, mutant IF that is degraded by acid/pepsin, mutant IF that cannot bind Cbl. Cbl remains bound to haptocorrin and not recognized by IFCR.
2. Defective IFCR?	(b) Abnormal IF that binds to Cbl but the IF-Cbl complex is not recognized by the IFCR. Transport incompetent; failure of expression at the apical plasma membrane; lack of endocytosis of IF-Cbl.
3. TC II deficiency	Cbl released from IF cannot exit the enterocytes due to lack of TC II.
4. Cbl F (defective lysosomal transporter?)	Cbl is retained in the lysosomes and is not transferred to TC II.

A. Digestion of Dietary Folates

Digestion and absorption of dietary folate has been the subject of intense investigations over the past three decades. In the diet, folate exists mainly in the form of folate polyglutamates (Butterworth *et al.*, 1963). Because of their molecular size and multiple negative charge, these forms of folate cannot be absorbed as such and must be converted to folate monoglutamates prior to absorption. This hydrolysis of polyglutamate occurs enzymatically by the action of the enzyme folate hydrolase (also called folylpoly-γ-glutamate carboxypeptidase) (Reisenauer *et al.*, 1971). Two forms of this enzyme have been identified in the small intestine of humans and pigs (Chandler *et al.*, 1986; Reisenauer *et al.*, 1971; Wang *et al.*, 1986a, 1986b). A brush-border form that is zinc-activated acts as an exopeptidase, that is, it cleaves the glutamate moieties of the folate polyglutamates sequentially, and has a pH optimum of 5.5 to 6.5 (Chandler *et al.*, 1986). The intracellular (lysosomal) form of folate conjugase acts as an endopeptidase, and has a pH optimum of 4.5 (Wang *et al.*, 1986b). Functional and immunological studies have shown the brush-border from of folate hydrolase to be mainly localized in the proximal part of the small intestine, with little or no activity in the ileum (Chandler *et al.*, 1991). On the other hand, distribution of the intracellular form of the folate hydrolase was found to be uniform along the entire length of the small intestine (Chandler *et al.*, 1991). The brush-border form of folate hydrolase has been recently cloned by Halsted *et al.* (1998) from the pig jejunum. Using an amplified homologous probe incorporating primer sequences from human prostate-specific membrane antigen (PSM), a protein that itself has been found to be capable of hydrolyzing folate polyglutamates (Pinto *et al.,* 1996), the full-length cDNA of the folate hydrolase of pig jejunum was isolated. The cDNA was found to contain an open reading frame (ORF) of 2253 bases encoding a polypeptide of 751 amino acid residues. At the nucleotide level, the ORF of the cloned pig intestinal folate hydrolase was found to have 88% and 83% identity with the human PSM and rat brain N-acetylated-?-linked acidic dipeptidase (NAALADase), respectively; however, very little homology was found at the 5' untranslated regions of these cDNAs. At the amino acid level, the pig folate hydrolase was found to be 91% identical to that of the human PSM, and 83% identical to that of the rat brain NAALADase. The Kyte Doolittle hydropathy plot (Kyte and Doolittle, 1982) of pig folate hydrolase showed a membrane topology identical to that of human PSM and rat NAALADase, and typified a type II membrane protein with a single hydrophobic transmembrane-spanning domain and a conserved short N-terminal cytoplasmic region. The encoded polypeptide of the cloned folate hydrolase appeared to have multiple putative glycosylation and zinc binding sites. Functional identity of the cloned cDNA was confirmed by expression in PC3 cells. It was shown that PC3 cells expressing the cloned pig jejunal cDNA exhibited activity of the membrane folate hydrolase, with functional and kinetic characteristics similar to those of the native pig intestinal membrane folate hydrolase. The expressed protein also was found to react with antibody prepared against the native folate hydrolase of pig intestine. Furthermore, mRNA transcripts corresponding to the cloned folate conjugase were identified by Northern blot analysis in mRNA from pig jejunal mucosa but not from ileal mucosa. The regional distribution pattern corresponds well with the functional and immunological distribution of the membrane form of the folate hydrolase in the native intestine (Chandler *et al.,* 1991). Furthermore, a 2.8-kb transcript also was found in human jejunum and in the human-derived prostate cancer cell line LCCap, as well as in the rat brain. From these studies, it was concluded that the pig folate hydrolase, the human PSM, and the rat NAALADase may represent a varied expression of the same gene in different species and tissues, and that their differences may reflect tissue differences in available substrate (Halsted *et al.,* 1998).

B. Mechanisms of Intestinal Absorption of Folate Monoglutamates

Early studies on the mechanism of absorption of folate monoglutamate in the small intestine have used a variety of intact intestinal tissue and cellular preparations (*in vitro* and *in vivo* perfusion techniques, everted sacs, isolated cells, etc.) (Rose, 1987, and references therein). These studies have identified important features of the intestinal absorption process of folate. These include identification of the proximal part of the small intestine as the preferential site of absorption of dietary folate monoglutamates; demonstration of the existence of a specific, carrier-mediated system for folate transport; and demonstration of pH dependence of the uptake process, with a higher uptake at acid pH compared with neutral or alkaline pH. However, these studies have not addressed the issue of polarity of carrier distribution in a functional enterocyte. This issue was subsequently addressed using purified BBM vesicles (BBMV) and BLM vesicles (BLMV) isolated from the small intestine of animals and humans (Said and Redha, 1987; Said *et al.*, 1987a; Schron *et al.*, 1985; Selhub and Rosenberg, 1981). Studies with BBMV isolated from rat, rabbit, and human jejunum have demonstrated the existence of a specific, carrier-mediated system for folate uptake (Said *et al.*, 1987a; Schron *et al.*, 1985; Selhub and Rosenberg, 1981). This system was found to be expressed mainly in the proximal part of the small intestine, and not in the ileum under normal physiological conditions (Said *et al.*, 1987a; Said *et al.*, 1988b; Schron *et al.*, 1985); following resection of the proximal part of the small intestine, however, this system adapts itself and is functionally expressed in the remaining ileal segment (Said *et al.*, 1988b). Uptake of folate by this system was found to be electroneutral in nature, as indicated by lack of effect of inducing a positive or a negative transmembrane electrical potential on the substrate uptake (Said *et al.*, 1987a). This folate uptake system was found to be shared (with similar affinity) by reduced (e.g., 5-methyltetrahydrofolate [5-MTHF]) and oxidized (e.g., folic acid) folate derivatives, as demonstrated by the mutual inhibition of uptake of these folate compounds and the similarity of the apparent K_m values of the inhibition constants (K_i) for the various folate derivatives (Said *et al.*, 1987a; Schron *et al.*, 1988; Selhub and Rosenberg, 1981). This characteristic is unique to the intestine and is different from the characteristics of the widely investigated cellular folate uptake system noted in the mouse leukemia L1210 cells, where the uptake has been shown to prefer reduced over oxidized folate derivatives (and thus it is referred to as the reduced folate carrier [RFC]) (Goldman *et al.*, 1963; Henderson and Zevely, 1984). It is also different from the folate transport systems described in other epithelia, where clear hierarchy with regard to the different folate derivatives has been reported (Selhub *et al.*, 1987; Sweiry and Yudilevich, 1988). The uptake process of folate in intestinal BBMV also was found to be sensitive to the inhibitory effect of the anion transport inhibitors 4,4'-diisothiocyano-2-2-disulfonic acid stilbene (DIDS), and acetamidoisothiocyanostilbene-2,2'disulfonic acid (SITS) (Said *et al.*, 1987; Schron *et al.*, 1985). Furthermore, as with the observation in intact intestinal preparations, folate uptake by intestinal BBMV was found to be highly pH dependent and increased with decreasing incubation buffer from pH 7.4 to 5.5 (Said *et al.*, 1987a; Schron *et al.*, 1985; Selhub and Rosenberg, 1981) (Fig. 5). Moreover, in the presence of a transmembrane pH gradient (i.e., pH_i = 7.4, pH_o = 5.0), a transient accumulation of folate inside the BBMV (i.e., an

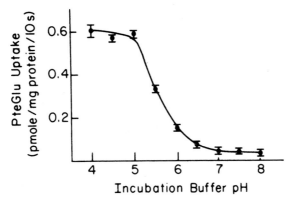

FIGURE 5 Dependence of folate uptake by human jejunal BBMV on incubation buffer pH. Reproduced from Said and Redha (1987) by permission.

overshoot) has been observed, indicating transport of the substrate against a concentration gradient (Said *et al.*, 1987a; Schron *et al.*, 1985). These observations have led to the conclusion that a folate:OH exchanger (or a folate:H^+ cotransport) is involved in intestinal folate transport (Said *et al.*, 1987a; Schron *et al.*, 1985). However, the observed stimulatory effect of an outside acidic pH on folate uptake could not be attributed solely to the existence of a transmembrane OH^- (or H^+) gradient. Rather, part of the stimulation in folate uptake by acidic incubation buffer pH appears to be a direct effect of the acidic pH on the folate carrier (for example, through alterations in the ionization state of certain groups in the carrier system). This conclusion was based on the observation that folate uptake by intestinal BBMV incubated at acid pH, but in the absence of a transmembrane pH gradient ($pH_i = pH_o = 5.0$) was significantly higher compared with uptake by vesicles incubated at an alkaline pH in the absence of a transmembrane pH gradient (i.e., $pH_i = pH_o = 7.4$) (Fig. 6) (Said *et al.*, 1987).

Unlike folate transport in renal and placental epithelia, where an alternative uptake mechanism through the so-called folate binding protein or the folate receptor (a glycosyl-phosphotidylinositol-linked membrane protein) also has been identified, no such mechanism was found in the normal small intestine, based on functional, molecular, and immunological studies (Said *et al.*, 1987; Said *et al.*, 1997; Said *et al.*, 1989; Witman *et al.*, 1992). As to the mechanism of folate exit out of the enterocyte, i.e., transport across the basolateral membrane (BLM), this process was investigated using purified BLMV isolated from rat small intestine. It was shown that folate transport across this membrane domain occurred via a specific, carrier-mediated system which that appears to be shared by oxidized and reduced folate derivatives and is DIDS sensitive (Said and Redha, 1987).

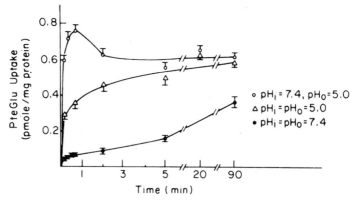

FIGURE 6   Effect of different pH gradient conditions on folate uptake by human jejunal BBMV. Reproduced from Said and Redha (1987) by permission.

Our understanding of the molecular identity of the intestinal folate transport process has been significantly increased since the identification of two cDNA clones that appear to be involved in folate transport. One clone was isolated from mouse small intestine and the other from human small intestine (Nguyen *et al.*, 1997; Said *et al.*, 1996). The mouse intestinal cDNA was found to have an open reading frame identical to that of the reduced folate carrier (RFC) of the mouse leukemia L1210 cells (Said *et al.*, 1996). The human intestinal cDNA clone was found to be similar to the cDNA clones of the human reduced folate transporters identified in other human tissues (Nguyen *et al.*, 1997). The two clones were found to share 74 and 66% homology with each other at the nucleotide and amino acid sequence levels, respectively. Both clones appeared to encode a protein with 12 putative transmembrane domains, with multiple potential glycosylation and protein-kinase phosphorylation sites. Both polypeptides appeared to carry net positive charge, which may be important for interaction of the carrier with its negatively charged substrate. Furthermore, when expressed in *Xenopus* oocytes, both clones were found to cause a significant and specific increase in the uptake of 5-MTHF (Nguyen *et al.*, 1997; Said *et al.*, 1996). Uptake by the expressed system was saturable as a function of substrate concentration with apparent K_m values similar to those reported with native intestinal preparations. However, the pH-dependence profile and sensitivity to inhibition by oxidized and reduced folate structural analogs, were found to be different when compared with characteristics of folate transport in the native intestine. Although folate uptake in native small intestinal preparations is significantly higher at acidic pH compared with neutral or alkaline pH (Said *et al.*, 1987a; Schron *et al.*, 1985; Selhub and Rosenberg, 1981), cRNA-induced folate uptake in *Xenopus* oocytes does not show such pH preponderance

(Nguyen *et al.,* 1997; Said *et al.,* 1996). Furthermore, although the folate transport system of the native small intestine handles both oxidized and reduced folate derivatives with similar affinity (Said *et al.,* 1987a; Schron *et al.,* 1988; Selhub and Rosenberg, 1981), the cRNA-induced folate uptake in *Xenopus* oocytes showed preferential inhibition by reduced over-oxidized folate derivatives. To further investigate these differences and to develop a better understanding of their possible cause(s), Kumar *et al.* (1998) determined the characteristics of the intestinal folate carrier RFC when stably expressed in epithelial cells of intestinal origin and compared the findings with its characteristics when expressed in *Xenopus* oocytes. The rat-derived, nontransformed small intestinal epithelial cells IEC-6 were used in the study. These cells were chosen because they reportedly have a folate uptake mechanism similar to that of the native small intestine, and like the native intestine they lack expression of the membrane folate receptor (Said *et al.,* 1997). Using these cells, it was shown that the expressed RFC carrier exhibits characteristics similar to those known for folate transport by the native small intestine (i.e., being pH-dependent with a higher uptake at acidic pH) and has similar affinities for reduced and oxidized folate derivatives. In contrast, the RFC expressed in *Xenopus* oocytes displayed the characteristics described earlier. From these studies, it was concluded that the characteristics of RFC vary according to the cell system in which they are expressed. Differences in membrane composition between cells, involvement of cell or tissue-specific posttranslational modification(s), and/or involvement of an auxiliary protein that modulates the activity of the expressed RFC leading to alterations in its functional characteristics, have been suggested as a possible mechanism(s) (Said *et al.,* 1997). In related studies the distribution of mRNA transcripts of the cloned intestinal folate carrier along the crypt-villus axis also has been examined by an *in situ* hybridization technique using human jejunal sections and by Northern blot analysis using poly(A^+) RNA isolated from rat crypt and villus cells (Nguyen *et al.,* 1997; Said *et al.,* 1996). In both cases, expression was significantly higher in the mature epithelial cells of the villus tip than in the immature epithelial cells of the crypt. Because mature cells of the villus are responsible for nutrient absorption, the latter findings on the distribution of mRNA transcripts of the folate carrier further suggest a role for this cloned carrier in normal intestinal absorption of dietary folate monoglutamates.

C. Regulation of Intestinal Folate Absorption

Certain aspects of intracellular regulation of the folate uptake process in the small intestine also have been examined (Said *et al.,* 1997). Using mature confluent monolayers of the intestinal epithelial cell line IEC-6 as an *in vitro* model system for enterocytes, together with specific modulators of certain protein kinase–mediated pathways, it has been demonstrated that agents that inhibit the activity of protein tyrosine kinase (PTK) significantly inhibit folate uptake by intestinal cells. This inhibitory effect was found to be mediated via inhibition in the activity (or number) and affinity of the folate uptake system, as indicated by the significant decrease in V_{max} and increase in apparent K_m of the folate uptake system. The mechanism by which PTK affects intestinal folate uptake is not clear. However, the finding from another study using the human-derived, normal NCM 460 colonic epithelial cells (which have a folate uptake mechanism similar to that of the small intestine; see later) that treatment with the tyrosine phosphatase inhibitor orthovanadate leads to significant stimulation in folate uptake raises the possibility that phosphorylation of the folate uptake system by PTK may be involved. Further support for possible involvement of phosphorylation in the PTK regulation of folate transport comes from the recent observations of Bhushan *et al.* (1996). These workers found that phosphorylation of the tyrosine moieties of the folate transporter of a subline of L1210 cells that are resistant to the antifolate compound methotrexate (due to impairment in the function of the membrane folate carrier, i.e., RFC), was significantly lower than phosphorylation of the tyrosine moieties of the folate transporter of the wild-type cells.

Increasing intracellular cAMP level also was found to lead to inhibition in folate uptake by IEC-6 cells. The mechanism of action of cAMP is not clear, but it appears not to be mediated via activation of protein kinase A (PKA). This is because inhibitors of PKA failed to affect folate uptake by these cells, or to reverse the inhibitory effect caused by cAMP. In contrast to the possible role of PTK and cAMP in the regulation of intestinal folate uptake, no roles for protein kinase C and Ca^+/calmodulin-mediated pathways in the regulation of folate uptake by IEC-6 cells were found (Said *et al.,* 1997). Further studies are needed to clarify the mechanism of action of PTK and cAMP on folate uptake, and to explore possible involvement of other unidentified cellular pathways in the regulation of the intestinal folate uptake process.

D. Absorption of Bacterially Synthesized Folate in the Large Intestine

As mentioned earlier, the intestine is exposed to a second exogenous source of folate, i.e., the bacterially synthesized folate in the large intestine. Studies have shown significant amounts of this folate to be in the form of folate monoglutamate, that is, available for absorption (Rong *et al.,* 1991). Using [^{3}H]p-aminobenzoic acid to label the newly synthesized folate by the normal microflora of rat large intestine, Rong *et al.* (1991) have shown that a portion of this folate is indeed absorbed and incorporated into various rat tissues. The mechanism of absorption of folate in the large intestine has been recently characterized using apical membrane vesicles (AMV) isolated from human native colon (Dudeja et al., 1997) and by studies utilizing the human-derived, non-transformed colonic epithelial cells (NCM 460) as a model system for colonocytes (Kumar et al., 1997). In the colonic AMV study, it has been shown that folate uptake takes place via a specific, carrier-mediated, pH-dependent, DIDS-sensitive uptake mechanism that is similar to that of the small intestine. A comparison of the V_{max} of folate uptake by human colonic AMV (V_{max} of 19.8 pmol/mg protein/10 s) (Dudeja et al., 1997) with that of the vitamin uptake by human small intestinal BBMV (V_{max} 4.7 pmol/mg protein/10 s) (Said et al., 1987a) indicated that the maximum velocity of the colonic folate transporter is in fact three to four times greater than that of the small intestine. However, the apparent K_m of the human colonic folate transporter (8 μM) was higher than that of the folate transporter of the human small intestine (1.69 μM), suggesting that the affinity of the former folate transporter is lower than that of the latter. It should be mentioned here that when comparing folate transport in the large intestine and the small intestine, one should take into consideration the longer residual time of luminal content in the large intestine.

The finding with the human-derived NCM 460 colonic epithelial cells has confirmed the observations noted with human colonic AMV and has demonstrated the involvement of a pH-dependent, carrier-mediated system for folate uptake (Kumar et al., 1997). Furthermore, these studies also have shown that the intracellular regulation of the colonic folate uptake process is similar to that of the small intestine, being regulated by PTK- and cAMP-mediated pathways. These similarities in the folate transport mechanism and regulation in colonic and small intestinal epithelial cells, together with the identification of a 3.3-kb transcript in Northern blot analysis of colonic mRNA with a cDNA probe of the folate transporter cloned from the small intestine, suggest that the folate transport systems in these two tissues are similar (or identical). The identification of an efficient carrier-mediated mechanism for folate absorption in the human colon suggests that this source of folate may contribute to the overall body folate homeostasis or may serve as a localized source of folate for the colonocytes. Furthermore, identification of a folate carrier in the human colon may lead to a better understanding of the cause of the localized folate deficiency described in this epithelia, which is believed to be associated with premalignant changes in the colonic mucosa (Cravo et al., 1991; Lashner et al., 1989).

VI. PANTOTHENIC ACID

Pantothenic acid is needed for the synthesis of coenzyme A and acyl carrier protein in mammalian cells, and thus is important in the metabolism of carbohydrate, fat, and to a lesser extent, protein. As with various other vitamins, the intestine is exposed to two sources of pantothenic acid: dietary and bacterial. Dietary pantothenic acid exists mainly in the form of CoA, which is hydrolyzed to free pantothenic acid in the intestinal lumen prior to absorption (Shibata *et al.,* 1983). The mechanism of absorption of pantothenic acid in the small intestine recently has been shown to involve the same carrier-mediated, Na^+-dependent system as that of biotin (see biotin section). Similarly, colonic absorption of pantothenic acid was found to involve the same biotin Na^+-dependent, carrier-mediated system (see biotin section). Interaction between biotin and pantothenic acid transport also has been observed in other tissues, such as the blood brain barrier (Spector and Mock, 1987), the heart (Beinlich *et al.,* 1990), and the placenta (Grassl, 1992).

VII. RIBOFLAVIN

Riboflavin (vitamin B_2), in its coenzyme forms riboflavin-5'-phosphate (FMN) and flavin adenine dinucleotide (FAD), is involved in key metabolic reactions including carbohydrate, amino acid, and lipid metabolism, and in the conversion of folic acid and pyridoxine into their coenzyme forms (Cooperman and Lopez, 1984; Rivlin, 1970). Vitamin B_2 is obtained from the diet, where it exists mainly in the forms of FMN and FAD. These forms are hydrolyzed to free riboflavin in the intestinal lumen prior to absorption by the action of intestinal phosphatases (Akiyama *et al.,* 1982; Campbell and Morrison, 1963; Daniel *et al.,* 1983a). Using a variety of intact intestinal preparations, the absorption process of riboflavin has been shown to occur by a specialized carrier-mediated system mainly in the proximal part of the small intestine (Akiyama *et al.,* 1982; Daniel *et al.,* 1983b; Said and Hollander, 1985). Using purified intestinal BBMV and BLMV, the transport process of riboflavin in these two preparations was found to be Na^+-independent, carrier-mediated, and electroneutral in nature (Said *et al.,* 1993a, 1993b). These findings have been confirmed in studies with human-derived intestinal epithelial Caco-2 cells (Said and Ma, 1994). The latter studies with Caco-2 cells have also shown that the inhibitor of the membrane Na^+/H^+ exchanger amiloride also inhibits riboflavin uptake and that this inhibition is competitive in nature (Said and Ma, 1994).

Regulation of the riboflavin intestinal uptake process by extracellular substrate levels and by specific intracellular protein kinase–mediated pathways has also been investigated (Said and Khani, 1993; Said and Ma, 1994; Said *et al.,* 1994). Over-supplementation of rats with pharmacological amounts of riboflavin was found to cause a significant and specific down-regulation in riboflavin uptake by intestinal BBMV compared with controls. On the other hand, riboflavin deficiency was found to cause a significant and specific up-regulation in the vitamin's uptake compared with pair-fed controls. These effects were found to be mediated through changes in the V_{max}, but not the apparent K_m, of the riboflavin uptake process, suggesting an increase in the number/activity of the functional riboflavin uptake carriers with no changes in their affinity (Said and Khani, 1993). Similar findings were observed in studies with Caco-2 cells (Said and Ma, 1994). In another study utilizing Caco-2 cells, compounds that increased intracellular cAMP levels down-regulated the riboflavin uptake process via an effect that is mediated through a decrease in the activity of riboflavin uptake carriers (Said et al., 1994). This led to the suggestion that a PKA-mediated pathway plays an important role in regulating riboflavin intestinal uptake process. With respect to the molecular nature of the riboflavin transport process across mammalian cell membranes, there has been no published report thus far on the subject.

VIII. THIAMINE

Thiamine (vitamin B_1), in its pyrophosphate form, acts as a cofactor in several vital reactions involved in metabolism and energy production. Specifically, it serves as a coenzyme in the decarboxylation of pyruvic acid and alpha-ketoglutaric acids; it also plays a role in the utilization of pentose in the hexose monophosphate shunt. Humans obtain thiamine from the diet. In the diet, thiamine exists mainly in phosphorylated forms (predominantly as thiamine pyrophosphate), which are hydrolyzed (by the action of a variety of intestinal phosphatases) to free thiamine in the intestinal lumen prior to absorption (Sklan and Trostler, 1977). Absorption of free thiamine then takes place predominantly in the proximal part of the small intestine (Komai et al., 1974; Rindi, 1984; Sklan and Trostler, 1977). The mechanism of absorption of thiamine (which exists as a monocation at physiological pHs of the intestinal lumen) has been studied using a variety of intestinal tissue preparations (Rindi, 1984). Animal studies at the intact intestinal tissue level have shown the uptake process of physiological concentrations of thiamine (< 2 μM) to be via a specific carrier-mediated system that is energy and Na^+ dependent and ouabain sensitive in nature (Ferrari et al., 1971; Hoyumpa et al., 1975). This carrier system appears to be the main mechanism responsible for absorption of the dietary thiamine that exists in the intestinal lumen, at an estimated concentration range of 0.2–2 μM (Hoyumpa et al., 1975). Uptake of high pharmacological concentrations of thiamine, on the other hand, was shown to take place by simple diffusion. It has also been shown that a significant amount of the thiamine that is taken up by the enterocyte is phosphorylated mainly to thiamine pyrophosphate by the action of cytoplasmic thiamine pyrophosphokinase (Cusaro et al., 1977; Komai et al., 1974; Rindi,

1984). This finding led to the early belief that phosphorylation is responsible for the saturable uptake mechanism of thiamine. However, it is well accepted now that intracellular phosphorylation of transported thiamine and the membrane up-take process are two different and separate cellular events, and that the role of phosphorylation is most probably to serve as a trapping mechanism for the trans-ported thiamine inside the absorptive cell. Other studies have shown that the thi-amine that exits out of the enterocyte across the basolateral membrane is in the form of free (i.e., nonphosphorylated) thiamine (Ferrari *et al.*, 1982; Rindi, 1984). The structural requirements for the efficient uptake of thiamine by a mem-brane transport system has also been studied by Komai and Shindo (1974), us-ing everted loops of rat small intestine. Evidence obtained has shown the im-portance of the amino group of the pyrimidine moiety, the size of the group at the 2′ position of the pyrimidine moiety, and the quaternary nitrogen in the thi-azole moiety of the thiamine molecule in the interaction and transport of the thi-amine molecule by the intestinal transport system.

Intestinal absorption of thiamine from the lumen to the blood represents move-ment of the vitamin across two structurally and functionally different membrane domains of the functionally polarized absorptive epithelial cells, that is, the brush border membrane (BBM) and the basolateral membrane (BLM). Thus, for an in-depth understanding of the mechanism of absorption of dietary thiamine in the small intestine, knowledge about the transport event at the individual mem-brane domain is critical. These issues were addressed with the use of purified small intestinal BBMV and BLMV (Casirola *et al.*, 1988; Hayashi *et al.*, 1981; Laforenza *et al.*, 1993 and 1998). Studies with purified BBMV isolated from guinea pig small intestine have shown that thiamine transport across the brush border membrane is linear as a function of concentration and is insensitive to in-hibition by unlabeled thiamine and structural analogs (Hayashi *et al.*, 1981). It was concluded that thiamine transport across the intestinal BBMV occurred by simple diffusion. Studies with purified BBMV isolated from rat intestine, on the other hand, presented evidence for the involvement of a specific thiamine: H^+ antiporter capable of transporting the vitamin against an intravesicular concen-tration gradient in the presence of an outwardly directed H^+ gradient (Ferrari *et al.*, 1982; Laforenza *et al.*, 1998). Furthermore, the process was reported to be competitively inhibited by certain unrelated organic cations. For example unla-beled quanidine reportedly inhibited the uptake of ^{3}H-thiamin, and unlabeled thi-amin inhibited the uptake of ^{3}H-quanidine by rat small intestinal BBMV (Laforenza *et al.*, 1998). The reported inhibition constants, however, were very high (10 and 7.8 mM for inhibition of ^{3}H thiamine and ^{3}H-quanidine uptake by unlabeled quanidine and thiamine, respectively). This raises the question of the physiological relevance of such observations with regard to absorption of dietary thiamine that exists in the low micromolar range. The cause of the discrepancy in the mechanism of thiamine uptake by intestinal BBMV between the guinea pig and the rat is not clear, but species variation (among other things) has been proposed as a possible explanation (Casirola *et al.*, 1988). The exit of thiamine out of the enterocyte via the BLM also has been examined using purified BLMV isolated from rat small intestine (Laforenza *et al.*, 1993). The results showed the involvement of a specialized carrier-mediated process. This system was reported to be Na^+ and ATP dependent and required the function of an active Na^+-K^+-ATPase (Laforenza *et al.*, 1993).

With regard to thiamine transport in the human small intestine, *in vivo* studies have suggested the involvement of a carrier-mediated system in the absorption process (Thomson and Leevy, 1972). This suggestion was confirmed by subsequent studies using human intestinal surgical and biopsy specimens (Hoyumpa *et al.*, 1982; Laforenza *et al.*, 1997; Rindi and Ferrari, 1977), and human-derived intestinal epithelial Caco-2 cells (Said *et al.*, 1999). Recent studies (Laforenza *et al.*, 1997; Said *et al.*, 1999) also have established that the intestinal thiamine uptake process is Na^+ independent in nature. As to the cellular regulation of the intestinal uptake process of thiamine, the studies with Caco-2 cells have shown that the process may be under the regulation of an intracellular Ca^{2+}/calmodulin-mediated pathway (Said *et al.*, 1999). However, to date, nothing is known about the molecular characteristics of the intestinal thiamine transport system(s). The recent simultaneous cloning of a thiamine transporter from human fetal brain, fibroblasts, and skeletal muscle cDNAs involved in the development of thiamine-responsive megaloblastic anemia (Labay *et al.*, 1999; Fleming *et al.*, 1999; Diaz *et al.*, 1999) should assist in the molecular identification and characterization of the intestinal thiamine absorption process.

IX. VITAMIN B$_6$

Vitamin B$_6$ consists of a group of structurally related nutrients that are both phosphorylated (pyridoxine phosphate, pyridoxyl phosphate, pyridoxamine phosphate) and nonphosphorylated (pyridoxine, pyridoxyl, and pyridoxamine). The compound with the most biological activity is pyridoxal phosphate, which functions as a cofactor in several enzymatic reactions including amino acid metabolism. The form frequently used for intestinal absorption studies is pyridoxine·HCl, which is also the active ingredient in medicinal preparations and food fortifications.

Hydrolysis of phosphorylated forms of the vitamin occurs in the intestinal lumen (Hamm *et al.*, 1979; Middleton, 1979 and 1985). Pyridoxal phosphate is taken up from the lumen of perfused rat jejunum; the process is saturable and dependent on the function of intraluminal alkaline phosphatase, thus suggesting intraluminal metabolism of the vitamin prior to absorption.

The intestine has not been shown to demonstrate net transepithelial transport or intracellular accumulation of vitamin B_6 in various intestinal preparations (Serebro *et al.*, 1966), in contrast to several other water-soluble vitamins. Brush border membrane vesicles of guinea pig jejunum showed no indication of a saturation process (Yoshida *et al.*, 1981). Isolated vascularly perfused rat intestine *in vivo* also indicated a nonsaturable process of pyridoxamine and pyridoxal uptake (Hamm *et al.*, 1979). Thus, the uptake process appeared to be by simple diffusion.

A feature of vitamin B_6 absorption is that the various components accumulate within the absorptive cell, primarily in the phosphorylated form (Middleton, 1978). In contrast to the indication of nonsaturable uptake in short-term studies, vitamin B_6 entry into the intestinal mucosa in prolonged studies on perfused intestine *in vivo* (Middleton, 1985) or on everted rings *in vitro* (Tsuji, 1973) involved saturation. This may be attributable to the metabolism of the vitamin intracellularly. The phosphate esters are dephosphorylated over time, with final transfer of the products to the circulation.

Roth-Maier *et al.* (1982) evaluated the possibility that dietary intake of vitamin B_6 influences the future absorptive capacity of the nutrient. They found that although the vitamin B_6 status of the rat is regulated homeostatically, this is not related to alterations in intestinal absorption.

X. CONCLUDING REMARKS

It is clear that significant progress has been made in recent years in our understanding of the absorption process of water-soluble vitamins in the small and large intestine, at both the cellular and the molecular levels. It is now well recognized that absorption of this diverse group of organic and essential micronutrients involves specialized and unique transport systems. Some of these transport systems have been cloned and characterized at the molecular level, and knowledge about their cellular and molecular regulation has begun to emerge. Others, however, await further investigation. It is believed that with the availability of powerful cellular/molecular biology tools, the coming years will witness a significant expansion in our understanding of the mechanism of absorption of these nutrients under normal physiological conditions and how aberration in their absorption occurs under certain pathophysiological conditions.

References

Agus, D. B., Gambhir, S. S., Pardridge, W. M., Spielholz, C., Baselga, J., Vera, J. C., and Golde, D. W. (1997). Vitamin C crosses the blood-brain barrier in the oxidized form through the glucose transporters. *J. Clin. Invest.* **100**, 2842–2848.

Akiyama, T., Selhub, J., and Rosenberg, I. H. (1982). FMN phosphatase and FAD pyrophosphatase in rat intestinal brush border membrane: Role in intestine absorption of dietary riboflavin. *J Nutr.* 112, 263–268.

Aminoff, M., Tahvanainen, E., Grasbeck, R., Weissenbach, J., Broach, H., and dela Chapelle, A. (1995). Selective intestinal malabsorption of vitamin B12 displays recessive mendelian inheritance: Assignment of a locus to chromosome 10 by linkage. *Am. I. Hum. Genet.* 57, 824–831.

Barth, C. A., Frigg, M., and Hogemeister, H. (1986). Biotin absorption from the hindgut of the pig. *J. Anim. Physiol. Anim. Nutr.* 55, 128–134.

Beinlich, C. J., Naumovitz, R. D., Song, W. O., and Neely, J. R. (1990). Myocardial metabolism of pantothenic acid in chronically diabetic rats. *J. Mol. Cell. Cardiol.* 22, 323–332.

Bhushan, A., Wroblewski, D., Xuan, Y., Tritton, T. R., and Hacker, M. P. (1996). Correlation of altered tyrosine phosphorylation with methotrexate resistance in a cisplatin-resistant subline of L1210 cells. *Biochem. Pharmacol.* 51, 477–482.

Bianchi, J., Wilson, F. A., and Rose, R. C. (1986). Dehydroascorbic acid and ascorbic acid transport in the guinea pig ileum. *Am. J. Physiol.* 250, G461–G468.

Birn, H., Verroust, PJ., Nexo, E., Hager, H., Jacobsen, C., Christensen, E. I., and Moestrup, S. (1997). Characterization of an epithelial 460 kDa protein that facilitates endocytosis of intrinsic factor-vitamin B12 and binds to receptor-associated protein. *J. Biol. Chem.* 272, 26497–26504.

Bork, P., and Beckmann, G. (1993). The CUB domain. A wide spread module in developmentally regulated proteins. *J. Mol. Biol.* 231, 539–545.

Bose, S., Chapin, S. J., Seetharam, S., Feix, J., Mostov, K. E., and Seetharam, B. (1998). Brefeldin A (BFA) inhibits the basolateral membrane delivery and dimerization of transcobalamin in human intestinal epithelial Caco-2 cells. *J. Biol. Chem.* 273, 16163–16169.

Bose, S., Dahms, N. M., Seetharam, S., and Seetharam B. (1997). Bipolar functional expression of transcobalamin II receptor in human intestinal polarized Caco-2 cells. *J. Biol. Chem.* 272, 11718–11725.

Bose, S., Feix, J., Seetharam, S., and Seetharam, B. (1996a). Dimerization of transcobalamin II-receptor: Requirement of a structurally ordered lipid bilayer. *J. Biol. Chem.* 271, 11718–11725.

Bose, S., Komorowski, R. A., Seetharam, S., Gilfix, B., Rosenblatt, D. S., and Seetharam, B. (1996b). *In vitro* and *in vivo* inactivation of transcobalamin II-receptor by its antiserum. *J. Biol. Chem.* 271, 4195–4200.

Bose, S., and Seetharam, B. (1997). Effect of disulfide bonds of transcobalamin II-receptor on its activity and basolateral targeting in human intestinal epithelial Caco-2 cells. *J. Biol. Chem.* 272, 20920–20928.

Bose, S., Seetharam, S., and Seetharam, B. (1995a). Membrane expression and interactions of human transcobalamin II receptor. *J. Biol. Chem.* 270, 8152–8157.

Bose, S., Seetharam, S., Hammond, T. G., and Seetharam, B. (1995b). Regulation of expression of transcobalamin II-receptor in the rat. *Biochem. J.* 310, 923–929.

Brown, B. B., and Rosenberg, J. H. (1987). Biotin absorption by distal rat intestine. *J. Nutr.* 117, 2121–2126.

Burkholder, P. R., and McVeigh, I. (1942). Synthesis of vitamins by intestinal bacteria. *Proc. Natl. Acad. Sci. (USA).* 28, 285–289.

Burman, J. F., Jenkins, W. J., Walker-Smith, J A., Philips, A. D., Sourial, N. A., Williams, C. B., and Mollin, D. L. (1985). Absent ileal uptake of IF-bound vitamin B12 *in vivo* in Immerslund-Grasbeck syndrome (Familial vitamin B12 malabsorption with proteinuria). *Gut.* 26, 311–314.

Butterworth, C. E., Santini, R., and Frommyer, W. B. (1963). The pteroylglutamate components of American diets as determined by chromatographic fractionation. *J. Clin. Invest.* 42, 1929–1939.

Campbell, J. A., and Morrison, C. A. (1963). Some factors affecting the absorption of vitamins. *Am. J. Clin. Nutr.* 12, 162–169.

Casirola, D., Ferrari, G., Gastaldi, G., Patrini, C., and Rindi, G. (1988). Transport of thiamin by brush border membrane vesicles from rat small intestine. *J. Physiol.* **398**, 329–339.

Chandler, C. J., Harrison, D. A., Buffington, C. A., Santiago, N. A., and Halsted, C. H. (1991). Functional specificity of jejunal brush-border pteroylpolyglutamate hydrolase in pig. *Am. J. Physiol.* **260**, G865–G872.

Chandler, C. J., Wang, T., and Halsted, C. H. (1986). Pteroylpolyglutamate hydrolase from human jejunal brush borders: purification and characterization. *J. Biol. Chem.* **261**, 928–933.

Chatterjee, N. S., Kumar, C. K., Ortiz, A., Rubin, S.A., and Said, H. M. (2000). Molecular mechanism of the intestinal biotin transport process. *Am. J. Physiol: Cell Physiol.* 277:C605–613.

Choi, J. L., and Rose, R. C. (1989). Regeneration of ascorbic acid by rat colon. *Proc. Soc. Exp. Biol. Med.* **190**, 369–374.

Christensen, E. I., Birn, H., Verroust, P. J., and Moestrup, S. (1998). Membrane receptors for endocytosis in the renal proximal tubule. *Int. Rev. Cytol.* **180**, 237–284.

Cole, H., Reynolds, T. R., Lockyer, J. M., Buck, G. A., Denson, T., Spence, J. E., Hymes, J., and Wolf, B. (1994). Human serum biotinidase: cDNA cloning, sequence, and characterization. *J. Biol. Chem.* **269**, 6566–6570.

Cooperman, J. M., and Lopez, R. (1984). Riboflavin. *In* "Handbook of Vitamins: Nutritional, Biochemical and Clinical Aspects" (L. J. Machlin, ed.), pp. 299–327. Dekker, New York.

Cravo, M. L., Mason, J. B., Selhub, J., and Rosenberg, I. H. (1991). Use of deoxyuridine suppression test to evaluate localized folate deficiency in rat colonic epithelium. *Am. J. Clin. Nutr.* **53**, 1450–1454.

Cusaro, G., Rindi, G., and Sciorelli, G. (1977). Subcellular distribution of thiamin-pyrophosphokinase and thiamin-pyrophosphatase activities in rat isolated enterocytes. *Int. J. Vitam. Nutr. Res.* **47**, 99–106.

Dan, N., and Cutler, D. F. (1994). Transcytosis and processing of intrinsic factor-cobalamin in Caco-2 cells. *J. Biol. Chem.* **269**, 18849–18855.

Daniel, H., Binninger, E., and Rehner, G. (1983a). Hydrolysis of FMN and FAD by alkaline phosphatase of the intestinal brush border membrane. *Int. J. Vitam. Nutr. Res.* **53**, 109–114.

Daniel, H., Wille, U., and Rehner, G. (1983b). *In vitro* kinetics of the intestinal transport of riboflavin in rats. *J. Nutr.* **113**, 636–643.

Diaz, G. A., Banikazemai, M., Oishi, K., Desnick, R. J., and Gelb, B. D. (1999). Mutations in a new gene encoding a thiamine transporter cause thiamine-responsive megaloblastic anaemia syndrome. *Nature Genet.* **22**, 309–312.

Dieckgraefe, B. K., Seetharam, B., and Alpers, D. H. (1988). Developmental regulation of rat intrinsic factor mRNA. *Am. J. Physiol.* **254**, G913–G919.

Dudeja, P. K., Torania, S. A., and Said, H. M. (1997). Evidence for the existence of an electroneutral, pH-dependent, DIDS-sensitive carrier-mediated folate uptake mechanism in the human colonic luminal membrane vesicles. *Am. J. Physiol.* **272**, G1408–G1415.

Dyer, D. L., and Said, H. M. (1997). Biotin uptake in cultured cell lines. *Methods Enzymol.* **279**, 393–405.

Eaton, D. M., Livingston, J. H., Seetharam, B., and Puntis, J. W. (1998). Overexpression of an unstable intrinsic factor-cobalamin receptor in Imerslund-Grasbeck syndrome. *Gastroenterology.* **115**, 173–176.

Fenton, W. A., and Rosenberg, L. E. (1995). Inherited disorders of cobalamin transport and metabolism. *In* "Metabolic basis of inherited disease" (C. R. Scriver, A. L. Beaudet, W. S. Sly, and D. Valley, eds.), Vol. 2, 7th ed., pp. 3111–3128. McGraw-Hill Information Service Co., New York.

Ferrari, B., Patrini, G., and Rindi, G. (1982). Intestinal thiamin transport in rats. Thiamin and thiaminphosphoester content in the tissue and serosal fluid of inverted jejunal sacs. *Pflugers Arch.* **39**, 37–41.

Ferrari, G., Ventura, U., and Rindi, G. (1971). The Na$^+$-dependence of thiamin intestinal transport *in vitro. Life Sci.* **10**, 67–75.

Fleming, J. C., Tartaglini, E., Steinkamp, M. P., Schorderet, D. F., Cohen, N., and Neufeld, E. J. (1999). The gene mutated in thiamine-responsive anaemia with diabetes and deafness (TRMA) encodes a functional thiamine transporter. *Nature Genet.* **22**, 305–308.

Fyfe, J. C., Giger, U., and Hall, C. A. (1991a). Inherited selective intestinal cobalamin malabsorption and cobalamin deficiency in dogs. *Pediatr. Res.* **29**, 24–31.

Fyfe, J. C., Ramanujam, K. S., Ramaswamy, K., Patterson, D. F., and Seetharam, B. (1991b). Defective brush border expression of intrinsic-factor cobalamin receptor in inherited selective intestinal cobalamin malabsorption. *J. Biol. Chem.* **266**, 4489–4494.

Goldman, I. D., Lichenstein, N. S., and Oliverio, V. T. (1968). Carrier-mediated transport of the folic acid analogue, methotrexate, in the L1210 leukemia cell. *J. Biol. Chem.* **243**, 5007–5017.

Gordon, M. M., Howard, T., Becich, M. J., and Alpers, D. H. (1995). Cathepsin L mediates intracellular ileal digestion of gastric intrinsic factor. *Am. J. Physiol.* **268**, G33–G40.

Gordon, M. M., Hu, C., Chokshi, H., Hewitt, J. E., and Alpers, D. H. (1991). Glycosylation is not required for ligand or receptor binding by expressed rat intrinsic factor. *Am. J. Physiol.* **260**, 736–742.

Grasbeck, R. (1969). *In* "Progress in Hematology" (E. B. Brown and C. V. Moore, eds.), Vol. 6, pp. 223–260. Grune and Stratton, Philadelphia.

Grassl, S. M. (1992). Human placental brush-border membrane Na$^+$-pantothenate cotransport. *J. Biol. Chem.* **267**, 22902–22906.

Gueant, J. L., Saunier, M., Gastin, I., Safi, A., Lamireau, T., Duclos, B., Bigard, M. A., and Grasbeck, R. (1995). Decreased activity of intestinal and urinary intrinsic factor in Grasbeck-Imerslund syndrome. *Gastroenterology* **108**, 1622–1628.

Halsted, C. H., Ling, E., Carter, R. L., Villanueva, J. A., Gardner, J. M., and Coyle, J. T. (1998). Folylpoly-γ-glutamate carboxypeptidase from pig jejunum: Molecular characterization and relation to glutamate carboxypeptidase II. *J. Biol. Chem.* **273**, 20417–20424.

Hamm, M. W., Hehansho, H., and Henderson, L. M. (1979). Transport and metabolism of pyridoxamine and pyridoxamine phosphate in the small intestine. *J. Nutr.* **109**, 1552–1559.

Hayashi, K., Yoshida, S., and Kawasaki, T. (1981). Thiamin transport in the brush border membrane vesicles of the guinea-pig jejunum. *Biochim. Biophys. Acta* **641**, 106–113.

Henderson, G. B., and Zevely, E. M. (1984). Transport routes utilized by L1210 cells for the influx and efflux of methotrexate. *J. Biol. Chem.* **259**, 1526–1531.

Hewitt, J. E., Gordon, M. M., Taggart, T., Mohandas T. K., and Alpers, D. H. (1991). Human gastric intrinsic factor: Characterization of cDNA and genomic clones and localization to human chromosome 11. *Genomics* **10**, 432–440.

Hoozen, C. M., Ling, E., and Halsted, C. H. (1996). Folate binding protein: Molecular characterization and transcript distribution in pig liver, kidney and jejunum. *Biochem. J.* **319**, 725–729.

Howard, T. A., Misra, D. N., Grove, M., Becich, M. J., Shao, J. S., Gordon, M., and Alpers, D. H. (1996). Human gastric intrinsic factor expression is not restricted to parietal cells. *J. Anat.* **189**, 303–313.

Hoyumpa, A. M., Strickland, R., Sheehan, J. J., Yarborough, G, and Nichols, S. (1982). Dual system of intestinal thiamin transport in humans. *J. Lab. Clin. Med.* **99**, 701–708.

Hoyumpa, A., Middleton, H. M., Wilson, F. A., and Schenker, S. (1975). Thiamin transport across the rat intestine. *Gastroenterology* **68**, 1218–1227.

Idriss, J. M., and Jonas, A. J. (1992). Vitamin B12 transport by rat liver lysosomal membrane vesicles. *J. Biol. Chem.* **266**, 9438–9441.

Karasov, W. H., Darken, B. W., and Bottum, M. C. (1991). Dietary regulation of intestinal ascorbate uptake in guinea pigs. *Am. J. Physiol.* **260**, G108–G118.

Komai, T., and Shindo, H. (1974). Structural specificities for the active transport system of thiamin in rat small intestine. *J. Nutr. Sci. Vitaminol.* **20,** 179–187.

Komai, T., Kawai, K., and Shindo, H. (1974). Active transport of thiamin from rat small intestine. *J. Nutr. Sci. Vitaminol.* **20,** 163–177.

Kozyraki, R., Kristiansen, M., Silahtaroglu, A., Hansen, C., Jacobsen, C., Tommerup, N., Verroust, P. J., and Moestrup, S. K. (1998). The human intrinsic factor-vitamin B12 receptor, cubulin: Molecular characterization and chromosomal mapping of the gene to 10p within the autosomal recessive megaloblastic anemia (MGA1) region. *Blood* **91,** 3593–3600.

Kumar, C. K., Moyer, M. P., Dudeja, P. K., and Said, H. M. (1997). A protein-tyrosine kinase regulated, pH-dependent carrier-mediated uptake system for folate by human normal colonic epithelial cell line NCM 460. *J. Biol. Chem.* **272,** 6226–6231.

Kumar, C. K., Nguyen, T. T., Gonzales, F. B., and Said, H. M. (1998). Comparison of intestinal folate carrier clone expressed in IEC-6 cells and in *Xenopus* oocytes. *Am. J. Physiol.* **274,** C289–C294.

Kyte, J., and Doolittle, R. F. (1982). A simple method for displaying the hydropathic character of a protein. *J. Mol. Biol.* **157,** 105–132.

Labay, V., Raz, T., Baron, D., Mandel, H., Williams, H., Barrett, T., Szargel, R., McDonald, L., Shalata, A., Nosaka, K., Gregory, S., and Cohen, N. (1999). Mutations in *SLC19A2* cause thiamine-responsive megaloblastic anaemia associated with diabetes mellitus and deafness. *Nature Genet.* **22,** 300–304.

Laforenza, U., Gastaldi, G., and Rindi, G. (1993). Thiamin outflow from the enterocyte: A study using basolateral membrane vesicles from rat small intestine. *J. Physiol.* **468,** 401–412.

Laforenza, U., Orsenigo, M. N., and Rindi, G. (1998). A thiamin: H^+ antiport mechanism for thiamin entry into brush border membrane vesicles from rat small intestine. *J. Memb. Biol.* **161,** 151–161.

Laforenza, U., Patrini, C., Alvisi, C., Faelli, A., Licandro, A., and Rindi, G. (1997). Thiamin uptake in human intestinal biopsy specimens, including observations from a patient with acute thiamin deficiency. *Am. J. Clin. Nutr.* **66,** 320–326.

Laframboise, R., Cooper, B. A., and Rosenblatt, D. S. (1992). Malabsorption of vitamin B12 from the intestine in a child with Cbl F disease: Evidence of lysosomal-mediated absorption. *Blood* **80,** 291–92.

Lampen, J., Hahler, G., and Peterson, W. (1942). The occurance of free and bound biotin. *J. Nutr.* **23,** 11–21.

Lashner, B. A., Heidenreich, P. A., Su, G. L., Kane, S. V., and Hanauer, S. B. (1989). The effect of folate supplementation on the incidence of displasia and cancer in chronic ulcerative colitis: a case-control study. *Gastroenterology* **97,** 255–259.

Li, N., Rosenblatt, D. S., Kamen, B. A., Seetharam, S., and Seetharam, B. (1994b). Identification of two mutant alleles of transcobalamin II in an affected family. *Hum. Mol. Genet.* **3,** 1835–1840.

Li, N., Rosenblatt, D. S., and Seetharam, B. (1994a). Nonsense mutations in human transcobalmin II deficiency. *Biochem. Biophys. Res. Commun.* **204,** 1111–1118.

Li, N., Seetharam, S., Lindemans, J., Alpers, D. H., Arwert, F., and Seetharam, B. (1993). Isolation and sequence analysis of variant forms of human transcobalamin II. *Biochim. Biophys. Acta.* **1172,** 21–30.

Li, N., Seetharam, S., Rosenblatt, D. S., and Seetharam, B. (1994c). Expression of transcobalamin II mRNA in human tissues and cultured fibroblasts form normal and TC II deficient patients. *Biochem. J.* **301,** 585–590.

Li, N., Seetharam, S., and Seetharam, B. (1995). Genomic organization of human transcobalamin II gene: Comparison to human intrinsic factor and transcobalamin I. *Biochem. Biophys. Res. Commun.* **208,** 756–764.

Li, N., Seetharam, S., and Seetharam, B. (1998). Characterization of the human transcobalamin II promoter: A proximal GC/GT box is a dominant negative element. *J. Biol. Chem.* **273,** 16104–16111.

Lobie, P. E., Garcia-Aragon, J., and Waters, M. J. (1992). Growth hormone (GH) regulation of gastric structure and function in the GH deficient rat: Up-regulation of intrinsic factor. *Endocrinology* **130,** 3015–3024.

Lorenz, R. G., and Gordon, J. I. (1993). Use of transgenic mice to study the regulation of gene expression in the parietal cell lineage of gastric units. *J. Biol. Chem.* **268,** 26559–26570.

Mellors, A. J., Nahrwold, D. L., and Rose, R. C. (1977). Ascorbic acid flux across the mucosal border of guinea pig and human ileum. *Am. J. Physiol.* **233,** E374–E379.

Middleton, H. M. (1978). Jejunal phosphorylation and dephosphorylation of absorbed pyridoxine HCl *in vitro. Am. J. Physiol.* **253,** E272–E278.

Middleton, H. M. (1979). Intestinal absorption of pyridoxal-5′phosphate disappearance from perfused segments of rat jejunum *in vivo. J. Nutr.* **109,** 975–981.

Middleton, H. M. (1985). Uptake of pyridoxine by *in vivo* perfused segments of rat small intestine: A possible role for intracellular vitamin metabolism. *J. Nutr.* **115,** 1079–1088.

Moestrup, S. K., Kozyarki, R., Kristiansen, M., Kaysen, J. H., Rasmussen, H. H., Brault, D., Pontilon, F., Goda, F. O., Christensen, E. I., Hammond, T. G., and Verrsout, P. J. (1998). The intrinsic factor-vitamin B12 receptor and target of teratogenic antibodies is a megalin-binding peripheral protein with homology to developmental proteins. *J. Biol. Chem.* **273,** 5235–5242.

Nguyen, T. T., Dyer, D. L., Dunning, D. D., Rubin, S. A., and Said, H. M. (1997). Human intestinal folate transport: Cloning, expression, and distribution of complementary RNA. *Gastroenterology* **1112,** 783–791.

Patterson, L. T., Nahrwold, D. L., and Rose, R. C. (1982). Ascorbic acid uptake in guinea pig intestinal mucosa. *Life Sci.* **31,** 2783–2791.

Pinto, J. T., Suffoletto, B. P., Berzin, T. M., Qiao, C. H., Lin, S., Tong, W. P., May, F., Mukherjee, B., and Heston, W. D. W. (1996). Prostate-specific membrane antigen: A novel folate hydrolase in human prostatic carcinoma cells. *Clin. Cancer Res.* **2,** 1445–1451.

Prasad, P. D., Wang, H. Huang, W., Fei, Y., Leibach, F. H., Devoe, L. D., and Ganapathy, V. (1999). Cloning and functional characterization of the intestinal Na^+-dependent multivitamin transporter. *Arch. Biochem. Biophys.* **366,** 95–106.

Prasad, P. D., Wang, H., Kekuda, R., Fujita, T., Feis, Y. J., Devoe, L. D., Beibach, F. H., and Ganapathy, V. (1998b). Cloning and functional expression of a cDNA encoding a mammalian-sodium dependent vitamin transporter mediating the uptake of pantothenate, biotin and lipoate. *J. Biol. Chem.* **273,** 7501–7506.

Qutob, S., Dixon, S. J., and Wilson, J. X. (1998). Insulin stimulates vitamin C recycling and ascorbate accumulation in osteoblastic cells. *Endocrinology* **139,** 51–56.

Ramanujam, K. S., Seetharam, S., Dahms, N., and Seetharam, B. (1991a). Functional expression of intrinsic factor-cobalamin receptor by renal proximal tubular epithelial cells. *J. Biol. Chem.* **266,** 13135–13140.

Ramanujam, K. S., Seetharam, S., Dahms, N. M., and Seetharam, B. (1994). Effect of processing inhibitors on cobalamin transcytosis in polarized opossum kidney cells. *Arch. Biochem. Biophys.* **315,** 8–15.

Ramanujam, K. S., Seetharam, S., Ramasamy, M., and Seetharam, B. (1991b). Cellular expression of cobalamin transport proteins and cobalamin transcytosis in human colon adenocarcinoma cells. *Am. J. Physiol.* **260,** G416–G422.

Ramanujam, K. S., Seetharam, S., and Seetharam, B. (1992). Leupeptin and ammonium chloride inhibit cellular transcytosis of cobalamin in opossum kidney cells. *Biochem. Biophys. Res. Commun.* **182,** 439–446.

Ramanujam, K. S., Seetharam, S., and Seetharam, B. (1993a). Regulated expression of intrinsic factor-cobalamin receptor by rat visceral yolk sac and placental membranes. *Biochim. Biophys. Acta* **1146,** 243–246.

Ramanujam, K. S., Seetharam, S., and Seetharam, B. (1993b). Intrinsic factor-cobalamin receptor activity in a marsupial, the American opossum (*Didelphis virginiana*). *Comp. Biochem. Physiol.* **104,** 771–775.

Reisenauer A. M., Krumdieck, C. L., and Halsted, C. H. (1971). Folate conjugase: Two separate activities in human jejunum. *Science* **198,** 196–197.

Rindi, G. (1984). Thiamin absorption b small intestine. *Acta Vitaminol. Enzymol.* **6,** 47–55.

Rindi, G., and Ferrari, G. (1977). Thiamin transport by human intestine *in vitro. Experentia* **33d,** 211–213.

Rivlin, R. S. (1970). Riboflavin metabolism. *N. Engl. J. Med.* **283,** 463–472.

Rong, N. I., Selhub, J., Goldin, B. R., and Rosenberg, I. (1991). Bacterially synthesized folate in rat large intestine is incorporated into host tissue folylpolyglutamates. *J. Nutr.* **121,** 1955–1959.

Rose, R. C. (1987). Intestinal absorption of water-soluble vitamins. *In* "Physiology of the Gastrointestinal Tract" (L. R. Johnson, ed.), pp. 1581–1596. Raven Press, New York.

Rose, R. C., and Bode, A. M. (1993). Biology of free radical scavengers: An evaluation of ascorbate. *FASEB J.* **7,** 1135–1142.

Rose, R. C., and Choi, J. L. (1990). Intestinal absorption and metabolism of ascorbic acid in rainbow trout. *Am. J. Physiol.* **258,** R1238–R1241.

Rose, R. C., Choi, J. L., and Koch, M. J. (1988). Intestinal transport and metabolism of oxidized ascorbic acid (dehydroascorbic acid). *Am. J. Physiol.* **254,** G824–G828.

Rose, R. C., and Nahrwold, D. L. (1978). Intestinal ascorbic acid transport following diets of high or low ascorbic acid content. *Int. J. Vitam. Nutr. Res.* **48,** 382–386.

Rosenblatt, D. S., Hosack, A., Matiazuk, N. V., Cooper, B. A., and Laframboise, R. (1985). Defect in vitamin B12 release from lysosomes: Newly described inborn error of vitamin B12 metabolism. *Science* **228,** 1319–1321.

Roth-Maier, D. A., Zinner, P. M., and Kerchgessner, M. (1982). Effect of varying dietary vitamin B_6 supply on intestinal absorption of vitamin B_6. *Int. J. Vitam. Nutr. Res.* **52,** 272–279.

Rumsey, S. C., Kwon, O., Xu, G. W., Burant, C. F., Simpson, I., and Levine, M. (1997). Glucose transporter isoforms GLUT1 and GLUT3 transport dehydroascorbic acid. *J. Biol. Chem.* **272,** 18982–18989.

Sahali, D., Mulliez, N., Chatlet, F., Dupuis, R., Ronco, P., and Verroust, PJ. (1988). Characterization of a 280 protein restricted to the coated pits of the renal brush border and the epithelial cells of the yolk sac. *J. Exp. Med.* **167,** 213–218.

Said, H. M. (1999). Cellular uptake of biotin: Mechanisms and regulation. *J. Nutr.* 129:4905–493.

Said, H. M., and Khani, R. (1993). Uptake of riboflavin across the brush border membrane of rat intestine: Regulation by dietary vitamin levels. *Gastroenterology* **105,** 1294–1298.

Said, H. M., and Ma, T. Y. (1994). Mechanism of riboflavin uptake by Caco-2 human intestinal epithelial cells. *Am. J. Physiol.* **266,** G15–G21.

Said, H. M., and Mohammedkhani, R. (1992). Involvement of histidine residues and sulfhydryl groups in the function of the biotin transport carrier of rabbit intestinal brush-border membrane. *Biochim. Biophys. Acta* **1107,** 238–244.

Said, H. M., and Redha, R. (1987). A carrier-mediated transport for folate in basolateral membrane vesicles of rat small intestine. *Biochem. J.* **247,** 141–146.

Said, H. M., and Redha, R. (1988). Ontogenesis of the intestinal transport of biotin in the rat. *Gastroenterology* **94,** 68–72.

Said, H. M., Ghishan, F. K., and Redha, R. (1987a). Folate transport by human intestinal brush-border membrane vesicles. *Am. J. Physiol.* **252,** G229–G236.

Said, H. M., and Hollander, D. (1985). A dual concentration dependent transport system for riboflavin in rat intestine *in vitro. Nutr. Res.* **5,** 1269–1279.

Said, H. M., Hollander, D., and Khani, R. (1993a). Uptake of riboflavin by intestinal basolateral membrane vesicles: A specialized carrier-mediated process. *Biochim. Biophys. Acta.* **1148,** 263–268.

Said, H. M., Khani, R., and McCloud, E. (1993b). Mechanism of transport of riboflavin in rabbit intestinal brush border membrane vesicles. *Proc. Soc. Exp. Biol. Med.* **202,** 428–434.

Said, H. M., Ma, T. Y., and Grant, K. (1994). Regulation of riboflavin intestinal uptake by protein kinase A: Studies with Caco-2 cells. *Am. J. Physiol.* **267,** G955–G959.

Said, H. M., Ma, T. Y., Ortiz, A., Tapia, A., and Valerio, C. K. (1997). Intracellular regulation of intestinal folate uptake: Studies with cultured IEC-6 epithelial cells. *Am. J. Physiol.* **272,** C729–C736.

Said, H. M., Mock, D. M., and Collins, J. (1989). Regulation of intestinal biotin transport in the rat: Effect of biotin deficiency and supplementation. *Am. J. Physiol.* **256,** G306–G311.

Said, H. M., Nguyen, T. T., Dyer, D. L., Cowan, K. H., and Rubin, S. A. (1996). Intestinal transport of folate: Identification of a mouse intestinal cDNA and localization of its mRNA. *Biochim. Biophys. Acta* **1281,** 164–172.

Said, H. M., Ortiz, A., McCloud, E., Dyer, D., Moyer, M. P., and Rubin, S. A. (1998). Biotin uptake by the human colonic epithelial cells NCM460: A carrier-mediated process shared with pantothenic acid. *Am. J. Physiol.* **275,** C1365–C1371.

Said, H. M., Ortiz, A., Kumar, C.K., Chartterjee, N., Dudeja, P.K., and Rubin, S.A. (1999). Transport of thiamine in the human intestine: Mechanism and regulation in intestinal epithelial cell model Caco-2. *Am. J. Physiol: Cell Physiol.* 277:C645–651.

Said, H. M., Redha, R., and Nylander, W. (1987b). A carrier-mediated Na$^+$-gradient dependent transport for biotin in human intestinal brush-border membrane vesicles. *Am. J. Physiol.* **253,** G631–G636.

Said, H. M., Redha, R., and Nylander, W. (1988a). Biotin transport in basolateral membrane vesicles of human intestine. *Gastroenterology* **94,** 1157–1163.

Said, H. M., Redha, R., Tipton, W., and Nylander, W. (1988b). Folate transport in ileal brush border membrane vesicles following extensive resection of proximal and middle small intestine in the rat. *Am. J. Clin. Nutr.* **47,** 75–79.

Said, H. M., Thuy, L. P., Sweetman, L., and Schatzman, B. (1993c). Transport of the biotin dietary derivative biocytin (*N*-biotinyl-L-lysine) in rat small intestine. *Gastroenterology* **104,** 75–79.

Schell, D. A., and Bode, A. M. (1993). Measurement of ascorbic acid and dehydroascorbic acid in mammalian tissue utilizing HPLC and electrochemical detection. *Biomed. Chromatography* **7,** 267–272.

Schron, C. M., Washington, C., and Blitzer, B. (1985). The trans-membrane pH gradient drives uphill folate transport in rabbit jejunum. *J. Clin. Invest.* **76,** 2030–2033.

Schron, C. M., Washington, C., and Blitzer, B. (1988). Anion specificity of the jejunal folate carrier: Effects of reduced folate analogues on folate uptake and effect. *J. Membr. Biol.* **102,** 175–183.

Seetharam, B. (1994). Gastric intrinsic factor and cobalamin absorption. *In* "Physiology of the Gastrointestinal Tract" (Leonard Johnson, ed.), Chap. 62, pp. 1997–2026. Raven Press, New York.

Seetharam, B., Bagur, S. S., and Alpers, D. H. (1981) Interaction of receptor for intrinsic factor-cobalamin complex with synthetic and brush border lipids. *J. Biol. Chem.* **256,** 9813–9815.

Seetharam, B., Christensen, I., Moestrup, S. K., Hammond, T. G., and Verroust, P. J. (1997). Identification of rat yolk sac teratogenic antibodies, gp280, as intrinsic factor-cobalamin receptor. *J. Clin. Invest.* **99,** 2317–2322.

Seetharam, B., Levine, J. S., Ramasamy, M., and Alpers, D. H. (1988). Purification, properties, and immunochemical localization of a receptor for intrinsic factor-cobalamin complex in the rat kidney. *J. Biol. Chem.* **263,** 4443–4449.

Seetharam, S., Ramanujam, K., and Seetharam, B. (1992). Biosynthesis and processing of intrinsic factor cobalamin receptor. *J. Biol. Chem.* **267,** 7421–7427.

Selhub, J, and Rosenberg, J. H. (1981). Folate absorption in isolated brush border membrane vesicles from rat intestine. *J. Biol. Chem.* **256,** 4489–4493.

Selhub, J., Emmanonel, D., Stauropoulons, T., and Arnold, R. (1987). Renal folate absorption and the kidney folate binding protein: 1–Urinary clearance studies. *Am. J. Physiol.* **252,** F750–F756.

Serebro, H. A., Solomon, H. M., Johnson, J. H., and Hendrix, T. R. (1966). The intestinal absorption of vitamin B$_6$ compounds by the rat and hamster. *Johns Hopkins Hosp. Bull.* **119,** 166–171.

Shao, J.-S., Schepp, W., and Alpers, D. H. (1998). Expression of intrinsic factor and pepsinogen in the rat stomach identifies a subset of parietal cells. *Am. J. Physiol.* **274,** G62–G70.

Shibata, K., Gross, C. J., and Henderson, L. M. (1983). Hydrolysis and absorption of pantothenate and its coenzymes in the rat small intestine. *J. Nutr.* **113,** 2107–2115.

Siliprandi, L., Vanni, P., Kessler, M., and Semenza, G. (1979). Na$^+$-dependent electroneutral L-ascorbate transport across brush-border membrane vesicles from guinea pig small intestine. *Biochim. Biophys. Acta.* **552,** 129–142.

Simpson, K. W., Alpers, D. H., DeWille, J., Swanson, P., Farmer, S., and Sherding, R. G. (1993). Cellular localization and hormonal regulation of pancreatic intrinsic factor secretion in the dog. *Am. J. Physiol.* **265,** G178–G188.

Siushansian, R., Tao, L., Dixon, S. J., and Wilson, J. X. (1997). Cerebral astrocytes transport ascorbic acid and dehydroascorbic acid through distinct mechanisms regulated by cyclic AMP. *J Neurochem.* **68,** 2378–2385.

Sklan, D., and Trostler, N. (1977). Site and extent of thiamin absorption in the rat. *J. Nutr.* **107,** 353–356.

Sorrell, M. F., Frank, O., Thomson, A. D., Aquino, A., and Baker, H. (1971). Absorption of vitamins from the large intestine. *Nutr. Res. Int.* **3,** 143–148.

Spector, R., and Mock, D. (1987). Biotin transport through the blood brain barrier. *J. Neurochem.* **48,** 400–404.

Stevenson, N., and Bush, M. (1969). Existence and characteristics of Na$^+$-dependent active transport of ascorbic acid in guinea pig. *Am. J. Clin. Nutr.* **22,** 318–326.

Sweiry, J. H., and Yudilevich, L. (1988). Characterization of folate uptake in guinea pig placenta. *Am. J. Physiol.* **254,** C735–C743.

Tang, L. H., Chokshi, H., Hu, C., Gordon, M. M., and Alpers, D. H. (1992). The intrinsic factor-cobalamin receptor binding site is located in the amino terminal portion of IF. *J. Biol. Chem.* **267,** 22982–22986.

Thomson, A. D., and Leevy, C. M. (1972). Observations on the mechanism of thiamin hydrochloride absorption man. *Clin. Sci.* **43,** 153–163.

Tsuji, T., Yamada, R., and Nose, Y. (1973). Intestinal absorption of vitamin B6. I. Pyridoxal uptake by rat intestinal tissue. *J. Nutr. Sci. Vitaminol.* **19,** 401–417.

Tsukaguchi, H, Tokui, T., Mackenzie, B., Berger, U. V., Chen, X., Wang, Y., Brubaker, R. F., and Hediger, M. A. (1999). A family of mammalian Na$^+$-dependent L-ascorbic acid transporters. *Nature* **399,** 70–75.

Vassiliadis, A., Rosenblatt, D. S., Cooper, B. A., and Bergeron, J. J. M. (1991). Lysosomal cobalamin accumulation in fibroblasts from a patient with an inborn error of cobalamin metabolism (Cbl F complementation group): Visualization by electronmicroscope autoradiography. *Exp. Cell Res.* **195,** 295–302.

Vera, J. C., Rivas, C. I., Fishbarg, J., and Golde, D. W. (1993). Mammalian facilitative hexose transporters mediates the transport of dehydroascorbic acid. *Nature* **364,** 79–82.

Wang, T., Reisenauer, A. M., and Halsted, C. H. (1986a). Comparison of folate conjugase activities in human, pig, rat, and monkey intestine. *J. Nutr.* **115,** 814–819.

Wang, T., Reisenauer, A. M., and Halsted, C. H. (1986b). Intracellular pteroylpolyglutamate hydrolase from human jejunal mucosa: Isolation and characterization. *J. Biol. Chem.* **261,** 13551–13555.

Wen, J., Gordon, M. M., and Alpers, D. H. (1997). A receptor binding site on intrinsic factor is located between amino acids 25–44 and interacts with other parts of the protein. *Biochem. Biophys. Res. Commun.* **231,** 48–51.

Witman, S.D., Lark, R.H., Coney, L.R., Fort, D. W., Frasca, V., Zurawski, V. R., and Kamen, B. A. (1992). Distribution of the folate receptor in normal and malignant cell lines and tissues. *Cancer Res.* **52,** 3396–3401.

Wolf, B., Heard, G. S., Secor-McVoy, J. R., and Raetz, H. M. (1984). Biotinidase deficiency: The possible role of biotinidase in the processing of dietary protein-bound biotin. *J. Inherited Metab. Dis.* **7,** 121–122.

Wrong, O. M., Edmonds, C. J., and Chadwich, V. S. (1981). Vitamins. *In* "The Large Intestine; Its Role in Mammalian Nutrition and Homeostasis" (Wrong, O. M., Edmonds, C. J., and Chadwick, V. S., eds.), pp. 157–166. Wiley, New York.

Yoshida, S., Hayashi, K., and Kawasaki, T. (1981). Pyridoxine transport in brush border membrane vesicles of guinea pig jejunum. *J. Nutr. Sci. Vitaminol.* **27,** 311–317.

CHAPTER 3

Specialized Properties of Colonic Epithelial Membranes: Apparent Permeability Barrier in Colonic Crypts

Satish K. Singh and Henry J. Binder
Departments of Internal Medicine and Cellular and Molecular Physiology, Yale University, New Haven, CT, 06520-8019

I. INTRODUCTION

The mammalian large intestine has several functions, and the epithelial cells lining the colon have specialized transport properties that reflect these functions. The colon plays a major role in fluid and electrolyte homeostasis. Further, the colon is a fermentation compartment that controls disposal of digestive waste material. In health, the colon absorbs substantial amounts of water, serving to

Current Topics in Membranes, Volume 50

dehydrate feces. The normal human colon absorbs Na^+, Cl^-, and water, and secretes K^+ and HCO_3^-. The characteristics of ion transport in the mammalian colon have been studied for more than 40 years and reflect, in part, the scavenger function of the colon, in that aldosterone substantially up-regulates Na^+ absorption. In recent studies, summarized in this chapter, we have established that the apical membrane of the colonic crypt appears to possess a diffusion barrier to weak acid, weak base, and gas movement that serves to maintain the intracellular pH (pH_i) of crypt epithelial cells relatively constant. In view of the high concentration of weak acids and bases produced by colonic microflora from nonabsorbed dietary material, a diffusion barrier would be crucial for the less mature crypt cells because proliferative functions are especially pH sensitive.

II. COLONIC ION TRANSPORT PROCESSES

A. Fluid and Ion Movements

Transepithelial water movement is the result of osmotic pressure gradients created by solute transport. Mammalian colon actively absorbs Na^+ against a significant Na^+ gradient (Devroede and Phillips, 1969) that serves as a scavenger to retain Na^+ at times of Na depletion and/or reduced dietary Na intake. Net water absorption is dependent on—and proportional to—the rate of Na^+ absorption (Curran and Schwartz, 1960). In rat distal colon, Na^+ uptake is primarily electroneutral and Cl^- dependent, representing parallel apical Na-H and Cl-HCO_3 exchange processes (Binder and Rawlins, 1973; Charney and Feldman, 1984; Binder et al., 1987; Rajendran and Binder, 1990; Rajendran and Binder, 1993). All colonic epithelia, like other epithelial cells, share a common Na,K-ATPase that extrudes Na^+ across the basolateral membrane. Under the influence of aldosterone, electroneutral Na^+ absorption becomes electrogenic, consistent with the down-regulation of Na-H exchange and induction of Na^+ channels (Foster et al., 1983; Halevy et al., 1986). Overall colonic Cl^- absorption is linked to HCO_3^- secretion (Smith et al., 1986; Feldman and Stephenson, 1990).

Extensive in vitro flux-chamber studies in rat, rabbit, and human small and large intestine have consistently demonstrated secretagogue-induced active Cl^- secretion. Cl^- secretion is largely driven by a basolateral electroneutral Na-K-2Cl cotransporter (Heintze et al., 1983; Dharmsathaphorn et al., 1985). In its simplest form, active Cl^- secretion is the result of second-messenger induced apical Cl^- conductances (Weymer et al., 1985). Secreted Cl^- causes an increase in membrane potential, the driving force for passive leakage of Na^+ into the lumen. Secreted Na^+ and Cl^- together create an osmotic pressure gradient that is the driving force for water movement into the lumen.

Colonic HCO_3^- secretion, when studied in intact tissue, is luminal Cl^- dependent, serosal Na^+ dependent, and stilbene sensitive (Frizzell et al., 1976; Smith et al., 1986; Feldman and Stephenson, 1990). Several enterotoxins and neurohumoral agents induce colonic fluid and electrolyte secretion. In vivo studies regularly show a HCO_3^--rich, plasma-like solution (Hubel, 1974; Donowitz and Binder, 1976), although definitive identification of the mechanisms of secretagogue-induced HCO_3^- secretion is lacking. However, when intestinal epithelia have been studied in vitro in flux chambers, secretagogues induced active Cl^- secretion usually with very little evidence for HCO_3^- secretion (Dietz and Field, 1973; Sheerin and Field, 1975; Frizzell et al., 1976); to date there is no clear explanation for this failure to consistently observe HCO_3^- secretion in vitro.

B. Functional Heterogeneity along the Crypt-Surface Axis

Within a given segment, the epithelium of the mammalian large intestine is not structurally or functionally homogeneous. The colonic crypt is a specialized "subset" of the colonic epithelium. Colonic glands or crypts of Lieberkuhn are numerous, $\sim 12,000$ per cm^2 mucosa in the distal colon of the rat (Mendizabal and Naftalin, 1992). Crypts are elongated tubular structures that protrude, in humans, approximately 500 μm into the lamina propria, where they are in close association with neural, vascular, immune, and connective tissue elements. Thus, a substantial fraction (roughly 40%) of the total epithelial surface area in the colon is contained within crypts.

Crypts also are the site of stem cell proliferation. Undifferentiated cells at the base of the crypt divide. Proliferative daughter cells then differentiate and mature into two predominant cell types: (1) Microvacuolated cells are columnar, with a basolaterally located nucleus, prominent apical Golgi apparatus, endoplasmic reticulum, and mitochondria. Microvacuolated cells have rudimentary apical microvilli, and are attached near their apical poles by junctional complexes. It is believed that microvacuolated cells transport ions. (2) Mucus-secreting goblet cells have a nucleus near the basolateral pole and are dominated by mucin-containing granules that compress cell organelles into a rim of cytoplasm at the cell periphery. Both microvacuolated and goblet cells migrate toward the surface and, in 3 to 4 days, are shed from "zones of extrusion" between the openings of adjacent crypts. Enteroendocrine cells exist in the crypt but are few in number, and migrate more slowly to the surface.

A model of a spatial distribution of specialized colonocyte function had been a physiological paradigm for more than 20 years (Welsh et al., 1982). Absorptive processes were believed to be restricted to surface cells, whereas secretory processes were believed to exist exclusively in the epithelial cells within the crypts. However, the data supporting this model of compartmentalized function were based primarily on indirect observations, because technical constraints permitted limited direct determinations of crypt function. The most direct support for heterogeneity of transport processes along the crypt-surface axis came from rabbit colon, and suggested that crypts are the site of secretagogue-induced salt and fluid secretion (Welsh *et al.*, 1982). Welsh *et al.*, found that luminal application of 0.1 mM amiloride, an inhibitor of apical Na^+ channels, increased apical membrane resistance of surface epithelial cells but had no effect in crypt cells. In contrast in crypt, but not in surface epithelial cells, prostaglandin E_2 increased apical membrane conductance presumed to be due to an increase in apical membrane Cl^- permeability. Further, PGE_2 caused microscopic fluid droplets to appear in areas overlying—but not between—crypt openings (Welsh *et al.*, 1982).

Based on these and other observations, a concept of separate, functionally distinct compartments for absorptive and secretory processes between surface and crypt cells came to be well accepted (Serebro *et al.*, 1969; Roggin *et al.*, 1972; Field *et al.*, 1980; Welsh *et al.*, 1982; Braaten *et al.*, 1988; Field, 1991). However, this model also came to imply that secretory processes did *not* exist in surface cells, and that absorptive processes did *not* exist in crypt cells. In fact, previous experimental observations had not established the *absence* of absorptive function in the crypt, nor the *absence* of secretory function in the surface cell. Some studies indicated that the spatial-distribution model may have been an oversimplification, in that surface epithelial cells in the rat distal colon exhibited a Cl^- conductance when exposed to DBcAMP, suggesting that surface cells are capable of secretion (Kockerling and Fromm, 1993). Additionally, work by Naftalin and colleagues indicated that colonic crypt cells possess the ability to absorb a hypertonic fluid (Bleakman and Naftalin, 1990; Naftalin and Pedley, 1990). However, none of the previous studies directly assessed absorptive function in crypt epithelial cells.

C. Direct Study of Transport Functions: The Isolated, Perfused, Colonic Crypt

Our group attempted to assess the absorptive and secretory functions of isolated colonic crypt epithelium by developing and employing a system to directly measure fluid and electrolyte movement in isolated colonic crypts. This system is based on established methods used to study isolated renal tubules for more than two decades. Microperfusion of the lumen of intact crypts permits isolation of solutions bathing the apical and basolateral membranes of an intact epithelium (Burg *et al.*, 1966; Burg, 1972) (see Fig. 1). Colonic crypts lend themselves to microperfusion because they are similar to renal tubules (Lohrmann and Greger, 1995). Both are hollow, tubular structures lined by polarized epithelial cells, the major difference being that the crypt has a blind end. A model system in which epithelial orientation is maintained has obvious advantages for studies of ion and fluid transport in polarized cells.

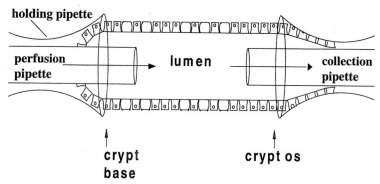

FIGURE 1 General method for microperfusion of isolated colonic crypts. An assembly of concentric glass micropipettes is fashioned to hold and cannulate the crypt lumen. Bell-shaped holding pipettes are used to stabilize the crypt immersed in the bath chamber. On the afferent side of the perfusion system, fluid is introduced into the lumen via a perfusion pipette that is passed through a bell-shaped holding pipette. The perfusion pipette is used to perforate the base of the crypt in order to cannulate the lumen of the crypt. Thus, the normally blind-ended crypt is rendered a perfusable tubular epithelial structure. On the efferent side of the perfusion system, a mated bell-shaped holding pipette is used to stabilize the crypt os. Fluid that has passed through the crypt lumen is collected with a calibrated collection pipette. Isolated crypts are perfused using techniques identical to those used for the perfusion of isolated nephron segments.

1. The Technique for Luminal Perfusion of Colonic Crypts

Techniques are adapted from the microperfusion of nephron segments, and rat colon is removed and immersed in ice-cold HEPES-buffered Ringer's solution. Particular segments are identified from anatomic features and stripped of the muscle layers. The remaining epithelial sheet is placed in a cooled (4°C) well on the stage of a dissecting microscope. Under direct visualization, sharpened forceps are used to tease out clusters of colonic crypts. Within these clusters, single, intact, nonbranched crypts are selected. With great care to avoid direct contact with epithelial cells, the connective tissue sheath surrounding the crypt is teased away with sharpened forceps. The isolated, free-floating crypt is sucked from the well into the tip of a glass transfer pipette and placed into a temperature-controlled, superfusable chamber on the stage of an inverted microscope capable of both epi- and trans-illumination. An assembly of concentric glass micropipettes mounted on V tracks (Fig. 1) is used to hold the blind end of the crypt. A perfusion pipette is driven into the blind end of the crypt to introduce perfusate into the lumen. A second set of micropipettes is used to cannulate the open end of the crypt and collect the effluent. Thus, solutions flow continuously through the lumen in a direction from the base to the os, with a collection rate of ~4 nl/min. Dead space is minimized by delivery of solutions near the tip of the perfusion pipette via a fluid exchange pipette system open to atmospheric pressure (Greger and Hampel, 1981). Simultaneously, the bath (i.e., serosal or blood side of the crypt) is superfused continuously in the chamber at a rate of ~3 ml/min. Both lumen and bath solutions are delivered without disrupting flow, using multiple-inflow single-outflow, sliding valves. Measured aliquots of effluent are sampled with a volume-calibrated pipette.

In this manner, we microperfused single isolated crypts from rat distal colon to assess fluid movement into and from the crypt lumen (Singh *et al.*, 1995). Studies with ^{3}H-inulin as a nonabsorbed marker demonstrated that crypts absorb fluid constitutively and that this absorptive process is Na^+ dependent (Fig. 2). Further, perfused crypts could be stimulated to secrete reversibly by increasing intracellular cAMP with the membrane-permeant form dibutyryl cAMP (Fig. 3).

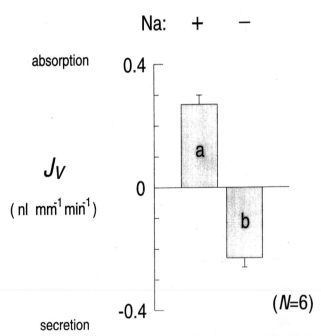

FIGURE 2 Net fluid movement (J_v) in isolated perfused rat distal colon crypts. J_v was determined in paired experiments when bath and lumen were perfused with a CO_2/HCO_3^--containing Ringer's solution (bar a), and with simultaneous perfusion of both bath and lumen with a Na^+-free solution in which the Na^+ was replaced with choline (bar b). A positive value for J_v indicates absorption; a negative value indicates secretion. The vertical hatch marks indicate standard errors of the mean; the number of paired observations appears in parentheses. Reprinted with permission from Singh, S. K., Binder, H. J., Boron, W. F., and Geibel, J. P. (1995). Fluid absorption in isolated perfused colonic crypts. (*J. Clin. Invest.* **96**, 2373-2379).

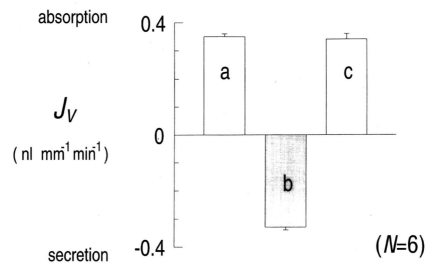

FIGURE 3   Effect of dibutyryl-cAMP on net fluid movement (J_v) in perfused rat distal colon crypts. J_v was measured in paired experiments on isolated perfused colonic crypts in the resting state (bar a), in the presence of 1 mM dibutyryl-cAMP (DBcAMP) added to the bath solution (bar b), and after washout of the drug (bar c). Both bath and lumen were perfused with CO_2/HCO_3^- Ringer's solution. Collections of luminal effluent were begun after 15 minutes of exposure to each solution. A positive value for J_v indicates net absorption; a negative value indicates net secretion. The vertical hatch marks indicate standard errors of the mean; the number of paired observations appears in parentheses. Reprinted with permission from Singh, S. K., Binder, H. J., Boron, W. F., and Geibel, J. P. (1995). Fluid absorption in isolated perfused colonic crypts. (*J. Clin. Invest.* **96**, 2373–2379).

Additionally, secretagogues that act as receptors localized to the basolateral domain and that induce secretion via distinct intracellular signaling pathways induced fluid secretion in perfused crypts. Vasoactive intestinal peptide, the hormone that stimulates adenylyl cyclase, and the neurotransmitter acetylcholine, which elevates cytosolic Ca^{2+}, both stimulated crypts to change reversibly from a state of net fluid absorption to secretion. Thus, it appears that crypt epithelium is capable of absorption as well as secretion, and that its function *in vivo* may be dictated largely by neurohumoral influences originating in the lamina propria. These findings directly challenge the long-standing bias that the crypt is capable of fluid secretion alone and lend support to the hypothesis that colonic crypts have a role in the dehydration of stool. However, since the isolated crypt preparation lacks nonepithelial elements, we have speculated that the observed absorption may result from a preparation devoid of tonic secretory stimulation from lamina propria cells, including the pericryptal myofibroblasts. Thus, crypts may very well secrete or perform no net fluid transport in their basal state *in situ* when they are potentially regulated by the several neurohumoral agonists released by multiple cells in the lamina propria (Bern *et al.,* 1989; Berschneider and Powell, 1992).

2. Fluorescence Imaging of Cells in Perfused Crypts

We also initiated studies using an intracellular pH (pH$_i$)-sensitive dye, BCECF-AM, to study ion and acid-base transport processes at the cellular level in crypt colonocytes. We combined microperfusion techniques with ratiometric fluorescence imaging, which provided two critical advantages over previous experimental methods: (1) acid-base transport could be studied in an intact polarized epithelial cell preparation at the apical (separate from the basolateral) membrane of individual cells, and (2) cells of interest could be positioned in situ along the maturational continuum of the crypt-surface axis.

We removed a segment of proximal colon, namely the fusus coli (Snipes *et al.,* 1982), from New Zealand white rabbits, dissecting single colonic crypts and perfusing them according to the same approach used with rat crypts (Fig. 4A). We chose the fusus coli because crypts in this region are numerous and of sufficient length (500–1000 μM) for optical studies. Moreover, the fusus coli has high rates of electrolyte and fluid transport *in vivo* (Snipes *et al.,* 1982). Light and transmission electron microscopy of sections of dissected crypts from the rabbit fusus coli (not shown) also confirmed that these crypts are composed primarily of two cell types: mucus-containing goblet cells and microvacuolated columnar-type cells. We noted no apparent extraepithelial cellular elements.

The chamber in which we perfused the crypts had a floor that consisted of a glass coverslip. We mounted the chamber on the stage of a Zeiss IM-35 inverted microscope, equipped for epi-illumination. We again used an assembly of concentric glass micropipettes to hold the blind end of the crypt, puncturing it with the perfusion pipette (i.e., the pipette used to deliver solutions to the lumen). We used a "holding" pipette to stabilize the open end of the crypt and collect effluent. We loaded crypt cells with the pH-sensitive dye BCECF by switching the bath to a HEPES-Ringer solution containing 10 μM BCECF-AM, a membrane permeant form of the dye, from which cytoplasmic esterases cleave the acetoxy methyl ester, thus trapping the dye within cells. As seen in Fig. 4B, dye uptake was not homogeneous. Examination by laser-scanning confocal microscopy (not shown) showed that BCECF localized to the microvacuolated cells. The goblet cells did not accumulate appreciable dye; large mucin aggregates displaced the cytoplasm to a thin rim of negligible fluorescence.

We used a solid-state intensified television system with digital image acquisition and analysis to study intracellular pH. We monitored fluorescence emission at 530 nm, while alternately exciting dye at 440 nm and 490 nm. Between excitation periods, the cells were in the dark to minimize photobleaching of dye and photodynamic damage. We acquired 440/490 data pairs as often as 1 per 2.5 s, although we matched the sampling rate (typically 1 per 5–30 s) to match the

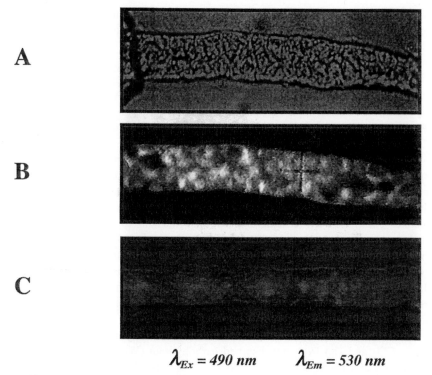

$$\lambda_{Ex} = 490\ nm \qquad \lambda_{Em} = 530\ nm$$

FIGURE 4   Perfusion and imaging of BCECF-loaded crypts. (A) 320× transmitted image of a perfused crypt from rabbit fusus coli. (B) The same crypt after a 10-min exposure to 10 μM BCECF-AM from the bath. BCECF was restricted to microvacuolated crypt cells. Goblet cells did not accumulate appreciable dye as their cytoplasm was filled with mucin. (C) Overlaid transmitted and fluorescent images of a crypt during luminal perfusion with cell-impermeant fluorescein dextran. Luminal dye pulses were observed quantitatively and reliably. Extravasation from lumen to bath was not detected in this experiment.

speed of pH_i changes. We used a program to outline the boundaries of individual crypt cells and grouped pixels by cells. For each cell in each sample, we obtained I_{490} and I_{440} values, and thus the I_{490}/I_{440} ratio. We converted the ratios to pH_i values using the nigericin/high-K^+ technique and averaged the pH_i values for all cells (typically 25) in the crypt to obtain an average pH_i record versus time for the experiment.

In other selected experiments, luminal perfusion was confirmed using the non-absorbable fluorescent marker fluorescein-dextran (40,000 Da, 6 fluorophores/mole; Molecular Probes, Eugene, OR). Fig. 4C is a composite transmitted and fluorescence image of a crypt perfused with 10 μM fluorescein-dextran in HEPES-Ringer solution. Dye filled and was confined to the lumen, and there was no evidence of lumen-to-bath leakage. After dye washout, fluorescence intensity decreased markedly, with only a faint rim of fluorescence remaining near the apical boundaries of the cells.

III. COLONIC LUMEN ECOLOGY—A UNIQUE ACID-BASE ENVIRONMENT

A. Colonic Microflora, Fermentation, and Energy Salvage

Colonic contents are dominated by the metabolic activity of more than 400 species of predominantly anaerobic colonizing bacteria that constitute 55% of fecal solids (Stephen and Cummings, 1980). Bacterial symbiosis permits colonic salvage of energy from dietary components that are not absorbed in the small intestine. The microflora ferments nonabsorbed nutrients to energy-rich products that are absorbed by the colon. This caloric salvage pathway is of major nutritional significance for many species, particularly ruminants that consume and ferment nonstarch polysaccharides from grasses (McBurney and VanSoest, 1991). Fermentation is a series of catabolic, energy-liberating reactions with several by-products. Because of the diversity of bacteria, different types of fermentation occur: (1) Most carbohydrate-fermenting bacteria utilize the Embden-Meyerhoff-Parnas pathway to produce short-chain fatty acids (as well as the gases CO_2 and H_2) from nonabsorbed carbohydrate (Wolin and Miller, 1983; McBurney and Thompson, 1990). About 95% of these short-chain fatty acids are absorbed by the colon, providing 3 to 11% of metabolizable energy intake in humans (Cummings et al., 1987; McBurney et al., 1988). Short-chain fatty acids produced in the colon are essential nutrients for colonocytes (Roediger, 1982) and are utilized by a wide range of tissues (Williamson, 1964; Lundquist et al., 1973). (2) CO_2, H_2, and CH_4 gases are produced by multiple fermentations in the colon (Cummings and Macfarlane, 1991; Macfarlane and Cummings, 1991). In addition, luminal acid can titrate secreted HCO_3^- to CO_2 (Levitt, 1972). (3) High concentrations of ammonia (NH_3/NH_4^+) also result from bacterial urea degradation and protein/amino acid deamination (Summerskill and Wolpert, 1970; Wolpert et al., 1971; Macfarlane et al., 1986; Macfarlane et al., 1988; Macfarlane and Cummings, 1991; Macfarlane et al., 1992). Ammonia either is used by bacteria as a nitrogen source or is absorbed, converted to urea by the liver, and then excreted in urine. This is a major nitrogen salvage pathway in animals, particularly rodents. (4) Other compounds, produced in small quantity from fermentation of carbohydrate and protein, include branched-chain fatty acids (isobutyric acid, isovaleric acid), phenols, and other organic acids.

Fermentation creates a unique acid-base environment in the colon. The pH of colonic contents is ~7.0 (Wrong et al., 1965; Down et al., 1972). In this range, at least three fermentation products could affect the pH$_i$ of colonic epithelial cells:

(1) Ammonia is a lipophilic weak base (pK$_a$ ~9.5) that exists at a high concentration (14–20 mM) in the colonic lumen (Wrong and Vince, 1984; Macfarlane et al., 1986). (2) Short-chain fatty acids are lipophilic, aliphatic, monocarboxylic weak acids (pK$_a$ ~4.8) that are the predominant luminal anion (80–130 mM). Acetic, propionic, and butyric acids constitute more than 95% of colonic short-chain fatty acids (Mitchell et al., 1985; Cummings et al., 1987). (3) Carbon dioxide effectively acts as a permeant weak acid (pK$_a$ 6.1). The CO_2 content of colonic gas is reportedly as high as 60% (Kirk, 1949; Steggerda and Dimmick, 1966).

B. Relationship of Colonic Epithelial Cells to Luminal Contents

All cells, whether polarized or not, regulate pH$_i$ via the coordinated activity of transporters located in the plasma membrane that serve to move acids and bases into and out of the cell. These transporters, among other functions, counteract acid produced from cellular metabolism and compensate for influences on pH$_i$ from the extracellular milieu (Roos and Boron, 1981). For the polarized epithelial cells that line the colonic lumen, two relatively unique phenomena are relevant to the regulation of pH$_i$. First, as described above, the colon is a fermentation compartment. As a result, the apical aspects of colonocytes are exposed to a unique acid-base environment where pH fluctuates and cell-permeant acids and bases (such as ammonia, short chain fatty acids, and CO_2) are abundant (Moore et al., 1978). Second, the colonic mucosa undergoes continuous and rapid renewal. Crypts are the site of epithelial cell proliferation, where stem cells at the base of the crypt proliferate, differentiate, migrate to the surface, and shed into the fecal stream. Therefore, at any given time, the epithelial cells that comprise the crypt represent a "snapshot" of a heterogeneous population of cells along a proliferative and maturational continuum. The "proliferative zone" in the bottom third of the crypt nearest the base contains a larger proportion of actively replicating cells than do the more superficial regions (Sunter et al., 1978; Goodlad et al., 1992). Cellular proliferation and the modulation of pH$_i$ are well recognized to be closely associated processes (Johnson and Epel, 1976; Webb and Nuccitelli, 1981; Busa and Nuccitelli, 1984; Pouyssegur et al., 1984; Aerts et al., 1985; Boonstra et al., 1988; Grandin and Charbonneau, 1990; Grandin et al., 1991; Ganz and Boron, 1994). Further, pH$_i$ homeostasis is vital to normal cell function (Busa, 1986). The activities of phosphofructokinase (the rate-limiting enzyme in glycolysis) (Trivedi and Danforth, 1966) and the ribosomal S6 protein (Pouyssegur et al., 1982) are exquisitely sensitive to changes in pH$_i$, as is the function of certain K^+ channels (Fitz et al., 1989; Copello et al., 1991).

1. Ion Transporters Located in the Plasma Membrane Regulate Cell pH
Acid-base transporters located in the plasma membrane that participate in the

regulation of pH_i can be broadly classified into two categories: (1) "Acid extruders," which tend to increase pH_i, can be subclassified into primary active transporters (e.g., the H^+-pump and gastric and colonic H-K exchanging ATPases), secondary active transporters (e.g., Na-H exchange and Na^+-dependent Cl-HCO_3 exchange), and tertiary monocarboxylic acid transporters (Boron, 1992). Some cells, including those of the rat distal colon, possess Na-HCO_3 cotransporters that appear to function as acid extruders (Boron and Boulpaep, 1983; Deitmer and Schlue, 1989; Rajendran et al., 1991). Na-H exchange and Na^+-dependent Cl-HCO_3 exchange are strongly pH_i-dependent, becoming more active at low pH_i and less active at high pH_i (Boron et al., 1979; Boyarsky et al., 1990). (2) "Acid loaders," which tend to decrease pH_i, include Cl-HCO_3 and Cl-OH exchange, as well as Na/HCO_3 cotransport. Cl-HCO_3 exchangers are pH_i-dependent as well, and are more active at alkaline pH_i. In addition, cellular metabolism as well as the environmental milieu tends to produce a tonic, "background" acid load (Boyarsky et al., 1990; Boron, 1992; Sjaastad et al., 1992). Thus, steady-state pH_i is determined by the balance between *all* acid-loading and acid-extruding processes.

2. Acid-Base Transport and pH_i Regulation in Colonic Crypt Cells

In isolated, dispersed human colonic crypt cells, Na-H exchangers and Na/HCO_3 cotransporters act as acid extruders, and an apparent Na^+-dependent Cl-HCO_3 exchange functions as an acid extruder. In superfused crypts from rat distal colon, Na-H exchange participates in recovery from SCFA-induced acid loads (Diener et al., 1993). Because both studies were performed on nonpolarized colonocyte preparations, the apical versus basolateral location of the transporters described is not known.

3. General Effect of Weak Bases and Weak Acids on pH_i

The extracellular environment has a major impact on pH_i via fluxes of permeant buffers across the plasma membrane. At physiological pH, a portion of any weak acid ($HA \leftrightarrow H^+ + A^-$) or base ($B + H^+ \leftrightarrow BH^+$) in solution is ionized, and thus water-soluble. The remainder of the compound is nonionized, and more lipid-soluble. Generally, lipid membranes are far more permeable to the lipid-soluble neutral species (HA and B) of a buffer pair than to the charged species (A^- and BH^+). The equilibrium ratio of nonionized to ionized species (HA/A^- and B/BH^+) is a function of the dissociation constant of the acid or base (K_a). The extent of association or dissociation of a weak acid or base is related to the pH of the solution and given by the Henderson-Hasselbalch equation, that is, $pH = pK_a + \log ([A^-]/[HA]$. Thus, at pH near 7.0 (as in the colon), this equation predicts that either a weak acid with a pK_a of 5 or a weak base with a pK_a of 9 will have a ratio of neutral to ionic forms of 1:100. Exposing a cell to a permeant buffer will result in pH_i changes over time that are a function of the total concentration of the buffer, the extracellular pH, the effective permeability of the membrane to the components of the buffer pair, the initial pH_i, and the cellular buffering capacity (β) (Roos and Boron, 1981).

4. General Effect of Neutral Weak Bases on pH_i

Because membrane permeability to BH^+ is much less than to B, BH^+ fluxes have a smaller effect on pH_i. When a cell is exposed to B/BH^+, B rapidly enters the cell across the plasma membrane. Most B combines with intracellular H^+ to form BH^+, resulting in a pH_i increase (B + H^+ $\leftrightarrow$ BH^+). Eventually extra- and intracellular [B] become equal, and net B influx ceases. When this happens, the pH_i increase ceases, and pH_i stabilizes (the "plateau phase"), provided that BH^+ does not enter and that the transport rate for other acids or bases is unchanged. Removal of extracellular B/BH^+ causes all these processes to reverse, so that in theory pH_i should return exactly to its initial value. However, this is not the case with NH_3/NH_4^+ because the influx of NH_4^+ leads to a slow decrease in pH_i during the plateau phase. After the NH_3/NH_4^+ is removed, pH_i decreases markedly, to a value that is less than the initial pH_i (an "undershoot"). This NH_3/NH_4^+ "prepulse technique" has wide utility among many cell types as a means of acid loading cells in order to study pH_i regulation, as well as a means to study the permeability of the plasma membrane to NH_3 and NH_4^+. In the colon, NH_3/NH_4^+ is present at 14 to 20 mM. Thus, handling of NH_3/NH_4^+ by crypt colonocytes is of interest because ammonia is part of their native milieu.

5. General Effect of Neutral Weak Acids and CO_2 on pH_i

The total short-chain fatty acid content of the colon exceeds 100 mM (Cummings et al., 1987), and P_{CO_2} can reach 60 mm Hg (Steggerda and Dimmick, 1966). Typically, if a cell is exposed to a monoprotic weak acid, the neutral HA molecule rapidly enters the cell and dissociates to form the conjugate weak base A^- as well as H^+ ([HA] $\leftrightarrow$ [H^+] + [A^-]). Thus, the influx of HA is accompanied by a decrease in pH_i. If A^- cannot cross the membrane, and if there is no compensatory acid-base transport, HA entry (and the pH_i decrease) continues until HA equilibrates across the membrane. Thus, under ideal conditions, one expects that exposing a cell to HA will cause pH_i to decrease to a new stable value.

When a cell is exposed to a CO_2/HCO_3^- solution, CO_2 influx dominates initially, causing a rapid pH_i decrease owing to the intracellular hydration of CO_2 ($CO_2 + H_2O(H_2CO_3 \leftrightarrow HCO_3^- + H^+$, overall $pK_a = 6.4$). The first of these two reactions is, in effect, catalyzed by carbonic anhydrase (CA). As CO_2 continues to diffuse into the cell, intracellular [CO_2], [HCO_3^-], and [H^+] increase until the overall equilibrium $CO_2 + H_2O \leftrightarrow HCO_3^- + H^+$ is achieved inside the cell, and until [CO_2] is the same inside and outside the cell. The magnitude of the CO_2-induced acidification is accentuated by the extracellular [CO_2], the initial pH_i, and by the total intracellular buffering power.

IV. EVIDENCE FOR A BARRIER TO NEUTRAL MOLECULES IN CRYPTS

It is widely accepted that the permeability of a biological membrane to a solute is directly related to the lipophilicity of the solute. Thus, as outlined earlier, cell membranes generally are far more permeable to neutral forms of weak acids and bases than to their complementary ionized species. However, as outlined, the colonic flora ferments nonabsorbed dietary nutrients (i.e., large carbohydrates and proteins) and produces large quantities of NH_3, short-chain fatty acids (neutral weak acids), and CO_2, which then are absorbed by the colon. Each of these acid-base species possesses a small, lipophilic, membrane permeant form that would be predicted to impact pH_i substantially and affect colonocyte function.

A. An Apical Barrier to a Permeant Weak Base in Perfused Crypts

Human colonic effluent contains typically 14 to 20 mM of NH_3/NH_4^+. This relatively high concentration is the result of the degradation of urea and the deamination of amino acids by colonic bacteria (Wrong and Vince, 1984; Macfarlane et al., 1986). Up to 4 g of NH_3/NH_4^+ is absorbed daily by the colon, making it by far the major site of ammonia production and absorption (Walser and Rodenlas, 1959; Wrong, 1978). Because NH_3/NH_4^+ has a pK_a of ~9.5, the majority of total ammonia is NH_4^+ at the neutral pH that prevails in the colonic lumen (Wrong and Metcalfe-Gibson, 1965; Down et al., 1972). Nevertheless, sufficient NH_3 is present that pH_i should be affected by diffusion of this weak base from the colonic lumen into colonocytes (Down et al., 1972). As such, if NH_3 were to cross the apical borders of these cells, the resultant alkali load could impair colonocyte function.

1. The Apical Membrane of Crypt Cells Has Low NH_3/NH_4^+ Permeability

In perfused crypts we found that, although exposing the basolateral surface of crypts to NH_3/NH_4^+ caused expected pH_i transients, exposing the apical surface to NH_3/NH_4^+ had no detectable effect on pH_i, even for very high concentrations of NH_3. This finding provided the first suggestion that the apical barrier of crypt colonocytes possesses specialized permeability properties. Figure 5A shows an experiment in which we examined the effect of applying 20 mM NH_3/NH_4^+ (extracellular pH fixed at 7.40) to either the lumen or the bath. The luminal exposure did not change pH_i, and similarly, removing NH_3/NH_4^+ from the lumen also had no effect on pH_i. In contrast, when we exposed the same cells to 20 mM NH_3/NH_4^+ from the bath, pH_i increased rapidly, presumably because of rapid diffusion of NH_3 into the cell, followed by protonation to form NH_4^+. A slower acidification followed, owing to NH_4^+ influx and/or other acidifying processes

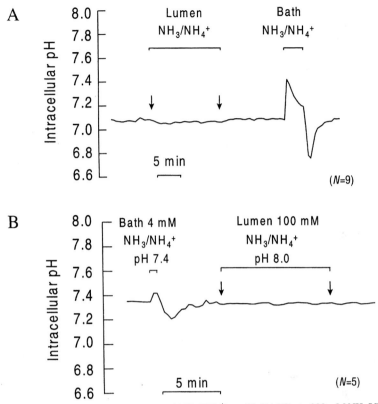

FIGURE 5   Effect of apical and basolateral NH_3/NH_4^+ on pH_i. (A) Effect of 20 mM NH_3/NH_4^+, pH 7.4, from lumen and bath. (B) Effect of 4 mM NH_3/NH_4^+, pH 7.4, from bath, followed by 100 mM NH_3/NH_4^+, pH 8.0, from lumen. The pairs of arrows indicate the times luminal solutions were changed. Adapted with permission from Singh, S. K., Binder, H. J., Geibel, J. P., and Boron, W. F. (1995). An apical permeability barrier to NH_3/NH_4^+ in isolated, perfused colonic crypts. (*Proc. Natl. Acad. Sci. USA* **92**, 11573–11577).

(Roos and Boron, 1981). Removing bath NH_3/NH_4^+ caused the opposite pH_i transients, that is, pH_i decreased rapidly, presumably because of NH_3 efflux, and then recovered slowly because of the action of homeostatic acid-extruding processes.

That lumen NH_3/NH_4^+ caused no pH_i change was consistent with the presence of an apical permeability barrier and/or absence of a carrier for NH_3 and NH_4^+. However, there were several alternative explanations that required consideration. The first possibility was that the apical membrane was so highly permeable to NH_4^+ that the parallel fluxes of the weak-base NH_3 and the conjugate weak-acid NH_4^+ produced no net change in pH_i under the imposed experimental conditions. This explanation is unlikely because the flux ratio of NH_3 (compared with that of NH_4^+) needed to produce no net pH_i change would vary exponentially with pH_i. Nonetheless, to test this hypothesis directly, we manipulated the $[NH_4^+]/[NH_3]$ ratio in the luminal fluid by increasing total luminal ammonia to 100 mM and by increasing the pH of the solution to 8.0. This represents a $\sim$20-fold increase in $[NH_3]$, but a $\sim$fivefold increase in $[NH_4^+]$. Despite this proportional increase in lumen $[NH_3]$ relative to $[NH_4^+]$, when this solution was introduced into the crypt lumen, no pH_i change was observed. On the other hand, when the same crypts were pulsed with 20 mM NH_4Cl at pH 7.4 from the bath, the expected pH_i transients occurred (Fig. 5B).

As a further test of the parallel NH_3/NH_4^+ influx hypothesis, and to better quantitate the differential permeability to NH_3 at the apical versus basolateral membranes, we performed a study in which a low $[NH_3]$ was presented to the bath and a high $[NH_3]$ was presented to the lumen. In this experiment (Fig. 5B), we first examined the effect on pH_i of exposing the basolateral surface of the crypt to 0.075 mM NH_3, achieved by decreasing total bath ammonia to 4 mM at a fixed extracellular pH of 7.4. Pulsing at this low level of bath NH_3/NH_4^+ caused the expected series of pH_i changes, the magnitudes of which were substantially smaller than when crypts were exposed from their basolateral aspect to 20 mM NH_4Cl. The bath pulse of 4 mM NH_3/NH_4^+ at pH 7.4 caused pH_i to increase by an average of 0.076 after 30 s. On the other hand, when the lumen of the same crypt was pulsed with a solution containing 100-fold greater $[NH_3]$, and 23-fold greater $[NH_4^+]$, there was no detectable effect on pH_i. Thus, the model of parallel influx of NH_3 and NH_4^+ was all but excluded. These data provided further indication that the *permeability* $\times$ *area* product for NH_3 is much higher at the basolateral cell border than at the apical cell border.

We used fluorescein-dextran in a poorly buffered solution to monitor the pH of the lumen to examine other possible explanations for the lack of effect of luminal NH_3/NH_4^+ on pH_i. These alternative scenarios were as follows: (1) Lumen perfusion solutions did not reach the lumen. However, fluorescein-dextran solution could be introduced reliably into the lumen (Fig. 4C). (2) Perfused crypts had an unusually high permeability to NH_3 such that it exited the lumen before reaching the apical borders of cells downstream. Such rapid exit of NH_3 would have caused the lumen to become progressively more acidic downstream from the perfusion pipette, a phenomenon that we did not observe. (3) Acid was secreted into the lumen, which titrated NH_3 to the less permeant NH_4^+. However, a progressively acidic lumen pH was not observed.

Figure 6 shows an experiment in which we monitored luminal pH during normal luminal flow. Switching solution pH from 7.4 to 8.0 caused the expected increase in the I_{490}/I_{440} ratio, consistent with an increase in luminal pH and delivery of solution to the lumen. When the solution was changed to one containing 100 mM NH_4Cl at pH 8.0, we observed no change in the I_{490}/I_{440} ratio. The

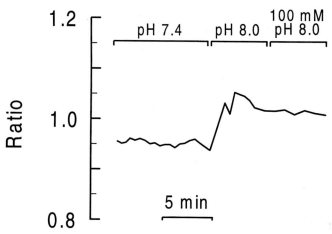

FIGURE 6   Effect on luminal pH of increasing the pH of luminal perfusate or adding $NH_3/NH_4{}^+$. Plotted on the y-axis are non-normalized I_{490}/I_{440} ratios for fluorescein dextran (40,000 MW). Switching the luminal perfusate pH from 7.4 to 8.0 could be observed to increase I_{490}/I_{440} ratios. Subsequent addition of 100 mM $NH_3/NH_4{}^+$, pH 8.0, did not change luminal pH along the length of the crypt, even when flow was stopped, which suggests that NH_3 and $NH_4{}^+$ do not exit the lumen appreciably. Adapted, with permission, from Singh, S. K., Binder, H. J., Geibel, J. P., and Boron, W. F. (1995). An apical permeability barrier to $NH_3/NH_4{}^+$ in isolated, perfused colonic crypts. *Proc. Natl. Acad. Sci. USA* **92**, 11573–11577).

record shown represents a portion of the lumen midway along the crypt, but there was no appreciable alkalinization along the length axis of the crypt lumen. Moreover, when lumen flow was stopped, the I_{490}/I_{440} ratio in the lumen remained uniform along the length of the lumen and did not change for up to 15 min. The results of these experiments argue against either a rapid exit of NH_3 from the lumen or an appreciable secretion of H^+ into the lumen.

2. The Effect of a Mucolytic on the Barrier to $NH_3/NH_4{}^+$ in Perfused Crypts

Unstirred mucus layers secreted by goblet cells serve as diffusion barriers to ions throughout the gastrointestinal tract (Forstner, 1978; Williams and Turnberg, 1980; Flemstrom and Kivilaakso, 1983; Silen, 1985). However, mucus has not been shown to prevent diffusion of neutral species. Nonetheless, we explored the possibility that chemically disaggregating the mucus barrier on the apical surface of colonic crypt cells might enhance their apical permeability to $NH_3/NH_4{}^+$. Before initiation of luminal perfusion, the crypt lumen was filled with mucus that is largely, though not entirely, washed away by the perfusion process. We used dithiothreitol (DTT), a mucolytic agent, to disaggregate further mucus within the crypt lumen to determine whether mucus has a role in the apical barrier to

NH_3/NH_4^+ permeation. In a series of experiments not shown, pulsing the lumen with 100 mM NH_3/NH_4^+ at pH 8.0 had no significant effect on pH_i; a second luminal pulse after visible disaggregation of luminal mucus with 2 mM DTT also had no effect on pH_i. Subsequent pulsing from the bath with 20 mM NH_3/NH_4^+ at pH 7.4 caused the usual series of pH_i changes (Singh *et al.*, 1995).

The most direct explanation for these observations is that there is indeed very low flux of NH_3/NH_4^+ across the apical barrier of the crypt colonocytes. This "permeability barrier" could have taken the form of a low-permeability (per se) or a low-apical surface area. That is, if the apical area were low enough, NH_3 entering via the apical membrane could exit via the basolateral membrane without measurably increasing pH_i. However, when we compared the effects of NH_3 in the lumen at a concentration 100-fold greater than that which elicited an easily detectable pH_i change from the bath, we observed no pH_i change. Based on transmission electron micrographs of random sections, and modeling the microvacuolated cell as a rectangular box, we estimated that no less than 10% of cell surface area is contained in the apical domain. Thus, even if the basolateral-to-apical area ratio were as high as 9:1, membrane area alone could not explain why luminal NH_3 did not alkalinize the cell.

B. An Apical Barrier to Weak Acids in Perfused Crypts

Approximately 300 mmol of short-chain fatty acids (SCFAs) are produced in the human colon each day, but fecal excretion is only 10 mmol per day, consistent with substantial colonic SCFA absorption (Cummings *et al.*, 1987; Macfarlane and Cummings, 1991). SCFAs (and especially butyrate) are important as the primary nutrient for colonocytes (Roediger, 1982; Roediger *et al.*, 1982; Roediger and Rae, 1982). Several mechanisms for colonic SCFA absorption have been proposed, including both passive, nonionic diffusion (VonEngelhardt and Rechkemmer, 1988; Sellin and DeSoignie, 1990; VonEngelhardt and Rechkemmer, 1992) and one or more active transport processes. Most transepithelial flux studies performed under voltage-clamp conditions studies have been consistent with nonionic diffusion as a dominant mechanism for colonic SCFA transport (Charney *et al.*, 1998; Sellin and DeSoignie, 1990; VonEngelhardt and Rechkemmer, 1992; Sellin *et al.*, 1993; von Engelhardt *et al.*, 1993). However, a series of studies in rat distal colon that used both transepithelial flux determinations (Binder and Mehta, 1989) and apical and basolateral membrane vesicles (Mascolo *et al.*, 1991; Reynolds *et al.*, 1993; Rajendran and Binder, 1994) yielded one model of both SCFA absorption and SCFA stimulation of electroneutral Na-Cl absorption. First, these studies identified distinct and different $SCFA^-/HCO_3^-$ exchanges in both apical (Mascolo *et al.*, 1991) and basolateral (Reynolds *et al.*, 1993) membranes. In these studies that used apical membranes primarily derived from surface and not crypt epithelial cells, limited nonionic uptake of SCFA was observed (Mascolo *et al.,* 1991). A role for both an apical Cl^--butyrate exchange and Na-H exchange was proposed for rat distal colon (Rajendran and Binder, 1994). Thus, in rat distal colon, SCFA uptake occurs as a result of an apical $SCFA-HCO_3$ exchange that, in theory, decreases pH_i, thus activating both Na-H and Cl^--butyrate exchanges; the net effect is electroneutral Na-Cl absorption with some recycling of SCFA across the apical membrane. SCFAs exit from the surface epithelial cell via a basolateral membrane $SCFA-HCO_3$ exchange. It is likely that carrier-mediated transport mechanisms explain, in part, colonic SCFA absorption; however, the relative contributions of carrier-mediated transport and nonionic diffusion are not definitively known. Less is known about how crypt cells, in particular, handle SCFAs, although studies of intracellular and extracellular pH indicate that both nonionic diffusion and carrier-mediated transport participate in acid-base movement in mouse colonic crypts (Chu and Montrose, 1996). More recent studies have shown that $SCFA-HCO_3$ exchange is virtually absent from crypt cell apical membranes (Rajendran, V.M., unpublished observations).

SCFAs have profound effects on pH_i and cell volume in colonic epithelia. Propionic acid transiently decreases pH_i and causes cells of the shark rectal gland to swell, probably by initial nonionic diffusion of SCFA with compensatory stimulation of Na-H exchange (Feldman *et al.,* 1989). Similarly, in a human colonic carcinoma line, propionic acid caused a rapid increase in pH_i (presumably from nonionic diffusion) and a subsequent increase in cell volume. A stilbene-sensitive, Na^+-dependent exit step was implicated in the ensuing regulatory volume decrease (Rowe *et al.,* 1993). The pH_i of isolated dispersed cells of rabbit proximal colon also decreased rapidly in the presence of propionic acid, but this SCFA-induced acidification was partially Na^+ dependent. Further, recovery from this acid load was sensitive to amiloride as well as to 4-α-OH-cinnamate (a blocker of H^+-monocarboxylate cotransport) but not to DIDS (DeSoignie and Sellin, 1994). Cells within superfused crypts from rat distal colon acidified and subsequently swelled upon exposure to butyric acid. This swelling and acid extrusion was mediated by compensatory activation of Na-H exchange, with the greatest activity found in cells at the opening of the crypt. A subsequent regulatory volume decrease was HCO_3^- dependent, and suppressed by inhibitors of K^+ and Cl^- channels (Diener *et al.,* 1993). None of these studies was performed on an intact, polarized, native colonic epithelium such as the perfused colonic crypt, and thus apical and basolateral permeability and transport processes were not separable.

1. Evidence for a Barrier to SCFA Permeation and Transport in the Crypt

Figure 7A shows the results of an experiment in which we monitored pH_i while exposing the cells of a colonic crypt, the lumen of which was pretreated

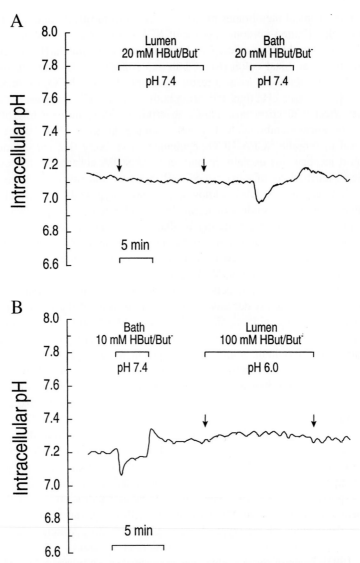

FIGURE 7 Effect of apical versus basolateral But$^-$/HBut on pH$_i$ in perfused colonic crypts. (A) Response of a single crypt colonocyte to application and removal of a pH-7.4 solution containing 20 mM But$^-$/HBut, first from the lumen and then from the bath. (B) Response of a single crypt colonocyte to application and removal of a pH-7.4 solution containing 10 mM But$^-$/HBut from the bath, followed by application and removal of a pH-6.0 solution containing 100 mM But$^-$/HBut from the lumen. In neither case did luminal exposure cause a significant change in pH$_i$, although both bath exposures did change pH$_i$ significantly.

with the mucolytic DTT, to a pH-7.4 solution containing 20 mM butyrate/butyric acid (HBut/But$^-$). The addition of this solution (which contains 0.08 mM HBut) to the lumen had no effect on pH$_i$; however, when this same solution was introduced into the bath, a rapid decrease in pH$_i$ was elicited because of the influx of HBut, followed by a slower increase presumably caused by the active extrusion of acid by homeostatic processes. Removing But$^-$/HBut caused the opposite series of changes: a rapid pH$_i$ increase owing to HBut efflux, followed by a slower recovery. As with NH$_3$/NH$_4{}^+$, the most straightforward explanation for our data is the presence of an apical barrier with extremely low permeabilities to But$^-$ and HBut that also lacks a carrier-mediated transport mechanism. However, as with NH$_3$/NH$_4{}^+$, six alternative explanations required exclusion. These are described herein.

The first possibility is that the apical membrane is permeable to both But$^-$ and HBut, but that the alkalinizing effect of But$^-$ influx exactly counteracts the acidifying effect of HBut influx. If true, then altering the ratio of HBut to But$^-$ influx would unmask a pH$_i$ change. However, as shown in Fig. 7B, when a pH-6.0 solution containing 100 mM But$^-$/HBut was introduced into the lumen, there was no effect on pH$_i$, even though the [HBut]/[But$^-$] ratio was 25-fold greater than that of the comparable luminal solution change in Fig. 7A. This result ruled out the parallel But$^-$/HBut influx model.

The second alternative is that homeostatic acid-extruding mechanisms prevented the pH$_i$ decrease expected from HBut influx across the apical membrane. In other work, we have observed that in the absence of CO_2/$HCO_3{}^-$, the primary acid-extruder in crypt colonocytes is a basolateral Na-H exchanger (Singh, S.K., unpublished observations). Blocking this exchanger with amiloride (1 mM) did not unmask acidification in experiments in which we added 100 mM But$^-$/HBut at pH 6.0 to the lumen.

The third possibility is that the apical border of crypt cells has normal HBut permeability, but that the apical surface area is substantially smaller than the basolateral area. However, the surface-area amplification of the apical membrane, at least of colonic crypt cells in the rat, is fivefold greater than the amplification of the basolateral membrane (Kashgarian et al., 1980), and our electron micrographs suggest that rabbit crypt cells do not differ substantially from rat colonocytes (not shown). Moreover, Fig. 7B shows that perfusing the bath with an [HBut] of 0.026 mM produced a pH$_i$ decrease, even though perfusing the lumen with an [HBut] 250-fold greater had no effect. Modeling this observation mathematically (Boron et al., 1994), this can be explained only if the *permeability* × *area* product of the apical membrane exceeds 4000-fold that of the basolateral membrane. Based on the surface-area amplification data, this 4000-fold difference must involve a difference in permeabilities.

The fourth alternative is that the solution with which we attempted to perfuse the lumen actually did not reach the lumen, but as shown in Figure 8A, lower-

A

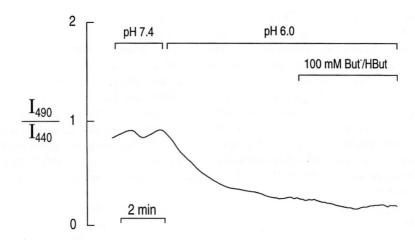

B

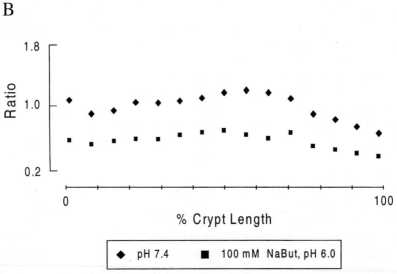

% Crypt Length

◆ pH 7.4   ■ 100 mM NaBut, pH 6.0

FIGURE 8 Effect on luminal pH of decreasing the pH of luminal perfusate or adding But⁻/HBut. (A) We monitored luminal pH in crypts perfused with fluorescein dextran; increasing values of I_{490}/I_{440} correspond with an increase in luminal pH. The lumen was first perfused with a poorly buffered pH-7.4 solution, then with a poorly buffered pH-6.0 solution, and finally with a similar solution containing 100 mM total butyrate at pH 6.0. This record was obtained midway along the crypt. (B) Records from 15 different positions along the length of the crypt lumen showed that I_{490}/I_{440} was essentially uniform down the length of the lumen. When flow was stopped with 100 mM Na-butyrate/pH 6.0 in the lumen, a uniform I_{490}/I_{440} ratio was observed in the lumen for up to 10 min. Thus, lumen pH does not vary appreciably as a function of distance along the crypt from the tip of the perfusion pipette, either in the basal state (A) or in the presence of 100 mM NaBut, pH 6.0 (B). If alkali were secreted into the lumen or if HBut had an unusually high rate of permeation out the crypt, then luminal pH should have increased distal to the perfusion pipette.

ing the pH of the luminal perfusate from 7.4 to 6.0 produced the expected decrease in luminal pH.

The fifth possibility is that crypts are so unusually permeable to HBut that it exits the lumen before reaching the cells of interest. However, for various reasons this is an unlikely explanation: (1) The cells closest to the fresh, incoming luminal solution exhibited the same lack of pH_i change as the cells further downstream. (2) Luminal pH did not change when the luminal perfusate was switched from a pH-6.0 solution containing no butyrate to a pH-6.0 solution containing 100 mM $But^-/HBut$ (Fig. 8A). Had HBut diffused from the lumen upstream of our measuring site, a more alkaline lumen should have been observed in the presence of $But^-/HBut$ than in its absence. (3) When the lumen was perfused with 100 mM $But^-/HBut$ at pH 6.0, luminal pH was uniform long the entire lumen (Fig. 8B). If HBut had exited the lumen, luminal pH should have become progressively more alkaline as the fluid flowed down the crypt. These observations exclude a significant loss of HBut from the lumen.

The sixth alternative is that crypts secreted enough alkali into the lumen to titrate all luminal HBut to the less permeant But^-. In this case, as in the previous one, the lumen should have become progressively more alkaline downstream along the crypt, which was not observed (Fig. 8B).

The observations of an apical barrier to SCFA must be viewed in the context that, at least in the rat distal colon, apical $SCFA-HCO_3$ exchange is restricted to surface cells, that is, absent from crypt cells. Thus, the absence of a carrier for $SCFA^-$ in rabbit proximal colon crypts, in combination with a membrane that possesses low inherent permeability to the nonionic form of SCFAs, would explain the apparent permeability barrier at the apical border of crypt cells.

2. Apical versus Basolateral Permeability to CO_2/HCO_3^-

Carbon dioxide is a small lipophilic molecule that permeates all known membranes, except those comprising the luminal borders of gastric-gland cells (Waisbren et al., 1994). Colonic fermentation of nonabsorbed nutrients produces a gas composed of up to 60% CO_2 (Steggerda and Dimmick, 1966; Strocchi and Levitt, 1998). This fact raises the issue of whether the apical borders of colonic crypt cells, which we have demonstrated have a negligible permeability to NH_3/NH_4^+ and $HBut/But^-$, have low permeability to CO_2/HCO_3^- as well. In recent experiments perfusing the lumen of a crypt with pH-7.4 bath solution containing 5% $CO_2/22$ mM HCO_3^- caused no change in crypt cell pH_i, whereas exposing the basolateral surface of the crypt to the same solution produced a rapid decrease in pH_i caused by CO_2 influx, followed by a slower increase presumably related to the active extrusion of acid (Singh, SK, unpublished observations). When the CO_2/HCO_3^- was removed, these effects were reversed. Even when the lumen was perfused with a solution containing 100% $CO_2/22$ mM HCO_3^- (pH 6.1), there was no effect on pH_i (Fig. 9). Perfusion of the lumen with this pH-6.1 solution produced the expected luminal acidification, which was uniform

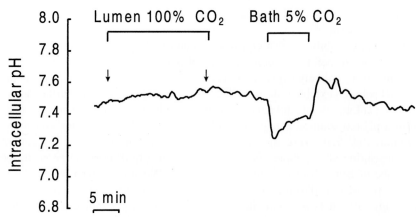

FIGURE 9 Lumen 100% CO_2 does not affect pH_i. Response of a single crypt colonocyte first to application and removal of a pH-6.1 luminal solution containing 100 % CO_2/22 mM HCO_3^-, and then to application and removal of a pH-7.4 basolateral solution containing 5% CO_2/22 mM HCO_3^-. Luminal exposure to CO_2/HCO_3^- did not cause a significant change in pH_i, although the bath exposure did. CO_2/HCO_3^- appears to follow a pattern similar to that observed with HBut/But$^-$.

along the entire lumen of the crypt (Singh, Sk, unpublished observations). Thus, the apical border of the microvacuolated colonic crypt cell has negligible permeability to both CO_2 and HCO_3^-.

C. Noncolonic Membranes with Specialized Permeability Characteristics

Although novel, the present observations are not without precedent, as low rates of permeation by NH_3 and CO_2 have been described in other specialized cells. Significant quantities of NH_3 do not cross either the apical membrane of the mouse renal medullary thick ascending limb (Kikeri *et al.*, 1989) or the plasma membrane of *Xenopus* oocytes (Burckhardt and Fromter, 1992). However, both these membranes have substantial pathways for NH_4^+ transport. The apical borders of gastric glands have low permeability to NH_3/NH_4^+ and CO_2/HCO_3^- (Waisbren *et al.*, 1994). Apical membrane endosomes from rabbit urinary bladder epithelium, which maintains large chemical (including pH) and osmotic gradients, also have exceptionally low NH_3 permeability (Chang *et al.*, 1994).

Colonic crypt cells are only the second example of cells with a CO_2-impermeable apical membrane, the other being the chief and parietal cells of the gastric gland (Waisbren *et al.*, 1994). In contrast, crypt cells are the first example of a membrane impermeable to an SCFA, a class of molecules that rapidly penetrate all other membranes that have been examined. We suspect that epithelial progenitor cells exposed to extremely inhospitable environments (e.g., the gastric gland and the colonic crypt) have apical borders with no demonstrable permeability to small, lipophilic molecules that rapidly penetrate other cell membranes. It is believed that NH_3, CO_2, and butyric acid penetrate most cell membranes by simply dissolving in the membrane lipid (Finkelstein, 1976; Gutknecht et al., 1977; Orbach and Finkelstein, 1980; Walter et al., 1982; Golchini and Kurtz, 1988). Because the composition and biophysical properties of the apical and basolateral membranes are substantially different in a variety of epithelial cells (Brasitus, 1983; Brasitus and Keresztes, 1984; Brasitus and Dudeja, 1986; Brasitus et al., 1986; van Meer and Simons, 1988; van't Hof et al., 1992; Tietz et al., 1995), it is possible that a difference in lipid composition underlies the difference in permeability properties in cells of the gastric gland and colonic crypt. Even though we pretreated our crypt lumens with dithiothreitol to remove mucus, we cannot exclude the possibility that an unusually adherent mucus layer might underlie the low permeability of the apical barrier. However, regardless of mechanism, the permeability barrier in crypts appears to protect proliferating cells from the harsh environment of the colonic fermentation compartment.

V. GAS AND WEAK-ELECTROLYTE PERMEABILITY OF SURFACE CELLS

That NH_3/NH_4^+, $HBut/But^-$, and CO_2/HCO_3^- do not appreciably traverse the apical membrane of colonic crypt cells must be reconciled with (1) the well-known fact that these three substances are absorbed by the colon (Castell, 1965; Down et al., 1972; Mitch et al., 1977; Charney and Haskell, 1984; Cummings et al., 1987), and (2) the fact that $HBut/But^-$ has been identified as a primary nutrient for colonocytes (Roediger, 1982). Certainly the presence of a barrier in the crypt presents a dilemma. How do crypt colonocytes "obtain" butyrate in the face of an apical permeability barrier? Assuming that the barrier observed in the crypt is present in vivo as well, it is necessary to postulate that all weak-electrolyte absorption occurs via surface epithelial cells. Indeed, to test this possibility, we measured the pH_i of polarized surface colonocytes in situ from the same proximal segment from which crypts were dissected. We measured the pH_i of surface colonocytes, adapting a method that permits control of solutions bathing the mucosal (apical) separate from the serosal aspects of an excised patch of colon (Fig. 10A) while imaging surface cells loaded with BCECF (Fig. 10B) (Rajendran et al., 1998). As shown in Figure 11, unlike their ancestral cells within the crypts, the pH_i of surface colonocytes is affected by apical exposure to NH_3/NH_4^+, $HBut/But^-$, and CO_2/HCO_3^-. The difference in apical permeability of surface and crypt cells was observed in somewhat different types of cell preparations. However, experimental methodology and design between preparations were quite similar. Thus, it is doubtful that different experimental conditions could explain fully the observed difference in apical permeability. In

A

B

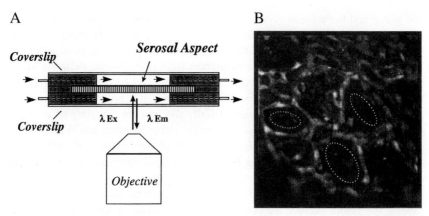

FIGURE 10   pH$_i$ records in colonic surface epithelial cells. (A) Schematic illustration of the chamber used to image individual surface epithelial cells in rabbit proximal colon (*fusus coli*). A series of retaining rings is used to hold and stretch an unstripped colonic sheet sandwiched between two glass coverslips. The chamber design permits independent superfusion of the mucosal and serosal aspects of the tissue during imaging of the mucosal aspect. Surface epithelial cells are loaded with dye by exposing the mucosal aspect to 10 μM BCECF-AM for approximately 15 min. (B) Fluorescence ($\lambda_{\text{emission}}$ at 530 nm, $\lambda_{\text{excitation}}$ of 490 nm) image of surface cells loaded with BCECF (64×). White dotted areas outline crypt openings.

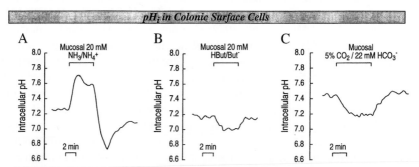

FIGURE 11   Proximal colon surface cells are permeable to neutral fermentation products. Effect on pH$_i$ of surface epithelial cells in rabbit proximal colon (*fusus coli*) to mucosal application of pH-7.4 solutions containing (A) 20 mM NH$_4$Cl, (B) 20 mM Na-butyrate, and (C) 22 mM HCO$_3^-$ / 5% CO$_2$. The apical aspect of proximal colon surface cells, unlike that of the ancestral crypt cells of the same segment, is permeable to these small, lipophilic neutral fermentation products. Thus, the surface epithelium provides a potential pathway for uptake of NH$_3$/NH$_4^+$, HBut/But$^-$, and CO$_2$/HCO$_3^-$ into portal blood.

addition, it must also be remembered that differences in permeability were observed at the apical and basolateral membrane in the same experiment.

Thus, surface cells provide an apparent pathway for movement of gases and
weak electrolytes into the lamina propria and the portal circulation. Because surface cells have high apparent permeability to neutral molecules, SCFA concentrations in the lamina propria could reach substantial levels as a result of free diffusion from the lumen to the serosal aspect of crypt cells, which, as we have
observed, possess basolateral uptake pathways (see Figs. 5, 7, and 9). However,
it should be noted that the total concentration of SCFA in the colonic lumen (of
~120 mM) is diluted 1000-fold in portal blood. We propose that *in vivo,* as crypt
cells function in relation to continuous blood flow, potentially cytotoxic levels
of SCFA (as are found in the lumen) do not occur at the basolateral aspect of
crypt cells. Such a paradigm would serve to blunt large effects on crypt-cell pH_i
from the high lumen-to-cell gradient of fermentation products while permitting
uptake of SCFA via the basolateral membrane (see Fig. 12).

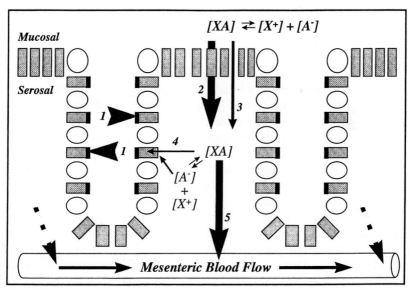

FIGURE 12 Proposed scheme for colonic absorption of acid-base fermentation products. The
arrangement of a crypt/gland barrier (1) in parallel with permeable surface epithelium (2,3) prevents
substantial influx of permeant weak electrolytes (XA) into proliferative gland cells as it permits their
movement via more mature surface cells into the lamina propria. In the lamina propria, equilibration
with plasma flow would dilute (XA) far below luminal levels. From the lamina propria, XA should
be able to enter crypt cells via their highly permeable basolateral borders (4); A⁻ also may be taken
up by ion transport processes.

To determine whether this paradigm can be observed in the stomach as well, we measured the pH_i of gastric surface epithelial cells to see if, by analogy to the colon, the barrier to NH_3 and CO_2 permeation in gastric glands (Waisbren et al., 1994) is absent from gastric surface cells. Indeed, despite a permeability barrier in gastric glands (Waisbren et al., 1994), the stomach also transports weak electrolytes and gases into blood, as witnessed by ^{14}C-urea breath testing for *Helicobacter pylori* (Dill et al., 1990; Henze et al., 1990). As with surface colonocytes, the pH_i of gastric surface cells alkalinizes on exposure to 20 mM NH_4Cl, pH 7.4 ($\Delta pH/30s \times 10^{-2} = +79.18 \pm 10.57$, $n = 35$ cells/4 mucosal preps/4 rabbits) and acidifies with 5% CO_2/22 mM HCO_3^-, pH 7.4 ($\Delta pH/30s \times 10^{-2} = -18.29 \pm 1.26$, $n = 35$ cells/4 mucosal preps/4 rabbits), consistent with high apical permeability to NH_3 and CO_2. Thus, gastric surface epithelial cells are a pathway for gas absorption from the stomach, much as surface cells are a pathway for absorption in the colon. Although we have identified uptake of NH_3/NH_4^+, HBut, and CO_2 into surface cells in the colon, we do not know whether they permeate the apical membrane per se and/or whether they take a paracellular route. What is clear is that the permeability barrier present in colonic crypts and gastric glands is not present in their respective surface epithelium.

VI. CONCLUDING REMARKS

This chapter has presented several observations that describe heterogeneity in the permeability of plasma membranes of the rabbit proximal colon and stomach. There is clear evidence that the apparent permeability of apical membranes of crypt cells to weak electrolytes and gases is lower than that of both basolateral membranes of crypt cells and apical membranes of surface cells. These observations of spatial and cellular heterogeneity must be viewed in context with other observations of apparent spatial and cellular heterogeneity in colonic epithelial cells. In addition to the well-described differences in absorptive and secretory processes in surface and crypt cells discussed previously, there have been several observations that ion transport processes are not distributed homogeneously along surface and crypt cells of rat distal colon. We have recently established a distinct distribution of anion exchanges in apical membranes of surface and crypt cells. Thus, $Cl^- - HCO_3^-$ and $SCFA-HCO_3$ exchanges are present exclusively on the apical membrane of surface cells and not on the apical membrane of crypt cells of the rat distal colon (Rajendran and Binder, 1999; Rajendran, V.M., unpublished observations). Whether the relative absence of $SCFA-HCO_3$ exchange in the apical membrane of crypt cells accounts, at least in part, for the low apparent permeability of SCFA across the apical membrane of crypt cells is not known. Also unknown is whether there are fundamental differences in the composition of the crypt apical membrane that result in the observed decrease in

SCFA (and NH_3 and CO_2) permeation. If, as with the rat distal colon, SCFA-HCO_3 exchange is not present in the apical membrane of crypt cells of rabbit proximal colon, then absence of this transporter likely represents a substantial component of the observed diffusion barrier. Clearly, membrane heterogeneity is more the rule than the exception.

References

Aerts, R. J., Durston, A. J. and Moolenaar, W. H. (1985). Cytoplasmic pH and the regulation of the Dictyostelium cell cycle. *Cell* **43**, 653–657.

Bern, M. J., Sturbaum, C. W., Karayalcin, S. S., Berschneider, H. M., Wachsman, J. T., and Powell, D. W. (1989). Immune system control of rat and rabbit colonic electrolyte transport. Role of prostaglandins and the enteric nervous system. *J. Clin. Invest.* **83**, 1810–1820.

Berschneider, H. M., and Powell, D. W. (1992). Fibroblasts modulate intestinal secretory responses to inflammatory mediators. *J. Clin. Invest.* **89**, 484–489.

Binder, H. J., Foster, E. S., Budinger, M. E., and Hayslett, J. P. (1987). Mechanism of electroneutral sodium chloride absorption in distal colon of the rat. *Gastroenterology* **93**, 449–455.

Binder, H. J., and Mehta, P. (1989). Short-chain fatty acids stimulate active sodium and chloride absorption *in vitro* in the rat distal colon. *Gastroenterology* **96**, 989–996.

Binder, H. J., and Rawlins, C. L. (1973). Electrolyte transport across isolated large intestinal mucosa. *Am. J. Physiol.* **225**, 1232–1239.

Bleakman, D., and Naftalin, R. J. (1990). Hypertonic fluid absorption from rabbit descending colon *in vitro*. *Am. J. Physiol.* **258**, G377–G390.

Boonstra, J., Tertoolen, L. G. J., Mumery, C. L., and de Laat, S. W. (1988). Regulation of intracellular pH during the G1/S-phase transition of the neuroblastoma cell cycle. *Exp Cell Res* **174**, 521–524.

Boron, W. F. (1992). Control of intracellular pH. *In* "The Kidney: Physiology and Pathophysiology" (2nd ed.) D. W. Seldin and G. Giebisch, eds.), pp 219–263, Raven, New York.

Boron, W. F., and Boulpaep, E. L. (1983). Intracellular pH regulation in the renal proximal tubule of the salamander. Basolateral HCO_3^- transport. *J. Gen. Physiol.* **81**, 53–94.

Boron, W. F., McCormick, W. C., and Roos, A. (1979). pH regulation in barnacle muscle fibers: Dependence on intracellular and extracellular pH. *Am. J. Physiol.* **237**, C185–C193.

Boron, W. F., Waisbren, S. J., Modlin, I. M., and Geibel, J. P. (1994). Unique permeability barrier of the apical surface of parietal and chief cells in isolated perfused gastric glands. *J. Exp. Biol.* **196**, 347–360.

Boyarsky, G., Ganz, M. B., Cragoe, E. J., Jr., and Boron, W. F. (1990). Intracellular-pH dependence of Na-H exchange and acid loading in quiescent and arginine vasopressin-activated mesangial cells. *Proc. Natl. Acad. Sci. USA* **87**, 5921–5924.

Braaten, B., Madara, J. L., and Donowitz, M. (1988). Age-related loss of nongoblet crypt cells parallels decreased secretion in rabbit descending colon. *Am. J. Physiol.* **255**, G72–G84.

Brasitus, T. A. (1983). Lipid dynamics and protein-lipid interactions in rat colonic epithelial cell basolateral membranes. *Biochim. Biophys. Acta* **728**, 20–30.

Brasitus, T. A., and Dudeja, P. K. (1986). Modulation of lipid fluidity of small- and large-intestinal antipodal membranes by Ca^{2+}. *Biochem. J.* **239**, 625–631.

Brasitus, T. A., Dudeja, P. K., and Dahiya, R. (1986). Premalignant alterations in the lipid composition and fluidity of colonic brush border membranes of rats administered 1,2 dimethylhydrazine. *J. Clin. Invest.* **77**, 831–840.

Brasitus, T. A., and Keresztes, R. S. (1984). Protein-lipid interactions in antipodal plasma membranes of rat colonocytes. *Biochim. Biophys. Acta* **773**, 290–300.

Burckhardt, B. C., and Fromter, E. (1992). Pathways of NH_3/NH_4^+ permeation across *Xenopus laevis* oocyte cell membrane. *Pflugers Arch.—Eur. J. Physiol.* **420**, 83–86.

Burg, M., Grantham, J., Abramow, M., and Orloff, J. (1966). Preparation and study of fragments of single rabbit nephrons. *Am. J. Physiol.* **210,** 1293–1298.

Burg, M. B. (1972). Perfusion of isolated renal tubules. *Yale J Biol. Med.* **45,** 321–326.

Busa, W. B. (1986). Mechanisms and consequences of pH-mediated cell regulation. *Ann. Rev. Physiol.* **48,** 389–402.

Busa, W. B., and Nuccitelli, R. (1984). Metabolic regulation via intracellular pH. *Am. J. Physiol.* **246,** R409–R438.

Castell, D. O. (1965). The ammonia tolerance test: An index of portal hypertension. *Gastroenterology* **49,** 539–543.

Chang, A., Hammond, T. G., Sun, T. T., and Zeidel, M. L. (1994). Permeability properties of the mammalian bladder apical membrane. *Am. J. Physiol.* **267,** C1483–C1492.

Charney, A. N., and Feldman, G. M. (1984). Systemic acid-base disorders and intestinal electrolyte transport. *Am. J. Physiol.* **247,** G1–G12.

Charney, A. N., and Haskell, L. P. (1984). Relative effects of systemic pH, P_{CO_2}, and H_{CO_3} concentration on colonic ion transport. *Am. J. Physiol.* **246,** G159–G165.

Charney, A. N., Micic, L., and Egnor, R. W. (1998). Nonionic diffusion of short chain fatty acids across rat colon. *Am. J. Physiol.* **274,** G518–G524.

Chu, S., and Montrose, M. H. (1996). Non-ionic diffusion and carrier-mediated transport drive extracellullar pH regulation of mouse colonic crypts. *J. Physiol.* **494,** 783–793.

Copello, J., Segal, Y., and Reuss, L. (1991). Cytosolic pH regulates maxi K^+ channels in Necturus gallbladder epithelial cells. *J. Physiol.* **434,** 577–590.

Cummings, J. H., and Macfarlane, G. T. (1991). The control and consequences of bacterial fermentation in the human colon. *J Appl. Bacteriol.* **70,** 443–459.

Cummings, J. H., Pomare, E. W., Branch, W. J., Naylor, C. P., and Macfarlane, G. T. (1987). Short chain fatty acids in human large intestine, portal, hepatic and venous blood. *Gut* **28,** 1221–1227.

Curran, P. F., and Schwartz, G. F. (1960). Na, Cl, and water transport by rat colon. *J Gen. Physiol.* **43,** 555–571.

Deitmer, J. W., and Schlue, W. R. (1989). An inwardly directed electrogenic sodium-bicarbonate cotransport in leech glial cells. *J Physiol. (Lond.)* **411,** 179–194.

DeSoignie, R., and Sellin, J. H. (1994). Propionate-initiated changes in intracellular pH in rabbit colonocytes. *Gastroenterology* **107,** 347–356.

Devroede, G. J., and Phillips, S. F. (1969). Conservation of sodium, chloride, and water by the human colon. *Gastroenterology* **56,** 101–109.

Dharmsathaphorn, K., Mandel, K. G., Masui, H., and McRoberts, J. A. (1985). Vasoactive intestinal polypeptide-induced chloride secretion by a colonic epithelial cell line. Direct participation of a basolaterally localized Na^+,K^+,Cl^- cotransport system. *J. Clin. Invest.* **75,** 462–471.

Diener, M., Helmle-Kolb, C., Murer, H., and Scharrer, E. (1993). Effect of short-chain fatty acids on cell volume and intracellular pH in rat distal colon. *Pflugers Arch.—Eur. J. Physiol.* **424,** 216–223.

Dietz, J., and Field, M. (1973). Ion transport in rabbit ileal mucosa. IV. Bicarbonate secretion. *Am. J. Physiol.* **225,** 858–861.

Dill, S., Payne-James, J. J., Misiewicz, J. J., Grimble, G. K., McSwiggan, D., Pathak, K., Wood, A. J., Scrimgeour, C. M., and Rennie, M. J. (1990). Evaluation of 13C-urea breath test in the detection of *Helicobacter pylori* and in monitoring the effect of tripotassium dicitratobismuthate in non-ulcer dyspepsia. *Gut* **31,** 1237–1241.

Donowitz, M., and Binder, H. J. (1976). Effect of enterotoxins of *Vibrio cholerae, Escherichia coli,* and *Shigella dysenteriae* type 1 on fluid and electrolyte transport in the colon. *J. Infect. Dis.* **134,** 135–143.

Down, P. F., Agostini, L., Murison, J., and Wrong, O. M. (1972). The interrelations of faecal ammonia, pH, and bicarbonate: Evidence of colonic absorption of ammonia by non-ionic diffusion. *Clin. Sci.* **43,** 101–114.

Down, P. F., Agostini, L., Murison, J., and Wrong, O. M. (1972). The relation between faecal pH, ammonia, and bicarbonate in man. *Clin. Sci.* **42**, 24P.

Feldman, G. M., and Stephenson, R. L. (1990). H^+ and HCO_3^- flux across apical surface of rat distal colon. *Am. J. Physiol.* **259**, C35–C40.

Feldman, G. M., Ziyadeh, F. N., Mills, J. W., Booz, G. W., and Kleinzeller, A. (1989). Propionate induces cell swelling and K^+ accumulation in shark rectal gland. *Am. J. Physiol.* **257**, C377–C384.

Field, M. (1991). Intestinal ion transport mechanisms. *In* "Diarrheal Diseases" M. Field, (ed), pp 3–21, Elsevier Science, New York.

Field, M., Smith, P. L., and Bolton, J. E. (1980). Ion transport across the isolated intestinal mucosa of the winter flounder, *Pseudopleuronectes americans:* II. Effects of cyclic AMP. *J. Membr. Biol.* **55**, 157–163.

Finkelstein, A. (1976). Water and nonelectrolyte permeability of lipid bilayer membranes. *J Gen. Physiol.* **68**, 127–135.

Fitz, J. G., Trouillot, T. E., and Scharschmidt, B. F. (1989). Effect of pH on membrane potential and K^+ conductance in cultured rat hepatocytes. *Am. J. Physiol.* **257**, G961–G968.

Flemstrom, G., and Kivilaakso, E. (1983). Demonstration of a pH gradient at the luminal surface of rat duodenum *in vivo* and its dependence on mucosal alkaline secretion. *Gastroenterology* **84**, 787–794.

Forstner, J. F. (1978). Intestinal mucins in health and disease. *Digestion* **17**, 234–263.

Foster, E. S., Zimmerman, T. W., Hayslett, J. P., and Binder, H. J. (1983). Corticosteroid alteration of active electrolyte transport in rat distal colon. *Am. J. Physiol.* **245**, G668–G675.

Frizzell, R. A., Koch, M. J., and Schultz, S. G. (1976). Ion transport by rabbit colon. I. Active and passive components. *J. Membr. Biol.* **27**, 297–316.

Ganz, M. B., and Boron, W. F. (1994). Long-term effects of growth factors on pH and acid-base transport in rat glomerular mesangial cells. *Am. J. Physiol.* **266**, F576–F585.

Golchini, K., and Kurtz, I. (1988). NH_3 permeation through the apical membrane of MDCK cells is via a lipid pathway. *Am. J. Physiol.* **255**, F135–F141.

Goodlad, R. A., Lee, C. Y., and Wright, N. A. (1992). Colonic cell proliferation and growth fraction in young, adult, and old rats. *Virchows Arch. B, Cell Pathol. Incl. Mol. Pathol.* **61**, 415–417.

Grandin, N., and Charbonneau, M. (1990). Cycling of intracellular pH during cell division of *Xenopus* embryos is a cytoplasmic activity depending on protein synthesis and phosphorylation. *J Cell Biol* **111**, 523–532.

Grandin, N., Rolland, J. P., and Charbonneau, M. (1991). Changes in intracellular pH following egg activation and during the early cell cycle of the amphibian *Pleurodeles waltlii* coincide with changes in MPF activity. *Biol Cell* **72**, 259–267.

Greger, R., and Hampel, W. (1981). A modified system for in vitro perfusion of isolated renal tubules. *Pflugers Arch.—Eur. J. Physiol.* **389**, 175–176.

Gutknecht, J., Bisson, M. A., and Tosteson, F. C. (1977). Diffusion of carbon dioxide through lipid bilayer membranes. *J. Gen. Physiol.* **69**, 779–794.

Ha, S. N., Giammona, A., Field, M., and Brady, J. W. (1988). A revised potential-energy surface for molecular mechanics studies of carbohydrates. *Carbohydr. Res.* **180**, 207–221.

Halevy, J., Budinger, M. E., Hayslett, J. P., and Binder, H. J. (1986). Role of aldosterone in the regulation of sodium and chloride transport in the distal colon of sodium-depleted rats. *Gastroenterology* **91**, 1227–1233.

Heintze, K., Stewart, C. P., and Frizzell, R. A. (1983). Sodium-dependent chloride secretion across rabbit descending colon. *Am. J. Physiol.* **244**, G357–G365.

Henze, E., Malfertheiner, P., Clausen, M., Burkhardt, H., and Adam, W. E. (1990). Validation of a simplified carbon-14-urea breath test for routine use for detecting *Helicobacter pylori* noninvasively. *J. Nucl. Med.* **31**, 1940–1944.

Hubel, K. A. (1974). The mechanism of bicarbonate secretion in rabbit ileum exposed to choleragen. *J. Clin. Invest.* **53**, 964–970.

Johnson, J. D., and Epel, D. (1976). Intracellular pH and activation of sea urchin eggs after fertilisation. *Nature* **262,** 661–664.

Kashgarian, M., Taylor, C. R., Binder, H. J., and Hayslett, J. P. (1980). Amplification of cell membrane surface in potassium adaptation. *Lab. Invest.* **42,** 581–588.

Kikeri, D., Sun, A., Zeidel, M. L., and Hebert, S. C. (1989). Cell membranes impermeable to NH_3. *Nature* **339,** 478–480.

Kirk, E. (1949). The quantity and composition of human colonic flatus. *Gastroenterology* **12,** 782–794.

Kockerling, A., and Fromm, M. (1993). Origin of cAMP-dependent Cl^- secretion from both crypts and surface epithelia of rat intestine. *Am. J. Physiol.* **264,** C1294–C1301.

Levitt, M. D. (1972). Intestinal gas production. *J. Am. Diet. Assoc.* **60,** 487–490.

Lohrmann, E., and Greger, R. (1995). The effect of secretagogues on ion conductances of *in vitro* perfused, isolated rabbit colonic crypts. *Pflugers Arch.* **429,** 494–502.

Lundquist, F., Sestoft, L., Damgaard, S. E., Clausen, J. P., and Trap-Jensen, J. (1973). Utilization of acetate in the human forearm during exercise after ethanol ingestion. *J. Clin. Invest.* **52,** 3231–3235.

Macfarlane, G. T., Allison, C., Gibson, S. A., and Cummings, J. H. (1988). Contribution of the microflora to proteolysis in the human large intestine. *J Appl. Bacteriol.* **64,** 37–46.

Macfarlane, G. T., and Cummings, J. H. (1991). The colonic flora, fermentation, and large bowel digestive function. *In* "The Large Intestine: Physiology, Pathophysiology, and Disease" (J. F. Phillips, R. G. Shorter and J. H. Pemberton, eds.), pp 51–92, Raven, New York.

Macfarlane, G. T., Cummings, J. H., and Allison, C. (1986). Protein degradation by human intestinal bacteria. *J. Gen. Microbiol.* **132,** 1647–1656.

Macfarlane, G. T., Gibson, G. R., and Cummings, J. H. (1992). Comparison of fermentation reactions in different regions of the human colon. *J Appl. Bacteriol.* **72,** 57–64.

Mascolo, N., Rajendran, V. M., and Binder, H. J. (1991). Mechanism of short-chain fatty acid uptake by apical membrane vesicles of rat distal colon. *Gastroenterology* **101,** 331–338.

McBurney, M., and VanSoest, P. (1991). Structure-function relationships: Lessons from other species. *In* "The Large Intestine: Physiology, Pathophysiology, and Disease" (J. F. Phillips, R. G. Shorter and J. H. Pemberton, eds.), pp 37–49, Raven, New York.

McBurney, M. I., and Thompson, L. U. (1990). Fermentative characteristics of cereal brans and vegetable fibers. *Nutr. Cancer* **13,** 271–280.

McBurney, M. I., Thompson, L. U., Cuff, D. J., and Jenkins, D. J. (1988). Comparison of ileal effluents, dietary fibers, and whole foods in predicting the physiological importance of colonic fermentation. *Am. J. Gastroenterol.* **83,** 536–540.

Mendizabal, M. V., and Naftalin, R. J. (1992). Effects of spermine on water absorption, polyethylene glycol 4000 permeability and collagenase activity in rat descending colon *in vivo. Clin. Sci.* **83,** 417–423.

Mitch, W. E., Lietman, P. S., and Walser, M. (1977). Effects of oral neomycin and kanamycin in chronic uremic patients: I. Urea metabolism. *Kidney Int.* **11,** 116–122.

Mitchell, B., Lawaon, M. J., Kerr-Grant, A., Roediger, W. E. W., Illman, R. J., and Topping, D. L. (1985). Volatile fatty acids in the human large bowel: Studies in surgical patients. *Nutr. Res.* **5,** 1089–1092.

Moore, W. E., Cato, E. P., and Holdeman, L. V. (1978). Some current concepts in intestinal bacteriology. *Am. J. Clin. Nutr.* **31,** S33–S42 (suppl 10).

Naftalin, R. J., and Pedley, K. C. (1990). Video enhanced imaging of the fluorescent Na^+ probe SBFI indicates that colonic crypts absorb fluid by generating a hypertonic interstitial fluid. *FEBS Lett* **260,** 187–194.

Orbach, E., and Finkelstein, A. (1980). The nonelectrolyte permeability of planar lipid bilayer membranes. *J. Gen. Physiol.* **75,** 427–436.

Pouyssegur, J., Chambard, J. C., Franchi, A., Paris, S., and Van Obberghen-Schilling, E. (1982). Growth factor activation of an amiloride-sensitive Na-H exchange system in quiescent fibroblasts: Coupling to ribosomal protein S6 phosphorylation. *Proc. Natl. Acad. Sci. USA* **79**, 3935–3939.

Pouyssegur, J., Sardet, C., Franchi, A., L'Allemain, G., and Paris, S. (1984). A specific mutation abolishing Na/H antiport activity in hamster fibroblasts precludes growth at neutral and acidic pH. *Proc. Natl. Acad. Sci. USA* **81**, 4833–4837.

Rajendran, V. M., and Binder, H. J. (1990). Characterization of Na-H exchange in apical membrane vesicles of rat colon. *J. Biol. Chem.* **265**, 8408–8414.

Rajendran, V. M., and Binder, H. J. (1993). Cl-HCO_3 and Cl-OH exchanges mediate Cl uptake in apical membrane vesicles of rat distal colon. *Am. J. Physiol.* **264**, G874–G879.

Rajendran, V. M., and Binder, H. J. (1994). Apical membrane Cl-butyrate exchange: mechanism of short chain fatty acid stimulation of active chloride absorption in rat distal colon. *J. Membr. Biol.* **141**, 51–58.

Rajendran, V. M., and Binder, H. J. (1999). Distribution and regulation of apical Cl-anion exchanges in surface and crypt cells of rat distal colon. *Am. J. Physiol.* **276**, G132–G137.

Rajendran, V. M., Oesterlin, M., and Binder, H. J. (1991). Sodium uptake across basolateral membrane of rat distal colon. Evidence for Na-H exchange and Na-anion cotransport. *J. Clin. Invest.* **88**, 1379–1385.

Rajendran, V. M., Singh, S. K., Geibel, J., and Binder, H. J. (1998). Differential localization of colonic H,K-ATPase isoforms in surface and crypt cells. *Am. J. Physiol.* **274**, G424–G429.

Reynolds, D. A., Rajendran, V. M., and Binder, H. J. (1993). Bicarbonate-stimulated [14C] butyrate uptake in basolateral membrane vesicles of rat distal colon. *Gastroenterology* **105**, 725–732.

Roediger, W. E. (1982). Utilization of nutrients by isolated epithelial cells of the rat colon. *Gastroenterology* **83**, 424–429.

Roediger, W. E., Heyworth, M., Willoughby, P., Piris, J., Moore, A., and Truelove, S. C. (1982). Luminal ions and short chain fatty acids as markers of functional activity of the mucosa in ulcerative colitis. *J Clin. Pathol.* **35**, 323–326.

Roediger, W. E., and Rae, D. A. (1982). Trophic effect of short chain fatty acids on mucosal handling of ions by the defunctioned colon. *Br J Surg* **69**, 23–25.

Roggin, G. M., Banwell, J. G., Yardley, J. H., and Hendrix, T. R. (1972). Unimpaired response of rabbit jejunum to cholera toxin after selective damage to villus epithelium. *Gastroenterology* **63**, 981–989.

Roos, A., and Boron, W. F. (1981). Intracellular pH. *Physiol. Rev.* **61**, 296–434.

Rowe, W. A., Blackmon, D. L., and Montrose, M. H. (1993). Propionate activates multiple ion transport mechanisms in the HT29-18-C1 human colon cell line. *Am. J. Physiol.* **265**, G564–G571.

Sellin, J. H., and DeSoignie, R. (1990). Short-chain fatty acid absorption in rabbit colon *in vitro*. *Gastroenterology* **99**, 676–683.

Sellin, J. H., DeSoignie, R., and Burlingame, S. (1993). Segmental differences in short-chain fatty acid transport in rabbit colon: Effect of pH and Na. *J. Membr. Biol.* **136**, 147–158.

Serebro, H. A., Iber, F. L., Yardley, J. H., and Hendrix, T. R. (1969). Inhibition of cholera toxin action in the rabbit by cycloheximide. *Gastroenterology* **56**, 506–511.

Sheerin, H. E., and Field, M. (1975). Ileal HCO_3 secretion: Relationship to Na and Cl transport and effect of theophylline. *Am. J. Physiol.* **228**, 1065–1074.

Silen, W. (1985). Pathogenetic factors in erosive gastritis. *Am. J. Med.* **79**, 45–48.

Singh, S. K., Binder, H. J., Boron, W. F., and Geibel, J. P. (1995). Fluid absorption in isolated perfused colonic crypts. *J. Clin. Invest.* **96**, 2373–2379.

Singh, S. K., Binder, H. J., Geibel, J. P., and Boron, W. F. (1995). An apical permeability barrier to NH_3/NH_4^+ in isolated, perfused colonic crypts. *Proc. Natl. Acad. Sci. USA* **92**, 11573–11577.

Sjaastad, M. D., Wenzl, E., and Machen, T. E. (1992). pH_i dependence of Na-H exchange and H delivery in IEC-6 cells. *Am. J. Physiol.* **262**, C164–C170.

Smith, P. L., Sullivan, S. K., and McCabe, R. D. (1986). Concentration-dependent effects of disulfonic stilbenes on colonic chloride transport. *Am. J. Physiol.* **250**, G44–G49.

Snipes, R. L., Clauss, W., Weber, A., and Hornicke, H. (1982). Structural and functional differences in various divisions of the rabbit colon. *Cell Tissue Res.* **225**, 331–346.

Steggerda, F. R., and Dimmick, J. F. (1966). Effect of bean diets on concentration of carbon dioxide in flatus. *Am. J. Clin. Nutr.* **19**, 120–124.

Stephen, A. M., and Cummings, J. H. (1980). The microbial contribution to human faecal mass. *J Med Microbiol* **13**, 45–56.

Strocchi, A., and Levitt, M. D. (1998). Intestinal gas. *In* "Gastrointestinal and Liver Disease: Pathophysiology, Diagnosis, and Management" (M. Feldman, B. F. Scharschmidt, and M. H. Sleisenger, eds.), pp 153–160, Saunders, Philadelphia.

Summerskill, W. H., and Wolpert, E. (1970). Ammonia metabolism in the gut. *Am. J. Clin. Nutr.* **23**, 633–639.

Sunter, J. P., Wright, N. A., and Appleton, D. R. (1978). Cell population kinetics in the epithelium of the colon of the male rat. *Virchows Arch. B. Cell Pathol.* **26**, 275–287.

Tietz, P. S., Holman, R. T., Miller, L. J., and LaRusso, N. F. (1995). Isolation and characterization of rat cholangiocyte vesicles enriched in apical or basolateral plasma membrane domains. *Biochemistry* **34**, 15436–15443.

Trivedi, B., and Danforth, W. H. (1966). Effect of pH on the kinetics of frog muscle phosphofructokinase. *J. Biol. Chem.* **241**, 4110–4112.

van Meer, G., and Simons, K. (1988). Lipid polarity and sorting in epithelial cells. *J. Cell. Biochem.* **36**, 51–58.

van't Hof, W., Silvius, J., Wieland, F., and van Meer, G. (1992). Epithelial sphingolipid sorting allows for extensive variation of the fatty acyl chain and the sphingosine backbone. *Biochem. J.* **283**, 913–917.

vonEngelhardt, W., Burmester, M., Hansen, K., Becker, G., and Rechkemmer, G. (1993). Effects of amiloride and ouabain on short-chain fatty acid transport in guinea-pig large intestine. *J. Physiol.* **460**, 455–466.

VonEngelhardt, W., and Rechkemmer, G. (1988). Concentration and pH-dependence of short-chain fatty acid absorption in the proximal and distal colon of the guinea pig. *Comp. Biochem. Physiol.* **91**, 959–963.

VonEngelhardt, W., and Rechkemmer, G. (1992). Segmental differences of short-chain fatty acid transport across guinea pig large intestine. *Exp. Physiol.* **77**, 491–499.

Waisbren, S. J., Geibel, J. P., Modlin, I. M., and Boron, W. F. (1994). Unusual permeability properties of gastric gland cells. *Nature* **368**, 332–335.

Walser, M., and Rodenlas, L. (1959). Urea metabolism in man. *J. Clin. Invest.* **53**, 1617–1626.

Walter, A., Hastings, D., and Gutknecht, J. (1982). Weak acid permeability through lipid bilayer membranes. Role of chemical reactions in the unstirred layer. *J. Gen. Physiol.* **79**, 917–933.

Webb, D. J., and Nuccitelli, R. (1981). Direct measurement of intracellular pH changes in *Xenopus* eggs at fertilization and cleavage. *J. Cell Biol.* **91**, 562–567.

Welsh, M. J., Smith, P. L., Fromm, M., and Frizzell, R. A. (1982). Crypts are the site of intestinal fluid and electrolyte secretion. *Science* **218**, 1219–1221.

Weymer, A., Huott, P., Liu, W., McRoberts, J. A., and Dharmsathaphorn, K. (1985). Chloride secretory mechanism induced by prostaglandin E_1 in a colonic epithelial cell line. *J. Clin. Invest.* **76**, 1828–1836.

Williams, S. E., and Turnberg, L. A. (1980). Retardation of acid diffusion by pig gastric mucosa: A potential role in mucosal protection. *Gastroenterology* **79**, 299–304.

Williamson, J. R. (1964). Effects of insulin and starvation on the metabolism of acetate and pyruvate by the perfused rat heart. *Biochem. J.* **93**, 97–106.

Wolin, M. J., and Miller, T. L. (1983). Interactions of microbial populations in cellulose fermentation. *Fed. Proc.* **42**, 109–113.

Wolpert, E., Phillips, S. F., and Summerskill, W. H. (1971). Transport of urea and ammonia production in the human colon. *Lancet* **2,** 1387–1390.

Wrong, O. (1978). Nitrogen metabolism in the gut. *Am. J. Clin. Nutr.* **31,** 1587–1593.

Wrong, O., and Metcalfe-Gibson, A. (1965). The electrolyte content of faeces. *Proc. R. Soc. Med.* **58,** 1007–1009.

Wrong, O. M., Metcalfe-Gibson, A., Morrison, R. B. I., Ing, T. S., and Howard, A. V. (1965). *In vivo* dialysis of faeces as a method of stool analysis. I. Technique and results in normal subjects. *Clin. Sci.* **28,** 357–375.

Wrong, O. M., and Vince, A. (1984). Urea and ammonia metabolism in the human large intestine. *Proc. Nutr. Soc.* **43,** 77–86.

CHAPTER 4

Genetic Regulation of Expression of Intestinal Biomembrane Transport Proteins in Response to Dietary Protein, Carbohydrate, and Lipid

Lon J. Van Winkle

Department of Biochemistry, Midwestern University, Downers Grove, Illinois 60515.

Current Topics in Membranes, Volume 50

I. INTRODUCTION AND SCOPE

Nutrients may act directly on enterocytes to influence biomembrane transport, or their effects may be indirect or even circuitous. In the latter two cases, regulation involves changes in other cells that eventually lead to alterations in enterocyte biomembrane transport. The results of most of the studies to be discussed herein are consistent with the conclusion that dietary constituents affect genetic regulation of expression of enterocyte biomembrane transport proteins directly, although it appears that indirect or circuitous effects occur in some cases.

Genetic regulation of monosaccharide, amino acid, peptide, and fatty-acid transport, in this order, will be discussed in the next three main sections of this chapter. The decision to focus on biomembrane transport of nutrients that supply free energy was made for several reasons. First, a principal research interest of the author is amino acid and peptide transport. Moreover, there has been a persistently unfulfilled need in the field of amino acid transport to apply classical as well as newer techniques to the study of amino acid and peptide transport (e.g., see chapters 4 and 6 of Van Winkle, 1999). Now that we have begun to study genetic regulation of this transport, it is imperative to characterize the transport systems in tissues well enough to know which transport activities should be attributed to the transport-related proteins under investigation. Similarly, there is some controversy about whether lipophilic substances such as long-chain-length fatty acids require transport proteins to cross biomembranes (e.g., chapter 4 of Van Winkle, 1999). For this reason, regulation of fatty acid transport into enterocytes is considered herein. In contrast, dietary regulation of monosaccharide transport across enterocyte plasma membranes has been studied thoroughly and reviewed recently (e.g., Ferraris and Diamond, 1997). Nevertheless, data published since that review have left previous conclusions about the regulation of monosaccharide transport in need of some revision.

II. REGULATION OF MONOSACCHARIDE TRANSPORT

A. Introduction

The diets of vertebrates contain mixtures of mono-, di-, and polysaccharides that may vary in the proportion of different carbohydrates and the proportion of total kilocalories supplied by this category of nutrients. Hydrolases catalyze digestion of dietary carbohydrates to monosaccharides in the intestinal lumen (Van Beers *et al.*, 1995), and transport proteins catalyze monosaccharide transport across the brush-border and basolateral membranes of the intestinal epithelium. Glucose, fructose, and galactose are the major products of digestion, and glucose and galactose are believed to be transported across intestinal epithelial cell membranes by the same proteins. For these reasons, the regulation of glucose and fructose transport is considered herein.

1. Mechanisms of Glucose and Fructose Absorption

Glucose is transported across the intestinal brush-border membrane principally by the sodium-dependent glucose transporter 1 (SGLT1). Thus, this monosaccharide may be driven into enterocytes by an electrochemical potential gradient of Na^+ depending on the luminal and intracellular Na^+ and glucose concentrations at various locations in the small intestine (Fig. 1). Recent data also are consistent with the interpretation that Na^+ and glucose gradients may

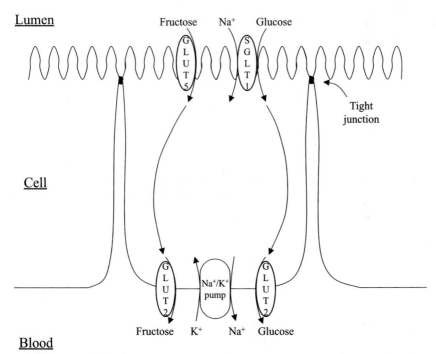

FIGURE 1 Absorption of dietary glucose and fructose occurs primarily via a transcellular pathway across the intestinal epithelium. In the model shown (Wright *et al.,* 1994), glucose is driven into enterocytes by SGLT1-catalyzed symport with Na^+ along the Na^+ total chemical potential gradient. For most models, it is proposed that glucose then leaves enterocytes via a GLUT2-catalyzed uniport. In contrast, fructose is both taken up and released by enterocytes via a uniport catalyzed by GLUT5 and GLUT2. Although transport of these monosaccharides across the brush border membrane by SGLT1 and GLUT5 is well established, their transport out of enterocytes by GLUT2 alone is less certain, especially in regard to the increased transport activity resulting from high-carbohydrate diets (see text).

drive the cotransport of water by SGLT1 in the intestine (e.g., Meinild *et al.,* 1998), although migration of water across the brush-border membrane via the SGLT1 protein remains difficult to demonstrate unambiguously. Glucose is believed to be transported out of enterocytes by glucose transporter 2 (GLUT2), after which it can migrate to the blood without passing across other biomembrane barriers. Alternatively, some dietary carbohydrate is converted to lactate in enterocytes, and the lactate is released to portal blood (Nicholls *et al.,* 1983; Holloway and Parsons, 1984). Similarly, the glucose leaving enterocytes may increase owing to conversion of some dietary fructose to glucose (Holloway and Parsons, 1984; Bismut *et al.,* 1993).

In contrast to the active transport with Na^+ that is possible for glucose across the brush-border membrane, it appears that fructose crosses this barrier via GLUT5, which is not known to concentrate fructose against a gradient (Bell *et al.,* 1993). Because glucose is not transported by GLUT5, at least in some species, fructose uptake from the intestinal lumen should not be slowed significantly by competition with lumenal glucose. In fact, glucose and amino acids facilitate fructose uptake possibly owing to the movement of water with their Na^+-dependent transport (Hoekstra and Van den Aker, 1996; Hoekstra *et al.,* 1996; Shi *et al.,* 1997). Such solvent drag could conceivably increase the rate of net fructose uptake via GLUT5.

Fructose exodus from enterocytes, however, is believed to occur via GLUT2, as is the case for glucose (Fig. 1). Owing to the potential for Na^+ gradient-driven active uptake of glucose, the intracellular glucose concentration may greatly exceed the fructose concentration in enterocytes after a carbohydrate-rich meal. Moreover, the K_m value for fructose transport by GLUT2 (67 mM), is, according to some reports, 5- to 11-fold higher than the K_m value for glucose transport (6 to 12 mM) (Bell *et al.,* 1993). If such is the case, one may wonder whether much intestinal transport of fructose actually occurs via GLUT2 *in vivo.* The detection of nearly equal concentrations of GLUT5 in basolateral and brush-border membranes by Blakemore and associates (1995) helps to resolve this question in humans. However, it is generally accepted that GLUT5 is expressed predominantly or exclusively in the intestinal brush-border membrane of rats and most other species that have been studied (e.g., Corpe *et al.,* 1996; Ferraris and Diamond, 1997; Corpe *et al.,* 1998). Regardless of their precise roles in intestinal monosaccharide absorption, each of the proteins, SGLT1, GLUT5, and GLUT2, appears to be involved in this process. Moreover, expression of each of these transporters is regulated in response to dietary carbohydrate.

2. Transcriptional and Posttranscriptional Regulation of Transporter Expression

Both brief (minutes) and more prolonged (hours to days) exposure of the intestinal tract to a change in carbohydrate concentration have been shown to alter enterocyte monosaccharide transport activity (Figs. 2 and 3). For example,

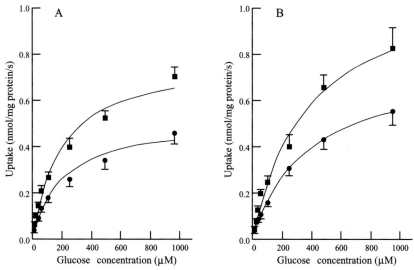

FIGURE 2   SGLT1-mediated glucose transport activity is increased by incubating (A) jejunal loops or (B) isolated jejunum with 25 mM glucose (squares) for 30 minutes relative to the same treatment with 25 mM mannitol (circles). Adapted from Sharp *et al.,* 1996, with permission from the British Society of Gastroenterology.

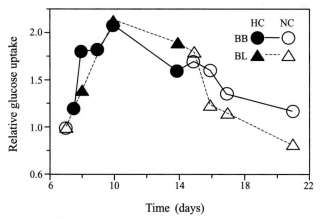

FIGURE 3   High dietary carbohydrate increases glucose transport across the brush-border (BB; circles) and basolateral (BL; triangles) membranes of rodent enterocytes. Transport activity increases when animals consume high-carbohydrate diets (HC; solid symbols) and decreases when they consume low- or no-carbohydrate diets (NC; open symbols). Adapted from Farraris, 1994, with permission from Raven Press.

incubation of the jejunum with D-glucose rapidly increases enterocyte SGLT1 transport activity in the brush-border membrane (Fig. 2). Much of this rapid increase *in vivo* occurs as a result of hexose-induced secretion of glucagon-like peptide-2 (e.g., Cheeseman and Tsang, 1996), which appears to cause trafficking of SGLT1 from an intracellular pool to the brush-border membrane (Cheeseman, 1997). Such acute enteric hormonal regulation of SGLT1 expression (Hirsh and Cheeseman, 1998) may be superimposed on long-term regulation of this expression (see section B.). While an understanding of both short-term and long-term regulation of transport is needed fully to understand how absorption of monosaccharides is influenced by nutritional factors, the principal focus here is on long-term regulation. Only in the case of long-term regulation do extant data indicate that the regulation may involve transcription of genes encoding hexose transport proteins.

B. Regulation of SGLT1 Expression

1. Which Enterocytes Undergo Regulatory Changes?

The amount of SGLT1 expressed in brush-border membranes in response to a long-term change in the quantity of dietary carbohydrate appears to be regulated only in enterocytes in the crypt, which then migrate along the villus to alter transport also at the tip (Fig. 4). In a series of experiments, Ferraris, Diamond, and associates (Ferraris and Diamond, 1992; Ferraris *et al.*, 1992; Ferraris and Diamond, 1993) showed that D-glucose–protectable phlorizin-specific binding to SGLT1 increases first in the crypt and only later at other positions along the villus upon switching mice from a diet containing no carbohydrate to one containing 55% sucrose (Fig. 5). Note that at 12 hours after the dietary change the phlorizin binding site density is highest in the crypt, and that it decreases about linearly along the villus to the tip, where no increase in phlorizin binding site density is yet realized. At 36 hours, nearly maximal binding density also occurs in the upper villus, and the binding density is at the same high level all along the villus after 156 hours of a high-carbohydrate diet (Fig. 5). These changes in D-glucose–protectable phlorizin-specific binding are reversed first in the crypt and later in the rest of the villus when mice are switched from a high-carbohydrate to a no-carbohydrate diet (Fig. 5).

It may be necessary to study development of enterocytes further in response to diet because transcription of the SGLT1 gene undergoes large diurnal changes in the small intestine (see later). Minimum and maximum SGLT1 mRNA and protein levels occur at about 12-hour intervals (Rhoads *et al.*, 1998). Consequently, the difference between the zero- and 12-hour phlorizin binding site density shown in Fig. 5 may be complicated by this diurnal effect. This daily rhythmicity is shifted by about half a day in diurnal rhesus monkeys relative to

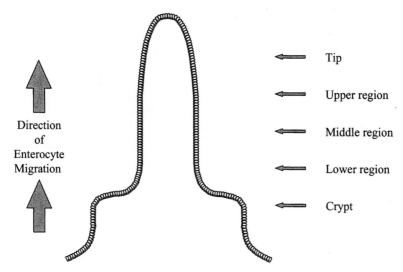

FIGURE 4 Enterocytes are produced in crypts and then migrate along intestinal villi toward the tips. Two or three days are required for the migration in mice (Ferraris *et al.*, 1992) after which the cells at the tip are sloughed off.

nocturnal rats. For these reasons, it is now important to study such rythmicity in each species used to investigate the effects of diet on transport. Moreover, investigators must remember to report the times of day at which measurements are made even in studies not designed explicitly to examine diurnal rhythmicity. Nevertheless, it appears unlikely that additional investigations will show a pattern of long-term regulation of SGLT1 expression in enterocytes in response to a change in the amount of dietary carbohydrate that is greatly different from that shown in Fig. 5.

2. Possible Transcriptional Regulation of the SGLT1 Gene in the Intestine

Ferraris and Diamond (1997) noted that the dramatic (more than 99%) reduction in D-glucose transport activity in lamb intestine upon weaning is accompanied by an equally dramatic decrease in SGLT1 protein concentration, but by only a 75% reduction in the SGLT1 mRNA level (Shirazi-Beechey *et al.*, 1991; Lescale-Matys *et al.*, 1993). Similarly, the time course for the increase in the intestinal SGTL1 mRNA level has been perceived (Ferraris and Diamond, 1997) to differ significantly from the time course for the increase in intestinal transport activity in rats placed on a high-glucose diet (Miyamoto *et al.*, 1993). Such data have been used to conclude that the SGLT1 protein level and transport activity are regulated primarily at a posttranscriptional (rather than a transcriptional) level of gene expression.

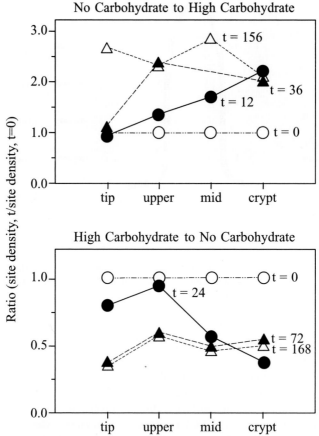

FIGURE 5   The high-affinity phlorizin binding site density (i.e., SGLT1 density) increases first in the crypts of villi and then migrates along villi toward the tips when mice are switched from a no-carbohydrate to a high-carbohydrate diet (top). Similarly, the SGLT1 density decreases first in and near the crypt when mice are switched from a high- to a no-carbohydrate diet (bottom). The levels (site densities) of high-affinity phlorizin binding sites in various positions along villi (relative to the levels at time = zero; open circles) are shown at 12 or 24 (solid circles), 36 or 72 (solid triangles), and 156 or 168 (open triangles) hours after switching mice from a no-carbohydrate to a high-carbohydrate diet (upper panel) or from a high-carbohydrate to a no-carbohydrate diet (lower panel). Adapted from Ferraris and Diamond, 1993, with permission from the National Academy of Sciences (USA).

A 75% reduction in the mRNA level could, however, account for a similar decrease in transport activity. If such is the case, then only about 25% of the reduction in activity in the intestine of weaning sheep might be attributable to a posttranscriptional mechanism, depending on whether mRNA stability also is regulated in this case. Moreover, statistically significant increases in transport activity and the SGLT1 mRNA level occur at about the same time in rat intestine 3 days after these animals begin to receive a high-glucose diet (Fig. 6), contrary to the interpretation of these same data elsewhere and as discussed earlier in this chapter. In fact, the correlation between transport activity and the SGLT1 mRNA level (Fig. 6) appears to be particularly good considering that transepithelial transport was measured (Miyamoto *et al.*, 1993), and SGLT1 accounts for only part of such transport (Fig. 1). Moreover, some increase in SGLT1 activity is expected to occur within minutes and before accumulation of SGLT1 mRNA and protein (Fig. 2).

Shirazi-Beechey and associates also concluded initially that expression of SGLT1 is regulated principally at the translational or even posttranslational level

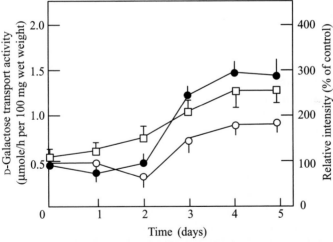

FIGURE 6 The D-galactose transport activity (open squares) and SGLT1 mRNA level (solid circles) increase on about the same time schedule in small intestine after rats begin to consume a high-glucose diet. Also shown are GLUT2 mRNA levels (open circles). Galactose transport across the intestinal epithelium, which is depicted here, may represent both SGTL1 and GLUT2 activity because SGLT1 transports this monosaccharide across the brush-border membrane whereas GLUT2 may be its primary route of migration across the basolateral membrane (see text). The increases in this D-galactose transport activity and the SGLT1 mRNA level are statistically significant 3 days after rats begin to consume the high-glucose diet ($P < .05$). Adapted from Miyamoto *et al.*, 1993, with permission from the Biochemical Society and Portland Press.

based primarily on the time course of disappearance of SGLT1 mRNA, protein, and transport activity from lamb intestine after weaning (Lescale-Matys *et al.*, 1993). However, the ability of dietary glucose to induce expression of SGLT1 mRNA, protein, and transport activity in 2- to 3-year-old sheep has led these same investigators to the revised conclusion that SGTL1 gene transcription contributes significantly to regulation of expression of glucose transport in the ovine intestine (Shirazi-Beechey *et al.*, 1996). Control of mRNA stability also could contribute to regulation in this case. Finally, diurnal regulation of SGLT1 mRNA and protein expression on about the same schedule in the rat intestine (Rhoads *et al.*, 1998) also supports the conclusion that intestinal SGLT1 glucose transport activity may be regulated at the level of gene transcription (see subsection b.), although regulation clearly is not limited to transcription alone (Ferraris and Diamond, 1997).

a. Nature of the Signal Leading to Stimulation of SGLT1 Expression by Dietary Monosaccharides. Long-term feeding of fructose as well as D-glucose induces expression of SGLT1 activity and protein in the intestines of mice, rats, and sheep (summarized in Table VIII of Ferraris and Diamond, 1997). In the case of rats, the level of SGLT1 mRNA also is known to increase in response to these carbohydrates. Because fructose and other monosaccharides such as 2-deoxy-D-glucose that are not transported by SGLT1 induce its expression, increased expression does not depend on transport by extant SGLT1 protein. Moreover, 2-deoxy-D-glucose remains in the lumen of the intestine (Hopfer, 1987), so it is not metabolized, and nonmetabolizable SGLT1 substrates also induce SGLT1 expression (Ferraris and Diamond, 1997). Consequently, metabolites of glucose and other metabolizable sugars appear not to form part of this signaling process. In fact, because 2-deoxy-D-glucose remains in the lumen, regulation of enterocyte SGLT1 expression by dietary carbohydrate must involve sensing of sugar molecules at the luminal surface (Minami *et al.*, 1993; Dyer *et al.*, 1994).

Furthermore, the lumenal sugar signal may affect SGTL1 expression locally and may not require systemic involvement. When D-glucose is infused into the ileal lumen of sheep for 4 days, SGTL1 protein expression in the duodenum is unaffected (Shirazi-Beechey, *et al.*, 1994). Similarly, vascular infusion of D-glucose into sheep for 4 days does not influence brush-border SGLT1 expression, although basolateral GLUT2 expression is increased by such treatment (Shirazi-Beechey *et al.*, 1994). Blood-to-lumen migration of glucose (Levine *et al.*, 1982) during weeks of diabetes may help to account for the sometimes increased brush-border glucose transport activity in this condition (reviewed by Ferraris and Diamond, 1997).

Although systemic effects may not be required for long-term regulation of SGLT1 expression, acute regulation of SGLT1 expression occurs through diet-induced enteric hormone secretion (Cheeseman, 1997; Hirsh and Cheeseman, 1998) (see section A.2). Indirect effects also are needed to account for the increase in SGLT1 expression in the rat jejunum following a 3- or 4-hour infusion of the ileum with glucose (Debnam, 1985). In light of all these studies, it is clear that direct, indirect, and even circuitous mechanisms are involved in dietary regulation of SGLT1 expression. However, each mechanism may be made to operate independently of the others by the right experimental protocol.

b. Possible Relationship between the Mechanism of Long-Term Dietary Regulation of SGLT1 Expression and the Genetic Mechanism of Its Diurnal Regulation. As mentioned above, SGLT1 mRNA and protein follow about the same pattern of diurnal regulation of expression in the intestines of rats and probably of other mammals (e.g., Fig. 7). Increases in the levels of SGLT1 mRNA and protein are directly attributable to an increase in the rate of transcription of the SGLT1 gene and concomitant translation of the accumulating SGLT1 mRNA. As shown in Fig. 8, the increase in the rate of transcription appears immediately to precede the daily increase in the level of SGLT1 mRNA and protein in the rat jejunum (Fig. 7). Surprisingly, this increase in the rate of transcription occurs in nuclei isolated from enterocytes at the tips of villi (Rhoads *et al.,* 1998). Hence, it appears that both transcriptional and posttranscriptional (e.g., Sharp *et al.,* 1996) regulation of expression of SGLT1 transport activity occur in enterocytes regardless of their position along villi. Perhaps cells in the crypt are programmed to be able to display a given level of transport sites in response to diet composition (see Fig. 5), and this level is further modulated depending on when during the day food is consumed.

Hence, transcriptional changes in response to diet composition could occur by some variation in the mechanism of diurnal regulation of SGLT gene transcription in enterocytes, especially because this diurnal periodicity appears to depend on the prior, light-cycle–related timing of food intake (Stevenson *et al.,* 1975; Stevenson and Fierstein, 1976). In this regard, Rhoads and associates (1998) found that the promoter region of the rat SGLT1 gene has a hepatocyte nuclear factor-1 (HNF-1) binding sequence immediately upstream to the TATA box (Fig. 9), and such is also the case for the ovine and human SLGT1 genes (Wood *et al.,* 1999). To date, α (Courtois *et al.,* 1987) and β (Mendel *et al.,* 1991a) isoforms of this atypical homeodomain protein (Gehring *et al.,* 1994) have been identified in liver, kidney, and intestine (Mendel and Crabtree, 1991). HNF-1 isoforms form both homo- and heterodimers in association with a third protein termed the dimerization cofactor of HNF-1 (DCoH). HNF-1 dimerization is stabilized by DCoH, which also helps to enhance gene transcription (Mendel *et al.,* 1991b). Interestingly, the HNF-1 complex at the HNF-1 binding sequence of the SGLT1 gene in enterocytes is composed primarily of HNF-1α dimers when transcription is relatively rapid (i.e., at 1000 to 1100 h), whereas both the α and β

A

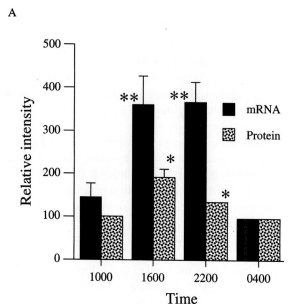

B   Monkey mRNA

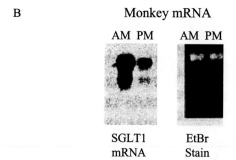

FIGURE 7 SGTL1 mRNA and protein levels vary diurnally in (A) rat and (B) monkey enterocytes. However, these schedules are reversed in the diurnal rhesus monkey relative to the nocturnal rat. Relative intensities in Northern and Western blots are shown for rat SGLT1 mRNA and protein (A), whereas Northern blots are shown (B) for monkey enterocyte mRNA obtained at 0900h (AM) or 2000h (PM). All animals were permitted free access to chow, but it is important to note that the daily variation in SGLT1 activity appears to depend on the previous day's feeding pattern. (See the beginning of the increases in transport activity relative to the beginning of feeding in Fig. 1 of Stevenson *et al.,* 1975 and in Fig. 3 of Stevenson and Fierstein, 1976. Also note that the SGLT1 gene transcription shown in Fig. 8 of this chapter occurs several hours before the increase in SGLT1 protein shown in this figure.) The present figure was adapted from Rhoads *et al.,* 1998, with permission from the American Society for Biochemistry and Molecular Biology.

A

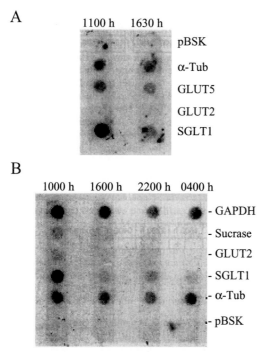

FIGURE 8   An increase in the rate of SGLT1 gene transcription in rat enterocytes precedes the increase in SGLT1 mRNA and protein levels shown to occur on a diurnal schedule in Fig. 7. Nuclei were isolated from villus tip enterocytes at the times indicated and used to assess the rates of transcription. The relative amounts of RNA produced from other genes at the indicated times of day also are shown for comparison with those of SGLT1. Adapted from Rhoads *et al.,* 1998, with permission from the American Society for Biochemistry and Molecular Biology.

isoforms are present in complexes when transcription is slower (i.e., at 1600 to 1630 h) (Rhoads *et al.,* 1998). Hence, it is possible that DCoH mediates exchange of the HNF-1 β isoform for the HNF-1 α isoform in order to form homodimers that increase the rate of transcription. Subsequent exchange of one of the HNF-1 α subunits in homodimers for an HNF-1 β subunit would slow transcription according to this hypothesis (Rhoads *et al.,* 1998).

In regard to the possibility that dietary carbohydrate might influence the rate of transcription of the SGLT1 gene by a related mechanism, it is interesting to note that expression of two other carbohydrate-responsive genes depends on HNF-1. Genes encoding sucrase-isomaltase (Wu *et al.,* 1994) and liver-type pyruvate kinase, which also is expressed in the intestine (Ogier *et al.,* 1987), contain HNF-1 responsive elements. In particular, the 5-terminal nucleotide residues

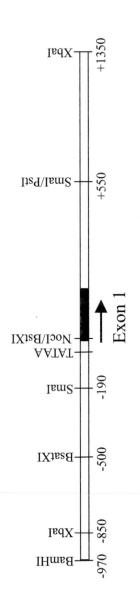

```
  1 GGATCCTGTT  AATATTGAAC  TTCAGGGCTT  TTACATACAC  GTGACATAAC  TATATATCCA  TAATACAACC  CACACGTCTT
 81 CTCGTGTGTG  GTTTAAAGCA  TCTCTAGATT  ACCTGTGACA  CCACCGTGAG  CAAGCTTATGA AACGTTACTC  ACTGTACTGC
161 ATGGCTTAGG  GAAACACAGG  GGAAATAGTC  TGTACAAGTT  CAGTACAAAA  CTATTTCCAT  CCGTGCGTGG  TGGGTGGAAT
241 CAGTTAGGTG  AAGACACAGG  ACAAAGAGAT  GGCTGTGGTA  GGACCCCCTT  CTCCTAGGCT  GGTTGTGGCT  CTCCTTACAG
321 TGGTGGGAAT  CGCGCAGTTC  CTAGGGATTG  CATGCTCTGG  GGAAGTCAGT  AACGGTGGCT  AAGGGATTTG  TATTGCGTTT
401 TCCCCAGCCC  CATCCCAGCAA TCTCATCCTC  CTGGAAGATT  CTCCAGGAGA  CCAAGAACTT  GGCTGAGTTC  TCTGCAAAGT
481 CCTGCACTCG  TTTTCCACA   GAGAAATCAG  TCCACAGAGA  TGCCAGGTCT  GCTTCAAAAA  GCACCTGCCA  CAGATAAGCC
561 TCAGCCTATA  GACCCTTGGC  TCCAGGTGCC  TCCGTCCCTC  CTGTGTAGCT  ACTCCTTTTC  TGCTTCACGT  GGACTCCGTG
641 CAAGCCAGCC  GTGGCAACTG  TGCGAGGTGC  CCAGGCGGGG  AGCACAGAGG  TCTTTTTGTG  CCAGGCGGTC  ACCACCCTTT
721 AGCCCGGCCC  CTCCCTGCTC  ACCGAACAGA  CTTTACTGCC  GGCCCGGGCG  CCCGCAGTCT  CTCTTGCAGC  TTTGCTGACT
801 TAGAGCACTG  CACACCTAGG  AGCTGCTTCC  CGACGGTGCG  GTGCAGCGCA  GCAGCCCAGG  CTCCTGCCGC  TCCTGCCGCG
881 CAGCCCCTCC  GTCGCGGTTC  CCCTCCCTCG  GGGCTGATCA  TTAACTCAAA  AGCAGTATAA  GGAGTGAGTG  GCCCCGGTGA

961 GCCGGTGAGC  C
```

MLTF/USF   HNF1   TATAA box

FIGURE 9   The proximal promoter region of the rat SGLT1 gene contains potential binding sites for MLTF/USF (major late transcription factor/upstream stimulatory factor) and HNF-1 (hepatocyte nuclear factor-1). A restriction map (top) and the nucleotide sequence (bottom) of this region are shown. Adapted from Rhoads et al., 1998, with permission from the American Society for Biochemistry and Molecular Biology.

of these 13-nucleotide elements are identical in all these genes (Rhoads *et al.*, 1998). In contrast, neither the GLUT2 (Ahn *et al.*, 1995) nor the GLUT5 (Mahraoui *et al.*, 1994) gene contains an HNF-1 binding sequence in the proximal region of its promoter. Hence, other mechanisms probably account for the diurnal and dietary regulation of transcription of the GLUT2 and GLUT5 genes.

C. GLUT5 Expression and Regulation of Fructose Transport

1. Regulatory Changes in GLUT5 Expression Occur Only after Enterocytes Leave the Crypt

Recall from section B.1 that the concentration of SGLT1 transport sites in the plasma membrane increases first in enterocytes in the crypt in response to long-term increases in dietary carbohydrate (see Fig. 5). Only later do increases also occur at other positions along the villus, eventually reaching the tip (Ferraris and Diamond, 1993). Similarly, villi lose their increased concentrations of SGLT1 transport sites first from enterocytes in the crypt when carbohydrate is removed from the diets of mice. Only in the case of diurnal regulation does SGLT1 protein production appear to occur in cells outside the crypt, apparently in anticipation of actual consumption of the food (Rhoads *et al.*, 1998).

In contrast to SGLT1, no increase in GLUT5 mRNA or protein is detected in crypt enterocytes of rats following gavage of their gastrointestinal tract with a high-fructose solution (Corpe *et al.*, 1998). Rather, increased GLUT5 mRNA appears first in the lower to mid regions of villi, in response to luminal fructose, whereas GLUT5 protein is detected mainly in the upper regions of the villi. It is important to emphasize that GLUT 5 mRNA is *never* detected in crypt cells even under control conditions, whereas some of this mRNA is always detected in the lower to mid regions of the villi (Corpe *et al.*, 1998). The simplest interpretation for these data appears to be that GLUT5 mRNA synthesis (or conceivably stabilization) begins after enterocytes leave the crypt, and this mRNA accumulation is followed by accumulation of GLUT5 protein as enterocytes migrate toward the villus tip. However, the specific position along the villus that enterocytes first detect the presence of fructose in the intestinal lumen is not clear.

2. Regulation of the GLUT5 mRNA Level May Control GLUT5 Protein Concentration and Transport Activity in the Intestine

In contrast to the imperfect correlation between mRNA level and protein concentration or transport activity described earlier for SGLT1 expression, diet-induced changes in these aspects of GLUT5 expression correlated well (Burant and Saxena, 1994; Shu *et al.*, 1998). In rats, fructose transport activity increases dramatically in the intestinal brush-border membrane at about the time of weaning, regardless of whether weaning to a milk-free diet actually occurs (Fig. 10).

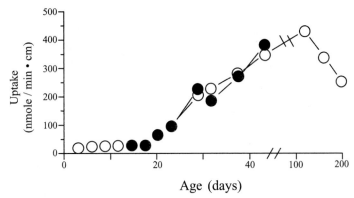

FIGURE 10   Brush-border fructose transport increases in weaning rats regardless of whether they are maintained on a milk-replacer diet (solid circles) or switched to a rat-chow diet (open circles). Weaning in rats occurs between about 15 and 28 days of age. Adapted form Toloza and Diamond, 1992, with permission from the American Physiological Society.

Nevertheless, both weaning and older rats increase expression of GLUT5 mRNA and transport activity several fold in response to dietary fructose (e.g., Fig. 11). In contrast, dietary fructose has no effect on glucose transport activity in weaning rats (Fig. 11), whereas dietary fructose dramatically increases glucose transport and the SGLT1 mRNA level in enterocytes of adult rats (Miyamoto *et al.*, 1993). This difference between weaning and adult rats may reflect differences in the length of high-fructose feeding in the two studies (i.e., 1 day vs 5 days), or it could relate to other differences in regulation between adult and weaning animals such as the presence of diurnal regulation of GLUT5 expression in adult rats (Corpe and Burant, 1996) but not in weaning rats (Shu *et al.*, 1998) (see below).

   a. Emerging Details of the Mechanism by which GLUT5 Gene Expression Is Regulated.   As for dietary regulation of SGLT1 expression, regulation of GLUT5 expression in response to dietary fructose requires signaling from the luminal side of the intestine, although a requirement for a combination of luminal and systemic signals has not been ruled out. In weaning rats fed a high-fructose diet, the GLUT5 mRNA level and fructose transport activity increase in enterocytes in contact with luminal fructose, but not in enterocytes surgically restricted from such contact (Fig. 11). The peak GLUT5 mRNA level precedes the peak increases in GLUT5 protein and fructose transport activity in both weaning (Shu *et al.*, 1998) and adult (Corpe and Burant, 1996) rats. These data are consistent with a model in which dietary fructose helps to stimulate its own transport by inducing transcription of the GLUT5 gene. This hypothesis was confirmed recently when nuclear run-on assays were used to show that dietary fructose and

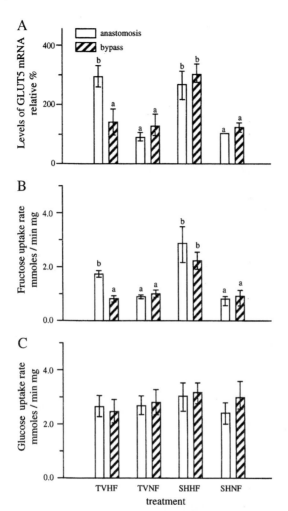

FIGURE 11 Dietary fructose induces expression of GLUT5 mRNA and transport in enterocytes that make direct contact with this monosaccharide. Weaning rats were surgically manipulated such that enterocytes were or were not in contact with dietary constituents. Only those enterocytes isolated from the region of the intestine that was bypassed (represented by hatched bars) in rats receiving a Thiry-Vella loop construct (TV) could not make contact with dietary constituents. TV and sham-operated (SH) animals received either a high-fructose (HF) or a no-fructose (NF) diet for 5 days prior to undergoing measurement of GLUT5 mRNA levels and fructose or glucose transport in their enterocytes. Adapted from Shu *et al.*, 1998, with permission from the American Physiological Society.

sucrose (but not glucose) increase the rate of GLUT5 gene transcription (Kishi *et al.,* 1999). It remains to be determined how transcription of the GLUT5 gene is regulated by dietary fructose. At least three binding sites for regulatory proteins are present in the 5' region of the GLUT5 gene, and one of these sites has homology to sites for hepatocyte nuclear factor-4 (HNF-4) (Klein *et al.,* 1998). However, little else is currently known about the details of regulation of this gene.

  b. Dietary Fructose Dampens Diurnal Regulation of GLUT5 Expression. As for SGLT1, GLUT5 mRNA and protein levels change in enterocytes depending on the time of day and apparently in anticipation of food consumption (Castelló *et al.,* 1995; Corpe and Burant, 1996). The increase in the GLUT5 protein level between 9:00 AM and 9:00 PM (control in Fig. 12B) is preceded by a peak in the GLUT5 mRNA level at 3:00 PM (control in Fig. 12A) and as anticipated from other data discussed earlier. Interestingly, these diurnal variations in GLUT5 mRNA and protein levels are greatly dampened by fructose-enriched feeding for 7 days (Fig. 12). Apparently, the free energy that would be needed to produce the wider diurnal variations in GLUT5 expression on a high fructose diet would partially reduce the advantage gained by increasing GLUT5 expression in order to take advantage of a high-fructose diet. That is, the more a high-fructose diet seems assured, the less the value in remodeling the brush border each day in anticipation of a diet in which the proportions of monosaccharides may vary. Moreover, the diurnal variation in expression of GLUT5 in adults was not present in weaning rats. Rat pups feed almost continuously, rather than mainly at night (Shu *et al.,* 1998), and such continuous eating may dampen diurnal variation of GLUT5 expression in pups like a high-dietary fructose does in adults.

D. Is GLUT2 Expression Involved in Regulation of Transepithelial Transport by Dietary Monosaccharides?

  In contrast to the data concerning regulation of GLUT5 and SGLT1 expression by dietary sugars, data from different laboratories on GLUT2 regulation by diet appear inconsistent. For example, basolateral glucose transport activity (see Fig. 3) and GLUT2 mRNA levels (but not necessarily GLUT2 protein) increase in enterocytes when rats are fed high-glucose or high-fructose diets (Cheeseman and Harley, 1991; Miyamoto *et al.,* 1993). More recently, however, Corpe and Burant (1996) observed no change in GLUT2 mRNA and a large decrease in GLUT2 protein upon feeding rats a high-fructose diet for 1 or 7 days (Fig. 13). Although the difference in the levels of GLUT2 mRNA detected in these studies in response to dietary fructose is difficult to explain, the meaning of the un-

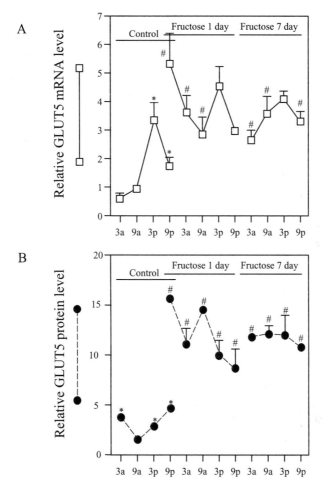

FIGURE 12   Dietary fructose induces GLUT5 (A) mRNA and (B) protein expression and dampens diurnal regulation in rat enterocytes. Rats were fed a control or fructose-enriched diet for the indicated times before measurement of GLUT5 mRNA and protein levels in their enterocytes at the indicated times of day. Asterisks indicate that levels at the specified times were significantly different from levels at 9:00 AM in control rats. Pound signs indicate that levels in rats fed a fructose-enriched diet differed significantly from levels in rats fed a control diet for the specified time of day ($P < .05$). Adapted from Corpe and Burant, 1996, with permission from the American Physiological Society.

equivocal decrease in the GLUT2 protein level in response to dietary fructose may have a simple explanation.

Perhaps GLUT2 is not the only transporter of monosaccharides in the basolateral membrane. The absence of GLUT2 in mice does not appear to prevent or even to impair absorption of dietary carbohydrate (Guillam *et al.*, 1997; 1998)

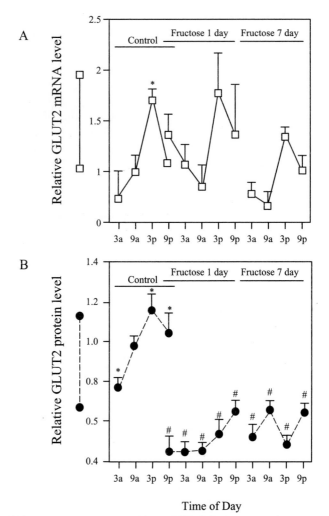

FIGURE 13 GLUT2 protein expression apparently cannot account for the increased monosaccharide transport in the basolateral membrane of enterocytes from rats fed a high-fructose diet (see text). Experiments were performed as described in the legend of Figure 12, except that GLUT2 (rather than GLUT5) mRNA and protein levels were measured. Adapted from Corpe and Burant, 1996, with permission from the American Physiological Society.

possibly because hexoses can be converted to lactate and released to portal blood and because other transporters are expressed to compensate for the GLUT2 deficiency. In this regard, expression of the GLUT5 protein may not be limited just to the brush-border membrane of enterocytes in humans (Blakemore *et al.,* 1995). Moreover, the use of cytochalasin B to study the well-documented expression of GLUT2 in the basolateral membrane may inadvertently include undetected changes in the expression of other cytochalasin B–sensitive monosaccharide transport proteins. Boyer and associates (1996) detected low levels of the cytochalasin B–sensitive protein (Bell *et al.,* 1993) GLUT1 in basolateral membranes of rat enterocytes, and the level of GLUT1 increases dramatically in diabetic rats. In fact, by using chemical means (such as cytochalasin B sensitivity) or molecular probes to detect only known monosaccharide transporters in tissues, we risk missing transport by as yet unidentified proteins. Although thorough kinetic studies of monosaccharide transport across rat basolateral membrane vesicles have produced no definitive evidence of such a protein (Cheeseman, 1993; Tsang *et al.,* 1993), a reproducible increase in the K_m as well as the V_{max} value for transport after exposure of enterocytes to luminal hexoses is consistent with the possibility that expression of more than one basolateral transporter is affected by dietary carbohydrate.

1. The Mechanism by which Dietary Carbohydrate Increases Basolateral Monosaccharide Transport Is Currently under Investigation

Increased levels of monosaccharides in blood plasma and the intestinal lumen increase monosaccharide transport in the basolateral membrane (reviewed by Ferraris and Diamond, 1997). Hence, it is unclear to what extent dietary carbohydrate acts directly on enterocytes to increase basolateral transport rather than indirectly through a diet-associated increase in the plasma glucose level. However, it is now known that short-term dietary regulation of basolateral transport occurs indirectly through secretion of the enteric hormones gastric inhibitory polypeptide and glucagon-like peptide-2 (Cheeseman and Tsang, 1996). As discussed earlier, it remains to be determined whether this increased transport can be attributed entirely to GLUT2 expression. A low level of GLUT1 also is expressed in the basolateral membrane, and its expression is increased greatly in diabetic rats (Boyer *et al.,* 1996) perhaps as a consequence of hyperglycemia. Thus, dietary regulation of monosaccharide transport may be more complex in the basolateral than in the brush-border membrane of enterocytes. In this regard, however, insulin did not reduce GLUT1 expression in the basolateral membrane of diabetic rats but rather increased it in the brush-border membrane, where this protein also is expressed in diabetes (Boyer *et al.,* 1996). The unanticipated expression of GLUT1 in the brush-border membranes of diabetic rats and its stimulation by insulin eventually may help to explain why the vast literature on the effects of diabetes on intestinal monosaccharide transport has been largely inconclusive and even contradictory (reviewed by Ferraris and Diamond, 1997). Similarly, signaling by C-peptide as well as insulin may be needed for more nearly normal gene expression and metabolism (e.g., Forst *et al.*, 1998).

2. Diurnal Regulation of GLUT2 Expression May Help to Supply Glucose to Enterocytes during Fasting

Although the basis for dietary regulation of monosaccharide transport across the basolateral membrane awaits a full description, it has been demonstrated that expression of GLUT2 in this membrane is regulated diurnally (Corpe and Burant, 1996). The increase in GLUT2 mRNA levels in enterocytes between 9:00 AM and 3:00 PM is associated with a concomitant increase in the level of GLUT2 protein (controls in Fig. 13). However, this increase in GLUT2 protein is better timed to serve to increase transport of glucose into enterocytes during fasting of rats during the day, rather than in anticipation of nocturnal feeding. Both glucose and glutamine serve as substrates to supply free energy for enterocyte metabolism (e.g., Fleming *et al.*, 1997), so a more-or-less continuous supply of these substrates may be needed by enterocytes. Uptake of glucose and amino acids from the intestinal lumen presumably supplies free energy to enterocytes during the fed state. As discussed earlier for dietary regulation, a full understanding of diurnal regulation of monosaccharide transport across the basolateral membrane probably will require consideration of transport proteins in addition to GLUT2. These other proteins conceivably could be as important as GLUT2 for transepithelial transport of dietary monosaccharides.

E. Summary

Good progress is being made in understanding how dietary carbohydrates influence monosaccharide uptake across the brush-border membrane of intestinal epithelial cells. Expression of SGLT1 is increased by elevations in a variety of dietary monosaccharides including fructose, although this protein transports primarily glucose and galactose. The sugars appear to act locally on intestinal cells and do not need to be taken up to influence long-term SGLT1 expression. Although much of this effect can be attributed to increased SGLT1 gene transcription, posttranscriptional changes also contribute significantly to regulation of SGLT1 expression. The details of how dietary constituents influence gene transcription have just begun to emerge, and they could involve some of the same nuclear proteins that bind to the SGLT1 gene during its diurnal regulation.

Similarly, expression of GLUT5 is increased in response to increases in dietary carbohydrate, but in this case the response is specific for the GLUT5 substrate fructose. The change in expression can be attributed entirely to changes in the GLUT5 mRNA level, and mRNA stability as well as GLUT5 gene transcription conceivably could contribute to these changes. GLUT5 mRNA and consequently changes in the amount of this RNA are detected only in enterocytes outside the villus crypt, in contrast to long-term regulation of SGLT1 expression in response to diet composition, which appears to occur only while cells are in the crypt. Expression of GLUT5 protein in brush-border membranes in response to more dietary fructose appears to increase as enterocytes migrate along villi.

In contrast to regulation of monosaccharide transport across the brush-border membrane, much less is known about long-term regulation of transport across the basolateral membrane in response to dietary carbohydrate. Expression of the major basolateral monosaccharide transport protein GLUT2 is decreased (rather than increased) in response to dietary carbohydrate. Moreover, its diurnal regulation seems better suited to supply glucose to enterocytes during fasting than for transepithelial monosaccharide transport during the fed state. Because basolateral membrane transport activity increases in response to dietary carbohydrate, its nature may warrant further investigation. Other cytochalasin B-sensitive and -insensitive GLUT proteins are expressed in the basolateral membrane, and their expression conceivably could be regulated by diet. However, more than one hexose transport activity in the basolateral membrane has not been detected in kinetic studies, although such studies do not rule out the possibility that multiple transport activities are present.

Although kinetic studies of hexose transport in enterocytes have been relatively thorough, amino-acid transport across both the basolateral and brush-border membranes has not been characterized completely. Nevertheless, interesting data on the regulation of expression of an accessory protein for amino-acid transport are consistent with the possibility that such regulation could contribute to diet-induced changes in amino-acid transport across both the basolateral and brush-border membranes (Segawa *et al.*, 1997).

III. REGULATION OF AMINO-ACID AND PEPTIDE TRANSPORT

A. Introduction

It has been known for some time that dietary amino acids and protein influence expression of amino-acid and peptide transport systems in the small intestine (e.g., Wolfram *et al.*, 1984; Stein *et al.*, 1987). Only in this decade, however, has it become possible to examine fully the mechanisms of this influence owing to the cloning of a growing number of cDNAs encoding mammalian transport-related proteins for amino acids (Van Winkle, 1993;1999; Malandro and Kilberg, 1996; Palacín *et al.*, 1998) and di- and tripeptides (Ganapathy and Leibach, 1996). Dietary regulation of expression of only a few of these proteins has been examined so far, and the discussion herein will be limited mainly to the two that have been studied in greatest detail: the amino acid transport-related protein BAT (Segawa *et al.*, 1997) and the intestinal peptide transport protein PepT1 (Walker *et al.*, 1998).

B. Proposed Mechanisms of Amino-Acid and Peptide Absorption Owing to BAT and PepT1 Expression

1. Transport Associated with BAT Expression

The BAT protein appears to be a component of several transport systems that are composed of more than one subunit. These systems may include those for pyruvate and uridine (Yao *et al.*, 1998a) as well as those that receive amino acids (reviewed by Van Winkle, 1999). Similarly, the activities of at least three distinct amino acid transport systems are increased in *Xenopus* oocytes expressing BAT (Van Winkle, 1993; Peter *et al.*, 1996; Ahmed *et al.*, 1997), so regulation of BAT expression could have numerous effects on the transport of nutrients and their metabolites. By most accounts, BAT spans the membrane only once, and it does not appear to form aggregates, so it is probably an activator of transport proteins rather than a transport protein *per se* (reviewed by Van Winkle, 1999). This activation may result, at least in part, from BAT-regulated trafficking of transport proteins to the plasma membrane, as is the case for the function of the BAT-related protein 4F2hc (Nakamura *et al.*, 1999). Moreover, a model in which BAT spans the membrane four times has been used to propose that BAT itself may form a pathway for amino acid transport across biomembranes (e.g., Mosckovitz *et al.*, 1994). We should probably remain open to such possibilities even though evidence is growing in support of the activator function of BAT (e.g., Estévez *et al.*, 1998). Many transport proteins have more than one function (Van Winkle, 1999) and BAT likely is no exception.

BAT function has been studied most frequently in regard to its conspicuous effect on amino acid transport system $b^{o,+}$. Although system $b^{o,+}$ was first well characterized in preimplantation mouse blastocysts (Van Winkle *et al.*, 1988), it now appears to transport amino acids in a wide variety of tissues including renal and intestinal epithelia (Furriols *et al.*, 1993; Pickel *et al.*, 1993). Because system $b^{o,+}$ may catalyze obligatory exchange of cationic amino acids for zwitterionic ones (reviewed by Van Winkle, 1999), it may help to concentrate cationic amino acids from the intestinal lumen against their concentration gradient. Zwitterionic amino acids that are driven into enterocytes against their gradient by symport with Na^+ could be exchanged for cationic amino acids via system $b^{o,+}$ (Fig. 14). In this model (Chillarón *et al.*, 1996), system B^o or system $B^{o,+}$ (Munck and Munck, 1994) catalyzes the Na^+-dependent uptake of zwitterionic amino acids, which then drive uptake of cationic amino acids via system $b^{o,+}$-catalyzed antiport. Cationic amino acid uptake against a concentration gra-

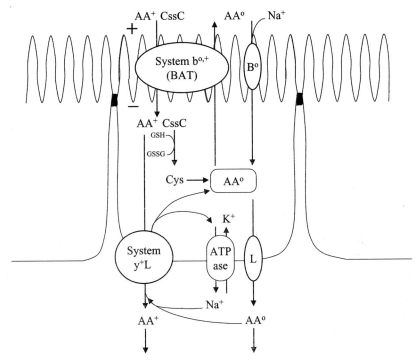

FIGURE 14 Model for the absorption and reabsorption of cationic amino acids (AA^+) and the tetrapolar amino acid cystine (CssC) via system $b^{o,+}$ in the brush-border membrane of intestinal and renal epithelia. In the model (Chillarón *et al.*, 1996), dipolar amino acids (AA^o) are taken up against their gradient via a Na^+/AA^o symporter termed B^o. The gradient of AA^o is then used to drive up-take of AA^+ and CssC against their gradients via the $b^{o,+}$ antiporter. AA^+ and AA^o (including Cys produced from CssC) are proposed to leave epithelial cells via systems y^+L and L (or similar systems), respectively, in the basolateral membrane. Such exodus of AA^+ would be driven by the Na^+ total chemical potential gradient (see text).

dient also is driven by the membrane electrical potential, which must be overcome for these amino acids to enter the blood. System y^+L may facilitate such exodus of cationic amino acids across the basolateral membrane through Na^+-gradient-driven, electrically neutral exchange (Fig. 14).

With the exception that system y^+L transports Na^+ in association with zwitterionic amino acids, its characteristics are very similar to those of system $b^{o,+}$. It may also be pertinent that system $b^{o,+}$ interacts with Na^+ (Van Winkle *et al.*, 1990a) although it does not appear to transport this cation with amino acids. Interestingly, a truncated version of the BAT protein stimulates amino-acid transport in *Xenopus* oocytes by a system that resembles y^+L in regard to its strong Na^+-dependent interaction with L-leucine (Miyamoto *et al.*, 1996)[1]. Although system y^+L transport is believed to be activated by expression of a related protein (i.e., 4F2hc; e.g., Yao *et al.*, 1998b), expression of BAT also could conceivably increase transport by system y^+L in the basolateral membrane under some circumstances. Similarly, 4F2hc expression appears in some cases to stimulate system $b^{o,+}$ activity (Bröer *et al.*, 1998), and it also appears to stimulate system L activity in nonepithelial cells (Bröer *et al.*, 1997).

In fact, several proteins with which 4F2hc associates were identified recently, and they display transport activities that resemble systems L (Kanai *et al.*, 1998; Mastroberardino *et al.*, 1998) and y^+L (Torrents *et al.*, 1998). Because the later proteins have significant homology with cationic amino acid transporters that correspond to system y^+ (Van Winkle, 1993; Malandro and Kilberg, 1996; Palacín *et al.*, 1998), the transport proteins with which 4F2hc, <u>BAT</u>, and related proteins associate may form the continuum of transport systems: L, $B^{o,+}$, y^+L, $b^{o,+}$, y^+, b^+_2 and b^+_1. As one proceeds along this continuum, the relative selectivities of the systems change from a virtually exclusive selection for zwitterionic substrates by system L, to a similarly strong selection for cationic amino acids by systems b^+_1 and b^+_2 (Van Winkle and Campione, 1990). Furthermore, regulation of expression of one or a few accessory proteins could conceivably help to regulate amino acid transport by several of these systems simultaneously in various tissues including the brush-border and basolateral membranes of the intestinal epithelium. (See also the possible effect of BAT on system $B^{o,+}$, discussed below.) The list of systems that may be regulated in this manner was expanded recently to include system x_c^- for anionic amino acids (Sato *et al.*, 1999).

Recently we found that chymotrypsin treatment of mouse blastocysts stimulates the activities of their $b^{o,+}$ and b^+_2 systems by about twofold. (Van Winkle and Campione, unpublished data). This simultaneous regulation of $b^{o,+}$ and b^+_2 activities could conceivably occur through regulation of expression of a single accessory protein, such as BAT (see above). We anticipate that the regulation may occur through a possibly novel proteinase-activated receptor in the epithelial cells (trophectoderm) of these embryos, although further studies are needed to test this hypothesis. Nevertheless, if such a mechanism of proteinase activation of system $b^{o,+}$ and possibly other amino acid transport systems also occurs in the intestinal epithelium, then it would provide a mechanism by which protein digestion might be coupled to absorption of its products. In this regard, pro-

---

[1]These investigators attributed this transport to system y^+, which does not interact strongly with L-leucine regardless of whether Na^+ is present (reviewed by Devés and Boyd, 1998; Devés *et al.*, 1998).

teinase-activated receptor 2 in the intestinal epithelium appears to respond to physiological concentrations of trypsin in the rat gastrointestinal tract (Kong *et al.*, 1997), although the relationship of this regulation to nutrient transport in general, and BAT expression in particular, remains to be determined.

2. PepT1-Catalyzed Transport

In contrast to the system $b^{o,+}$ antiport that is associated with <u>BAT</u> expression, PepT1 catalyzes symport of di- and tripeptides with protons (e.g., Mackenzie *et al.*, 1996). In this case, the actions of Na^+ K^+-ATPase and a Na^+/H^+ exchanger create an acidic extracellular microenvironment (McEwen *et al.*, 1990) that drives peptide uptake (Fig. 15). The peptides are hydrolyzed inside enterocytes, and the resultant amino acids can leave the cells across the basolateral membrane as described earlier (Fig. 14). Although PepT1 and related proteins

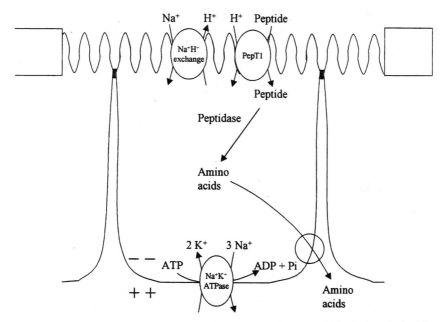

FIGURE 15 Model for proton-driven transport of peptides by PepT1 in the intestinal epithelium. In this model (Ganapathy *et al.*, 1994), the combined actions of $Na^+K^+ATPase$ and a Na^+/H^+ exchanger concentrate protons in the microenvironment on the luminal side of the brush-border membrane. This total chemical potential gradient of protons then drives transport of peptides in the lumen of the intestine into enterocytes against their total chemical potential gradients. The peptides are hydrolyzed intracellularly, and the resultant amino acids leave the cell across the basolateral membrane (e.g., Fig. 14).

generally are believed to catalyze a single component of peptide transport, some published data are consistent with the interpretation that their expression is associated with peptide transport by two or more processes (e.g., Fig. 16). Although such kinetic data may seem insufficient alone to prove that more than one component of mediated peptide transport is present, such data for BAT (Van Winkle, 1993) led to more thorough studies to demonstrate that BAT expression is associated with transport by at least two systems for amino acids (Peter *et al.*, 1996).

C. Dietary Regulation of the BAT mRNA Level and the Amino Acid Transport Activities Associated with BAT Expression

Dietary amino acids induce amino acid transport activity in the brush-border membrane of the intestinal epithelium (Wolfram *et al.*, 1984; Stein *et al.*, 1987). Moreover, the characteristics of much of this transport resemble those of the Na^+-independent system $b^{o,+}$, which is associated with BAT expression. Although some intestinal transport is Na^+ dependent, and hence dissimilar from system $b^{o,+}$, data reported recently are consistent with the possibility that expression of BAT is associated with an increase in both Na^+-dependent and Na^+-

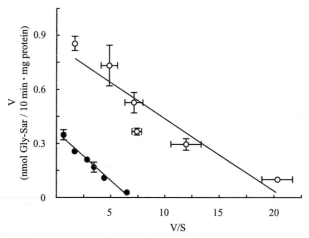

FIGURE 16 PepT protein expression appears to be associated with more than one component of dipeptide transport. The data shown are for substrate-saturable transport by PepT2 in a rat renal proximal tubular epithelial cell line in the presence (solid circles) or absence (open circles) of the histidyl-reactive chemical diethylpyrocarbonate (DEP). After linear transformation, the data do not fit well on the straight line in the Hofstee plot shown, especially in the absence of DEP. A curved line would better fit the data and would be consistent with the presence of more than one transport activity. Adapted from Bandsch *et al.*, 1997, with permission from Elsevier Science.

independent transport of cationic and zwitterionic amino acids across the brush-border membrane (see later). Such transport is catalyzed by systems $B^{o,+}$ (Van Winkle *et al.*, 1985) and $b^{o,+}$ (Van Winkle *et al.*, 1988), respectively, so an increase in BAT expression could help to explain why the activities of these transport systems increase dramatically at the same time during development of preimplantation mouse embryos (Van Winkle, 1992).

The levels of BAT mRNA and $b^{o,+}$-like transport activity in mouse small intestine increase together in response to a high-protein diet (Segawa *et al.*, 1997). Of the individual amino acids tested (Table I), only dietary L-aspartate increases the V_{max} value for total Na^+-independent L-cystine transport in the brush-border membrane of enterocytes. Similarly, dietary L-aspartate increases the level of BAT mRNA in these cells (Fig. 17A), although other amino acids apparently were not tested in this regard. In mice, therefore, a dietary signal for increasing BAT expression appears to reside largely if not entirely in aspartate released during protein digestion. In an earlier study, aspartate was found to be the most potent amino acid to increase lysine transport activity in mouse small intestine (Stein *et al.*, 1987). However, other amino acids may serve such a function in different species. For example, $b^{o,+}$-like L-lysine transport increases in the intestinal brush-border membrane of chickens consuming a lysine-enriched diet

TABLE I

An Aspartate–Enriched Diet Increases L-Cystine Transport
Activity in the Brush-Border Membranes of Murine Enterocytes

Diet	V_{max} (pmol/15 s per mg of protein)	K_m (mM)
Untreated	144 ± 48	0.12 ± 0.03
Protein-free	128 ± 32	0.11 ± 0.02
Glycine	151 ± 31	0.13 ± 0.02
Alanine	134 ± 41	0.14 ± 0.03
Lysine	157 ± 25	0.11 ± 0.02
Arginine	169 ± 18	0.14 ± 0.04
Aspartate	228 ± 34*	0.12 ± 0.03
Glutamate	127 ± 32	0.11 ± 0.02

In most cases (except untreated controls), mice consumed a protein-free diet for 2 weeks before receiving diets enriched in the indicated amino acid (5%) for 4 days.

*$P < .05$ for the V_{max} value relative to the value in brush-border membrane vesicles from mice that continued to receive the protein-free diet.

Adapted from Segawa *et al.*, 1997, with permission from the Biochemical Society and Portland Press.

(Torras-Llort *et al.,* 1998). In contrast, a high-protein diet did not increase expression of BAT mRNA in rat enterocytes, although it did raise expression of mRNAs encoding PepT1 and a glutamate and aspartate transport protein in this species (Erickson *et al.,* 1995). In the human Caco-2 intestinal cell model, incubation of cells with extracellular lysine reduced their system $b^{o,+}$ transport activity gradually over a 2-day period (Satsu *et al.,* 1998). Such possible differences among species make it imperative to study regulation of their amino acid and peptide transport processes in detail for each case rather than attempting to extrapolate results from one species to another.

Moreover, it remains essential to carefully characterize all the systems responsible for the observed transport. However, careful and complete characterization of each system contributing to transport often is overlooked, especially in the case of transport of amino acids. In the present case, mouse BAT expression in *Xenopus* oocytes leads to a small (about 20% of the total at 50 μM substrate concentration) but consistent and statistically significant increase in Na^+-dependent transport of cationic and zwitterionic amino acids (Table II). Similarly, a significant (about 25% of the total) Na^+-dependent component of L-cystine uptake is present in brush-border membrane vesicles (BBMVs) from the small intestine of mice fed an aspartate-enriched diet (Fig. 17B). A similar Na^+-dependent component of lysine transport appears in BBMVs from the intestine of chickens fed a lysine-enriched diet, but this component of transport is not observed when the diet is not enriched with lysine (Torras-Llort *et al.,* 1998). Characterization of this Na^+-dependent component of L-lysine transport in chicken intestine was abandoned after it was learned that the amino acid analog

TABLE II

A Component of Mouse BAT Induced
Amino Acid Transport is Na^+-Dependent

Amino Acid	Xenopus oocyte injected with	Uptake (pmol/min · oocyte)	
		$-Na^+$	$+Na^+$
L-Arginine	Water	0.24 ± 0.11	0.48 ± 0.14
	BAT cRNA	13.4 ± 2.3	16.3 ± 1.5
L-Leucine	Water	0.14 ± 0.05	0.39 ± 0.12
	BAT cRNA	11.1 ± 2.6	14.8 ± 3.5
L-Cystine	Water	0.10 ± 0.04	0.16 ± 0.04
	BAT cRNA	6.6 ± 1.5	9.0 ± 1.2

Although transport is too variable among xenopus oocytes to conclude that a component of it is Na^+-dependent for each amino acid tested (50 μM), the average transport rate is consistently higher in the presence of Na^+ in oocytes expressing BAT ($P < .02$).

Adapted from Segawa *et al.,* 1997, with permission from the Biochemical Society and Portland Press.

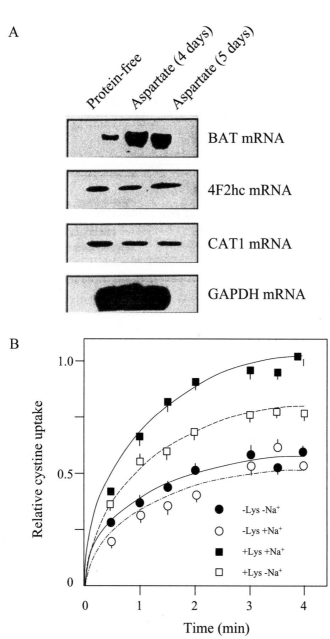

FIGURE 17 (A) Feeding mice an L-aspartate–enriched diet for 4 or 5 days induces BAT mRNA expression in their enterocytes. In contrast, expression of several other mRNAs is not affected by dietary aspartate. (B) A component of L-cystine uptake is Na^+ dependent in brush-border membrane vesicles from mice fed an L-aspartate–enriched diet for 4 days. The Na^+-dependent component of L-cystine transport is observed only in vesicles loaded with L-lysine (+L-Lys), so it could conceivably occur owing to exchange. Adapted from Segawa *et al.*, 1997, with permission from the Biochemical Society and Portland Press.

2-aminoendobicyclico(2.2.1)heptane-2-carboxylic acid (BCH) does not selectively inhibit the component (Torras-Llort *et al.,* 1998). Similarly, a component of transport attributed to the Na^+-independent system y^+ becomes in part Na^+ stimulated in the brush-border membrane of chickens fed a lysine-enriched diet, although this Na^+-dependent transport has not yet been characterized. Hence, not only may high-lysine and high-aspartate diets induce expression of both Na^+-dependent and Na^+-independent components of transport in intestines of chickens and mice, but each of these components may contain more than one transport system (recall discussion in section B above).

To comprehend completely the effects of diet on transport, the various transport system activities must be isolated and characterized, as described in detail elsewhere (e.g., appendix to chapter 4 of Van Winkle, 1999). Best-fit analyses of kinetic data to determine whether two or more transport processes are present for a particular substrate may be insensitive because different systems may have nearly the same K_m values for transport of the substrate. However, analog inhibition and ABC testing can be used more efficiently to identify, isolate, and characterize each component of transport. If multiple components of transport are present, each must be studied independently to fully understand how transport is influenced by diet. Concomitant correlations between transport activity and transporter mRNA and protein levels are more likely to be drawn appropriately if all the potentially pertinent transport activities have been identified. In spite of the need for more complete studies, the data currently under discussion are convincing enough to conclude that diet-induced changes in the expression of BAT in mouse and probably chicken enterocytes lead to changes in the rates of transport of cationic and zwitterionic amino acids in the brush-border membrane of the intestinal epithelium.

It also remains to be determined whether dietary amino acids regulate BAT expression in enterocytes at the level of transcription, although the data available so far are consistent with this possibility, especially in the mouse (see discussion herein). However, it is possible that dietary aspartate somehow acts to stabilize BAT mRNA in mouse enterocytes, and that aspartate leads to an increase in amino acid transport activity in other ways in addition to increasing BAT mRNA levels. If regulation of BAT expression occurs at least in part at the level of transcription, such regulation could occur through a direct effect of dietary amino acids on the BAT gene in enterocytes, or indirectly or even circuitously through other signaling molecules. The rat BAT gene promoter region has been characterized, and positive and negative regulatory elements of it have been detected upon expression in pig LLC-PK1 cells (Yan *et al.,* 1994). However, it may be necessary to characterize the promoter region of the BAT gene in each species to understand its regulation by dietary amino acids. For example, a high-protein diet has been found to induce BAT expression in mice (Segawa *et al.,* 1997) but not in rats (Erickson *et al.,* 1995), and thus, even among rodents, the genes may be regulated somewhat differently.

D. Regulation of PepT1 Expression by Its Substrates

Expression of PepT1 mRNA is increased in enterocytes of rats when the concentration of protein in their diets is increased (Erickson *et al.,* 1995). The increase in the PepT1 mRNA level occurs selectively in the distal region of the small intestine. Hence, it appears that this region plays a particularly important role in the transport of peptides produced during digestion of dietary protein (Erickson *et al.,* 1995).

To attempt to learn whether such induction of PepT1 mRNA expression might occur without hormonal or neuronal control, Walker and associates (1998) used the Caco-2 human intestinal cell line as a model for enterocytes. Replacement of 4.0 mM glutamine with 4.0 mM glycylglutamine in the medium for 3 days led to a 64% increase in the Na^+-independent peptide transport activity in Caco-2 cells and to parallel increases in lysine and leucine uptake. The increase in peptide transport activity probably results directly from an increase in the PepT1 protein because about 1.7-fold greater binding of antibody specific for this protein is observed in cells incubated for 3 days in medium containing glycylglutamine (Walker *et al.,* 1998). Similarly, the level of PepT1 mRNA increases by about 1.9-fold in Caco-2 cells exposed to 4.0 mM glycylglutamine for 3 days. This increase in mRNA levels is attributable at least in part to increased mRNA stability, although these investigators concluded that the greater stability could not account fully for the greater level of PepT1 mRNA observed. However, the effect of glycylglutamine exposure on the rate of transcription of the PepT1 gene in Caco-2 cells should be measured to verify their conclusion that mRNA levels also increase owing to more rapid mRNA synthesis.

Similar results were reported for PepT1 mRNA levels in the rat ileal epithelium itself, and an amino acid and peptide responsive element was detected in the promoter region of the PepT1 gene (Shiraga *et al.,* 1999). It remains to be determined, however, whether the rate of transcription of the PepT1 gene actually increases in the intestinal epithelium in response to dietary protein, peptides, and amino acids. Moreover, it was reported recently that expression of the PepT1 protein itself is reduced by dietary amino acids and increased by starvation in rat jejunal epithelium (Oginara *et al.,* 1999). Hence, regulation of intestinal peptide transport by diet may be more complex than previously anticipated.

E. Summary

The cloning of cDNAs encoding amino acid and peptide transporters in this decade now allows the study of genetic regulation of their expression in enterocytes in response to dietary protein, peptides, and amino acids. Such studies already have revealed interesting details of this regulation that may vary among species. For example, aspartate appears to be one of the most potent components of dietary protein in determining regulation of BAT-associated amino acid transport in mice, but it appears not to have the same effect in rats, at least as a component of dietary protein. The regulation in mice appears to occur at the level of transcription, although posttranscriptional mechanisms also may contribute to the regulation of transport activity. BAT expression may affect several Na^+-dependent and Na^+-independent amino acid transport systems in the brush-border and possibly the basolateral membrane, although the precise nature of all the systems affected by BAT expression remains to be determined.

Such detailed characterization of each amino acid transport system in the brush-border and basolateral membranes of the intestine seems particularly important now that the opportunity to study the mechanisms of their dietary regulation is at hand. Indeed, such regulation is difficult to study definitively if one does not know for sure which transport systems are present and which are affected by dietary protein. Although the amino acid transport systems present in the small intestine have been listed in numerous reviews, the effects of BAT expression on the still poorly defined transport activities discussed earlier in this chapter indicate that these lists may be incomplete or even incorrect. Interestingly, the activity of some of the same transport systems for cationic and zwitterionic amino acids may be increased upon exposure of human Caco-2 cells to 4.0 mM glycylglutamine for 3 days, although it remains to be determined whether BAT expression underlies these changes in amino acid transport. Similarly, the mechanism by which lysine in the medium suppresses system $b^{o,+}$-like activity in Caco-2 cells is still under investigation.

In contrast to suppression of system $b^{o,+}$ transport activity by its substrate lysine, peptide transport activity increases in Caco-2 cells after exposure to glycylglutamine. Both an increased rate of PepT1 gene transcription and a decreased rate of PepT1 mRNA degradation may contribute to this increase in expression of PepT1 transport activity, although the pertinent transcription rates have not been measured. Because PepT1 expression may be associated with more than one transport activity, this transport may need to be characterized more thoroughly. Moreover, regulation of PepT1 expression by dietary peptides and amino acids may differ at different positions along the intestine. Hence, its expression needs to be studied more thoroughly at each position in order to fully understand the dietary regulation of intestinal peptide transport.

IV. REGULATION OF FATTY ACID TRANSPORT

A. Introduction

Integral membrane proteins provide a pathway for the migration of hydrophilic peptides, amino acids, and monosaccharides across the hydrophobic interior region of the membrane phospholipid bilayer. However, the theory that such proteins also are needed to transport fatty acids across biomembranes has gained acceptance only recently (chapter 4 of Van Winkle, 1999). Moreover, cytosolic fatty-acid-binding proteins appear to influence the biomembrane transport of these substances (Fig. 18A), as do extracellular proteins such as albumin in blood plasma in the case of cells that take up fatty acids from the circulation. Although all these factors and additional processes (e.g., Fig. 18B) should be considered in discussing completely the regulation of absorption of dietary fatty acids from the lumen of the small intestine, here the focus is primarily on dietary regulation of expression of biomembrane fatty-acid transport proteins and on regulation of expression of the liver isoform of the cytosolic fatty-acid-binding protein (L-FABP$_c$). This isoform of FABP$_c$ also is expressed in the intestine, and its regulation appears to parallel that of an intestinal biomembrane fatty-acid transporter (Poirier *et al.*, 1996).

At least three families of fatty-acid transport proteins have been identified so far, and two of them are expressed conspicuously in enterocytes (Besnard *et al.*, 1996).[2] One of these proteins is known as the plasma membrane fatty-acid-binding protein (FABP$_{pm}$). Interestingly, FABP$_{pm}$ is identical to the protein that in mitochondria serves as aspartate amino transferase (mAspAT) (see Berk *et al.*, 1996 for review). Such multiple functions of biomembrane transport proteins are not uncommon (Van Winkle, 1999), although no well-documented case is as dramatic as mAspAT/FABP$_{pm}$. The other conspicuous fatty-acid transport protein in enterocytes is the fatty-acid transporter (FAT). FAT is an 88-kDa plasma membrane glycoprotein with two predicted transmembrane segments (Abumrad *et al.*, 1993). As for FABP$_{pm}$, FAT catalyzes fatty-acid biomembrane transport with K$_m$ values in the nanomolar range (Ibrahimi *et al.*, 1996), so this transport is likely to be physiologically significant (Van Winkle, 1999; Sfeir *et al.*, 1999). Because FAT, but not FABP$_{pm}$, has been studied in enterocytes in regard to regulation by diet, FAT and L-FABP$_c$ are the main focus of the discussion herein.

B. <u>FAT</u> and L-FABP$_c$ Function in Context

Absorption of dietary triacyglycerols involves five major steps: hydrolysis of triacylglycerols in the intestinal lumen, transport of the products of hydrolysis across the brush-border membrane into enterocytes, reesterification of fatty acids and monoacylglycerols into triacylglycerols and other lipids, incorporation of the triacylglycerols into chylomicrons (and to a much smaller extent, intestinal VLDLs), and secretion of the chylomicrons (and VLDLs) by exocytosis for

---

[2]Relatively weak expression of the third family of long-chain fatty-acid transport proteins (FATP) has also been detected in mouse intestine (Motojima *et al.*, 1998).

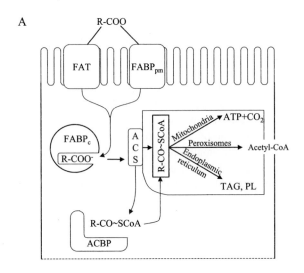

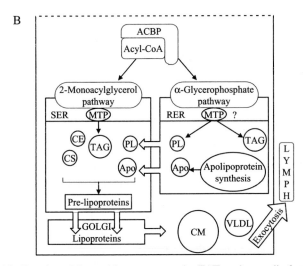

FIGURE 18 Function of fatty-acid transport protein (FAT) and cytosolic fatty-acid-binding protein (FABP$_c$) in absorption of dietary fatty acids. FAT catalyzes migration of fatty acids across the brush-border membrane, whereas FABP$_c$ binds intracellular fatty acids until they are activated to form acyl-CoA. (A) The principal fate of dietary fatty acids in enterocytes is the formation of tri-acylglycerols (TAG) and phospholipids (PL). (B) TAG and PL are then incorporated into lipoproteins and secreted into the lymph. Other abbreviations: ACBP, acyl-CoA-binding protein; ACS, long-chain acyl-CoA synthetase; Apo, apoprotein; CE, cholesterol esters; CM, chylomicrons; CS, cholesterol; FABP$_{pm}$, plasma membrane fatty-acid-binding protein (also a fatty-acid transporter); MTP, microsomal triacylglycerol transfer protein; SER, smooth endoplasmic reticulum; RER, rough endoplasmic reticulum; VLDL, very low density lipoproteins. Adapted form Besnard *et al.*, 1996, with permission from the Proceedings of the Nutrition Society.

transport to blood via the lymph (Figs. 18A and 18B). FAT and $FABP_{pm}$ catalyze the second of these two processes, whereas several binding proteins including $FABP_c$ facilitate intracellular processing of fatty acids and fatty acyl-CoA (reviewed by Besnard *et al.*, 1996). What is known about the genetic regulation of expression of the liver isoform (but not the intestinal isoform) of $FABP_c$ in intestine resembles what is known about this regulation of expression of FAT in response to diet (Poirier *et al.*, 1996).

C. Genetic Regulation of FAT and L-FABP_c Expression by Dietary Lipid

FAT mRNA and protein are expressed primarily in mature enterocytes near the tips of villi (Poirier *et al.*, 1996). Moreover, FAT expression is limited to the small intestine of the rat gastrointestinal tract, as expected for a protein involved in dietary fatty-acid transport. Consumption of a diet enriched with sunflower oil for 3 days results in dramatic increases in FAT mRNA levels, and to a lesser degree FAT protein concentrations, in all segments of the small intestine, whereas only small increases in FAT mRNA levels are observed when rats consume a diet enriched with medium-chain-length fatty acids (Fig. 19). The effect of di-

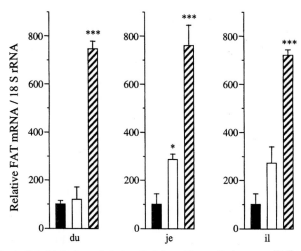

FIGURE 19 A diet rich in long-chain-length fatty acids greatly increases FAT mRNA levels in rat enterocytes. Such a diet, containing sunflower oil as the source of fatty acids, was fed to rats for 3 days before measurement of the FAT mRNA levels in their enterocytes. In contrast, dietary medium-chain-length fatty acids (MCFA) increased FAT mRNA levels only moderately. Solid bars, control rats; open bars, rats fed MCFA; hatched bars, rats fed sunflower oil; du, duodenum; je, jejunum; il, ileum. Adapted from Poirier *et al.*, 1996, with permission from Springer-Verlag.

etary lipid seems likely to be directly on enterocytes rather than through neu-ronal, paracrine, or endocrine signaling, because expression of the liver isoform of $FABP_c$ in Caco-2 cells is induced by long-chain fatty acids (Poirier et al., 1997) and expression of this protein is regulated coordinately with FAT expres-sion in enterocytes (Poirier et al., 1996).

Although it remains to be verified that the level of FAT mRNA increases in enterocytes as a result of transcription, such data already are available for ex-pression of L-$FABP_c$. Consumption of a diet enriched with sunflower oil in-creases both L-$FABP_c$ mRNA levels and the rate of transcription of the L-$FABP_c$ gene in mouse enterocytes (Mallordy et al., 1995). Similarly, transcription of the gene encoding L-$FABP_c$ is induced by a peroxisone-proliferator hypolipidemic drug, and this drug also increases the level of FAT mRNA in enterocytes. For these reasons, it has been suggested that long-chain fatty acids may induce ex-pression of both of these genes by a mechanism similar to that of peroxisome proliferators (Poirier et al., 1996). In fact, peroxisome proliferators have been shown to increase the levels of mRNAs encoding members of all three families of long-chain fatty-acid transport proteins in intestine (Motojima et al., 1998), so all these genes may be regulated by a similar mechanism.

In this regard, exposure of preadipocytes to long-chain fatty acids induces ex-pression of a putative fatty-acid-activated receptor (FAAR) that is related to per-oxisome proliferator-activated receptors (PPARs) (Amri et al., 1995). FAAR is expressed in several tissues including intestine, and it binds to peroxisome pro-liferator-response elements. However, the otherwise limited similarity of FAAR to PPARs, and differences among the tissues in which they are most highly ex-pressed, indicates that FAAR and related proteins form a distinct subtype of the PPAR family (Amri et al., 1995). Amri and associates (1995) argue convincingly that FAAR helps to mediate regulation of transcription of the FAT gene, and they have recently begun to characterize the pertinent binding site in the FAT gene promoter region (Grimaldi et al., 1999)

D. Summary

Dietary regulation of fatty-acid biomembrane transport per se has been stud-ied much less extensively in enterocytes than has regulation of monosaccharide, amino acid, and peptide transport in these cells. Nevertheless, more is known about the possible mechanism of dietary regulation of the genes encoding FAT and L-$FABP_c$ than of those encoding transporters of these other nutrients. This circumstance has arisen, in part, because only recently has it been documented that fatty-acid biomembrane transport requires catalysis by transport proteins and because fatty-acid absorption from the lumen of the small intestine is more com-plex than absorption of other kilocalorie-containing nutrients. However, it is also known now that biomembrane transport of all of these nutrients may be influenced significantly by the experimental procedures used to study the transport. Hence, it continues to be necessary to study both nutrient transport and the mechanisms of dietary regulation of the transport in enterocytes.

V. PERSPECTIVE: WHY WE STILL NEED TO PERFORM STUDIES AT THE PHYSIOLOGICAL AS WELL AS THE MOLECULAR LEVEL OF INVESTIGATION

Considerable progress is being made toward understanding dietary regulation of intestinal biomembrane transport of monosaccharides, amino acids, peptides, and fatty acids. Some of this regulation occurs at the level of gene transcription, whereas other components are posttranscriptional. Similarly, some regulation of transport occurs through direct effects of nutrients only on enterocytes, and other aspects of transport regulation involve neuronal, paracrine, and endocrine effects. In all cases, the details of the molecular mechanisms by which the regulation occurs are presently emerging. In regard to the main purpose for this chapter, we are now beginning to learn how the promoter regions of genes encoding biomembrane transport proteins interact with regulator proteins. The activities of these regulatory proteins may in turn be influenced directly or indirectly by components of the diet.

Despite this important progress, there is good reason to believe that the most important variable measured in transport studies, namely the quantity of transport activity itself, may not be measured accurately in most studies. If such inaccuracies in measurements occur consistently to the same degree in all cases, then little misunderstanding should occur in regard to regulation of transport activity, because we need only adjust all our measurements of transport activity by the proper proportion. All our comparisons should correctly reflect regulation of transport, and the relative amounts of transport activity should vary in their physiologically pertinent directions even if they are not absolutely accurate. However, the question still to be answered is whether measurement of transport in preparations from the intestine of animals in different treatment groups is always disturbed in equal proportions upon isolating the cells or membrane vesicles, or even just disturbing the intestine and all its tissues with a surgical procedure. Studies by Uhing and associates (1995 and 1997) raise the question of the impact of surgery on transport, whereas other studies indicate that transport in cell membranes may sometimes change dramatically when the cells are isolated from their environments *in vivo*.

Uhing and associates (1995 and 1997) showed that monosaccharide and amino acid transport are altered significantly by disturbing the small intestine surgically (e.g., Fig. 20). Transport activity appears to be reduced by more than 80% immediately after surgery and remains somewhat reduced for about 24 hours, at least in

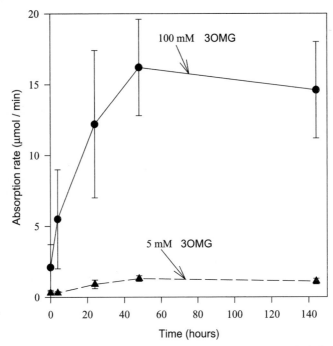

FIGURE 20 Laparotomy and anesthesia appear greatly to reduce intestinal absorption of the
D-glucose analog 3-O-methyl-glucose (3OMG). The rate of 3OMG absorption was measured at the
indicated times after surgical placement of catheters in the duodenums and pertinent blood vessels
of adult male rats. The rate of transport appears to be suppressed by anesthesia and surgery for at
least 4 and perhaps as long as 24 hours. Moreover, acute anesthesia administration to rats 4 or more
days after the initial surgery reduced absorption by about 26% (data not shown). Mediated transport
constituted at least 71% of the total shown immediately after surgery and 95% of the total 6 days
later. Data from Uhing and Kimura, 1995.

the case of amino acid transport (Uhing and Arango, 1997). Monosaccharide transport also is reduced about 26% by anesthesia alone (Uhing and Kimura, 1995). Of
the total 3-O-methyl-D-glucose transport shown in Fig. 20, at least 71% is transporter mediated immediately after surgery, and 95% is mediated 6 days later.
Because most studies on transport across the gut are performed near the time of
surgical disturbance, it needs to be determined whether the effect of surgery influences transport by the same proportion in animals consuming different diets.

Similarly, separating epithelial cells from their environment *in vivo* alters their
transport activities dramatically in some cases (Van Winkle, 1999). Although
such data have not been obtained for intestinal epithelia, they have been for rat
and mouse trophectoderm in preimplantation blastocysts (Van Winkle and
Campione, 1987; Van Winkle *et al.*, 1990b; 1990c). The activities of the Na$^+$-

dependent amino acid transport systems $B^{o,+}$, X_{AG}^{-} and β increase dramatically in blastocysts when they are removed from the reproductive tract a few hours prior to implantation, but not when they are removed a day before this event (e.g., Fig. 21, and unpublished data). The increase is complete in less than 1 hour, and it occurs in blastocysts near implantation, regardless of whether they are actually removed from the uterine lumen or simply disturbed by massaging the uterus with a blunt instrument (Van Winkle and Campione, 1987). In contrast, lysine transport by the amino acid transport system b^{+}_{2} decreases significantly in activity at the same time system $B^{o,+}$ activity increases, whereas lysine uptake by system $b^{o,+}$ remains unchanged (Van Winkle *et al.*, 1990b). Because all three systems appear to catalyze net uptake of positive charge under the conditions of the transport assays, the observed changes in transport activity cannot be attributable to a change in the membrane potential alone upon isolation of blastocysts from the reproductive tract. In addition, because the transport activities of blastocysts inside the uterine lumen are difficult to measure accurately, the transport activities could conceivably change immediately upon disturbing these embryos as well as between about 6 minutes and 1 hour after they are disturbed (Fig. 21).

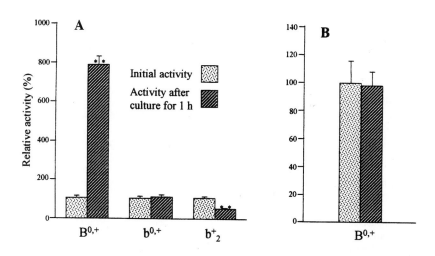

Transport system

FIGURE 21 (A) The amino acid transport activities of systems $B^{o,+}$ and b^{+}_{2} (but not $b^{o,+}$) change upon removal of blastocysts form the uterus just prior to implantation. (B) No such change is observed in system $B^{o,+}$ activity when blastocysts are removed from the reproductive tract 24 hours before implantation. The changes are unlikely to result from changes in membrane electrical potential alone because each system appears to catalyze the uptake of net positive charge under the conditions of its assay. Double asterisks indicate statistically significant differences ($P < .01$). Data from Van Winkle *et al.*, 1990b.

Regardless of when or how many times the transport activity changes, it is clear that such changes in activity complicate interpretation of experiments designed to determine the effects of various treatments of animals on transport activity in blastocysts. For example, for many years it was believed that estrogen treatment of mice ovariectomized after mating increases the amino acid transport activity in the blastocysts within their reproductive tracts. Now it is known that either a decrease or an increase in this activity may be observed 24 hours after estrogen administration, depending on whether transport activity is measured immediately after isolating the blastocysts or 30 to 60 minutes thereafter (Van Winkle and Campione, 1987). Similarly, it needs to be determined in what manner the transport activities in the intestinal epithelium may change upon disturbing it to isolate cells or membrane vesicles after the consumption of various diets. Until such measurements are made, we will not know for certain the effects of dietary factors on the activities of intestinal nutrient transport proteins. Perhaps there will be increased interest in obtaining such measurements or even in measuring intestinal transport under the most physiological conditions achievable (e.g., see Uhing, 1998) as experiments accumulate in which the measured transport activities do not match those anticipated from molecular studies.

Acknowledgements

The author thanks Drs. Nalini Chandar, Chris Cheeseman, Maria Hatzoglou, and Susan Viselli for commenting on earlier versions of the manuscript, and Barb Le Breton, Mike Moore, and Eileen Suarez for helping prepare it.

References

Abumrad, N. A., el-Maghrabi, M. R., Amri, E. Z., Lopez, E., and Grimaldi, P. A. (1993). Cloning of a rat adipocyte membrane protein implicated in binding or transport of long-chain fatty acids that is induced during preadipocyte differentiation. Homology with human CD36. *J. Biol. Chem.* **268**, 17665–17668.

Ahmed, A., Yao, P.-C., Brant, A. M., Peter, G. J., and Harper, A. A. (1997). Electrogenic L-histidine transport in neutral and basic amino acid transporter (NBAT)–expressing *Xenopus laevis* oocytes. *J. Biol. Chem.* **272**, 125–130.

Ahn, Y.-h., Kim, J.-w., Han, G.-s., Lee, B.-g., and Kim, Y.-s. (1995). Cloning and characterization of rat pancreatic beta-cell/liver type glucose transporter gene: A unique exon/intron organization. *Arch. Biochem. Biophys.* **323**, 387–396.

Amri, E.-Z., Bonino, F., Ailhaud, G., Abumrad, N. A., and Grimaldi, P. A. (1995). Cloning of a protein that mediates transcriptional effects of fatty acids in preadipocytes. *J. Biol. Chem.* **270**, 2367–2371.

Bell, G. I., Burant, C. F., Takeda, J., and Gould, G. W. (1993). Structure and function of mammalian facilitative sugar transporters. *J. Biol. Chem.* **268**, 19161–19164.

Berk, P. D., Bradbury, M., Zhou, S.-L., Stump, D., and Han, N.-I. (1996). Characterization of membrane transport processes: Lessons from the study of BSP, bilirubin, and fatty acid uptake. *Semin. Liver Dis.* **16**, 107–120.

Besnard, P., Niot, I., Bernard, A., and Carlier, H. (1996). Cellular and molecular aspects of fat metabolism in the small intestine. *Proc. Nutr. Soc.* **55**, 19–37.

Bismut, H., Hers, H. G., and Van Schaftingen, E. (1993). Conversion of fructose to glucose in the rabbit small intestine. A reappraisal of the direct pathway. *Eur. J. Biochem.* **213,** 721–726.

Blakemore, S. J., Aledo, J. C., James, J., Campbell, F. C., Lucocq, J. M., and Hundal, H. S. (1995). The Glut5 hexose transporter is also localized in the basolateral membrane of the human jejunum. *Biochem. J.* **309,** 7–12.

Boyer, S., Sharp, P. A., Debnam, E. S., Baldwin, S. A., and Srai, S. K. S. (1996). Streptozotocin diabetes and the expression of GLUT1 at the brush border and basolateral membranes of intestinal enterocytes. *FEBS Lett.* **396,** 218–222.

Brandsch, M., Brandsch, C., Ganapathy, M. E., Chew, C. S., Ganapathy, V., and Leibach, F. H. (1997). Influence of proton and essential histidyl residues on the transport kinetics of the H^+/peptide cotransport systems in intestine (PEPT1) and kidney (PEPT2). *Biochim. Biophys. Acta* **1324,** 251–262.

Bröer, S., Bröer, A., and Hamprecht, B. (1997). Expression of the surface antigen 4F2hc affects system-L-like neutral-amino-acid transport activity in mammalian cells. *Biochem. J.* **324,** 535–541.

Bröer, A., Hamprecht, B., and Bröer, S. (1998). Discrimination of two amino acid transport activities in 4F2 heavy chain–expressing *Xenopus laevis* oocytes. *Biochem. J.* **333,** 549–554.

Burant, C. F., and Saxena, M. (1994). Rapid reversible substrate regulation of fructose transporter expression in rat small intestine and kidney. *Am. J. Physiol.* **265,** G71–G79.

Burant, C. F., Takeda, J., Brot-Laroche, E., Bell, G. I., and Davidson, N. O. (1992). Fructose transporter in human spermatozoa and small intestine is GLUT5. *J. Biol. Chem.* **267,** 14523–14526.

Castelló, A., Gumá, A., Sevilla, L., Furriols, M., Testar, X., Palacín, M., and Zorzano, A. (1995). Regulation of GLUT5 gene expression in rat intestinal mucosa: Regional distribution, circadian rhythm, perinatal development and effect of diabetes. *Biochem. J.* **309,** 271–277.

Cheeseman, C. I. (1993). Glut2 is the transporter for fructose across the rat intestinal basolateral membrane. *Gastroenterology* **105,** 1050–1056.

Cheeseman, C. I. (1997). Upregulation of SGLT-1 transport activity in rat jejunum induced by GLP-2 infusion *in vivo. Am. J. Physiol.* **273,** R1965–R1971.

Cheeseman, C. I., and Harley, B. (1991). Adaptation of glucose transport across rat enterocyte basolateral membrane in response to altered dietary carbohydrate intake. *J. Physiol.* **437,** 563–575.

Cheeseman, C. I., and Tsang, R. (1996). The effect of GIP and glucagon-like peptides on intestinal basolateral membrane hexose transport. *Am. J. Physiol.* **271,** G477–G482.

Chillarón, J., Estévez, R., Mora, C., Wagner, C. A., Suessbrich, H., Lang, F., Gelpí, J. L., Testar, X., Busch, A. E., Zorzano, A., and Palacín, M. (1996). Obligatory amino acid exchange via systems $b^{o,+}$-like and y^+-like. *J. Biol. Chem.* **271,** 17761–17770.

Corpe, C. P., Basaleh, M. M., Affleck, J., Gould, G., Jess, T. J., and Kellett, G. L. (1996). The regulation of GLUT5 and GLUT2 activity in the adaptation of intestinal brush-border fructose transport in diabetes. *Pflugers Arch.* **432,** 192–201.

Corpe, C. P., Bovelander, F. J., Hoekstra, J. H., and Burant, C. F. (1998). The small intestinal fructose transporters: Site of dietary perception and evidence for diurnal and fructose sensitive control elements. *Biochim. Biophys. Acta* **1402,** 229–238.

Corpe, C. P., and Burant, C. F. (1996). Hexose transporter expression in rat small intestine: Effect of diet on diurnal variations. *Am. J. Physiol.* **271,** G211–G216.

Debnam, E. S. (1985). Adaptation of hexose uptake by the rat jejunum induced by the perfusion of sugars into the distal ileum. *Digestion* **31,** 25–30.

Devés, R., Angelo, S., and Rojas, A. M. (1998). System y^+L: The broad scope and cation modulated amino acid transporter. *Exp. Physiol.* **83,** 211–220.

Devés, R., and Boyd, C. A. R. (1998). Transporters for cationic amino acids in animal cells: Discovery, structure and function. *Physiol. Rev.* **78,** 487–545.

Dyer, J., Scott, D., Beechey, R. B., Care, A. D., Abbas, S. K., and Shirazi-Beechey, S. P. (1994). Dietary regulation of intestinal glucose transport. *In* "Mammalian Brush-Border Membrane Proteins" (M. J. Lentze, R. J. Grand, and H. Y. Naim, eds.), Vol. 2, Pt. II, pp. 65–72, Thieme Verlag, Stuttgart, Germany.

Erickson, R. H., Gum, J. R., Lindstrom, M. M., McKean, D., and Kim, Y. S. (1995). Regional expression and dietary regulation of rat small intestinal peptide and amino acid transporter mRNAs. *Biochem. Biophys. Res. Comm.* **216,** 249–257.

Estévez, R., Camps, M., Rojas, A. M., Testar, X., Devés, R., Hediger, M. A., Zorzano, A., and Palacín, M. (1998). The amino acid transport system $y^+L/4F2hc$ is a heteromultimeric complex. *FASEB J.* **12,** 1319–1329.

Ferraris, R. P. (1994). Regulation of intestinal nutrient transport. *In* "Physiology of Gastrointestinal Tract" (L. Johnson, ed.), Vol. 2, pp. 1821–1844, Raven, New York.

Ferraris, R. P., and Diamond, J. (1992). Crypt-villus site of glucose transporter induction by dietary carbohydrate in mouse intestine. *Am. J. Physiol.* **262,** G1069–G1073.

Ferraris, R. P., and Diamond, J. M. (1993). Crypt/villus site of substrate-dependent regulation of mouse intestinal glucose transporters. *Proc. Natl. Acad. Sci. USA* **90,** 5868–5872

Ferraris, R. P., and Diamond, J. (1997). Regulation of intestinal sugar transport. *Physiol. Rev.* **77,** 257–302.

Ferraris, R. P., Villenas, S. A., and Diamond, J. M. (1992). Regulation of brush-border enzyme activities and enterocyte migration rates in mouse small intestine. *Am. J. Physiol.* **262,** G1047–G1059.

Fleming, S. E., Zambell, K. L., and Fitch, M. D. (1997). Glucose and glutamine provide similar proportions of energy to mucosal cells of rat small intestine. *Am. J. Physiol.* **273,** G968–G978.

Forst, T., Kunt, T., Pfützner, A., Beyer, J., and Wahren, J. (1998). New aspects on biological activity of C-peptide in IDDM patients. *Exp. Clin. Endocrinol. Diabetes* **106,** 270–276.

Furriols, M., Chillarón, J., Mora, C., Castelló, A., Bertran, J., Camps, M., Testar, X., Vilaró, S., Zorzano, A., and Palacín, M. (1993). rBAT, related to L-cysteine transport, is localized to the microvilli of proximal straight tubules, and its expression is regulated in kidney by development. *J. Biol. Chem.* **268,** 27060–27068.

Ganapathy, V., Brandsch, M., and Leibach, F. H. (1994). Intestinal transport of amino acids and peptides. *In* "Physiology of the Gastrointestinal Tract" (Leonard R. Johnson, ed.), Chap. 52, pp. 1773–1794. Raven, New York.

Ganapathy, V., and Leibach, F.H. (1996). Peptide transporters. *Curr. Opin. Nephrol. Hypertension* **5,** 395–400.

Gehring, W. J., Affolter, M., and Burglin, T. (1994). Homeodomain proteins. *Ann. Rev. Biochem.* **63,** 487–526.

Green, S. (1995). PPAR: A mediator of peroxisome proliferator action. *Mutat. Res.* **333,** 101–109.

Grimaldi, P. A., Teboul, L., Gaillard, D., Armengod, A. V., and Amri, E. Z. (1999). Long chain fatty acids as modulators of gene transcription in preadipose cells. *Mol. Cell. Biochem.* **192,** 63–68.

Guillam M.-T., Hümmler, E., Schaerer, E., Yeh, J. I., Birnbaum, M. J., Beermann, F., Schmidt, A., Deriaz, N., and Thorens, B. (1997). Early diabetes and abnormal postnatal pancreatic islet development in mice lacking Glut-2. *Nature Genet.* **17,** 327–330.

Guillam, M.-T., Burcelin, R., and Thorens, B. (1998). Normal hepatic glucose production in the absence of GLUT2 reveals an alternative pathway for glucose release from hepatocytes. *Proc. Natl. Acad. Sci. USA* **95,** 12317–12321.

Hirsh, A. J., and Cheeseman, C. I. (1998). Cholecystokinin decreased intestinal hexose absorption by a parallel reduction in SGLT1 abundance in the brush-border membrane. *J. Biol. Chem.* **273,** 14545–14549.

Hoekstra, J. H., and van der Akar, J. H. L. (1996). Facilitating effect of amino acids on fructose and sorbitol absorption in children. *J. Pediatr. Gastroenterol. Nutr.* **23,** 118–124.

Hoekstra, J. H., van der Akar, J. H. L., Kneepkens, C. M., Stellaard, F., Geypens, B., and Ghoos, Y. F. (1996). Evaluation of $^{13}CO_2$ breath tests for the detection of fructose malabsorption. *J. Lab. Clin. Med.* **127**, 303–309.

Holloway, P. A., and Parsons, D. S. (1984). Absorption and metabolism of fructose by rat jejunum. *Biochem. J.* **222**, 57–64.

Hopfer, U. (1987). Membrane transport mechanisms for hexoses and amino acids in the small intestine. *In* "Physiology of the Gastrointestinal Tract" (L. Johnson, ed.), 2nd ed., pp. 1499–1526. Raven, New York.

Ibrahimi, A., Sfeir, Z., Magharaie H., Amri, E. Z., Grimaldi, P., and Abumrad, N. A. (1996). Expression of the CD36 homolog (FAT) in fibroblast cells: Effects on fatty acid transport. *Proc. Natl. Acad. Sci. (USA)* **93**, 2646–2651.

Kanai, Y., Segawa, H., Miyamoto, K.-I., Uchino, H., Takeda, E., and Endou, H. (1998). Expression cloning and characterization of a transporter for large neutral amino acids activated by the heavy chain of 4F2 antigen (CD98). *J. Biol. Chem.* **273**, 23629–23632.

Kishi, K., Tanaka, T., Igawa, M., Takase, S., and Goda, T. (1999). Sucrase-isomaltase and hexose transporter gene expressions are coordinately enhanced by dietary fructose in rat jejunum. *J. Nutr.* **129**, 953–956.

Klein, S., Alpers, D. H., Grand, R. J., Levin, M. S., Lin, H. C., Mansbach, C. M., Burant, C., Reeds, P., and Rombeau, J. L. (1998). Advances in nutrition and gastroenterology: Summary of the 1997 ASPEN research workshop. *J. Parenter. Enter. Nutr.* **22**, 3–13.

Kong, W., McConalogue, K., Khitin, L. M., Hollenberg, M. D., Payan D. G., Böhm, S. K., and Bunnett, N. W. (1997). Luminal trypsin may regulate enterocytes through proteinase-activated receptor 2. *Proc. Natl. Acad. Sci. (USA)* **94**, 8884–8889.

Lescale-Matys, L., Dyer, J., Scott, D., Freeman, T. C., Wright, E. M., and Shirazi-Beechey, S. P. (1993). Regulation of the bovine intestinal Na$^+$/glucose co-transporter (SGLT) is dissociated from mRNA abundance. *Biochem. J.* **291**, 435–440.

Levine, G. M., Shiau, Y. F., and Deren, J. A. (1982). Characteristics of intestinal glucose secretion in normal and diabetic rats. *Liver Physiol.* **5**, G455–G459.

MacKenzie, B., Loo, D. D. F., Fei, Y.-J., Ganapathy, V., Leibach, F. H., and Wright, E. M. (1996). Mechanism of the human intestinal H$^+$-coupled oligopeptide transporter hPEPT1. *J. Biol. Chem.* **271**, 5430–5437.

Mahraoui, L., Takeda, J., Mesonero, J., Chantret, I., Dussaulx, E., Bell, G. I., and Brot-Laroche, E. (1994). Regulation of expression of the human fructose transporter (GLUT5) by cyclic AMP. *Biochem. J.* **301**, 169–175.

Malandro, M. S., and Kilberg, M. S. (1996). Molecular biology of mammalian amino acid transporters. *Ann. Rev. Biochem.* **65**, 305–336.

Mallordy, A., Poirier, H., Besnard, P., Niot, I., and Carlier, H. (1995). Evidence for transcriptional induction of the liver fatty-acid-binding-protein gene by benzafibrate in the small intestine. *Eur. J. Biochem.* **227**, 801–807.

Mastroberardino, L., Spindler, B., Pfeiffer, R., Skelly, P. J., Loffings, J., Shoemaker, C. B., and Verrey, F. (1998). Amino-acid transport by heterodimers of 4F2hc/CD98 and members of a permease family. *Nature* **395**, 288–291.

McEwan, G. T. A., Schousboe, B., and Skadhauge, R. (1990). Direct measurement of mucosal surface pH of pig jejunum *in vivo. Zentralbl Veterinarmed [A]* **37**, 439–444.

Meinild, A.-K., Klaerke, D. A., Loo, D. D. F., Wright, E. M., and Zeuthen, T. (1998). The human Na$^+$-glucose cotransporter is a molecular water pump. *J. Physiol.* **508**, 15–21.

Mendel, D. B., and Crabtree, G. R. (1991). HNF-1, a member of a novel class of dimerizing homeodomain proteins. *J. Biol. Chem.* **266**, 677–680.

Mendel, D. B., Hansen, L. P., Graves, M. K., Conley, P. B., and Crabtree, G. R. (1991a). HNF-1 alpha and HNF-1 beta (vHNF-1) share dimerization and homeodomains, but not activation domains, and form heterodimers *in vitro. Genes Dev.* **5**, 1042–1056.

Mendel, D. B., Khavari, P. A., Conley, P. B., Graves, M. K., Hansen, L. P., Admon, A., and Crabtree, G. R. (1991b). Characterization of a cofactor that regulates dimerization of a mammalian homeodomain protein. *Science* **254,** 1762–1767.

Minami, H., Kim, J. R., Tada, K., Takahashi, F., Miyamoto, K, Nakabou, Y., Sakai, K., and Hagihira, H. (1993). Inhibition of glucose absorption by phlorizin affects intestinal function in rats. *Gastroenterology* **105,** 692–697.

Miyamoto, K., Hase, K., Takagi, T., Fujii, T., Taketani, Y., Minami, H., Oka, T., Nakabou, Y. (1993). Differential responses of intestinal glucose transporter mRNA transcripts to levels of dietary sugars. *Biochem. J.* **295,** 211–215.

Miyamoto, K., Segawa, H., Tatsumi, S., Katai, K., Yamamoto, H., Taketani, Y., Haga, H., Morita, K., and Takeda, E. (1996). Effects of truncation of the COOH-terminal region of Na^+-independent neutral and basic amino acid transporter on amino acid transport in *Xenopus* oocytes. *J. Biol. Chem.* **271,** 16758–16763.

Mosckovitz, R., Udenfriend, S., Felix, A., Heimer, E., and Tate, S. S. (1994). Membrane topology of the rat kidney neutral and basic amino acid transporter. *FASEB J.* **8,** 1069–1074.

Motojima, K., Passilly, P., Peters, J. M., Gonzalez, F. J., and Latruffe, N. (1998). Expression of putative fatty acid transporter genes are regulated by peroxisome proliferator-activated receptor α and γ activators in a tissue- and inducer-specific manner. *J. Biol. Chem.* **273,** 16710–16714.

Munck, L. K., and Munck, B. G. (1994). Amino acid transport in the small intestine. *Physiol. Res.* **43,** 335–346.

Nakamura, E., Sato, M., Yang, H., Miyagawa, F., Harasaki, M., Tomita K., Matsuoka, S., Noma, A., Iwai, K., and Minato, N. (1999). 4F2 (CD98) heavy chain is associated covalently with an amino acid transporter and controls intracellular trafficking and membrane topology of 4F2 heterodimer. *J. Biol. Chem.* **274,** 3009–3016.

Nicholls, T. J., Leese, H. J., and Bronk, J. R. (1983). Transport and metabolism of glucose by rat small intestine. *Biochem. J.* **212,** 183–187.

Ogier, H., Munnich, A., Lyonnet, S., Vaulong, S., Reach, G., and Kahn, A. (1987). Dietary and hormonal regulation of L-type pyruvate kinase gene expression in rat small intestine. *Eur. J. Biochem.* **166,** 365–370.

Ogihara, H., Suzuki, T., Nagamachi, Y., Inui, K., and Takata K. (1999). Peptide transporter in the rat small intestine: Ultrastructural localization and the effect of starvation and administration of amino acids. *Histochem. J.* **31,** 169–174.

Palacín, M., Estévez, R., Bertran, J., and Zorzano, A. (1998). Molecular biology of mammalian plasma membrane amino acid transporters. *Physiol. Rev.* **78,** 696–1054.

Peter, G. J., Davidson, I. G., Ahmed, A., McIlroy, L., Forrester, A. R., and Taylor, P. M. (1996). Multiple components of arginine and phenylananine transport induced in neutral and basic amino acid transporter-cRNA-injected *Xenopus* oocytes. *Biochem. J.* **318,** 915–922.

Pickel, V. M., Nirenberg, M. J., Chan, J., Mosckovitz, R., Undenfriend, S., and Tate, S. S. (1993). Ultrastructural localization of a neutral and basic amino acid transporter in rat kidney and intestine. *Proc. Natl. Acad. Sci. (USA)* **90,** 7779–7783.

Poirier, H., Degrace, P., Niot, I., Bernard, A., and Besnard, P. (1996). Localization and regulation of the putative membrane fatty-acid transporter (FAT) in the small intestine: Comparison with fatty acid–binding proteins (FABP). *Eur. J. Biochem.* **238,** 368–373.

Poirier, H., Niot, I., Degrace, P., Monnot, M.-C., Bernard, A., and Besnard, P. (1997). Fatty acid regulation of fatty acid–binding protein expression in the small intestine. *Am. J. Physiol.* **273,** G289–G295.

Rand, E. B., Depaoli, A. M., Davidson, N. O., Bell, G. I., and Burant, C. F. (1993). Sequence, tissue distribution, and functional characterization of the rat fructose transporter GLUT5. *Am. J. Physiol.* **264,** G1169–G1176.

Rhoads, D. B., Rosenbaum, D. H., Unsal, H., Isselbacher, K. J., and Levitsky, L. L. (1998). Circadian periodicity of intestinal Na$^+$/glucose cotransporter 1 mRNA levels is transcriptionally regulated. *J. Biol. Chem.* **273,** 9510–9516.

Sato, H., Tamba, M., Ishii, T., and Bannai, S. (1999). Cloning and expression of a plasma membrane cystine/glutamate exchange transporter composed of two distinct proteins. *J. Biol. Chem.* **274,** 11455–11458.

Satsu, H., Watanabe, H., Arai, S., and Shimizu, M. (1998). System b$^{o,+}$-mediated regulation of lysine transport in Caco-2 human intestinal cells. *Amino Acids* **14,** 379–384.

Segawa, H., Miyamoto, K., Ogura, Y., Haga, H., Morita, K., Katai, K., Tatsumi, S., Nii, T., Taketani, Y., and Takeda, E. (1997). Cloning, functional expression and dietary regulation of the mouse neutral and basic amino acid transporter (NBAT). *Biochem. J.* **328,** 657–664.

Sfeir, Z., Ibrahimi, A., Amri, E., Grimaldi, P., and Abumrad, N. (1999). CD36 antisense expression in 3T3-F442A preadipocytes. *Mol. Cell. Biochem.* **192,** 3–8.

Sharp, P. A., Debnam, E. S., and Srai, S. K. S. (1996). Rapid enhancement of brush border glucose uptake after exposure of rat jejunal mucosa to glucose. *Gut* **39,** 545–550.

Shi, X., Shedl, H. P., Summers, R. M., Lambert, G. P., Chang, R. T., Xia, T., and Gisolfi, C. V. (1997). Fructose transport mechanisms in humans. *Gastroenterology* **113,** 1171–1179.

Shiraga, T., Miyamoto, K.-I., Tanaka, H., Yamamoto, H., Taketani, Y., Morita, K., Tamai, I., Tsuji, A., and Takeda, E. (1999). Cellular and molecular mechanisms of dietary regulation on rat intestinal H$^+$/peptide transporter PepT1. *Gastroenterology* **116,** 354–362.

Shirazi-Beechey, S. P., Hirayama, B. A., Wang, Y., Scott, D., Smith, M. W., and Wright, E. M. (1991). Ontogenic development of lamb intestinal sodium-glucose co-transporter is regulated by diet. *J. Physiol. (Lond.)* **437,** 699–708.

Shirazi-Beechey, S. P., Gribble, S. M., Wood, I. S., Tarpey, P. S., Beechey, R. B., Dyer, J., Scott, D., and Barker, P.J. (1994). Dietary regulation of the intestinal sodium-dependent glucose cotransporter (SGLT1). *Biochem. Soc. Trans.* **22,** 655–658.

Shirazi-Beechey, S. P., Dyer, J., Allison, G., and Wood, I. S. (1996). Nutrient regulation of intestinal sugar-transporter expression. *Biochem. Soc. Trans.* **24,** 389–392.

Shu, R., David, E. S., and Ferraris, R. P. (1998). Luminal fructose modulates fructose transport and GLUT5 expression in small intestine of weaning rats. *Am. J. Physiol.* **274,** G232–G239.

Stein, E. D., Chang, S. D., and Diamond, J. M. (1987). Comparison of different dietary amino acids as inducers of intestinal amino acid transport. *Am. J. Physiol.* **252,** G626–G635.

Stevenson, N. R., Ferrigni, F., Parnicky, K., Day, S., and Fierstein, J. S. (1975). Effect of changes in feeding schedule on the diurnal rhythms and daily activity levels of intestinal brush border enzymes and transport systems. *Biochim. Biophys. Acta* **406,** 131–145.

Stevenson, N. R., and Fierstein, J. S. (1976). Circadian rhythms of intestinal sucrase and glucose transport: Cued by time of feeding. *Am. J. Physiol.* **230,** 731–735.

Toloza, E. M., and Diamond, J. M. (1992). Ontogenetic development of nutrient transporters in rat intestine. *Am. J. Physiol.* **263,** G593–G604.

Torras-Llort, M., Soriano-Garcia, J. F., Ferrer, R., and Moretó, M. (1998). Effect of a lysine-enriched diet on L-lysine transport by the brush border membrane of the chicken jejunum. *Am. J. Physiol.* **274,** R69–R75.

Torrents, D., Estévez, R., Pineda, M., Fernández, E., Lloberas, J., Shi, Y.-B., Zorzano, A., and Palacín, M. (1998). Identification and characterization of a membrane protein (y$^+$L amino acid transporter-1) that associates with 4F2hc to encode the amino acid transport activity y$^+$L. *J. Biol. Chem.* **273,** 32437–32445.

Tsang, R., Ao, Z., and Cheeseman, C. (1994). Influence of vascular and lumenal hexoses on rat intestinal basolateral glucose transport. *Can. J. Physiol. Pharmacol.* **72,** 317–326.

Uhing, M. R. (1998). Effect of sodium ion coupled nutrient transport on intestinal permeability in chronically catheterized rats. *Gut* **43,** 22–28.

Uhing, M. R., and Arango, V. (1997). Intestinal absorption of proline and leucine in chronically catheterized rats. *Gastroenterology* **113**, 865–874.

Uhing, M. R., and Kimura, R. E. (1995). The effect of surgical bowel manipulation and anesthesia on intestinal glucose absorption in rats. *J. Clin. Invest.* **95**, 2790–2798.

Van Beers, E. H., Büller, H. A., Grand, R. J., Einerhand, A. W. C., and Dekker, J. (1995). Intestinal brush border glycohydrolases: Structure, function, and development. *Crit. Rev. Biochem. Mol. Biol.* **30**, 197–262.

Van Winkle, L. J. (1992). Amino acid transport during embryogenesis. *In* "Mammalian Amino Acid Transport: Mechanisms and Control" (M. S. Kilberg and D. Häussinger, eds.), pp. 75–87. Plenum, New York.

Van Winkle, L. J. (1993). Endogenous amino acid transport systems and expression of mammalian amino acid transport proteins in *Xenopus* oocytes. *Biochim. Biophys. Acta* **1154**, 157–172.

Van Winkle, L. J. (1999). Biomembrane Transport. Academic Press, San Diego.

Van Winkle, L. J., and Campione, A. L. (1987). Development of amino acid transport system $B^{o,+}$ in mouse blastocysts. *Biochim. Biophys. Acta* **925**, 164–174.

Van Winkle, L. J., and Campione, A. L. (1990). Functional changes in cation-preferring amino acid transport during development of preimplantation mouse conceptuses. *Biochim. Biophys. Acta* **1028**, 165–173.

Van Winkle, L. J., Campione, A. L., and Gorman, J. M. (1990a). Inhibition of transport system $b^{o,+}$ in blastocysts by inorganic and organic cations yields insight into the structure of its amino acid receptor site. *Biochim. Biophys. Acta* **1025**, 215–224.

Van Winkle, L. J., Campione, A., and Gorman, J. M. (1988). Na^{+}-independent transport of basic and zwitterionic amino acids in mouse blastocysts by a shared system and by processes which distinguish between these substrates. *J. Biol. Chem.* **263**, 3150–3163.

Van Winkle, L. J., Campione, A. L., and Farrington, B. H. (1990b). Development of system $B^{o,+}$ and a broad-scope Na^{+}-dependent transporter of zwitterionic amino acids in preimplantation mouse conceptuses. *Bichim. Biophys. Acta* **1025**, 225–233.

Van Winkle, L. J., Christensen, H. N., and Campione, A. L. (1985). Na^{+}-dependent transport of basic, zwitterionic, and bicyclic amino acids by a broad-scope system in mouse blastocysts. *J. Biol. Chem.* **260**, 12118–12123.

Van Winkle, L. J., Iannaccone, P. M., Campione, A. L., and Garton, R. L. (1990c). Transport of cationic and zwitterionic amino acids in preimplantation rat conceptuses. *Dev. Biol.* **142**, 184–193.

Walker, D., Thwaites, D. T., Simmons, N. L., Gilbert, H. J., and Hirst, B. H. (1998). Substrate upregulation of the human small intestinal peptide transporter, hPepT1. *J. Physiol.* **507**, 697–706.

Wolfram, S., Giering, H., and Scharrer, E. (1984). Na^{+}-gradient dependence of basic amino acid transport into rat intestinal brush border membrane vesicles. *Biochem. Physiol.* **78A**, 475–480.

Wood, I. S., Allison, G. G., and Shirazi-Beechey, S. P. (1999). Isolation and characterization of a genomic region upstream from the ovine Na^{+}/D-glucose cotransporter (SGLT1) cDNA. *Biochem. Biophys. Res. Comm.* **257**, 533–537.

Wright, E. M., Hirayama, B. A., Loo, D. D., Turk, E., and Hager, K. (1994). Intestinal sugar transport. *In* "Physiology of the Gastrointestinal Tract" (L. Johnson, ed.), Vol. 2, pp. 1751–1772, Raven, New York.

Wu, G. D., Chen, L., Forslund, K., and Traber, P. G. (1994). Hepatocyte nuclear factor-1 alpha (HNF-1 alpha) and HNF-1 beta regulate transcription via two elements in an intestine-specific promoter. *J. Biol. Chem.* **269**, 17980–17085.

Yan, N., Mosckovitz, R., Gerber, L. D., Mathew, S., Murty, V. V. V. S., Tate, S. S., and Udenfriend, S. (1994). Characterization of the promoter region of the gene for the rat neutral and basic amino acid transporter and chromosomal localization of the human gene. *Proc. Natl. Acad. Sci. USA* **91**, 7548–7552.

Yao, S. Y. M., Muzyka, W. R., Cass, C. E., Cheeseman, C. I., and Young, J. D. (1998a). Evidence that the transport-related proteins BAT and 4F2hc are not specific for amino acids: Induction of Na^+-dependent uridine and pyruvate transport activity by recombinant BAT and 4F2hc expressed in *Xenopus* oocytes. *Biochem. Cell Biol.* **76,** 859–865.

Yao, S. Y. M., Muzyka, W. R., Elliott, J. F., Cheeseman, C. I., and Young, J. D. (1998b). Cloning and functional expression of a cDNA from rat jejunal epithelium encoding a protein (4F2hc) with system y^+L amino acid transport activity. *Biochem. J.* **330,** 745–752.

CHAPTER 5

Molecular Structure and Regulation of Tight Junctions

Christina M. Van Itallie and James Melvin Anderson
Section of Digestive Diseases and Department of Cell Biology, Yale School of Medicine,
New Haven, CT 06520

I. INTRODUCTION

Solute and water transport in the intestine requires both active transcellular transport mechanisms and the presence of a paracellular seal between epithelial cells; tight junctions provide this seal. The tight junction was defined physiologically and microscopically during the 1960s and 1970s, long before any biochemical components were identified. This lack of biochemical insight impeded rationalization of the junction's regulated barrier properties. In 1986, ZO-1, the first tight junction protein, was identified (Stevenson *et al.*, 1986); now there are more than two dozen known junction-associated proteins. This has allowed recent attempts to describe the biochemical basis of tight-junction function. The first goal of this review is to critically assess methods commonly used to study tight-junction physiology. Second, we will review known protein components, with special attention to those suspected of having a role in the organization or regulation of the junction. Third, we will briefly review issues that are specifically relevant to tight junctions in the gastrointestinal tract, such as the molecular mechanisms of bacterial toxins that contribute to diarrhea by disrupting the paracellular barrier. Junction physiology has been the subject of many reviews (Spring, 1998; Reuss, 1991).

II. CHARACTERIZING TIGHT-JUNCTION PROPERTIES

A hallmark of epithelia is the ability to accomplish vectoral transport between distinct apical and basal tissue compartments. The tight junction is located as a continuous gasket-like seal at the most apical region of the lateral cell membranes. The interactions of tight junctions between adjacent cells form a barrier to the paracellular diffusion of water and solutes and maintain gradients generated by transcellular transport mechanisms (Figs. 1 and 2). The tight junction

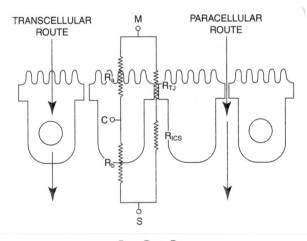

$$R_c = R_a + R_b$$

$$R_S = R_{TJ} + R_{ICS}$$

$$R_T = \frac{(R_C)(R_S)}{R_C + R_S}$$

FIGURE 1 Two permeation routes across epithelia, the transcellular resistance (R_C) and resistance of the paracellular shunt path (R_S), operate in parallel. R_C and R_S each contain two resistances in series: R_a and R_b are resistances of apical and basolateral cell membranes, respectively, whereas R_{TJ} and R_{ICS} are resistances of tight-junction proper and intercellular space, respectively. Reprinted with permission from Powell, D. W., *Am. J. Physiol.* **241**, G275–G288, (1981).

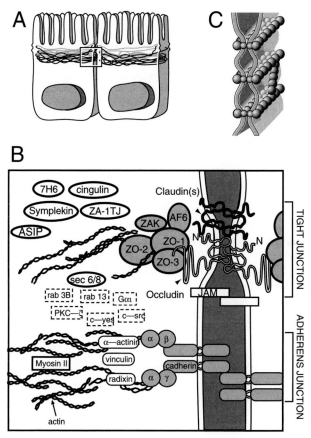

FIGURE 2 Model depicting the protein components of the apical junction complex, the tight junction, and the adherens junctions, in a highly polarized epithelial cell. (A) Continuous cell-cell contact is made at the apical end of the lateral interspace, where a ring of bidirectional actin filaments is concentrated. (B) The boxed region is depicted at higher magnification and shows identified proteins and their protein associations. Some proteins are specific to one junction; others may be shared (dashed boxes). (C) Interpretation of freeze-fracture EM images (see Fig. 3) shows barriers are formed by the intercellular contacts of branching linear polymers of occludin and claudin. Adapted with permission from Mitic, L. L., and Anderson, J. M., *Ann. Rev. Physiol.* **60**, 121–42 (1998).

also is the site that demarcates the apical and basolateral membrane domains, although its role in the generation or maintenance of cell polarity remains unclear. For decades the transport properties of intestine were studied on isolated pouches and intact sheets of whole tissue mounted in Ussing chambers. Microelectrode studies permitted electrical circuit modeling of transcellular and paracellular components (Fig. 1). These approaches provided insight into the tight junction's basic barrier properties but provided limited potential for direct experimental manipulation and investigation of its molecular structure. Since the early 1970s, with the advent of epithelial cell culture models that form sealed tight junctions, there has been a shift to the use of monolayers grown on porous filter surfaces. As gene knockout and transgenic mouse models for junction proteins are developed, the use of whole-tissue mounts likely will become popular again. A number of different methods have been used to study tight-junction physiology, all with some limitations. Widely used techniques will be briefly reviewed because they form the basis for our present understanding of the behavior of this structure, and also have begun to be applied in the analysis of the roles that specific tight-junction proteins play in the physiology of the junction. Further reviews of methods can be found elsewhere (Madara, 1998; Yap *et al.,* 1998; Spring, 1998; Reuss, 1991).

The most commonly used technique in analysis of tight junctions has been measurement of transepithelial electrical resistance (TEER). With this method, a brief current pulse of known amplitude is applied to a cell or tissue layer placed on a permeable support, and the voltage deflection across the layer is measured. The TEER is then calculated using Ohm's law, treating the monolayer as a linear circuit of parallel and series resistors. Resistance to the flow of current is determined by parallel components within both the transcellular (R_c) and paracellular or shunt (R_s) routes (Fig. 1). R_C is composed of the serial resistance of the apical (R_a) and basal (R_b) plasma membranes and R_S from the serial resistances of the tight junction (R_{TJ}) and the intercellular space (R_{ICS}). Because R_C is typically several orders of magnitude higher than R_S (Table I), current pre-

TABLE I

Electrical Resistances of Various Epithelia

Tissue	Species	Resistances (ohm cm^2)					
		R_a	R_b	R_c	R_s	R_t	R_c/R_s
Gallbladder	Rabbit	156	143	299	21	20	14
	Goose	625	378	1003	30	29	33
Small intestine							
Duodenum	Rat				67	51	
Ileum	Rat					88	
	Rabbit				115	100	
Colon	Rabbit	1570	100	1670	345	286	5
Gastric mucosa							
Antrum	Necturus	4343	3354	7697	2241	1730	3
Fundus	Necturus	1779	1047	2826	10573	2230	0.3
Urinary bladder	Rabbit	4–150 k	5–10 k	9–160 k	6–300 k	5–10 k	<0.1

dominantly flows through, and the measured total resistance (R_T) reflects barrier properties of the paracellular route.

Numerous qualifications must be considered before assuming that TEER describes the tight junction's barrier; unfortunately, in practice they often are not. The measured voltage drop is normalized to the area exposed to current; however, the true surface area often is difficult to measure. For example, when studying tissues with micro- or macroscopic folds, the TEER should be corrected for the real surface area by expressing values relative to the capacitance (Lewis and Diamond, 1976; Clausen et al., 1979). Because TEER is related to the length of junction available for electrolyte movement, it is dependent on cell size and the contour of cell-cell contacts. An epithelial monolayer filled with small cells will have more junction length per area than will the same area containing larger cells (Gonzales-Mariscal, 1992). This is one possible explanation for the finding that cells in culture often show decreases in TEER (from peak values), over time because often the cells will continue to divide and pack more tightly after they have formed a resistive monolayer. Interdigitating cell-cell contours also increases the area available for paracellular shunting. Both these confounding factors can be overcome by histological characterization of junction length and contour per unit area.

Defects in the cell monolayer create shunts that lower the TEER and interfere with characterization of the junction's barrier. Defects result from nonconfluent growth, the presence of dividing or apoptotic cells in the monolayer, and edge effects related to mounting the tissue or growing cell layers in permeable chambers. In practice, TEER is even sensitive to changes in temperature (Gonzalez-Mariscal et al., 1984). In addition, the contribution of the lateral intercellular space to the paracellular resistance is usually ignored (Fig. 2). However, several studies have demonstrated that changes in the geometry of the intercellular space related to changes in cell volume substantially effect TEER, especially in tissues with low paracellular resistance (Pappenheimer and Reiss, 1987; Stoddard and Reuss, 1988). Finally, using TEER as a way of detecting changes in the state of tight junctions in response to drugs or other treatments is complicated by the fact that many treatments also may change the activity of ion channels and the resistance of basolateral or apical membranes (Barrett, 1993). TEER using both direct and alternating current (impedance) and pitfalls associated with its use have been extensively reviewed by Reuss (1991) and Madara (1998). There are several infrequently used methods that theoretically are not subject to the same artifacts as TEER. For example, in the scanning conductance method, a microelectrode pipette is passed across a monolayer and current sinks are quantified over a small isolated length of junction; unfortunately this method requires special skill and equipment (Gitter et al., 1997).

Changes in permeability of tight junctions also are assessed by measurement of changes in the paracellular flux of impermeant uncharged compounds of varying sizes, including radiolabeled mannitol and inulin as well as size-graded fluorescent dextrans (Balda *et al.,* 1996). Flux is proportional to the cross-sectional area available to the tracer, and consequently, like TEER, is confounded by defects in the cell monolayer. However, since the intact junction does not permit passage of molecules above a distinct size, if a monolayer has lost size discrimination it must contain defects, and neither flux nor TEER will reflex properties of the tight junction. Although flux measurements seem prone to fewer problems in interpretation than TEER measurements, flux of any of these tracers across a confluent monolayer is minimal. This means that local changes in concentration of these solutes in the lateral intercellular space may be much higher than would be represented in an aliquot sampled from the media chamber; Madara *et al.,* (1992) have suggested that both upper and lower chambers be continuously stirred during the sampling.

Recently, Spring and coworkers (1998) have pioneered techniques for measurement of ionic composition and water flux through the lateral intercellular space by using optical microscopic methods. This technique has allowed determination not only of the permeability of the tight junction to Na^+ but also of the concentration of this ion in the lateral intercellular space between cultured MDCK cells. More recent work by this group has used the ability of confocal microscopy to generate deblurred images of the lateral intercellular space of MDCK cells filled with a tight-junction impermeant fluorescent dextran (70,000 M_R). These methods can be used to map concentration differences in the lateral intercellular space over time and also after application of osmotically active solutes to the apical and basolateral surfaces. One interesting outcome of these studies is the observation that dilution of the fluorescent dextran is never seen immediately at the tight junction; it is detected only further down the lateral intercellular space. This suggests that water does not flow across the tight junction, but enters from the lateral surfaces by osmosis. Optical imaging is a potentially promising technique to differentiate between effects at the level of the tight junction and those at the lateral membrane.

III. PHYSIOLOGICAL CHARACTERISTICS OF THE TIGHT JUNCTION

Early studies of epithelial physiology in frog skin, which has extremely high paracellular resistance, led to the conclusion that tight junctions were impermeable structures. This conclusion was reinforced by the demonstration that tight junctions formed barriers to proteins and large colloidal tracers (Miller, 1960; Kaye and Pappas, 1962a; 1962b). So-called leaky epithelia, such as those of small intestine, were believed to be fragile and easily damaged in preparation (Cerijido *et al.,* 1998). Eventually, it came to be realized that although the electrical resistance of small intestine and similar tissues was low, these tissues maintained the ability to transport water and solutes when studied in isolated preparations. Analysis of the transepithelial electrical resistance in these tissues was unknowingly complicated by the presence of gap junctions. Current injected into one cell to measure cell membrane resistance spread into adjacent cells. Analysis also was made more difficult by variable tissue geometry and the presence of multiple cell types in most isolated epithelial cell sheets (Diamond, 1977). However, in the early 1970s, Fromter and Diamond used a simple flat epithelium, the *Necturus* gallbladder, and concluded that the resistance of the cell membranes was 23 times higher that the transepithelial electrical resistance (Fromter and Diamond, 1972; Fromter, 1972). This meant that a paracellular shunt accounted for more than 95% of the transepithelial ion conductance. The location of the paracellular shunts at the tight junctions was confirmed by voltage scanning experiments. During transepithelial current passage, extracellular microelectrodes located local current sinks above the cellular junctions. Results from these and similar experiments led to the realization that barriers actually vary widely among epithelia, giving rise to the functional distinction between "leaky" and "tight" epithelia (see Table I) (Powell, 1981). The former are defined as having a total resistance (R_T) of less than 1000 ohm cm^2, or a R_S of less than 50% of R_T. Epithelia within the gastrointestinal tract can be subdivided into three major classes: leaky (gallbladder, liver, and small intestine), moderately leaky or moderately tight (colon and gastric antrum), and tight (gastric fundus and esophagus) (Powell, 1981). The esophagus contains a nonkeratinizing squamous epithelia and lacks tight junctions. The magnitude of the paracellular barrier defines the degree of passive back-diffusion after transcellular mechanisms have generated electrolyte and solute gradients. As a consequence, leaky epithelia, such as those in the liver, allow rapid paracellular movement of counter ions and secrete isosmotic fluids. Tight epithelial, such as those in the renal collecting ducts, maintain high electro-osmotic gradients and can secrete either diluted or concentrated products.

The variable permeability of tight junctions led Diamond (1977) to describe this role of the junction as a "gate." Experiments aimed at characterizing tight-junction permeability were extensively reviewed by Reuss (1991) and will be only briefly summarized here. When nonionic hydrophilic tracers were used to assess junction permeability, the paracellular pathway behaved as an aqueous path containing pores with radii of 30 to 40 Å (Van Os *et al.*, 1974). Experiments using apical or lateral ion substitutions to examine changes in voltage generated at the junction (diffusion potentials) demonstrated that most junctions are somewhat cation selective (Powell, 1981). These data suggest that the proposed tight-junction pores were lined with negative charges (Stevenson *et al.*, 1986). As Reuss (1991) noted, the high ion permeability of the junctions of leaky epithelia would suggest high water permeability. Surprisingly, this is not the case. For example, in the ascending loop of Henle there is high paracellular electrical conductance (Herbert *et al.*, 1984) but very low if any transepithelial water permeability (Rocha and Kokko, 1973). In addition, as cited above, the recent work by Spring and coworkers, using a specially developed optical microscopic technique, demonstrated that although the tight junctions of MDCK cells are leaky to ions, there is no significant transjunctional water flow (Kovbasnjuk *et al.*, 1998). The amount of water flow across the tight junctions of other leaky epithelia, particularly those specialized for fluid reabsorption (such as in the small intestine and renal proximal tubule), remains to be determined. Perhaps the aquaporin water channels are more significant than the paracellular pathway to water movement. Considerable controversy remains regarding the ability of junctions to permit water movement, and the molecular basis for discriminating between water and ions.

The tight junction is the site that marks the distinction between apical and basolateral membranes, and it is tempting to ascribe it a role in generating polarity. It has been shown to restrict diffusion of membrane lipids in the outer but not inner leaflet of the plasma membrane (van Meer and Simons, 1986). However, the role of the tight junction in regulating the polarized distribution of proteins remains unclear, and local protein-protein interactions with plasma membrane domains likely play major roles in determining their distribution. It is clear, in terms of the function of epithelia, that the existence of the appropriately polarized distribution of proteins in relationship to the tight junction is essential; what is not understood is the role, if any, of the tight junction in setting up or maintaining this polarized distribution. In this vein, two proteins localized to the tight junction, the Sec 6/8 complex (Grindstaff *et al.*, 1998) and ASIP, a mammalian homolog of a *Caenorhabditis elegans* polarity protein (Izumi *et al.*, 1998), eventually may provide some insight (see Fig. 2). These proteins will be discussed more thoroughly later in this chapter.

IV. STRUCTURE OF THE TIGHT JUNCTION

Although the tight junction is difficult to distinguish at the light microscopic level, well over 100 years ago vital stains allowed recognition of dark "terminal bars" between adjacent columnar epithelial cell near the apical surface. This contact was assumed to contain local accumulations of "intercellular cement" (Bloom and Fawcett, 1968). With the arrival of the transmission electron microscope, Farquhar and Palade (1963) recognized that the tight junction was characterized by a series of discrete sites of membrane contact between adjacent cells. There are from one to many spots or kisses, which form a continuous seal around the apical border of epithelial cells and measure between 0.1 and 0.7 μm in depth (Schneeberger and Lynch, 1992). A small amount of filamentous material is associated with the area containing the membrane contacts, although much more is associated with the adherens junction immediately below the tight junction (see Fig. 2). This filamentous material is actin rich (Meza *et al.,* 1980) and forms a perijunctional ring that may be important in the regulation of paracellular permeability (Madara *et al.,* 1987). The tight junction, adherens junction, and desmosomes are together considered the apical junctional complex. Also included with these junctions should be gap junctions, which mediate cellular communication. A subset of the various types of gap junction proteins, connexins, sometimes are observed within the tight junction.

The structure of tight junctions is best revealed by freeze-fracture electron micrographs (Fig. 3). This technique exposes structures embedded in the bilayer, some of which can be identified by their characteristic shapes (gap junction protein plaques, for example). Membrane fractures through the region of epithelial cell tight junctions demonstrate a complex meshwork of strands. Under appro-

FIGURE 3   A freeze-fracture electron micrograph of rat small intestine absorptive cells. The characteristic image of the tight junction is visible, showing branching and anastomosing fibrils in the fracture face. The junction fibrils demarcate the boundary between microvilli on the apical surface (above) and the lateral cell surface domain (below). Reprinted with permission from Stevenson, B. R, Anderson, J. M., and Bullivant, S., *Mol. Cell. Biochem.* **83,** 129–145 (1988).

priate fixation methods, each strand appears composed of a continuous string of particles, usually embedded in the cytoplasmic half of the plasma membrane with complementary grooves observed in the exoplasmic half of the lipid bilayer (Bullivant, 1977). Because tight junctions are shared between cells, it is likely that each cell contributes to strand formation, but the nature of the coordination of this interaction is not known (Schneeberger and Lynch, 1984). One obvious possibility is that strand components form adhesive contacts with strands on an adjacent cell.

The tight-junction strands revealed by freeze-fracture analysis are presumed to form the physical basis for the observed physiological barrier. In general, the number and complexity of the strands correlate positively with the degree of tightness of the paracellular barrier (Claude and Goodenough, 1973). An increase in the number of tight-junction strands among different tissues is related to an exponential increase in transepithelial electrical resistance. The logarithmic (rather than linear) relationship between ion permeation through the tight junction was hypothesized to result from pores that would alternate between open and closed states (Claude, 1978). There are some discrepancies in the relationship between the strand number and the transepithelial electrical resistance in a number of instances; some of these may be related to heterogeneity of cell types in the tissue in question or to local areas of heterogeneity within the junction of a single cell. These issues are discussed in a review by Schneeberger and Lynch (1984).

The morphology of tight junctions along the small intestine was assessed by Madara et al., (1980). These studies used freeze-fracture electron microscopy to examine tight-junction depth, strand number, and organization. They found that tight junctions were deeper and more complex in villus cells than in crypt cells, which suggests that the crypt epithelium has greater permeability than the villus epithelium. They also found that tight junctions were deeper and that there are more strands in monkey ileum than jejunum, which correlates with the known differences in permeability. These investigators also noted that there was great heterogeneity in the tight junctions in the intestine and further analyzed this by comparing tight-junction structure in absorptive cells and goblet cells from rat ileal mucosa (Madara et al., 1980). Absorptive cells had uniform, well-organized tight junctions, whereas goblet cell junctions were extremely variable. Analyses of structure and permeability to tracers suggest that the tight junction of villus goblet cells might contribute to the overall low tissue resistance of the small intestine.

This heterogeneity emphasizes the difficulty in interpreting transepithelial paracellular resistances and flux across tissues containing multiple cell types. The basis for the heterogeneity in the magnitude of the paracellular barriers may be a quantitative difference in the amount of tight junction material or may be attributable to qualitative differences in the protein components expressed in the different cells. The next section describes some of the identified tight-junction components and their characteristics.

V. COMPONENTS OF THE TIGHT JUNCTION

The identified protein components of the tight junction fall into four major groups: the transmembrane proteins; the peripheral membrane proteins associated with the cytoplasmic plaque including ZO-1, -2, and -3 among others; signaling proteins including kinases and GTP-binding proteins; and perijunctional actin and myosin. These proteins have been the subject of a number of reviews (Fanning *et al.*, 1998; Itoh *et al.*, 1997). An overview will be presented here, with an effort to emphasize newly available information and minimize information reviewed in detail elsewhere.

A. Transmembrane Proteins

Three families of transmembrane proteins have been identified in the tight junction, but their precise roles in forming the paracellular barrier have not been determined. These proteins are occludin, the claudin multigene family, and junctional adhesion molecule (JAM).

1. Occludin

Occludin was the first tight-junction transmembrane protein described (Furuse *et al.*, 1993). The amino acid sequence of occludin from several different species (Ando-Akatsuka *et al.*, 1996) predicts four hydrophobic transmembrane-spanning domains and two extracellular loops; this topology has been confirmed by antibody accessibility studies (Van Itallie and Anderson, 1997). The extracellular loops of occludin are presumed to play a role in forming or regulating the paracellular barrier, because forced expression of occludin increases the transepithelial electrical resistance of cultured MDCK cells (Balda *et al.*, 1996; McCarthy *et al.*, 1996). When expressed under an inducible promoter, occludin expression correlates with enhanced TEER in a reversible manner (Fig. 4) (McCarthy *et al.*, 1996). Paradoxically, overexpression of occludin induces an increase in paracellular flux to uncharged solutes. The physical basis by which occludin simultaneously increases TEER and flux remains unexplained. This paradox is similar to the dissociation of water movement and TEER, and suggests the barrier may contain channels or pores that can discriminate.

The carboxy terminal cytoplasmic portion of occludin is important in formation of the tight-junction barrier (Balda *et al.*, 1996; Chen *et al.*, 1997) because deletion of this region results in increased permeability of the tight junction. The

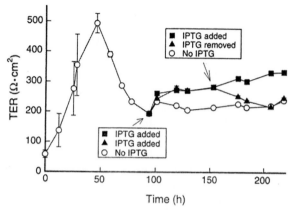

FIGURE 4   Forced expression of occludin under an IPTG-inducible promoter in MDCK cells results in reversible increases in TEER. Three sets of occludin-transfected MDCK cells were incubated for 96 h, at which time IPTG was added to two of the three sets. TEER increased as occludin was overexpressed. When IPTG was withdrawn from one set, its TEER returned to the level of that for uninduced cells. Reprinted with permission from McCarthy, K. M., *et al.*, *J. Cell Sci.* **109**, 2287–2298 (1996).

nature of the interactions involved in this regulatory region of occludin is not clear, although it is known that the last 150 amino acids of occludin contain a binding site for ZO-1, one of the cytoplasmic plaque proteins associated with the tight junction (Furuse *et al.*, 1994). ZO-1 binding is not required for occludin localization in cells that contain endogenous occludin, and Chen and coworkers (1997) demonstrated that mutant occludin (lacking the ZO-1 binding region) oligomerizes with endogenous occludin. However, the interaction of occludin with ZO-1 or other tight-junction proteins may serve important regulatory functions in paracellular permeability.

There is evidence that phosphorylation may regulate assembly of occludin into the junction. Occludin is phosphorylated on serine and threonine residues. A more highly phosphorylated form is both inextractable in nonionic detergents and physically localized to the tight junction. Less phosphorylated forms of occludin are more extractable and located along the lateral surface of cultured epithelial cells as well as in intact intestine (Sakakibara *et al.*, 1997). The importance of occludin phosphorylation in regulating movement in and out of the junction remains to be determined (Cordenonsi *et al.*, 1997; Wong and Gumbiner, 1997).

Although there is a well-demonstrated role for occludin in tight-junction function, gene knockout experiments unexpectedly demonstrated that occludin-deficient cells still formed a resistive paracellular barrier, and an extensive network of the typical tight junction strands was observed by freeze-fracture electron microscopy (Saitou *et al.*, 1998). These observations led to a search for additional transmembrane components and identification of the claudins.

2. Claudins

At least 20 claudins exist. The first two claudins identified were isolated by direct biochemical fractionation of junction-enriched membranes from chicken liver (Furuse *et al.*, 1998a). Peptide microsequencing allowed cloning of two closely related proteins named claudin-1 and -2. The amino acid sequence of these small (22 kDa) proteins predicted four transmembrane-spanning regions, two extracellular loops and very short cytoplasmic N and C termini. Although the predicted topology of the proteins is similar to that of occludin, they share no sequence homology. Both claudins were localized to the tight-junction fibrils, as determined by immunogold freeze-fracture labeling (Furuse *et al.*, 1998a). In addition, it was demonstrated that claudin-1 and -2 have the ability to induce the formation of tight junction-like strands when transfected into fibroblasts that lack tight junctions (Furuse *et al.*, 1998b). Thus, they appear to be the primary structural and functional components of the barrier, and occludin presumably modulates some property of the barrier. Because occludin overexpression increases flux out of proportion to TEER, perhaps occludin actually introduces channels.

Six more claudins have been reported recently (Morita *et al.*, 1999), and sequence data bases reveal at least eight more, indicating the existence of a multigene family (Table II). Three of these were identified by similarity searches of sequence databases; these are RVP1 (a cDNA elevated in regressing ventral prostate, identified by subtraction cloning) (Briehl and Miesfeld, 1991), the *Clostridium perfringens* enterotoxin-receptor (CPE-R) (Katahira *et al.*, 1997a; 1997b), and the human TMVCF protein (transmembrane protein deleted in velo-cardio-facial syndrome) (Sirotkin *et al.*, 1997). Although previously characterized with other properties, all three of these claudins are located within junction fibrils, as determined by immunoelectron microscopy. Northern blot analysis shows that relative message levels for the claudin family members are differentially expressed among tissues (Table II).

Claudins may be responsible for the variable properties of paracellular permeability. This is suggested by their differential expression patterns and by the phenotype of a heritable human disease called recessive renal hypomagnesemia, which results from mutations in paracellin (Simon, *et al.*, 1999). Paracellin is clearly a member of the claudin family and the human ortholog of bovine claudin-16 (GenBank, AB030082). Paracellin is expressed exclusively in the kidney and predominantly in the thick ascending loop of Henle. This is the site where magnesium and calcium ions are resorbed from the tubule through the paracellular pathway driven by a lumen-positive electrical gradient. Patients with mutations in paracellin fail to resorb these cations, which are lost in the urine. Apparently both magnesium and calcium are lost; however, for magnesium, un-

TABLE II

Distribution of the Claudin Family

Claudin	Protein or mRNA tissue distribution					
	Lung	Liver	Kidney	Testis	Heart	Myelin
1	+	++++	+++	+	+	0
2	0	+	++++	0	0	0
3 (RVP1)	+++	++++	+	+	0	0
4 (CPE-R)	+	0	++++	0	0	0
5 (TMVCF)	++++	+	+	+	+	0
6	0	0	0	0	0	0
7	++++	+	++++	++	0	0
8	+	0	++++	+	0	0
9						
10			+			
11 (OSP)	0	0	+	++++	0	++++
12						
13						
14		+				
15	+					
16 (Paracellin)	0	0	++++	0	0	0

Relative mRNA or protein expression pattern of claudins 1–16 among various tissues (Morita *et al.*, 1999). ++++ = highest, 0 = none, blank = unknown. Three claudins were previously cloned and characterized under different names: RVP1 (rat ventral prostate-1), CPE-R (*Clostridium perfringens* enterotoxin receptor), and TMVCF (transmembrane protein deleted in velo-cardio-facial syndrome). Paracellin is the human ortholog of bovine caludin-16. Other claudins and their distributions are inferred from GenBank (NCBI) entries.

like calcium, there are no counterbalancing hormonal mechanisms to restore it. Profound hypomagnesemia and seizures result. Although much work remains to be done, these results suggest that paracellin/claudin-16 may be required to allow divalent cations to pass through the tight junction. This is the first major insight about the basis for selective paracellular permeability. Conceivably, other claudins could confer anion and perhaps size selectivity.

3. Junctional Adhesion Molecule

JAM is a novel member of the immunoglobulin superfamily that localizes to the tight junctions of epithelial and endothelial cells by both immunofluorescence and immunoelectron microscopy (Martin-Padura *et al.*, 1998). When expressed in COS cells, which lack tight junctions and do not normally express JAM, the protein accumulated at sites of cell-cell contact and induced adhesion.

A monoclonal antibody to JAM inhibited both spontaneous and chemokine-induced monocyte transmigration through endothelial cell monolayers. This suggests that JAM might play a role in the regulation of monocyte transmigration. JAM is not exclusively localized to tight junctions, and any potential contribution to barrier formation is presently unknown.

B. Peripheral Membrane Proteins

A plaque of peripheral membrane proteins is associated with the cytoplasmic surface of the tight junction, and growing evidence suggests these are involved in assembly, scaffolding, and regulation of the transmembrane proteins (see Fig. 2). These proteins include ZO-1 (Stevenson *et al.,* 1986) and two coprecipitating proteins, ZO-2 (Gumbiner *et al.,* 1991; Jesaitis and Goodenough, 1994) and ZO-3 (Haskins *et al.,* 1998). ZO-1, -2, and -3 are members of the MAGUK (Membrane Associated Guanylate Kinase) family of proteins (Willott *et al.,* 1993; Jesaitis and Goodenough, 1994), some members of which are responsible for coupling transmembrane proteins within specific cell membrane domains. This coupling typically is based on the interaction of the extreme carboxy terminus of the transmembrane protein, with a PDZ domain found within the MAGUK proteins. However, although occludin binds to ZO-1 (and to ZO-2 and ZO-3), this interaction is not via any of the three PDZ domains in ZO-1. Therefore, the PDZ domains in these three tight junction MAGUKs are presumably available for interaction with other proteins. Of note, all claudins display a PDZ-binding consensus sequence at their C termini. Both ZO-1 and ZO-3 also bind actin (Fanning *et al.,* 1998), although it is unclear at present whether this interaction is direct or though an intermediate actin-binding protein. Other tight junction proteins about which less is known include cingulin (Citi *et al.,* 1988), 7H6 (Zhong *et al.,* 1993), and symplekin (Keon *et al.,* 1996).

C. Perijunctional Actin and Signaling Proteins

A thick ring of actin filaments forms a belt under the apical junction complex (see Fig. 1). Although most filaments are associated with the adherens junction, electron microscopic images demonstrate individual filaments from this band terminate at cell-cell contacts, where claudin and occludin are located. Virtually all intracellular signaling pathways have been reported to affect paracellular permeability, and in many cases the changes in paracellular permeability correlate with changes in the organization of this actin band. This information has been reviewed (Mitic and Anderson, 1998; Madara, 1998), and only new material will be included here. One interesting new development is that ZO-1, which cosediments with actin (Itoh *et al.,* 1997; Fanning *et al.,* 1998), has been demonstrated to bind the actin-binding protein cortactin. This protein is a known substrate for the *src* kinase, and its phosphorylation correlates with reorganization of the cortical actin cytoskeleton (Katsube *et al.,* 1998). This provides a mechanism whereby the actin cytoskeleton might be linked through ZO-1 to the transmembrane protein occludin. It also might provide a mechanism for how these interactions are regulated by tyrosine phosphorylation. The roles of signaling proteins in modulating this or other linkages between the cytoskeleton and the paracellular seal remain to be determined.

Along with the large number of putative signaling proteins localized to the apical junction that have been reviewed elsewhere (Mitic and Anderson, 1998), several new tight junction–associated proteins have been identified. The mammalian homolog of the yeast Sec6/8 proteins is localized at the tight junction, and there is some evidence that it is involved in membrane trafficking at the tight junction or in the establishment or maintenance of cell polarity (Grindstaff *et al.,* 1998). ASIP (Atypical PKC isotype specific interacting protein), which interacts with PKCλ (Izumi *et al.,* 1998), also was recently found to be associated with the tight junction in intestinal epithelial cells, hepatocytes, and MDCK cells. ASIP and its *C. elegans* homolog, polarity protein PAR-3, are PDZ-domain–containing proteins, and Izumi and coworkers suggest that ASIP may function as a scaffold to hold a subset of the cellular PKCλ (and perhaps PKCζ) at the tight junction. The *ras* target AF-6 was demonstrated to interact with the tight junction protein ZO-1 and was localized to the tight junction by immunoelectron microscopy (Yamamoto *et al.,* 1997). Transformation of epithelial cells with activated *ras* results in disruption of cell contacts, and these investigators suggest that one mechanism for this disorganization might be disruption of the AF-6/ZO-1 complex.

Although these proteins are, in this context, identified as being associated with the tight junction, almost all the proteins described herein also are found elsewhere in the cell and in many cases are found in cells that lack tight junctions. However, it seems reasonable to assume that some of the same protein interactions are preserved in different cellular contexts. ZO-1 is concentrated at tight junctions in epithelial cells, but it also is found in intercalated disks in the heart and in astrocytes in the brain, among other sites. However, in analogous fashion to its role at the tight junction, ZO-1 in other settings is still likely to act as a cortical membrane scaffold and may interact with a subset of the same proteins with which it interacts at the tight junction. The ZO-1 homolog in *Drosophila tamou* is not a tight-junction protein, and plays a role in regulating levels of transcription factors controlling bristle number (Takahisa *et al.,* 1996). Like β-catenin, another junction protein (see Fig. 2), perhaps ZO-1 in mammals also will be found to have a second role, in interacting with transcription factors. Similarly, occludin has been identified in activated T lymphocytes, where it may be involved in lymphocyte adhesion and trafficking (Alexander *et al.,* 1998). Other tight-junction components are found elsewhere and are likely to serve multiple roles in a variety of cellular activities.

VI. TIGHT JUNCTIONS IN THE GASTROINTESTINAL TRACT

The large surface area of the gastrointestinal tract means that this epithelium has an unusual degree of exposure to the external environment. It is the major surface across which nutrients must be absorbed, but it also must form a barrier to pathologic organisms. The tight junction is an important component regulating both nutrient absorption and barrier function. The role of the intestinal tight junction in the regulation of solute transport was reviewed recently; therefore, only a brief overview will be presented here. Pathologic states in which there are alterations to the structure or function of the intestinal tight junction will be covered in somewhat greater detail.

A. Solute Absorption

Tight junctions in the intestine are subject to the same regulatory influences observed elsewhere in the body, but in addition may be regulated by nutrients in the intestinal lumen. Pappenheimer and coworkers (Pappenheimer and Renieri, 1987; Madara and Pappenheimer, 1987) have demonstrated that activation of Na^+-glucose (or Na^+-coupled amino acid) transport in the small intestine enhances tight-junction permeability. They suggest that increased permeability is caused by contraction of the apical actin-myosin ring. Consistent with this idea is the finding that increased myosin light chain phosphorylation coincides with solute-induced changes in permeability (Madara, 1998). However, this finding is controversial (Ballard *et al.,* 1995) because others have been unable to detect this in human intestine (Fine *et al.,* 1993) and there continues to be debate on the role of the tight junction in nutrient absorption.

B. Intestinal Tight Junctions in Disease

Disruption of tight junctions is a feature of several pathologic states. The best described of these disruptions fall into two main classes: first, responses to inflammation or specific cytokines; and second, effects of several different bacterial toxins.

During inflammation or other immunologically mediated responses, immune cells migrate through the paracellular barrier between endothelial and epithelial cells. The experimental intestinal model system in which this has been most studied is T84 colon carcinoma cells, which form a resistive monolayer when grown on semipermeable filters. The migration of polymorphonuclear leukocytes across this monolayer can be induced by chemotactic peptides (Madara, 1989). Low rates of transmigration (18 neutrophils/mm/40 min) have no detectable effects on TEER (Parsons et al., 1987). In contrast, higher rates result in transient and reversible decreases in transepithelial electrical resistance and in increases in solute flux (Nash et al., 1987). This transmigration effect may be mediated through a cytosolic Ca^{2+}-dependent signaling pathway (Huang et al., 1993).

In vivo, inflammation also is associated with the release of a large number of inflammatory mediators, including reactive oxygen species, cytokines, and proteases. Many of these compounds can induce cytoskeletal reorganizations, which may have secondary effects on tight-junction permeability; a well-studied example is the effect of interferon-γ on intestinal tight-junction permeability (Madara and Stafford, 1989). Another example is that of histamine, which increases the paracellular permeability of endothelial cells by stimulating phosphorylation of myosin light chain with a resulting increase in perijunctional cytoskeletal tension (Lum and Malik, 1994). Structural and functional changes in the tight junction also have been documented in chronic inflammatory states of the bowel. For example, in patients with ulcerative colitis, the epithelial surface resistance decreases to approximately 25% of control values, determined from biopsy samples in which alternating current methods were used (Schmitz et al., 1999). This defect correlates with a simplification in junctional strand number from about 7 to 5, similar to the effects of ZOT toxin displayed in Figure. 5. This and similar studies suggest that inflammatory mediators induce a defect in the junctional barrier and contribute to diarrhea by permitting ionic back-leak and decreased net absorption. Experimental colitis in rats has been shown to cause cholestasis and very similar structural and functional changes in hepatocyte tight junctions (Lora et al., 1997). The defect in bile transport presumably is secondary to bacterial toxins or inflammatory mediators released from the inflamed colon.

Several bacterial toxins increase tight-junction permeability and through this mechanism may contribute to diarrhea. Perhaps the best studied of these is the ZOT produced by Vibrio cholerae. ZOT induces increases in small intestinal tight-junction permeability through a protein kinase C–dependent reorganization of the actin cytoskeleton (Fasano et al., 1995). Enchanced permeability correlates with a simplification of the strand network observed in freeze-fracture electron micrographic images (Fig. 5). Consequently, actin rearrangements are coupled through ZO-1 to the barrier-forming transmembrane protein (see Fig. 2). Fasano and coworkers (1997) have demonstrated that use of this toxin can enhance intestinal absorption of insulin and other macromolecules into the bloodstream of intact animals, and speculated on its potential application to enhance

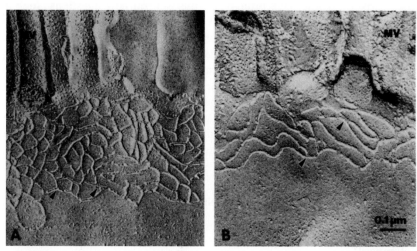

FIGURE 5   Freeze-fracture studies of rabbit ileal tissue exposed to culture supernatants of *V. cholerae* for 60 min. (A) An intact ZO with numerous intersections (arrowheads) between junctional strands. MV, microvilli. (B) An affected ZO from ileal tissue exposed to *V. cholerae* ZO-toxin containing media; the reticulum appears simplified owing to greatly decreased incidence of strand intersections. Reprinted with permission from Fasano, A., *et al.*, *Proc. Natl. Acad. Sci. USA* **88**, 5242–5246 (1991).

intestinal drug absorption. *Clostridium difficile* toxin A also increases tight permeability (Moore *et al.*, 1990), as does the *Bacteroides fragilis* toxin fragilysin (Obiso *et al.*, 1997). Toxin A induces perijunctional actin rearrangements dependent on the small GTP-binding protein rho. The suggested mechanism of action of the fragilysin is direct proteolytic degradation of tight-junction proteins. This is a mechanism similar to that described for other bacterial protease toxins, but that act elsewhere in the body, such as the *Pseudomonas aeruginosa*-produced metalloproteinase that degrades tight junctions and basement membrane in the lung (Azghani, 1996). Finally, enteropathic *Escherichia coli* also increases paracellular permeability (Philpott *et al.*, 1996), as does the vacuolating toxin released by pathogenic strains of *Helicobacter pylori* (Papini *et al.*, 1998), although specific actions on the tight junction are not documented. It appears more likely that some toxins enhance paracellular permeability by inducing cell death or global changes in cell shape.

VII. FUTURE RESEARCH

Recent advances in enumerating and characterizing tight-junction proteins is beginning to provide a molecular explanation for the junction's physiological characteristics. These have brought to light several new questions, such as whether electrical resistance, solute flux, and water flux are regulated independently. Junctions appear to be biochemically different among cell types, in terms of protein expression patterns (e.g., claudins 1 through 16) and alternatively spliced protein forms. These insights should lead to an understanding of differential physiological properties and their regulation. We can predict that mutations in junction proteins will cause inherited differences in junction properties. Paracellin/claudin-16 is the first such example. Finally, inherited differences may underlie differential predisposition to antigen exposure, infection, allergies, and chronic inflammatory disease of the intestine.

Acknowledgments

The authors are supported by grants from the NIH (DK 45134, DK 38979, CA 66263), and thank D. Franco and L. Mitic for help with manuscript in preparation.

References

Alexander, J. S., Dayton, T., Davis, C., Hill, S., Jackson, T. H., Blaschuk, O., Symonds, M., Okayama, N., Kevil, C. G., Laroux, F. S., Berney, S. M., and Kimpel, D. (1998). Activated T-lymphocytes express occludin, a component of tight junctions. *Inflammation* **22**, 573–582.

Ando-Akatsuka, Y., Saitou, M., Hirase, T., Kishi, M., Sakakibara, A., Itoh, M., Yonemura, S., Furuse, M., and Tsukita, S. (1996). Interspecies diversity of the occludin sequence: cDNA cloning of human, mouse, dog, and rat-kangaroo homologues. *J. Cell Biol.* **133**, 43–47.

Azghani, A. O. (1996). *Pseudomonas aeruginosa* and epithelial permeability: role of virulence factors elastase and exotoxin A. *Am. J. Respir. Cell Mol. Biol.* **15**, 132–140.

Balda, M. S., Whitney, J. A., Flores, C., Gonzalez, S., Cereijido, M., and Matter, K. (1996). Functional dissociation of paracellular permeability and transepithelial electrical resistance and disruption of the apical-basolateral intramembrane diffusion barrier by expression of a mutant tight junction membrane protein. *J. Cell Biol.* **134**, 1031–1049.

Ballard, S. T., Hunter, J. H., and Taylor, A. E. (1995). Regulation of tight junction permeability during nutrient absorption across the intestinal epithelium. *Ann. Rev. Nutr.* **15**, 35–55.

Barrett, K. E. (1993). Positive and negative regulation of chloride secretion in T84 cells. *Am. J. Physiol.* **265**, C859–C868.

Bloom, W., and Fawcett, D. W. (1968). "A Textbook of Histology." Saunders, Philadelphia.

Briehl, M. M., and Miesfeld, R. L. (1991). Isolation and characterization of transcripts induced by androgen withdrawal and apoptotic cell death in the rat ventral prostate. *Mol. Endocrinol.* **10**, 1381–1388.

Bullivant, S. (1977). Evaluation of membrane structure facts and artefacts produced during freeze-fracturing. *J Microsc* **111**, 101–106.

Cerijido, M., Valdes, J., Shoshani, L., and Contreras, R. G. (1998). Role of tight junctions in establishing and maintaining cell polarity. *Ann. Rev. Physiol.* **60**, 161–177.

Chen, Y., Merzdorf, C., Paul, D. L., and Goodenough, D. A. (1997). COOH terminus of occludin is required for tight junction barrier function in early *Xenopus* embryos. *J. Cell Biol.* **138**, 891–899.

Citi, S., Sabannay, H., Jakes, R., Geiger, B., and Kendrich-Jones, J. (1988). Cingulin, a new peripheral component of tight junctions. *Nature* **333**, 272–275.

Claude, P. (1978). Morphological factors influencing transepithelial permeability: A model for the resistance of the zonula occludens. *J. Membr. Biol.* **39**, 219–232.

Claude, P., and Goodenough, D. A. (1973). Fracture faces of zonulae occludentes from "tight" and "leady" epithelia. *J. Cell Biol.* **58,** 390–400.

Clausen, C., Lewis, S. A., and Diamond, J. M. (1979). Impedance analysis of a tight epithelium using a distributed resistance model. *Biophys. J.* **26,** 291–317.

Cordenonsi, M., Mazzon, E., DeRigo, L., Baraldo, S., Meggio, F., and Citi, S. (1997). Occludin dephosphorylation in early development of *Xenopus* laevis. *J. Cell Sci.* **110,** 3131–3139.

Diamond, J. (1977). The epithelial junction: Bridge, gate, and fence. *Physiologist* **20,** 10–18.

Fanning, A. S., Jameson, B., Jesaitis, L. A., and Anderson, J. M. (1998). The tight junction protein ZO-1 establishes a link between the transmembrane protein occludin and the actin cytoskeleton. *J. Biol. Chem.* **273,** 29745–29753.

Farquhar, M., and Palade, G. E. (1963). Junctional complexes in various epithelia. *J. Cell Biol.* **17,** 375–412.

Fasano, A., Florentini, C., Gianfranco, D., Kaper, J. B., Margaretten, K., Ding, X., Guandalini, S., Comstock, L., and Goldblum, G. E. (1995). Zonula occludens toxin (ZOT) modulates tight junctions through protein kinase C-dependent actin reorganization *in vitro. J. Clin. Invest.* **96,** 710–720.

Fasano, A., and Uzzau, S. (1997). Modulation of intestinal tight junctions by zonula occludens toxin permits enteral administration of insulin and other macromolecules in an animal model. *J. Clin. Invest.* **99,** 1158–1164.

Fine, K. D., Santa Ana, C. A., Porter, J. L., and Fortran, J. S. (1993). Effects of D-glucose on intestinal permeability and its passive absorption in human small intestine *in vivo. Gastroenterology* **105,** 1117–1125.

Fromter, E. (1972). The route of passive ion movement through the epithelium of *Necturus* gallbladder. *J. Membr. Biol.* **8,** 259–301.

Fromter, E., and Diamond, J. (1972). Route of passive ion permeation in epithelia. *Nat. Struct. Biol.* **235,** 9–13.

Furuse, M., Fujita, K., Hiiragi, T., Fujimoto, K., and Tsukita, S. (1998a). Claudin-1 and -2: Novel integral membrane proteins localizing at tight junctions with no sequence similarity to occludin. *J. Cell Biol.* **141,** 1539–1550.

Furuse, M., Hirase, T., Itoh, M., Nagafuchi, A., Yonemura, S., and Tsukita, S. (1993). Occludin: A novel integral membrane protein localizing at tight junctions. *J. Cell Biol.* **123,** 1777–1788.

Furuse, M., Itoh, M., Hirase, T., Nagafuchi, F., Yonemura, S., and Tsukita, S. (1994). Direct association between occludin and ZO-1 and its possible involvement in the localization of occludin at tight junctions. *J. Cell Biol.* **127,** 1617–1626.

Furuse, M., Sasaki, H., Fujimoto, K., and Tsukita, S. (1998b). A single gene product, claudin-1 or -2, reconstitutes tight junction strands and recruits occludin in fibroblasts. *J. Cell Biol.* **143,** 391–401.

Gitter, A. H., Bertog, M., Schulzke, J.-D., and Fromm, M. (1997). Measurement of paracellular epithelial conductivity by conductance scanning. *Pflugers Arch.* **434,** 830–840.

Gonzales-Mariscal, L. (1992). The relationship between structure and function of tight junctions. *In* "Tight Junctions" (M. Cereijido, ed.), pp. 67–76, CRC Press, Boca Raton, FL.

Gonzalez-Mariscal, L., Chavez de Ramirez, B., and Cereijido, M. (1984). Effect of temperature on the occludin junctions of monolayers of epitheloid cells. *J. Membr. Biol.* **79,** 175–184.

Grindstaff, K. K., Yeaman, C., Anandasabapathy, N., Hsu, S.-C., Rodriguez-Boulan, E., Scheller, R. H., and Nelson, W. J. (1998). Sec6/8 complex is recruited to cell-cell contacts and specifies transport vesicle delivery to the basal-lateral membrane in epithelial cells. *Cell* **93,** 731–740.

Gumbiner, B., Lowenkopf, T., and Apatira, D. (1991). Identification of 160-kDa polypeptide that binds to the tight junction protein ZO-1. *Proc. Natl. Acad. Sci. USA* **88,** 3460–3464.

Haskins, J., Gu, L., Wittchen, E. S., Hibbard, J., and Stevenson, B. R. (1998). ZO-3, a novel member of the MAGUK protein superfamily found at the tight junction, interacts with ZO-1 and occludin. *J. Cell Biol.* **141,** 199–208.

Herbert, S.C., Friedman, P. A., and Andreoli, T. E. (1984). Effects of antidiuretic hormone on cellular conductive pathways in mouse medullary thick ascending loops of Henle. *J. Membr. Biol.* **80,** 201–210.

Huang, A.J., Manning, J. E., Bandak, T. M., Ratau, M. C., Hauser, K. R., and Silverstein, S. C. (1993). Endothelial cell cytosolic free calcium regulates neutrophil migration across monolayers of endothelial cells. *J. Cell Biol.* **120,** 1371–1380.

Itoh, M., Nagafuchi, A., Moroi, S., and Tsukita, S. (1997). Involvement of ZO-1 in cadherin-based adhesion through its direct binding to α catenin and actin filaments. *J. Cell Biol.* **138,** 181–192.

Izumi, Y., Hirose, T., Tamai, Y., Hirai, S., Nagashima, Y., Fujimoto, T., Tabuse, Y., Kemphues, K. J., and Ohno, S. (1998). An atypical PKC directly associates and colocalizes at the epithelial tight junction with ASIP, a mammalian homologue of *Caenorhabditis elegans* polarity protein PAR-3. *J. Cell Biol.* **143,** 95–106.

Jesaitis, L. A., and Goodenough, D. A. (1994). Molecular characterization and tissue distribution of ZO-2, a tight junction protein homologous to ZO-1 and the *Drosophila* discs–large tumor suppressor protein. *J. Cell Biol.* **124,** 949–961.

Katahira, J., Inoue, N., Horiguchi, Y., Matsuda, M., and Sugimoto, N. (1997a). Molecular cloning and functional characterization of the receptor for *Clostridium perfringens* enterotoxin. *J. Cell Biol.* **136,** 1239–1247.

Katahira, J., Sugiyama, N., Inoue, N., Horiguchi, Y., Matsuda, M., and Sugimoto, N. (1997). *Clostridium perfringens* enterotoxin utilizes two structurally related membrane proteins as functional receptors *in vivo. J. Biol. Chem.* **272,** 26652–26658.

Katsube, T., Takahisa, M., Ueda, R., Hashimoto, N., Kobayashi, M., and Togashi, S. (1998). Cortactin associates with the cell-cell junction protein ZO-1 in both *Drosophila* and mouse. *J. Biol. Chem.* **273,** 29672–29677.

Kaye, G. I., and Pappas, G. D. (1962a). Studies on the cornea I. The uptake and transport of colloidal particles by the living rabbit cornea *in vivo. J. Cell Biol.* **12,** 457–479.

Kaye, G. I., and Pappas, G. D. (1962b). Studies on the cornea II. The fine structure of the rabbit cornea and the uptake and transport of colloidal particles by the cornea *in vitro. J. Cell Biol.* **12,** 481–501.

Keon, B. H., Schafer, S., Kuhn, C., Grund, C., and Franke, W. W. (1996). Symplekin, a novel type of tight junction plaque protein. *J. Cell Biol.* **134,** 1003–1018.

Kovbasnjuk, O., Leader, J. P., Weinstein, A. M., and Spring, K. R. (1998). Water does not flow across the tight junctions of MDCK cell epithelium. *Proc. Natl. Acad. Sci. USA* **95,** 6526–6530.

Lewis, S. A., and Diamond, J. M. (1976). Na$^+$ transport by rabbit urinary bladder, a tight epithelium. *J. Membr. Biol.* **28,** 1–40.

Lora, L., Mazzon, E., Martines, D., Fries, W., Muraca, M., Martin, A., d'Odorico, A., Naccarato, R., and Citi, S. (1997). Hepatocyte tight-junctional permeability is increased in rat experimental colitis. *Gastroenterology* **113,** 1347–1354.

Lum, H., and Malik, A. B. (1994). Regulation of vascular endothelial barrier function. *Am. J. Physiol.* **267,** L223–L241.

Madara, J. L. (1989). Loosening tight junctions. Lessons from the intestine. *J. Clin. Invest.* **83,** 1089–1094.

Madara, J. L. (1998). Regulation of the movement of solutes across tight junctions. *Ann. Rev. Physiol.* **60,** 143–159.

Madara, J. L., Moore, R., and Carlson, S. (1987). Alteration of intestinal tight junction structure and permeability by cytoskeletal contraction. *Am. J. Physiol.* **253,** C854–C861

Madara, J. L., and Pappenheimer, J. R. (1987). Structural basis for physiological regulation of paracellular pathways in intestinal epithelia. *J. Membr. Biol.* **100,** 149–164.

Madara, J. L., Parkos, C., Colgan, S., Nusrat, A, Atisook, K., Kaoutzani, K., and Kaoutzani, P. (1992). The movement of solutes and cells across tight junctions. *Ann. NY Acad. Sci.* **664,** 47–60.

Madara, J. L., and Stafford, J. (1989). Interferon-gamma directly affects barrier function of cultured intestinal epithelial monolayers. *J. Clin. Invest.* **83,** 724–727.

Madara, J. L., Trier, J. S., and Neutra, M. R. (1980). Structural changes in the plasma membrane accompanying differentiation of epithelial cells in human and monkey small intestine. *Gastroenterology* **78,** 963–975.

Martin-Padura, I., Lostaglio, S., Schneemann, M., Williams, L., Romano, M., Fruscella, P., Panzeri, C., Stoppacciaro, A., Ruco, L., Villa, A., Simmons, D., and Dejana, E. (1998). Junctional adhesion molecule, a novel member of the immunoglobulin superfamily that distributes at intercellular junctions and modulates monocyte transmigration. *J. Cell Biol.* **142,** 117–127.

McCarthy, K. M., Skare, I. B., Stankewich, M. D., Furuse, M., Tsukita, S., Rogers, R. A., Lynch, R. D., and Schneeberger, E. E. (1996). Occludin is a functional component of the tight junction. *J. Cell Sci.* **109,** 2287–2298.

Meza, I., Ibarra, G., Sabanero, M., Martinez-Palomo, A., and Cereijido, M. (1980). Occludin junctions and cytoskeletal components in a cultured transporting epithelium. *J. Cell Biol.* **87,** 746–754.

Miller, F. (1960). Hemoglobin absorption by the cells of the proximal convoluted tubule in mouse kidney. *J. Biophys. Biochem. Cytol.* **8,** 689–718.

Mitic, L., and Anderson, J. M. (1998). Molecular architecture of tight junctions. *Ann. Rev. Physiol.* **60,** 121–141.

Moore, R., Pothoulakis, G., LaMont, J. T., Carlson, S., and Madara, J. L. (1990). *C. difficile* toxin A increases intestinal permeability and induces Cl⁻ secretion. *Am. J. Physiol.* **259,** G165–G179.

Morita, K., Furuse, M., Fujimoto, K., and Tsukita, S. (1999). Claudin multigene family encoding four-transmembrane domain protein components of tight junction strands. *Proc. Natl. Acad. Sci. USA* **96,** 511–516.

Nash, S., Stafford, J., and Madara, J. L. (1987). Effects of polymorphonuclear leukocytes transmigration on the barrier function of cultured intestinal epithelial monolayers. *J. Clin. Invest.* **80,** 1104–1113.

Obiso, R. J., Azghani, A. O., and Wilkins, T. D. (1997). The *Bacteroides fragilis* toxin fragilysin disrupts the paracellular barrier of epithelial cells. *Infect. Immun.* **65,** 1431–1439.

Papini, E., Satin, B., Norais, N., de Bernard, M., Telford, J. L., Rappuoli, R., and Montecucco, C. (1998). Selective increase of the permeability of polarized epithelial cell monolayers by Helicobacter pylori vacuolating toxin. *J. Clin. Invest.* **102,** 813–820.

Pappenheimer, J. R., and Reiss, K. Z. (1987). Contribution of solvent drag through intercellular junctions to absorption of nutrients by the small intestine of the rat. *J. Membr. Biol.* **100,** 123–136.

Pappenheimer, J. R., and Renieri, A. (1987). Contribution of solvent drag through intercellular junctions to absorption of nutrients by the small intestine of the rat. *J. Membr. Biol.* **100,** 123–136.

Parsons, P. E., Sugahara, K., Cott, G. R., Mason, R. J., and Henson, P. M. (1987). Effect of neutrophil migration and prolonged neutrophil contact in epithelial permeability. *Am. J. Pathol.* **129,** 302–312.

Philpott, D. J., McKay, D. M., Sherman, P. M., and Perdue, M. H. (1996). Infection of T84 cells with enteropathogenic *Escherichia coli* alters barrier and transport functions. *Am. J. Physiol.* **270,** G634–G645

Powell, D. W. (1981). Barrier function of epithelia. *Am. J. Physiol.* **241,** G275–G288

Reuss, L. (1991). Tight junction permeability to ions and water. *In* "Tight Junctions" (M. Cereijido, ed.), pp. 49–66, CRC Press, Boca Raton, FL.

Rocha, A. S., and Kokko, J. P. (1973). Sodium chloride and water transport in the medullary thick ascending limb of Henle. Evidence for active chloride transport. *J. Clin. Invest.* **52,** 612–623.

Saitou, M., Fujimoto, K., Doi, Y., Itoh, M., Fujimoto, T., Furuse, M., Takano, H., Noda, T., and Tsukita, S. (1998). Occludin-deficient embryonic stem cells can differentiate into polarized epithelial cells bearing tight junctions. *J. Cell Biol.* **141,** 397–408.

Sakakibara, A., Furuse, M., Saitou, M., Ando-Akatsuka, Y., and Tsukita, S. (1997). Possible involvement of phosphorylation of occludin in tight junction formation. *J. Cell Biol.* **137,** 1393–1401.

Schmitz, H., Barmeyer, C., Fromm, M., Runkel, N., Foss, H.-D., Bentzel, C. J., Riecken, E.-O., and Schulzke, J.-D. (1999). Altered tight junction structure contributes to the impaired epithelial barrier function in ulcerative colitis. *Gastroenterology* **116,** 301–309.

Schneeberger, E. E., and Lynch, R. D. (1984). Tight junctions: Their structure, composition, and function. *Circ. Res.* **55,** 723–733.

Schneeberger, E. E., and Lynch, R. D. (1992). Structure, function, and regulation of cellular tight junctions. *Am. J. Physiol.* **262,** L647–L661

Simon, D. B., Lu, Y., Choate, K. A., Velazquez, H., Al-Sabban, E., Praga, M., Casari, G., Bettinelli, A., Colussi, G., Rodriguez-Soriano, J., McCredie, D., Milford, D., Sanjad, S., Lifton, R. P. (1999). Paracellin-1, a renal tight junction protein required for paracellular $Mg2^+$ resorption. *Science* **265,** 103–106.

Sirotkin, H., Morrow, B., Saint-Jore, B., Puech, A., Gupta, R. D., Patanjali, S. R., Skoultchi, A., Weissman, S., and Kucherlapati, R. (1997). Identification, characterization and precise mapping of a human gene encoding a novel membrane spanning protein from the 22q11 region deleted in velo-cardio-facial syndrome. *Genomics* **42,** 245–251.

Spring, K. R. (1998). Routes and mechanism of fluid transport by epithelia. *Ann. Rev. Physiol.* **60,** 105–119.

Stevenson, B. R., Silicano, J. D., Mooseker, M., and Goodenough, D. A. (1986). Identification of ZO-1: A high molecular weight polypeptide associated with the tight junction (zonula occludens) in a variety of epithelia. *J. Cell Biol.* **103,** 755–766.

Stoddard, J. S., and Reuss, L. (1988). Voltage- and time-dependence of apical membrane conductance during current clamp in *Necturus* gallbladder epithelium. *J. Membr. Biol.* **103,** 191–204.

Takahisa, M., Togashi, S., Suzuki, T., Kobayashi, M., Murayama, A., Kondo, K., Miyake, T., and Ueda, R. (1996). The *Drosophila tamou* gene, a component of the activating pathway of *extramacrochaetae* expression, encodes a protein homologous to mammalian cell-cell junction-associated protein ZO-1. *Genes. Dev.* **10,** 1783–1795.

Van Itallie, C. M., and Anderson, J. M. (1997). Occludin confers adhesiveness when expressed in fibroblasts. *J. Cell Sci.* **110,** 1113–1121.

van Meer, G., and Simons, K. (1986). The function of tight junctions in maintaining differences in lipid composition between the apical and the basolateral cell surface domains of MDCK cells. *EMBO J.* **5,** 1455–1464.

Van Os, C. H., de Jong, M. D., and Slegers, J. F. G. (1974). Dimensions of polar pathways through the rabbit gallbladder epithelium. The effect of phloretin on nonelectrolyte permeability. *J. Membr. Biol.* **15,** 363–370.

Willott, E., Balda, M. S., Fanning, A. S., Jameson, B., Van Itallie, C., and Anderson, J. M. (1993). The tight junction protein ZO-1 is homologous to the *Drosophila* discs–large tumor suppressor protein of septate junctions. *Proc. Natl. Acad. Sci. USA* **90,** 7834–7838.

Wong, V., and Gumbiner, B. (1997). Synthetic peptide corresponding to the extracellular domain of occludin perturbs the tight junction permeability barrier. *J. Cell Biol.* **136,** 399–409.

Yamamoto, T., Harada, N., Kano, K., Taya, S., Canaani, E., Matsuura, Y., Mizoguchi, A., Ide, C., and Kaibuchi, K. (1997). The *ras* target AF-6 interacts with ZO-1 and serves as a peripheral component of tight junctions in epithelial cells. *J. Cell Biol.* **139,** 785–795.

Yap, A.S., Mullin, J. M., and Stevenson, B. R. (1998). Molecular analyses of tight junction physiology: Insights and paradoxes. *J. Membr. Biol.* **163,** 159–167.

Zhong, Y., Saitoh, T., Minase, T., Sawada, N., Enomoto, K., and Mori, M. (1993). Monoclonal antibody 7H6 reacts with a novel tight junction-associated protein distinct from ZO-1, cingulin, and ZO-2. *J. Cell Biol.* **120,** 477–483.

CHAPTER 6

The Cystic Fibrosis Transmembrane Conductance Regulator in the Gastrointestinal System

Erik M. Schwiebert* and Richard Rozmahel†

*Departments of Physiology, Biophysics, and Cell Biology, and The Gregory Fleming James Cystic Fibrosis Research Center, University of Alabama, Birmingham, Alabama 35294; †Department of Genetics & Genomic Biology, The Hospital for Sick Children, and the Department of Pharmacology, University of Toronto, Toronto, Ontario

Current Topics in Membranes, Volume 50

I. INTRODUCTION

Cystic fibrosis (CF) commonly presents with pancreatic, intestinal, and hepatobiliary lesions (reviewed in Welsh *et al.,*1995, and Eggermont, 1996). Thus, the gastrointestinal (GI) system has been the focus of intensive investigations into the function of the cystic fibrosis transmembrane conductance regulator (*CFTR*) gene, and its dysfunction in CF.

In this review, *CFTR* expression, function, and dysfunction in the GI system will be addressed anatomically throughout the GI system. Because a large body of information relevant to this subject is derived from nonhuman sources, including rodents and sharks, data from such sources also are included, where applicable. With this construction, the authors wish to achieve several goals through this review. First, it is hoped that no contributions to the field will be neglected, and the authors apologize in advance for any oversights or omissions owing to length constraints. Second, published work from all aspects of the GI system will be highlighted. The literature was researched on several levels, from the specific organ of the GI system to the CF researcher, and these key words were cross-referenced to CF and *CFTR*. Third, through this construction, the review will expose regions of the GI system where our knowledge regarding expression and/or function of *CFTR* remains incomplete. Fourth, simultaneous discussion of *CFTR* expression, function, and dysfunction in the context of an anatomic analysis of the GI system will be covered to provide an integrative analysis of the role of *CFTR* at these sites. Because CF disease presents commonly in the pancreas, intestine, liver, and gallbladder, these regions of the GI system will be the focus of the discussion of *CFTR* dysfunction. The integrative nature of the discussion is intended to provide a clear picture of *CFTR* in the context of normal GI function. Finally, the genetics of CF GI disease will be discussed. In particular, studies demonstrating the existence of genes that "modify" CF intestinal disease severity (so-called modifier genes) will be covered. Where appropriate, pharmacological, biological, and genetic interventions with foundations in the GI system will be addressed.

II. FUNCTION AND DYSFUNCTION OF CFTR

A. Cystic Fibrosis Before CFTR or "BC"

Before the cloning of *CFTR* by Tsui and colleagues (Kerem *et al.,* 1989; Riordan *et al.,* 1989; Rommens *et al.,* 1989) and its characterization, studies had suggested the basic CF defect to be impaired epithelial ion conductance. Physiological studies had shown absence of a cyclic-AMP (cAMP)-activated Cl^- conductance (Knowles *et al.,* 1983; Quinton, 1983; Widdicombe *et al.,* 1985; Frizzell *et al.,* 1986; Welsh, 1986; Li *et al.,* 1988) and heightened Na^+ permeability (Knowles *et al.,* 1983) across CF epithelia.

B. Member of the ABC Transporter Family

Cloning and characterization of *CFTR* revealed it to be a member of a family of ATP-binding cassette (ABC) transporters that includes the multidrug resistance proteins (MDR-1, MRP), the sulfonylurea receptor (SUR), and STE-6 (a yeast α-mating factor transporter). Since the cloning of *CFTR,* hundreds of additional ABC transporters have been cloned from mammalian cells, yeast, and bacteria (Bartosz, 1998; Croop, 1998; Fath and Kolter, 1993; Higgins, 1992; Nikaido and Hall, 1998; van Veen and Konings, 1998). Because these proteins generally are believed to be transporters and not channels, the fact that CFTR is a member of this family was, and continues to be, perplexing.

C. Functions Attributed to CFTR

Regardless of the fact that *CFTR* is classified as a member of the ABC transporter family, numerous studies have conclusively demonstrated its function as a cAMP-activated Cl^- channel (Bear *et al.,* 1991; 1992; Kartner *et al.,* 1991; Rommens *et al.,* 1991). However, it should be noted that the Cl^- channel activity of *CFTR* does not preclude it from having transport functions similar to those of other members of the ABC transporter family. In keeping with this possibility, it has been suggested that overexpression of *CFTR* in heterologous cells confers a multidrug-resistance phenotype (Wei *et al.,* 1995). Several laboratories have suggested that CFTR also facilitates transport of ATP (Reisin *et al.,* 1994; Prat *et al.,* 1996; Pasyk and Foskett, 1997) and other adenine nucleotides (Pasyk *et al.,* 1997). Nevertheless, the potential ATP transport function of *CFTR* has been disputed (Grygorczyk *et al.,* 1996; Li *et al.,* 1996; Reddy *et al.,* 1996; Grygorczyk and Hanrahan, 1997), and the various aspects of this controversy have been summarized (Schwiebert, 1999b). In addition to these potential functions, Linsdell and Hanrahan have shown that *CFTR* transports anions much larger than Cl$^-$ (Linsdell *et al.*, 1997; Linsdell and Hanrahan, 1998a; 1998b). Principal among these larger anions is glutathione, which suggests that *CFTR* affects cellular metabolism (Linsdell *et al.*, 1998b).

CFTR traffics to the apical membrane within trans-Golgi–derived secretory vesicles that are yet to be fully defined (reviewed in Bradbury, 1999b). The recycling of *CFTR* from the apical membrane has been shown to occur through clathrin-coated pits (Lukacs *et al.*, 1997; Bradbury *et al.*, 1999a). Because *CFTR* resides for periods within cytoplasmic organelles and vesicles, it is possible that it also has a function at these sites. Indeed, Pasyk and Foskett have reported that *CFTR* maintains a Cl$^-$ channel activity within the endoplasmic reticulum (ER) membrane (Pasyk and Foskett, 1995). In addition, they showed that both ATP and the physiological sulfate donor PAPS can be transported by *CFTR* across ER membranes, suggesting that *CFTR* facilitates the loading of vesicles and organelles with these molecules (Pasyk *et al.*, 1997).

The *CFTR* Cl$^-$ channel also is required to facilitate acidification of intracellular compartments. To maintain electroneutrality when H$^+$ is loaded into vesicles and organelles via the H$^+$-ATPase pump, a Cl$^-$ channel must open simultaneously. In this regard, Al-Awqati and coworkers showed that organelles from CF epithelial cells fail to acidify as readily as their non-CF counterparts (Barasch and Al-Awqati, 1993; Barasch *et al.*, 1991). It was suggested that such a difference in acidification within biosynthetic organelles might underlie the changes in posttranslational modification of proteins (in particular, the carbohydrate moieties of glycoproteins) observed in CF. Studies documenting differences in mucus rheology, mucin composition, and GI secretions that provide support for such a biosynthetic defect are described in section III.

Numerous studies have demonstrated the regulation of other conductances by *CFTR*. This concept has been reviewed recently in different contexts by the author (Schwiebert *et al.*, 1999a). Therefore, this review will focus only on specific studies of *CFTR* regulation of conductance related to GI models. In addition, because Na$^+$ absorption in the GI tract is largely through Na$^+$-coupled cotransporters, similar to kidney proximal tubule (reviewed in Wright *et al.*, 1997), CFTR's regulation of these channels will not be discussed. However, the regulation of other Cl$^-$ and K$^+$ channels by *CFTR* will be described briefly.

The emerging role for *CFTR* as an effector of inflammatory function will be addressed throughout this review. Because the primary cause of morbidity in CF patients is recurrent and increasingly virulent bacterial lung infection and inflammation, most of these studies have been performed in lung-derived cell models and tissues. Numerous studies have clearly demonstrated an overactive inflammatory response in CF compared with non-CF infants, even in the absence of detectable pathogens (Balough *et al.*, 1995; Khan *et al.*, 1995). These studies have indicated elevated levels of the pro-inflammatory cytokine IL-8 and a decrease of anti-inflammatory IL-10 in CF lungs (Balough *et al.,* 1995; Bonfield *et al.,* 1995a; 1995b; Khan *et al.,* 1995; Dosanjh *et al.,* 1998; Tabary *et al.,* 1998). Meanwhile, other studies have demonstrated significantly increased numbers of neutrophils in bronchoalveolar lavage fluid (BALF) from the lungs of CF infants (Konstan *et al.,* 1994; Khan *et al.,* 1995) as well as elevated levels of primed inflammatory cells (Roberts and Stiehm, 1989; Koller *et al.,* 1995). Furthermore, even during pulmonary infection, CF individuals were found to have an abnormally heightened inflammatory response (Koller *et al.,* 1995). Nevertheless, other studies have suggested that the observed CF inflammatory response is the consequence of earlier infections (Armstrong *et al.,* 1997).

Although it has been argued that CF infants have infections that cannot be readily detected, Prince and coworkers found heightened NFκB activity in CF versus non-CF cultured epithelial cells following *Pseudomonas aeruginosa* infection (DiMango *et al.,* 1998). Furthermore, studies of CF and non-CF epithelial cell lines and primary cultures by Schwiebert and coworkers found that RANTES, a chemokine that promotes transmigration of inflammatory cells from blood to tissue, was lacking in CF cells (Schwiebert *et al.,* 1999c). Complementation of the CF cells with wild-type *CFTR* restored RANTES gene expression and secreted protein levels (Schwiebert *et al.,* 1999c). Interestingly, the transfection of *CFTR* was directly associated with increased RANTES gene transcription. Moreover, mutation of the putative NFκB recognition site in the RANTES promoter eliminated *CFTR*'s regulation of its activity. Additional studies also demonstrated a similar lack of RANTES secretion in a CF pancreatic cell model (CFPAC-1). These observations are consistent with the work of Prince and colleagues (DiMango *et al.,* 1998) in suggesting that *CFTR* modulates expression of inflammatory (and possibly other) genes. Hence, a defect of the inflammatory process, causing some of the CF manifestations, warrants consideration. The numerous functions that have been attributed to CFTR thus far are summarized in Table I.

D. CFTR Mutations Associated with CF Disease

At the time of writing, more than 800 sequence variations implicated as CF-causing have been identified in the *CFTR* gene. (See http://www.genet. sickkids.on.ca/cftr/ for the CF Mutation Database supported by the CF Genetic Analysis Consortium.) The mutations can be grouped into two distinct categories corresponding to their effect on pancreatic function (reviewed in Welsh, 1995). Patients with adequate levels of exocrine pancreatic enzymes (designated as less than 7% fecal excretion of dietary fat intake) to maintain basic nourishment (approximately 15% of CF patients) are classified as pancreatic sufficient (PS). On the other hand, patients with inadequate pancreatic function are classified as pan-

TABLE I

Reported Functions of CFTR

Reported CFTR activities	References
1. Chloride channel	Bear *et al.*, 1991; Rommens *et al.*, 1991; Anderson *et al.*, 1991; Tabcharani *et al.*, 1991
2. Regulation of epithelial conductances	Gabriel *et al.*, 1993; Jovov et al., 1995b; Schwiebert *et al.*, 1995; Stutts *et al.*, 1995a; Kunzelmann, *et al.*, 1997
3. Vesicle trafficking	Bradbury *et al.*, 1992; Bradbury and Bridges, 1992
4. ATP transport	Schwiebert *et al.*, 1995; Reisin *et al.*, 1994; Prat *et al.*, 1996; Pasyk and Foskett, 1997
5. Regulation of cytosolic organellar pH	Barasch *et al.*, 1991; Barasch and Al-Awqati, 1993
6. Regulation of cellular energy levels	Quinton and Reddy, 1992
7. Transport of sulfate donors	Pasyk and Foskett, 1997
8. Facilitator of multidrug resistance	Wei *et al.*, 1995
9. Bicarbonate transporter or channel	Clarke and Harline, 1998
10. Glutathione transport	Linsdell and Hanrahan, 1999b
11. Chemokine regulation	Schwiebert *et al.*, 1999b

See text (section II) for details on each function of CFTR.

creatic insufficient (PI). Patients with at least one copy of a milder (PS) *CFTR* mutation normally present with a PS phenotype, whereas those with two severe mutations are PI (Welsh, 1995). Although a correlation between PS and milder disease has been observed, the relationship is not absolute. In addition, no association between CFTR genotype and severity of respiratory disease is evident (Kerem *et al.*, 1990b; Santis *et al.*, 1990a; 1990b). Nevertheless, *CFTR* mutations can be classified at the molecular level into several categories by their effect on gene expression, protein truncation, protein processing, and channel function and regulation (Tsui, 1992; Welsh and Smith, 1993).

One category of mutations is defined by its detrimental effect on gene expression. However, owing to *CFTR*'s poorly defined regulatory sequences, the difficulties in proving its effect on expression, and its obscure nature, the ascertainment of CFTR is highly problematic. To date, several candidates for disease-causing promoter mutations are known; these include -741T→G (Bienvenu *et al.*, 1994), -816C→T, -895T→G (Bienvenu *et al.*, 1995), -471delAGG, -102T→A, -94G→T, and -33G→A (unpublished data; see http://www. genet.sickkids.on.ca/cftr/). Although their nature as disease-causing mutations remains uncertain, the lack of identification of additional mutations on the corresponding CF chromosomes supports their involvement.

A second class of mutations results in absence or truncation of the CFTR protein. This category includes mutations that eliminate the translation initiation codon, frameshift, nonsense, and aberrant splicing mutations. Examples of the first type include M1V, M1K, M1I(ATA), and M1I(ATT) (see http://www.genet.sickkids.on.ca/cftr/ for the CF Mutation Database), all of which are predicted to abolish translation initiation. Numerous frameshift and nonsense mutations also have been identified throughout the *CFTR* gene (see http://www.genet.sickkids.on.ca/cftr/). Interestingly, several of the mutations leading to *CFTR* truncations cause mRNA instability and degradation (Hamosh *et al.*, 1992; Will *et al.*, 1995) before translation (Bienvenu *et al.*, 1996; Will *et al.*, 1995). An example of a nonsense mutation causing mRNA instability is W1282X (Hamosh *et al.*, 1992; Will *et al.*, 1995). The W1282X mutation has been identified as common in the Israeli Ashkenazi population (Shoshani *et al.*, 1992). Without any confirmed exception, *CFTR* truncation mutations result in severe CF disease. Both intron and exon sequence variations leading to aberrant splicing and diminished levels of intact *CFTR* mRNA make up a less prevalent group of CF-causing mutations. Examples of this class are the 621 + 1G→T mutation in intron 4 and the 1898 + 1G→T mutation of intron 12 (Crawford *et al.*, 1995; Hull *et al.*, 1993). Variations in length of a polypyrimidine tract within CFTR intron 8, conferring differences in exon 9 splicing efficiency, can modify the severity of milder CF alleles (i.e., R117H, discussed later) (Chu *et al.*, 1993; Kiesewetter *et al.*, 1993). The length of the polypyrimidine tract normally varies (i.e., 5, 7, or 9 contiguous thymidines—(so-called 5T, 7T, and 9T variants), and the efficiency of correct exon 9 splicing is directly related to its length. *CFTR* alleles with 5T produce significantly less exon 9^+ mRNA than do 9T alleles. Hence, *CFTR* mutations normally resulting in mild pancreatic disease on a 9T-containing chromosome are more severe if present on a 5T chromosome.

The third category of CF mutations affects CFTR protein trafficking. This group includes ΔF508, the most common mutation, making up approximately 70% of all Caucasian CF chromosomes (Kerem *et al.*, 1989). The ΔF508 mutation is a phenylalanine deletion at position 508, within the first ATP-binding domain of *CFTR*. Homozygosity for ΔF508 results in severe CF disease, including PI. Expression and functional analysis of the ΔF508 CFTR protein shows only slightly diminished chloride channel function (Dalemans *et al.*, 1991; Drumm *et al.*, 1991; Li *et al.*, 1993; Pasyk and Foskett, 1995). However, ΔF508 CFTR protein is unstable (Lukacs *et al.*, 1993; Ward and Kopito, 1994) and undergoes incorrect processing leading to biosynthetic arrest and degradation by the ubiquitin-proteosome pathway (reviewed in Kopito, 1999) in the endoplasmic reticulum (ER) (Cheng *et al.*, 1990; Kartner *et al.*, 1992; reviewed in Welsh *et al.*, 1993). Because posttranslational processing of *CFTR* is normally inefficient (an estimated 25% of wild-type CFTR protein is correctly translated and inserted into the plasma membrane (Ward and Kopito, 1994), the additional reduction conferred by the ΔF508 mutation is disastrous. A second mutation in the first ATP-binding domain allelic to ΔF508 can complement the protein maturation defect (Dork *et al.*, 1991; Teem *et al.*, 1993; 1996), which suggests correction of ΔF508 CFTR by secondary structural changes. Such structural changes of ΔF508 CFTR can be introduced by pharmaceutical intervention giving rise to functional *CFTR* (Thomas *et al.*, 1992). The labile nature of the ΔF508 aberration also is demonstrated by partial correction of processing and trafficking by temperature reduction (Denning *et al.*, 1992). Cells expressing ΔF508 *CFTR* grown at reduced temperatures produce a correctly processed and trafficked protein whose activity in the plasma membrane is maintained following temperature upshift. These studies suggest that improper cellular trafficking of ΔF508 *CFTR* results in the disease, rather than incapacitation of Cl$^-$ transport. Furthermore, because ΔF508 carriers are normal, loss of apical membrane transport, rather than gain of misdirected intracellular transport, causes the disease. Characterization of different CFTR missense mutant proteins indicates that many severe disease mutations have similar biosynthetic impairments (Cheng *et al.*, 1990).

A fourth category comprises *CFTR* missense mutations coding for proteins with proper maturation and trafficking, but impaired regulation. Such mutations (which include G551D) (Cutting *et al.*, 1990) generally result in a severe CF phenotype. Hence, although these mutant CFTR proteins are targeted correctly to the plasma membrane, once there, their activation by phosphorylation and/or ATP binding/hydrolysis is defective.

Another category of *CFTR* missense mutations codes for correctly processed and trafficked *CFTR* that gives rise to clinically mild CF. Patients harboring this class of mutation in a homozygous or heterozygous state display the less severe pancreatic disease. These milder mutations can be grouped into two sets according to their location in the gene. The first set (A455E and P574H) is located in the first ATP-binding domain (Kerem *et al.*, 1990a). This set of mild mutations causes only minor perturbations of protein structure leading to slightly reduced processing and/or anion transport capacity. The second set (R117H, R334W, R347E, and R347P) consists of missense mutations resulting in substitution of basic charged residues in the first transmembrane domain of the protein (Dean *et al.*, 1990; Gasparini *et al.*, 1991). The three mutants R334W, R347E, and R347P are situated in membrane-spanning segment 6, which likely participates in the *CFTR* channel pore and thereby contributes to its conductance properties. The R117H mutation is located in *CFTR*'s first extracellular loop, and thus interacts with the pore. Because all these mutations are substitutions into residues with decreased positive charge under physiological conditions, the anion conductance properties of the mutant *CFTR* would likely be altered (Sheppard *et al.*, 1993). The different classes of *CFTR* mutations and their contribution to CF have been reviewed (Tsui, 1992; Welsh and Smith, 1993).

It is of interest that the identification and characterization of disease-causing *CFTR* mutations exposes a poor correlation between *CFTR* genotype and severity

TABLE II

Classification of CFTR Gene Variations Associated with CF Disease

Mutation class	Effect on CFTR	Pancreatic phenotype	Examples
1. Gene regulation*	Abolish or diminish CFTR gene expression	PI	−741T>G
2. Abnormal protein			
Translation initiation	No protein made	PI	M1V
Frameshift	mRNA instability, exon skipping, truncated protein	PI	441delA
Nonsense	mRNA instability, exon skipping, truncated protein	PI	W1282X
Splicing	mRNA instability, exon skipping, truncated protein	PI	1898+1G>T
5T Variant	Absense of exon 9	PS > PI	Exon 9−
3. Trafficking	CFTR misfolding, biosynthetic arrest, degradation	PI	ΔF508,A455E
4. Regulation	Defective channel activation	PI	G551D
5. Conductance	Reduced conductance	PS	R117H

See details in text (section II) on mutations and their effects.
*Note that the effect of these gene regulation mutations on CFTR expression and function remains unclear.

of CF intestinal and respiratory disease. Patients harboring identical *CFTR* mutations do not necessarily present with similar disease onset and/or progression, which suggests a strong influence from secondary factors. Such secondary factors can be environmental or genetic, but it is more likely that a combination of the two is involved. Identification of these factors and elucidation of their contribution to the CF phenotype could provide a greater understanding of the pathophysiology of the disease and provide novel therapeutic interventions. Studies aimed at their identification, specifically with regard to CF-associated intestinal disease, are discussed in section VI. Table II summarizes the different categories of *CFTR* mutations.

III. CYSTIC FIBROSIS AND *CFTR* EXPRESSION IN THE GASTROINTESTINAL SYSTEM

A. Oral Cavity

1. Expression, Function, and Dysfunction in CF

The expression and function of *CFTR* in the buccal epithelium is not well defined. However, Klinger and coworkers (Richards *et al.*, 1993) have clearly demonstrated *CFTR* mRNA in cells collected by buccal swabs. Despite this report, however, the level and specific cell types expressing *CFTR,* as well as its physiological function at this site, remain unclear. Similarly, the CF phenotype of buccal epithelium has not been investigated.

B. Salivary Glands

1. Expression

The expression and function of *CFTR* in salivary glands has been investigated extensively. Using Northern blot analysis, Kelley *et al.,* (1992) demonstrated relatively high levels of *CFTR* expression in murine salivary glands. In addition, cell-specific studies of *CFTR* in rat salivary glands using RNA *in situ* hybridization showed intense expression in ductal epithelial cells (Trezise and Buchwald, 1991). A similar expression pattern was demonstrated subsequently in human fetal salivary glands (Tizzano *et al.,* 1993; Trezise *et al.,* 1993). In other studies, Zeng *et al.* (1997) used immunocytochemical and functional analyses to show CFTR in apical membranes of submandibular ductal and acinar epithelial cells of wild-type, but not $Cftr^{-/-}$ and ΔF508 mice.

2. Function

Until recently, CFTR function in submandibular glands was ill defined. However, Cook and colleagues (Dinudom *et al.,* 1995) showed that mandibular salivary gland granular cells contain forskolin-stimulated Cl^- currents. These Cl^- currents displayed *CFTR* characteristics, including (1) a linear current-voltage relationship, (2) time- and voltage-independent activation kinetics, (3) a halide permselectivity sequence of $Br^- > Cl^- > I^-$; and (4) inhibition with 1 mM diphenylamine carboxylate (DPC). The detected Cl^- current was stimulated by β-adrenergic agonists, providing additional support for it being conferred by *CFTR,* and not a previously observed hyperpolarization-activated ClC-2-like current (Komwatana *et al.,* 1994). Additionally, similar to *CFTR,* the observed current was not inhibited by DIDS. In other studies by Thomas and colleagues (Zeng *et al.,* 1997), a cAMP-activated Cl^- channel with biophysical and pharmacological properties consistent with CFTR was observed in acinar and ductal cells lining submandibular glands. Because Cl^- secretion regulates transepithelial fluid movement, *CFTR*'s activity at these sites supports its role in modulating the rheology of salivary gland secretions.

3. Dysfunction in CF

Numerous studies have documented abnormalities of salivary glands and their secretions in CF (Lieberman and Littenberg, 1969; Ceder, 1983; Bardon, 1987; Rigas *et al.,* 1989). However, despite the growing knowledge of *CFTR*'s expression and function in this tissue, little is known regarding the physiological consequences of its dysfunction at this site. It has been speculated that absence of *CFTR* from salivary gland acinar and ductal epithelia complicates the disease by decreasing salivary fluid secretion and thereby altering its composition (Zeng *et al.*, 1997). Clearly, additional investigations of CF salivary glands are warranted to understand the molecular and physiological consequences.

C. Esophagus

1. Expression, Function, and Dysfunction in CF
The expression and function of *CFTR* in the esophageal epithelium is not well characterized; consequently, the effects of its dysfunction at this site have not been investigated and will not be covered here.

D. Stomach

1. Expression
Kozak and colleagues (Kelley *et al.*, 1992) and Strong *et al.* (1994) demonstrated *CFTR* mRNA in the stomach by Northern blot analysis and RNA *in situ* hybridization, respectively. Relative to other GI tissues, however, expression of *CFTR* throughout the gastric mucosa is relatively low (Strong *et al.*, 1994), and it remains unclear as to which specific cell types express *CFTR*. Thus, additional investigations of cell-specific expression of *CFTR* among gastric parietal cells, chief cells, goblet cells, and gastrin-, secretin-, or somatostatin-secreting endocrine cells are warranted.

2. Function
As with the cell-specific expression of *CFTR,* its function in the gastric mucosa is not well defined. However, a role for another epithelial Cl^- channel in gastric parietal cells has emerged from the work of Cuppoletti and colleagues (Malinowska *et al.*, 1995; Sherry *et al.*, 1997). These studies demonstrated that, in rabbit gastric parietal cells, a PKA-activated Cl^- conductance is required for acid secretion by the H^+, K^+-ATPase pump following histamine stimulation (Cuppoletti *et al.*, 1993; Malinowska *et al.*, 1995). To maintain electroneutrality during robust acid secretion, Cl^- must follow H^+ and K^+ movement across the apical membrane of the acid-secreting cells. Isolation of apical membrane vesicles from gastric parietal cells and fusion with lipid bilayers revealed a Cl^- channel activated at low pH (3.0), located on the extracellular side of the bilayer (Cuppoletti *et al.*, 1993). The single-channel conductance of this pH-activated Cl^- channel was 28 pS in 800 mM CsCl solutions (supraphysiologic $[Cl^-]$) and

6.9 pS in 150 mM CsCl solutions (physiological [Cl$^-$]). Although a conductance of 6.9 pS is similar to *CFTR,* additional characterization showed it to have a permselectivity of I$^-$>Cl$^-$≥Br$^-$ and stimulation by hyperpolarization (0 to −80 mV shift in voltage), unlike *CFTR.* Studies of wild-type and ΔF508 *CFTR* showed reduced Cl$^-$ channel activity of both forms, with diminished extracellular pH (Sherry *et al.,* 1994), further evidence against *CFTR* being the principal Cl$^-$ channel observed in the H$^+$, K$^+$-ATPase-enriched gastric membranes. Taken together, these characteristics suggest that *CFTR* is not responsible for maintaining electroneutrality during gastric parietal cell secretions.

Other investigations showed that ClC-2 was the Cl$^-$ channel whose activity in parallel with H$^+$, K$^+$-ATPase was critical for acid secretion from gastric parietal cells. Malinowska *et al.* (1995) cloned a ClC-2 Cl$^-$ channel [designated ClC-2G(2α)] from rabbit gastric parietal cells with high amino acid identity to the rat ClC-2 cloned by Jentsch and coworkers (Thiemann *et al.,* 1992) and the human ClC-2 identified by Cid *et al.* (1995). Expression of ClC-2G(2α) in *Xenopus* oocytes showed high activity with an extracellular pH of 3.0 and a single-channel conductance of 29 pS in 800 mM CsCl (Malinowska *et al.,* 1995). Additional biophysical properties of ClC-2G(2α) were similar to the Cl$^-$ chan-

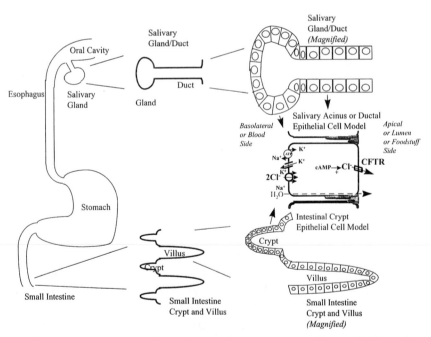

FIGURE 1 Schematic of *CFTR* expression and function in the upper GI tract. The illustrations show increasing magnification of the salivary gland and duct and the intestinal villus and crypt. An epithelial cell model is included, showing *CFTR* expression and function in the apical membrane.

nel native to the gastric parietal cells. Both channels were potentiated by cAMP and cAMP-dependent protein kinase and stimulated by a "pH or acid sensor" identified in the extracellular region of the protein encoded by the residues EELE at position 416-419. Taken together, these results prove that ClC-2, and not *CFTR,* maintains electroneutrality during gastric acid secretion. However, as described below, *CFTR* confers Cl^-, HCO_3^-, and fluid secretion (neutral or alkaline) in distal intestinal segments.

Thus, the significance of *CFTR* expression and function in the gastric mucosa is not known. Its function and the consequences of dysfunction in normal and CF gastric glands need to be examined further.

E. Intestine

1. Expression

Meconium ileus (MI), an obstructive lesion of the small intestine, is a major clinical feature of CF (occurring in approximately 15% of CF newborns). Thus, numerous laboratories have undertaken in-depth investigations of *CFTR* expression in the intestinal mucosa of humans, rats, mice, and pigs (Trezise *et al.,* 1991; 1993; Kelley *et al.,* 1992; Tizzano *et al.,* 1993; Gaillard *et al.,* 1994; Strong *et al.,* 1994; Ameen *et al.,* 1995; Hayden and Carey, 1996). These studies showed relatively higher *CFTR* expression in the proximal intestine compared with the more distal segments, and in the crypt relative to the villus. This gradient of expression is consistent with *CFTR*'s role in intestinal Cl^- and fluid secretion (Trezise *et al.,* 1991; Strong *et al.,* 1994). Hayden and Carey (1996) used immunolocalization to demonstrate high levels of *CFTR* in intestinal goblet cells of pigs and mice. It was found that a subpopulation of villus cells and superficial crypt cells in human and rat intestine displayed intense *CFTR* staining (Ameen *et al.,* 1995), which suggests a distinct transport function for these cells. Similarly, a scattered population of villi epithelial cells with unusually high *CFTR* expression was observed by Strong and colleagues (1994) in the human duodenum and jejunum but not in the ileum. Other intestinal cells expressing high levels of *CFTR* were observed in the Brunner's gland and mucosal epithelial cells near lymph nodules (Strong *et al.,* 1994). Several laboratories have shown that intestinal expression of *CFTR* changes little during development, in contrast to the lung epithelium, in which expression is high in the fetus and then gradually lower throughout development (Trezise *et al.,* 1993; Gaillard *et al.,* 1994).

Investigation of *CFTR* in human intestinal epithelial cell models (HT-29, Caco-2, T84, etc.) showed its expression to be tightly coupled to cellular differentiation (Buchwald *et al.,* 1991; Montrose-Rafizadeh *et al.,* 1991; 1992; Sood *et al.,* 1992). As intestinal epithelial monolayers developed greater transepithelial resistance and expression of differentiation markers (sucrase-isomaltase, villin, α1-antitrypsin), *CFTR* mRNA increased from 8-fold to 20-fold (Buchwald *et al.,* 1991; Montrose-Rafizadeh *et al.,* 1991; 1992; Sood *et al.,* 1992), while protein increased roughly 5-fold (Montrose-Rafizadeh *et al.,* 1991). However, an important caveat to be considered is that the experimental cell lines were derived from intestinal carcinomas, which could have abnormally high levels of *CFTR* expression. In agreement with unnatural expression of *CFTR* in cells with altered phenotypes, Strong and Collins have observed enhanced *CFTR* expression at the site of intestinal polyps and masses (T. V. Strong, personal communication).

The HT-29 intestinal epithelial cell line (which constitutively expresses high levels of CFTR) has been used to evaluate the effects of cytokines, protein kinase C agonists, cAMP, and calcium agonists on *CFTR* expression (Trapnell *et al.,* 1991; Bargon *et al.,* 1992a; 1992b; 1998; Nakamura *et al.,* 1992). TNFα, a potent proinflammatory cytokine, had no effect on *CFTR* transcription in these cells but decreased the half-life of its mRNA by 65% (Nakamura *et al.,* 1992). Because earlier work had demonstrated that phorbol ester stimulated intestinal Cl^- secretion through CFTR and a Ca^{2+}-dependent conductance pathway, the effect of these two agonists on *CFTR* expression also was investigated (Trapnell *et al.,* 1991; Bargon *et al.,* 1992a; Bargon *et al.,* 1992b). Both agonists were found to down-regulate *CFTR* mRNA expression in intestinal cell lines. Additionally, studies of the calcium ionophores ionomycin and A23187 showed decreased *CFTR* mRNA and protein in HT-29 cells (Bargon *et al.,* 1992a). Taken together, these results show that *CFTR* expression in intestinal epithelial cell models is highly regulated by agonists and second-messenger systems, particularly kinases.

Surprisingly, it has been reported that intestinal *CFTR* expression is mutually exclusive to another member of the ABC transporter family, P-glycoprotein (product of the *Mdr*-1 gene) (Trezise *et al.,* 1992). These two genes were later shown to be coordinately regulated (Trezise *et al.,* 1997). P-glycoprotein has a physiological function in the canalicular secretion of organic cations as well as the transport of exogenous drugs. As stated earlier, normal expression of *CFTR* shows a decreasing gradient from the crypt to the villus (Trezise *et al.,* 1992). In contrast, expression of *Mdr*-1 is highest at the distal portion of the villus and decreases with proximity to the crypt (Trezise *et al.,* 1992). These results imply a switching of gene expression from *CFTR* to *Mdr*-1 as the epithelial cells differentiate during migration toward the villus tip. In other studies, it was found that at the onset of pregnancy, the expression of *CFTR* in the uterine epithelium decreases while *Mdr*-1 increases (Trezise *et al.,* 1992), which supports coordinate regulation of the two genes. These findings also suggest that *CFTR* and P-glycoprotein have related activities, and cells require only one of the proteins to maintain normal function. A relationship between *CFTR* and *Mdr*-1 also was demonstrated by Breuer *et al.* (1993) in a different manner. It was observed that upon acquiring a multidrug-resistance phenotype (induced by growth in colchicine-containing medium), HT-29 cells elevated Mdr-1 but markedly reduced *CFTR* expression (Breuer *et al.*, 1993). Upon reversal of the multidrug-resistance phenotype by removal from drug selection, *Mdr*-1 expression decreased and normal CFTR levels were restored (Breuer *et al.*, 1993). An *in vivo* demonstration that *CFTR* and *Mdr*-1 expression are coordinately regulated was provided by Trezise and colleagues (1997) utilizing CF knockout mice. In the intestines of the CF mice, *CFTR* expression was reduced fourfold while *Mdr*-1 was increased fourfold (Trezise *et al.*, 1997). In older animals, intestinal *CFTR* expression was absent, and *Mdr*-1 also decreased (Trezise *et al.*, 1997). In contrast to *Mdr*-1, there was no difference in *Mdr*-2 expression.

Numerous additional ABC transporters have now been identified in GI tissue. The epithelial basolateral conductance regulator (EBCR) (van Kuijck *et al.*, 1996), multidrug-resistance–associated protein (MRP) (Cole *et al.*, 1992; Makhey *et al.*, 1998), *Mdr*-2 (Buschman *et al.*, 1992), and many other *Mdr* isoforms are expressed in the GI system. The physiological function of EBCR is unknown; however, its expression is not limited to the GI system. Reportedly it has been expressed along the nephron of the kidney. MRP is involved in secretion of non-bile acid organic anions, whereas *Mdr*-2 and *Mdr*-3 act on biliary phospholipid secretion (reviewed in Ling, 1997). Although one function of CFTR is that of an inorganic anion channel, it also might transport other substrates. This concept will be revisited below.

2. Function

Early patch-clamp studies of rodent and human intestinal epithelial cell models identified the functional activity of CFTR, even before its gene had been identified. Tabcharani *et al.* (1990) recorded two types of Cl^- channels in T84 human colon carcinoma cells. A linear and small-conductance Cl^- channel was stimulated by cAMP, while a second, larger conductance channel with an outwardly rectifying I-V relationship was observed (Tabcharani *et al.*, 1990). Upon characterization of CFTR, it was determined that the former conductance was consistent with CFTR. In hindsight, this report was the first documented observation of the CFTR Cl^- channel in an intestinal epithelial cell model. Subsequently, it was found that CFTR regulates the outwardly rectifying Cl^- channel (ORCC) (Gabriel *et al.*, 1993; Jovov *et al.*, 1995a; Schwiebert *et al.*, 1995; 1998; Julien *et al.*, 1999); however, this channel has not been cloned.

Confirmation of CFTR's anion channel function has been replicated numerous times by many different laboratories. Much of this work was designed to differentiate between the cAMP- and Ca^{2+}-stimulated Cl^- conductance pathways. To provide a resource to differentiate the two activities, Bijman *et al.* (1993) identified a rat embryonic intestinal cell line, IEC-6, that lacked *CFTR* expression and the 8 pS Cl^- channel. Stable transfection of IEC-6 cells with *CFTR* produced an intestinal epithelial cell model (IEC-CF7) with a cAMP-stimulated Cl$^-$ channel, along with *CFTR* mRNA and protein (Bijman *et al.*, 1993). These cells provided an important foundation to characterize both the biophysical and physiological properties of *CFTR*. For a similar reason, Valverde *et al.* (1995) used *CFTR*-deficient mouse models to show that *CFTR* is indeed responsible for cAMP-stimulated Cl$^-$ conductance.

Dharmsathaphorn, Barrett, and colleagues showed early on that both cAMP- and Ca^{2+}-dependent Cl$^-$ secretion pathways exist in intestinal epithelia (McRoberts *et al.*, 1985; Dharmsathaphorn and Pandol, 1986; Mandel *et al.*, 1986). Later it was shown that these Cl$^-$ secretory pathways were mediated by *CFTR* and a recently cloned member of the Ca^{2+}-activated Cl$^-$ channel (CaCC) family (Cunningham *et al.*, 1995; Gruber *et al.*, 1998). Studies also showed that agonists stimulating each of these two secretory mechanisms were additive or synergistic in their ability to stimulate Cl$^-$ secretion (Bajnath *et al.*, 1993a). The synergism of these two Cl$^-$ channels has been shown numerous times in HT-29 clone 19A cells by Bijman, de Jonge, Groot, and coworkers (Bajnath *et al.*, 1993b; 1995; Kansen *et al.*, 1993). Groot and colleagues showed that forskolin, (adenylyl cyclase agonist), and phorbol esters (protein kinase C agonists) had synergistic stimulatory effects on CFTR Cl$^-$ channel activity (Bajnath *et al.*, 1993a; 1993b). This finding was further validated by studies of heterologous CFTR Cl$^-$ channel activity in Chinese hamster ovary (CHO) cells by Tabcharani *et al.* (1991).

Other studies have examined the effects of agonists that drive second-messenger systems on intestinal Cl$^-$ secretion. Purinergic agonists, especially adenosine or adenosine precursors, have been shown to stimulate Cl$^-$ secretion via cyclic nucleotides (Barrett *et al.*, 1990; Cantiello *et al.*, 1994; Stutts *et al.*, 1995b; O'Reilly *et al.*, 1998). Histamine also is a potent intestinal Cl$^-$ and fluid secretagogue (Traynor *et al.*, 1993; Barrett, 1995; Homaidan *et al.*, 1997). Gastrointestinal peptides (such as vasoactive intestinal peptide and brain natriuretic peptide) also stimulate Cl$^-$ secretion (Traynor *et al.*, 1993; Barrett, 1995). Carbachol stimulates Cl$^-$ secretion via Ca^{2+}-mediated pathways (Dharmsathaphorn *et al.*, 1986; Bohme *et al.*, 1991; Chandan *et al.*, 1991; Diener *et al.*, 1991; Vaandrager *et al.*, 1991; Warhurst *et al.*, 1991; Devor and Duffey, 1992; Walters *et al.*, 1992; McEwan *et al.*, 1994). Prostaglandins as well as arachidonic acid and its metabolites have been found to promote Cl$^-$ secretion (Musch *et al.*, 1983; Martinez and Barker, 1987; Diener and Rummel, 1990; Sakai *et al.*, 1992; Hyun and Binder, 1993; Kachur *et al.*, 1995). In short, the list of agonists that influence intestinal Cl$^-$ secretion is endless. However, teasing out *CFTR*'s contribution to these secretagogue effects is important for the development of CF therapies using "alternative" Cl$^-$ conductances; this will be revisited below.

Although the various and diverse functions attributed to *CFTR* remain contentious, its activity as a cAMP-activated Cl$^-$ channel of the intestinal mucosa is firmly established. This activity of *CFTR* is best illustrated by the fact that two pathogens that induce diarrhea, *Escherichia coli* and *Vibrio cholera,* act through CFTR to promote intestinal Cl⁻ and fluid secretion. The mechanism of action of the *E. coli* heat-stable enterotoxin (STa) to stimulate Cl⁻ secretion has been investigated (Tien *et al.,* 1994; Seidler *et al.,* 1997). STa has been shown to bind to the guanylin receptor (whose natural ligands are guanylin and uroguanylin) and activate its intrinsic guanylyl cyclase activity. The resultant cGMP then activates the cGMP-dependent protein kinases and cross-activates cAMP-dependent protein kinases. Both cyclic nucleotide-dependent kinases activate *CFTR,* leading to chronic Cl⁻ and fluid secretion (see Fig. 2). Further support for this mechanism comes from studies performed on mouse models (Cuthbert *et al.,* 1994a), CF patients, and heterozygous carriers (Goldstein *et al.,* 1994). Investigation of mouse models deficient of the guanylyl cyclase C receptor conferred resistance to STa (Mezoff *et al.,* 1992; Swenson *et al.,* 1996; Mann *et al.,* 1997). Although there is strong support for the direct role of STa in stimulating intestinal secretion, it is also possible that STa causes an immune response and the stimulation of intestinal secretion through inflammatory mediators (Barrett *et al.,* 1990; Perdue *et al.,* 1990; Madara *et al.,* 1992; Cooke *et al.,* 1993; Traynor *et al.,* 1993; Colgan *et al.,* 1994; Kandil *et al.,* 1994; Kachur *et al.,* 1995; Schmitz *et al.,* 1996; O'Loughlin *et al.,* 1997; Benya *et al.,* 1998). In

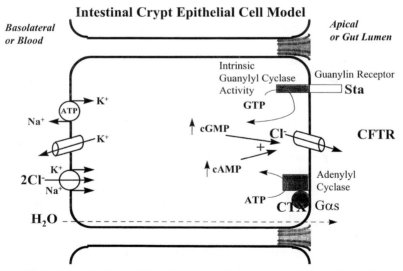

FIGURE 2   Schematic of modulation of *CFTR* activity by secretory diarrhea-inducing bacterial toxins. The epithelial cell model illustrates the cellular mechanism of action of *Escherichia coli* heat-stable enterotoxin (Sta) and *Vibrio cholera* cholera toxin (CTX). Each toxin stimulates *CFTR* by increasing the cellular concentration of cyclic nucleotides.

other studies, it has been shown that guanylin, uroguanylin, and/or STa stimulate CFTR Cl⁻ secretion through the type II cGMP–dependent protein kinase (Chao *et al.*, 1994; Markert *et al.*, 1995; Guba *et al.*, 1996; Lohmann *et al.*, 1997; Mann *et al.*, 1997; Seidler *et al.*, 1997). Additional studies by Clarke and coworkers (Joo *et al.*, 1998) showed that uroguanylin stimulates Cl⁻, HCO₃⁻, and fluid secretion across intestinal epithelia derived from all intestinal segments. Although all intestinal segments were responsive, the duodenum and cecum appeared to be stimulated most robustly. In other investigations, Madara and colleagues showed that STa-induced Cl⁻ and fluid secretion require an intact actin cytoskeleton (Matthews *et al.*, 1992; Matthews *et al.*, 1993); however, the significance to *CFTR* remains unclear.

It has been demonstrated that cholera toxin stimulates *CFTR* by activation of the heterotrimeric G protein (Gₛ) of adenylyl cyclase, which leads to constitutive production of cAMP. In turn, cyclic-AMP, through its activation of PKA, stimulates *CFTR* to cause chronic intestinal Cl⁻ and fluid secretion (see Fig. 2).

Because intestinal pathogens causing Cl⁻ induced diarrhea may do so by exploiting *CFTR,* Gabriel *et al.* (1994) have suggested a heterozygous advantage for mutant *CFTR.* This hypothesis stems from the fact that the frequency of mutant *CFTR* carriers (particularly ΔF508) within the Caucasian population is significantly greater than among the general population. This observation prompted the speculation that *CFTR* mutations are selected for by conferring an advantageous phenotype, such as resistance to bacterial pathogens causing diarrhea (Gabriel *et al.*, 1994). This hypothesis has been supported by comparing the tolerance of homozygous CF, heterozygous CF, and wild-type mice to cholera toxin-induced diarrhea (Gabriel *et al.*, 1994). The results, which showed STa stimulation of CFTR-dependent Cl⁻ and fluid secretion, prompted Cuthbert *et al.* (1994) to reach similar conclusions in justifying a heterozygous advantage. In follow-up studies, however, Cuthbert *et al.* (1995) found no evidence of a heterozygous advantage in ileal or colonic epithelia of CF mice, at least with respect to acute responses. They reasoned that if a heterozygous advantage for CF mutations did exist, it would involve either (1) non-*CFTR*-dependent transport processes, (2) prolonged exposure to secretagogues, or (3) a mutant *CFTR* contributing a protective effect (Cuthbert *et al.*, 1995). Along the same lines, Colledge and colleagues (Pier *et al.*, 1998) investigated whether *CFTR* mutations could protect against typhoid fever, which is initiated when *Salmonella typhi* infects intestinal epithelial cells. Their results suggest that *S. typhi,* but not the related pathogen *S. typhimurium,* gains cellular entry through *CFTR.* It was found that cells expressing wild-type *CFTR* internalized more of the pathogen than cells expressing ΔF508 *CFTR.* In addition, antibodies and synthetic peptides containing a sequence corresponding to the first extracellular domain of *CFTR* blocked *S. typhi* internalization. Mice heterozygous for ΔF508 *CFTR* were found to translocate 86% less *S. typhi* than wild-type mice, whereas homozygous

ΔF508 *CFTR* animals did not translocate the pathogen. In contrast, *CFTR* geno-type had no effect on translocation of *S. typhimurium*. Furthermore, immuno-electron microscopy showed that more *CFTR* bound to *S. typhi* in the submu-cosa of the wild-type than of the heterozygous ΔF508 mice. As in the previous studies, it was suggested that diminished *CFTR* in heterozygotes provides resis-tance to typhoid.

CFTR also may be required for intestinal HCO_3^- secretion. It was observed by Ainsworth and colleagues that both basal and acid-stimulated HCO_3^- and fluid secretion in the proximal duodenum were impaired in CF mice compared with normal mice (Hogan *et al.*, 1997a; 1997b). In addition, they showed that *CFTR* mediates basal, cAMP-, and Ca^{2+}-induced duodenal epithelial HCO_3^- se-cretion. In other studies, Clarke and Harline (1998) demonstrated that cAMP-stimulated HCO_3^- secretion across the duodenum occurred by two mechanisms. First, electrogenic HCO_3^- secretion occurs via *CFTR*, implying that *CFTR* also is permeable to HCO_3^- (Clarke and Harline, 1998). The second mechanism in-volved potentiation of Cl^-/HCO_3^- exchange by *CFTR*. This mechanism was found to be closely associated with an unknown carbonic anhydrase activity (Clarke and Harline, 1998). Seidler *et al.* (1997) also showed that cAMP, cGMP, and Ca^{2+} stimulated intestinal HCO_3^- secretion. Taken together, these results suggest that *CFTR* mediates HCO_3^- secretion via multiple mechanisms and that both Cl^- and HCO_3^- efflux promote fluid secretion across intestinal mucosa in an osmotically facilitated manner. The latter conclusion is borne out by the work of Sinaasappel and colleagues, who observed that water secretion mirrored Cl^- secretion in human jejunum (Teune *et al.*, 1996), which was defective in CF in-testines (reviewed in Eggermont, 1996). *CFTR*'s regulation of other ion con-ductances and transporters/exchangers will be revisited below.

3. Dysfunction in CF

Although bacterial toxins utilize *CFTR* to promote intestinal disease, the ab-sence of *CFTR* from this system also results in disease. As mentioned, approxi-mately 15% of CF newborns present with obstructive MI (reviewed in Eggermont, 1996, and Welsh, 1995). MI is believed to result from the dehydration of in-testinal secretions owing to diminished Cl^- and fluid secretion. In additional to MI, a postnewborn, distal intestinal obstructive syndrome (DIOS) also is a com-plication of CF. Although the pathogenesis of DIOS is unclear, inefficient di-gestion of food products, related to insufficiency of exocrine pancreatic enzymes, is believed to be responsible. Decreased HCO_3^- secretion also may lead to in-testinal dehydration and an inability to neutralize luminal acidity, thereby also causing impaired digestive enzyme activity (pancreatic lipase in particular). Consequently, malabsorption leading to steatorrhea and failure to thrive is typi-cal of CF. Liver and pancreatic disease (see later), also common in CF, com-pound the intestinal complications.

F. Rectum

1. Expression

Similar to the more proximal intestine, rectal Cl^- and fluid secretion is impaired in both CF patients and mouse models (Bijman *et al.*, 1991; Goldstein *et al.*, 1991; 1994; Gowen *et al.*, 1991; Veeze *et al.*, 1991; Rao *et al.*, 1994; Wilschanski *et al.*, 1996). Potential difference (PD) measurements to examine CFTR function have been performed in the mouse by Wilschanski *et al.* (1996) and in the human rectum by Veeze *et al.* (1991; 1994). Because of the success of these clinical tests, their use in confirming a CF diagnosis was proposed (Veeze *et al.*, 1991; 1994; Wilschanski *et al.*, 1996). In fact, PD studies in which CF mice with mild and severe intestinal disease were compared suggest a correlation between CF intestinal disease severity and function of alternative Cl^- conductances (Wilschanski *et al.*, 1996). The functional expression of this alternative Cl^- conductance was determined by CF modifier genes (Rozmahel *et al.*, 1996). In support of the mouse results, differences in rectal PD measurements of patients also correlated with disease severity (i.e., residual anion secretion was observed in patients with milder disease) (Veeze *et al.*, 1991; 1994).

G. Shark Rectal Gland as a Model for the Intestinal Epithelia

While CFTR in the rodent and human intestine were being investigated, studies of Cl^- secretion in the rectal gland of the dogfish shark, *Squalus acanthias*, also were underway. The shark rectal gland (SRG) provides a convenient and well-characterized model of intestinal salt and fluid secretion (reviewed in Forrest, 1996; Silva *et al.*, 1997). It was found that agonists that increase cAMP levels (such as adenosine, C-type natriuretic peptide [CNP], and vasoactive intestinal peptide [VIP]) and tyrosine-kinase inhibitors (genistein) stimulated Cl^- secretion from SRG. These properties suggested that CFTR, or a CFTR-like Cl^- channel, was active in this tissue (Forrest, 1996). Hence, shortly after cloning of the human *CFTR* gene, its shark homolog (*sCFTR*) was cloned and found to be 72% identical to the human gene (Marshall *et al.*, 1991). Studies by Kelley, Forrest, and coworkers showed that production of adenosine by SRG closely coupled Cl^- secretion (Kelley *et al.*, 1990; 1991; Forrest, 1996). This work also identified SRG receptors that were homologous to mammalian intestinal peptide receptors (Forrest, 1996; Aller *et al.*, 1999).

Recent work by Forrest's group focused on *CFTR* trafficking, expression, and function in a thorough and integrative manner (Lehrich *et al.*, 1998). In these studies, the effects of VIP, forskolin, and genistein on Cl^- secretion and immunostaining for CFTR in SRG were conducted in parallel (Lehrich *et al.*, 1998). Although each agonist stimulated Cl^- secretion, the diffuse staining of *CFTR* at the apical pole of SRG epithelial cells shifted to more intense and concentrated staining at or near the apical membrane (Lehrich *et al.*, 1998). This result suggested that agonist-induced Cl^- secretion from SRG epithelium involved, at least in part, trafficking and insertion of *CFTR* into apical membranes. This work also solidified *CFTR*'s role in Cl^- secretion in SRG intestinal epithelia.

H. Liver and Gallbladder

1. Expression

Because liver disease is a common clinical manifestation of CF, *CFTR* expression and function have been examined extensively in hepatic tissue. In parallel with the study of other GI tissues, Northern blot analysis and *in situ* hybridization have shown *CFTR* expression in the liver and gallbladder (Kelley *et al.*, 1992; Tran *et al.*, 1992; Cohn *et al.*, 1993; Tizzano *et al.*, 1993; Trezise *et al.*, 1993). Strong *et al.* (1994) demonstrated uniform expression of *CFTR* along the gallbladder epithelium. More detailed characterization of liver expression by immunolocalization showed that *CFTR* resides predominantly at the apical pole of biliary duct epithelial cells; relatively little staining was observed in hepatocytic epithelial cells (Cohn *et al.*, 1993). The expression of *CFTR* and *Mdr*-1 also was examined during rat liver regeneration following partial hepatectomy (Tran *et al.*, 1992). It was found that *CFTR* mRNA expression increased after surgery, peaking at 2 and 24 hours. In contrast, *Mdr*-1 expression did not change (Tran *et al.*, 1992). Similar to the regulation of *CFTR* expression in human intestinal epithelial cells, Lascols and colleagues (Kang-Park *et al.*, 1998) showed a PKC-α and PKC-ϵ mediated phorbol ester-induced down-regulation of CFTR mRNA in BC1 human liver epithelial cells. Together, these results show robust *CFTR* expression throughout gallbladder and intrahepatic bile duct epithelia, and regulation similar to that of the intestine. Although *CFTR* function at these sites is not clear, a role in mediating or neutralizing hepatic bile acid secretion is possible.

2. Function in Liver

An important function of the liver is the biotransformation of xenobiotics. Interestingly, numerous studies have shown CF livers to exhibit abnormally high rates of xenobiotic metabolism (Kearns *et al.*, 1990; Spino, 1991; Colombo *et al.*, 1992; Kearns, 1993; Kearns *et al.*, 1996; Parker *et al.*, 1997; Rey *et al.*, 1998). Other members of the ABC transporter family, including the multidrug-resistance transporters (*Mdr*-1, MRP, *Mdr*-3, etc.), also are expressed in the liver and likely are involved in xenobiotic transport and bile acid and phospholipid secretions (Lomri *et al.*, 1996; Roman *et al.*, 1997). *Mdr*-1 participates in biliary secretion of cationic molecules, and *Mdr*-2 or "MRP" secretes phospholipids. *Mdr*-3, or canalicular multiple organic anion transporter (cMOAT), secretes organic anions such as bilirubin (reviewed in Erlinger, 1996a; 1996b). With respect to *CFTR*, it has been found that *CFTR* overexpression in heterologous cells confers a multidrug-resistance phenotype, which is distinct from MDR-related proteins (Wei *et al.*, 1995). As previously suggested by their coordinate regulation, *CFTR* may share some functions of the MDR proteins, such as drug, phospholipid, or organic anion transport. Alternatively, the MDR proteins may share some of *CFTR*'s function. In accordance, it has been suggested that MDR proteins enhance ATP release and potentiate ion transport involved in regulatory volume decrease (RVD) (Roman *et al.*, 1997). Similarly, the facilitation of ATP transport by *CFTR* could participate in autocrine regulation of cell volume (Schwiebert *et al.*, 1995; and reviewed in Schwiebert, 1999a).

The transport processes of different hepatic cell types, particularly hepatocytic and cholangiocytic epithelial cells, are essential for normal liver function and bile formation (reviewed in Strazzabosco, 1997). Hepatocytic epithelial cells primarily secrete biliary osmolytes and bile acids, whereas cholangiocytic epithelial cells secrete Cl^- and HCO_3^-, thereby promoting fluid secretion and biliary acid neutralization. By secreting osmolytes, hepatocytic epithelial cells also participate in establishing an osmotic gradient that provides a driving force for cholangiocytic fluid secretion. Some of the first indications of *CFTR* expression and function in bile duct epithelia came from the work of Cohn and coworkers (Fitz *et al.*, 1993). Similar to intestinal and pancreatic epithelia, their study showed both Ca^{2+}- and cAMP-stimulated Cl^- secretion in bile duct epithelial cells (Fitz *et al.*, 1993). The observed cAMP-stimulated Cl^- currents had a linear I-V relationship, were time- and voltage-independent, had $Cl^- > I^-$ permselectivity, and were insensitive to 150 μM DIDS, identical to *CFTR* (Fitz *et al.*, 1993). Both biochemical and immunocytochemical analyses confirmed *CFTR* in cholangiocytic, but not hepatocytic, epithelial cells (Fitz *et al.*, 1993). From these studies it was concluded that defective secretion into bile ducts might contribute to CF cholestatic liver disease (Fitz *et al.*, 1993). LaRusso and colleagues subsequently confirmed *CFTR*'s presence in apical membranes of medium and large (but not small) cholangiocytic epithelial cells (Tietz *et al.*, 1995; Alpini *et al.*, 1997).

3. Function in the Gallbladder

Ductal obstructions are common features of CF. Thus, gallbladder ductal epithelium has been used extensively as a model to study secretion of mucins, extracellular matrix, Cl^-, HCO_3^-, and fluid. Scholte and coworkers (Peters *et al.*, 1996) demonstrated *CFTR* expression and function in murine primary gallbladder epithelial cells. Housset and colleagues (Dray-Charier *et al.*, 1995) were among the first to show functional, biochemical, and immunocytochemical evidence of *CFTR* in human gallbladder epithelial cells. High levels of *CFTR* mRNA and glycosylated protein were detected in primary cultures of human gallbladder epithelial cells, where *CFTR* was immunolocalized to the apical membrane and subapical poles (Dray-Charier *et al.*, 1995). The human cells showed both Ca^{2+}- and cAMP-stimulated Cl^- efflux, and the cAMP-activated current was inhibitable by DPC (Dray-Charier *et al.*, 1995). Both VIP and secretin, agonists that signal through cAMP, also stimulated Cl^- efflux in these cells (Dray-Charier *et al.*, 1995). Similar results were found in *Necturus* gallbladder epithelium by Reuss and coworkers (Torres *et al.*, 1996). As with the other GI-derived cells, a comparison of CF and non-CF cells showed that *CFTR* mediates cAMP-stimulated Cl^- and fluid secretion in gallbladder epithelia (Peters *et al.*, 1997). A link between *CFTR* function and mucus secretion has been suggested by Osborne and colleagues (Kuver *et al.*, 1994). In these studies, transfected gallbladder epithelial cells with a fivefold increase in CFTR protein showed a concomitant fourfold enhancement of mucin labeling and secretion compared with cells that had normal levels of CFTR (Kuver *et al.*, 1994). These results suggest that CFTR might regulate mucin synthesis and secretion. However, in studies by Scholte and colleagues (Peters *et al.*, 1996), no difference in mucus secretion was detected between the CF and non-CF cells.

4. Dysfunction in Liver and Gallbladder

CF liver disease is considered the first inherited disorder affecting the transport mechanism of cholangiocytic epithelial cells (Colombo *et al.*, 1998). Although little is known regarding the underlying defect, impaired secretion

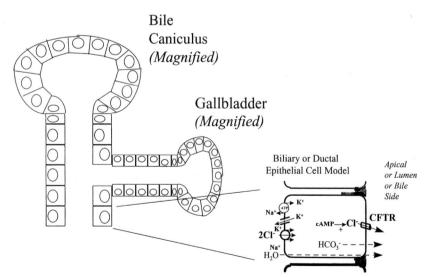

FIGURE 3 Schematic of *CFTR* expression and function in the liver. The illustrations show increasing magnification of the bile caniculus, bile duct, and gallbladder. Also shown is *CFTR* expression and function in the apical membrane of the biliary duct.

from the biliary epithelium likely results in dehydrated and acidic luminal contents and damage to bile ducts through buildup of cytotoxic compounds and infectious agents (Colombo *et al.*, 1998). Because of defective Cl^- and HCO_3^- transport from the biliary epithelia, secretin-stimulated choleresis (similar to primary sclerosing cholangitis) also is associated with CF (McGill *et al.*, 1996). The CF biliary disease may have a pathogenesis similar to that of the small intestine and may contribute to impaired nutritional status.

Interestingly, Hopwood and coworkers (Hill *et al.*, 1997) showed that gallbladder epithelial cells of CF mice have oversulfation of glycosaminoglycans and altered extracellular matrix. The significance of this finding with respect to underlying basis and contribution to CF pathophysiology is not known; however, similar oversulfation defects have been observed in other CF tissues.

I. Pancreas

1. Expression

The most common CF gastrointestinal disease is exocrine pancreatic insufficiency (PI). PI manifests in approximately 85% of CF patients and requires clinical intervention consisting of continuous enzyme replacement therapy.

CFTR RNA *in situ* hybridization studies of the pancreas by Strong *et al.* (1994) demonstrated high expression in the small intercalated ducts and lower expression in the interlobular ducts. Similar expression was found in other studies (Trezise *et al.*, 1991; 1993; Tizzano *et al.*, 1993; Kopelman *et al.*, 1995; Hyde *et al.*, 1997). Interestingly, RNA *in situ* hybridization studies detected little, if any, *CFTR* expression in pancreatic acinar cells. Immunostaining of the pancreas with a *CFTR*-specific antibody by Cohn and coworkers (Marino *et al.*, 1991) detected *CFTR* at the apical pole of proximal duct epithelial cells, confirming the expression pattern detected by RNA *in situ* hybridization experiments. More recent investigations using second-generation *CFTR* antibodies and functional and pharmacological identification shows *CFTR* in pancreatic acini and ducts (Zeng *et al.*, 1997). In fact, because of its early and widespread expression in pancreatic ducts, *CFTR* is an important marker to follow pancreatic duct development (Hyde *et al.*, 1997). Because *CFTR* expression is present in pancreatic acinar and ductal cells, abnormal components of their secretions, as well as dehydration, likely underlie CF pancreatic disease.

Sweezey and colleagues (1996) have used the pancreatic epithelial cell line PANC-1 to investigate the effects of different steroid hormones on the regulation of CFTR. These studies demonstrated that the female hormones progesterone and estradiol inhibited *CFTR*-mediated ion transport. Extrapolation of these findings to other epithelial cells and their significance to CF disease warrant further investigation.

2. Function

To our knowledge, the first definitive patch-clamp studies showing functional evidence for a cAMP-activated, low-conductance Cl^- channel with *CFTR* characteristics were by Gray *et al.* (1988; 1989). These Cl^- channels were observed in apical membranes of duct cells cultured from fetal pancreas (Gray *et al.*, 1989). It was found that channel activity was potentiated about threefold by secretin, cAMP analogs, and forskolin (Gray *et al.*, 1988). Similar to the observations made by Tabcharani *et al.* (1990) in T84 colon carcinoma cells, an outwardly rectifying Cl^- channel also was observed, although rarely. From these studies, Gray *et al.* (1990) proposed that the CFTR Cl^- channel worked in concert with Cl^-/HCO_3^- exchange to promote fluid secretion in pancreatic ducts. This hypothesis was especially relevant because the low-conductance Cl^- channel was only slightly permeable to HCO_3^-. Their hypothesis was supported by subsequent patch-clamp studies of pancreatic duct cells (Gray *et al.*, 1993). Unequivocal confirmation of the identity of this low-conductance Cl^- channel as CFTR was provided by Frizzell and colleagues (Cliff *et al.*, 1992). In these studies, a CF pancreatic adenocarcinoma epithelioid cell line (CFPAC-1) was stably transfected with CFTR. The CFTR-expressing cells showed a 7–10 pS Cl^- channel, which was absent in the nontransfected cells (Cliff *et al.*, 1992). The newly expressed Cl^- channel (1) was stimulated by cAMP agonists, (2) was activated by PKA and ATP, (3) was insensitive to the Cl^- channel inhibitors DNDS and DIDS, and (4) had permselectivity of $Br^- > Cl^- > I^-$, consistent with CFTR (Cliff *et al.*, 1992). In other studies, Kopelman *et al.* (1993) eliminated the cAMP- but not the Ca^{2+}-activated Cl^- conductance by introduction of CFTR antisense oligonucleotides to the pancreatic epithelial cell model PANC-1. Similar to intestinal and biliary epithelial cells, separate pathways for Ca^{2+}- and cAMP-stimulated Cl^- conductance are present in pancreatic ducts (Cliff *et al.*, 1992; Kopelman *et al.*, 1993; Gray *et al.*, 1994; Nguyen *et al.*, 1997). Kopelman *et al.* (1995) provided evidence for CFTR in rabbit pancreatic acinar cells, and Thomas and colleagues (Zeng *et al.*, 1997) confirmed CFTR in pancreatic acinar and ductal cells.

Impaired ductal HCO_3^- secretion is believed to be a major contributor to CF pancreatic disease. Soleimani and colleagues (Shumaker *et al.*, 1999) demonstrated expression of NBC-1, an electrogenic Na^+/HCO_3^- cotransporter, in CF (CFPAC-1 cells) and normal cultured pancreatic duct cells (CAPAN-1 cells and CFTR transfected CFPAC-1 cells). It was shown that NBC-1 mediates HCO_3^- uptake across the basolateral membrane of these cells and that this activity is potentiated by cAMP stimulation of CFTR. They suggested that defective pancreatic ductal HCO_3^- secretion in CF results from decreased basolateral HCO_3^- entry mediated by NBC-1, secondary to a diminished driving force owing to CFTR absence.

In summary, CFTR functions in pancreatic acini and ducts as a Cl^- and HCO_3^- channel and to facilitate fluid and HCO_3^- secretion.

3. Dysfunction in CF

Not all CF patients are PI. Durie (1992) correlated *CFTR* genotype to pancreatic disease in 538 patients to classify mutations as "severe" or "mild" with respect to pancreatic function. It was found that "compromised conductance" and "partial trafficking" (see section II.D and Table II) mutations on one or both CF chromosomes conferred a mild (PS) phenotype. On the other hand, the PI phenotype was observed when nonsense, splicing, or severe trafficking mutations were present on both CF chromosomes.

Another common clinical manifestation of the CF pancreas is pancreatitis. For this reason, Cohn *et al.* (1998) investigated *CFTR* mutations in idiopathic pancreatitis patients without CF-associated lung disease. In their study, 17 common *CFTR* mutations and the length of the intron 8 polypyrimidine variant (see section II.D and Table II) were examined. In total, eight *CFTR* mutations were detected in the patients with chronic idiopathic pancreatitis; 10 of the 27 patients had at least one abnormal *CFTR* allele. Interestingly, three of the 10 patients had both *CFTR* alleles affected, but did not have a typical CF diagnosis, even after reassessment. However, patients with two *CFTR* mutations were found to have abnormal nasal cAMP-mediated Cl⁻ transport. Thus, an association between *CFTR* mutations and chronic manifestation of idiopathic pancreatitis was concluded. Moreover, the *CFTR* mutations in the patients with pancreatitis were similar to those associated with male infertility.

Thus, loss of *CFTR* in the pancreas has a several-fold effect on CF disease. First, the acidic environment owing to compromised Cl^-, HCO_3^-, and fluid secretion is unfavorable to pancreatic enzyme activity and nutrient absorption.

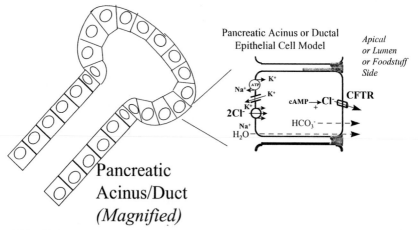

FIGURE 4 Schematic of *CFTR* expression and function in the pancreas. Increasing magnification of the pancreatic acinus and duct shows *CFTR* expression and function in the apical membrane of epithelial cells.

Second, *CFTR* dysfunction in the pancreas leading to ductal obstructions, pancreatic deterioration, and pancreatitis further impairs digestion by diminishing digestive enzyme production and secretion.

IV. INTEGRATION OF CFTR AND CF GASTROINTESTINAL PATHOPHYSIOLOGY

CFTR has been associated with a plethora of physiological functions, including transport (Cl^-, ATP, PAPS, bicarbonate, glutathione, and drugs), regulation (epithelial conductance; pH and cellular energy), membrane trafficking, protein processing, and inflammation. Although evidence of CFTR's role as a cAMP-stimulated Cl^- channel is uncontestable, this activity does not preclude it from having additional and diverse activities. Indeed, compared with other Cl^- channels, the relatively complex structure of CFTR and its equally intricate regulation support the possibility of collateral functions. In addition, with regard to CF pathophysiology, not all disease presentations can be readily attributed to the loss of Cl^- channel activity. In this section, CFTR-mediated processes will be reviewed in context of the GI system, and the latest hypotheses as to their modulation by CFTR will be presented.

A. CFTR: cAMP-Activated Cl^- Channel and Facilitation of Fluid Secretion

These two CFTR functions have been reviewed extensively herein. CFTR acts as a cAMP-activated Cl^- channel in apical membranes of salivary gland acinus and duct, pancreatic acinus and duct, gallbladder epithelium, biliary duct, and intestinal crypt epithelium. CFTR-dependent Cl^- secretion across these epithelia establishes an osmotic driving force for fluid secretion. This activity was illustrated clearly by the requirement of CFTR for heat-stable enterotoxin and cholera toxin-induced secretory diarrhea. The most pronounced pathological lesions observed in the CF intestinal tract likely relate to loss of this function. As discussed previously, the CF intestinal tract is prone to obstructive lesions, including MI in newborns and DIOS later in life. Both these obstructive diseases are believed to relate to dehydration and ensuing inspissation of luminal contents. Because fluid secretion into the intestinal lumen is directly coupled to Cl^- secretion, loss of CFTR-mediated Cl^- secretion will give rise to both pathological conditions.

B. CFTR: Bicarbonate Secretion

Confirmation of CFTR as a Cl^- channel provided the impetus for investigating its ability to transport other anions. From such studies, evidence has accumulated that CFTR conducts HCO_3^- in a cAMP-dependent manner, and regulates its secretion through other processes.

Studies have demonstrated that numerous mechanisms contribute to intestinal HCO_3^- secretion, including cAMP, cGMP, and Ca^{2+}-mediated pathways (Seidler et al., 1997). Two mechanisms relevant to CF are electrogenic HCO_3^- secretion directly through CFTR and potentiation of Cl^-/HCO_3^- exchange by CFTR, both of which are cAMP-dependent (Clarke et al., 1998). Direct CFTR mediation of basal, cAMP, and Ca^{2+}-induced HCO_3^- secretion has been demonstrated in mice (Hogan et al., 1997a; Hogan et al., 1997b). Defective HCO_3^- secretion in CF mice diminished duodenal fluid secretion. Because HCO_3^- secretion contributes to osmotically facilitated fluid secretion, its impairment would have an impact on CF intestinal obstructive lesions. Moreover, acidic pH owing to decreased HCO_3^- secretion is detrimental to digestive enzyme activity, resulting in malabsorption and an aggravation of the intestinal disease.

C. CFTR: Modulation of Cytosolic and Organellar pH

CFTR is an intracellular Cl^- channel. Using patch-clamp analysis of intracellular membranes, Pasyk and Foskett (1995) demonstrated that CFTR acts as a Cl^- channel in intracellular organelles such as the ER. Its function at these internal sites is hypothesized to provide a parallel anion shunt to facilitate H^+ transport into organelles for maintenance of physiological pH. Accordingly, Al-Awqati, Barasch, and coworkers (Barasch et al., 1991) showed that ER, trans-Golgi, and lysosomal compartments of CF cells failed to acidify as readily as normal cells. In addition, Elgavish (1991) showed that cytosolic pH was elevated almost 0.4 pH units in mock- versus CFTR-transfected CFPAC-1 cells. It was proposed that abnormal organellar pH resulted in defective protein processing because the optimum pH for enzymes responsible for glycosylation was not maintained. Indeed, numerous studies of CF have shown impaired glycosylation (including diminished sialylation and oversulfation) of proteins such as mucins. Whether or not such defects contribute to CF pathophysiology remains unclear; however, it is likely that they would impact on the physiological properties of proteins, including mucins.

CFTR is a regulator of the Cl^-/HCO_3^- and Na^+/H^+ exchangers. As mentioned earlier, CFTR not only directly participates in HCO_3^- secretion but also stimulates the Cl^-/HCO_3^- exchanger. This dual role of CFTR in HCO_3^- secretion was demonstrated by Clarke and Harline (1998) in mouse duodenum. Mastrocola et al. (1998) also found that C127 cells lacking CFTR or expressing $\Delta F508$ CFTR had similar Cl^-/HCO_3^- exchanger-dependent cytosolic acidification or alkalinization. However, only cells expressing wild-type CFTR had cAMP-dependent HCO_3^- secretion, suggesting that CFTR is a HCO_3^- channel. Poulson and Machen (1996) showed a similar role for CFTR in airway epithelial cells.

Along the same lines, Roepe and coworkers (Wei et al., 1995) showed that *CFTR* overexpression in NIH3T3 fibroblast cells caused decreased or depolarized membrane potentials. A similar effect was observed in fibroblasts overexpressing *Mdr*-1 (Wei *et al.*, 1995). However, in contrast to the *Mdr*-1–transfected cells that were alkaline with respect to controls, *CFTR* transfected cells had mildly acidified intracellular pH (Wei *et al.*, 1995; 1997). An explanation for this observation is that Cl^-/HCO_3^- exchange was significantly reduced during *Mdr*-1 expression, whereas the pH optima for Na^+/H^+ exchanger activity was lower in *CFTR*-expressing cells (Wei *et al.*, 1997).

Rajendran *et al.* (1999) also uncovered a role for CFTR and ORCC in Cl^--dependent Na^+/H^+ exchange across intestinal crypts. It was found that Na^+/H^+ exchange was inhibited by the Cl^- channel blockers NPPB and DIDS, and by an antibody to CFTR (Rajendran *et al.*, 1999). A dependence on parallel Cl^-/HCO_3^- exchange was not apparent. These results suggest that CFTR also modulates apical membrane Na^+/H^+ exchange in intestinal crypt cells.

Finally, because CF cells are inept at acidifying their cytosol, and Cl^- and HCO_3^- secretion is impaired, Gottlieb and Dosanjh (1996) studied the relationship between *CFTR*, cytosolic pH, and apoptosis. It was found that C127 cells expressing $\Delta F508$ *CFTR* failed to acidify, in contrast to cells with wild-type CFTR (Gottlieb and Dosanjh, 1996). Consequently, the cells also failed to undergo apoptosis. These investigators concluded that failure to acidify the cytoplasm and undergo apoptosis might contribute to CF pathogenesis.

In contrast to the above studies, McGraw and colleagues as well as Verkman and coworkers failed to observe perturbations of cytoplasmic, vesicular, or organellar pH in CF cells (Dunn *et al.*, 1994; Seksek *et al.*, 1996).

D. CFTR: Regulation of Other Ion Channels

A recent review has focused on CFTR's role in conductance regulation (Schwiebert *et al.*, 1999a). We will briefly summarize studies performed in GI epithelial cells to demonstrate this function of CFTR.

1. ORCC

Much of the work showing regulation by CFTR of the outward rectifying chloride channel (ORCC) has been performed in airway epithelial cells by Guggino and colleagues, Gabriel *et al.*, and Benos *et al.* (Egan *et al.*, 1992; Gabriel *et al.*, 1993; Jovov *et al.*, 1995a; 1995b; Schwiebert *et al.*, 1995). Several studies, however, also have investigated this regulatory interaction in GI epithelial cell models. Simmons and colleagues (McEwan *et al.*, 1994) showed that several different Cl^- conductances were present in the apical membranes of T84 colon carcinoma cells. Carbachol-induced Cl^- secretion (via intracellular Ca^{2+}) correlated with basal Cl^- secretory I_{SC} (McEwan *et al.*, 1994). However, at low basal I_{SC}, the Cl^- channel blocker DIDS inhibited both carbachol- and hypo-tonicity-induced I_{SC}, but not VIP-stimulated Cl^- secretion (via cAMP) (McEwan *et al.*, 1994). On the other hand, at high basal I_{SC}, the carbachol-induced increase in Cl^- secretion was insensitive to DIDS. Taken together, these results suggest the presence of at least three Cl^- conductance pathways in the T84 cells, a VIP (cAMP)-stimulated channel (likely CFTR) and both a DIDS-sensitive and -insensitive Ca^{2+}-activated conductance.

Recently, Madara and colleagues (Merlin *et al.*, 1998) showed that T84 cells possess two different cAMP-stimulated Cl^- secretory pathways and a Ca^{2+}-stimulated pathway. Thapsigargin-induced Cl^- secretion was completely abolished by DIDS, whereas forskolin-induced Cl^- secretion was only partially inhibited. The DIDS-sensitive component of the forskolin-induced apical-membrane Cl^- secretion had a permselectivity sequence of $I^- > Cl^-$, which was unlike *CFTR* but resembled the ORCC (Merlin *et al.*, 1998). In other studies, Okada and coworkers showed that guinea pig Paneth cells (intestinal crypt cells that secrete antimicrobial agents such as colicidins, lysozyme, and defensins [Ouellette, 1997]) displayed Cl^- currents consistent with *CFTR*. However, mRNA and protein studies failed to detect *CFTR* in these cells (Tsumura *et al.*, 1998). The reason for the discrepancy is not known, but as with other studies, *CFTR* function does not always correlate with expression. Although the standard for establishing *CFTR* expression is immunoprecipitation, function may prove more accurate and sensitive.

2. CaCC

CF mouse models normally do not present with lung disease. As an explanation, Clarke *et al.* (1994) suggested that the absence of lung disease was related to expression of a Ca^{2+}-activated Cl^- conductance (CaCCs) in mouse airways. Functional expression of a Ca^{2+}-dependent Cl^- conductance also has been demonstrated in mouse intestine, and its activity at this site correlated with milder intestinal disease in CF mice (Rozmahel *et al.*, 1996; Wilschanski *et al.*, 1996). If this alternative Cl^- conductance could indeed circumvent the loss of CFTR, its manipulation could be used in CF therapeutics. This potential strategy will be revisited in the "Therapies" section herein.

3. ENaC

Much of the work showing *CFTR* inhibition of Na^+ absorption via the amiloride-sensitive epithelial Na^+ channel (ENaC) has been performed in lung epithelia, heterologous cell systems, and planar lipid bilayers. Seminal work by various groups (Stutts *et al.*, 1995a; Benos *et al.*, 1996; Ismailov *et al.*, 1996a; Stutts *et al.*, 1997; Schreiber *et al.*, 1999) has provided insight into this regulatory relationship. Possible mechanisms of the CFTR-ENaC regulatory interactions are the focus of a recent report (Schreiber *et al.,* 1999).

With regard to the intestinal tract, Clarke and Harline (1996) studied the relationship of CFTR to ENaC in the jejunum of wild-type heterozygous and homozygous CF mice. These studies showed absence of basal Cl^- secretion in CF versus heterozygous and wild-type mice and, as expected, the CF animals lacked cAMP-stimulated Cl^- secretion. Under basal conditions, Na^+ was absorbed similarly in all three groups of animals. However, while cAMP stimulation abolished NaCl absorption and induced Cl^- secretion in wild-type and heterozygous mice, it had no effect on either property in the CF animals (Clarke and Harline, 1996b). These results suggest that in the intestine, as in the airway, *CFTR* is required for both Cl^- secretion and the modulation of cAMP inhibition of Na^+ absorption. In related studies, Grubb and Boucher (1997a) compared the effects of different Na^+ diets on proximal and distal colonic and cecal electrophysiological properties of CF and wild-type mice. When wild-type or CF mice consumed a high-Na^+ diet, little or no amiloride-sensitive Na^+ absorption was observed. In contrast, a low-Na^+ diet resulted in significant Na^+ absorption in all three intestinal regions of CF mice, whereas wild-type animals had less Na^+ absorptive current and only in the distal colon and cecum (Grubb *et al.,* 1997a). The animals on a low Na^+ diet had elevated levels of aldosterone. When treated with exogenous aldosterone, the Na^+ absorption in the CF distal colon was greater than that in the wild-type colon (Grubb *et al.,* 1997a). These results support *CFTR*'s role in modulating absorptive Na^+ channels in the mouse intestine.

Grubb and Gabriel (1997b) reviewed the intestinal pathophysiology of CF mouse models. Consistent between CF patients and mice was the absence of cAMP-stimulated Cl^- secretion and HCO_3^- secretion, defective cAMP regulation of Na^+ absorption, and elevated electrogenic Na^+ absorption. It was concluded that the intestinal disease of CF mice is similar to its counterpart in humans, but of greater severity.

4. ROMK

Egan and coworkers showed that *CFTR* confers sulfonylurea sensitivity and inhibition to renal potassium (K^+) channels (ROMK) (McNicholas *et al.,* 1996; 1997), extending the studies of Welling and colleagues (Ruknudin *et al.,* 1998). A recent review focused on the possible mechanisms of CFTR-ROMK regulatory interactions in comparison with the SUR-BIR (K_{ATP} channel) studies of pancreatic β cells (Ishida-Takahashi *et al.,* 1998). Similar to the sulfonylurea receptor (SUR) regulation of pancreatic β cell K^+ channels, *CFTR* modulates the activity of the ROMK channel by forming of a 1:1 stoichiometric complex (Aguilar-Bryan *et al.,* 1998). Through such studies, *CFTR*'s regulations of K^+ channel function in airways and the GI tract have come under investigation. Cuthbert *et al.,* (1994b) identified a difference in colonic K^+ secretion between

CF and wild-type mice. It was found that forskolin and carbachol caused a reduction in I_{SC} that was antagonized by apical application of the K^+ channel inhibitors barium chloride (Ba^{2+}) and tetraethylammonium (TEA) chloride. These investigators showed that the forskolin-induced K^+ secretory response was greater in normal mice than in CF mice. In addition, Greger and colleagues as well as Cuthbert *et al.* emphasized the importance of basolateral K^+ channels for triggering Cl^- secretion (Cuthbert, 1994b; Lohrmann and Greger, 1993; Devor *et al.*, 1996; MacVinish *et al.*, 1998). It was found that a K^+ channel agonist (1-EBIO) stimulates Cl^- secretion effectively in mouse and human intestinal epithelia (Devor *et al.*, 1996; MacVinish *et al.*, 1998), but only poorly in airway epithelia (MacVinish *et al.*, 1998). Moreover, Ba^{2+} and another K^+ channel inhibitor, 293B, attenuate both basolateral K^+ channels and Cl^- secretion in colonic epithelia (Warth *et al.*, 1996; MacVinish *et al.*, 1998). Ba^{2+} but not 293B inhibited basolateral K^+ channels in nasal epithelia (MacVinish *et al.*, 1998). These results suggest that basolateral K^+ channels are critical for maintaining epithelial Cl^- and fluid secretion and that the types of basolateral K^+ channels differ between the GI tract and airways. Along the same lines, Valverde *et al.* (1995) compared cell-volume regulation in normal and CF mouse intestinal crypts by using crypt cell height as a measure of cell volume. They found that regulatory volume decrease (RVD) in response to hypotonic challenge was defective in CF. This defect was attributable to impaired stimulation of K^+ conductance important for K^+ efflux during RVD, rather than lack of Cl^- secretion. These results suggest that CFTR modulates K^+ channels and cell volume. Whether defects in these processes contribute to CF disease requires clarification.

5. ATP Release and ATP Channels

Schwiebert and coworkers (Taylor *et al.*, 1998) demonstrated that ATP release under isotonic and hypotonic conditions is up-regulated in the presence of CFTR. It was hypothesized that CFTR facilitates ATP release for autocrine purinergic control of cell volume regulation (Taylor *et al.*, 1998). Indeed, ATP, via purinergic receptors, may regulate K^+ and Cl^- channels in RVD; thus, the impaired K^+ channels in CF could relate to loss of this release. The mechanism and physiology of CFTR-facilitated ATP release have been reviewed (Schwiebert, 1999a; Schwiebert *et al.*, 1999b).

The importance of ATP release for cell-volume regulation was suggested by Fitz and coworkers through a study of channel and cell-volume regulation in hepatocytic and cholangiocytic epithelial cells (Lomri *et al.*, 1996; Roman *et al.*, 1997; 1998; Feranchak *et al.*, 1998). It was found that forskolin and ionomycin stimulated ATP release from cholangiocytic epithelial cells, giving rise to micromolar amounts of ATP and its metabolites in bile (Chari *et al.*, 1996). Similarly, nucleotides were detected in the effluent of *in vitro* models of hepatocytic (HTC rat hepatoma cells) and cholangiocytic (Mz-Cha-1 "Witt" cholangiocarcinoma cells) epithelial cells (Chari *et al.,* 1996). In parallel studies, Fitz and Sostman (1994) showed that ATP acts as an agonist to stimulate cation, K^+, and Cl^- currents (in that order and time course) in whole-cell patch-clamp recordings of HTC cells. Cation and K^+ currents were activated transiently, and Cl^- currents were sustained and showed outward rectification (Fitz and Sostman, 1994). Moreover, flux studies of biliary epithelial cells showed similar results with respect to ATP stimulation of both K^+ and Cl^- permeability (McGill *et al.,* 1994). It was also shown that ATP, via purinergic receptors, promotes biliary secretion (Schlenker *et al.,* 1997). Because ATP stimulated these three specific conductances in a time-dependent manner, Fitz and Sostman hypothesized that ATP may be released by hepatocytic and/or cholangiocytic epithelial cells to activate ionic conductances for RVD during cell-volume regulation in the liver. In HTC cells, Wang *et al.,* (1996) showed that hypotonic challenge stimulated an ATP whole-cell conductance (where ATP was the only charge carrier). There was little or no ATP conductance under isotonic conditions (Wang *et al.,* 1996). Stimulation of RVD and swelling-activated Cl^- channels was prevented by ATP receptor antagonists or apyrase (Wang *et al.,* 1996). More recently, Roman *et al.* (1997) showed that expression of multiple MDR isoforms augments the autocrine ATP control of liver cell-volume regulation. Because *CFTR* also is expressed in cholangiocytic epithelial cells, it is possible that these ABC transporters may facilitate ATP release into bile for cell-volume regulation during secretion and resorption of bile acids in the bile duct and gallbladder. In keeping with this hypothesis, collaborations between Fitz's and Schwiebert's groups could uncover physiological roles for ABC transporter-facilitated ATP release and signaling.

E. CFTR: Regulation of Vesicle Trafficking

In the early 1990s, collaborations between Jilling and Kirk, together with Bradbury and Bridges, described defects in vesicle trafficking in CF pancreatic epithelial cell models (Bradbury and Bridges, 1992a; Bradbury *et al.,* 1992b). In particular, the lack of cAMP-regulated exocytosis and the presence of enhanced endocytosis (which could be corrected by CFTR) were observed in CFPAC-1 cells (Bradbury and Bridges, 1992a; Bradbury *et al.,* 1992b). In fact, as early as 1989, it was proposed that CF was a disease of abnormal Cl^- permeability, protein secretion, and vesicle traffic (Kirk, 1989). These early studies resulted in seminal reports in which it was suggested that CFTR regulated plasma membrane recycling (Bradbury *et al.,* 1992b; Bradbury and Bridges, 1994a). The latter studies emerged from the conjecture that cAMP stimulated Cl^- secretion from GI epithelial cells by inserting more CFTR Cl^- channels into the plasma membrane—a hypothesis that is still widely held. In keeping with this hypothesis, CFTR has been shown in endosomes (Lukacs *et al.,* 1992) and clathrin-coated vesicles (Bradbury *et al.,* 1994b), but not in caveolae (Bradbury *et al.,* 1999a) of airway, GI, or heterologous cell systems. Thus, stimulation of Cl⁻ secretion by cAMP could be achieved by activating CFTR resident in the plasma membrane or by adding additional channels to the membrane. Both mechanisms likely contribute to CFTR-mediated Cl⁻ secretion, depending on the epithelial cell type, its origin, and its conductive state.

Several laboratories have continued to study CFTR's role in membrane trafficking. First, Fuller and Benos showed a dependence on vesicle trafficking for CFTR Cl⁻ secretion in T84 cells. In these studies, it was found that nocodazol and brefeldin A, inhibitors of vesicle trafficking, blocked CFTR Cl⁻ secretion (Tousson *et al.,* 1996). Second, Frizzell and coworkers showed polarization-dependent and cAMP-mediated insertion of vesicles and a concomitant increase in Cl⁻ permeability in multiple systems (T84 colonic epithelial cells, heterologous cell systems, *Xenopus* oocytes) (Morris *et al.,* 1994; Takahashi *et al.,* 1996). Thirdly, Schwiebert *et al.* (1994) showed that the G protein and cAMP effects on Cl⁻ permeability in airway epithelial cells involved exocytosis of CFTR or its regulators. Fourth, Collawn and coworkers investigated rates of endocytosis of wild-type and C-terminal mutant forms of CFTR in T84 cells (Prince *et al.,* 1994; Prince *et al.,* 1999). By following the rate and magnitude of internalization, Prince *et al.,* showed that cAMP inhibited the internalization of CFTR, consistent with general inhibition of endocytosis in colonic epithelial cells by cAMP. More recently, Collawn and colleagues (Prince *et al.,* 1999) showed efficient endocytosis of CFTR requires a tyrosine-based signal, at Y1424. Finally, although no role in intracellular compartment acidification had been indicated (Biwersi and Verkman, 1994), Biwersi et al (1996) showed CFTR's function in endosome fusion in recycling pathways.

Frizzell and coworkers (Howard *et al.,* 1996) used epitope-tagged CFTR to show that cAMP-stimulation caused insertion of CFTR into the plasma membrane, particularly in the *Xenopus* oocyte expression system. Similar data suggesting that cAMP-dependent insertion of CFTR into plasma membrane were provided by Forrest (1996) using the shark rectal gland (described earlier in this chapter). Recently, Hollande *et al.* demonstrated CFTR protein and its targeting to the apical membrane in human pancreatic duct cells (Capan-1 cells), although the role of cAMP in this process was not addressed (Hollande *et al.,* 1998). In contrast, however, Stanton and coworkers (Moyer *et al.,* 1998) did not find cAMP-induced movement of GFP-tagged CFTR in MDCK epithelial cells. In their studies, GFP-tagged CFTR was resident in the apical plasma membrane with or without cAMP treatment. In general, these studies showed that CFTR is present and functional in endosomal compartments, which suggests that its insertion into the plasma membrane provides a mechanism to enhance Cl^- secretion. Nevertheless, it remains possible that exocytosis inserts cofactors or regulators of CFTR to enhance Cl^- secretion.

Studies by Stanton and colleagues suggest that a PDZ-binding motif at the C-terminus of CFTR is responsible for its anchoring in the apical membrane. Mutagenesis of this motif deflected some CFTR to the lateral membrane, whereas cleavage of the C-terminus distal to serine 1455 caused most CFTR to reside in lateral membranes (unpublished data). In intestinal epithelial and heterologous cell models, various studies by Kirk and coworkers provided compelling evidence that vesicle adapter proteins syntaxin 1A and Munc18 associate with and regulate CFTR (Naren *et al.,* 1997; 1998). Syntaxin 1A and Munc18 are involved in vSNARE/tSNARE complexes, as well as vesicle docking and fusion. These investigators showed that epithelial cells express multiple syntaxin isoforms; however, only syntaxin 1A (which resides in the apical membrane of colonic epithelial cells) had affinity for CFTR (Naren *et al.,* 1997). On the other hand, Munc18 displaced syntaxin 1A from CFTR (Naren *et al.,* 1997). Naren *et al.* (1998) extended this work to show that the syntaxin-CFTR interactions were domain-specific such that the H3 domain of syntaxin 1A bound with high affinity to the N-terminus of CFTR. Because SNARE and PDZ-binding proteins interact with the N-termini and C-termini of CFTR, respectively, a model is beginning to evolve whereby a dynamic "push-pull" of CFTR at each of its intracellular termini could have functional ramifications. In this regard, CFTR N-terminus interactions with vesicle adapter and receptor proteins may mediate membrane trafficking, whereas interactions with PDZ binding proteins at the C-terminus may anchor or cluster CFTR in plasma membrane microdomains. In addition, these interactions could participate in CFTR function, and their loss may prevent Cl^- channel activity or regulatory interactions with other transport proteins via cytoskeletal scaffolding complexes. However, CFTR localization, vesicle trafficking, and regulation of trafficking may differ between epithelial and heterologous cell systems, and their state of differentiation. Moreover, the types of vesicles in which CFTR resides also may differ among cell types. Thus, much remains to be understood regarding CFTR's intracellular targeting, regulation, and function.

F. CFTR: Post-Translational Protein Processing

Barasch and coworkers (1991) reported CF-associated hypersulfation and diminished sialic acid addition to glycoproteins. Along the same lines, sulfate incorporation into glycosaminoglycans of altered extracellular matrix (ECM) was reported in the liver and pancreas of CF mice (Hill *et al.,* 1997). Similarly, LaMont and coworkers (Bhaskar *et al.,* 1998) showed dysregulation of proteoglycan processing and secretion by comparing normal and CF biliary epithelial cells. In addition, increased incorporation of glucosamine was shown to underlie enhanced glycosylation of chondroitin sulfate in CF cells (Bhaskar et al., 1998). Finally, differences in sulfate transport have been identified between CF pancreatic cells (CFPAC-1) and their CFTR complemented counterpart (PANC-1 cells) (Elgavish and Meezan, 1992).

Glycosylation defects of glycoproteins are observed in CF cells (Barasch et al., 1991), thus altered mucus properties and rheology may to be central to intestinal, bile duct, and pancreatic duct blockages. Parmley and Gendler (1998) showed that MUC1 protein and mRNA were upregulated twofold and sixfold, respectively, in CF mice. Moreover, in CF mice where the MUC1 gene also was deleted, intestinal obstructive lesions (similar to MI in humans) were significantly diminished and the animals survived on normal diets. In other studies, De Lisle et al. (1997) showed that gp300 (MUCLIN), a sulfated mucin-like glycoprotein, was secreted by non-CF epithelial cells lining the intestine, pancreas, and gallbladder. In CF tissues, however, MUCLIN expression was increased in the pancreas and duodenum, but not in the gallbladder. Luminal surface labeling of MUCLIN along the villus also was elevated in CF (De Lisle et al., 1997). Additional studies showed that MUCLIN's electrophoretic mobility was retarded, indicating altered processing (De Lisle et al., 1997). In keeping with the contribution of mucins to CF pathophysiology, MUC6 was a major constituent of the tenacious secretions blocking pancreatic ducts (Reid et al., 1997). Taken together, these results suggest that abnormalities in proteoglycans (particularly mucins) contribute to the CF obstructive lesions. However, not all studies show correlation between CFTR and abnormal protein processing. Peters et al. (1996) found no direct link between CFTR expression and mucin secretion in gallbladder epithelial cells. De Lisle et al. also found no abnormality in the gallbladder mucus and ECM, which suggests that lack of CFTR is not associated with defective protein processing throughout the GI system. In related experiments of airway epithelial cells, Wiesz and coworkers (Jiang et al., 1997) reported that differences in glycosylation between CF and CFTR-complemented CF cells were independent of CFTR. Distinct glycosylation of surface proteins was observed in CF cells with or without wild-type CFTR expression (Jiang et al., 1997). This suggests that secondary factors contribute to the CF processing defects.

G. CFTR: Inflammation and Apoptosis

Recent focus of CF research has been on the cause of excessive inflammation. Although overactive inflammation in CF airways is regarded as the primary cause of morbidity, its potential contribution to GI disease is not ruled out. Indeed, reports of increased intestinal wall thickness (Haber et al., 1997) and goblet cell hyperplasia emphasize the possibility that inflammatory events also may contribute to CF GI disease.

Altered chemokine expression and release have been observed in CF cells. Although most of this work was performed in airway epithelial cell models, Schwiebert *et al.* (1999c) recently showed altered chemokine response in CFPAC-1 cells. They found that CFPAC-1 cells did not express the chemokine RANTES; however, CFTR complementation restored RANTES expression and release. Similar results were found in airway epithelial cells, which indicates a global inflammatory response defect mediated by CF epithelia.

It is of interest that several studies also have suggested a role for CFTR in apoptosis. Auricchio and coworkers (Maiuri *et al.*, 1997) observed that DNA fragmentation (an indicator of apoptosis) was higher in CF enterocytes and tracheocytes than in controls. In contrast, however, Gottlieb and Dosanjh (1996) showed that heterologous cells expressing ΔF508 CFTR failed to enter into apoptosis, whereas wild-type CFTR mediated cytosolic acidification and initiated programmed cell death.

It should be stressed that inflammatory and/or apoptotic defects should not be discounted as contributing to CF GI disease. Madara and colleagues used T84 cells as a model of intestinal epithelial immune effectors. In different studies, they showed that chemokines, bacterial-derived peptides, 5'-AMP, interleukins, and activated eosinophils or neutrophils stimulated Cl^- secretion across T84 monolayers (Madara *et al.*, 1992; Matthews *et al.*, 1993; Colgan *et al.*, 1994; Zund *et al.*, 1996). Finally, Pier and coworkers (1998) suggested that *Salmonella typhi* infects intestinal epithelia through CFTR. As described earlier, cholera toxin and heat-stable enterotoxin, derived from different pathogens, stimulate cyclic nucleotide-driven signaling in intestinal epithelial cells to activate CFTR. The fact that multiple pathogens target CFTR emphasizes that combined microbiologic and inflammatory mechanisms contribute to CF GI disease.

H. Synopsis: Possible Role of CFTR in Pancreatitis, MI, and Cholangitis

It is likely that various factors contribute to the global abnormality observed in the CF GI system. Diminished Cl^-, HCO_3^-, and fluid secretion in the pancreatic acini and duct, intestinal crypt, bile duct, and gallbladder are considered the most pronounced physiological lesions of the CF GI system. However, it remains unclear how these abnormalities underlie the global defects observed in CF intestinal disease. Although the tenacious nature of luminal secretions caused by dehydration is considered the primary source of the disease, the possible role of the other dysfunctions also should be considered. Other possible contributing factors include abnormal processing and overexpression of mucins and extracellular matrix proteins. An obvious link between colonic wall thickening and interstitial fibrosis with diminished fluid secretion likewise remains elusive. It is important for the CF research community to remain responsive to other possibilities, including epithelial remodeling, posttranslational protein modification, inflammation, and apoptosis.

V. THE GI SYSTEM AS A MODEL FOR CF THERAPEUTICS

A severe intestinal phenotype and easy accessibility have made the intestinal tract an important model for CF therapeutic studies. In this regard, the major focus of CF treatment is on correcting the intestinal disease in CF mouse models.

A. Alternative Conductances to Circumvent the CFTR Defect

The severe intestinal and mild airway disease found in CF mice are not characteristic of the disease in patients. A potential explanation for this difference is proposed by Boucher and colleagues (Clarke *et al.,* 1994). In CF mice, it was found that CFTR is the primary Cl^- channel in organs that demonstrate a severe phenotype. On the other hand, tissues with less severe disease also expressed a Ca^{2+}-regulated Cl^- conductance (CaCC). Thus, both the airways and the pancreas of mice expressed CaCC, but the intestines did not. Furthermore, human lung and pancreas express low levels of CaCC, whereas the intestine has high levels. Studies by Rozmahel *et al.* (1996) and Wilschanski *et al.* (1996) support this concept by showing that severity of intestinal disease in CF mice correlates indirectly with function of a Ca^{2+}-activated Cl^- channel. Taken together, these results suggest that manipulation of alternative Cl^- conductance may be of therapeutic value in CF.

Argent and coworkers (Winpenny *et al.,* 1995) studied the relationship between CFTR and the Ca^{2+}-activated Cl^- conductance elevated in CF mouse pancreatic duct cells. To elucidate whether elevation of this Cl^- current is attributable to increased activity of Ca^{2+}-activated chloride channels, the number of these channels was compared among pancreatic duct cells harvested from wild-type, heterozygous, and homozygous CF animals. The study showed no difference between steady-state Ca^{2+}-activated current densities among the three groups of animals, which suggests that absence of CFTR was not responsible for elevation of this current.

A potential route for manipulating CaCC activity is suggested by studies of Barrett and colleagues (Uribe *et al.,* 1996a) that show two inhibitory mechanisms of EGF in T84 cells. One means of inhibition was shared with carbachol, whereas the other was independent of intracellular Ca^{2+} levels. The first type of EGF-dependent CaCC inhibition was mediated through phosphatidylinositol

3-kinase (PI 3-kinase) (Uribe *et al.*, 1996b). It was found that carbachol inhibited CaCC by transactivating the EGF receptor and MAP kinase, limiting carbachol-stimulated Cl^- secretion in T84 cells (Keely *et al.*, 1998). It also was determined that a second CaCC inhibitory mechanism along this pathway relies on the inositol phosphate, D-Ins (3,4,5,6)P4), acting on calmodulin kinase II phosphory-lated channels (Ismailov *et al.*, 1996b). Other studies determined that addition of the flavanoid quercetin to the apical surface of T84 cells activated Cl^- secretion through CaCC (Sanchez de Medina *et al.*, 1997). In contrast, the addition of quercetin to the basolateral surface of these cells inhibited this secretion (Sanchez de Medina *et al.*, 1997). In other studies, Argent and colleagues (Lingard *et al.*, 1994) used the pancreatic ductal adenocarcinoma cell line BxPC-3 to demonstrate that acetylcholine (10 μM), ATP (100 μM), ADP (100 μM), bombesin (10 nM), or cholecystokinin (10 nM) stimulated Cl^- secretion triggered by increased intracellular Ca^{2+}. In studies of pancreatic duct cells, Verdon *et al.*, (1995) showed a volume-activated Cl^- current that required Ca^{2+} as a trigger for activity. Further investigation suggested that cell swelling led to Ca^{2+} influx that activated PKC phosphorylation of the channel or a regulator causing its opening (Verdon *et al.*, 1995).

Treatment of PI by activation of CaCC has been proposed by Gray and colleagues (Winpenny *et al.*, 1998). Using whole-cell patch-clamping of freshly isolated human pancreatic duct cells, CaCC activity was observed after addition of 1 μM ionomycin to the bath solution or 1 μM calcium to the pipette solution. The elicited conductances had similar characteristics, suggestive of their identity. Because this channel already had been implicated as compensatory for CFTR loss in murine pancreas, it was proposed that its activation in the human tissue could be of therapeutic benefit in CF (Winpenny *et al.*, 1998).

The Paneth cell defensins, cryptdins, which form anion conductive pores in phospholipid bilayers, may provide an alternative intestinal therapy for CF. Because cryptdins are homologs of molecules known to form anion pores, Madara and colleagues (Lencer *et al.*, 1997) tested their ability to confer channel activity when applied to apical membranes of intestinal epithelial cells in culture. These studies showed that cryptdins 2 and 3 (but not 1, 4, 5, and 6) stimulated reversible and dose-dependent Cl^- secretion from polarized T84 monolayers (Lencer *et al.*, 1997). The secretory response was not inhibited by pretreatment with 8-phenyltheophyline, or dependent on intracellular levels of cAMP or cGMP, which indicates lack of involvement of apical adenosine and guanylin receptors. Hence, cryptdins 2 and 3 can selectively permeate apical cell membranes of epithelial cells in culture to cause Cl^- secretion, thus potentially acting as intestinal paracrine secretagogues. Thus, CF intestinal disease therapy based on their activity may be feasible.

Another approach for activating intestinal Cl^- secretion by neutrophil-derived secretagogue has been proposed by Mrsny and colleagues (Madara *et al.*, 1992;

Madara *et al.,* 1993). Activated neutrophils in the intestinal lumen have been shown to release 5'-AMP, to which the epithelia respond with Cl⁻ secretion and secretory diarrhea (Madara *et al.,* 1993). Additional studies have suggested that 5'-AMP is converted to adenosine at the epithelial cell surface by ecto-5'-nucleotidase, which then activates the intestinal secretion through adenosine receptors on the apical membranes. These observations suggest that ATP or its metabolites may serve as intestinal paracrine mediators of Cl⁻ secretion.

From such studies of the physiological mechanisms underlying the activity and regulation of Cl⁻ channels and their manipulation through pharmacological approaches, information necessary to formulate potential CF therapeutics is rapidly being acquired.

B. Ursodeoxycholic Acid as a Therapy for CF Liver Disease

Liver disease is a serious problem for CF patients. Several studies have demonstrated improvements in CF liver function following administration of ursodeoxycholic acid (UDCA) (O'Brien *et al.,* 1996; Lepage *et al.,* 1997; Scher *et al.,* 1997; Lindblad *et al.,* 1998). To investigate the clinical and biochemical effect of UDCA, Colombo *et al.* (1990) administered 10 to 15 mg/kg body weight per day to 9 CF patients with liver disease. Effectiveness of the therapy was assayed by liver function tests, serum bile acid concentration, and biliary bile composition before and during therapy. Two months after initiation of therapy, UDCA in biliary bile acid was roughly 20% higher than the pretreatment level. This enrichment was at the expense of cholic and chenodeoxycholic acids, thus decreasing bile hydrophobicity. The concentration of UDCA in serum also was elevated, but its increase was more variable. UDCA also became the major fecal bile acid (12%–67% of total), indicating highly variable absorption. Importantly, it was found that liver function improved significantly in all treated patients after 2 to 6 months. Both fat absorption and growth rate appeared unaffected by UDCA; however, some improvement was observed in patients with more severe steatorrhea (Colombo *et al.,* 1990). In a follow-up study, the efficacy and safety of UDCA with respect to improvements in clinical and nutritional factors were established (Colombo *et al.,* 1996). Supplementation with taurine also was investigated for patients who received UDCA. In all, 55 CF patients with liver disease received 1 year of either UDCA (15 mg/kg body weight daily) plus taurine (30 mg/kg body weight daily), UDCA plus placebo, placebo plus taurine, or double placebo. Clinical deterioration occurred in patients who received placebo but not in those who had UDCA. UDCA also improved gamma-glutamyl transpeptidase and 5'-nucleotidase levels in the patients. The taurine treatment resulted in a significant increase of serum prealbumin and a reduction in fat malabsorption, but there was no effect on the liver biochemistry. Severe side effects were not observed in any treatment group. It was concluded that UDCA improved clinical and biochemical parameters of CF liver disease and that taurine supplementation benefited patients with severe PI and compromised nutritional status (Colombo *et al.*, 1996). Strandvik and colleagues (Lindblad *et al.*, 1998) showed efficacy of 2-year UCDA treatment on CF liver disease. Eight of their 10 patients showed normalization of liver function, all had decreased serum levels of immunoglobulin G, and 7 showed improved morphology and diminished inflammation and/or bile duct proliferation. However, one patient showed progression of the liver disease after treatment. The concentration of UDCA in urine varied, but increased during treatment from approximately 4% to 40% of total bile acids. In contrast, the concentration of secondary bile acids did not increase significantly. This study showed that UDCA reduced CF liver inflammation and improved its morphology.

Because CF patients with liver disease are more likely to have essential fatty acid deficiency, and UDCA alters lipid profiles, Lepage *et al.* (1997) investigated lipid profiles and status of fat-soluble vitamins after UDCA treatment. Nineteen CF children with liver dysfunction received 15 mg/kg UDCA per day, which was increased to 30 mg/kg in the absence of a 50% decrease in alanine transaminase or aspartate transaminase within 2 months. Initially, all patients showed essential fatty acid deficiency. However, during a period of 25 months, significant decreases in triglycerides, cholesterol, and total fatty acids were evident. In addition, UDCA improved essential fatty acid status and reduced the 20:3n-9/20:4n-6 fatty acid ratio. Although no change in vitamin E levels was evident, there was an increase in the molar ratio of unesterified retinol to retinol-binding protein, but no difference in retinol-binding protein concentrations. This study confirmed that UDCA alters lipoprotein metabolism and improves essential fatty acid and retinol status in CF liver disease. Geddes and colleagues (Thomas *et al.*, 1995) showed improved vitamin E absorption in CF liver disease after UDCA treatment.

Hegarty and coworkers (O'Brien *et al.*, 1996) investigated whether increased circulating hepatotoxic bile acids contributed to the cholestasis characteristic of CF liver disease. Serum bile acid profiles of CF patients with and without liver disease were compared, and the effect of UDCA (a nonhepatotoxic bile acid) on liver biochemistry and serum bile acids was evaluated. Fasting and postprandial serum bile acid levels were analyzed for 15 patients with liver disease and compared with levels for 18 patients without liver disease and 10 control subjects. The same clinical parameters also were measured in six CF patients with liver disease before and 6 months after 20 mg/kg/day UDCA treatment and compared with those of six control patients with CF-associated liver disease. It was found that the total fasting and postprandial serum bile acid levels were significantly higher in patients with liver disease compared with patients without liver disease and controls. Following 6 months of treatment, the serum was saturated with

UDCA, and improved liver biochemistry was observed in the treatment group, without any reduction of hepatotoxic serum bile acids. Therefore, although circulating hepatotoxic serum bile acids are elevated in CF liver disease, improvements in liver disease after UDCA treatment cannot be attributed solely to their decrease.

The basis for the clinical benefit of UDCA in CF liver disease is not clear. One explanation is provided by Fitz and coworkers (Shimokura *et al.*, 1995), who demonstrated that UDCA stimulates a bicarbonate-rich choleresis. Because ductular secretion also increases biliary bicarbonate levels, the effect of UDCA on intracellular Ca^{2+} levels and membrane Cl^- permeability was measured in human cholangiocarcinoma cells. Exposure to UDCA (2.5 mmol) increased intracellular Ca^{2+} from 180 ± 25 to 639 ± 84 nM. This change stimulated ^{125}I efflux (indicator of cellular anion permeability) approximately threefold over basal. Furthermore, patch clamping showed that exposure to extracellular (1.25 mM) or intracellular (100 μM) UDCA activated Cl^- currents. Intracellular administration of UDCA increased current density, which had been inhibited by intracellular Ca^{2+} chelation. These results suggest that UDCA increased intracellular Ca^{2+} to activate Cl^- channels in biliary cells. It was speculated that UDCA increased bile flow by stimulating ductular secretion. Thus, UDCA could be therapeutic to CF patients through circumvention of CFTR-dependent biliary secretion.

C. Dietary Manipulation

The earliest success in extending the lifespan of CF mice by circumventing their intestinal disease was achieved by weaning them onto a low-residue liquid nutrient diet (Eckman *et al.*, 1995; Clarke *et al.*, 1996a; Kent *et al.*, 1996). The underlying conjecture of this therapy was that a solid mouse-chow diet aggravated the intestinal obstructive lesions that resulted in the death of the animals. Indeed, after dietary manipulation, the majority of CF animals had a greatly extended lifespan, with only minor pathological changes in their intestinal tracts (Eckman *et al.*, 1995; Clarke *et al.*, 1996a; Kent *et al.*, 1996).

D. Transgenic Studies

Whitsett and colleagues (Zhou *et al.*, 1994) showed that heterologous expression of human *CFTR* in transgenic mouse models could correct the CF intestinal defect. By expressing *CFTR* along the intestinal villi epithelia, under the control of the rat intestinal fatty-acid binding protein (FABP) promoter, the investigation provided the first evidence of the therapeutic potential of *CFTR* cDNA. In these studies, high levels of *CFTR* were demonstrated throughout the intestinal epithelia by RNA *in situ* hybridization analysis. However, in contrast to endogenous *CFTR* expression, which is primarily in the intestinal crypts, FABP expression was predominantly in villi epithelial cells, with significantly less in the crypt epithelia. Electrophysiological analysis of the transgene expressing CF animals confirmed partial correction of the cAMP-stimulated Cl⁻ secretory response. Morphologically, the transgene-expressing CF mice showed correction of the CF-associated ileal goblet cell and crypt cell hyperplasia (Zhou *et al.*, 1994). Importantly, the transgene-expressing CF mice demonstrated a normal survival profile. A noteworthy observation of this study was that although the transgene expression did not mimic endogenous expression, the animals had correction of the intestinal disease. This finding has important repercussions for CF gene therapy, where therapeutic gene expression would be different from endogenous. Overall, these results suggest that endogenous expression levels and patterns might not be needed to correct the CF defect.

In contrast to this report, however, other investigations of heterologous *CFTR* expression in intestinal epithelia of CF mice have not been so positive. In particular, Buchwald and colleagues showed that expression of CFTR under control of the villin promoter conferred only minor amelioration of the intestinal disease in CF mice (Lu *et al.*, 1996, and personal communication). Only slight improvement in electrophysiological properties and survival of the CF animals expressing the transgene was observed. In view of the FABP-CFTR mouse studies, this result was somewhat unexpected because the villin promoter-driven expression of *CFTR* more closely resembled the endogenous gene. Although the basis for the discrepancy between the two studies is not known, a possible explanation is that *CFTR* expression is significantly higher in the FABP-CFTR animals (Lu *et al.*, 1996). Nevertheless, additional studies would be needed to confirm this explanation.

In related studies, Rozmahel *et al.* (1997) reported that low-level *CFTR* expression conferred only minor correction of intestinal disease in CF mice. In these studies, gene targeting was used to introduce human *CFTR* cDNA into exon 1 of the murine CFTR gene, conferring expressional replacement. Characterization of the homozygous mice confirmed expression of the introduced cDNA similar to endogenous CFTR, albeit at lower levels. Importantly, these animals showed minor improvement in intestinal electrophysiological properties but no significant improvement in survival compared with homozygous mutant animals (Rozmahel *et al.*, 1997). Although the basis of the incomplete rescue of these animals is not clear, insufficient expression of the transgene (roughly 25% of normal), combined with differences in regulation of human and mouse *CFTR*, was suggested (Rozmahel *et al.*, 1997).

An elaborate *CFTR* YAC transgenic experiment was performed by Huxley and coworkers (Manson *et al.*, 1997). To circumvent the possible loss of genomic sequences required for proper expression, a 320-kb YAC spanning the human *CFTR* locus was used to generate a transgenic mouse model. Characterization of *CFTR* null mice carrying the YAC transgene showed expression identical to that of the endogenous gene, and correction of all intestinal lesions including the bioelectric defects. It was found that expression of the *CFTR* transgene was slightly higher than that of the cognate gene (Manson *et al.*, 1997).

Taken together, the above transgenic studies suggest that the level of *CFTR* expression is crucial for normal intestinal function in mice. In addition, expression of *CFTR* in cells not normally expressing it appears to have no deleterious consequences. These findings are of particular importance for *CFTR*-based gene therapy studies.

Porteous and colleagues (Dorin *et al.*, 1996) investigated the level of *CFTR* required for normal intestinal function in mice by intercrossing mice expressing different CFTR levels. Using this approach, *CFTR* expression levels ranging from 0% to 100% could be achieved. The results showed that 5% of normal CFTR levels largely corrected the electrophysiological defect (50% of normal) and completely rescued the intestinal disease (100% survival) (Dorin *et al.*, 1996). An important lesson from this study was that low CFTR expression and only partial correction of its activity provided significant pathological correction in mice. However, correlation of these results with the therapeutic levels needed to correct the disease in patients has not been examined. Importantly, the level of *CFTR* required for normal intestinal function in this study was inconsistent with the transgenic experiments, which required significantly higher expression for complete correction. The reason for this discrepancy is not clear, but it might reflect differences between human and mouse *CFTR*.

E. Gene Therapy

The most commonly used tissue for *CFTR* gene therapy investigations has been the lung. CF mouse intestine also has been used as a model for gene therapy studies. One of the earliest studies of *CFTR* gene therapy in CF mice was reported by Alton *et al.* (1993). By directly instilling *CFTR*-containing liposomes into different sections of the intestinal tract, the authors demonstrated (through PD measurements) *CFTR* expression and slight correction of forskolin-activated current. Importantly, these studies provided encouraging data regarding the safety of DNA liposome complexes (Alton *et al.*, 1993).

Cuthbert and colleagues (Curtis *et al.*, 1998) examined electrogenic anion transport in mouse gallbladders after intratracheal transfer of *CFTR*-containing liposomes. Gallbladders from untreated CF animals did not show *CFTR* mRNA or electrogenic anion transport in response to forskolin. However, intratracheal

TABLE III

Therapies with Foundations in the GI System

Treating the Symptoms

  Pancrease™ (humans)

  Liquid diet and oral osmotic laxatives (CF knockout mice)

Circumventing the Lack of CFTR

  Identify other "alternative" chloride channels, define their normal expression and normal activity in CF epithelial cell models, and determine how to augment their activity

  Upregulate other ABC transporters to substitute for CFTR

Fixing Broken CFTR

  Ursodeoxycholic acid (liver disease in humans)

  Phosphodiesterase inhibitors, phosphatase inhibitors, flavonoids, genistein (augment mutant CFTR activity

Adding Back Wild-Type or Normal CFTR

  Viral-based gene therapy

  Cationic lipid-based gene therapy

See text (section V) for details on each putative therapy.

gene transfer restored the response to wild-type levels. Moreover, *CFTR* mRNA also was detected at this site following treatment (Curtis *et al.*, 1998). In contrast, no correction was observed when the liposome-*CFTR* complexes were administered orally. The efficiency of this unusual route for delivering *CFTR* to the gallbladder, and possibly other intestinal tissues, clearly warrants further investigation. A summary of putative CF therapies is provided in Table III.

VI. GENETICS OF CF INTESTINAL DISEASE

The severity of CF intestinal disease is highly variable. This variability is indicated by the fact that MI occurs in only 15% of patients (Welsh *et al.*, 1995). This clinical manifestation has little correlation with CFTR genotype, which implies that a contribution from secondary factors contributes its presentation. Using CF mice, Rozmahel *et al.* (1996) demonstrated that secondary genetic factors do influence CF intestinal disease.

Severe intestinal obstruction occurs in the majority of CF mice and leads to death soon after birth or weaning. However, Rozmahel *et al.* (1996) showed that some CF animals derived from the 129/Sv × CD1 strains survived for more than 6 weeks. The CF animals with prolonged survival were designated Class III CF mice, the ones that died by age 10 days were Class I, and those that died at weaning (transition to solid food) were considered Class II (Rozmahel *et al.*, 1996).

Histopathological investigation of the Class III CF mice showed increased intestinal mucosal thickness, suggestive of hypertrophic growth and an increased goblet cell population, but little mucin accumulation. The absence of intestinal lumen accumulations in the Class III CF mice suggested that prolonged survival resulted from decreased mucus viscosity and adequate clearance (Rozmahel *et al.*, 1996). Because all three classes of CF animals were maintained in same environments, the prolonged survival of Class III mice likely was related to an unlinked modifier gene. To confirm the contribution of a modifier gene, CF mice were crossed with different inbred strains to generate animals with various genetic backgrounds. It was found that intestinal disease of the CF animals varied significantly among the distinct genetic backgrounds. The mode of inheritance appeared to be either incomplete dominance or recessiveness at an unlinked locus (Rozmahel *et al.*, 1996). To identify the genetic locus or loci responsible for variable CF intestinal disease severity in the mice, a genome scan was performed with polymorphic DNA markers. One marker, *D7Mit56,* located at proximal mouse chromosome 7, showed significant deviation from expected random segregation (Rozmahel *et al.*, 1996). Further analysis with closely linked markers supported the linkage, indicating the presence of a modifier gene near *D7Mit152.* It was also observed that the severe intestinal disease of Class I CF mice was determined by this locus. These results were highly suggestive ($P < .001$) of a genetic modifier of CF intestinal disease on mouse chromosome 7. It was also determined that the chromosome 7 modifier did not contribute solely to disease variability, implicating additional factors (Rozmahel *et al.*, 1996).

Further genetic studies determined that a second CF intestinal disease modifier locus was located at mouse medial chromosome 5 ($P < .001$) (Rozmahel *et al.*, manuscript in preparation). Importantly, linkage of the identified modifiers to intestinal disease severity was consistent with a gender-biased model. Interestingly, the two implicated modifier loci contain the gender-dependent steroid sulfotransferase genes *Std* and *Ste,* which conjugate steroid hormones and bile acids in the intestinal tract and liver. Additional studies are underway to confirm and clarify the role of these genes as CF intestinal disease modifiers.

In follow-up studies, Tsui and colleagues (Zielenski *et al.*, 1999) evaluated the contribution of the human locus 19q13 (syntenic to mouse proximal chromosome 7) to CF severity in patients. Two CF clinical phenotypes were evaluated: intestinal disease was scored by presence or absence of MI at birth, and pulmonary phenotype was measured by a standard deviation score of age-adjusted forced expiratory volume in one second. Nine polymorphic microsatellite markers, spanning the region 19q13.2 to 19q13.4, were analyzed for 197 CF sibling pairs from 161 families. For each pair, the number of alleles identical by descent at each of the nine marker loci was derived using multipoint analysis. It was found that the sibling-pair difference for MI correlated significantly with the calculated estimate of shared alleles at all points across the locus. The strongest statistical support was observed for marker *D19S112* ($P = 6.3 \times 10^{-6}$). In contrast, no correlation was found between this locus and the pulmonary phenotype ($P > .28$) (Zielenski *et al.,* 1999). The detection of a modifier of intestinal but not pulmonary disease was consistent with the CF mouse studies.

These studies show that other genetic factors have an impact on CF intestinal disease. The identification and characterization of these genetic factors will provide a better understanding of CF heterogeneity and provide insight into prognosis and therapeutics for CF.

VII. QUESTIONS AND FUTURE DIRECTIONS

Ten years after identification of the genetic basis of CF, a comprehensive understanding of the underlying physiological defects remains elusive. Although important knowledge regarding the normal function and disease-associated dysfunction of *CFTR* has been gathered over the years, our understanding of the most fundamental cause of CF has progressed slowly. Of particular importance is a thorough comprehension of the numerous and diverse functions of *CFTR* in different organelles and cell types; this must be determined before consensus regarding the basic CF defect can be achieved. Historically, the most obvious physiological lesion constituting the CF phenotype has been the loss of a cAMP-activated, Cl^- channel from epithelial cells, and accompanying up-regulated Na^+ absorption. However, as more insight into CF pathophysiology is gained, it is becoming more obvious all facets of the disease cannot be explained by these defects alone. As indicated in this review of CFTR in the intestinal system, its primary role in different cells and at different times may vary. Basic roles in Cl^- secretion and in regulation of Na^+ absorption have been proven unequivocally for CFTR; however, other functions attributable to the protein remain contentious. Yet, CFTR's regulation of other conductances, transcription, inflammation, or apoptosis, and its contribution to the protein processing milieu or to transport of other molecules could be of greater importance to the disease than altered Cl^- and Na^+ conductance. Thus, correcting CF Cl^- or Na^+ transport defects alone may not provide global and efficacious therapy. Future research into CF should focus on gaining a more complete understanding of CFTR function and lack thereof in the disease. Much knowledge has been gained over the last 10 years; nevertheless, many questions remain. With the current pace of CF research and the diligence of scientists, it is likely that many of these questions will be answered in the near future.

References

Aguilar-Bryan, L., Clement, J. P. T. (1998). Toward understanding the assembly and structure of KATP channels. *Physiol. Rev.* **78,** 227–245.

Aller, S. G., Lombardo, I. D. (1999). Cloning, characterization, and functional expression of a CNP receptor regulating CFTR in the shark rectal gland. *Am. J. Physiol.* **276,** C442–C449.

Alpini, G., Ulrich, C. (1997). Molecular and functional heterogeneity of cholangiocytes from rat liver after bile duct ligation. *Am. J. Physiol.* **272**, G289–G297.

Alton, E. W., Middleton, P. G. (1993). Non-invasive liposome-mediated gene delivery can correct the ion transport defect in cystic fibrosis mutant mice. *Nature Genet.* **5**, 135–142.

Ameen, N. A., Ardito, T. (1995). A unique subset of rat and human intestinal villus cells express the cystic fibrosis transmembrane conductance regulator. *Gastroenterology* **108**, 1016–1023.

Anderson, M. P., Rich, D. P. (1991). Generation of cAMP-activated chloride currents by expression of CFTR. *Science* **251**, 679–682.

Armstrong, D. S., Grimwood, K. (1997). Lower airway inflammation in infants and young children with cystic fibrosis. *Am. J. Respir. Crit. Care. Med.* **156**, 1197–1204.

Bajnath, R. B., Dekker, K. (1995). Chloride secretion induced by phorbol dibutyrate and forskolin in the human colonic carcinoma cell line HT-29Cl.19A is regulated by different mechanisms. *Pflugers Arch.* **430**, 705–712.

Bajnath, R. B., Groot, J. A. (1993a). Synergistic activation of non-rectifying small-conductance chloride channels by forskolin and phorbol esters in cell-attached patches of the human colon carcinoma cell line HT-29cl.19A. *Pflugers Arch.* **425**, 100–108.

Bajnath, R. B., van den Berghe, N. (1993b). Activation of ion transport by combined effects of ionomycin, forskolin, and phorbol ester on cultured HT-29cl.19A human colonocytes. *Pflugers Arch.* **425**, 90–99.

Balough, K., McCubbin, M. (1995). The relationship between infection and inflammation in the early stages of lung disease from cystic fibrosis. *Pediatr. Pulmonol.* **20**, 63–70.

Barasch, J., and Al-Awqati, Q. (1993). Defective acidification of the biosynthetic pathway in cystic fibrosis. *J. Cell Sci. Suppl.* **17**, 229–233.

Barasch, J., Kiss, B. (1991). Defective acidification of intracellular organelles in cystic fibrosis. *Nature* **352**, 70–73.

Bardon, A. (1987). Cystic fibrosis. Carbohydrate metabolism in CF and in animal models for CF. *Acta. Paediatr. Scand. Suppl.* **332**, 1–30.

Bargon, J., Loitsch, S. (1998). Modulation of cystic fibrosis transmembrane conductance regulator gene—Expression by elevation of intracellular cyclic AMP. *Eur. J. Med. Res.* **3**, 256–262.

Bargon, J., Trapnell, B. C. (1992a). Down-regulation of cystic fibrosis transmembrane conductance regulator gene expression by agents that modulate intracellular divalent cations. *Mol. Cell Biol.* **12**, 1872–1888.

Bargon, J., Trapnell, B. C. (1992b). Expression of the cystic fibrosis transmembrane conductance regulator gene can be regulated by protein kinase C. *J. Biol. Chem.* **267**, 16056–16060.

Barrett, K. E. (1995). Effect of the diglyceride lipase inhibitor, RG80267, on epithelial chloride secretion induced by various agents. *Cell Signal* **7**, 225–233.

Barrett, K. E., Cohn, J. A. (1990). Immune-related intestinal chloride secretion. II. Effect of adenosine on T84 cell line. *Am. J. Physiol.* **258**, C902–C912.

Bartosz, G. (1998). ABC transporters in human cells. *Postepy Biochem* **44**, 136–151.

Bear, C. E., Duguay, F. (1991). Cl-channel activity in *Xenopus* oocytes expressing the cystic fibrosis gene. *J. Biol. Chem.* **266**, 19142–19145.

Bear, C. E., Li, C. H. (1992). Purification and functional reconstitution of the cystic fibrosis transmembrane conductance regulator (*CFTR*). *Cell* **68**, 809–818.

Benos, D. J., Awayda, M. S. (1996). Diversity and regulation of amiloride-sensitive Na^+ channels. *Kidney Int.* **49**, 1632–1637.

Benya, R. V., Matkowskyj, K. A. (1998). Galanin causes Cl^- secretion in the human colon. Potential significance of inflammation-associated NF-kappa B activation on galanin-1 receptor expression and function. *Ann. NY Acad. Sci.* **863**, 64–77.

Bhaskar, K. R., Turner, B. S. Dysregulation of proteoglycan production by intrahepatic biliary epithelial cells bearing defective (delta-f508) cystic fibrosis transmembrane conductance regulator. *Hepatology* **27**, 7–14.

Bienvenu, T., Beldjord, C. (1996). Analysis of alternative splicing patterns in the cystic fibrosis transmembrane conductance regulator gene using mRNA derived from lymphoblastoid cells of cystic fibrosis patients. *Eur. J. Hum. Genet.* **4,** 127–134.

Bienvenu, T., Lacronique, V. (1994). A potential CF mutation at position −741 upstream from the *CFTR* gene induces altered interaction with transactivating factors. *Am. J. Hum. Genet.* **55,** 1236 (Abstr., Suppl.).

Bienvenu, T., Lacronique, V. (1995). Three novel sequence variations in the 5′ upstream region of the cystic fibrosis transmembrane conductance regulator (*CFTR*) gene: Two polymorphisms and one putative molecular defect. *Hum Genet* **95,** 698–702.

Bijman, J., Dalemans, W. (1993). Low-conductance chloride channels in IEC-6 and CF nasal cells expressing *CFTR*. *Am. J. Physiol.* **264,** L229–L235.

Bijman, J., Veeze, H. (1991). Chloride transport in the cystic fibrosis enterocyte. *Adv. Exp. Med. Biol.* **290,** 287–294.

Biwersi, J., Emans, N. (1996). Cystic fibrosis transmembrane conductance regulator activation stimulates endosome fusion *in vivo*. *Proc. Natl. Acad. Sci. USA* **93,** 12484–12489.

Biwersi, J., Verkman, A. S. (1994). Functional CFTR in endosomal compartment of *CFTR*-expressing fibroblasts and T84 cells. *Am. J. Physiol.* **266,** C149–C156.

Bohme, M., Diener, M. (1991). Calcium- and cyclic-AMP-mediated secretory responses in isolated colonic crypts. *Pflugers Arch.* **419,** 144–151.

Bonfield, T. L., Konstan, M. W. (1995a). Normal bronchial epithelial cells constitutively produce the anti- inflammatory cytokine interleukin-10, which is downregulated in cystic fibrosis. *Am. J. Respir. Cell Mol. Biol.* **13,** 257–261.

Bonfield, T. L., Panuska, J. R. (1995b). Inflammatory cytokines in cystic fibrosis lungs. *Am. J. Respir. Crit. Care. Med.* **152,** 2111–2118.

Bradbury, N. A. (1999b). Intracellular *CFTR:* Localization and function. *Physiol. Rev.* **79,** S175–S191 (Suppl. 1).

Bradbury, N. A., and Bridges, R. J. (1994a). Role of membrane trafficking in plasma membrane solute transport. *Am. J. Physiol.* **267,** C1–C24.

Bradbury, N. A., and Bridges, R. J. (1992a). Regulated membrane vesicle trafficking: A defect in cystic fibrosis corrected by gene transfer. *Biochem. Soc. Trans.* **20,** 124S.

Bradbury, N. A., Clark, J. A. (1999a). Characterization of the internalization pathways for the cystic fibrosis transmembrane conductance regulator. *Am. J. Physiol.* **276,** L659–L668.

Bradbury, N. A., Cohn, J. A. (1994b). Biochemical and biophysical identification of cystic fibrosis transmembrane conductance regulator chloride channels as components of endocytic clathrin-coated vesicles. *J. Biol. Chem.* **269,** 8296–8302.

Bradbury, N. A., Jilling, T. (1992b). Regulation of plasma membrane recycling by CFTR. *Science* **256,** 530–532.

Breuer, W., Slotki, I. N. (1993). Induction of multidrug resistance downregulates the expression of *CFTR* in colon epithelial cells. *Am. J. Physiol.* **265,** C1711–C1715.

Buchwald, M., Sood, R. (1991). Regulation of expression of *CFTR* in human intestinal epithelial cells. *Adv. Exp. Med. Biol.* **290,** 241–250.

Buschman, E., Arceci, R. J. (1992). *mdr2* encodes P-glycoprotein expressed in the bile canalicular membrane as determined by isoform-specific antibodies. *J. Biol. Chem.* **267,** 18093–18099.

Cantiello, H. F., Prat, A. G. (1994). External ATP and its analogs activate the cystic fibrosis transmembrane conductance regulator by a cyclic AMP-independent mechanism. *J. Biol. Chem.* **269,** 11224–11232.

Ceder, O. (1983). Cystic fibrosis. *In vitro* and *in vivo* studies on the biochemical background to the pathogenesis. *Acta. Paediatr. Scand. Suppl* **309,** 1–47.

Chandan, R., O'Grady, S. M. (1991). Modulation of Na^+, Cl^- and HCO_3^- transport by carbachol in pig distal jejunum. *Eur. J. Pharmacol.* **193,** 257–264.

Chao, A. C., de Sauvage, F. J. (1994). Activation of intestinal CFTR Cl^- channel by heat-stable enterotoxin and guanylin via cAMP-dependent protein kinase. *Embo. J.* **13,** 1065–1072.

Chari, R. S., Schutz, S. M. (1996). Adenosine nucleotides in bile. *Am. J. Physiol.* **270,** G246–G252.

Cheng, S. H., Gregory, R. J. (1990). Defective intracellular transport and processing of CFTR is the molecular basis of most cystic fibrosis. *Cell* **63,** 827–834.

Chu, C. S., Trapnell, B. C. (1993). Genetic basis of variable exon 9 skipping in cystic fibrosis transmembrane conductance regulator mRNA. *Nature Genet.* **3,** 151–156.

Cid, L. P., Montrose-Rafizadeh, C. (1995). Cloning of a putative human voltage-gated chloride channel (CIC-2) cDNA widely expressed in human tissues. *Hum. Mol. Genet.* **4,** 407–413.

Clarke, L. L., Gawenis, L. R. (1996a). Increased survival of *CFTR* knockout mice with an oral osmotic laxative. *Lab. Anim. Sci.* **46,** 612–618.

Clarke, L. L., Grubb, B. R. (1994). Relationship of a non-cystic fibrosis transmembrane conductance regulator-mediated chloride conductance to organ-level disease in Cftr(-/-) mice. *Proc. Natl. Acad. Sci. USA* **91,** 479–483.

Clarke, L. L., and Harline, M. C. (1996b). *CFTR* is required for cAMP inhibition of intestinal Na$^+$ absorption in a cystic fibrosis mouse model. *Am. J. Physiol.* **270,** G259–G267.

Clarke, L. L., and Harline, M. C. (1998). Dual role of *CFTR* in cAMP-stimulated HCO$_3^-$ secretion across murine duodenum. *Am. J. Physiol.* **274,** G718–G726.

Cliff, W. H., Schoumacher, R. A. (1992). cAMP-activated Cl channels in CFTR-transfected cystic fibrosis pancreatic epithelial cells. *Am. J. Physiol.* **262,** C1154–C1160.

Cohn, J. A., Friedman, K. J. (1998). Relation between mutations of the cystic fibrosis gene and idiopathic pancreatitis. *N. Engl. J. Med.* **339,** 653–658.

Cohn, J. A., Strong, T. V. (1993). Localization of the cystic fibrosis transmembrane conductance regulator in human bile duct epithelial cells. *Gastroenterology* **105,** 1857–1864.

Cole, S. P., Bhardwaj, G. (1992). Overexpression of a transporter gene in a multidrug-resistant human lung cancer cell line. *Science* **258,** 1650–1654.

Colgan, S. P., Resnick, M. B. (1994). IL-4 directly modulates function of a model human intestinal epithelium. *J. Immunol.* **153,** 2122–2129.

Colombo, C., Apostolo, M. G. (1992). Liver disease in cystic fibrosis. *Neth. J. Med.* **41,** 119–122.

Colombo, C., Battezzati, P. M. (1996). Ursodeoxycholic acid for liver disease associated with cystic fibrosis: A double-blind multicenter trial. The Italian group for the study of ursodeoxycholic acid in cystic fibrosis. *Hepatology* **23,** 1484–1490.

Colombo, C., Battezzati, P. M. (1998). Liver and biliary problems in cystic fibrosis. *Semin. Liver Dis.* **18,** 227–235.

Colombo, C., Setchell, K. D. (1990). Effects of ursodeoxycholic acid therapy for liver disease associated with cystic fibrosis. *J. Pediatr.* **117,** 482–489.

Cooke, H. J., Wang, Y. Z. (1993). Coordination of Cl$^-$ secretion and contraction by a histamine H2$^-$ receptor agonist in guinea pig distal colon. *Am. J. Physiol.* **265,** G973–G978.

Crawford, J., Labrinidis, A. (1995). A splicing mutation (1898 + 1G → T) in the *CFTR* gene causing cystic fibrosis. *Hum. Mutat.* **5,** 101–102.

Croop, J. M. (1998). Evolutionary relationships among ABC transporters. *Methods Enzymol.* **292,** 101–116.

Cunningham, S. A., Awayda, M. S. (1995). Cloning of an epithelial chloride channel from bovine trachea. *J. Biol. Chem.* **270,** 31016–31026.

Cuppoletti, J., Baker, A. M. (1993). Cl$^-$ channels of the gastric parietal cell that are active at low pH. *Am. J. Physiol.* **264,** C1609–C1618.

Curtis, C. M., Martin, L. C. (1998). Restoration by intratracheal gene transfer of bicarbonate secretion in cystic fibrosis mouse gallbladder. *Am. J. Physiol.* **274,** G1053–G1060.

Cuthbert, A. W., Halstead, J. (1995). The genetic advantage hypothesis in cystic fibrosis heterozygotes: A murine study. *J. Physiol. (Lond.)* **482,** 449–454.

Cuthbert, A. W., Hickman, M. E. (1994a). Chloride secretion in response to guanylin in colonic epithelial from normal and transgenic cystic fibrosis mice. *Br. J. Pharmacol.* **112,** 31–36.

Cuthbert, A. W., MacVinish, L. J. (1994b). Ion-transporting activity in the murine colonic epithelium of normal animals and animals with cystic fibrosis. *Pflugers Arch.* **428**, 508–515.

Cutting, G. R., Kasch, L. M. (1990). A cluster of cystic fibrosis mutations in the first nucleotide-binding fold of the cystic fibrosis conductance regulator protein. *Nature* **346**, 366–369.

Dalemans, W., Barbry, P. (1991). Altered chloride ion channel kinetics associated with the delta F508 cystic fibrosis mutation. *Nature* **354**, 526–528.

De Lisle, R. C., Petitt, M. (1997). MUCLIN expression in the cystic fibrosis transmembrane conductance regulator knockout mouse. *Gastroenterology* **113**, 521–532.

Dean, M., White, M. B. (1990). Multiple mutations in highly conserved residues are found in mildly affected cystic fibrosis patients. *Cell* **61**, 863–870.

Denning, G. M., Anderson, M. P. (1992). Processing of mutant cystic fibrosis transmembrane conductance regulator is temperature-sensitive. *Nature* **358**, 761–764.

Devor, D. C., and Duffey, M. E. (1992). Carbachol induces K^+, Cl^-, and nonselective cation conductances in T84 cells: A perforated patch-clamp study. *Am. J. Physiol.* **263**, C780–C787.

Devor, D. C., Singh, A. K. (1996). Modulation of Cl^- secretion by benzimidazolones. II. Coordinate regulation of apical GCl and basolateral GK. *Am. J. Physiol.* **271**, L785–L795.

Dharmsathaphorn, K., and Pandol, S. J. (1986). Mechanism of chloride secretion induced by carbachol in a colonic epithelial cell line. *J. Clin. Invest.* **77**, 348–354.

Diener, M., Egleme, C. (1991). Phospholipase C-induced anion secretion and its interaction with carbachol in the rat colonic mucosa. *Eur. J. Pharmacol.* **200**, 267–276.

Diener, M., and Rummel, W. (1990). Distension-induced secretion in the rat colon: Mediation by prostaglandins and submucosal neurons. *Eur. J. Pharmacol.* **178**, 47–57.

DiMango, E., Ratner, A. J. (1998). Activation of NF-kappaB by adherent *Pseudomonas aeruginosa* in normal and cystic fibrosis respiratory epithelial cells. *J. Clin. Invest.* **101**, 2598–2605.

Dinudom, A., Komwatana, P. (1995). A forskolin-activated Cl^- current in mouse mandibular duct cells. *Am. J. Physiol.* **268**, G806–G812.

Dorin, J. R., Farley, R. (1996). A demonstration using mouse models that successful gene therapy for cystic fibrosis requires only partial gene correction. *Gene Ther.* **3**, 797–801.

Dork, T., Wulbrand, U. (1991). Cystic fibrosis with three mutations in the cystic fibrosis transmembrane conductance regulator gene. *Hum. Genet.* **87**, 441–446.

Dosanjh, A. K., Elashoff, D. (1998). The bronchoalveolar lavage fluid of cystic fibrosis lung transplant recipients demonstrates increased interleukin-8 and elastase and decreased IL-10. *J. Interferon Cytokine Res.* **18**, 851–854.

Dray-Charier, N., Paul, A. (1995). Expression of cystic fibrosis transmembrane conductance regulator in human gallbladder epithelial cells. *Lab. Invest.* **73**, 828–836.

Drumm, M. L., Wilkinson, D. J. (1991). Chloride conductance expressed by delta F508 and other mutant CFTRs in *Xenopus* oocytes. *Science* **254**, 1797–1799.

Dunn, K. W., Park, J. (1994). Regulation of endocytic trafficking and acidification are independent of the cystic fibrosis transmembrane regulator. *J. Biol. Chem.* **269**, 5336–5345.

Durie, P. R. (1992). Pathophysiology of the pancreas in cystic fibrosis. *Neth. J. Med.* **41**, 97–100.

Eckman, E. A., Cotton, C. U. (1995). Dietary changes improve survival of *CFTR* S489X homozygous mutant mouse. *Am. J. Physiol.* **269**, L625–L630.

Egan, M., Flotte, T. (1992). Defective regulation of outwardly rectifying Cl^- channels by protein kinase A corrected by insertion of *CFTR*. *Nature* **358**, 581–584.

Eggermont, E. (1996). Gastrointestinal manifestations in cystic fibrosis. *Eur. J. Gastroenterol. Hepatol.* **8**, 731–738.

Elgavish, A. (1991). High intracellular pH in CFPAC: A pancreas cell line from a patient with cystic fibrosis is lowered by retrovirus-mediated *CFTR* gene transfer. *Biochem. Biophys. Res. Commun.* **180**, 342–348.

Elgavish, A. and Meezan, E. (1992). Altered sulfate transport via anion exchange in CFPAC is corrected by retrovirus-mediated *CFTR* gene transfer. *Am. J. Physiol.* **263**, C176–C186.

Erlinger, S. (1996a). Mechanisms of hepatic transport and bile secretion. *Acta Gastroenterol. Belg.* **59,** 159–162.

Erlinger, S. (1996b). Review article: New insights into the mechanisms of hepatic transport and bile secretion. *J. Gastroenterol. Hepatol.* **11,** 575–579.

Fath, M. J., and Kolter, R. (1993). ABC transporters: Bacterial exporters. *Microbiol. Rev.* **57,** 995–1017.

Feranchak, A. P., Roman, R. M. (1998). Phosphatidylinositol 3-kinase contributes to cell volume regulation through effects on ATP release. *J. Biol. Chem.* **273,** 14906–14911.

Fitz, J. G., Basavappa, S. (1993). Regulation of membrane chloride currents in rat bile duct epithelial cells. *J. Clin. Invest.* **91,** 319–328.

Fitz, J. G., and Sostman, A. H. (1994). Nucleotide receptors activate cation, potassium, and chloride currents in a liver cell line. *Am. J. Physiol.* **266,** G544–G553.

Forrest, J. N., Jr. (1996). Cellular and molecular biology of chloride secretion in the shark rectal gland: Regulation by adenosine receptors. *Kidney Int.* **49,** 1557–1562.

Frizzell, R. A., Rechkemmer, G. (1986). Altered regulation of airway epithelial cell chloride channels in cystic fibrosis. *Science* **233,** 558–560.

Gabriel, S. E., Brigman, K. N. (1994). Cystic fibrosis heterozygote resistance to cholera toxin in the cystic fibrosis mouse model. *Science* **266,** 107–109.

Gabriel, S. E., Clarke, L. L. (1993). CFTR and outward rectifying chloride channels are distinct proteins with a regulatory relationship. *Nature* **363,** 263–268.

Gaillard, D., Ruocco, S. (1994). Immunohistochemical localization of cystic fibrosis transmembrane conductance regulator in human fetal airway and digestive mucosa. *Pediatr. Res.* **36,** 137–143.

Gasparini, P., Nunes, V. (1991). The search for south European cystic fibrosis mutations: Identification of two new mutations, four variants, and intronic sequences. *Genomics* **10,** 193–200.

Goldstein, J. L., Sahi, J. (1994). *Escherichia coli* heat-stable enterotoxin-mediated colonic Cl^- secretion is absent in cystic fibrosis. *Gastroenterology* **107,** 950–956.

Goldstein, J. L., Shapiro, A. B. (1991). *In vivo* evidence of altered chloride but not potassium secretion in cystic fibrosis rectal mucosa. *Gastroenterology* **101,** 1012–1019.

Gottlieb, R. A., and Dosanjh, A. (1996). Mutant cystic fibrosis transmembrane conductance regulator inhibits acidification and apoptosis in C127 cells: Possible relevance to cystic fibrosis. *Proc. Natl. Acad. Sci. USA* **93,** 3587–3591.

Gowen, C. J., Gowen, M. A. (1991). Colonic transepithelial potential difference in infants with cystic fibrosis. *J. Pediatr.* **118,** 412–415.

Gray, M. A., Greenwell, J. R. (1988). Secretin-regulated chloride channel on the apical plasma membrane of pancreatic duct cells. *J. Membr. Biol.* **105,** 131–142.

Gray, M. A., Harris, A. (1989). Two types of chloride channel on duct cells cultured from human fetal pancreas. *Am. J. Physiol.* **257,** C240–C251.

Gray, M. A., Plant, S. (1993). cAMP-regulated whole cell chloride currents in pancreatic duct cells. *Am. J. Physiol.* **264,** C591–C602.

Gray, M. A., Pollard, C. E. (1990). Anion selectivity and block of the small-conductance chloride channel on pancreatic duct cells. *Am. J. Physiol.* **259,** C752–C761.

Gray, M. A., Winpenny, J. P. (1994). CFTR and calcium-activated chloride currents in pancreatic duct cells of a transgenic CF mouse. *Am. J. Physiol.* **266,** C213–C221.

Grubb, B. R., and Boucher, R. C. (1997a). Enhanced colonic Na^+ absorption in cystic fibrosis mice versus normal mice. *Am. J. Physiol.* **272,** G393–G400.

Grubb, B. R., and Gabriel, S. E. (1997b). Intestinal physiology and pathology in gene-targeted mouse models of cystic fibrosis. *Am. J. Physiol.* **273,** G258–G266.

Gruber, A. D., Elble, R. C. (1998). Genomic cloning, molecular characterization, and functional analysis of human CLCA1, the first human member of the family of Ca^{2+}-activated Cl^- channel proteins. *Genomics* **54,** 200–214.

Grygorczyk, R., and Hanrahan, J. W. (1997). CFTR-independent ATP release from epithelial cells triggered by mechanical stimuli. *Am. J. Physiol.* **272**, C1058–C1066.

Grygorczyk, R., Tabcharani, J. A. (1996). CFTR channels expressed in CHO cells do not have detectable ATP conductance. *J. Membr. Biol.* **151**, 139–148.

Guba, M., Kuhn, M. (1996). Guanylin strongly stimulates rat duodenal HCO_3^- secretion: Proposed mechanism and comparison with other secretagogues. *Gastroenterology* **111**, 1558–1568.

Haber, H. P., Benda, N. (1997). Colonic wall thickness measured by ultrasound: Striking differences in patients with cystic fibrosis versus healthy controls. *Gut* **40**, 406–411.

Hamosh, A., Rosenstein, B. J. (1992). *CFTR* nonsense mutations G542X and W1282X associated with severe reduction of *CFTR* mRNA in nasal epithelial cells. *Hum. Mol. Genet.* **1**, 542–544.

Hayden, U. L., and Carey, H. V. (1996). Cellular localization of cystic fibrosis transmembrane regulator protein in piglet and mouse intestine. *Cell Tissue Res.* **283**, 209–213.

Higgins, C. F. (1992). ABC transporters: From microorganisms to man. *Ann. Rev. Cell Biol.* **8**, 67–113.

Hill, W. G., Harper, G. S. (1997). Organ-specific over-sulfation of glycosaminoglycans and altered extracellular matrix in a mouse model of cystic fibrosis. *Biochem. Mol. Med.* **62**, 113–122.

Hogan, D. L., Crombie, D. L. (1997a). Acid-stimulated duodenal bicarbonate secretion involves a CFTR-mediated transport pathway in mice. *Gastroenterology* **113**, 533–541.

Hogan, D. L., Crombie, D. L. (1997b). CFTR mediates cAMP- and Ca^{2+}-activated duodenal epithelial HCO_3^- secretion. *Am. J. Physiol.* **272**, G872–G878.

Hollande, E., Fanjul, M. (1998). Targeting of CFTR protein is linked to the polarization of human pancreatic duct cells in culture. *Eur. J. Cell. Biol.* **76**, 220–227.

Homaidan, F. R., Tripodi, J. (1997). Regulation of ion transport by histamine in mouse cecum. *Eur. J. Pharmacol.* **331**, 199–204.

Howard, M., Jilling, T. (1996). cAMP-regulated trafficking of epitope-tagged CFTR. *Kidney Int.* **49**, 1642–1648.

Hull, J., Shackleton, S. (1993). Abnormal mRNA splicing resulting from three different mutations in the *CFTR* gene. *Hum. Mol. Genet.* **2**, 689–692.

Hyde, K., Reid, C. J. (1997). The cystic fibrosis transmembrane conductance regulator as a marker of human pancreatic duct development. *Gastroenterology* **113**, 914–919.

Hyun, C. S., and Binder, H. J. (1993). Mechanism of leukotriene D4 stimulation of Cl^- secretion in rat distal colon *in vitro. Am. J. Physiol.* **265**, G467–G473.

Ishida-Takahashi, A., Otani, H. (1998). Cystic fibrosis transmembrane conductance regulator mediates sulphonylurea block of the inwardly rectifying K^+ channel Kir6.1. *J. Physiol. (Lond.)* **508**, 23–30.

Ismailov, I. I., Awayda, M. S. (1996a). Regulation of epithelial sodium channels by the cystic fibrosis transmembrane conductance regulator. *J. Biol. Chem.* **271**, 4725–4732.

Ismailov, I. I., Fuller, C. M. (1996b). A biologic function for an "orphan" messenger: D-myo-inositol 3,4,5,6- tetrakisphosphate selectively blocks epithelial calcium-activated chloride channels. *Proc. Natl. Acad. Sci. USA* **93**, 10505–10509.

Jiang, X., Hill, W. G. (1997). Glycosylation differences between a cystic fibrosis and rescued airway cell line are not CFTR dependent. *Am. J. Physiol.* **273**, L913–L920.

Joo, N. S., London, R. M. (1998). Regulation of intestinal Cl^- and HCO_3^- secretion by uroguanylin. *Am. J. Physiol.* **274**, G633–G644.

Jovov, B., Ismailov, II, (1995a). Cystic fibrosis transmembrane conductance regulator is required for protein kinase A activation of an outwardly rectified anion channel purified from bovine tracheal epithelia. *J. Biol. Chem.* **270**, 1521–1528.

Jovov, B., Ismailov, II, (1995b). Interaction between cystic fibrosis transmembrane conductance regulator and outwardly rectified chloride channels. *J. Biol. Chem.* **270**, 29194–29200.

Julien, M., Verrier, B. (1999). Cystic fibrosis transmembrane conductance regulator (CFTR) confers glibenclamide sensitivity to outwardly rectifying chloride channel (ORCC) in Hi-5 insect cells. *J. Membr. Biol.* **168**, 229–239.

Kachur, J. F., Won-Kim, S. (1995). Eicosanoids and histamine mediate C5a-induced electrolyte secretion in guinea pig ileal mucosa. *Inflammation* **19**, 717–725.

Kandil, H. M., Berschneider, H. M. (1994). Tumour necrosis factor alpha changes porcine intestinal ion transport through a paracrine mechanism involving prostaglandins. *Gut* **35**, 934–940.

Kang-Park, S., Dray-Charier, N. (1998). Role for PKC alpha and PKC epsilon in down-regulation of *CFTR* mRNA in a human epithelial liver cell line. *J. Hepatol.* **28**, 250–262.

Kansen, M., Bajnath, R. B. (1993). Regulation of chloride channels in the human colon carcinoma cell line HT29.cl19A. *Pflugers Arch.* **422**, 539–545.

Kartner, N., Augustinas, O. (1992). Mislocalization of delta F508 *CFTR* in cystic fibrosis sweat gland. *Nature Genet.* **1**, 321–327.

Kartner, N., Hanrahan, J. W. (1991). Expression of the cystic fibrosis gene in non-epithelial invertebrate cells produces a regulated anion conductance. *Cell* **64**, 681–691.

Kearns, G. L. (1993). Hepatic drug metabolism in cystic fibrosis: Recent developments and future directions. *Ann. Pharmacother.* **27**, 74–79.

Kearns, G. L., Crom, W. R. (1996). Hepatic drug clearance in patients with mild cystic fibrosis. *Clin. Pharmacol. Ther.* **59**, 529–540.

Kearns, G. L., Mallory, G. B., Jr., (1990). Enhanced hepatic drug clearance in patients with cystic fibrosis. *J. Pediatr.* **117**, 972–979.

Keely, S. J., Uribe, J. M. (1998). Carbachol stimulates transactivation of epidermal growth factor receptor and mitogen-activated protein kinase in T84 cells. Implications for carbachol-stimulated chloride secretion. *J. Biol. Chem.* **273**, 27111–27117.

Kelley, G. G., Aassar, O. S. (1991). Endogenous adenosine is an autacoid feedback inhibitor of chloride transport in the shark rectal gland. *J. Clin. Invest.* **88**, 1933–1939.

Kelley, G. G., Poeschla, E. M. (1990). A1 adenosine receptors inhibit chloride transport in the shark rectal gland. Dissociation of inhibition and cyclic AMP. *J. Clin. Invest.* **85**, 1629–1636.

Kelley, K. A., Stamm, S. (1992). Expression and chromosome localization of the murine cystic fibrosis transmembrane conductance regulator. *Genomics* **13**, 381–388.

Kent, G., Oliver, M. (1996). Phenotypic abnormalities in long-term surviving cystic fibrosis mice. *Pediatr. Res.* **40**, 233–241.

Kerem, B., Rommens, J. M. (1989). Identification of the cystic fibrosis gene: Genetic analysis. *Science* **245**, 1073–1080.

Kerem, B. S., Zielenski, J. (1990a). Identification of mutations in regions corresponding to the two putative nucleotide (ATP)-binding folds of the cystic fibrosis gene. *Proc. Natl. Acad. Sci. USA* **87**, 8447–8451.

Kerem, E., Corey, M. (1990b). The relation between genotype and phenotype in cystic fibrosis— Analysis of the most common mutation (delta F508). *N. Engl. J. Med.* **323**, 1517–1522.

Khan, T. Z., Wagener, J. S. (1995). Early pulmonary inflammation in infants with cystic fibrosis. *Am. J. Respir. Crit. Care. Med.* **151**, 1075–1082.

Kiesewetter, S., Macek, M. J. (1993). A mutation in *CFTR* produces different phenotypes depending on chromosomal background. *Nature Genet.* **5**, 274–278.

Kirk, K. L. (1989). Defective regulation of epithelial Cl$^-$ permeability and protein secretion in cystic fibrosis: The putative basic defect. *Am. J. Kidney Dis.* **14**, 333–338.

Knowles, M. R., Stutts, M. J. (1983). Abnormal ion permeation through cystic fibrosis respiratory epithelium. *Science* **221**, 1067–1070.

Koller, D. Y., Urbanek, R. (1995). Increased degranulation of eosinophil and neutrophil granulocytes in cystic fibrosis. *Am. J. Respir. Crit. Care. Med.* **152**, 629–633.

Komwatana, P., Dinudom, A. (1994). Characterization of the Cl$^-$ conductance in the granular duct cells of mouse mandibular glands. *Pflugers Arch.* **428**, 641–647.

Konstan, M. W., Hilliard, K. A. (1994). Bronchoalveolar lavage findings in cystic fibrosis patients with stable, clinically mild lung disease suggest ongoing infection and inflammation. *Am. J. Respir. Crit. Care. Med.* **150**, 448–454.

Kopelman, H., Ferretti, E. (1995). Rabbit pancreatic acini express *CFTR* as a cAMP-activated chloride efflux pathway. *Am. J. Physiol.* **269**, C626–C631.

Kopelman, H., Gauthier, C. (1993). Antisense oligodeoxynucleotide to the cystic fibrosis transmembrane conductance regulator inhibits cyclic AMP-activated but not calcium-activated cell volume reduction in a human pancreatic duct cell line. *J. Clin. Invest.* **91**, 1253–1257.

Kopito, R. R. (1999). Biosynthesis and degradation of *CFTR. Physiol. Rev.* **79**, S167–S173 (Suppl 1).

Kunzelmann, K., Kiser, G. L. (1997). Inhibition of epithelial Na$^+$ currents by intracellular domains of the cystic fibrosis transmembrane conductance regulator. *FEBS Lett.* **400**, 341–344.

Kuver, R., Ramesh, N. (1994). Constitutive mucin secretion linked to *CFTR* expression. *Biochem. Biophys. Res. Commun.* **203**, 1457–1462.

Lehrich, R. W., Aller, S. G. (1998). Vasoactive intestinal peptide, forskolin, and genistein increase apical *CFTR* trafficking in the rectal gland of the spiny dogfish, *Squalus acanthias.* Acute regulation of *CFTR* trafficking in an intact epithelium. *J. Clin. Invest.* **101**, 737–745.

Lencer, W. I., Cheung, G. (1997). Induction of epithelial chloride secretion by channel-forming cryptdins 2 and 3. *Proc. Natl. Acad. Sci. USA* **94**, 8585–8589.

Lepage, G., Paradis, K. (1997). Ursodeoxycholic acid improves the hepatic metabolism of essential fatty acids and retinol in children with cystic fibrosis. *J. Pediatr.* **130**, 52–58.

Li, M., McCann, J. D. (1988). Cyclic AMP-dependent protein kinase opens chloride channels in normal but not cystic fibrosis airway epithelium. *Nature* **331**, 358–360.

Li, C., Ramjeesingh, M. (1993). The cystic fibrosis mutation (delta F508) does not influence the chloride channel activity of CFTR. *Nature Genet.* **3**, 311–316.

Li, C., Ramjeesingh, M. (1996). Purified cystic fibrosis transmembrane conductance regulator (CFTR) does not function as an ATP channel. *J. Biol. Chem.* **271**, 11623–11626.

Lieberman, J., and Littenberg, G. D. (1969). Increased kallikrein content of saliva from patients with cystic fibrosis of the pancreas. A theory for the pathogenesis of abnormal secretions. *Pediatr. Res.* **3**, 571–578.

Lindblad, A., Glaumann, H. (1998). A two-year prospective study of the effect of ursodeoxycholic acid on urinary bile acid excretion and liver morphology in cystic fibrosis–associated liver disease. *Hepatology* **27**, 166–174.

Ling, V. (1997). Multidrug resistance: molecular mechanisms and clinical relevance. *Cancer Chemother. Pharmacol.* **40**, S3–S8 (Suppl.).

Lingard, J. M., al-Nakkash, L. (1994). Acetylcholine, ATP, bombesin, and cholecystokinin stimulate 125I efflux from a human pancreatic adenocarcinoma cell line (BxPC-3). *Pancreas* **9**, 599–605.

Linsdell, P., and Hanrahan, J. W. (1998a). Adenosine triphosphate–dependent asymmetry of anion permeation in the cystic fibrosis transmembrane conductance regulator chloride channel. *J. Gen. Physiol.* **111**, 601–614.

Linsdell, P., and Hanrahan, J. W. (1998b). Glutathione permeability of *CFTR. Am. J. Physiol.* **275**, C323–C326.

Linsdell, P., Tabcharani, J. A. (1997). Permeability of wild-type and mutant cystic fibrosis transmembrane conductance regulator chloride channels to polyatomic anions. *J. Gen. Physiol.* **110**, 355–364.

Lohmann, S. M., Vaandrager, A. B. (1997). Distinct and specific functions of cGMP-dependent protein kinases. *Trends Biochem. Sci.* **22**, 307–312.

Lohrmann, E., and Greger, R. (1993). Isolated perfused rabbit colon crypts: Stimulation of Cl$^-$ secretion by forskolin. *Pflugers Arch.* **425**, 373–380.

Lomri, N., Fitz, J. G. (1996). Hepatocellular transport: Role of ATP-binding cassette proteins. *Semin. Liver Dis.* **16**, 201–210.

Lu, Z., Auerbach, W. (1996). Introduction of human *CFTR* into CF mice under the control of a human villin gene promoter by transgenic techniques. Tenth Annual North American Cystic Fibrosis Conference, Orlando, FL, Wiley-Liss.

Lukacs, G. L., Chang, X. B. (1992). The cystic fibrosis transmembrane regulator is present and functional in endosomes. Role as a determinant of endosomal pH. *J. Biol. Chem.* **267**, 14568–14572.

Lukacs, G. L., Chang, X. B. (1993). The delta F508 mutation decreases the stability of cystic fibrosis transmembrane conductance regulator in the plasma membrane. Determination of functional half-lives on transfected cells. *J. Biol. Chem.* **268**, 21592–21598.

Lukacs, G. L., Segal, G. (1997). Constitutive internalization of cystic fibrosis transmembrane conductance regulator occurs via clathrin-dependent endocytosis and is regulated by protein phosphorylation. *Biochem. J.* **328**, 353–361.

MacVinish, L. J., Hickman, M. E. (1998). Importance of basolateral K^+ conductance in maintaining Cl^- secretion in murine nasal and colonic epithelia. *J. Physiol. (Lond.)* **510**, 237–247.

Madara, J. L., Parkos, C. (1992). Cl^- secretion in a model intestinal epithelium induced by a neutrophil-derived secretagogue. *J. Clin. Invest.* **89**, 1938–1944.

Madara, J. L., Patapoff, T. W. (1993). 5′-adenosine monophosphate is the neutrophil-derived paracrine factor that elicits chloride secretion from T84 intestinal epithelial cell monolayers. *J. Clin. Invest.* **91**, 2320–2325.

Maiuri, L., Raia, V. (1997). DNA fragmentation is a feature of cystic fibrosis epithelial cells: A disease with inappropriate apoptosis? *FEBS Lett.* **408**, 225–231.

Makhey, V. D., Guo, A. (1998). Characterization of the regional intestinal kinetics of drug efflux in rat and human intestine and in Caco-2 cells. *Pharm. Res.* **15**, 1160–1167.

Malinowska, D. H., Kupert, E. Y. (1995). Cloning, functional expression, and characterization of a PKA-activated gastric Cl^- channel. *Am. J. Physiol.* **268**, C191–C200.

Mandel, K. G., Dharmsathaphorn, K. (1986). Characterization of a cyclic AMP-activated Cl^- transport pathway in the apical membrane of a human colonic epithelial cell line. *J. Biol. Chem.* **261**, 704–712.

Mann, E. A., Jump, M. L. (1997). Mice lacking the guanylyl cyclase C receptor are resistant to STa-induced intestinal secretion. *Biochem. Biophys. Res. Commun.* **239**, 463–466.

Manson, A. L., Trezise, A. E. (1997). Complementation of null CF mice with a human *CFTR YAC* transgene. *Embo. J.* **16**, 4238–4249.

Marino, C. R., Matovcik, L. M. (1991). Localization of the cystic fibrosis transmembrane conductance regulator in pancreas. *J. Clin. Invest.* **88**, 712–716.

Markert, T., Vaandrager, A. B. (1995). Endogenous expression of type II cGMP-dependent protein kinase mRNA and protein in rat intestine. Implications for cystic fibrosis transmembrane conductance regulator. *J. Clin. Invest.* **96**, 822–830.

Marshall, J., Martin, K. A. (1991). Identification and localization of a dogfish homolog of human cystic fibrosis transmembrane conductance regulator. *J. Biol. Chem.* **266**, : 22749–22754.

Martinez, J. R., and Barker, S. (1987). Effect of prostaglandins on Cl and K transport in rat submandibular salivary acini. *Arch. Oral Biol.* **32**, 843–847.

Mastrocola, T., Porcelli, A. M. (1998). Role of *CFTR* and anion exchanger in bicarbonate fluxes in C127 cell lines. *FEBS Lett.* **440**, 268–272.

Matthews, J. B., Awtrey, C. S. (1992). Microfilament-dependent activation of $Na^+/K^+/2Cl^-$ cotransport by cAMP in intestinal epithelial monolayers. *J. Clin. Invest.* **90**, 1608–1613.

Matthews, J. B., Awtrey, C. S. (1993). $Na^{(+)}$-$K^{(+)}$-$2Cl^-$ cotransport and Cl^- secretion evoked by heat-stable enterotoxin is microfilament dependent in T84 cells. *Am. J. Physiol.* **265**, G370–G378.

McEwan, G. T., Hirst, B. H. (1994). Carbachol stimulates Cl^- secretion via activation of two distinct apical Cl^- pathways in cultured human T84 intestinal epithelial monolayers. *Biochim. Biophys. Acta* **1220**, 241–247.

McGill, J. M., Basavappa, S. (1994). Adenosine triphosphate activates ion permeabilities in biliary epithelial cells. *Gastroenterology* **107**, 236–243.

McGill, J. M., Williams, D. M. (1996). Survey of cystic fibrosis transmembrane conductance regulator genotypes in primary sclerosing cholangitis. *Dig. Dis. Sci.* **41**, 540–542.

McNicholas, C. M., Guggino, W. B. (1996). Sensitivity of a renal K^+ channel (ROMK2) to the inhibitory sulfonylurea compound glibenclamide is enhanced by coexpression with the ATP-binding cassette transporter cystic fibrosis transmembrane regulator. *Proc. Natl. Acad. Sci. USA* **93**, 8083–8088.

McNicholas, C. M., Nason, Jr., M. W., (1997). A functional CFTR-NBF1 is required for ROMK2-CFTR interaction. *Am. J. Physiol.* **273**, F843–F848.

McRoberts, J. A., Beuerlein, G. (1985). Cyclic AMP and Ca^{2+}-activated K^+ transport in a human colonic epithelial cell line. *J. Biol. Chem.* **260**, 14163–14172.

Merlin, D., Jiang, L. (1998). Distinct Ca^{2+}- and cAMP-dependent anion conductances in the apical membrane of polarized T84 cells. *Am. J. Physiol.* **275**, C484–C495.

Mezoff, A. G., Giannella, R. A. (1992). *Escherichia coli* enterotoxin (STa) binds to receptors, stimulates guanyl cyclase, and impairs absorption in rat colon. *Gastroenterology* **102**, 816–822.

Montrose-Rafizadeh, C., Blackmon, D. L. (1992). Regulation of cystic fibrosis transmembrane conductance regulator (CFTR) gene transcription and alternative RNA splicing in a model of developing intestinal epithelium. *J. Biol. Chem.* **267**, 19299–19305.

Montrose-Rafizadeh, C., Guggino, W. B. (1991). Cellular differentiation regulates expression of Cl⁻ transport and cystic fibrosis transmembrane conductance regulator mRNA in human intestinal cells. *J. Biol. Chem.* **266**, 4495–4499.

Morris, A. P., Cunningham, S. A. (1994). Polarization-dependent apical membrane CFTR targeting underlies cAMP-stimulated Cl⁻ secretion in epithelial cells. *Am. J. Physiol.* **266**, C254–C268.

Moyer, B. D., Loffing, J. (1998). Membrane trafficking of the cystic fibrosis gene product, cystic fibrosis transmembrane conductance regulator, tagged with green fluorescent protein in madin-darby canine kidney cells. *J. Biol. Chem.* **273**, 21759–21768.

Musch, M. W., Kachur, J. F. (1983). Bradykinin-stimulated electrolyte secretion in rabbit and guinea pig intestine. Involvement of arachidonic acid metabolites. *J. Clin. Invest.* **71**, 1073–1083.

Nakamura, H., Yoshimura, K. (1992). Tumor necrosis factor modulation of expression of the cystic fibrosis transmembrane conductance regulator gene. *FEBS Lett.* **314**, 366–370.

Naren, A. P., Nelson, D. J. (1997). Regulation of CFTR chloride channels by syntaxin and Munc18 isoforms. *Nature* **390**, 302–305.

Naren, A. P., Quick, M. W. (1998). Syntaxin 1A inhibits CFTR chloride channels by means of domain-specific protein-protein interactions. *Proc. Natl. Acad. Sci. USA* **95**, 10972–10977.

Nguyen, T. D., Koh, D. S. (1997). Characterization of two distinct chloride channels in cultured dog pancreatic duct epithelial cells. *Am. J. Physiol.* **272**, G172–G180.

Nikaido, H., and Hall, J. A. (1998). Overview of bacterial ABC transporters. *Methods Enzymol.* **292**, 3–20.

O'Brien, S. M., Campbell, G. R. (1996). Serum bile acids and ursodeoxycholic acid treatment in cystic fibrosis–related liver disease. *Eur. J. Gastroenterol. Hepatol.* **8**, 477–483.

O'Loughlin, E. V., Zhe, L. (1997). Colonic structural and ion transport abnormalities in suckling rabbits infected with *Escherichia coli* K12. *J. Pediatr. Gastroenterol. Nutr.* **25**, 394–399.

O'Reilly, C. M., O'Farrell, A. M., (1998). Purinoceptor activation of chloride transport in cystic fibrosis and CFTR-transfected pancreatic cell lines. *Br. J. Pharmacol.* **124**, 1597–1606.

Ouellette, A. J. (1997). Paneth cells and innate immunity in the crypt microenvironment. *Gastroenterology* **113**, 1779–1784.

Parker, A. C., Pritchard, P. (1997). Enhanced drug metabolism in young children with cystic fibrosis. *Arch. Dis. Child* **77**, 239–241.

Parmley, R. R., and Gendler, S. J., (1998). Cystic fibrosis mice lacking *Muc1* have reduced amounts of intestinal mucus. *J. Clin. Invest.* **102**, 1798–1806.

Pasyk, E. A., and Foskett, J. K. (1995). Mutant (delta F508) cystic fibrosis transmembrane conductance regulator Cl⁻ channel is functional when retained in endoplasmic reticulum of mammalian cells. *J. Biol. Chem.* **270**, 12347–12350.

Pasyk, E. A., and Foskett, J. K. (1997). Cystic fibrosis transmembrane conductance regulator-associated ATP and adenosine 3'-phosphate 5'-phosphosulfate channels in endoplasmic reticulum and plasma membranes. *J. Biol. Chem.* **272**, 7746–7751.

Perdue, M. H., Marshall, J. (1990). Ion transport abnormalities in inflamed rat jejunum. Involvement of mast cells and nerves. *Gastroenterology* **98**, 561–567.

Peters, R. H., French, P. J. (1996). *CFTR* expression and mucin secretion in cultured mouse gall-bladder epithelial cells. *Am. J. Physiol.* **271**, G1074–G1083.

Peters, R. H., van Doorninck, J. H. (1997). Cystic fibrosis transmembrane conductance regulator mediates the cyclic adenosine monophosphate–induced fluid secretion but not the inhibition of resorption in mouse gallbladder epithelium. *Hepatology* **25**, 270–277.

Pier, G. B., Grout, M. (1998). *Salmonella typhi* uses CFTR to enter intestinal epithelial cells. *Nature* **393**, 79–82.

Poulsen, J. H., and Machen, T. E. (1996). HCO_3^- dependent pHi regulation in tracheal epithelial cells. *Pflugers Arch.* **432**, 546–554.

Prat, A. G., Reisin, I. L. (1996). Cellular ATP release by the cystic fibrosis transmembrane conductance regulator. *Am. J. Physiol.* **270**, C538–C545.

Prince, L. S., Peter, K. (1999). Efficient endocytosis of the cystic fibrosis transmembrane conductance regulator requires a tyrosine-based signal. *J. Biol. Chem.* **274**, 3602–3609.

Prince, L. S., Workman, R. J. (1994). Rapid endocytosis of the cystic fibrosis transmembrane conductance regulator chloride channel. *Proc. Natl. Acad. Sci. USA* **91**, 5192–5196.

Quinton, P. M. (1983). Chloride impermeability in cystic fibrosis. *Nature* **301**, 421–422.

Quinton, P. M., and Reddy, M. M. (1992). Control of CFTR chloride conductance by ATP levels through non-hydrolytic binding. *Nature* **360**, 79–81.

Rajendran, V. M., Geibel, J. (1999). Role of Cl channels in Cl-dependent Na/H exchange. *Am. J. Physiol.* **276**, G73–G78.

Rao, M. C., Bissonnette, G. B. (1994). Rectal epithelial expression of protein kinase A phosphorylation of cystic fibrosis transmembrane conductance regulator. *Gastroenterology* **106**, 890–898.

Reddy, M. M., Quinton, P. M. (1996). Failure of the cystic fibrosis transmembrane conductance regulator to conduct ATP. *Science* **271**, 1876–1879.

Reid, C. J., Hyde, K. (1997). Cystic fibrosis of the pancreas: involvement of MUC6 mucin in obstruction of pancreatic ducts. *Mol. Med.* **3**, 403–411.

Reisin, I. L., Prat, A. G. (1994). The cystic fibrosis transmembrane conductance regulator is a dual ATP and chloride channel. *J. Biol. Chem.* **269**, 20584–20591.

Rey, E., Treluyer, J. M. (1998). Drug disposition in cystic fibrosis. *Clin. Pharmacokinet.* **35**, 313–329.

Richards, B., Skoletsky, J. (1993). Multiplex PCR amplification from the *CFTR* gene using DNA prepared from buccal brushes/swabs. *Hum. Mol. Genet.* **2**, 159–163.

Rigas, B., Korenberg, J. R. (1989). Prostaglandins E_2 and E_2 alpha are elevated in saliva of cystic fibrosis patients. *Am. J. Gastroenterol.* **84**, 1408–1412.

Riordan, J. R., Rommens, J. (1989). Identification of the cystic fibrosis gene: Cloning and characterization of complementary DNA. *Science* **245**, 1066–1073.

Roberts, R. L., and Stiehm, E. R. (1989). Increased phagocytic cell chemiluminescence in patients with cystic fibrosis. *Am. J. Dis. Child.* **143**, 944–950.

Roman, R. M., Bodily, K. O. (1998). Activation of protein kinase Calpha couples cell volume to membrane Cl^- permeability in HTC hepatoma and Mz-ChA-1 cholangiocarcinoma cells. *Hepatology* **28**, 1073–1080.

Roman, R. M., Wang, Y. (1997). Hepatocellular ATP-binding cassette protein expression enhances ATP release and autocrine regulation of cell volume. *J. Biol. Chem.* **272**, 21970–21976.

Rommens, J. M., Dho, S. (1991). cAMP-inducible chloride conductance in mouse fibroblast lines stably expressing the human cystic fibrosis transmembrane conductance regulator. *Proc. Natl. Acad. Sci. USA* **88**, 7500–7504.

Rommens, J. M., Iannuzzi, M. C. (1989). Identification of the cystic fibrosis gene: Chromosome walking and jumping. *Science* **245**, 1059–1065.

Rozmahel, R., Gyomorey, K. (1997). Incomplete rescue of cystic fibrosis transmembrane conductance regulator deficient mice by the human *CFTR* cDNA. *Hum. Mol. Genet.* **6**, 1153–1162.

Rozmahel, R., Wilschanski, M. (1996). Modulation of disease severity in cystic fibrosis transmembrane conductance regulator deficient mice by a secondary genetic factor. *Nature Genet.* **12,** 280–287.

Ruknudin, A., Schulze, D. H. (1998). Novel subunit composition of a renal epithelial KATP channel. *J. Biol. Chem.* **273,** 14165–14171.

Sakai, H., Okada, Y. (1992). Arachidonic acid and prostaglandin E$_2$ activate small-conductance Cl$^-$ channels in the basolateral membrane of rabbit parietal cells. *J. Physiol. (Lond.)* **448,** 293–306.

Sanchez de Medina, F., Galvez, J. (1997). Effects of quercetin on epithelial chloride secretion. *Life Sci.* **61,** 2049–2055.

Santis, G., Osborne, L. (1990a). Genetic influences on pulmonary severity in cystic fibrosis. *Lancet* **335.**

Santis, G., Osborne, L. (1990b). Independent genetic determinants of pancreatic and pulmonary status in cystic fibrosis. *Lancet* **336,** 1081–1084.

Scher, H., Bishop, W. P. (1997). Ursodeoxycholic acid improves cholestasis in infants with cystic fibrosis. *Ann. Pharmacother.* **31,** 1003–1005.

Schlenker, T., Romac, J. M. (1997). Regulation of biliary secretion through apical purinergic receptors in cultured rat cholangiocytes. *Am. J. Physiol.* **273,** G1108–G1117.

Schmitz, H., Fromm, M. (1996). Tumor necrosis factor-alpha induces Cl$^-$ and K$^+$ secretion in human distal colon driven by prostaglandin E$_2$. *Am. J. Physiol.* **271,** G669–G674.

Schreiber, R., Hopf, A. (1999). The first-nucleotide binding domain of the cystic-fibrosis transmembrane conductance regulator is important for inhibition of the epithelial Na$^+$ channel. *Proc. Natl. Acad. Sci. USA* **96,** 5310–5315.

Schwiebert, E. M. (1999b). ABC transporter-facilitated ATP conductive transport. *Am. J. Physiol.* **276,** C1–C8.

Schwiebert, E. M., Benos, D. J. (1999a). CFTR is a conductance regulator as well as a chloride channel. *Physiol. Rev.* **79,** S145–S166 (Suppl. 1).

Schwiebert, E. M., Egan, M. E. (1995). CFTR regulates outwardly rectifying chloride channels through an autocrine mechanism involving ATP. *Cell* **81,** 1063–1073.

Schwiebert, L. M., Estell, K. (1999c). Chemokine expression in CF epithelia: Implications for the role of *CFTR* in RANTES expression. *Am. J. Physiol.* **276,** C700–C710.

Schwiebert, E. M., Gesek, F. (1994). Heterotrimeric G proteins, vesicle trafficking, and CFTR Cl$^-$ channels. *Am. J. Physiol.* **267,** C272–C281.

Schwiebert, E. M., Morales, M. M. (1998). Chloride channel and chloride conductance regulator domains of CFTR, the cystic fibrosis transmembrane conductance regulator. *Proc. Natl. Acad. Sci. USA* **95,** 2674–2679.

Seidler, U., Blumenstein, I. (1997). A functional CFTR protein is required for mouse intestinal cAMP-, cGMP-, and Ca$^{(2+)}$-dependent HCO$_3^-$ secretion. *J. Physiol. (Lond.)* **505,** 411–423.

Seksek, O., Biwersi, J. (1996). Evidence against defective trans-Golgi acidification in cystic fibrosis. *J. Biol. Chem.* **271,** 15542–15548.

Sheppard, D. N., Rich, D. P. (1993). Mutations in *CFTR* associated with mild-disease form channels with altered pore properties. *Nature* **362,** 160–164.

Sherry, A. M., Stroffekova, K. (1997). Characterization of the human pH- and PKA-activated ClC-2G(2 alpha) Cl$^-$ channel. *Am. J. Physiol.* **273,** C384–C393.

Shimokura, G. H., McGill, J. M. (1995). Ursodeoxycholate increases cytosolic calcium concentration and activates Cl$^-$ currents in a biliary cell line. *Gastroenterology* **109,** 965–972.

Shoshani, T., Augarten, A. (1992). Association of a nonsense mutation (W1282X), the most common mutation in the Ashkenazi Jewish cystic fibrosis patients in Israel, with presentation of severe disease. *Am. J. Hum. Genet.* **50,** 222–228.

Shumaker, H., Amlal, H. (1999). CFTR drives Na$^+$-nHCO^{-3} cotransport in pancreatic duct cells: A basis for defective HCO^{-3} secretion in CF. *Am. J. Physiol.* **276,** C16–C25.

Silva, P., Solomon, R. J. (1997). Transport mechanisms that mediate the secretion of chloride by the rectal gland of *Squalus acanthias*. J. Exp. Zool. **279,** 504–508.

Sood, R., Bear, C. (1992). Regulation of *CFTR* expression and function during differentiation of intestinal epithelial cells. *Embo. J.* **11,** 2487–2494.

Spino, M. (1991). Pharmacokinetics of drugs in cystic fibrosis. *Clin. Rev. Allergy* **9,** 169–210.

Strazzabosco, M. (1997). Transport systems in cholangiocytes: Their role in bile formation and cholestasis. *Yale J. Biol. Med.* **70,** 427–434.

Strong, T. V., Boehm, K. (1994). Localization of cystic fibrosis transmembrane conductance regulator mRNA in the human gastrointestinal tract by *in situ* hybridization. *J. Clin. Invest.* **93,** 347–354.

Stutts, M. J., Canessa, C. M. (1995a). CFTR as a cAMP-dependent regulator of sodium channels. *Science* **269,** 847–850.

Stutts, M. J., Lazarowski, E. R. (1995b). Activation of CFTR Cl⁻ conductance in polarized T84 cells by luminal extracellular ATP. *Am. J. Physiol.* **268,** C425–C433.

Stutts, M. J., Rossier, B. C. (1997). Cystic fibrosis transmembrane conductance regulator inverts protein kinase A–mediated regulation of epithelial sodium channel single channel kinetics. *J. Biol. Chem.* **272,** 14037–14040.

Sweezey, N. B., Gauthier, C. (1996). Progesterone and estradiol inhibit CFTR-mediated ion transport by pancreatic epithelial cells. *Am. J. Physiol.* **271,** G747–G754.

Swenson, E. S., Mann, E. A. (1996). The guanylin/STa receptor is expressed in crypts and apical epithelium throughout the mouse intestine. *Biochem. Biophys. Res. Commun.* **225,** 1009–1014.

Tabary, O., Zahm, J. M. (1998). Selective up-regulation of chemokine IL-8 expression in cystic fibrosis bronchial gland cells *in vivo* and *in vitro. Am. J. Pathol.* **153,** 921–930.

Tabcharani, J. A., Chang, X. B. (1991). Phosphorylation-regulated Cl⁻ channel in CHO cells stably expressing the cystic fibrosis gene. *Nature* **352,** 628–631.

Tabcharani, J. A., Low, W. (1990). Low-conductance chloride channel activated by cAMP in the epithelial cell line T84. *FEBS Lett.* **270,** 157–164.

Takahashi, A., Watkins, S. C. (1996). CFTR-dependent membrane insertion is linked to stimulation of the CFTR chloride conductance. *Am. J. Physiol.* **271,** C1887–C1894.

Taylor, A. L., Kudlow, B. A. (1998). Bioluminescence detection of ATP release mechanisms in epithelia. *Am. J. Physiol.* **275,** C1391–C1406.

Teem, J. L., Berger, H. A. (1993). Identification of revertants for the cystic fibrosis delta F508 mutation using STE6-CFTR chimeras in yeast. *Cell* **73,** 335–346.

Teem, J. L., Carson, M. R. (1996). Mutation of R555 in CFTR-delta F508 enhances function and partially corrects defective processing. *Recept. Channels* **4,** 63–72.

Teune, T. M., Timmers-Reker, A. J. (1996). *In vivo* measurement of chloride and water secretion in the jejunum of cystic fibrosis patients. *Pediatr. Res.* **40,** 522–527.

Thiemann, A., Grunder, S. (1992). A chloride channel widely expressed in epithelial and non-epithelial cells. *Nature* **356,** 57–60.

Thomas, P. J., Ko, Y. H. (1992). Altered protein folding may be the molecular basis of most cases of cystic fibrosis. *Febs Lett.* **312,** 7–9.

Thomas, P. S., Bellamy, M. (1995). Malabsorption of vitamin E in cystic fibrosis improved after ursodeoxycholic acid. *Lancet* **346,** 1230–1231.

Tien, X. Y., Brasitus, T. A. (1994). Activation of the cystic fibrosis transmembrane conductance regulator by cGMP in the human colonic cancer cell line, Caco-2. *J. Biol. Chem.* **269,** 51–54.

Tietz, P. S., Holman, R. T. (1995). Isolation and characterization of rat cholangiocyte vesicles enriched in apical or basolateral plasma membrane domains. *Biochemistry* **34,** 15436–15443.

Tizzano, E. F., Chitayat, D. (1993). Cell-specific localization of *CFTR* mRNA shows developmentally regulated expression in human fetal tissues. *Hum. Mol. Genet.* **2,** 219–224.

Torres, R. J., Altenberg, G. A. (1996). Polarized expression of cAMP-activated chloride channels in isolated epithelial cells. *Am. J. Physiol.* **271,** C1574–C1582.

Tousson, A., Fuller, C. M. (1996). Apical recruitment of CFTR in T-84 cells is dependent on cAMP and microtubules but not Ca^{2+} or microfilaments. *J. Cell Sci.* **109**, 1325–1334.

Tran-Paterson, R., Davin, D. (1992). Expression and regulation of the cystic fibrosis gene during rat liver regeneration. *Am. J. Physiol.* **263**, C55–C60.

Trapnell, B. C., Zeitlin, P. L. (1991). Down-regulation of cystic fibrosis gene mRNA transcript levels and induction of the cystic fibrosis chloride secretory phenotype in epithelial cells by phorbol ester. *J. Biol. Chem.* **266**, 10319–10323.

Traynor, T. R., Brown, D. R. (1993). Effects of inflammatory mediators on electrolyte transport across the porcine distal colon epithelium. *J. Pharmacol. Exp. Ther.* **264**, 61–66.

Trezise, A. E., and Buchwald, M. (1991). *In vivo* cell-specific expression of the cystic fibrosis transmembrane conductance regulator. *Nature* **353**, 434–437.

Trezise, A. E., Chambers, J. A. (1993). Expression of the cystic fibrosis gene in human foetal tissues. *Hum. Mol. Genet.* **2**, 213–218.

Trezise, A. E., Ratcliff, R. (1997). Co-ordinate regulation of the cystic fibrosis and multidrug resistance genes in cystic fibrosis knockout mice. *Hum. Mol. Genet.* **6**, 527–537.

Trezise, A. E., Romano, P. R. (1992). The multidrug resistance and cystic fibrosis genes have complementary patterns of epithelial expression. *Embo. J.* **11**, 4291–4303.

Tsui, L. C. (1992). The spectrum of cystic fibrosis mutations. *Trends Genet.* **8**, 392–398.

Tsumura, T., Hazama, A. (1998). Activation of cAMP-dependent Cl^- currents in guinea-pig Paneth cells without relevant evidence for *CFTR* expression. *J. Physiol. (Lond.)* **512**, 765–777.

Uribe, J. M., Gelbmann, C. M. (1996a). Epidermal growth factor inhibits $Ca(2^+)$-dependent Cl^- transport in T84 human colonic epithelial cells. *Am. J. Physiol.* **271**, C914–C922.

Uribe, J. M., Keely, S. J. (1996b). Phosphatidylinositol 3-kinase mediates the inhibitory effect of epidermal growth factor on calcium-dependent chloride secretion. *J. Biol. Chem.* **271**, 26588–26595.

Vaandrager, A. B., Bajnath, R. (1991). Ca^{2+} and cAMP activate different chloride efflux pathways in HT-29.cl19A colonic epithelial cell line. *Am. J. Physiol.* **261**, G958–G965.

Valverde, M. A., O'Brien, J. A. (1995). Impaired cell volume regulation in intestinal crypt epithelia of cystic fibrosis mice. *Proc. Natl. Acad. Sci. USA* **92**, 9038–9041.

van Kuijck, M. A., van Aubel, R. A. (1996). Molecular cloning and expression of a cyclic AMP-activated chloride conductance regulator: A novel ATP-binding cassette transporter. *Proc. Natl. Acad. Sci. USA* **93**, 5401–5406.

van Veen, H. W., and Konings, W. N. (1998). The ABC family of multidrug transporters in microorganisms. *Biochim. Biophys. Acta* **1365**, 31–36.

Veeze, H. J., Halley, D. J. (1994). Determinants of mild clinical symptoms in cystic fibrosis patients. Residual chloride secretion measured in rectal biopsies in relation to the genotype. *J. Clin. Invest.* **93**, 461–466.

Veeze, H. J., Sinaasappel, M. (1991). Ion transport abnormalities in rectal suction biopsies from children with cystic fibrosis. *Gastroenterology* **101**, 398–403.

Verdon, B., Winpenny, J. P. (1995). Volume-activated chloride currents in pancreatic duct cells. *J. Membr. Biol.* **147**, 173–183.

Walters, R. J., O'Brien, J. A. (1992). Membrane conductance and cell volume changes evoked by vasoactive intestinal polypeptide and carbachol in small intestinal crypts. *Pflugers Arch.* **421**, 598–605.

Wang, Y., Roman, R. (1996). Autocrine signaling through ATP release represents a novel mechanism for cell volume regulation. *Proc. Natl. Acad. Sci. USA* **93**, 12020–12025.

Ward, C. L., and Kopito, R. R. (1994). Intracellular turnover of cystic fibrosis transmembrane conductance regulator. Inefficient processing and rapid degradation of wild-type and mutant proteins. *J. Biol. Chem.* **269**, 25710–25718.

Warhurst, G., Higgs, N. B. (1991). Stimulatory and inhibitory actions of carbachol on chloride secretory responses in human colonic cell line T84. *Am. J. Physiol.* **261**, G220–G228.

Warth, R., Riedemann, N. (1996). The cAMP-regulated and 293B-inhibited K^+ conductance of rat colonic crypt base cells. *Pflugers Arch.* **432,** 81–88.

Wei, L. Y., Hoffman, M. M. (1997). Altered pHi regulation in 3T3/CFTR clones and their chemotherapeutic drug-selected derivatives. *Am. J. Physiol.* **272,** C1642–C1653.

Wei, L. Y., Stutts, M. J. (1995). Overexpression of the cystic fibrosis transmembrane conductance regulator in NIH 3T3 cells lowers membrane potential and intracellular pH and confers a multidrug resistance phenotype. *Biophys. J.* **69,** 883–895.

Welsh, M. J. (1986). An apical-membrane chloride channel in human tracheal epithelium. *Science* **232,** 1648–1650.

Welsh, M. J., Denning, G. M. (1993a). Dysfunction of *CFTR* bearing the delta F508 mutation. *J. Cell. Sci. Suppl.* **17,** 235–239.

Welsh, M. J., and Smith, A. E. (1993b). Molecular mechanisms of CFTR chloride channel dysfunction in cystic fibrosis. *Cell* **73,** 1251–1254.

Welsh, M. J., L.-C. Tsui, F. T. Boat, and A. L. Beaudet (1995). Cystic fibrosis. *In* "The Metabolic and Molecular Bases of Inherited Disease," C. R. Scriver, A. L. Beaudet, W. S. Sly, and D. Valle, eds., pp. 3799–3877, Chapter 127, McGraw-Hill, New York.

Widdicombe, J. H., Welsh, M. J. (1985). Cystic fibrosis decreases the apical membrane chloride permeability of monolayers cultured from cells of tracheal epithelium. *Proc. Natl. Acad. Sci. USA* **82,** 6167–6171.

Will, K., Dork, T. (1995). Transcript analysis of *CFTR* nonsense mutations in lymphocytes and nasal epithelial cells from cystic fibrosis patients. *Hum. Mutat.* **5,** 210–220.

Wilschanski, M. A., Rozmahel, R. (1996). *In vivo* measurements of ion transport in long-living CF mice. *Biochem. Biophys. Res. Commun.* **219,** 753–759.

Winpenny, J. P., Harris, A. (1998). Calcium-activated chloride conductance in a pancreatic adenocarcinoma cell line of ductal origin (HPAF) and in freshly isolated human pancreatic duct cells. *Pflugers Arch.* **435,** 796–803.

Winpenny, J. P., Verdon, B. (1995). Calcium-activated chloride conductance is not increased in pancreatic duct cells of CF mice. *Pflugers Arch.* **430,** 26–33.

Wright, E. M., Hirsch, J. R. (1997). Regulation of Na^+/glucose cotransporters. *J. Exp. Biol.* **200,** 287–293.

Zeng, W., Lee, M. G. (1997). Immuno and functional characterization of CFTR in submandibular and pancreatic acinar and duct cells. *Am. J. Physiol.* **273,** C442–C455.

Zhou, L., Dey, C. R. (1994). Correction of lethal intestinal defect in a mouse model of cystic fibrosis by human CFTR. *Science* **266,** 1705–1708.

Zielenski, J., Corey, M. (1999). Detection of a cystic fibrosis modifier locus for meconium ileus on human chromosome 19q13. *Nature Genet.* **22,** 128–129.

Zund, G., Madara, J. L. (1996). Interleukin-4 and interleukin-13 differentially regulate epithelial chloride secretion. *J. Biol. Chem.* **271,** 7460–7464.

CHAPTER 7

Integrated Signaling Mechanisms that Regulate Intestinal Chloride Secretion

S. J. Keely and K. E. Barrett

Department of Medicine, University of California, San Diego, School of Medicine, San Diego, CA 92103

I. INTRODUCTION

The intestinal epithelium is a continuous monolayer of cells that line the intestinal tract. Primary functions of this epithelium are the absorption and secretion of ions and water to and from the intestinal lumen. Under normal circumstances, net absorption predominates, which allows for conservation of the large amounts of fluid that pass through the intestine each day. The efficiency with which the epithelial layer carries out this function is remarkable considering only approximately 2% of this 9-liter daily fluid load is lost to the stool. However, in addition to its great capacity for absorption, the intestine is capable of secretion. Intestinal secretions normally are required to maintain an appropriate fluid environment within the lumen. This allows for transiting of ingested material through the alimentary canal and mixing of ingested food particles with digestive enzymes, and provides a medium for the diffusion of digested nutrients to the epithelium, where absorption may occur. The importance of ongoing secretory processes to normal intestinal function can be inferred from pathological conditions in which these processes are impaired. For example, in cystic fibrosis, a genetic disorder of epithelial chloride secretion, intestinal obstruction and malabsorption are common (Welsh *et al.,* 1995). The secretion of ions into the gut may be altered by many chemical and environmental stimuli. For example, the presence of toxins and pathogens within the intestinal lumen may elicit reflex secretory responses that constitute an innate defense, known as "the enteric tear mechanism" (Cooke, 1998), by which such harmful substances may be flushed from the body before they gain access to the systemic circulation. However, certain pathological conditions may result in excessive stimulation of ion and fluid secretion which, if they exceed the absorptive capacity of the intestine, may result in the onset of secretory diarrhea. The most dramatic example of this is cholera, where up to 20 liters of fluid may be lost from the body each day.

The ability of the intestine to rapidly convert from a net absorptive to a secretory organ suggests that epithelial function is closely regulated. Not only must it be able to respond appropriately to environmental, chemical, and hormonal cues, but also mechanisms must exist by which secretory responses may be terminated, or downregulated, to prevent excessive fluid loss from the body. Recent years have seen much progress in our understanding of epithelial biology and how chloride secretion is regulated in both health and disease. The cloning and sequencing of several transport proteins involved in the chloride secretory process also have provided new insight into how chloride secretion is regulated at the molecular level. Significant advances also have been made in understanding the biochemistry of epithelial signal transduction pathways and how different second messenger systems interact with each other, and ultimately with transport proteins, to alter chloride secretion in response to changes in the extracellular environment. At the same time, a greater understanding of intercellular communication between epithelial cells and regulatory cells of the intestinal mucosa, such as those of the mucosal immune system and enteric nervous system, has led to new insight into how epithelial physiology may be regulated at the whole-tissue level. Through the combined efforts of researchers in these diverse disciplines of science, we are coming to a more complete appreciation of the complexity of epithelial cell biology and how the net expression of epithelial chloride secretory responses represents the integrated sum of myriad endogenous and exogenous regulatory influences.

II. ORGANIZATION, STRUCTURE, AND FUNCTION OF THE INTESTINAL EPITHELIUM

The cellular composition of the epithelium is heterogeneous, although the predominant cell type is the transporting cell that carries out the absorptive and secretory functions. Other cells within the epithelial monolayer perform functions that modulate or are complementary to those of the transporting cells. For example, Paneth cells and goblet cells, which are found within the crypts, contribute to the protective function of the epithelium by secreting antibacterial defensins and mucus, respectively (LaMont, 1992; Selsted *et al.*, 1992). Similarly, enteroendocrine cells act as sensors and have the ability to initiate neuronal reflex arcs that ultimately modify chloride secretion across the epithelium in response to changes within the luminal environment (Goyal and Hirano, 1996; Cooke, 1998; Hofer *et al.*, 1999). However, despite their diverse phenotypes, all epithelial cells arise from pluripotent stem cell precursors found at the base of epithelial crypts (Babyatsky and Podolsky, 1999; Bjerknes *et al.*, 1999). For many years it was believed that the absorptive and secretory functions of the intestinal epithelium were spatially distinct processes, with absorption occurring across the villus cells (or surface cells of the colon) and secretion occurring in the crypts (Welsh and Smith, 1982). However, with the development of more precise experimental techniques, it is becoming increasingly evident that this scheme is an oversimplification and that both villus and crypt cells are capable of absorption and secretion (Powell, 1995).

Two essential features of epithelial cells that enable them to conduct the vectorial transport of ions, such as chloride, are their functional polarity and their ability to form tight intercellular junctions with one another (Cereijido *et al.*, 1998). Functional polarity refers to the asymmetric distribution of proteins between the apical and basolateral domains of epithelial cells. Asymmetry in the localization of the channels, pumps, and cotransporters that constitute ion transport mechanisms allows electrochemical gradients to be established between the apical and basolateral side of epithelia which, in turn, permits the vectorial transport of ions to occur. In addition to ion transport proteins, epithelial cells also differentially express receptors for various secretagogues on their apical and basolateral surfaces. Receptors for neurotransmitters and immune cell mediators, such as acetylcholine and histamine, are expressed predominantly on the basolateral side of the cell, where they are in close contact with elements of the mucosal immune system and the enteric nervous system. Other receptors are expressed only on the apical surface of intestinal epithelia and mediate the actions of various factors found within the lumen, such as bacterial enterotoxins and guanylin (Forte *et al.*, 1993; Kuhn *et al.*, 1994). Some receptors, such as those for adenosine and kinins, are distributed bilaterally and may have a role in mediating secretory responses to activated phagocytes which, under conditions of inflammation, migrate through the epithelial layer and release their mediators on the apical side (Cuthbert et al., 1985; Barrett et al., 1989; Strohmeier et al., 1995; Keely et al., 1997).

Tight junctions are found toward the apical pole of epithelial cells and function not only to regulate the movement of molecules within the plane of the lipid bilayer but also as selective pores that restrict the paracellular movement of substances, including ions. Tight junctions consist of a dense network of proteins around the circumference of the cell, which impinges on sites of intercellular membrane fusion, known as "kiss" sites, to form gasket-like seals between adjacent cells. Many protein components of the tight junctions have been identified, although the functions of several remain to be determined. Occludin, and the recently identified claudins (Denker and Nigam, 1998; Mitic and Anderson, 1998; Morita et al., 1999), are transmembrane components of the tight junction and are likely to act as the actual sealing proteins. Another junctional protein, paracellin, which displays structural and sequence homology to the claudins, recently has been identified in renal epithelia. Mutations of paracellin result in a rare autosomal recessive disease characterized by severe Mg^{2+} and Ca^{2+} wasting. Therefore, it has been proposed that this protein, either alone or in partnership with other junctional components, confers the cationic selectivity that is characteristic of tight junction pores (Simon et al., 1999). On the cytoplasmic side, tight junctions are associated with many other proteins that most likely are involved in regulation of junctional permeability. These include the membrane-associated guanylate kinase (MAGUK) proteins ZO-1 and ZO-2, cingulin, and the 7H6 antigen (Zhong et al., 1993; Anderson, J. M., 1996; Fanning et al., 1998). Also associated with the tight junctions are several signaling proteins including low-molecular-weight G proteins, heterotrimeric G proteins, and tyrosine kinases (Zahraoui et al., 1994; Nusrat et al., 1995; Denker et al., 1996; Hecht et al., 1996; Nybom and Magnusson, 1996). Tight junctions also are linked to the cellular cytoskeleton by way of actin microfilaments that extend from the "kiss sites" to the perijunctional ring, a dense band of actin filaments that underlies the tight junction. Many intracellular second messengers, including cyclic nucleotides, protein kinases, calcium, and phospholipids, are believed to regulate tight junction permeability through mechanisms involving alterations in the cytoskeleton and contraction of the perijunctional ring (Fasano et al., 1995; Van Itallie et al., 1995; Tai et al., 1996). The opening of tight junctions in response to alterations in the levels of intracellular second messengers is a crucial event in determining the extent of transepithelial ion and water fluxes because, as tight junctions open, the paracellular barrier to ion and fluid transport may decrease drastically.

III. BASIS OF EPITHELIAL CHLORIDE SECRETION

The secretion of water across epithelial cells is ultimately a passive process driven by the active transport of ions. In the small intestine and colon, chloride is the primary ion that drives fluid secretion. Typically, chloride ions are secreted transcellularly into the intestinal lumen, thereby creating an electrochemical gradient for the passive diffusion of cations, mainly sodium, via a paracellular route through the tight junctions. The resulting accumulation of salt on the apical side of the epithelium is followed by water flowing down its osmotic gradient through the tight junctions in order to maintain isotonicity.

A. Components of the Chloride Secretory Mechanism

As with all other ion transport mechanisms, chloride secretion across intestinal epithelia occurs via the concerted actions of several ion transport pathways, which together constitute the chloride secretory mechanism. Four main ion transport pathways are involved: the $Na^+/K^+/2Cl^-$ cotransporter, the Na^+/K^+ ATPase pump, K^+ channels, and Cl^- channels.

1. $Na^+/K^+/2Cl^-$ Cotransporter (NKCC)

The basolateral uptake step for chloride is mediated by a cotransporter. Cotransporters are carrier proteins that bind multiple solutes on one side of a plasma membrane and shuttle them across to the other side. Typically, cotransporters utilize favorable electrochemical gradients for the movement of one or more cotransported solutes to facilitate the movement of another solute against an unfavorable gradient. Because the direction of solute movement depends on existing electrochemical gradients, the activity of cotransporters is largely governed by the activity of ATPase pumps (see later). NKCCs not only have an important role in the chloride secretory mechanism but also are involved in regulation of cell volume. Two distinct isoforms of NKCC have been identified. NKCC2 has a restricted expression and, to date, has been identified only in absorptive kidney epithelia (Haas *et al.*, 1998). In contrast, NKCC1 is widely expressed and is localized to the basolateral membrane of most secretory epithelia, including those in the intestine.

The tertiary structure of NKCC has not been fully elucidated, but biochemical studies and amino acid analysis indicate it has a large central domain of 12 membrane-spanning regions and long cytoplasmic carboxy and amino termini. Ion binding occurs in the central domain in a highly ordered fashion, and full occupancy of the ion-binding domains is required for cotransport to occur (Haas and Forbush, 1998). Because two cations are transported into the cell along with two chloride ions, transport via NKCC1 is electroneutral that is, there is no net transfer of charge.

It appears that NKCC activity is regulated by a dynamic equilibrium of kinase and phosphatase activities that alter the phosphorylation state of the cytoplasmic carboxy and amino termini of the protein. Both osmotic stress and hormonal agonists that elevate intracellular cAMP, such as vasoactive intestinal peptide (VIP), have been demonstrated to increase phosphorylation of the protein (Lytle and Forbush, 1992; Matthews *et al.,* 1997; 1998). Although the specific protein kinases responsible for NKCC phosphorylation have not been identified, it appears that, to some degree, there is a convergence of the signaling pathways by which hormonal and osmotic stimuli activate the protein in intestinal epithelia. Studies of shark rectal gland and T_{84} cells demonstrate that hormonally stimulated cotransporter activation is, at least in part, mediated by opening of apical chloride channels (Lytle and Forbush, 1996; Haas and Forbush, 1998; Matthews *et al.,* 1998). The subsequent decrease in cytosolic Cl^- concentration is believed to activate an osmosensitive protein kinase, which phosphorylates, and activates, NKCC. Such direct cross-talk between apical chloride channels and basolateral NKCC helps to maintain intracellular electrolyte composition by ensuring that chloride efflux and influx are closely synchronized. In addition to phosphorylation, it has been demonstrated that the cellular cytoskeleton may play an important role in regulation of NKCC1 function. In cultured T_{84} colonic epithelial cells, agonists that stimulate chloride secretion via elevations in cAMP also induce extensive remodeling of the cytoskeleton. In addition, stabilization of the cytoskeleton with agents such as phalloidin has been demonstrated to inhibit both NKCC activity and cyclic nucleotide-dependent chloride secretion (Goldstein *et al.,* 1991; Matthews *et al.,* 1992; 1993b). It has been proposed that NKCC function may be regulated by F-actin assemblies at the basolateral side of the cell. Such assemblies may represent a barrier preventing NKCC activation, either by inhibition of ion translocation or by membrane-fusion events that regulate insertion of the protein into the membrane. Disassembly of the actin cytoskeleton, which may occur in response to agonists that activate protein kinase A, would bring about disruption of these F-actin barriers, thus favoring NKCC activation (Matthews *et al.,* 1997). Protein kinase C (PKC) also may have a role in regulation of NKCC function because exposure of T_{84} cells to phorbol esters, which activate classical and novel PKC isoforms, results in down-regulation of NKCC function (Matthews *et al.,* 1993a; Farokzhad *et al.,* 1998; Song and Matthews, 1999), perhaps by modulating the trafficking of the cotransporter to and from the membrane.

2. Na^+/K^+ ATPase

For ion and fluid transport to occur across intestinal epithelia, energy is required to create and maintain the appropriate osmotic gradients. This energy usually is derived from ATPase pumps. As their name suggests, these transport proteins utilize energy from the hydrolysis of ATP to pump ions against electrochemical gradients. The driving force for intestinal chloride secretion is derived from the activity of the Na^+/K^+ ATPase pump located on the basolateral side of all intestinal epithelial cells. The Na^+/K^+ ATPase functions to maintain a low Na^+ ion concentration within the cell by pumping $3Na^+$ ions from the cell in exchange for the movement of $2K^+$ ions into the cell. Because the pump activity results in a net loss of cations from the cell, it creates an electronegative potential within the cell with respect to the outside. The activity of the Na^+/K^+ ATPase pump has two main consequences with respect to chloride secretion. First, it maintains a low intracellular Na^+ concentration and thereby creates a gradient for Na^+ uptake across the basolateral membrane via NKCC1. Thus, it is an important part of the Cl^- secretory mechanism because it creates the appropriate gradient for loading cells with Cl^- above its electrochemical equilibrium. Second, the negative electrical potential generated inside the cell by Na^+ extrusion via the Na^+/K^+ ATPase is important in maintaining the appropriate electrical gradient for Cl^- exit to occur.

The Na^+/K^+ ATPase has been cloned and been found to consist of a heterodimer of α and β subunits. Although three isoforms of the α and two isoforms of the β subunits have been identified, only the $\alpha 1$ and $\beta 1$ subunits are expressed in mammalian intestine (Wild and Thompson, 1996). The α subunit is a 110-kDa protein and contains all the active sites necessary for Na^+ and K^+ binding and for ATP hydrolysis. It also contains phosphorylation sites that, like NKCC1, are likely involved in regulation of pump activity (Blanco et al., 1995; Ewart and Klip, 1995; Beguin et al., 1996). The β subunit is a highly glycosylated 55-kDa protein and is thought to be important in directing and anchoring the assembled pump to the membrane.

3. K^+ Channels

K^+ channels exist on both the apical and basolateral membranes of intestinal epithelial cells and have been studied most extensively in tissues or cell lines derived from the colon (Greger et al., 1997; Mayol et al., 1997b; Schultheiss and Diener, 1998). K^+ channels in the basolateral membrane are important in the Cl^- secretory mechanism, and at least two subtypes of this channel are known to exist in intestinal epithelia: one is activated by cAMP and is sensitive to blockade by Ba^{2+} and the other is activated by intracellular Ca^{2+} and is relatively insensitive to Ba^{2+} (Dharmsathaphorn and Pandol, 1985; McRoberts et al., 1985; Mandel et al., 1986). The latter Ca^{2+}-activated K^+ channel shares several properties with the recently cloned intermediate conductance, inwardly rectifying K^+ channel, hIK1 (Devor and Frizzell, 1993; Jensen et al., 1998). These channels serve as an exit pathway for K^+, which is accumulated within the cell by way of the basolateral Na^+/K^+ ATPase pump and NKCC1. Extrusion of K^+ via basolateral channels not only increases the driving force for Cl^- uptake via NKCC1 but also helps maintain the cell in a hyperpolarized state, thus increasing the gradient for chloride secretion. A third type of K^+ channel has been identified in the basolateral membrane of T_{84} colonic epithelial cells, which is distinct from the cAMP-activated K^+ conductance and independent of changes in intracellular Ca^{2+} (Devor and Frizzell, 1998a). This large-conductance channel is activated by arachidonic acid, and may play a role in mediating chloride secretion in response to adenosine, an inflammatory mediator that appears to stimulate intestinal epithelial chloride secretion, at least in part, by a mechanism independent of intracellular Ca^{2+} and cAMP (Barrett and Bigby, 1993).

Little is known about how K^+ channel activity is regulated on the molecular level, but studies employing both kinase and phosphatase inhibitors suggest that alterations in the phosphorylation/dephosphorylation state of protein are likely involved (Schultheiss and Diener, 1998). In addition, similar to NKCC, K^+ channel activity appears to be regulated by PKC because treatment of epithelial cells with phorbol esters results in inhibition of K^+ channel conductance (Matthews *et al.*, 1993a; Reenstra, 1993).

4. Cl⁻ Channels

Channels in the apical membrane of intestinal epithelia provide the exit pathway for chloride and are essential for vectorial secretion to occur. Several types of chloride channel are known to exist in secretory epithelia including cAMP-activated channels, Ca^{2+}-activated channels, outwardly rectifying chloride channels (ORCC), and ClC channels. ORCC and members of the ClC family of chloride channels are known to be expressed in intestinal epithelia where they are believed to play a role in regulation of cell volume (Sakamoto *et al.*, 1996; Bond *et al.*, 1998; Foskett, 1998; Hagos *et al.*, 1999). However, to date there are few data regarding regulation of these channels or their contribution to transepithelial Cl⁻ and fluid secretion. In contrast, the cystic fibrosis transmembrane conductance regulator (CFTR) channel, which mediates Cl⁻ efflux in response to cAMP-dependent hormones, is probably the most intensely studied of ion channels. Mutations in the gene encoding this protein are responsible for cystic fibrosis, an inherited disorder that causes defective chloride transport in numerous organs including the intestine. Normal CFTR channels have an ion selectivity of Cl>Br>1 with a linear conductivity of 8–10 pS. The structure of CFTR has been well elucidated (Sheppard and Welsh, 1999). It consists of a 170-kDa polypeptide chain of 1480 amino acids with two membrane-spanning domains, each consisting of six transmembrane-spanning sequences. These regions appear to form the actual channel pore because point mutations within this domain alter the ion selectivity of the channel. CFTR also possesses two cytosolic nucleotide binding domains (NBDs) that have led it to be classified in a family of proteins known as the ATP-binding cassette (ABC) proteins. Binding and hydrolysis of ATP by these domains is necessary for channel gating (Anderson and Welsh, 1992). However, because the energy required for Cl⁻ extrusion via CFTR is derived from the electrochemical gradients established by the basolateral Na^+/K^+ ATPase, the requirement of energy derived from ATP hydrolysis by CFTR, per se for chloride secretion, has not been defined (Foskett, 1998). In addition to regulatory NBDs, CFTR also possesses a long cytoplasmic R domain, the deletion of which renders the channel constitutively active (Rich *et al.,* 1991). It is believed that the R domain acts by blocking the channel pore on the cytoplasmic face in a manner similar to that of the ball-and-chain models for cation channels. The R domain contains several consensus sites for phosphorylation by protein kinase A (Townsend *et al.,* 1996). Once it is phosphorylated, there is a change in conformation of the R domain that, in the presence of ATP, leads to an opening of the channel (Dulhanty and Riordan, 1994; Gadsby and Nairn, 1999). Closure of the channel is brought about by phosphatases that dephosphorylate the R domain, thus returning the channel to its quiescent state (Luo *et al.,* 1998).

Although activation of PKA by elevations in intracellular levels of cAMP is the main stimulus for activation of CFTR, several other factors appear to be involved in regulation of the channel. For example, the R domain contains sites for phosphorylation by PKC and it is believed that phosphorylation of these sites may facilitate phosphorylation of CFTR by PKA (Jia *et al.,* 1997). However, prolonged activation of PKC may lead to inhibition of epithelial chloride secretion caused by down-regulation of CFTR transcription (Trapnell *et al.,* 1991; Matthews *et al.,* 1993a). CFTR also may be activated by cGMP-dependent protein kinases (Vaandrager *et al.,* 1997b), and a further level of channel regulation is achieved through regulated insertion of the channel into the apical membrane by a mechanism that appears to involve the cytoskeleton (Prat *et al.,* 1995; Lehrich *et al.,* 1998; Prince *et al.,* 1999) and membrane fusion regulators, known as syntaxins (Naren *et al.,* 1997; 1998).

Finally, in addition to its role in conducting chloride, CFTR has been demonstrated to have a role in regulating the conductance of other ion channels, including ORCC. This hypothesis originally was based on the observation that, although ORCCs are present in the membranes of cystic fibrosis epithelia, they are insensitive to activation by PKA (Hwang *et al.,* 1989). It is unclear how CFTR regulates ORCC function, but it has been proposed to occur by an autocrine mechanism that may involve CFTR-dependent ATP release across the apical membrane. ATP may then act on purinergic receptors to stimulate ORCC opening (Jovov *et al.,* 1995; Schwiebert *et al.,* 1995). The mechanism by which CFTR mediates ATP release remains unclear, but it has been shown that specific and separate domains of CFTR are involved in its ability to act as a chloride channel and its ability to regulate ORCC (Schwiebert *et al.,* 1998).

The possible existence of a specific Ca^{2+}-dependent chloride channel in intestinal epithelia has been a point of debate for some years. In airway epithelia, chloride secretory responses to Ca^{2+}-dependent agonists are not affected by mutations in CFTR, which indicates that these agonists activate a chloride conductance that is distinct from the cAMP-dependent chloride channel (Willumsen and Boucher, 1989). However, in intestinal epithelia, Ca^{2+}-dependent chloride secretion also is impaired in cystic fibrosis, which suggests there is a functional lack of a Ca^{2+}-dependent Cl^- conductance (Taylor *et al.*, 1988; Hardcastle *et al.*, 1991). This idea is supported by the findings of Anderson and Welsh (1991), who failed to detect a Ca^{2+}-dependent Cl^- conductance in cultured monolayers of colonic epithelial cells. However, other investigators have detected anion conductances, distinct from those elicited by cAMP, in response to Ca^{2+}-dependent secretagogues (Vaandrager *et al.*, 1991; McEwan *et al.*, 1994; Arreola *et al.*, 1998; Merlin *et al.*, 1998), and recent studies have shown that a cloned Ca^{2+}-dependent chloride channel is expressed in the crypt epithelia of human small and large intestine (Gruber *et al.*, 1998). The reasons for such discrepancies in detecting Ca^{2+}-dependent Cl^- channels in intestinal epithelial cells, and the reason why Ca^{2+}-dependent intestinal epithelial Cl^- secretion is impaired in cystic fibrosis, remain unclear, but may be related to the multitude of negative influences to which Ca^{2+}-dependent Cl^- secretion is subjected (see later).

B. Summary

As described earlier, the individual transport proteins that comprise the intestinal Cl^- secretory mechanism have been well elucidated. It is the concerted and synchronized actions of these proteins that allow the electrogenic transfer of Cl^- ions from the basolateral to the apical side of the epithelium (Fig. 1). In summary,

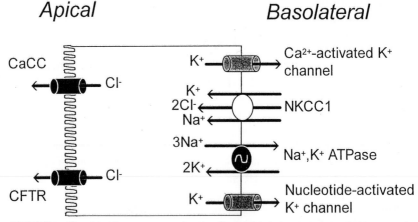

FIGURE 1 The chloride secretory mechanism. Transepithelial secretion of chloride across intestinal epithelial cells is accomplished by the coordinated activity of several transport proteins arranged asymmetrically. On the basolateral membrane, chloride uptake is accomplished by NKCC1. This is a secondary active transport, driven by the low intracellular sodium concentration established by the Na^+, K^+ ATPase. Potassium is recycled across basolateral potassium channels, which have not been definitively identified at the molecular level. At the apical membrane, chloride can exit the cell via cAMP-regulated CFTR chloride channels or via calcium-activated chloride channels (CaCC).

Cl^- is accumulated into the cell by way of a basolateral $Na^+/K^+/2Cl^-$ cotransporter (NKCC1) that moves two Cl^- ions into the cell along with $1K^+$ and $1Na^+$. The energy for Cl^- uptake is derived from the basolateral Na^+/K^+ ATPase pump that transports Na^+ out of the cell, against an electrochemical gradient, in exchange for K^+ coming into the cell. The action of this transporter keeps Na^+ concentrations low within cytoplasm with respect to the extracellular environment and thus creates a gradient for Na^+ uptake via the cotransporter. Because three Na^+ ions are pumped from the cell in exchange for two K^+ ions into the cell, the actions of the Na^+/K^+ ATPase also make the inside of the cell more electronegative with respect to the outside. Potassium, which enters the cell via NKCC1 and the Na^+/K^+ ATPase, does not accumulate in the cytosol because it may exit via regulated K^+ channels in the basolateral membrane. Thus, the concerted actions of NKCC1, the Na^+/K^+ ATPase, and the K^+ channels serve to increase Cl^- within the cell, above its electrochemical equilibrium, and to make the inside of the cell more electronegative whith respect to the outside. This creates a favorable gradient for Cl^- to be extruded from the cell when regulated chloride channels within the apical membrane are opened. When Cl^- is secreted across the cell, Na^+ and water follow by passive paracellular flux through the tight junctions.

IV. REGULATION OF INTESTINAL Cl^- SECRETION

Since Cl^- secretion across the intestinal epithelium is dependent on the concerted activity of multiple transport proteins, the overall process may be regulated by factors that alter the functional activity of any given component of this transport mechanism. Acute changes in transporter function may be brought about by signals that interact directly with the protein or that induce alterations in its phosphorylation state or its level of membrane insertion. Longer term alterations may be brought about by factors that increase or decrease transporter protein expression. In general, mechanisms involved in the regulation of Cl^- secretion can be divided into two classes. The first class of regulatory mechanisms is intercellular in nature and includes an array of chemical mediators that may be released from neurons, immunocytes, enteroendocrine cells, and fibroblasts, all of which reside within the intestinal mucosa in close proximity to the epithelial layer. The second class of regulatory mechanisms exists within the epithelial cells themselves and consists of the second messengers and signal transduction pathways that are activated when extracellular stimuli interact with epithelial cells, usually via specific receptors within the plasma membrane. Thus, receptors can be considered as the point at which information regarding the state of the extracellular environment is received by epithelial cells, while receptor-coupled second-messenger systems represent the mechanisms by which this information is processed and disseminated into the cells.

Inter- and intracellular mechanisms of regulation can be subdivided further into positive and negative regulatory mechanisms. Positive mechanisms can be considered as those that act to promote epithelial Cl^- secretion, and negative mechanisms are those that serve to down-regulate or inhibit the Cl^- secretory process. For the sake of clarity, positive and negative regulation of Cl^- secretion will be discussed in separate sections herein. However, it is important to bear in mind that the intestinal epithelium, and the component transport proteins of the chloride secretory mechanism, are under the combined influence of both positive and negative intra- and intercellular modes of regulation at all times. Furthermore, the extent of intestinal Cl^- secretion at any given time represents the finely balanced, integrated product of these opposing influences.

A. Intercellular Regulation of Intestinal Cl^- Secretion

Intestinal Cl^- transport is regulated in health and disease by a wide variety of factors that are extrinsic to the epithelial layer. Such factors include luminal stimuli, blood-borne hormones, and neuronally or immunologically derived mediators (Fig. 2; Table I). Luminal factors that elicit chloride secretion include normal physiological stimuli, such as a food bolus moving through the intestine, which trigger secretory reflexes in order to lubricate the epithelium as it passes. They also include pathophysiological stimuli such as enteric bacteria that can evoke excessive secretory responses resulting in conditions of secretory diarrhea. Although some effects of luminal stimuli on chloride secretion can be brought about by a direct action on the epithelial cells themselves, these responses usually are largely mediated by cellular elements residing within the lamina propria, a layer of loose connective tissue underlying the epithelium. Histological examination of the lamina propria shows a variety of cell types including immune cells (the mucosal immune system [MIS]) and an extensive neuronal network (the enteric nervous system [ENS]). The proximity of these cell types to each other and to the epithelial layer immediately suggests their important role in regulation of epithelial function. Also found within the lamina propria is the myofibroblastic sheath that directly underlies the epithelium and is believed to play a complementary role to that of the ENS and MIS in regulation of Cl^- secretion (Berschneider and Powell, 1992; Hinterleitner et al., 1996). Finally, blood vessels, which bring oxygen and nutrients to the intestinal mucosa, also appear to have the ability to release soluble factors that can influence epithelial secretion (Blume et al., 1998).

1. The Enteric Nervous System

The enteric nervous system (ENS) is made up of several nerve plexi, the major ones being the myenteric and submucosal plexi. Whereas the myenteric nerve

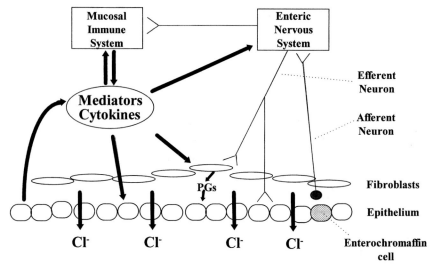

FIGURE 2 Intercellular regulation of intestinal chloride secretion. The secretion of chloride across the intestinal epithelium is regulated by cells of the immune system and the enteric nervous system that reside within the attendant lamina propria. Through the release of soluble mediators and neurotransmitters (see Table I), these cells communicate with the epithelium to evoke changes in chloride, and subsequently fluid secretion. By virtue of bidirectional communication, the enteric nervous and mucosal immune systems can recruit one another to amplify prosecretory signals. Further signal amplification may be achieved through stimulation of eicosanoid synthesis from the myofibroblastic sheath, which directly underlies the epithelial layer. Intercellular communication also can occur in the opposite direction, as epithelial and enterochromaffin cells continuously sense the luminal environment and send appropriate signals to the ENS and MIS, via cytokines and afferent neurons, respectively.

plexis is primarily involved in regulation of intestinal smooth muscle motility (Furness *et al.,* 1999; Hasler, 1999), the submucosal plexus is involved in regulation of epithelial function. However, neural connections exist between the two plexi, which allow for coordination of intestinal motor and secretory functions (Kirchgessner and Gershon, 1988). Bidirectional communication exists between the central nervous system (CNS) and the ENS through neural pathways that synapse within the plexi of the ENS. These include both sympathetic and parasympathetic pathways, which have opposing effects on ion transport. Activation of sympathetic neurons stimulates ion and water absorption, whereas parasympathetic pathways predominantly promote secretion (Greenwood *et al.,* 1987; Goyal and Hirano, 1996). However, complex reflex arcs enable the ENS to function independently of the CNS, which has led several investigators to refer to it as the "little brain" (Wood, 1991; Surprenant, 1994; Goyal and Hirano, 1996; Cooke, 1998). These reflex arcs are important in sensing changes within the intestinal environment and in elucidating an appropriate secretory response

TABLE I

Intercellular Regulators of Intestinal Chloride Secretion

Class[a]	Prosecretory[b]	Antisecretory[b]
Immunological mediators	Histamine (Rangachari, 1992)	Interleukin-13 (Zund et al., 1996)
	5-hydroxytryptamine (Scott et al., 1992; Budhoo et al., 1996)	Interleukin-10 (Madsen et al., 1996; 1997)
	Eicosanoids (Bern et al., 1989; Berschneider and Powell, 1992; Eberhart and Dubois, 1995)	Interferon-γ (Madsen et al., 1997)
	Kinins (Rangachari and Prior, 1993; Teather and Cuthbert, 1997; Cuthbert and Huxley, 1998)	
	Adenosine (Barrett et al., 1990; Madara et al., 1993; Strohmeier et al., 1995)	
	Platelet activating factor (Bern et al., 1989; Berschneider and Powell, 1992; Hinterleitner et al., 1997; Borman et al., 1998)	
	Reactive oxygen metabolites (Gaginella et al., 1995)	
	TNF-α (Kandil et al., 1994; Schmitz et al., 1996)	
	Interleukin-1 (Chang et al., 1990; Chissone et al., 1990; Theodorou et al., 1994)	
	Interleukin-3 (Chang et al., 1990)	
Neurotransmitters	Acetylcholine (Cooke, 1984; Warhurst et al., 1991; O'Malley et al., 1995)	Acetylcholine (Warhurst et al., 1991; Vajanaphanich et al., 1994; Keely et al., 1998b)
	Vasoactive intestinal peptide (Cox and Cuthbert, 1989a; Reddix et al., 1994; Martin et al., 1996)	Noradrenaline (Greenwood et al., 1987)
	5-hydroxytryptamine (Scott et al., 1992; Budhoo et al., 1996)	Somatostatin (Hurst et al., 1995; McKeen et al., 1995; Warhurst et al., 1996)
	Substance P (Reddix and Cooke, 1992; Cooke et al., 1997a; Goldhill and Angel, 1998)	γ-Aminobutyric acid (McNaughton et al., 1996)
		Neuropeptide Y (Cox et al., 1988; Bouritius et al., 1998)

Category	Mediators[a,b]	
	Calcitonin gene-related peptide (Cox et al., 1989b; Yarrow et al., 1991)	Nitric oxide (Reddix et al., 1998)
	Nitric oxide (McNaughton, 1994; Stack, et al., 1996)	
	PACAP (Cox, 1992; Kuwahara et al., 1993)	
	Neurotensin (Castagliuolo et al., 1999)	
Exogenous and/or luminally-derived factors	STa (Levine et al., 1991; Kuhn et al., 1994; Greenberg et al., 1997; Vaandrager et al., 1997a)	Ammonia (Mayol et al., 1997a)
	Bile acids (Gelbmann et al., 1995)	
	Enterotoxins (Beubler and Horina, 1990; Mourad et al., 1995; Lundgren and Jodal, 1997; Pothoulakis et al., 1998; Turvill et al., 1998)	
Growth factors and hormones	Guanylin (Forte et al., 1993; Kuhn et al., 1994; Greenberg et al., 1997)	EGF (Uribe et al., 1996a; 1996b)
	Endothelins (Kuhn et al., 1997; Reddix et al., 1998)	TGF-α (Beltinger et al., 1999)
	Melatonin (Chan et al., 1998)	Growth hormone (Guarino et al., 1995)
		Heregulin (Keely and Barrett, 1998)
		Peptide YY (Cox et al., 1998; Liu et al., 1995; Quin et al., 1995)

[a]Mediators are subdivided into major regulatory classes. However, it should be noted that some agents listed fall into more than one category.

[b]Note that some mediators can exert both positive and negative effects on chloride secretion.

(Cooke, 1998). Stimuli that trigger such intestinal secretory reflexes include mucosal stroking (which models the passage of food through the intestine), distension, and bacterial toxins (Cooke, 1998; Pothoulakis *et al.,* 1998).

The involvement of the ENS in regulation of intestinal Cl^- secretion has been extensively studied using voltage-clamped mucosal tissues that retain functional intrinsic nerve plexi. Treatment of such tissues with neurotoxins, such as tetrodotoxin (TTX), typically reduces basal levels of ion transport, implying a tonic influence of the ENS in regulation of epithelial function (Carey and Cooke, 1989). Treatment of intestinal preparations with TTX also has been demonstrated to reduce Cl^- secretion in response to a variety of agents including hormones, immunological stimuli, and enteric bacteria, which indicates that the ENS is likely to play an important role in intestinal pathophysiology by amplifying epithelial secretory responses to inflammatory (or other) stimuli (Cooke, 1994; Goyal and Hirano, 1996; Pothoulakis *et al.,* 1998). Electrical field stimulation of voltage-clamped intestinal tissues also has been employed to study neuronal regulation of epithelial secretion. This method, along with immunohistochemical techniques, has proven very useful in identifying neurotransmitters involved in regulation of Cl^- secretion. Acetylcholine appears to be a major prosecretory neurotransmitter within the ENS and is released from both interneurons, within reflex arcs, and from secretomotor neurons that impinge closely on the epithelial layer. However, many other conventional and peptide transmitters have been identified that are involved in regulation of intestinal Cl^- secretion (Table I). Investigations into the relative roles of individual transmitters in the regulation of ion transport are complicated by the fact that many nerves contain multiple neurotransmitters, some of which are pro-secretory and some inhibitory. The functional significance of multiple cotransmitters is unclear, but it is believed that, of the transmitters in an individual nerve, one is a primary transmitter while the others are subsidiary neurotransmitters or neuromodulators (McConalogue and Furness, 1994; Furness *et al.,* 1999).

2. The Mucosal Immune System

The gastrointestinal tract is extremely rich in immune cells, which is not surprising considering the continuous exposure of the intestine to bacterial, viral, and dietary antigens. Indeed, the high degree of ongoing immunological activity within the intestinal mucosa has led to the idea that, even under normal conditions, the intestine is an organ that is in a constant state of mild inflammation. Immune cells found within the mucosal layer include T and B lymphocytes, mast cells, neutrophils, eosinophils, and macrophages, all of which have been demonstrated to have a role in the regulation of epithelial function. The functional structure of the MIS can broadly be considered to consist of afferent and efferent components. The afferent component is involved in detecting antigenic material and toxins within the intestinal lumen and in priming the effector limb of the MIS. The cells primarily involved in detection of luminal antigens are lymphocytes, which exist in multiple forms within the intestinal mucosa. They may be found as a diffuse population of cells within the lamina propria or within distinct areas, known as Peyer's patches. A large proportion of mucosal lymphocytes may be found interspersed between epithelial cells, where they are known as intraepithelial lymphocytes (Shanahan, 1997; Kagnoff, 1998; Shaw *et al.,* 1998). Activation of lymphocytes by luminal antigens results in the synthesis and release of a variety of mediators, including cytokines. Cytokines not only serve to stimulate the chemotaxis and activation of the effector cells of the MIS but also have been shown to have a direct role in regulating epithelial Cl^- secretion. Some cytokines have the ability to stimulate Cl^- secretion, whereas others have been shown to have the opposite effect (see Table I). Recently, it has become apparent that the epithelial layer is not just an innocent bystander to these events; rather, it may play an active role in initiating mucosal immune responses. In this regard, the epithelium behaves in a fashion complementary to lymphocytes in that it also may take part in antigen recognition, presentation, and production of proinflammatory cytokines (Eckmann *et al.,* 1997; Yang *et al.,* 1997; Gewirtz *et al.,* 1998; Seydel *et al.,* 1998).

Inflammatory conditions of the gut are characterized by the influx of immune effector cells to the submucosa, which occurs in response to the chemotactic cytokines, or chemokines, that are released from afferent cells. These phagocytic cells, along with a resident population of mast cells in the lamina propria, have a profound influence on the secretory state of the epithelium. Mast cells are found throughout the gut wall, along the entire length of the intestinal tract. They represent a rich source of biologically active mediators, some of which are stored within cytoplasmic granules and others are synthesized and released upon appropriate stimulation. Mast cells appear to play an important role in intestinal pathophysiology because changes in mucosal mast cell numbers and mediator release have been observed in various inflammatory conditions of the gut including food allergic disease, inflammatory bowel disease, and systemic mastocytosis (Ferguson *et al.,* 1987; Cherner *et al.,* 1988; Nolte *et al.,* 1990; Austin, 1992; Crowe and Perdue, 1992; Crowe *et al.,* 1997).

A variety of approaches have been employed to study the role of mast cells in regulation of intestinal function. For example, activation of mast cells by anti-IgE, which cross-links cell surface IgE molecules, has been demonstrated to stimulate Cl^- secretion across small and large intestinal tissues derived from humans and experimental animals (Crowe and Perdue, 1993; Crowe *et al.,* 1997; Stack *et al.,* 1995; Hinterleitner *et al.,* 1997). Models of gut hypersensitivity also have proven useful in elucidating the role of mast cells in mediating allergic responses in the intestine. In such experiments, animals are first sensitized to food proteins, such as cows' milk or ovalbumin, or to enteric parasites (Baird *et al.,* 1984; 1985; Russell, 1986; Castro *et al.,* 1987; Crowe *et al.,* 1990; O'Malley *et al.*, 1993; Santos *et al.*, 1998a). Subsequent challenge of voltage-clamped intestinal preparations from these animals with the appropriate antigen typically results in rapid-onset Cl⁻ secretory responses, which can be inhibited by pretreatment of the tissues with mast-cell stabilizers, such as cromoglycate and doxantrozole (Perdue and Gall, 1985; Crowe *et al.*, 1990). That mast cells are important in mediating these responses is further supported by studies in antigen-sensitized mast-cell–deficient mice, where Cl⁻ secretory responses to subsequent antigen challenge are reduced by approximately 70% (Perdue *et al.*, 1991). Models of gut hypersensitivity also have been useful in identifying potential mast-cell mediators involved in promoting Cl⁻ secretion by showing the effects of specific receptor antagonists on secretory responses to antigen challenge. Another approach to identify mast-cell mediators that likely are involved in regulation of Cl⁻ secretion has been to directly examine the effects of potential mediators on ion transport across voltage-clamped intestinal tissues. Histamine, 5-hydroxytryptamine, eicosanoids, adenosine, and cytokines have been implicated in mediating intestinal Cl⁻ secretion in response to mast-cell activation (See Table I).

Phagocytes, a group of cells consisting of neutrophils, eosinophils, and macrophages, migrate into the lamina propria in response to chemotactic gradients of bacterially-derived products and cytokines released from activated immunocytes. Much evidence exists to suggest a role for phagocytic cells in regulation of intestinal ion transport. Direct activation of phagocytic cells with the bacterially derived tripeptide f-met-leu-phe has been shown to stimulate chloride secretion across intestinal tissues from both animals and humans (Bern *et al.*, 1989; Finlay and Smith, 1989; Barrett and Chang, 1990; Stack *et al.*, 1995; Hinterleitner *et al.*, 1997). Similar to mast-cell activation, Cl⁻ secretion in response to phagocyte activation is brought about by an array of mediators, some of which overlap with those released from activated mast cells, such as eicosanoids and adenosine. However, a significant portion of phagocyte-stimulated Cl⁻ secretion appears also to be mediated by generation of reactive oxygen metabolites such as nitric oxide and hydrogen peroxide.

Interestingly, when phagocytes are co-cultured with intestinal epithelial cells, they have been found to be more effective in stimulating Cl⁻ secretion when they are present on the apical side rather than the basolateral side. This finding is of relevance in the clinical setting of inflammatory bowel disease, where neutrophils migrate through the epithelial layer and into the crypts to form crypt abscesses (Nash *et al.*, 1987; Meenan *et al.*, 1996). The primary mediator of secretory responses to apical phagocytes appears to be 5′AMP, which is converted to adenosine at the apical surface by the membrane-bound enzyme 5′-ectonucleotidase (Madara *et al.*, 1992; 1993; Resnick *et al.*, 1993). How neutrophils traverse the epithelial layer has been a subject of considerable study. It appears to involve a regulated unsealing of the tight junctions in a manner dependent on the CD11b/CD18 adhesion molecule complex expressed on the surface of activated neutrophils (Parkos *et al.,* 1991; Parkos, 1997). Although mechanisms exist within epithelial cells by which tight junctions may be rapidly sealed after neutrophil migration, if migration occurs at a high density it can lead to loss of the barrier function of the epithelium and may contribute to the formation of mucosal ulcers that often are associated with inflammatory bowel diseases (Nash *et al.,* 1987; Nusrat *et al.,* 1997).

3. Interactions between Regulatory Cells of the Lamina Propria

Over the past two decades, it has become apparent that the ENS and MIS do not operate independently of one another but rather function as an integrated unit in the regulation of intestinal Cl^- secretion (see Fig. 2). Evidence supporting neuroimmune interactions in the gut submucosal layer comes, in part, from anatomic studies of the lamina propria which show that there are intimate associations between cells of the immune system and enteric neurons, suggesting that there is intercellular communication between the ENS and MIS (Stead *et al.,* 1989; Arizono *et al.,* 1990). These findings are further supported by studies which demonstrate that pharmacological denervation of intestinal tissues with TTX results in attenuated secretory responses to immune cell activation (Bern *et al.,* 1989; Finlay and Smith, 1989; Crowe and Perdue, 1993; O'Malley *et al.,* 1993; Cooke, 1994). Furthermore, many putative immune cell mediators, including histamine, 5-hydroxytryptamine, eicosanoids, cytokines, reactive oxygen metabolites, and kinins, stimulate secretory responses that are partially sensitive to TTX (Bern *et al.,* 1989; Gaginella and Kachur, 1989; 1995; Chissone *et al.,* 1990; Karayalcin *et al.,* 1990; Rangachari, 1992; Siriwardena *et al.,* 1993; Budhoo *et al.,* 1996; Borman *et al.,* 1998; Kuwahara *et al.,* 1998). Thus, stimulation of the MIS by dietary or bacterial antigens results in the production of various inflammatory mediators that can have direct effects on the epithelium itself, as evidenced by studies with cultured epithelial cells (Chissone *et al.,* 1990; Barrett, 1992), or can stimulate secretion indirectly via activation of the ENS. Thus, it is believed that the ENS serves to amplify the Cl^- secretory response to a specific immunological stimulus.

Just as immunological stimuli can recruit the ENS to modulate epithelial secretion, it appears that the reverse also is true, and that functionally relevant communication between the MIS and ENS is bidirectional. The presence of receptors for neuropeptides, such as substance P, on mast cells and phagocytes suggests that enteric neurons may regulate intestinal ion transport via modulation of immunocyte function (Shanahan *et al.,* 1985; Stead *et al.,* 1987; Mathison *et al.,* 1993; Pothoulakis *et al.,* 1998). Functional studies support this idea because secretory responses to neuronal activation by electrical field stimulation are attenuated by agents that inhibit mast cell degranulation and antagonists of inflammatory mediators (Cooke, 1994; Perdue, 1996; Pothoulakis *et al.,* 1998;

Santos *et al.,* 1998a). Responses to electrical field stimulation also have been demonstrated to be diminished in mice that are genetically mast-cell deficient (Perdue *et al.,* 1991).

4. Paracrine Regulation of Epithelial Secretion

In addition to the MIS and ENS, other cell types exist within (or in close proximity to) the epithelium, which produce factors that act in a paracrine fashion to modify the "responsiveness" of the epithelium to neuroimmunologically-derived substances. For example, Paneth cells secrete defensins, which not only play a role in host defense against luminal pathogens but also appear to insert into the apical membrane of epithelial cells, forming anion selective pores and thereby increasing chloride secretion in response to conventional secretagogues (Lencer *et al.,* 1997; Ouellette, 1997).

Another key cell type involved in paracrine regulation of epithelial function is the myofibroblast. These cells form a continuous layer underlying the basement membrane of the intestinal epithelium (see Fig. 2), and, through the synthesis of a wide variety of cytokines and growth factors, are believed to play an important role in regulation of epithelial cell growth, differentiation, and repair (Powell *et al.,* 1999). However, more recently it has been found that they also appear to have a role in regulation of intestinal epithelial Cl^- secretion. In reductionist studies, employing co-cultures of myofibroblasts with T_{84} colonic epithelial cells, it has been found that secretory responses to several immune cell mediators such as histamine, 5-hydroxytryptamine, hydrogen peroxide, and kinins are enhanced when epithelial cells are cultured in the presence of fibroblasts. The potentiating effect of epithelial co-culture with fibroblasts was blocked by the cyclooxygenase inhibitor indomethacin, implying that immune cell mediators stimulate the synthesis and release of prostaglandins from subepithelial fibroblasts. In turn, these prostaglandins may act in a paracrine fashion to stimulate Cl^- secretion across the epithelium (Berschneider and Powell, 1992). Subsequent studies indicate that the potentiating effect of fibroblasts may be increased further by cytokines (such as IL-1 and TNF-α) that stimulate cyclooxygenase expression in myofibroblasts and thereby enhance their ability to synthesize prostaglandins upon exposure to immunological mediators (Hinterleitner *et al.,* 1996; Kim *et al.,* 1998). In addition to prostaglandins, other factors released by fibroblasts likely play a role in regulation of chloride secretion. For example, activated fibroblasts also release other cytokines and mediators such as TNF-α, IL-1, platelet activating factor, and free oxygen radicals (Powell *et al.,* 1999), all of which have been demonstrated to have prosecretory effects across intestinal epithelia (see Table I). Thus, similar to the ENS, subepithelial fibroblasts may be recruited by the mucosal immune system to serve as a mechanism by which secretory responses elicited by immunological stimuli may be amplified. In a manner similar to that of myofibroblasts, endothelial cells may alter epithelial function by a paracrine mechanism involving prostaglandin release. In this case, the PGI_2 metabolite 6-ketoPGF2α has been implicated as the endothelial-derived factor that alters epithelial cell secretion (Blume *et al.,* 1998).

Thus, when one considers the multiplicity of regulatory cell types present within the intestinal mucosa (see Fig. 2), and the myriad mediators they release upon activation (see Table I), the task of understanding how intestinal epithelia are regulated at the extracellular level is a daunting one—and even more so when it comes to understanding extracellular communication within the intestinal mucosa in the setting of inflammatory bowel diseases. However, the ever-increasing research interest that this challenging field of epithelial physiology receives, and the ongoing development of new tools and techniques that enable the molecular dissection of the mechanisms by which epithelial cells are influenced by regulatory cells of the lamina propria, promises exciting advances in our overall knowledge of epithelial physiology in the coming years.

B. Intracellular Regulation of Chloride Secretion: Positive Regulation

Generally, neuroimmune and hormonal agonists that bring about alterations in intestinal chloride secretion do so by binding to specific receptors on the surface of epithelial cells. Thus, binding of intercellular mediators to their specific receptors on the epithelial plasma membrane represents a critical step in the chain of events that links alterations in the extracellular environment to the illicitation of an appropriate physiological response. Most receptors for neuroimmune mediators exist within the basolateral domain of the cell; however, some receptors, particularly those for bacterial endotoxins and mediators released from migrating immune cells, exist in addition to, or exclusively on, the apical domain. Most receptors that are coupled to positive stimulation of chloride secretion belong to the G-protein–coupled class of receptors (GPCRs). Ligand binding to these receptors results in the activation of associated heterotrimeric G proteins which, in turn, leads to the initiation of intracellular signaling cascades. These signaling pathways induce the production of second messengers that interact with, and alter the activity of, one or more of the transport proteins of the chloride secretory mechanism (Table II). The two major second-messenger systems that have been characterized for their roles in regulation of epithelial Cl^- secretion are those that involve the production of cyclic nucleotides and those that elevate intracellular levels of free Ca^{2+}.

1. Cyclic Nucleotides

Agonists that elevate intracellular levels of cyclic 3'-5' adenosine monophosphate (cAMP) are known to stimulate Cl^- secretion across intestinal epithelial cells (see Fig. 3; Table II) (Cartwright *et al.,* 1985; Donowitz and Welsh, 1986;

Basolateral

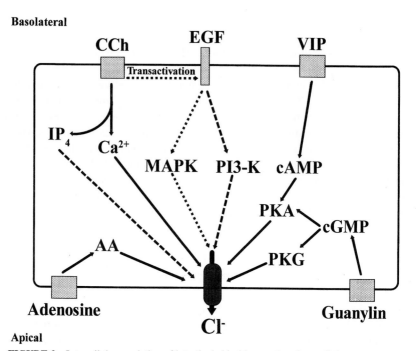

Apical

FIGURE 3 Intracellular regulation of intestinal chloride secretion. Intercellular messengers exert their effects on epithelial secretion via activation of specific receptors and stimulation of intracellular signal transduction pathways. Prosecretory signaling pathways (depicted by solid arrows) may be elicited by Ca^{2+}-dependent, cAMP-dependent, cGMP-dependent, and arachidonic acid–dependent agonists. Typically these second messengers increase chloride secretion via alterations in transport protein phosphorylation and/or insertion into the epithelial membrane. For simplicity, all regulatory influences are shown to have an impact on an apical chloride channel, although basolateral transport pathways, such as potassium channels, may be alternative or additional targets and therefore the figure is intended only to convey that the overall process of chloride secretion can be regulated by the signals shown. The chloride secretory process, particularly Ca^{2+}-dependent chloride secretion, also is subject to inhibitory signals (depicted by broken arrows). Signals that downregulate Ca^{2+}-dependent secretion may be generated by growth factors, such as EGF, which reduce chloride secretion via stimulation of PI3-K. Inhibitory signals also may be intrinsic to the Ca^{2+}-dependent signaling mechanism in that, in addition to elevating intracellular Ca^{2+}, Ca^{2+}-dependent agonists such as carbachol, also stimulate generation of Ins(3,4,5,6)P$_4$. Carbachol can evoke transactivation of the EGFr (pathway indicated by arrows with short dashes), with subsequent activation of MAPK. Note that the pathways activated by EGFr activation by EGF itself, versus EGFr transactivation via the receptor for carbachol, diverge at the level of the EGFr (as indicated by the different arrow types). The relationship of the ability of carbachol to evoke increases in InsP$_4$ and MAPK activity is unknown at present. It is the integrated influence of positive and negative intracellular signaling pathways that ultimately governs the extent of epithelial secretory responses.

TABLE II

Intracellular Regulators of Intestinal Chloride Secretion

	Prosecretory		Antisecretory		
Second messenger	Activating agonist[a]	References	Second messenger	Activating agonist[a]	References
cAMP	Vasoactive intestinal polypeptide	Mandel et al., 1986; Rodgers et al., 1990	Ins(3,4,5,6)P$_4$	Acetylcholine	Kachintorn et al., 1993b; Vajanaphanich et al., 1994; Xiu et al., 1998
Intracellular Ca^{2+}	Acetylcholine, histamine	Kachintorn et al., 1993a; Cartwright et al., 1985; Dharmsathaphorn and Pandol, 1985	PI3-kinase	EGF, HRG[b]	Uribe et al., 1996b; Keely and Barrett, 1998
cGMP	Guanylin, STa	Kuhn et al., 1994; Greenberg et al., 1997; Vaandrager et al., 1997a	MAPK[b]	Acetylcholine	(Keely et al., 1998b)
Arachidonic acid	Adenosine	Barrett and Bigby, 1993; Devor and Frizzell, 1998; Madara et al., 1992	Protein kinase C	Acetylcholine, histamine	Reenstra et al., 1993; Matthews et al., 1993; Kachintorn et al., 1992

[a]Representative examples are provided for each signaling pathway. Others also exist.
[b]MAPK, mitogen-activated protein kinase; EGF, epidermal growth factor; HRG, heregulin.

271

Mandel *et al.*, 1986; Rodgers *et al.*, 1990). cAMP is generated when neurohumoral mediators, such as vasoactive intestinal polypeptides, bind to their cell surface receptors and activate associated heterotrimeric G proteins of the G_s subtype. The G protein then disassociates into the $G\alpha_s$ and $G\beta\gamma$ subunits. $G\alpha_s$ subunits go on to activate adenylate cyclase, which catalyzes the synthesis of cAMP from ATP. The primary downstream effector of cAMP is protein kinase A (PKA), a multimeric protein consisting of two regulatory and two catalytic subunits. When two molecules of cAMP bind to each of the regulatory subunits, the catalytic subunits are released and may phosphorylate various subtrates on serine or threonine residues. Two subtypes of PKA are known to exist, based on their type of regulatory subunit and whether they are localized to the cytosolic or membrane compartment, and recent data suggest that both type I and type II isoforms of PKA exist in intestinal epithelial cells and that both are capable of mediating the effects of cAMP on Cl^- secretion (Singh *et al.*, 1998). Elevations in intracellular cAMP can have many actions on the Cl^- secretory process, including direct phosphorylation of CFTR, activation of basolateral K^+ channels and NKCC1, and remodeling of the actin cytoskeleton with associated alterations in transporter abundance at the plasma membrane. However, although cAMP alters the activity of many ion transport proteins involved in the Cl^- secretory mechanism, it is believed that the primary regulatory step for cAMP-dependent Cl^- secretion is at the level of Cl^- channel opening. It was originally believed that CFTR was the only Cl^- channel activated in response to cAMP; however, recent data suggest that at least part of the cAMP-dependent Cl^- secretory response is mediated by a Cl^- channel that also may be activated by Ca^{2+}-dependent agonists (Merlin *et al.*, 1998).

Exogenous factors also have the ability to stimulate intestinal chloride secretion via elevations in cytosolic cAMP. For example, cholera toxin exerts its pro-secretory effects by a mechanism that involves sustained production of cAMP. It does so by binding to specific receptors on the apical membrane of intestinal epithelia, from where it is translocated into the cell and causes irreversible activation of adenylate cyclase by ADP ribosylation. The resulting prolonged increase in cAMP production is accompanied by excessive chloride secretion, which manifests clinically as severe secretory diarrhea (Beubler and Horina, 1990; Nath *et al.*, 1994).

Another cyclic nucleotide known to have pro-secretory effects in the intestine is cyclic 3'-5' guanosine monophosphate (cGMP) (see Fig. 3; Table II). Mediators that stimulate intracellular production of cGMP include the heat-stable enterotoxin of *Escherichia coli* (STa) and its endogenous counterparts guanylin and uroguanylin. Binding of these agonists to their apically localized receptor activates an intrinsic guanylate cyclase activity, thereby increasing intracellular cGMP concentrations and producing a chloride secretory response qualitatively similar to that stimulated by cAMP (Kuhn *et al.*, 1994; Greenberg *et al.*, 1997; Vaandrager *et al.*, 1997a). As with PKA, two isoforms of cGMP-dependent protein kinase are known to exist, and it is the type II isoform that appears predominantly to mediate the pro-secretory effects of cGMP in intestinal epithelial cells (Vaandrager et al., 1997b). Also similar to cAMP-mediated secretion, the primary regulatory step in cGMP-mediated secretion appears to be at the level of CFTR channel phosphorylation. Indeed, cross-activation of PKA by cGMP has been reported in T_{84} epithelial cells, which may represent an additional pathway mediating chloride secretion in response to cGMP-mediated agonists in the intestine (Forte et al., 1992; Chao et al., 1994; Tien et al., 1994).

2. Intracellular Calcium

Agonists that elevate intracellular levels of Ca^{2+} also are known to stimulate Cl^- secretion across intestinal epithelial cells (see Fig. 3; Table II) (Cartwright et al., 1985; Dharmsathaphorn and Pandol, 1985; Donowitz et al., 1986; Kachintorn et al., 1993b). Such agonists, which are typified by the neurotransmitter acetylcholine and the mast-cell mediator histamine, act in a similar fashion to that of cyclic nucleotide-dependent agonists in that they initiate intracellular signaling by first binding to specific cell-surface receptors and activating receptor-coupled G proteins. However, it is there that the similarities in signaling mechanisms involved in regulation of Cl^- secretion stimulated by cyclic nucleotides and Ca^{2+} diverge. Calcium-dependent agonists bind to receptors that are linked to G_q. Thus, G-protein activation in response to Ca^{2+}-mediated agonists leads to stimulation of phospholipase C, which then cleaves membrane phospholipids to yield inositol trisphosphate (InsP$_3$) and diacylglycerol (DAG). InsP$_3$ stimulates the release of Ca^{2+} from intracellular stores, thereby increasing intracellular free Ca^{2+} concentrations. How elevations in intracellular Ca^{2+} lead to increases in Cl^- secretion has not been defined fully. Previously it was believed that, unlike airway epithelia, intestinal epithelia do not express Ca^{2+}-activated chloride channels and that the major point of control for Ca^{2+}-dependent Cl^- secretion is the opening of calcium-sensitive basolateral K^+ channels. The resulting efflux of K^+ from the cell is believed to bring about hyperpolarization of the cell, thus increasing the driving force for Cl^- through constitutively open apical Cl^- channels (Cartwright et al., 1985; Dharmsathaphorn and Pandol, 1985). However, evidence is accumulating that a specific Ca^{2+}-activated Cl^- channel does exist in intestinal epithelial cells and that this channel most likely represents an important point for regulation of Ca^{2+}-dependent secretion (Ismailov et al., 1996; Gruber et al., 1998; Merlin et al., 1998). The protein kinases that act downstream of elevations in intracellular Ca^{2+} also are not well understood. However, since inhibitors of Ca^{2+}-calmodulin-dependent kinase have been demonstrated to have antidiarrheal effects (Shook et al., 1989; Aikawa and Karasawa, 1998), and because calcium-activated chloride channels are substrates for this kinase (Fuller et al., 1994; Cunningham et al., 1995), it appears likely that this enzyme is involved in some way.

Intracellular regulation of Ca^{2+}-dependent Cl^- secretion appears to be much more complex than that involved in cyclic nucleotide-stimulated secretion. Whereas elevations in intracellular cAMP are quantitatively and temporally associated with changes in Cl^- secretion in response to cAMP-mediated agonists, the same is not true for Ca^{2+}-dependent hormones. Although elevations in Ca^{2+} alone are sufficient to stimulate Cl^- secretion (Kachintorn et al., 1993a), there is a poor correlation between changes in cytosolic Ca^{2+} levels and the magnitude of the resulting Cl^- secretory response (Dharmsathaphorn et al., 1989). Indeed, unlike cyclic nucleotide-dependent agonists, which may stimulate chloride secretory responses across cultured epithelial cells for as long as the agonist remains in contact with the cells, Ca^{2+}-dependent agonists stimulate secretory responses that are transient in nature, even if intracellular levels of Ca^{2+} remain elevated (Vajanaphanich et al., 1995). This suggests that inhibitory signals exist within epithelial cells, which serve to limit, or switch off, secretory responses to Ca^{2+}-mediated secretagogues (see later).

3. Arachidonic Acid

There are some stimuli of intestinal epithelial chloride secretion whose effects do not appear to be mediated by either cyclic nucleotides or intracellular Ca^{2+}. For example, the purinergic agonist adenosine, which is released from activated immune cells, induces Cl^- secretory responses that are qualitatively similar to those stimulated by cAMP-dependent agonists but that may not be accompanied by measurable increases in the intracellular levels of either cAMP, cGMP, or Ca^{2+} (Barrett et al., 1989; Madara et al., 1992). This has led to the speculation that additional pro-secretory messengers exist within intestinal epithelia. Studies in T_{84} cells implicate arachidonic acid (AA) as one candidate second messenger in mediating secretory responses to adenosine (Barrett and Bigby, 1993). The mechanism by which AA mediates stimulation of Cl^- secretion is not fully understood, but it may involve activation of a basolateral K^+ conductance distinct from those activated by either cAMP or Ca^{2+} (Devor and Frizzell, 1998a). Furthermore, the observation that AA, in addition to activating a subset of K^+ channels, can inhibit Ca^{2+}-dependent K^+ channels, implies this second messenger has a potentially pivotal role in regulation of intestinal secretion (Devor and Frizzell, 1998b).

4. Positive Interactions between Signaling Pathways

Cultured epithelial cells provide a useful tool for studying the mechanisms by which putative secretagogues, and their associated intracellular signaling pathways, interact with the Cl^- secretory mechanism in the absence of confounding extrinsic factors. However, in vivo, the situation is likely to be much more complex because the intestinal epithelium is constantly exposed to an array of neurohormonal mediators acting via different signaling mechanisms. Such a scenario also can be modeled crudely in cultured cells where combinations of different agonists may be added simultaneously to monolayers of epithelial cells and their combined effects on Cl^- secretion studied. In this way, it has been shown that when combinations of cAMP- and Ca^{2+}-dependent agonists are added to intestinal epithelial cells, synergistic Cl^- secretion results; that is, an amplified Cl^- secretory response occurs that is much greater than would be predicted by simply adding the two individual responses together (Cartwright et al., 1985). The same also has been shown to be true for combinations of either cGMP-dependent agonists and Ca^{2+}-dependent secretagogues (Levine et al., 1991), or adenosine and Ca^{2+}-dependent agonists (Barrett et al., 1990). This phenomenon of synergistic secretion most likely is relevant in conditions of intestinal inflammation, when levels of both cyclic nucleotide- and Ca^{2+}-dependent agonists may be elevated simultaneously within the intestinal mucosa.

The underlying basis of synergistic interactions between second messengers has been the subject of considerable research. An original hypothesis for the phenomenon was based on the idea that Cl^- secretion stimulated by cyclic nucleotides or Ca^{2+} is governed by different rate-limiting steps. It was believed that the major rate-limiting step for Ca^{2+}-dependent Cl^- secretion lies in the opening of apical Cl^- channels, whereas for cAMP-dependent secretion it is in the opening of basolateral K^+ channels (Barrett, 1993). When both pathways are activated simultaneously, the rate-limiting step for each type of secretion is removed and synergistic secretion results. However, more recent studies indicate that synergistic Cl^- secretion also may arise from interactions between signaling pathways at the level of the second-messenger generation. Although release of Ca^{2+} from intracellular stores promotes Cl^- secretion, it has been demonstrated that influx of Ca^{2+} from the extracellular milieu may serve as, or parallel, an inhibitory signal that limits the extent of Ca^{2+}-dependent Cl^- secretion, by an as yet undefined mechanism. cAMP has been shown to inhibit this Ca^{2+} influx step in intestinal epithelial cells, perhaps thereby removing its inhibitory influence and thus contributing to a greater overall secretory response (Vajanaphanich et al., 1995).

C. Intracellular Regulation of Chloride Secretion: Negative Regulation

As alluded to above, the characteristics of calcium-mediated chloride secretion suggest that negative mechanisms intrinsic to the epithelium serve to limit the overall extent of the secretory response. To reiterate, calcium-mediated chloride secretion is transient, even in the face of continued agonist presence and when elevations in cytosolic calcium concentrations persist. Furthermore, the extent of the chloride secretory response correlates poorly with the absolute magnitude of the increase in cytosolic calcium when responses to a range of agonists are examined (Barrett, 1997). Both of these points are in marked contradistinction to the secretory response induced by agonists that stimulate an increase in cellular cyclic nucleotides (Montrose *et al.*, 1999). Some time ago, these features of calcium-dependent chloride secretion led to the hypothesis that calcium-mediated agonists also produce additional second messengers within stimulated cells that interact with calcium and modify its effects on intracellular targets, either directly or indirectly (Barrett, 1993) (see Table II). Since that time, two major pathways whereby calcium-mediated chloride secretion is limited have been delineated (Vajanaphanich *et al.*, 1994; Uribe *et al.*, 1996a). Moreover, recent work has suggested that there are points of intersection between these negative signaling mechanisms (Keely *et al.*, 1998b).

1. Negative Regulation Mediated by Inositol Tetrakisphosphate

A starting point for the identification of negative signaling pathways that regulate chloride secretion in colonic epithelial cells came from the observation that certain calcium-dependent agonists, such as carbachol, rendered cells refractory to subsequent stimulation with a non-receptor-dependent means to elevate cytosolic calcium, such as thapsigargin. In contrast, other calcium-dependent agents, such as histamine, appeared not to evoke such negative signaling events (Kachintorn *et al.*, 1993b). This negative influence on secretion did not result from the failure of thapsigargin to increase cytosolic calcium in cells that had been pretreated with carbachol. Rather, the inhibition apparently reflected the "uncoupling" of increases in calcium from the downstream response of chloride secretion. A consideration of biochemical events that correlated with this "uncoupling" phenomenon then showed that carbachol, but not histamine, led to a sustained increase in an inositol phosphate metabolite, inositol (3,4,5,6) tetrakisphosphate [$Ins(3,4,5,6)P_4$], that had hitherto been considered an "orphan" messenger (Kachintorn *et al.*, 1993b; Menniti *et al.*, 1993). Subsequent studies indicated that the production of $Ins(3,4,5,6)P_4$ in cells stimulated with carbachol depended on tyrosine kinase activity, and that exogenous addition of a cell-permeant analogue of this molecule, resulting in biologically relevant intracellular concentrations, could reproduce the inhibitory effect of carbachol on subsequent calcium-mediated chloride secretion (Vajanaphanich *et al.*, 1994).

We and other authors went on to define the membrane target acted on by $Ins(3,4,5,6)P_4$ to exert its inhibitory effect on chloride secretion. As alluded to above, the positive signaling arm that is elicited by carbachol and other calcium-dependent stimuli to evoke chloride secretion is believed to impinge predominantly on a basolateral potassium channel as a primary target, although more recently a role for apical calcium-activated chloride conductance also has been suggested (Devor and Frizzell, 1993; Merlin *et al.*, 1998; Montrose *et al.*, 1999). Both patch-clamp and nuclide-efflux studies suggested that pretreatment of epithelial cells with carbachol could alter the subsequent activity of a chloride, but not a potassium, channel (Xie *et al.*, 1996; 1998; Barrett *et al.*, 1998). This effect was reproduced by both Ins(3,4,5,6)P_4 and a cell-permeant analogue of this messenger (Xie *et al.*, 1996; Barrett *et al.*, 1998). The involvement of a calcium-activated chloride channel (CaCC) as a target for Ins(3,4,5,6)P_4 was further substantiated by studies in a lipid bilayer model, using a cloned CaCC that was exogenously expressed in oocytes (Ismailov *et al.*, 1996). These experiments indicated that Ins(3,4,5,6)P_4, but not other isomers of this molecule nor Ins(1,4,5)P_3 or InsP$_5$, even at higher concentrations, was able to induce a profound reduction in CaCC open probability. However, the effect of Ins(3,4,5,6)P_4 in this system was complex, and dependent on both the ambient calcium concentration and the phosphorylation status of the channel. When unphosphorylated, CaCC opens only at very high concentrations of calcium that exceed the biological range (Fuller *et al.*, 1994). Following phosphorylation by calcium/calmodulin kinase II, the dose-response curve for calcium shifts to the left, such that channel opening occurs within the range of calcium concentrations that might be expected in stimulated cells, and actually is inhibited at higher calcium concentrations. In the presence of Ins(3,4,5,6)P_4, the activation curve shifts even further to the left, such that the channel is open only at calcium concentrations below those observed in resting cells, and is closed within the calcium range seen in stimulated cells (Fuller *et al.*, 1994; Ismailov *et al.*, 1996). These data, in total, indicate that Ins(3,4,5,6)P_4 likely targets an apical CaCC to exert its inhibitory effect on calcium-dependent chloride secretion, and probably does so via a direct interaction with the channel rather than by intermediary signaling events. Thus, an agonist such as carbachol initiates a positive effect on chloride secretion initially, mediated by an increase in Ins(1,4,5)P_3, a consequent increase in cytosolic calcium, and the activation of both CaCC and a basolateral potassium channel (Barrett, 1993). However, subsequently the response is turned off, at least in part because of the influence of the increasing concentrations of Ins(3,4,5,6)P_4 that rise progressively after carbachol addition (Vajanaphanich *et al.*, 1994). The fact that Ins(1,4,5)P_3 levels are increased only transiently after carbachol treatment, whereas the increase in Ins(3,4,5,6)P_4 is very sustained, suggests that the long-term effect of carbachol and related agonists on chloride secretion is likely to be inhibitory, unless coordinated with a response induced by cAMP (Vajanaphanich *et al.*, 1994; 1995).

2. Negative Regulation Mediated by Phosphatidylinositol 3-Kinase

The finding that the inhibitory effect of carbachol was somehow dependent on tyrosine kinase activity (Vajanaphanich *et al.*, 1993), and the knowledge that intestinal epithelial cells express growth factor receptors that contain intrinsic tyrosine kinase activity (Coffey *et al.*, 1995; Uribe and Barrett, 1997), also led to the delineation of a second pathway for the limitation of calcium-dependent chloride secretion. The prototypic peptide growth factor, epidermal growth factor (EGF), was shown to reduce epithelial chloride secretion evoked by calcium-dependent agonists (Uribe *et al.*, 1996a). However, unlike the dual effect of agonists such as carbachol on secretory responses, EGF itself did not serve as an agonist of secretion; rather it only inhibited responses to calcium-dependent stimuli such as thapsigargin, or carbachol itself. On some levels, the inhibitory effect of EGF appeared somewhat similar to that induced by carbachol. It reflected an "uncoupling" of changes in cytosolic calcium from the downstream response of secretion, and was associated with an increase in cellular levels of $Ins(3,4,5,6)P_4$, albeit to a lesser extent than that induced by carbachol (Vajanaphanich *et al.*, 1994; Uribe *et al.*, 1996b). However, several pieces of data then showed that the inhibitory effect of EGF must employ a mechanism that is at least partially independent of that evoked by carbachol and $Ins(3,4,5,6)P_4$. First, an inhibitory effect of EGF on chloride secretion could be observed as early as 1 minute after growth factor addition, a time when no detectable increase in $Ins(3,4,5,6)P_4$ was observed. Second, the inhibitory effects of carbachol and EGF were additive, even at maximally effective concentrations of each agent, implying that they must use independent pathways. Finally, nuclide-efflux studies showed that the target of EGF's inhibitory effect on secretion was most likely a basolateral potassium conductance (Barrett *et al.*, 1998). *In toto*, these various observations mandated a search for an alternative signaling mechanism whereby EGF and related growth factors might inhibit chloride secretion.

One alternate signaling cascade that has been linked to EGF receptor occupancy is that involving the tyrosine-kinase–dependent activation of the lipid kinase phosphatidylinositol 3-kinase (PI 3-kinase) (Toker and Cantley, 1997; Krymskaya *et al.*, 1999). EGF stimulates a rapid increase in the levels of two lipid products of this enzyme, phosphatidylinositol (PtdIns) (3,4) P_2 and $PtdIns(3,4,5)P_3$ in T_{84} cells, within 1 minute of addition of the growth factor (Uribe *et al.*, 1996b). These kinetics compare favorably with the rapidity of onset of the inhibitory effect of EGF on chloride secretion. Furthermore, EGF induces the recruitment of both the p85 regulatory and p110 catalytic subunits of PI 3-kinase to antiphosphotyrosine immunoprecipitates, and an increase in the *in vitro* PI 3-kinase activity of such immunoprecipitates. Importantly, the ability of EGF to stimulate PI 3-kinase activity is not shared by carbachol (Uribe *et al.*, 1996b). Finally, wortmannin, a relatively specific inhibitor of PI 3-kinase, blocked the accumulation of lipid products of this enzyme in T_{84} cells treated with EGF and also reversed the inhibitory effects of EGF on calcium-dependent chloride secretion. As expected, wortmannin and related inhibitors had no effect on the ability of carbachol to inhibit subsequent calcium-dependent chloride secretion. In sum, these findings imply that EGF recruits a tyrosine-kinase–dependent signaling cascade that involves activation of PI 3-kinase, and that the products of this enzyme in turn interact directly or indirectly with a basolateral potassium channel. However, the precise mechanism by which these lipids can alter channel activity is still a topic for investigation.

It is conceivable that at least part of the ability of PI 3-kinase to regulate intestinal chloride secretion could be at the level of the cytoskeleton and/or active delivery and retrieval of transport proteins to and from the plasma membrane. PI 3-kinase has been recognized as an important contributor to the regulation of vesicular transport (Jones and Clague, 1995; 1998; Toker and Cantley, 1997; Burd et al., 1998), which is increasingly understood as a mechanism that may etermine the plasma membrane abundance, and thus capacity, of some (but not all) membrane transport proteins, such as CFTR and NKCC1 (D'Andrea et al., 1996; Takahashi et al., 1996; Matthews et al., 1997; 1998). For example, Matthews et al., have shown that protein kinase C-ε, a putative target of the 3-phosphorylated lipids that are the products of PI 3-kinase (Toker, 1998), appears to participate in the endocytosis of NKCC1 from the basolateral membrane of T_{84} cells (Song and Matthews, 1999). It is also of interest that other peptide growth factors and hormones that share the ability of EGF to cause PI 3-kinase activation, such as transforming growth factor-α (TGF-α), insulin, and insulin-like growth factor, also can inhibit calcium-dependent chloride secretion without serving as agonists of this process (Uribe and Barrett, 1997). These findings may have implications for the control of chloride secretion in various physiological and pathophysiological settings, such as during mucosal injury and in the postprandial period. These topics will be discussed in greater detail in the following section.

3. Convergence and Divergence of Negative Signaling Mechanisms

As outlined earlier, two pathways appear to limit the overall extent of the calcium-dependent chloride secretory mechanism in intestinal epithelial cells. The first is activated by carbachol, and perhaps by additional ligands active at G-protein–coupled receptors, such as UTP (Barrett et al., 1998), and is mediated at least in part by the soluble messenger Ins(3,4,5,6)P_4. The second mechanism is activated by growth factors such as EGF, and is mediated by membrane-delimited products of PI 3-kinase. However, emerging evidence also implies some convergence in the signaling cascades used by carbachol and EGF to limit calcium-dependent chloride secretion. Thus, we observed that the ability of carbachol to reduce chloride secretion apparently was dependent on tyrosine kinase signaling, in that both inhibition of chloride secretion and elevation in Ins(3,4,5,6)P_4 induced by carbachol were coordinately reversed by the general tyrosine kinase inhibitor genistein (Vajanaphanich et al., 1993). More recently, it has emerged that at least part of the requirement for tyrosine kinase activation in signaling pathways evoked by carbachol represents transactivation of the EGF receptor (Keely et al., 1998b). In turn, this results in the recruitment of downstream components of the mitogen-activated protein (MAP) kinase cascade, with stimulation of the ERK 1 and 2 isoforms of MAP kinase. The significance of these signaling events for the regulation of chloride secretion was revealed by the observation that either an EGF kinase inhibitor (tyrphostin AG1478) or an inhibitor of MAP kinase activation reduced ERK phosphorylation and potentiated carbachol-stimulated chloride secretory responses. These data imply that recruitment of the MAP kinase cascade via EGF receptor transactivation represents a mechanism whereby the response of intestinal epithelial cells to calcium-dependent agonists is intrinsically limited. However, the downstream events that link MAP kinase activation to an inhibitory effect on the chloride secretory machinery have not been established. It is possible that the ability of carbachol to stimulate an increase in $Ins(3,4,5,6)P_4$ is regulated by such a mechanism, but this as well as other possible independent pathways to the inhibition of chloride secretion have not been explored. Moreover, transactivation of the EGFr and recruitment of the MAP kinase cascade could represent an "auto-inhibitory" influence of carbachol on its own secretory response, whereas the $Ins(3,4,5,6)P_4$-dependent mechanism might be more important in the longer term desensitization of the epithelium to restimulation by additional calcium-dependent secretagogues.

The findings described above, where at least a portion of the inhibitory effect of carbachol appears to be secondary to EGF receptor transactivation, do raise a paradox because of the differences in the inhibitory effects of carbachol and EGF on T_{84} cell secretory function. As described previously, there is substantial evidence that these inhibitory effects involve mechanisms that are at least partially independent. Thus, it is not clear how independent pathways can arise if the two ligands evoke signaling that in fact appears to converge at the level of the EGF receptor, as an early event. Therefore, it would be of interest to understand how the signal conferred by EGF receptor transactivation in response to carbachol, as opposed to that evoked by EGF binding, is appropriately interpreted. At least part of the explanation for subsequent signaling divergence downstream of the EGF receptor may lie in interactions with other receptor subtypes related to the EGF receptor. The EGF receptor is the prototypic member of a family of four receptors known as the ErbB receptors (Gullick and Srinivasan, 1998). Following ligand binding, EGF receptors can either homodimerize or heterodimerize with other family members if these also are present in the same cell. Because there is variation in the homo- or heterodimers that are formed, the components of downstream signaling cascades that are recruited also could be different.

We showed that stimulation of T_{84} cells with EGF, but not with carbachol, leads to the apparent recruitment of ErbB2 to the EGF receptor and thus the formation of EGF receptor/ErbB2 heterodimers (Keely et al., 1998a). This is of interest because the ErbB2 receptor is a so-called orphan receptor with no known ligand, but apparently also with the highest level of catalytic activity of any ErbB receptor (Huang et al., 1998). EGF also stimulated the recruitment of the PI 3-kinase p85 regulatory subunit to ErbB2, but carbachol was inactive in this regard, consistent with the inability of carbachol to stimulate PI 3-kinase activity in T_{84} cells (Uribe *et al.*, 1996b; Keely *et al.*, 1998a). Thus, signal coding at the level of the basolateral membrane of T_{84} cells, and presumably other intestinal epithelial cells, may depend on the relative formation of EGF receptor homodimers versus recruitment of other *Erb*B family members. In turn, different sets of downstream effectors may be activated by the different receptor complexes that form.

V. SUMMARY

A. Model for Integrated Regulation of Intestinal Chloride Secretion

Hopefully, it is apparent from the foregoing that the chloride secretory function of the intestinal epithelium is the subject of a plethora of regulatory influences. At the extracellular level, the epithelium is acted on by mediators derived from the endocrine, neurocrine, and mucosal immune systems, as well as by paracrine pathways, with the relative contributions of each component likely to vary according to the precise physiological state of the intestine (e.g., fed vs fasted) and the underlying presence of any pathology (e.g., the activation of mucosal immune effector cells in the setting of intestinal inflammation). However, with regard to the latter point, it is important to remember that the mucosal immune system is likely significant as a regulator of secretory function not only in pathological settings but also as a contributor to responses to normal "physiological" agonists. In part, the immune system may be recruited by the enteric nervous system (Frieling *et al.*, 1994; Wang *et al.*, 1995; Santos *et al.*, 1998b), and mast cells may be activated directly by endogenous chloride secretagogues (such as bile acids) to stimulate chloride secretion via the release of mediators such as histamine (Gelbmann *et al.*, 1995). It is also important to consider that paracrine, juxtacrine, and autocrine regulatory pathways may contribute to the control of chloride secretion in the intestine. For example, bidirectional interactions with underlying intestinal myofibroblasts almost certainly can stimulate the chloride secretory process (Hinterleitner *et al.*, 1996). Similarly, mediators produced by epithelial cells themselves, such as prostaglandins generated in response to bacterial invasion, or growth factors such as TGF-α, may act in the local environment to alter the function of neighboring cells (Babyatsky *et al.*, 1996; Eckmann *et al.*, 1997; Uribe and Barrett, 1997; Geoke *et al.*, 1998).

B. Physiological Implications

At this point, we understand relatively little about the minute-to-minute regulation of epithelial chloride secretion that is required to subserve the normal physiological function of the gastrointestinal tract. However, the fact that measurable active secretion occurs throughout the intestinal system (presumably to provide for an appropriately fluid environment for normal digestion and absorption), the observation that this is normally carefully counterbalanced with absorptive processes such that net absorption predominates overall, and the knowledge that significant disease can result when secretory processes are overexpressed, suggest that there is tight regulation of the overall secretory process (Montrose *et al.,* 1999). It is presumed that the level of secretory function, in common with other physiological processes in the gastrointestinal system, is coordinated with meal intake by the combined influence of central and local regulatory pathways. However, the extent to which the luminal presence of specific components of the meal can signal to induce secretory function in local or distant segments of the intestine is still not well understood. There is some evidence that the release of VIP from enteric nerve endings, mediated at least in part by central input, is responsible for at least a portion of secretory function in the postprandial period (Martin and Shuttleworth, 1996; Cooke *et al.,* 1997b; Guandalini, 1997). Similarly, both basal and stimulated secretion is affected by cholinergic blockade, perhaps reflecting vagal influences (Hogenauer *et al.,* 1999).

The potential for synergistic interactions between combinations of chloride secretagogues acting through different signaling mechanisms likely has physiological import. Thus, calcium-dependent chloride secretory responses are limited in their extent and transient unless coordinated with responses induced by cyclic nucleotides. This implies that there may be circumstances in which calcium- and cAMP-dependent agonists (such as acetylcholine and VIP) are released simultaneously by enteric nerve endings to result in a greater secretory response than that which could be evoked by a cAMP-dependent agonist acting alone. Furthermore, the fact that there are negative signaling mechanisms intrinsic to the epithelium that normally limit the extent of calcium-dependent secretagogues may underpin the utility of the calcium-dependent mechanism to subserve a need for short-term secretory responses. For example, Sidhu and Cooke (1995) have described a reflex arc whereby mucosal stoking evokes a transient secretory response of the mucosa, involving cholinergic mechanisms. Because mucosal stroking can be considered to model the passage of a bolus of food along the length of the gastrointestinal tract, the ability of this stimulus to evoke a short burst of secretion may provide for lubrication of the epithelium in a localized fashion as the bolus passes, without evoking a more extensive and sustained (and perhaps inappropriate) secretory response that could lead to excessive losses of fluid and electrolytes.

The mechanisms that regulate chloride secretion over a longer time scale, presumably involving changes in the level of transcription and/or translation of membrane transport proteins and regulatory elements, are even less well understood. By analogy with changes in colonic sodium absorptive processes that accompany salt deprivation or overfeeding (Barbry and Hofman, 1997), the chloride secretory process is likely modulated by alterations in whole-body chloride homeostasis over a prolonged period. However, the precise details of any such regulation remain to be elucidated. However, some data indicate that at least one regulatory system can be chronically regulated; guanylin expression is significantly depressed in animals subjected to salt restriction, which perhaps implies an important role for this novel hormone in salt homeostasis (Li *et al.,* 1996). It is to be expected that our burgeoning database on the molecular basis of the chloride secretory mechanism will spawn additional insight into how this process is regulated under both physiological and pathophysiological circumstances.

C. Disease Implications

A full understanding of the mechanisms that regulate intestinal chloride secretion may have implications for the pathogenesis and treatment of disease states that involve epithelial dysfunction. Significant morbidity can result from both over- and under-expression of chloride secretion, such as in diarrheal disorders and cystic fibrosis, respectively (Montrose *et al.,* 1999). It follows that a full understanding of the regulatory processes that set the overall extent of chloride secretion will likely have therapeutic implications.

The regulatory pathways described herein may be subverted to result in excessive secretory responses in the settings of intestinal infections and inflammation. For example, the heat-stable enterotoxin of *Escherichia coli* acts as a "super agonist" of the apical guanylin receptor to evoke sustained cGMP-dependent chloride secretion. The structure of the toxin confers relative resistance to degradation by luminal proteases, thus prolonging the ability of the toxin to induce a secretory response. Secretion induced by exogenous toxins may involve recruitment of other regulatory systems in addition to being mediated by a direct effect of the toxin on epithelial cells themselves. For example, the full expression of a secretory response to cholera toxin clearly involves activation of the enteric nervous system, and secretory effects of the toxin can be observed even in remote segments of the intestine that are not in direct contact with the offending microorganism or its toxin product (Nocerino *et al.,* 1995; Lundgren *et al.,* 1997). Similarly, toxins elaborated by *Clostridium difficile,* the major causative agent in antibiotic-associated diarrhea (Hogenauer *et al.,* 1998), can activate a prosecretory cascade that involves stimulation of sensory nerves and subepithelial mast cells, and the eventual recruitment of circulating inflammatory cells such as neutrophils that release additional substances with stimulatory effects on chloride secretion.

The synergistic chloride secretory responses that result when the epithelium is exposed to combinations of secretory agonists acting through different second messenger pathways are likely of paramount importance in the setting of intestinal inflammation. There, the epithelium is subjected to a barrage of secretagogues arising from both resident and infiltrating immunologic effector cells. A similar onslaught of pro-secretory factors is implicated in the setting of infectious diarrheal disorders. With the latter, the ability to call on a rapidly increased rate of secretory function, by virtue of synergistic interactions, can be considered a primitive host defense response designed to "flush out" the offending microorganism. However, this defensive reaction also carries the liability for severe diarrheal symptoms in inflammatory settings that do not share a microbial pathogenesis. The observation of synergistic chloride secretory responses also implies that there may be some potential for antidiarrheal therapies in these settings that target one or a few key secretagogues, because the corollary of synergism is that if one signaling pathway is no longer operative, then secretion will no longer be potentiated.

The negative signaling pathways that limit the extent of calcium-dependent chloride secretion may have implications in the setting of cystic fibrosis. In this disease, cAMP- (and likely cGMP-) dependent chloride secretion is either absent or at least markedly impaired in a variety of organs, including the intestinal tract and airways. In the airways, studies have shown that the calcium-dependent chloride secretory mechanism remains patent (Chan *et al.*, 1992), and perhaps is a suitable therapeutic target to bypass the secretory defect (Rubin, 1999; Zeitlin, 1999). The extent to which calcium-dependent chloride secretion is expressed has been shown to be a determinant of disease severity in mice that have been engineered to display the cystic fibrosis defect (Grubb and Boucher, 1999). However, in the intestine, calcium-dependent chloride secretion is defective in cystic fibrosis in common with the cyclic nucleotide-dependent secretory mechanism (Goldstein *et al.*, 1991; Ohmichi *et al.*, 1994). Initially this was believed to represent a lack of a calcium-activated chloride channel in intestinal epithelial cells, and thus a dependence of the calcium-dependent chloride secretory mechanism on the primary activation of basolateral potassium channels, ultimately providing a driving force for chloride exit across CFTR channels in normal tissues. However, more recent studies have suggested that calcium-activated chloride channels are likely present in intestinal as well as airway epithelia (Gruber *et al.*, 1998), and thus the failure to detect a chloride secretory response in intestinal tissues from patients with cystic fibrosis might instead reflect negative regulatory mechanisms of the type discussed in this chapter. It follows that a full understanding of the nature of these inhibitory signaling mechanisms might ultimately provide for optimization of therapies that target calcium-activated chloride secretion as a backup pathway in cystic fibrosis.

VI. CONCLUDING REMARKS

Intestinal chloride secretion is a process that is subject to a large number of regulatory influences at both the intra- and intercellular levels, and is likely regulated on a minute-to-minute basis, as well as in the longer term, in response to changes in whole-body fluid and electrolyte status. The plethora of regulatory influences on this process emphasizes the pivotal physiological importance of this transport mechanism for normal intestinal function. Moreover, the implications of knowledge in this area are underscored by the knowledge that substantial morbidity can result from both under- and over-expression of active chloride secretion. The last few years have witnessed an explosive growth in our understanding of the molecular basis of the chloride secretory mechanism and its regulation. Hopefully, such knowledge ultimately will allow for the development of more effective therapies for conditions such as cystic fibrosis, infectious diarrhea, and inflammatory bowel diseases.

Acknowledgments

The authors thank Glenda Wheeler-Loessel for assistance with manuscript submission. Studies from the authors' laboratory have been supported by grants from the National Institutes of Health (DK28305, DK53480, and DK35108 [Project 5]) to K.E.B., and by a Career Development Award from the Crohn's and Colitis Foundation of America to S.J.K.

References

Aikawa, N., and Karasawa, A. (1998). Effects of KW-5617 (zaldaride maleate), a potent and selective calmodulin inhibitor, on secretory diarrhea and on gastrointestinal propulsion in rats. *Jpn. J. Pharmacol.* **76**, 199–206.

Anderson, J. M. (1996). Cell signalling: MAGUK magic. *Curr. Biol.* **6**, 382–384.

Anderson, M. P., and Welsh, M. (1991). Calcium and cAMP activate different chloride channels in the apical membrane of normal and cystic fibrosis epithelia. *Proc. Natl. Acad. Sci. USA* **88**, 603–607.

Anderson, M. P., and Welsh, M. J. (1992). Regulation by ATP and ADP of CFTR chloride channels that contain mutant nucleotide-binding domains. *Science* **257**, 1701–1704.

Arizono, N., Matsuda, S., Hattor, T., Kojima, Y., Maeda, T., and Galli, S. J. (1990). Anatomical variation in mast cell nerve associations in the rat small intestine, heart, lung, and skin: Similarities of distances between neural processes and mast cells, eosinophils, or plasma cells in the jejunal lamina propria. *Lab. Invest.* **62**, 626–634.

Arreola, J., Melvin, J. E., and Begenisich, T. (1998). Differences in regulation of Ca^{2+}-activated Cl channels in colonic and parotid secretory cells. *Am. J. Physiol.* **274**, C161–C166.

Austin, K. F. (1992). Systemic mastocytosis. *N. Engl. J. Med.* **326**, 639–640.

Babyatsky, M. W., and Podolsky, D. K. (1999). Growth and development of the gastrointestinal tract. *In* "Textbook of Gastroenterology" (T. Yamada, ed.), pp. 547–584. Lippincott, Williams, and Wilkins, Philadelphia.

Babyatsky, M. W., Rossiter, G., and Podolsky, D. K. (1996). Expression of transforming growth factors α and β in colonic mucosa in inflammatory bowel disease. *Gastroenterology* **110**, 975–984.

Baird, A. W., Coombes, R. R. A., McLaughlin, P., and Cuthbert, A. W. (1984). Immediate hypersensitivity reactions to cows' milk proteins in isolated epithelium from ileum of milk-drinking guinea pigs: Comparisons with colonic epithelia. *Int. Arch. Allergy Appl. Immunol.* **75,** 255–263.

Baird, A. W., Cuthbert, A. W., and Pearce, F. L. (1985). Immediate hypersensitivity reactions in epithelia from rats infected with *Nippostrongylus brasiliensis. Br. J. Pharmacol.* **85,** 787–795.

Barbry, P., and Hofman, P. (1997). Molecular biology of Na^+ absorption. *Am. J. Physiol.* **273,** G571–G585.

Barrett, K. E. (1992). Effects of histamine and other mast cell mediators on chloride secretion by a human colonic epithelial cell line. *J. Immunol.* **147,** 959–964.

Barrett, K. E. (1993). Positive and negative regulation of chloride secretion in T_{84} cells. *Am. J. Physiol.* **265,** C859–C868.

Barrett, K. E. (1997). Integrated regulation of intestinal epithelial transport: Intercellular and intracellular pathways. *Am. J. Physiol.* **272,** C1069–C1076.

Barrett, K. E., and Bigby, T. D. (1993). Involvement of arachidonic acid in the chloride secretory response of intestinal epithelial cells. *Am. J. Physiol.* **264,** C446–C452.

Barrett, K. E., Cohn, J. A., Huott, P. A., Wasserman, S. I., and Dharmsathaphorn, K. (1990). Immune-related intestinal chloride secretion. II. Effect of adenosine on T_{84} cell line. *Am. J. Physiol.* **258,** C902–C912.

Barrett, K. E., Huott, P. A., Shah, S. S., Dharmsathaphorn, K., and Wasserman, S. I. (1989). Differing effects of apical and basolateral adenosine on the colonic epithelial cell line, T_{84}. *Am. J. Physiol.* **256,** C197–C203.

Barrett, K. E., Smitham, J., Traynor-Kaplan, A. E., and Barrett, K. E. (1998). Inhibition of Ca^{2+}-dependent chloride secretion in T_{84} cells: Membrane target(s) of inhibition is agonist specific. *Am. J. Physiol.* **274,** C958–C965.

Barrett, T. A., Musch, M. W., and Chang, E. B. (1990). Chemotactic peptide effects on intestinal electrolyte transport. *Am. J. Physiol.* **259,** G947–G954.

Beguin, P., Beggah, A., Cotecchia, S., and Geering, K. (1996). Adrenergic, dopaminergic, and muscarinic receptor stimulation leads to PKA phosphorylation of the Na-K-ATPase. *Am. J. Physiol.* **270,** C457–C464.

Beltinger, J., Hawkey, C. J., and Stack, W. A. (1999). TGF-α reduces bradykinin-stimulated ion transport and prostaglandin release in human colonic epithelial cells. *Am. J. Physiol.* **276,** C848–C855.

Bern, M. J., Sturbaum, C. W., Karayalcin, S. S., Berschneider, H. M., Wachsman, J. T., and Powell, D. W. (1989). Immune system control of rat and rabbit electrolyte transport: Role of prostaglandins and enteric nervous system. *J. Clin. Invest.* **83,** 1810–1820.

Berschneider, H. M., and Powell, D. W. (1992). Fibroblasts modulate intestinal secretory responses to inflammatory mediators. *J. Clin. Invest.* **89,** 484–489.

Beubler, E., and Horina, G. (1990). $5-HT_2$ and $5-HT_3$ receptor subtypes mediate cholera toxin-induced intestinal fluid secretion in the rat. *Gastroenterology* **99,** 83–89.

Bjerknes, M., and Cheng, H. (1999). Clonal analysis of mouse intestinal epithelial progenitors. *Gastroenterology* **116,** 7–14.

Blanco, G., Sanchez, G., and Mercer, R. W. (1995). Comparison of the enzymatic properties of the Na,K-ATPase α3 β1 and α3 β2 isozymes. *Biochemistry* **34,** 9897–9903.

Blume, E. D., Taylor, C. T., Lennon, P. F., Stahl, G. L., and Colgan, S. P. (1998). Activated endothelial cells elicit paracrine induction of epithelial chloride secretion: 6-keto-PGF1α is an epithelial secretagogue. *J. Clin. Invest.* **102,** 1161–1172.

Bond, T. D., Ambikapathy, S., Mohammad, S., and Valverde, M. A. (1998). Osmosensitive Cl^- currents and their relevance to regulatory volume decrease in human intestinal T_{84} cells: Outwardly vs. inwardly rectifying currents. *J. Physiol.* **511,** 45–54.

Borman, R. A., Jewell, R., and Hillier, K. (1998). Investigation of the effects of platelet-activating factor (PAF) on ion transport and prostaglandin synthesis in human colonic mucosa *in vitro*. *Br. J. Pharmacol.* **123**, 231–236.

Bouritius, H., Oprins, J. C. J., Bindels, R. J. M., Hartog, A., and Groot, J. A. (1998). Neuropeptide Y inhibits ion secretion in the intestinal epithelium by reducing chloride and potassium conductance. *Pflugers Arch. Eur. J. Physiol.* **435**, 219–226.

Budhoo, M. R., Harris, R. P., and Kellum, J. M. (1996). 5-hydroxytryptamine–induced Cl⁻ transport is mediated by 5-HT$_3$ and 5-HT$_4$ receptors in the rat distal colon. *Eur. J. Pharmacol* **298**, 137–144.

Burd, C. G., Babst, M., and Emr, S. D. (1998). Novel pathways, membrane coats, and PI kinase regulation in yeast lysosomal trafficking. *Semin. Cell Dev. Biol.* **9**, 527–533.

Carey, H. V., and Cooke, H. J. (1989). Tonic activity of submucosal neurons influences basal ion transport. *Life Sci.* **3**, 127–138.

Cartwright, C. A., McRoberts, J. A., Mandel, K. G., and Dharmsathaphorn, K. (1985). Synergistic action of cyclic adenosine monophosphate- and calcium-mediated chloride secretion in a colonic epithelial cell line. *J. Clin. Invest.* **76**, 1837–1842.

Castagliuolo, I., Wang, C.-C., Valenick, L., Pasha, A., Nikulasson, S., Carrraway, R. E., and Pothoulakis, C. (1999). Neurotensin is a proinflammatory neuropeptide in colonic inflammation. *J. Clin. Invest.* **103**, 843–849.

Castro, G. A., Harari, Y., and Russell, D. (1987). Mediators of anaphylaxis-induced ion transport changes in small intestine. *Am. J. Physiol.* **253**, G540–G548.

Cereijido, M., Valdes, J., Shoshani, L., and Contreras, R. G. (1998). Role of tight junctions in establishing and maintaining cell polarity. *Ann. Rev. Physiol.* **60**, 161–177.

Chan, H. C., Goldstein, J., and Nelson, D. J. (1992). Alternate pathways for chloride conductance activation in normal and cystic fibrosis airway epithelial cells. *Am. J. Physiol.* **262**, C1273–C1283.

Chan, H. C., Lui, K. M., Wong, W. S., and Poon, A. M. S. (1998). Effect of melatonin on chloride secretion by human colonic T$_{84}$ cells. *Life Sci.* **62**, 2151–2158.

Chang, E. B., Musch, M. W., and Mayer, L. (1990). Interleukins 1 and 3 stimulate anion secretion in guinea pig intestine. *Gastroenterology* **98**, 1518–1524.

Chao, A. C., Sauvage, F. J., Dong, Y.-J., Wagner, J. A., Goeddel, D. V., and Gardner, P. (1994). Activation of intestinal CFTR Cl⁻ channel by heat-stable enterotoxin and guanylin via cAMP-dependent protein kinase. *EMBO J.* **13**, 1065–1072.

Cherner, J. A., Jensen, R. T., Dubois, A., O'Dorisio, T. M., Gardner, J. D., and Metcalfe, D. D. (1988). Gastrointestinal dysfunction in systemic mastocytosis. *Gastroenterology* **95**, 657–667.

Chissone, D. C., Simon, P. L., and Smith, L. (1990). Interleukin 1: Effects on rabbit ileal mucosal ion transport *in vitro*. *Eur. J. Pharmacol* **180**, 217–228.

Coffey, R. J., Gangrosa, L. M., Damstrup, L., and Dempsey, P. J. (1995). Basic actions of transforming growth factor-α and related peptides. *Eur. J. Gastroenterol. Hepatol.* **7**, 923–927.

Cooke, H. J. (1984). Influence of enteric cholinergic neurons on mucosal transport in guinea pig ileum. *Am. J. Physiol.* **246**, G263–G267.

Cooke, H. J. (1994). Neuroimmune signalling in regulation of intestinal ion transport. *Am. J. Physiol.* **266**, G167–G178.

Cooke, H. J. (1998). 'Enteric tears': Chloride secretion and its neural regulation. *News Physiol. Sci.* **13**, 269–274.

Cooke, H. J., Sidhu, M., Fox, P., Wang, Y.-Z., and Zimmermann, E. M. (1997a). Substance P as a mediator of colonic secretory reflexes. *Am. J. Physiol.* **272**, G238–G245.

Cooke, H. J., Sidhu, M., and Wang, Y. Z. (1997b). Activation of 5-HT$_{1P}$ receptors on submucosal afferents subsequently triggers VIP neurons and chloride secretion in the guinea-pig colon. *J. Auton. Nerv. Syst.* **66**, 105–110.

Cox, H. M. (1992). Pituitary adenylate cyclase activating peptides, PACAP-27, and PACAP-38: Stimulators of electrogenic ion secretion in the rat small intestine. *Br. J. Pharmacol.* **106**, 498–502.

Cox, H. M., and Cuthbert, A. W. (1989a). Secretory actions of VIP, PHI, and helodermin in rat small intestine: The effects of putative VIP antagonists upon VIP-induced ion secretion. *Regul. Pept.* **26,** 127–135.

Cox, H. M., Cuthbert, A. W., Hakanson, R., and Wahlestedt, C. (1988). The effect of neuropeptide Y and peptide YY on electrogenic ion transport in rat intestinal epithelia. *J. Physiol.* **398,** 65–80.

Cox, H. M., Ferrar, J. A., and Cuthbert, A. W. (1989b). Effects of α- and β-CGRPs upon ion transport in rat descending colon. *Br. J. Pharmacol.* **97,** 996–998.

Crowe, S. E., Luthra, G. K., and Perdue, M. H. (1997). Mast cell mediated ion transport in intestine from patients with and without inflammatory bowel disease. *Gut* **41,** 785–792.

Crowe, S. E., and Perdue, M. H. (1992). Gastrointestinal food hypersensitivity: Basic mechanisms of pathophysiology. *Gastroenterology* **103,** 1075–1095.

Crowe, S. E., and Perdue, M. H. (1993). Anti-immunoglobulin E-stimulated ion transport in human small and large intestine. *Gastroenterology* **105,** 764–772.

Crowe, S. E., Sestini, P., and Perdue, M. H. (1990). Allergic reactions of rat jejunal mucosa: Ion transport reponses to luminal antigen and inflammatory mediators. *Gastroenterology* **99,** 74–82.

Cunningham, S. A., Awayda, M. S., Bubien, J. K., Ismailov, I. I., Arrate, M. P., Berdiev, B. K., Benos, D. K., and Fuller, C. M. (1995). Cloning of an epithelial chloride channel from bovine trachea. *J. Biol. Chem.* **270,** 31016–31026.

Cuthbert, A. W., and Huxley, C. (1998). The primary and final effector mechanisms required for kinin-induced epithelial chloride secretion. *Am. J. Physiol.* **274,** G578–G583.

Cuthbert, A. W., Kirkland, S. C., and MacVinish, L. J. (1985). Kinin effects on ion transport in monolayers of HCA-7 cells, a line from human colonic adenocarcinoma. *Br. J. Pharmacol.* **86,** 3–5.

D'Andrea, L., Lytle, C., Matthews, J. B., Hofman, P., Forbush, B., and Madara, J. L. (1996). Na:K:2Cl cotransport (NKCC) of intestinal epithelial cells. Surface expression in response to cAMP. *J. Biol. Chem.* **271,** 28969–28976.

Denker, B. M., and Nigam, S. K. (1998). Molecular structure and assembly of the tight junction. *Am. J. Physiol.* **274,** F1–F9.

Denker, B. M., Saha, C., Khawaja, S., and Nigam, S. K. (1996). Involvement of a heterotrimeric G protein in tight junction biogenesis. *J. Biol. Chem.* **271,** 25750–25753.

Devor, D. C., and Frizzell, R. A. (1993). Calcium-mediated agonists activate an inwardly rectified K^+-channel in colonic secretory cells. *Am. J. Physiol.* **265,** C1271–C1280.

Devor, D. C., and Frizzell, R. A. (1998a). Modulation of K^+ channels by arachidonic acid in T_{84} cells. I. Activation of a Ca^{2+}-independent K^+ channel. *Am. J. Physiol.* **274,** C149–C160.

Devor, D. C., and Frizzell, R. A. (1998b). Modulation of K^+ channels by arachidonic acid in T_{84} cells. I. Inhibition of the Ca^{2+}-dependent K^+ channel. *Am. J. Physiol.* **274,** C138–C148.

Dharmsathaphorn, K., Cohn, J., and Beuerlein, G. (1989). Multiple calcium-mediated effector mechanisms regulate chloride secretory responses in T_{84} cells. *Am. J. Physiol.* **256,** C1224–C1230.

Dharmsathaphorn, K., and Pandol, S. J. (1985). Mechanism of chloride secretion induced by carbachol in a colonic epithelial cell line. *J. Clin. Invest.* **77,** 348–354.

Donowitz, M., and Welsh, M. J. (1986). Ca^{2+} and cyclic AMP in regulation of intestinal Na, K, and Cl transport. *Ann. Rev. Physiol.* **48,** 135–150.

Dulhanty, A. M., and Riordan, J. R. (1994). Phosphorylation by cAMP-dependent protein kinase causes a conformational change in the R-domain of the cystic fibrosis transmembrane conductance regulator. *Biochemistry* **33,** 4072–4079.

Eberhart, C. E., and Dubois, R. N. (1995). Eicosanoids and the gastrointestinal tract. *Gastroenterology* **109,** 285–301.

Eckmann, L., Stenson, W. F., Savidge, T. C., Lowe, D. C., Barrett, K. E., Fierer, J., Smith, J. R., and Kagnoff, M. F. (1997). Role of intestinal epithelial cells in the host secretory response to infection by invasive bacteria. Bacterial entry induces epithelial prostaglandin H synthase-2 expression and prostaglandin E_2 and $F_{2\alpha}$ production. *J. Clin. Invest.* **100,** 296–309.

Ewart, H. S., and Klip, A. (1995). Hormonal regulation of the Na(+)-K(+)-ATPase: Mechanisms underlying rapid and sustained changes in pump activity. *Am. J. Physiol.* **269**, C295–C311.

Fanning, A. S., Jameson, B. J., Jesaitis, L. A., and Anderson, J. M. (1998). The tight junction protein ZO-1 establishes a link between the transmembrane protein occludin and the actin cytoskeleton. *J. Biol. Chem.* **273**, 29745–29753.

Farokzhad, O., Mun, E., Smith, J., Sicklick, J., Song, J., and Matthews, J. B. (1998). Regulation of intestinal epithelial Na-K-2Cl cotransporter expression by protein kinase C. *Gastroenterology* **114**, A368.

Fasano, A., Fiorentini, C., Donelli, G., Uzzau, S., Kaper, J. B., Margeretten, K., Ding, X., Guandalini, S., Comstock, L., and Goldblum, S. E. (1995). Zonula occludens toxin modulates tight junctions through protein kinase C-dependent actin reorganization, *in vitro. J. Clin. Invest.* **96**, 710–720.

Ferguson, A., Cummins, A. G., and Munroe, G. H. (1987). Roles of mucosal mast cells in intestinal cell mediated immunity. *Ann. Allergy* **59**, 40–43.

Finlay, R. B., and Smith, P. L. (1989). Stimulation of chloride by N-formyl-methionyl-leucyl-phenylalanine (FMLP) in rabbit ileal mucosa. *J. Physiol.* **417**, 403–419.

Forte, L. R., Eber, S. L., Turner, J. T., Freeman, R. H., Fok, K. F., and Currie, M. G. (1993). Guanylin stimulation of Cl^- secretion in human intestinal T_{84} cells via cyclic guanosine monophosphate. *J. Clin. Invest.* **91**, 2423–2428.

Forte, L. R., Thorne, P. K., Eber, S. L., Krause, W. J., Freeman, R. H., Francis, S. H., and Corbin, J. D. (1992). Stimulation of intestinal Cl^- transport by heat-stable enterotoxin: Activation of cAMP-dependent protein kinase by cGMP. *Am. J. Physiol.* **263**, C607–C615.

Foskett, J. K. (1998). ClC and CFTR chloride channel gating. *Ann. Rev. Physiol.* **60**, 689–717.

Frieling, T., Cooke, H. J., and Wood, J. D. (1994). Neuroimmune communication in the submucous plexus of guinea pig colon after sensitization to milk antigen. *Am. J. Physiol.* **267**, G1087–G1093.

Fuller, C. M., Ismailov, I. I., Keeton, D. A., and Benos, D. J. (1994). Phosphorylation and activation of a bovine tracheal anion channel by Ca^{2+}/calmodulin-dependent protein kinase II. *J. Biol. Chem.* **269**, 26642–26650.

Furness, J. B., Bornstein, J. C., Kunze, W. A., and Clerc, N. (1999). The enteric nervous system and its extrinsic connections. *In* "Textbook of Gastroenterology" (T. Yamada, ed.), pp. 11–35. Lippincott, Williams, and Wilkins, Philadelphia.

Gadsby, D. C., and Nairn, A. C. (1999). Control of CFTR channel gating by phosphorylation and nucleotide hydrolysis. *Physiol. Rev.* **79**, S77–S107.

Gaginella, T. S., and Kachur, J. F. (1989). Kinins as mediators of intestinal secretion. *Am. J. Physiol.* **256**, G1–G15.

Gaginella, T. S., Kachur, J. F., Tamai, H., and Keshavarzin, A. (1995). Reactive oxygen and nitrogen metabolites as mediators of secretory diarrhea. *Gastroenterology* **109**, 2019–2028.

Gelbmann, C. M., Schteingart, C. D., Thompson, S. M., Hofmann, A. F., and Barrett, K. E. (1995). Mast cells and histamine contribute to bile acid–stimulated secretion in the mouse colon. *J. Clin. Invest.* **95**, 2831–2839.

Geoke, M., Kanai, M., Lynch-Devaney, K., and Podolsky, D. (1998). Rapid mitogen-activated protein kinase activation by transforming growth factor α in wounded rat intestinal epithelial cells. *Gastroenterology* **114**, 697–705.

Gewirtz, A. T., McCormick, B., Neish, A. S., Petasis, N. A., Gronert, K., Serhan, C. N., and Madara, J. L. (1998). Pathogen-induced chemokine secretion from model intestinal epithelium is inhibited by lipoxin A_4 analogs. *J. Clin. Invest.* **101**, 1860–1869.

Goldhill, J., and Angel, I. (1998). Mechanism of tachykinin NK3 receptor-mediated colonic transport in the guinea pig. *Eur. J. Pharmacol* **363**, 161–168.

Goldstein, J. L., Shapiro, A. B., Rao, M. C., and Layden, T. J. (1991). *In vivo* evidence of altered chloride but not potassium secretion in cystic fibrosis rectal mucosa. *Gastroenterology* **101**, 1012–1019.

Goyal, R. J., and Hirano, I. (1996). The enteric nervous system. *N. Engl. J. Med.* **334,** 1106–1115.

Greenberg, R. N., Hill, M., Crytzer, J., Krause, W. J., Eber, S. L., Hamra, F. K., and Forte, L. R. (1997). Comparison of effects of uroguanylin, guanylin, and *Escherichia coli* heat-stable enterotoxin ST_a in mouse intestine and kidney: Evidence that uroguanylin is an intestinal natriuretic hormone. *J. Invest. Med.* **45,** 276–283.

Greenwood, B. L., Tremblay, L., and Davidson, J. S. (1987). Sympathetic control of motility, fluid transport, and transmural potential difference in the rabbit ileum. *Am. J. Physiol.* **253,** G726–G729.

Greger, R., Bleich, M., and Warth, R. (1997). New types of K^+ channels in the colon. *Weiner Klin. Wochenschr.* **109,** 497–498.

Grubb, B. R., and Boucher, R. C. (1999). Pathophysiology of gene-targeted mouse models for cystic fibrosis. *Physiol. Rev.* **79,** S193–S214.

Gruber, A. D., Elble, R. C., Ji, H. L., Schreur, K. D., Fuller, C. M., and Pauli, B. U. (1998). Genomic cloning, molecular characterization, and functional analysis of human CLCA1, the first human member of the family of Ca^{2+}-activated channel proteins. *Genomics* **54,** 200–214.

Guandalini, S. (1997). Enteric nervous system: Intestinal absorption and secretion. *J. Pediatr. Gastroenterol. Nutr.* **25,** S5–S6.

Guarino, A., Canani, R. B., Iafusco, M., Casola, A., Russo, R., and Rubino, A. (1995). *In vivo* and *in vitro* effects of human growth hormone on rat intestinal ion transport. *Pediatr. Res.* **37,** 576–580.

Gullick, W. J., and Srinivasan, R. (1998). The type 1 growth factor receptor family: New ligands and receptors and their role in breast cancer. *Br. Cancer Res. Treatment* **52,** 43–53.

Haas, M., and Forbush, B. (1998). The Na-K-Cl cotransporters. *J. Bioenerg. Biomembr.* **30,** 161–172.

Hagos, Y., Krick, W., and Burckhardt, G. (1999). Chloride conductance in HT29 cells: Investigations with apical membrane vesicles and RT-PCR. *Pflugers Arch. Eur. J. Physiol.* **437,** 724–730.

Hardcastle, J., Hardcastle, P. T., Taylor, C. J., and Goldhill, J. (1991). Failure of cholinergic stimulation to induce a secretory response from the rectal mucosa in cystic fibrosis. *Gut* **32,** 1035–1039.

Hasler, W. L. (1999). Motility of the small intestine and colon. *In* "Textbook of Gastroenterology" (T. Yamada, ed.), pp. 215–245. Lippincott, Williams, and Wilkins, Philadelphia.

Hecht, G., Pestic, L., Nikcevic, G., Koutsouris, A., Tripuraneni, J., Lorimer, D. D., Nowak, G., Guerriero, V., Ellson, E. L., and De Lanerolle, P. (1996). Expression of the catalytic domain of myosin light chain kinase increases paracellular permeability. *Am. J. Physiol.* **271,** C1678–C1684.

Hinterleitner, T. A., Saada, J. I., Nerschneider, H. M., Powell, D. W., and Valentich, J. D. (1996). IL-1 stimulates intestinal myofibroblast COX gene expression and augments activation of Cl^- secretion in T_{84} cells. *Am. J. Physiol.* **271,** C1262–C1268.

Hinterleitner, T. A., Valentich, J. D., Cha, J. H., Will, P., Welton, A., and Powell, D. W. (1997). Platelet-activating factor contributes to immune cell and oxidant-mediated intestinal secretion. *J. Pharmacol. Exp. Ther.* **281,** 1264–1271.

Hofer, D., Asan, E., and Drenckhahn, D. (1999). Chemosensory perception in the gut. *News Physiol. Sci.* **14,** 18–23.

Hogenauer, C., Aichbichler, B. W., Porter, J. L., and Fordtran, J. S. (1999). Effect of atropine and octreotide on normal active chloride secretion by the human jejunum *in vivo*. *Gastroenterology* **116,** A880.

Hogenauer, C., Hammer, H. F., Krejs, G. J., and Reisinger, E. C. (1998). Mechanisms and management of antibiotic associated diarrhea. *Clin. Infect. Dis.* **27,** 702–710.

Huang, G. C., Ouyang, X., and Epstein, R. J. (1998). Proxy activation of protein *Erb*B2 by heterologous ligands and receptors implies a heterotetrameric mode of receptor tyrosine kinase interaction. *Biochem. J.* **331,** 113–119.

Hurst, R. D., Ballantyne, G. H., and Modlin, I. M. (1995). Octreotide inhibition of serotonin-induced ileal chloride secretion. *J. Surg. Res.* **59,** 631–635.

Hwang, T. C., Lu, L., Zeitlin, P. L., Gruenert, D. C., Huganir, R., and Guggino, W. B. (1989). Cl^- channels in CF: Lack of activation by protein kinase C and cAMP-dependent protein kinase. *Science* **244**, 1351–1353.

Ismailov, I. I., Fuller, C. M., Berdiev, B. K., Shlyonsky, V. G., Benos, D. J., and Barrett, K. E. (1996). A biological function for an "orphan" messenger: D-myo-inositol (3,4,5,6) tetrakisphosphate selectively blocks epithelial calcium-activated chloride channels. *Proc. Natl. Acad. Sci. USA* **93**, 10505–10509.

Jensen, B. S., Stroebaek, D., Christophersen, P., Joergensen, T. D., Hansen, C., Silahtaroglu, A., Olesen, S. P., and Ahring, P. K. (1998). Characterization of the cloned human intermediate-conductance Ca^{2+}-activated K^+ channel. *Am. J. Physiol.* **275**, C848–C856.

Jia, Y. L., Mathews, C. J., and Hanrahan, J. W. (1997). Phosphorylation by protein kinase C is required for acute activation of cystic fibrosis transmembrane conductance regulator by protein kinase. *Am. J. Biol. Chem.* **272**, 4978–4984.

Jones, A. T., and Clague, M. J. (1995). Phosphatidylinositol 3-kinase activity is required for early endosome fusion. *Biochem. J.* **311**, 31–34.

Jones, A. T., Mills, I. G., Scheidig, A. J., Alexandrov, K., and Clague, M. J. (1998). Inhibition of endosome fusion by wortmannin persists in the presence of activated Rab5. *Mol. Cell Biol.* **9**, 323–332.

Jovov, B., Ismailov, I. I., Berdiev, B. K., Fuller, C. M., Sorscher, E. J., Dedman, J. R., Kaetzel, M. A., and Benos, D. J. (1995). Interaction between cystic fibrosis transmembrane conductance regulator and outwardly rectified chloride channels. *J. Biol. Chem.* **270**, 29194–29200.

Kachintorn, U., Vajanaphanich, M., Barrett, K. E., and Traynor-Kaplan, A. E. (1993b). Elevation of inositol tetrakisphosphate parallels inhibition of Ca^{2+}-dependent Cl^- secretion in T_{84} cells. *Am. J. Physiol.* **264**, C671–C676.

Kachintorn, U., Vajanaphanich, M., and Traynor-Kaplan, A. E., Dharmsathaphorn, K, and Barrett, K. E., (1993a). Activation by calcium alone of chloride secretion in T_{84} epithelial cells. *Br. J. Pharmacol.* **109**, 510–517.

Kachintorn, U., Vongkovit, P, Vajanaphancih, M., Dinh, S., and Barrett, K. E. (1992). Dual effects of a phorbol ester on calcium-dependent chloride secretion by T_{84} epithelial cells. *Am. J. Physiol.* **262**, C15–C22.

Kagnoff, M. F. (1998). Current concepts in mucosal immunity. III. Ontogeny and function of $\gamma\delta$ T cells in the intestine. *Am. J. Physiol.* **274**, G455–G458.

Kandil, H. M., Berschneider, H. M., and Argenzio, R. A. (1994). Tumor necrosis factor α changes porcine intestinal ion transport through a paracrine mechanism involving prostaglandins. *Gut* **35**, 934–940.

Karayalcin, S. S., Sturbaum, C. W., Wachsman, J. T., Cha, J. H., and Powell, D. W. (1990). Hydrogen peroxide stimulates rat colonic prostaglandin production and alters electrolyte transport. *J. Clin. Invest.* **86**, 60–68.

Keely, S. J., and Barrett, K. E. (1998). Heregulin inhibits calcium-dependent chloride secretion: A role for *Erb*B2 and *Erb*B3. *Gastroenterology* **114**, A385.

Keely, S. J., Halverson, M. J., and Barrett, K. E. (1998a). Colonic epithelial *Erb*B2 receptors: A possible role in diversification of inhibitory signaling via the EGF receptor. *Gastroenterology* **114**, A385.

Keely, S. J., Skelly, M. M., O'Donoghue, D. P., Baird, A. W., and Barrett, K. E. (1997). Bilateral expression of functional bradykinin receptors in the human colonic epithelium. *Gastroenterology* **112**, A375.

Keely, S. J., Uribe, J. M., and Barrett, K. E. (1998b). Carbachol stimulates transactivation of epidermal growth factor receptor and mitogen-activated protein kinase in T_{84} cells. *J. Biol. Chem.* **273**, 27111–27117.

Kim, E. C., Zhu, Y., Andersen, V., Sciaky, D., Cao, H. J., Meekins, H., Smith, T. J., and Lance, P. (1998). Cytokine-mediated PGE_2 expression in human colonic fibroblasts. *Am. J. Physiol.* **275**, C988–C994.

Kirchgessner, A. L., and Gershon, M. D. (1988). Projections of submucosal neurons to the myenteric plexus of the guinea pig intestine: *In vitro* tracing of microcircuits by retrograde and anterograde transport. *J. Comp. Neurol.* **236**, 487–498.

Krymskaya, V. P., Hoffman, R., Eszterhas, A., Kane, S., Ciocca, V., and Panettieri, R. A. (1999). EGF activates ErbB-2 and stimulates phosphatidylinositol 3-kinase in human airway smooth muscle cells. *Am. J. Physiol.* **276**, L246–L255.

Kuhn, M., Adermann, K., Jahne, J., Forssmann, W. G., and Rechkemmer, G. (1994). Segmental differences in the effects of guanylin and *Escherichia coli* heat-stable enterotoxin on Cl^- secretion in human gut. *J. Physiol.* **479**, 433–440.

Kuhn, M., Fuchs, M., Beck, F. X., Martin, S., Jahne, J., Klempnauer, J., Kaever, V., Rechkemmer, G., and Forssmann, W. G. (1997). Endothelin-1 potently stimulates chloride secretion and inhibits Na^+-glucose absorption in human intestine *in vitro*. *J. Physiol.* **499**, 391–402.

Kuwahara, A., Kuramoto, H., and Kadowaki, M. (1998). 5-HT activates nitric oxide generating neurons to stimulate chloride secretion in guinea pig distal colon. *Am. J. Physiol.* **1998**, G829–G834.

Kuwahara, A., Kuwahara, Y., Mochizuki, T., and Yanaihara, N. (1993). Action of pituitary adenylate cyclase activating polypeptide on ion transport in guinea pig distal colon. *Am. J. Physiol.* **264**, G433–G441.

LaMont, J. T. (1992). Mucus: The front line of intestinal defense. *Ann. NY Acad. Sci.* **664**, 190–201.

Lehrich, R. W., Aller, S. G., Webster, P., Marino, C. R., and Forrest, J. N. (1998). Vasoactive intestinal peptide, forskolin, and genistein increase apical CFTR trafficking in the rectal gland of the spiny dogfish, *Squalus acanthias*. *J. Clin. Invest.* **101**, 737–745.

Lencer, W. I., Cheung, G., Stroheimer, G. R., Currie, M. G., Ouellette, A. J., Selsted, M. E., and Madara, J. L. (1997). Induction of epithelial chloride secretion by channel-forming cryptdins 2 and 3. *Proc. Natl. Acad. Sci. USA* **94**, 8585–8589.

Levine, S. A., Donowitz, M., Watson, J. M., Sharp, W. G., Crane, J. K., and Weikel, C. S. (1991). Characterization of the synergistic interaction of *Escherichia coli* heat-stable enterotoxin and carbachol. *Am. J. Physiol.* **261**, G592–G601.

Li, Z., Knowles, J. W., Goyeau, D., Prabhakar, S., Short, D. B., Perkins, A. G., and Goy, M. F. (1996). Low salt intake downregulates the guanylin signaling pathway in rat distal colon. *Gastroenterology* **111**, 1714–1721.

Liu, C. D., Hines, O. J., Whang, E. E., Balasubramaniam, A., Newton, T. R., Zinner, M. J., Ashley, S. W., and McFadden, D. W. (1995). A novel synthetic analog of peptide YY, BIM-43004, given intraluminally, is proabsorptive. *J. Surg. Res.* **59**, 80–84.

Lundgren, O., and Jodal, M. (1997). The enteric nervous system and cholera toxin-induced secretion. *Comp. Biochem. Physiol.* **118**, 319–327.

Luo, J., Pato, M. D., Riordan, J. R., and Hanrahan, J. W. (1998). Differential regulation of single CFTR channels by PP2C, PP2A, and other phosphatases. *Am. J. Physiol.* **274**, C1397–C1410.

Lytle, C., and Forbush, B. (1992). The Na-K-Cl cotransport protein of shark rectal gland. II. Regulation by direct phosphorylation. *J. Biol. Chem.* **267**, 25438–25443.

Lytle, C., and Forbush, B. (1996). Regulatory phosphorylation of the secretory Na-K-Cl cotransporter: Modulation by cytoplasmic Cl. *Am. J. Physiol.* **270**, C437–C448.

Madara, J. L., Parkos, C., Colgan, S., MacLeod, R. J., Nash, S., Matthews, J., Delp, C., and Lencer, W. (1992). Cl^- secretion in a model intestinal epithelium induced by a neutrophil-derived secretagogue. *J. Clin. Invest.* **89**, 1938–1944.

Madara, J. L., Patapoff, T. W., Gillece-Castro, B., Colgan, S. P., Parkos, C., Delp, C., and Mrsny, R. J. (1993). 5'-Adenosine monophosphate is the neutrophil-derived paracrine factor that elicits chloride secretion from T_{84} intestinal epithelial cell monolayers. *J. Clin. Invest.* **91**, 2320–2325.

Madsen, K. L., Lewis, S. A., Tavernini, M. M., Hibbard, J., and Fedorak, R. N. (1997). Interleukin 10 prevents cytokine-induced disruption of T_{84} monolayer barrier integrity and limits chloride secretion. *Gastroenterology* **113**, 151–159.

Madsen, K. L., Tavernini, M. M., Mosmann, T. R., and Fedorak, R. N. (1996). Interleukin 10 modulates ion transport in rat small intestine. *Gastroenterology* **111,** 936–944.

Mandel, K. G., Dharmsathaphorn, K., and McRoberts, J. A. (1986). Characterization of a cyclic AMP-activated Cl$^-$ transport pathway in the apical membrane of a human colonic epithelial cell line. *J. Biol. Chem.* **261,** 704–712.

Martin, S. C., and Shuttleworth, T. J. (1996). The control of fluid-secreting epithelia by VIP. *Ann. NY Acad. Sci.* **805,** 133–147.

Mathison, R., Davidson, J. S., and Befus, A. D. (1993). Neural regulation of neutrophil involvement in pulmonary inflammation. *Comp. Biochem. Physiol.* **106C,** 39–48.

Matthews, J. B., Awtrey, C. S., Hecht, G., Tally, K. J., Thompson, R. S., and Madara, J. L. (1993a). Phorbol ester sequentially downregulates cAMP-regulated basolateral and apical Cl$^-$ transport pathways in T$_{84}$ cells. *Am. J. Physiol.* **265,** C1109–C1117.

Matthews, J. B., Awtrey, C. S., and Madara, J. L. (1992). Microfilament-dependent activation of Na$^+$/K$^+$/2Cl$^-$ cotransport by cAMP in intestinal epithelial monolayers. *J. Clin. Invest.* **90,** 1608–1613.

Matthews, J. B., Awtrey, C. S., Thompson, R., Hung, T., Tally, K. J., and Madara, J. L. (1993b). Na/K/2Cl cotransport and chloride secretion evoked by heat-stable enterotoxin is microfilament-dependent in T$_{84}$ cells. *Am. J. Physiol.* **265,** C370–C378.

Matthews, J. B., Smith, J. A., and Hrnjez, B. J. (1997). Effects of F-actin stabilization or disassembly in epithelial Cl$^-$ secretion and Na-K-2Cl cotransport. *Am. J. Physiol.* **272,** C254–C262.

Matthews, J. B., Smith, J. A., Mun, E. C., and Sicklick, J. K. (1998). Osmotic regulation of intestinal epithelial Na$^+$-K$^+$-2Cl$^-$ cotransport: Role of Cl$^-$ and F-actin. *Am. J. Physiol.* **274,** C697–C706.

Mayol, J. M., Hrnjez, B. J., Akbarali, H. I., Song, J. C., Smith, J. A., and Matthews, J. B. (1997a). Ammonia effect on calcium-activated chloride secretion in T$_{84}$ intestinal epithelial monolayers. *Am. J. Physiol.* **273,** C634–C642.

Mayol, J. M., Mun, E. C., Hrnjez, B. J., Song, J. C., Hassan, I., Smith, J. A., Akbarali, H. I., and Matthews, J. B. (1997b). Opposing effects of the tyrosine kinase inhibitor genistein on apical and basolateral transport sites of Cl$^-$ secretory epithelia. *Gastroenterology* **112,** A384.

McConalogue, K., and Furness, J. B. (1994). Gastrointestinal neurotransmitters. *Baillieres Clin. Endocrinol. Metab.* **8,** 51–76.

McEwan, G. T., Hirst, B. H., and Simmons, N. L. (1994). Carbachol stimulates Cl$^-$ secretion via activation of two distinct apical Cl$^-$ pathways in cultured human T$_{84}$ intestinal epithelial monolayers. *Biochim. Biophys. Acta* **1220,** 241–247.

McKeen, E. S., Feniuk, W., and Humphrey, P. P. (1995). Somatostatin receptors mediating inhibition of basal and stimulated electrogenic ion transport in rat isolated distal colonic mucosa. *Naunyn Schmiedebergs Arch. Pharmacol.* **352,** 402–411.

McNaughton, W. K. (1994). Nitric oxide donating compounds stimulate electrolyte transport in the guinea pig intestine *in vitro. Life Sci.* **53,** 585–593.

McNaughton, W. K., Pineau, B. C., and Krantis, A. (1996). ψ-Aminobutyric acid stimulates electrolyte transport in the guinea pig ileum *in vitro. Gastroenterology* **110,** 498–507.

McRoberts, J. A., Beuerlein, G., and Dharmsathaphorn, K. (1985). Cyclic AMP and Ca^{2+} activated K$^+$ transport in a human colonic epithelial cell line. *J. Biol. Chem.* **260,** 14163–14172.

Meenan, J., Mevissen, M., Monajemi, H., Radema, S. A., Soule, H. R., Moyle, M., Tytgat, G. N. J., and van Deventer, S. J. H. (1996). Mechanism underlying neutrophil adhesion to apical epithelial membranes. *Gut* **38,** 201–205.

Menniti, F. S., Oliver, K. G., Putney, J. W., and Shears, S. B. (1993). Inositol phosphates and cell signaling: New views of InsP$_5$ and InsP$_6$. *Trends Biochem. Sci.* **18,** 53–56.

Merlin, D., Jiang, L., Strohmeier, G. R., Nusrat, A., Alper, S. L., Lencer, W. I., and Madara, J. L. (1998). Distinct Ca^{2+}- and cAMP-dependent anion conductances in the apical membrane of polarized T$_{84}$ cells. *Am. J. Physiol.* **275,** C484–C495.

Mitic, L. L., and Anderson, J. M. (1998). Molecular architecture of tight junctions. *Ann. Rev. Physiol.* **60,** 121–142.

Montrose, M. H., Keely, S. J., and Barrett, K. E. (1999). Electrolyte secretion and absorption: Small intestine and colon. *In* "Textbook of Gastroenterology" (T. Yamada, ed.), pp. 320–355. Lippincott, Williams, and Wilkins, Philadelphia.

Morita, K., Furuse, M., Fujimoto, K., and Tsukita, S. (1999). Claudin multigene family encoding four-transmembrane domain protein components of tight junction strands. *Proc. Natl. Acad. Sci. USA* **96,** 511–516.

Mourad, F. H., O'Donnell, L. J. D., Dias, J. A., Ogutu, E., Andre, E. A., Turvill, J. L., and Farthing, M. J. G. (1995). Role of 5-hydroxytryptamine type 3 receptors in rat intestinal fluid and electrolyte secretion induced by cholera and *Escherichia coli* enterotoxins. *Gut* **37,** 340–345.

Naren, A. P., Nelson, D. J., Xie, W., Jovov, B., Pevsner, J., Bennett, M. K., Benos, D. J., Quick, M. W., and Kirk, K. L. (1997). Regulation of CFTR chloride channels by syntaxin and Munc18 isoforms. *Nature* **390,** 302–305.

Naren, A. P., Quick, M. W., Collawn, J. F., Nelson, D. J., and Kirk, K. L. (1998). Syntaxin 1A inhibits CFTR channels by means of domain-specific protein-protein interactions. *Proc. Natl. Acad. Sci.* **95,** 10972–10977.

Nash, S., Stafford, J., and Madara, J. L. (1987). Effects of polymorphonuclear leukocyte migration on the barrier function of cultured intestinal epithelial monolayers. *J. Clin. Invest.* **80,** 1104–1113.

Nath, S. K., Huang, X., L'helgoualc'h, A., Rautureau, M., Bisalli, A., Heyman, M., and Desjeux, J. F. (1994). Relation between chloride secretion and intracellular cyclic adenosine monophosphate in a cloned intestinal cell line HT-29cl 19A. *Gut* **35,** 631–636.

Nocerino, A., Iafusco, M., and Guandalini, S. (1995). Cholera toxin-induced small intestinal secretion has a secretory effect on the colon of the rat. *Gastroenterology* **108,** 34–39.

Nolte, H., Spjeldnaes, N., Kruse, A., and Windleborg, B. (1990). Histamine release from gut mast cells in patients with inflammatory bowel diseases. *Gut* **31,** 791–794.

Nusrat, A., Parkos, C. A., Liang, T. W., Carnes, D. K., and Madara, J. L. (1997). Neutrophil migration across model intestinal epithelia: Monolayer disruption and subsequent events in epithelial repair. *Gastroenterology* **113,** 1489–1500.

Nusrat, A., Turner, J. R., Colgan, S. P., Parkos, C. A., Carnes, D., Lemichez, E., Boquet, P., and Madara, J. L. (1995). Rho protein regulates tight junctions and perijunctional actin organization in polarized epithelia. *Proc. Natl. Acad. Sci. USA* **92,** 10629–10633.

Nybom, P., and Magnusson, K. E. (1996). Studies with wortmannin and cytochalasins suggest a pivotal role of phosphatidylinositols in the regulation of tight junction integrity. *Biosci. Rep.* **16,** 265–272.

Ohmichi, M., Sawada, T., Kanda, Y., Koike, K., Hirota, K., Miyake, A., and Saltiel, A. R. (1994). Thyrotropin-releasing hormone stimulates MAP kinase activity in GH3 cells by divergent pathways. *J. Biol. Chem.* **269,** 3783–3788.

O'Malley, K. E., Farrell, C. B., O'Boyle, K., and Baird, A. W. (1995). Cholinergic activation of chloride secretion in rat colonic epithelia. *Eur. J. Pharmacol* **275,** 83–89.

O'Malley, K. E., Sloan, T., Joyce, P., and Baird, A. W. (1993). Type I hypersensitivity reactions in intestinal mucosae from rats infected with *Fasciola hepatica*. *Parasite Immunol.* **15,** 449–453.

Ouellette, A. J. (1997). Paneth cells and innate immunity in the crypt microenvironment. *Gastroenterology* **113,** 1779–1784.

Parkos, C. A. (1997). Molecular events in neutrophil transepithelial migration. *Bioessays* **19,** 865–873.

Parkos, C., Delp, C., Arnaout, M. A., and Madara, J. L. (1991). Neutrophil migration across a cultured intestinal epithelium: Dependence on a CD11b/CD18-mediated event and enhanced efficiency in the physiologic direction. *J. Clin. Invest.* **88,** 1605–1612.

Perdue, M. H. (1996). Immunomodulation of epithelium. *Can. J. Gastroenterol.* **10,** 243–248.

Perdue, M. H., and Gall, D. G. (1985). Transport abnormalities during intestinal anaphylaxis in the rat: Effect of antiallergic agents. *J. Allergy Clin. Immunol.* **76**, 498–503.

Perdue, M. H., Masson, S., Wershil, B. K., and Galli, S. J. (1991). Role of mast cells in ion transport abnormalities associated with intestinal anaphylaxis. Correction of diminished secretory response in genetically mast cell deficient *W/Wv* mice by bone marrow transplantation. *J. Clin. Invest.* **87**, 687–693.

Pothoulakis, C., Castagliuolo, I., and LaMont, J. T. (1998). Nerves and intestinal mast cells modulate responses to enterotoxins. *News Physiol. Sci.* **13**, 58–63.

Powell, D. W. (1995). Dogma destroyed: Colonic crypts absorb. *J. Clin. Invest.* **96**, 2102–2103.

Powell, D.W., Mifflin, R.C., Valentich, J.D., Crowe, S.E., Saada, J.I., West, A.B. (1999) Myofibroblasts. I. Paracrine cells important in health and disease. *Am. J. Physiol.* **277**, C1–C19.

Prat, A. G., Xiao, Y.-F., Ausiello, D. A., and Cantiello, H. F. (1995). cAMP-dependent regulation of CFTR by the actin cytoskeleton. *Am. J. Physiol.* **268**, C1552–C1561.

Prince, L. S., Peter, K., Hatton, S. R., Zaliauskiene, L., Cotlin, L. F., Clancey, J. P., Marchase, R. B., and Collawn, J. F. (1999). Efficient endocytosis of the cystic fibrosis transmembrane conductance regulator requires a tyrosine-based signal. *J. Biol. Chem.* **274**, 3602–3609.

Quin, J. A., Sgambati, S. A., Goldenring, J. R., Basson, M. D., Fielding, L. P., Modlin, I. M., and Ballantyne, G. H. (1995). PYY inhibition of VIP-stimulated ion transport in the rabbit distal ileum. *J. Surg. Res.* **58**, 111–115.

Rangachari, P. K. (1992). Histamine: Mercurial messenger in the gut. *Am. J. Physiol.* **262**, G1–G13.

Rangachari, P. K., and Prior, T. (1993). Effects of bradykinin on the canine proximal colon. *Regul. Pept.* **46**, 511–522.

Reddix, R. A., and Cooke, H. J. (1992). Neurokinin 1 receptors mediate substance P-induced changes in ion transport in guinea pig ileum. *Regul. Pept.* **39**, 215–225.

Reddix, R. A., Kuwahara, A., Wallace, L., and Cooke, H. J. (1994). Vasoactive intestinal polypeptide: A transmitter in submucous neurons mediating secretion in guinea pig distal colon. *J. Pharmacol. Exp. Ther.* **269**, 1124–1129.

Reddix, R. A., Mullet, D., Fertel, R., and Cooke, H. J. (1998). Endogenous nitric oxide inhibits endothelin-1–induced chloride secretion in guinea pig colon. *Nitric Oxide* **2**, 28–36.

Reenstra, W. W. (1993). Inhibition of cAMP and Ca^{2+}-dependent chloride secretion by phorbol esters: Inhibition of basolateral K^+ channels. *Am. J. Physiol.* **264**, C161–C168.

Resnick, M. B., Colgan, S. P., Patapoff, T. W., Mrsny, R. J., Awtrey, C. S., Delp-Archer, C., Weller, P. F., and Madara, J. M. (1993). Activated eosinophils evoke chloride secretion in model intestinal epithelia primarily via regulated release of 5′AMP. *J. Immunol.* **151**, 5716–5723.

Rich, D. P., Gregory, R. J., Andersen, M. P., Manavalan, P., Smith, A. E., and Welsh, M. J. (1991). Effect of deleting the R-domain on CFTR-generated chloride channels. *Science* **253**, 205–207.

Rodgers, K. V., Goldman, P. S., Frizzell, R. A., and McKnight, G. S. (1990). Regulation of Cl^- transport in T_{84} cell clones expressing a mutant regulatory subunit of cAMP-dependent protein kinase. *Proc. Natl. Acad. Sci. USA* **87**, 8975–8979.

Rubin, B. K. (1999). Emerging therapies for cystic fibrosis lung disease. *Chest* **115**, 1120–1126.

Russell, D. A. (1986). Mast cells in regulation of intestinal electrolyte transport. *Am. J. Physiol.* **256**, G396–G403.

Sakamoto, H., Kawasaki, M., Uchida, S., Sasaki, S., and Marumo, F. (1996). Identification of a new outwardly rectifying Cl^- channel that belongs to a subfamily of the ClC Cl^- channels. *J. Biol. Chem.* **271**, 10210–10216.

Santos, J., and Perdue, M. H. (1998a). Immunological regulation of intestinal epithelial transport. *Digestion* **59**, 404–408.

Santos, J., Saperas, E., Nogueiras, C., Mourelle, M., Antolain, M., Cadahia, A., and Malagelada, J. R. (1998b). Release of mast cell mediators into the jejunum by cold pain stress in humans. *Gastroenterology* **114**, 640–648.

Schmitz, H., Fromm, M., Bode, P., Scholz, P., Riecken, E. O., and Schulzke, J. D. (1996). Tumor necrosis factor-α induces Cl^- and K^+ secretion in human distal colon driven by prostaglandin E_2. *Am. J. Physiol.* **271**, G669–G674.

Schultheiss, G., and Diener, M. (1998). K^+ and Cl^- conductances in the distal colon of the rat. *Gen. Pharmacol.* **31**, 337–342.

Schwiebert, E. M., Egan, M. E., Hwang, T. H., Fulmer, S. B., Allen, S. S., Cutting, G. R., and Guggino, W. B. (1995). CFTR regulates outwardly rectifying chloride channels through an autocrine mechanism involving ATP. *Cell* **81**, 1063–1073.

Schwiebert, E. M., Morales, M. M., Devidas, S., Egan, M. E., and Guggino, W. B. (1998). Chloride channel and chloride conductance regulator domains of CFTR, the cystic fibrosis transmembrane conductance regulator. *Proc. Natl. Acad. Sci. USA* **95**, 2674–2679.

Scott, C. M., Bunce, K. T., and Spraggs, C. F. (1992). Investigation of the 5-hydroxytryptamine receptor mediating the maintained short-circuit current response in guinea pig ileal mucosa. *Br. J. Pharmacol.* **106**, 877–882.

Selsted, M. E., Miller, S. I., Henschen, A. H., and Ouellette, A. J. (1992). Enteric defensins: Antibiotic peptide components of intestinal host defense. *J. Cell. Biol.* **118**, 929–936.

Seydel, K. B., Li, E., Zhang, Z., and Stanley, S. L. (1998). Epithelial cell-initiated inflammation plays a crucial role in early tissue damage in amebic infection of human intestine. *Gastroenterology* **115**, 1440–1453.

Shanahan, F. (1997). A gut reaction: Lymphoepithelial communication in the intestine. *Science* **275**, 1897–1898.

Shanahan, F., Denburg, J. A., Fox, J., Bienenstock, J., and Befus, A. D. (1985). Mast cell heterogeneity: Effects of neuroenteric peptides on histamine release. *J. Immunol.* **135**, 1331–1337.

Shaw, S. K., Hermanowski-Vosatka, A., Shibahara, T., McCormick, B. A., Parkos, C. A., Carlson, S. L., Ebert, E. C., Brenner, M. B., and Madara, J. L. (1998). Migration of intestinal intraepithelial lymphocytes into a polarized epithelial monolayer. *Am. J. Physiol.* **275**, G584–G591.

Sheppard, D. N., and Welsh, M. J. (1999). Structure and function of the CFTR chloride channel. *Physiol. Rev.* **79**, S23–S45.

Shook, J. E., Burks, T. F., Wasley, J. W., and Norman, J. A. (1989). Novel calmodulin antagonist CGS9343B inhibits secretory diarrhea. *J. Pharmacol. Exp. Ther.* **251**, 247–252.

Sidhu, M., and Cooke, H. J. (1995). Role for 5-HT and Ach in submucosal reflexes mediating colonic secretion. *Am. J. Physiol.* **269**, G346–G351.

Simon, D.B., Lu, Y., Choate, K.A., Heino, V., Al-Sabban, E., Praga, M., Casari, G., Bettinelli, A., Colussi, G., Rodriguez-Soriano, J., McCredie, D., Milford, D., Sanjad, S., Lifton., R.P. (1999). Paracellin-1, a renal tight junction protein required for paracellular Mg^{2+} resorption. *Science* **285**, 103–106.

Singh, A. K., Taksen, K., Walker, W., Frizzell, R. A., Watkins, S. C., Bridges, R. J., and Bradbury, N. A. (1998). Characterization of PKA isoforms and kinase-dependent activation of chloride secretion in T_{84} cells. *Am. J. Physiol.* **275**, C562–C570.

Siriwardena, A. K., Budhoo, M. R., Smith, E. P., and Kellum, J. M. (1993). A 5-HT-3 receptor agonist induces neurally mediated chloride transport in rat distal colon. *J. Surg. Res.* **55**, 55–59.

Song, J. C., and Matthews, J. B. (1999). Protein kinase Cε regulates basolateral membrane endocytosis and MARCKS dynamics in human intestinal epithelia. *Gastroenterology* **116**, A932.

Stack, W. A., Filipowicz, B., and Hawkey, C. J. (1996). Nitric oxide donating compounds stimulate human colonic ion transport *in vitro*. *Gut* **39**, 93–99.

Stack, W. A., Keely, S. J., O'Donoghue, D. P., and Baird, A. W. (1995). Immune regulation of human colonic electrolyte transport *in vitro*. *Gut* **36**, 395–400.

Stead, R. H., Dixon, M. F., Bramwell, N. H., Ridell, R. H., and Bienenstock, J. (1989). Mast cells are closely opposed to nerves in the human gastrointestinal mucosa. *Gastroenterology* **97**, 575–585.

Stead, R. H., Tomioka, M., Quinonez, G., Simon, G. T., Felten, S. Y., and Bienenstock, J. (1987). Intestinal mucosal mast cells in normal and nematode-infected rat intestines are in intimate contact with peptidergic nerves. *Proc. Natl. Acad. Sci. USA* **84**, 2975–2979.

Strohmeier, G. R., Reppert, S., Lencer, W., and Madara, J. (1995). The A_{2b} adenosine receptor mediates cAMP responses to adenosine agonists in human intestinal epithelia. *J. Biol. Chem.* **270**, 2387–2394.

Surprenant, A. (1994). Control of the gastrointestinal tract by enteric neurons. *Ann. Rev. Physiol.* **56**, 117–140.

Tai, Y.-T., Flick, J., Levine, S. A., Madara, J. L., Sharp, G. W. G., and Donowitz, M. (1996). Regulation of tight junction resistance in T_{84} monolayers by elevation in intracellular Ca^{2+}: A protein kinase C effect. *J. Membr. Biol.* **149**, 71–79.

Takahashi, A., Watkins, S. C., Howard, M., and Frizzell, R. A. (1996). CFTR-dependent membrane insertion is linked to stimulation of the CFTR chloride conductance. *Am. J. Physiol.* **271**, C1887–C1894.

Taylor, C. J., Baxter, P. S., Hardcastle, J., and Hardcastle, P. T. (1988). Failure to induce secretion in jejunal biopsies from children with cystic fibrosis. *Gut* **29**, 957–962.

Teather, S., and Cuthbert, A. W. (1997). Induction of bradykinin B_1 receptors in rat colonic epithelium. *Br. J. Pharmacol.* **121**, 1005–1011.

Theodorou, V., Eutamene, H., Fioramonti, J., Junien, J. L., and Bueno, L. (1994). Interleukin-1 induces a neurally mediated colonic secretion in rats: Involvement of mast cells and prostaglandins. *Gastroenterology* **106**, 1493–1500.

Tien, X.-Y., Brasitus, T. A., Kaetzel, M. A., Dedman, J. R., and Nelson, D. J. (1994). Activation of the cystic fibrosis transmembrane conductance regulator by cGMP in the human colon cancer cell line, Caco-2. *J. Biol. Chem.* **269**, 51–54.

Toker, A. (1998). Signaling through protein kinase C. *Frontiers Biosci.* **3**, D1134–D1147.

Toker, A., and Cantley, L. C. (1997). Signalling through the lipid products of phosphoinositide-3-OH kinase. *Nature* **387**, 673–676.

Townsend, R. R., Lipniunas, P. H., Tulk, B. M., and Verkman, A. S. (1996). Identification of protein kinase A phosphorylation sites on NBD1 and R domains of CFTR using electrospray mass spectrometry with selective phosphate ion monitoring. *Protein Sci.* **5**, 1865–1873.

Trapnell, B. C., Zeitlin, P. L., Chu, C.-S., Yoshimura, K., Nakamura, H., Guggino, W. B., Bargon, J., Banks, T. C., Dalemans, W., Pavirani, A., Lecocq, J.-P., and Crystal, R. C. (1991). Downregulation of cystic fibrosis gene mRNA transcript levels and induction of the cystic fibrosis phenotype in epithelial cells by phorbol ester. *J. Biol. Chem.* **266**, 10319–10323.

Turvill, J. L., Mourad, F. H., and Farthing, M. J. G. (1998). Crucial role for 5-HT in cholera toxin but not *Escherichia coli* heat-labile enterotoxin-intestinal secretion in rats. *Gastroenterology* **115**, 883–890.

Uribe, J. M., and Barrett, K. E. (1997). Nonmitogenic actions of growth factors: An integrated view of their role in intestinal physiology and pathophysiology. *Gastroenterology* **112**, 255–268.

Uribe, J. M., Gelbmann, C. M., Traynor-Kaplan, A. E., and Barrett, K. E. (1996a). Epidermal growth factor inhibits Ca^{2+}-dependent Cl^- transport in T_{84} human colonic epithelial cells. *Am. J. Physiol.* **271**, C914–C922.

Uribe, J. M., Keely, S. J., Traynor-Kaplan, A. E., and Barrett, K. E. (1996b). Phosphatidylinositol 3-kinase mediates the inhibitory effect of epidermal growth factor on calcium-dependent chloride secretion. *J. Biol. Chem.* **271**, 26588–26595.

Vaandrager, A. B., Bajnath, R., Groot, J. A., Bot, A. G. M., and DeJonge, H. R. (1991). Ca^{2+} and cAMP activate different chloride efflux pathways in HT-29.cl19A colonic epithelial cell line. *Am. J. Physiol.* **261**, C958–C965.

Vaandrager, A. B., Bot, A. G. M., and DeJonge, H. R. (1997a). Guanosine 3′-5′-cyclic monophosphate-dependent protein kinase II mediates heat-stable enterotoxin-provoked chloride secretion in rat intestine. *Gastroenterology* **112**, 437–443.

Vaandrager, A. B., Tilly, B. C., Smolenski, A., Schneider-Rasp, S., Bot, A. G. M., Edixhoven, M., Scholte, B. J., Jarchau, T., Walter, U., Lohmann, S. M., Poller, W. C., and de Jonge, H. R. (1997b). cGMP stimulation of cystic fibrosis transmembrane conductance regulator Cl$^-$ channels co-expressed with cGMP-dependent protein kinase type II but not type Iβ. *J. Biol. Chem.* **272,** 4195–4200.

Vajanaphanich, M., Schultz, C., Rudolf, M. T., Wasserman, M., Enyedi, P., Craxton, A., Shears, S. B., Tsien, R. Y., Barrett, K. E., and Traynor-Kaplan, A. E. (1994). Long-term uncoupling of chloride secretion from intracellular calcium levels by Ins(3,4,5,6)P$_4$. *Nature* **371,** 711–714.

Vajanaphanich, M., Schultz, C., Tsien, R. Y., Traynor-Kaplan, A. E., Pandol, S. J., and Barrett, K. E. (1995). Cross-talk between calcium and cAMP-dependent intracellular signalling pathways. *J. Clin. Invest.* **96,** 386–393.

Vajanaphanich, M., Wasserman, M., Buranawuti, T., Barrett, K. E., and Traynor-Kaplan, A. E. (1993). Carbachol uncouples chloride secretion from [Ca^{++}]: Association with tyrosine kinase activity. *Gastroenterology* **104,** A286.

Van Itallie, C. M., Balda, M. S., and Anderson, J. M. (1995). Epidermal growth factor induces tyrosine phosphorylation and reorganization of the tight junction protein ZO-1 in A431 cells. *J. Cell. Sci.* **108,** 1735–1742.

Wang, L., Stanisz, A. M., Wershil, B. K., Galli, S. J., and Perdue, M. H. (1995). Substance P induces ion secretion in mouse small intestine through effects on enteric nerves and mast cells. *Am. J. Physiol.* **269,** G85–G92.

Warhurst, G., Higgs, N. B., Fakoury, H., Warhurst, A. C., and Coy, D. H. (1996). Somatostatin receptor subtype 2 mediates somatostatin inhibition of ion secretion in rat distal colon. *Gastroenterology* **111,** 325–333.

Warhurst, G., Higgs, N. B., Tonge, A., and Turnberg, L. A. (1991). Stimulatory and inhibitory actions of carbachol on chloride secretory responses in human colonic cell line T$_{84}$. *Am. J. Physiol.* **261,** G220–G228.

Welsh, M. J., and Smith, P. L. (1982). Crypts are the site of intestinal fluid and electrolyte secretion. *Science* **218,** 1219–1221.

Welsh, M. J., Tsui, L.-C., Boat, T. F., and Beaudet, A. L. (1995). Cystic fibrosis. *In* "The Metabolic and Molecular Basis of Inherited Disease" (C.R. Scriver, ed.), pp. 3799–3876. McGraw–Hill, New York.

Wild, G. E., and Thompson, A. B. R. (1996). Na$^+$-K$^+$-ATPase α1 and β1-mRNA and protein levels in rat small intestine in experimental ileitis. *Am. J. Physiol.* **269,** G666–G675.

Willumsen, N. J., and Boucher, R. C. (1989). Activation of an apical Cl$^-$ conductance by Ca^{2+} ionophores in cystic fibrosis airway epithelia. *Am. J. Physiol.* **256,** C226–C235.

Wood, J. D. (1991). Communication between minibrain in gut and enteric immune system. *News Physiol. Sci.* **6,** 64–69.

Xie, W., Kaetzel, M. A., Bruzik, K. S., Dedman, J. R., Shears, S. B., and Nelson, D. J. (1996). Inositol 3,4,5,6-tetrakisphosphate inhibits the calmodulin-dependent protein kinase II–activated chloride conductance in T$_{84}$ colonic epithelial cells. *J. Biol. Chem.* **271,** 14092–14097.

Xie, W., Solomons, K. R., Freeman, S., Kaetzel, M. A., Bruzik, K. S., Nelson, D. J., and Shears, S. B. (1998). Regulation of Ca^{2+}-dependent chloride conductance in a human colonic epithelial cell line (T$_{84}$): Cross-talk between Ins(3,4,5,6)P$_4$ and protein phosphatases. *J. Physiol.* **510,** 661–673.

Yang, S.-K., Eckmann, L., Panja, A., and Kagnoff, M. F. (1997). Differential and regulated expression of C-X-C, C-C, and C-chemokines by human colon epithelial cells. *Gastroenterology* **113,** 1214–1223.

Yarrow, S., Ferrar, J. A., and Cox, H. M. (1991). The effects of capsaicin upon electrogenic ion transport in rat descending colon. *Naunyn Schmiedebergs Arch. Pharmacol.* **344,** 557–563.

Zahraoui, A., Joberty, G., Arpin, M., Fontaine, J.-J., Hellio, R., Tavitian, A., and Louvard, D. (1994). A small rab GTPase is distributed in cytoplasmic vesicles in non-polarized cells but colocalizes with the tight junction marker ZO-1 in polarized epithelial cells. *J. Cell. Biol.* **124,** 101–115.

Zeitlin, P. L. (1999). Novel pharmacologic therapies for cystic fibrosis. *J. Clin. Invest.* **103,** 447–452.

Zhong, Y., Saitoh, T., Minase, T., Sawada, K., Enomoto, K., and Mori, M. (1993). Monoclonal antibody 7H6 reacts with a novel tight junction-associated protein distinct from ZO-1, cingulin, and ZO-2. *J. Cell. Biol.* **120,** 477–483.

Zund, G., Madara, J. L., Dzus, A. L., Awtrey, C. S., and Colgan, S. P. (1996). Interleukin-4 and interleukin-13 differentially regulate epithelial chloride secretion. *J. Biol. Chem.* **271,** 7460–7464.

CHAPTER 8

Anion Absorption in the Intestine: Anion Transporters, Short-Chain Fatty Acids, and Role of the *DRA* Gene Product

Marshall H. Montrose* and Juha Kere†
*Department of Physiology and Biophysics, Indiana University School of Medicine, Indianapolis, Indiana, and the †Finnish Genome Center, University of Helsinki, Finland

I. INTRODUCTION

The goal of this chapter is to review current understanding about mechanisms that mediate intestinal absorption of water-soluble inorganic and organic anions. The chapter will mostly focus on absorption of the predominant anions of the mammalian small and large intestine. Chloride is the most abundant anion in the small intestinal lumen, and absorption of chloride is required to maintain body levels of this major electrolyte. It is less widely appreciated that short-chain fatty acids (SCFAs; monocarboxylates such as acetate, propionate, and butyrate) are the predominant anions in the colonic lumen. SCFAs are produced by bacterial fermentation, and once they are absorbed provide metabolic substrates that yield a significant portion of the body's energy. By virtue of the mechanisms that mediate their absorption, both luminal chloride and luminal SCFAs have additional physiological roles in stimulating water absorption and in transporting other ions such as sodium, bicarbonate, and protons. The first part of this chapter will discuss transport mechanisms and membrane proteins that mediate anion absorption. The second half of the chapter will emphasize the pathophysiology and genetics of congenital chloride diarrhea.

Disruption of anion absorption and anion metabolism can lead to diarrheal disease and acid/base disturbances, and may predispose to inflammatory bowel diseases. Congenital chloride diarrhea is a rare autosomal-recessive disease caused by mutations in the DRA (downregulated in adenoma) protein, which is predominantly expressed in the small intestine and colon. The recent expansion of studies about the function and dysfunction of the DRA protein is stretching our understanding of intestinal function.

II. INORGANIC ANION ABSORPTION

The body must absorb a number of inorganic ions to sustain life. Although chloride will be discussed at length in section II.D, it is worthwhile to summarize facts and controversies relating to absorption of other inorganic anions.

A. Inorganic Sulfate

Inorganic sulfate anions (SO_4) can be absorbed from the small intestinal lumen by active transport (Anast et al., 1965; Smith et al., 1981). Evidence suggests that the mechanism of transcellular absorption is via a Na/SO_4 cotransporter in the apical membrane and a SO_4/anion exchanger in the basolateral membrane (Langridge-Smith et al., 1983; Langridge-Smith and Field, 1981; Knickelbein and Dobbins, 1990). There is one report of an apical SO_4/OH exchange reaction in ileal membrane vesicles, but the physiological relevance remains unknown (Schron et al., 1985). The apical Na/SO_4 cotransporter has been cloned, and designated *ileal NaSi-1* (Norbis et al., 1994). The ileal NaSi-1 is identical in amino acid sequence to *renal NaSi-1*, although the mRNA transcript of ileal NaSi-1 has an additional 600 bp of 3' untranslated sequence (Norbis et al., 1994; Markovich et al., 1993). Because the long and short transcripts are found in both small intestine and kidney cortex (Murer et al., 1994), the *renal* and *ileal* designations are useful for identifying the source of mRNA used for expression cloning, but they do not indicate tissue-specific expression of each transcript. Initial results from rabbit and rat ileum suggested that the Na/SO_4 cotransport was electroneutral (Langridge-Smith *et al.,* 1983; Lucke *et al.,* 1981); however, more recent results suggest that NaSi-1 is electrogenic, with a probable stoichiometry of 3 Na:1 SO_4 (Busch *et al.,* 1994). The low electrical currents produced by such a stoichiometry may have been below the detection limit of earlier methods.

Less information is available about sulfate transport in the large intestine. The only pertinent information we can identify suggests that the colon actively secretes sulfate anions (Freel *et al.,* 1997). NaSi-1 is not expressed in the (rat) large intestine (Murer *et al.,* 1994), and there is no evidence of sulfate-coupled sodium transport in the colon (Edmonds and Mackenzie, 1984; Sandle, 1989). Some information implicates another gene product, *DRA,* as a potential colonic sulfate transporter. DRA protein is observed in the apical membrane of colonocytes (Antalis *et al.,* 1998; Haila *et al.,* in press), and DRA can mediate sulfate transport when heterologously expressed (Byeon *et al.,* 1998; Moseley *et al.,* 1999; Silberg *et al.,* 1995). However, the physiological importance of DRA as a sulfate transporter remains unsure. Naturally occurring mutations in *DRA* lead to diminished chloride absorption and disturbed acid/base absorption (Moseley *et al.,* 1999; Bieberdorf *et al.,* 1972; Darrow, 1945; Gamble *et al.,* 1945; Holmberg *et al.,* 1975; Turnberg, 1971; Höglund *et al.,* 1996), but there have been no reports of changes in intestinal sulfate handling as a consequence of *DRA* mutation.

B. Inorganic Phosphate

Inorganic phosphate (PO_4) also is actively absorbed from the small intestine, and this absorption can be increased either by vitamin D_3 or by restrictions in dietary phosphate (Walling, 1977; Lee *et al.,* 1986). It has been shown that an apical membrane Na/PO_4 cotransporter is a component of the transcellular absorption of phosphate, and is the major site for regulation of absorption (Caverzasio *et al.,* 1987; Hildmann *et al.,* 1982; Quamme, 1985). Within the three known gene families of Na/PO_4 cotransporters in vertebrates (types I, II, and III), several isoforms are expressed in the intestine. Type-III transporters are widely distributed in mammalian tissues, and both *PiT-1* (also known as *GLVR*) and *PiT-2* (also known as *Ram-1*) transcripts are found in the intestine (Katai *et al.,* 1999; Kavanaugh and Kabat, 1996). Among the type-II transporters, *NaPi-IIb* is the isoform expressed in murine and human intestine (Feild *et al.,* 1999; Hilfiker *et al.,* 1998).

It is controversial which isoform(s) contribute to intestinal phosphate absorption. Restrictions in dietary phosphate were shown to have no effect on mRNA abundance of *PiT-1*, *PiT-2*, or *NaPi-IIb* transcripts, but did increase the mRNA abundance of *PiUS* (an accessory protein likely to have a role in regulating phosphate transport), and did increase the amount of NaPi-IIb protein in the apical membrane of enterocytes (Katai et al., 1999; Hattenhauer et al., 1999). In vitamin D–deficient animals, administration of calcitriol did not increase mRNA abundance of PiT-1, NaPi-IIb, or PiUS (Katai et al., 1999). However, vitamin D_3 repletion increased mRNA abundance of PiT-2 transcripts and increased NaPi-IIb immunostaining at the enterocyte brush border (Katai et al., 1999; Hattenhauer et al., 1999). At this stage, the strongest case has been made for NaPi-IIb as the major intestinal Na/PO₄ cotransporter; it is expressed in the apical membrane, it transports phosphate with the substrate affinity and pH dependence observed in native tissue, and it is functionally regulated by vitamin D and dietary phosphate (Hilfiker et al., 1998; Hattenhauer et al., 1999).

C. Halide Anions Other than Cl

Halide anions other than Cl (I, F, or Br) are rarely studied with respect to their intestinal absorption. Evidence suggests that fluoride and iodide can be absorbed from the small intestine, although active iodide secretion predominates in most gut segments (Ilundain et al., 1987; Parkins, 1971). In intestinal brush-border membrane vesicles, fluoride uptake is stimulated by inwardly directed proton gradients and fluoride may compete for some of the same transport substrate sites as Cl anions (He et al., 1998). This tentatively suggests that $F/OH(HCO_3)$ exchange may be an alternative function of $Cl/OH(HCO_3)$ exchangers in the apical membrane. The DRA protein is a candidate to transport other halide anions besides chloride, because its highly homologous sister protein PDS (encoding the Pendred syndrome gene) transports iodide and chloride (but not sulfate) (Scott et al., 1999). The anion specificity of these closely related proteins merits further study because it may shed light on the molecular mechanisms that control ion selectivity.

D. Chloride

A significant portion of the Na^+ and Cl^- absorbed by the intestinal tract does not move net charge across the epithelium (i.e., it is electroneutral), and it is mutually dependent on the alternate ion (i.e., electroneutral Cl^- absorption requires the presence of Na^+ and vice versa) (Duffey et al., 1978; Frizzell et al., 1975; 1979; Nellans et al., 1973). Electroneutral NaCl absorption is a major route for small intestinal Cl^- absorption postprandially, after absorption of nutrients is complete, or in the fasting state (Maher et al., 1997). This NaCl absorptive mechanism is a principal route for absorption in the proximal colon, with less prominence in the distal colon (Hubel et al., 1987; Binder and Sandle, 1994). To explain these coupled fluxes, two models of NaCl uptake in the apical membrane have been proposed: (1) an electroneutral Na/Cl cotransporter and (2) the combined action of Na/H and Cl/HCO$_3$ exchangers.

Abundant information supports the existence of electroneutral NaCl absorption comprised of a Na/H exchange working in parallel with a Cl/HCO$_3$ exchange pathway (Knickelbein et al., 1983; 1985; Lubcke et al., 1986; Binder et al., 1987; Foster et al., 1986; 1990). Cl/HCO$_3$ exchange is present in vesicles prepared from the apical membranes of rabbit ileal villus cells (Knickelbein et al., 1985; 1988) and rat colon (Rajendran et al., 1993; 1999). In the dual exchanger model, coupling of the paired exchangers occurs through changes in intracellular pH (pH$_i$). The Na$^+$ gradient drives Na$^+$ uptake and H$^+$ efflux by means of Na/H exchange, which alkalinizes the cytoplasm and increases activity of the Cl/HCO$_3$ exchanger. The action of carbonic anhydrase produces HCO$_3^-$, which then leaves the cell in exchange for uptake of luminal Cl$^-$. The net reaction is Na$^+$ and Cl$^-$ uptake in exchange for H$^+$ and HCO$_3^-$ efflux. Once in the cell, Na$^+$ is pumped out by Na,K-ATPase, and Cl$^-$ follows by way of an electroneutral transport protein that has yet to be identified, but is likely to be K/Cl cotransport (Reuss, 1983). Alternatively, a basolateral Cl/HCO$_3$(OH) exchange may facilitate Cl$^-$ efflux, if prevailing ion gradients permit.

Genetic disorders provide evidence that the dual exchanger model is responsible for the majority of intestinal NaCl absorption in humans. Two distinct genetic disorders exist that selectively eliminate chloride or sodium absorption, namely congenital chloride diarrhea and congenital sodium diarrhea, respectively (Booth et al., 1985; Holmberg and Perheentupa, 1985). In each case, the disorder has been localized to an alteration in ion transport at the apical membrane of enterocytes (Haila et al., in press; Booth et al., 1985; Byeon et al., 1996; Kere et al., 1999). Because both diseases are autosomal recessive and apparently nonallelic, two distinct molecular mechanisms must be responsible for at least part of the sodium and chloride absorption.

Less evidence is available to support a functional role for directly coupled Na/Cl cotransport in the apical membrane of mammalian intestinal epithelial cells. Similar to the kidney medulla (Payne and Forbush, 1995), a directly coupled cotransporter has been shown to be important for Na$^+$ absorption in the intestine of winter flounder (Musch et al., 1982), but directly coupled Na/Cl cotransport has not been routinely observed in membrane vesicles or in intact tissue preparations from mammals. However, tissue from the small intestine expresses an apical Na/Cl cotransporter, so we may simply be lacking knowledge about the signals that activate this protein (Chang et al., 1996).

1. Cl/HCO$_3$(OH) Exchange

Cl/HCO$_3$ exchange acts in concert with Na/H exchange to mediate electroneutral NaCl uptake at the apical membrane of both colonocytes and small intestinal cells (Knickelbein et al., 1983; 1985; Lubcke et al., 1986; Binder et al., 1987;

Foster *et al.,* 1986; 1990). Unlike apical membrane Na/H exchange, which is expressed predominantly in villus or surface cells (in the small intestine or colon, respectively) (Knickelbein *et al.,* 1988; Hoogerwerf *et al.,* 1996; Bookstein *et al.,* 1994), apical Cl/HCO$_3$ exchange in the small intestine has little or no gradient along the villus-crypt axis (Knickelbein *et al.,* 1988). With the possible exception that the NHE2 apical Na/H exchanger isoform may be expressed in colonic crypts (Dudeja *et al.,* 1996), these results suggest that Cl/HCO$_3$ exchange may contribute to other physiological mechanisms in addition to NaCl absorption. In the colon, apical Cl/HCO$_3$ exchange is observed predominantly at the surface cells, although Cl/OH exchange has a less steep gradient such that it is still relatively active in crypt membranes (Rajendran and Binder, 1999). This and other kinetic evidence suggests that apical Cl/OH exchange is not mediated by the same protein as Cl/HCO$_3$ exchange, in either the small intestine or the colon (Rajendran and Binder, 1993; 1999; Vaandrager and deJonge, 1988; Liedtke and Hopfer, 1982). Molecular biologic approaches have identified many candidate proteins that may perform apical Cl$^-$/anion exchange, and it is unclear which, if any, are important for electroneutral Cl$^-$ absorption.

2. AE Gene Family

Multiple isoforms of the anion exchanger (*AE*) gene family are expressed in the small and large intestine. Most recently, a detailed study of intestinal AE2 expression has convincingly shown that among three known AE2 transcripts (AE2a, AE2b, AE2c1), it is the AE2a and AE2b transcripts that are abundant in the intestine (Alper *et al.,* 1999). In the same study, careful comparison among immunostaining with several AE2-specific antibodies confirmed a basolateral localization of most AE2 epitopes, although some staining conditions and antibodies produced a less stringent localization (Alper *et al.,* 1999). It appears likely that variations among the sequences of the AE transcripts, and the reliance on a single antibody, may have misled earlier analyses of AE2 expression (Chow *et al.,* 1992).

Evidence of the intestinal presence of AE1 and AE3 has been presented in preliminary form (Rajendran *et al.,* 1997; Alrefai *et al.,* 1997). Apical Cl/HCO$_3$ exchange in ileal brush-border membranes has been informally cited as matching the characteristics of AE3 in terms of a common pH dependence and lack of osmotic activation, hinting that AE3 may be a good candidate for the apical transporter in the small intestine (Alper *et al.,* 1999; Nader *et al.,* 1994). However, this conflicts with preliminary results in which AE3 was assigned to the basolateral membrane (Alrefai *et al.,* 1997). The function and localization of AE isoforms remains to be firmly established.

It also should be noted that mutations in a non-AE family member, the DRA protein, can lead to severely compromised chloride absorption in colon and ileum (see section IV). Based on the known ability of an apically localized DRA to mediate chloride transport (see section II.D.4), this is compelling evidence that *AE* family members may play only a minor role in chloride absorption across the apical membrane.

3. *KCC* Gene Family

Electroneutral K/Cl cotransport has been suggested as a major route of basolateral Cl efflux in NaCl absorptive epithelia (Reuss, 1983), and there is a rapidly expanding gene family of such cotransporters, designated *KCC*. There are currently four known KCC isoforms (Payne *et al.*, 1996; Hiki *et al.*, 1999; Holtzman *et al.*, 1998; Mount *et al.*, 1999; Race *et al.*, 1999). Only the initially discovered member of the gene family (*KCC1*) is known to be expressed in intestinal tissue, but its role in Cl absorption remains untested (Gillen *et al.*, 1996).

4. DRA

In contrast to the candidates listed in the previous sections, which currently have only a tentative link to intestinal physiology, there is strong evidence that DRA protein is intimately involved with Cl absorption. As described in section IV, *DRA* has been identified as the disease gene that causes congenital chloride diarrhea (CLD).

As described in section IV, the phenotype of CLD suggests that the disease is caused by a defective Cl/HCO$_3$ exchange in the apical membrane of colonic cells. DRA protein is found on the apical brush-border membrane of epithelial cells in the colon (Haila *et al.*, in press; Byeon *et al.*, 1996; Kere *et al.*, 1999). *In situ* hybridization has shown expression in the colon to be restricted to the uppermost third along the crypt-surface axis, whereas in the ileum, expression is seen deeper in the crypts as well (Höglund *et al.*, 1996; Yang *et al.*, 1998). This protein distribution is fully compatible with a role for DRA as a major absorptive chloride transporter. Besides the intestine from ileum to rectum, in other parts of the body DRA expression is seen only in the seminal vesicle epithelium and eccrine sweat glands in the skin (Haila *et al.*, in press).

DRA is a membrane transport protein. *DRA* was first cloned as a candidate tumor suppressor gene (Schweinfest *et al.*, 1993), but its high homology with a human sulfate transporter DTDST and the rat *sat-1* (for sulfate anion transporter-1) suggested a likely transporter function. Based on this homology, *DRA* was first shown to transport sulfate and oxalate in a sodium-independent and DIDS-sensitive manner when expressed in *Xenopus* oocytes (Silberg *et al.*, 1995). Subsequently, the list of substrates has grown to include at least chloride, iodide, hydroxyl, and bicarbonate (Byeon *et al.*, 1998; Maseley *et al.*, 1999; Melvin *et al.*, 1999). The chloride and sulfate transport mediated by *DRA* can be eliminated by the presence of CLD mutation (V317del) in the *DRA* sequence (Moseley *et al.*, 1997; 1999). It has been shown that an inwardly directed proton gradient (alkaline inside) will stimulate chloride uptake by DRA (Moseley *et al.*, 1997;

1999). Conversely, measurements of intracellular pH have shown that DRA mediates a Cl^- dependent and electroneutral change in acid/base flux consistent with Cl/OH and/or Cl/HCO_3 exchange (Melvin *et al.*, 1999).

The protein most closely related to DRA, named pendrin, also is encoded by a disease gene (*PDS*) associated with defective membrane transport. *PDS* is responsible for Pendred syndrome, a form of recessive hearing loss and thyroid disease (Everett *et al.*, 1997). Pendrin specifically does not transport sulfate, but at least chloride and iodine are transported (Scott *et al.*, 1999), and speculatively, a possible function for pendrin might be in the regulation of endolymph pH through Cl/HCO_3 exchange in the inner ear. These results confirm that the family of proteins earlier named as sulfate transporters may have much broader transport specificity. Some highly homologous family members such as DRA and PDS are clearly encoding functional anion exchangers that are structurally distinct from the AE (anion exchanger) family of genes and proteins. The protein domains regulating the transport specificity have not been experimentally determined. A third, highly homologous member of this gene family is *DTDST*, encoding a sulfate transporter defective in the recessive cartilage diseases diastrophic dysplasia (Hästbacka *et al.*, 1994), atelosteogenesis type II (Hästbacka *et al.*, 1996), and achondrogenesis type IB (Superti-Furuga *et al.*, 1996). Of interest for future physiological studies is the fact that this gene family is further expanding with the isolation of new members in humans (Lohi *et al.*, in press).

The number of known CLD mutations in the *DRA* gene now exceeds 20 (Table I). The mutations show slight clustering at three locations but also occur elsewhere in the protein. The mutations include small deletions and insertions as well as amino acid substitutions. Splice site mutations and nonsense mutations give rise to truncated or stucturally abnormal proteins. Functionally interesting is the observation that many mutations localize in the suggested intracellular C-terminal tail in the 12-transmembrane segment model of the protein (Kere *et al.*, 1999). It is perhaps less likely that these mutations would affect the synthesis or integration of the protein in the cell membrane. Alternatively, this C-terminal domain might have a distinct regulatory function. The presence of several putative protein-kinase C and casein-kinase II substrate sites in the C-terminal segment supports this suggestion, although their functional role remains untested. It has not been possible to establish any genotype-phenotype correlations because the success of the substitution therapy (Holmberg, 1986) mostly determines the clinical course of patients. However, all the known mutations have been detected in patients with full blown clinical phenotype and large excretion of chloride in stools. Thus, these mutations are likely to represent functionally completely or almost-completely deficient protein forms. As shown in Figure 1, the site of

TABLE I

Naturally Occurring Mutations in *DRA* Associated with Congenital Chloride Diarrhea

Nucleotide change	Codon	Protein change	No. of patients	References
Missense mutations, single amino acid insertions and deletions				
358G>A	120	Glycine → Serine, G120S	2	(Höglund et al., 1998a; 1998b)
371A>T	124	Histidine → Leucine, H124L	3	(Höglund et al., 1996; 1998a)
392C>G	131	Proline → Arginine, P131R	2	(Höglund et al., 1998b)
951-953delGGT	317	In-frame loss of Valine, V317del	32	(Höglund et al., 1996)
1487T>G	496	Leucine → Arginine, L496R	3	(Höglund et al., 1998a)
1578-1580delTTA	527	In-frame loss of Tyrosine, Y527del	2	(Höglund et al., 1998b)
2025-2026insATC	676	In-frame addition of Isoleucine, I675-676ins	12	(Höglund et al., 1998a)
Nonsense mutations, splice site mutations				
177-178insC	60	Frameshift at codon 60, STOP at codon 70	1	(Höglund et al., 1998b)
268-269insAA	90	Frameshift at codon 91, STOP at condon 93	3	(Höglund et al., 1998a)
344delT	115	Frameshift at codon 115, STOP at codon 133	3	(Höglund et al., 1996; 1998a)
559G>T	187	Glycine → STOP, G187X	9	(Höglund et al., 1998a)
IVS5-1G>T	—	Loss of intron acceptor site AG at nt −1 of intron 5	1	(Höglund et al., 1998a)
IVS5-2A>G	—	Loss of intron acceptor site AG at nt −2 of intron 5	1	(Höglund et al., 1998a)
915C>A	305	Tyrosine → STOP, Y305X	2	(Höglund et al., 1998b)
IVS11-1G>A	—	Loss of intron acceptor site AG at nt −1 of intron 11	1	(Höglund et al., 1998a)
1516delC	505	Frameshift at codon 505, STOP at codon 534	1	(Höglund et al., 1998b)
1526-1527delTT	509	Frameshift at codon 509, STOP at codon 517	1	(Etani et al., 1998)
1548-1551delAACC	516	Frameshift at codon 518, STOP at codon 534	1	(Höglund et al., 1998b)
1609delA	537	Frameshift at codon 537, STOP at codon 575	1	(Höglund et al., 1998a)
2116delA	706	Frameshift at codon 706, STOP at codon 711	1	(Höglund et al., 1998b)

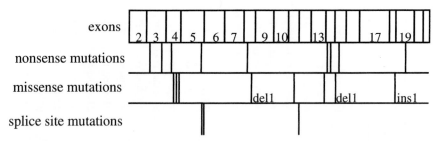

FIGURE 1 Naturally occurring mutations in DRA. The portion in black illustrates the exon structure of DRA with the spacing approximating the length of the sequence in each exon. Below the exon structure are vertical bars which indicate the position of naturally occurring mutations along this sequence. Results are presented separately for nonsense (blue bars), missense (red bars), and splice site (green) mutations which are found in CLD patients. del1 = in-frame deletion of single amino acid, ins1 = in-frame insertion of a single amino acid. (See color insert)

CLD mutations is widely dispersed along the DRA protein. Analogous to those *CFTR* mutations that associate with pancreatic sufficiency or vas deferens agenesis, functionally less-impaired mutations also are likely to exist in the CLD gene. However, the phenotype associated with those remains unknown.

Figure 2 displays current knowledge about the transporters just discussed. It depicts a tentative model that includes the most likely constellation of transport proteins that mediate chloride absorption. Although further research may prove the model wrong, it provides a working hypothesis for further work.

III. ORGANIC ANIONS

Other chapters in this volume discuss absorption of amino acids and peptides (Leibach, chapter 10) and nucleosides (Young, chapter 9); thus, these organic anions will not be discussed here. We will focus on the mechanisms of carboxylate (COOH) transport.

A. Di- and Tri-Carboxylate Transporters

Succinate, citrate, and α-ketoglutarate can be absorbed across the brush-border membrane of intestinal cells by Na-dependent transport in both the small and large intestine (Browne *et al.*, 1978; Kimmich *et al.*, 1991; Moe *et al.*, 1998; Wolffram *et al.*, 1990; 1992; 1994a; 1994b). There is strong evidence for at least two distinct Na-dependent dicarboxylate cotransporters, because the small in-

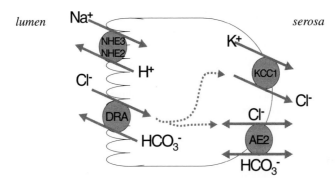

FIGURE 2 Mechanisms of chloride absorption. This diagram of an intestinal epithelial cell is shown with the most likely transporters involved. The apical membrane also is shown, with the Na/H exchangers that mediate the other half of the electroneutral NaCl absorption. Individual carrier proteins are indicated by shaded circles, and the name of the cloned transporter is indicated in the center of each circle.

testinal, but not colonic, transport can be inhibited by lithium (Wolffram *et al.,* 1994a). Molecular cloning has recently confirmed the existence of multiple Na-dependent dicarboxylate cotransporter isoforms (NaDC-1 and NaDC-2) in *Xenopus* (Pajor, 1995; Bai and Pajor, 1997). Although NaDC-1 and NaDC-2 express differential sensitivity to lithium, it is not clear whether these proteins are orthologs of the two anticipated intestinal forms in mammals, because only NaDC-2 is expressed in frog intestine (Pajor, 1995; Bai and Pajor, 1997). Identification of mammalian isoforms should be forthcoming, as a rat cDNA closely related to NaDC sequences has been reported (Khatri *et al.,* 1996). Ana Pajor has recently reviewed the functional and molecular aspects of epithelial dicarboxylate transport (Pajor, 1999).

B. SCFAs and other Monocarboxylates

Short-chain fatty acids (SCFAs) are monocarboxylates produced in the colon by bacterial catabolism of unabsorbed carbohydrate and protein. The combined concentration of the predominant luminal SCFAs (i.e., acetate, propionate, and butyrate) is 100–150 mM, so they are the major anions and osmolytes in the colonic lumen (Burgaut, 1987). Absorbed SCFA accounts for 7–10% of ingested calories and serves as an important energy source for colonic epithelial cells (Burgaut, 1987; Roediger, 1982; Bond *et al.,* 1980).

The mechanisms of SCFA transport across the epithelium are controversial, but it appears that there is a role for both nonionic diffusion (i.e., flux of the protonated form of these weak acids without intervention of a transport protein) and carrier-mediated transport of SCFA anions. One point is clear: SCFA transport is intimately involved with acid/base fluxes in the gut. Before delving into more complex systems, it is worthwhile to review the acid/base behavior of the SCFAs alone.

The abundant SCFAs in the colonic lumen all have pK_a values of 4.8. Because the normal colonic contents have a pH ranging from 6–8, SCFAs in the lumen exist predominantly as anions, with only 0.1–10% of SCFAs being in the uncharged (protonated) form. This skewed distribution of molecular forms does not mean *a priori* that anionic transport will be quantitatively more important because nonionized compounds can be 10^4- to 10^5-fold more permeable than their charged counterparts across most membranes (Montrose and Kimmich, 1986; Roos and Boron, 1981). The existence of two molecular forms does imply that when one form is preferentially transported across the plasma membrane, it will unbalance the equilibrium between the two molecular forms both inside and outside the cell. This will lead to a shift in the chemical balance between the two forms, which consequently changes pH. For example, selective uptake of nonionized SCFA molecules into an environment of pH 7 (the usual intracellular pH) will require that 99% of the nonionized SCFA molecules give up their proton to reestablish chemical equilibrium between the molecular forms, leading to cellular acidification. If the extracellular environment does not have an infinite supply of SCFAs, then cellular uptake of the nonionized form also will lead to a redistribution of molecular forms in the extracellular environment, tending to alkalinize the space as protons are consumed to regenerate the nonionized SCFA. This is not just a feature of nonionized SCFA fluxes. If cells absorb only anionic SCFA, then qualitatively opposite effects on pH are predicted, albeit of different magnitude.

1. Nonionic Diffusion of SCFAs

There are several predictions that can distinguish nonionic diffusion from carrier-mediated transport. Nonionic diffusion should be dependent on the concentration and hydrophobicity of the nonionized form, and diffusional flux should not saturate as the concentration of SCFA is increased. Despite these clear criteria, it has been difficult to rigorously distinguish the relative role of nonionic and ionized SCFA fluxes in the colon.

In isolated mouse colonocytes, the initial acidification in intracellular pH caused by SCFAs was shown to be proportional to the concentration of extracellular protonated SCFA, not the concentration of SCFA anion (Chu and Montrose, 1995). This is strong evidence that nonionic fluxes can predominate versus other routes of SCFA in terms of initial influence of intracellular pH. In isolated mouse and rabbit colonocytes, a wide variety of monocarboxylates elicited an initial acidification, with more rapid acidification occurring with unsubstituted aliphatic molecules (Chu and Montrose, 1995; DeSoignie and Sellin, 1994). In these studies it was not possible to assign membrane localization to the nonionic fluxes, but at a minimum either the apical or the basolateral membrane would be predicted to utilize nonionic diffusion for SCFA flux.

In Ussing chamber experiments using guinea pig, rat, or rabbit colon, the mucosal-to-serosal flux of SCFA is proportional to luminal SCFA concentration (Sellin *et al.,* 1993; Charney *et al.,* 1998; Von Engelhardt and Rechkemmer, 1992). Even at 100 mM SCFA, there was no evidence of flux saturation with either of two SCFAs having widely divergent rates of metabolism (Charney *et al.,* 1998). These results are important confirmation of the nonsaturability measured under less controlled circumstances in earlier studies (Umesaki *et al.,* 1979; Rechkemmer and Von Engelhardt, 1988). These results strongly suggest that nonionic diffusion is the rate-limiting step in the mucosal-to-serosal flux of SCFA, but other predictions of nonionic diffusion have yielded ambiguous results. For instance, although acidification of the medium stimulates SCFA uptake, the increased transport is not proportional to the predicted change in concentration of the nonionized SCFA (Sellin *et al.,* 1993; Charney *et al.,* 1998). It appears that this poor correlation may be related to changes in extracellular pH that (like intracellular pH changes) are a consequence of SCFA flux. Early pH microelectrode studies demonstrated that SCFAs were able to change pH near the tissue surface (Holtug *et al.,* 1992). Using a fluorescent dye to image extracellular pH within the colonic crypts, the extent of alkalinization in the crypt lumen was found to be linearly related to the predicted concentration of nonionized luminal SCFA at this local pH (Chu and Montrose, 1996). Such results show that events near the apical membrane can dynamically alter the quantitative relationship between flux and bulk medium pH, in a way that is difficult to account for without knowing pH in microdomains of the tissue. Adding yet another layer of complexity, it has been suggested in native tissue (and demonstrated in a tissue culture model) that SCFAs also generate intracellular pH gradients that can locally affect membrane transport events (Dagher *et al.,* 1996; Gonda *et al.,* 1999).

As described earlier, a compelling body of evidence favors nonionic diffusion as a major route of SCFA flux and of SCFA effects on cell and tissue pH. However, numerous observations suggest that other routes of SCFA transport also are present in the colon. Measurements of intracellular pH have shown that the prompt acidification induced by SCFAs is followed by other mechanisms that compensate for the acidification. Some of these mechanisms may include SCFA transporters. In isolated mouse colonocytes, the mechanism of Na-independent pH_i recovery was unrelated to differences in metabolism or hydrophobicity among the tested monocarboxyates, and was electroneutral (Chu and Montrose, 1995). Although the anion transport inhibitor DIDS had no effect on this Na-independent pH_i recovery mechanism, measurements of extracellular pH in mouse colonic mucosa identified a DIDS-sensitive and SCFA-dependent mechanism that contributed to alkalinization in the crypt lumen (Chu and Montrose, 1996). In isolated rabbit colonocytes (but not mouse colonocytes), pH_i recovery from an SCFA-induced acid load was sensitive to α-cyano-hydroxycinnamate, an inhibitor of MCT1 (see section III.B.4) (Chu and Montrose, 1995; DeSoignie and Sellin, 1994). Most suggestions of carrier-mediated SCFA flux have originated from studies of isolated membrane vesicles. As described in the following sections, when isotopic tracers are used to study SCFA transport in isolated membrane vesicles, the predominant SCFA transport mechanisms have characteristics that are difficult to reconcile with nonionic diffusion. Most notably, the extravesicular pH dependence of SCFA uptake into vesicles does not coincide with the pKa of the SCFAs (Mascolo et al., 1991; Ritzhaupt et al., 1998a; Harig et al., 1991; Reynolds et al., 1993). As described in more detail in the following sections, vesicular SCFA fluxes also saturate as a function of SCFA concentration, and the rate of SCFA uptake is dependent on the presence of alternate ions.

2. SCFA/HCO$_3$ Exchange

SCFA/HCO$_3$ exchange was identified in apical membrane vesicles from human ileum and colon (Harig et al., 1991; 1996), as well as apical and basolateral membrane vesicles from rat colon (Mascolo et al., 1991; Reynolds et al., 1993). The transport reaction has been most commonly defined as a stimulated uptake of isotopic SCFA in response to an outwardly directed bicarbonate/CO_2 gradient (usually in conjunction with an inwardly directed proton gradient). If bicarbonate/CO_2 loss is rapid, with consequent changes in intravesicular and/or extravesicular pH, many effects assigned to SCFA/HCO$_3$ exchange could be equally well explained by nonionic diffusion of SCFAs. This ambiguity of interpretation also applies to the basic observations that luminal SCFA stimulates bicarbonate secretion in intact tissue (Umesaki et al., 1979; Dohgen et al., 1994) and that removal of bicarbonate/CO_2 diminishes transepithelial SCFA flux (Engelhardt et al., 1994). However, some observations from vesicles provide strong evidence that carrier-mediated SCFA/HCO$_3$ exchange exists. In the absence of bicarbonate, inward proton gradients drive isotopic SCFA uptake that is not saturable at higher SCFA concentrations (consistent with nonionic diffusion) (Harig et al., 1991; 1996). However, in the presence of bicarbonate and a similar pH gradient, the stimulated SCFA uptake saturates (Mascolo et al., 1991; Ritzhaupt et al., 1998a; Harig et al., 1991; 1996; Reynolds et al., 1993). The bicarbonate-stimulated SCFA uptake into apical membrane vesicles from the colon is insensitive to the disulfonic stilbene DIDS (Mascolo et al., 1991; Ritzhaupt et al., 1998a; Harig et al., 1996; Stein et al., 1995), although it can be inhibited by niflumic acid (Harig et al., 1996) or MCT1 inhibitors (RitzHaupt et al., 1998a) (see section III.B.4). In contrast, the equivalent stimulated uptake into ileal apical vesicles or colonic basolateral vesicles can be inhibited by mM concentrations of DIDS (Harig *et al.,* 1991; Reynolds *et al.,* 1993), which suggests that functionally distinct pathways mediate the same transport reaction at the apical and basolateral membranes. Finally, uptake of SCFA is *cis*-inhibited by other monocarboxylates, but the rank potency of inhibition does not correlate with hydrophobicity of the *cis* inhibitor (Harig *et al.,* 1991; 1996).

Evidence from native tissue experiments also indicates a potential role for SCFA/HCO$_3$ exchange. Using isolated cecal tissue, measurements of extracellular pH near the basolateral membrane have shown a dramatic transient alkalinization in response to serosal butyrate that occurred in the presence, but not absence, of bicarbonate/CO$_2$ (Genz *et al.,* 1999). Measurements of extracellular pH near the luminal membrane did not detect a similar response after addition of luminal SCFA (Chu and Montrose, 1996; Genz *et al.,* 1999), which suggests that serosal SCFA/HCO$_3$ exchange is functionally the most active. Using pH stat titration of luminal contents, addition of SCFA was found to elicit a threefold greater addition of base equivalents to the lumen in the presence (versus absence) of bicarbonate/CO$_2$ (Dohgen *et al.,* 1994).

3. SCFA/Cl Exchange

SCFA/Cl exchange was originally identified in rat colonic apical membrane vesicles (Rajendran and Binder, 1994). SCFA/Cl exchange plays a negligible role in SCFA uptake across the apical membrane (Harig *et al.,* 1996). However, this anion exchange reaction has been proposed as a route for SCFA anion efflux across the apical membrane, potentially as a route for recycling SCFA anions across the apical membrane to fuel ongoing SCFA uptake via SCFA/HCO$_3$ exchange (Binder and Mehta, 1989), or as an alternative to HCO$_3$/Cl exchange in stimulating electroneutral Cl absorption (Rajendran and Binder, 1994; Binder and Mehta, 1990). It was not possible to detect SCFA/Cl exchange in human colonic luminal membrane vesicles, which suggests that observations may be species specific (Harig *et al.,* 1996). Evidence of a physiological role of SCFA/Cl exchange in SCFA transport also remains elusive in intact tissue studies. As far as we are aware, the only evidence of such a route is that removal of luminal chloride reduced serosal-to-mucosal flux of SCFAs in rat colon, although it should be noted that a similar effect was not detected in rabbit colon (Sellin *et al.,* 1993; Charney *et al.,* 1998). Most studies have suggested that SCFA/Cl exchange is not an important route for SCFA flux. Using a lipophilic pH sensor that reports from the apical membrane of guinea-pig colonocytes, butyrate was shown to induce similar changes in surface pH in the presence or absence of luminal chloride (Genz *et al.,* 1999). Similarly, removal of chloride did not affect mucosal-to-serosal SCFA fluxes across rabbit or rat distal colon (Sellin *et al.,* 1993; Charney *et al.,* 1998). These results suggest that absorptive transepithelial SCFA fluxes are chloride independent. However, SCFAs affect other ion transport events that clearly do require chloride. For example, chloride removal diminishes SCFA-stimulated sodium absorption in rabbit proximal colon and rat distal colon (Binder and Mehta, 1989; Sellin and DeSoignie, 1998), and surface pH changes induced by chloride removal are greater in the presence of butyrate (Genz *et al.,* 1999). These and other observations suggest that SCFAs may affect other transporters (e.g., via changes in pH) or alternatively SCFA anions may substitute for bicarbonate on an apical Cl/HCO_3 exchanger, with any resulting backflux of SCFA anions being too small to disturb the dominant mechanisms of SCFA flux.

4. MCT1

MCT is the designation for a gene family of monocarboxylate transporters. Structural and functional properties among *MCT* family members have been reviewed recently (Halestrap and Price, 1999). Among the seven members of this gene family, only the first cloned monocarboxylate transporter (*MCT1*) has confirmed expression in the intestines (Halestrap and Price, 1999; Garcia *et al.,* 1994). MCT1 is a proton/monocarboxylate cotransporter found in many tissues of the body. The MCT1 protein mediates an electroneutral transport reaction that functionally mimics nonionic diffusion by 1:1 coupling of proton and anion transport (Halestrap and Price, 1999). MCT1 is inhibited by α-cyano-hydroxycinnamates (CHC), and transports a broad range of monocarboxylates including pyruvate, L-lactate, propionate, and butyrate (Garcia *et al.,* 1994). MCT1 is certainly suited for transport of monocarboxylates that are too hydrophilic for efficient flux by nonionic diffusion through the lipid bilayer (e.g., L-lactate), but there is active debate concerning the role of MCT1 in transport of the more hydrophobic SCFAs found in the colonic lumen. There are conflicting reports about the ability of α-cyano-hydroxycinnamates to affect fluxes of hydrophobic SCFA in isolated colonocytes (Chu and Montrose, 1995; DeSoignie and Sellin, 1994).

The site of MCT1 expression in the intestinal epithelium is controversial. MCT1 protein was localized to the basolateral membrane of hamster colonocytes by using immunofluorescence staining (Garcia *et al.,* 1994). Predominant basolateral staining also has been confirmed in the small and large intestines of rat (Tamai *et al.,* 1999). Basolateral membrane vesicles from the small intestine contain a CHC-sensitive lactate flux, with evidence suggesting that the transporter may be shared by a wide variety of monocarboxylate, Cl, and bicarbonate anions (Orsenigo *et al.,* 1997; Cheeseman *et al.,* 1994; Storelli *et al.,* 1980). Based on these results, MCT1 has the potential to mediate basolateral SCFA efflux from cells as part of transcellular SCFA absorption. However, multiple reports identify robust proton/lactate cotransport in brush-border membrane vesicles from rabbit, rat, and human small intestine (Storelli *et al.,* 1980; Tiruppathi *et al.,* 1988; Friedrich *et al.,* 1991; 1992). MCT1 protein has been detected in

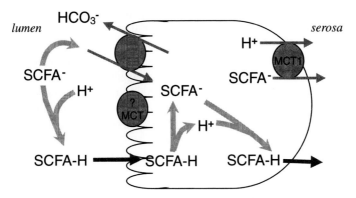

FIGURE 3 Mechanisms of SCFA absorption. This diagram of a colonic epithelial cell depicts the likely site and action of the different mechanisms mediating transcellular SCFA transport. Involvement of a carrier protein is indicated by shaded circles. The name of the cloned tranporter is in the center of the circle, if it is known. Nonionic diffusion is shown as SCFA-H directly traversing the membrane without benefit of a carrier (black arrows).

apical membrane vesicles of pigs and humans (Ritzhaupt *et al.,* 1998b), and minor immunostaining of the apical membrane by polyclonal anti-MCT1 antisera also has been reported (Tamai *et al.,* 1999). In isolated colonic luminal membrane vesicles from pigs, 20% of butyrate transport was inhibited by CHC (Ritzhaupt *et al.,* 1998a). Although some results may be influenced by contamination of apical membrane vesicles with basolateral membrane vesicles, these results raise the possibility that MCT1 (or a closely related protein) may have a role in apical uptake of SCFA from the colonic lumen. Such a possibility is enhanced by the proposal that unusual permeability properties of the colonic apical membrane could limit nonionic diffusion (H. Binder, chapter in this volume).

Figure 3 displays an attempt to build a consensus model of the various routes that allow SCFAs to traverse the intestinal epithelial cell. It includes the transport events that have been most strongly supported to date, but of course it is likely to change as we learn more about the mechanisms of SCFA transport.

IV. DISORDERS OF ANION ABSORPTION

This section will focus on congenital chloride diarrhea (CLD). This genetic disorder provides a unique opportunity to explore a failed experiment of nature that has a direct impact on anion absorption in the intestines.

Although genetic disorders can provide unique insight into the underlying molecular physiology of anion transport, the most common diseases of fluid and electrolyte absorption are not caused by structural changes in specific anion transporters. For a balanced perspective of clinical impact, it should be mentioned that stimulation of anion (usually choride) secretion is the most frequent mechanism that clinically compromises *net* anion and water absorption. For this reason, *diarrheal diseases* evoked by infectious agents are a major killer in non-industrialized nations, responsible for about 5 million deaths per year. In addition, it appears that some forms of *inflammatory bowel disease* involve disorders in anion handling. Rectal irrigation with SCFAs has been found to be a useful therapy for ulcerative colitis (Scheppach *et al.*, 1992; Breuer *et al.*, 1991), which suggests that restriction of anion absorption and metabolism can be a crucial factor in the progression of disease.

A. Congenital Chloride Diarrhea

1. Clinical Synopsis

CLD was identified as a distinct disease by Gamble *et al.* (1945) and Darrow (1945), who named it congenital alkalosis with diarrhea. Voluminous diarrhea develops *in utero,* leading to dilated bowel loops and distended abdomen for the baby and polyhydramnios for the mother, often associated with premature birth. An ultrasound diagnosis is possible, based on the typical bowel finding. Metabolic alkalosis, dehydration, hypochloremia, hyponatremia, and hypokalemia develop quickly in the newborn baby. Hyperbilirubinemia is an associated finding. Stools are watery and can be confused with urine. Without active correction of hydration and electrolyte balance, the baby is at risk of dying. Diagnosis, if only suspected, is simple and based on stool chloride measurement. With appropriately corrected electrolyte balance, the fecal chloride concentration always exceeds 90 mmol/L in CLD (compared with 10–15 mmol/L in normal individuals). It is important to note that an untreated child will not excrete chloride any more because of the depletion of body chloride contents. Thus, the diagnostic test should be administered only after the correction of losses. In newborns, treatment consists of intravenous administration of electrolyte solution. In older children and adults, simply drinking appropriately balanced sodium/potassium chloride solution ensures close-to-normal life. Stool volume tends to diminish with age, and growing children usually learn to control their urgency in a socially satisfying manner. Detailed instructions for therapy have been given by Holmberg (1986); in brief, it is essential to titrate the dose so that chloride is excreted in urine, but to avoid overdose of salts that would lead to the worsening of diarrhea by osmosis. Undersubstitution of water and salt losses may lead to secondary hyperaldosteronism, chronic contraction, and early kidney failure. Patients having good balance can grow and develop normally and may bear children.

2. Intestinal Pathophysiology of CLD

The mechanism of electrolyte absorption defect has been studied directly by intestinal catheterization with the aid of volunteer patients (Bieberdorf et al., 1972; Holmberg et al., 1975). These studies have established a defect in chloride absorption with concomitant luminal acidification and body alkalinization. There are also changes in sodium absorption, but when luminal pH is restored to a normal range, the changes in sodium absorption revert to normal, whereas chloride absorption remains defective (Holmberg et al., 1975). The results have been interpreted based on the conventional model of paired Na/H and Cl/HCO₃ exchangers mediating apical uptake of NaCl. From this viewpoint, results implicated a defective Cl/HCO₃ exchange mechanism as the primary cause of CLD, with secondary inhibition of the Na/H exchange through the changes in luminal pH. Diarrhea is caused by the compromise in salt absorption, which leads to the movement of water to the lumen by osmosis. Hyponatremia and contraction lead to secondary hyperaldosteronism and potassium wastage through the kidneys.

3. Familial Genetics to *CLD*-Disease Gene Candidates

Norio et al. (1971) found that CLD was inherited by autosomal recessive genetic mechanism. This finding was facilitated by the relatively high frequency of CLD in Finland, estimated at 1 in 20,000 births, especially in the eastern provinces. Patients volunteering for functional studies made it possible to determine the functional defect directly in Cl/HCO₃ exchange (Holmberg et al., 1975). The molecular defect became known first through positional cloning studies (Höglund et al., 1995). Considering *CFTR* as a candidate gene, a genetic linkage study in eight families led to the fortuitous mapping of CLD a few centimorgans proximal to the *CFTR* gene, and the simultaneous exclusion of *AE1* and *AE2* as candidate genes (Kere et al., 1993). Population-wide studies of rare genetic diseases have allowed the mapping and identification of numerous other disease genes in Finland (de la Chapelle, 1993).

4. Discovery of *DRA* as *CLD*-Disease Gene

During the process of positional cloning of the CLD gene, Höglund et al. (1996) found that the position of the CLD gene mapped close to a gene designated *DRA*. *DRA* (for *downregulated in adenoma*) had been previously isolated by subtraction cloning experiments in the search for carcinogenesis associated genes (Schweinfest et al., 1993). *DRA* was found to be highly expressed in normal colonic and ileal epithelium, but strongly downregulated in adenomatous tissue and adenocarcinoma of the colon. It was only later that its amino acid sequence was noted to be homologous to another human protein, the diastrophic dysplasia sulfate transporter, encoded by the *DTDST* gene. That gene was the target of a positional cloning project, and the observation that finally led to its cloning was high homology of a short stretch of genomic sequence to the rat sat-1 liver canalicular sulfate transporter (Hästbacka *et al.*, 1994). Based on this homology, other investigators tested and proved the hypothesis that heterologously expressed DRA was a sulfate transport (Silberg *et al.*, 1995). However, even prior to that publication, the position information, homology of DRA with sat-1 and DTDST (both known anion transporters), and the high expression level of DRA in the intestine had prompted a search for mutations in patients with CLD. Two coding changes were found in Finnish patients, and later, numerous more in patients from all over the world (see Table I). The geography of the rare disease and very different carrier frequencies of both sequence changes in Finland helped to implicate one of them, V317del, as the CLD mutation, whereas the other, C307W, was suggested to be a functionally silent polymorphism (Hoglund *et al.*, 1996). That conclusion was subsequently verified by functional sulfate and chloride transport analysis of DRA proteins in which the two mutant sequences were analyzed separately (Moseley *et al.*, 1999).

The role of the *DRA* (or *CLD*) gene in carcinogenesis was studied directly among CLD families. Neither patients themselves (recessive mutation homozygotes) nor their parents (obligate heterozyogus carriers) had a convincingly elevated number of cancer cases (Hemminki *et al.*, 1998). Furthermore, the mutations that have been associated with CLD in patients worldwide comprise a wide spectrum of types and structural consequences for the protein product (Kere *et al.*, 1999). Finally, DRA also has been shown to be downregulated, although to a lesser extent, in inflammatory bowel (Haila *et al.*, in press; Yang *et al.*, 1998). These observations suggest that any direct role for the DRA protein in carcinogenesis is unlikely.

Very recently we identified additional members of the CLD/PDS/DTDST gene family in humans. Several new members (PAT1 to PAT6, for *putative anion transporter 1–6*) have been identified based on high-sequence homology (Lohi *et al.*, in press). Interestingly, the new genes show highly specific tissue distribution patterns, suggesting them as candidates for important physiological anion transport functions. One of them, named PAT1, is expressed in the kidney and pancreas, and specifically in pancreatic ductal cell lines. These features make this new gene a plausible candidate to encode a ductal apical Cl/HCO_3 exchanger of the pancreas (Lohi *et al.*, in press). The secretion of HCO_3 to neutralize gastric juice entering the gut is the major function of the ductal cells, and Cl/HCO_3 exchanger activity has been found at the basolateral and luminal membranes of mouse pancreatic ducts (Zhao *et al.*, 1994).

Acknowledgments

Studies from the authors' laboratories have been supported by grants DK42457 and DK54940 from the National Institutes of Health (M.H.M.), as well as the Foundation for Pediatric Research, Ulla Hjelt Fund, the Sigrid Juselius Foundation, the Helsinki University Hospital research funds, and the Academy of Finland (J.K.).

References

Alper, S. L., Rossmann, H., Wilhelm, S., Stuart-Tilley, A. K., Shmukler, B. E., and Seidler, U. (1999). Expression of AE2 anion exchanger in mouse intestine. *Am. J. Physiol.* **277,** G321–G332.

Alrefai, W. A., Barakat, J., Tyagi, S., Layden, T. J., Ramaswamy, K., and Dudeja, P. K. (1997). AE-2 and AE-3 anion exchanger isoforms are localized to the basolateral but not the apical membranes in the human intestine. *Gastroenterology* **112,** A344.

Anast, C., Kennedy, P., Volk, G., and Adamson, L. (1965). *In vitro* studies of sulfate transport by the small intestine of the rat. *J Lab. Clin. Med.* **65,** 903–911.

Antalis, T. M., Reeder, J. A., Gotley, D. C., Byeon, M. K., Walsh, M. D., Henderson, K. W., Papas, T. S., and Schweinfest, C. W. (1998). Down-regulation of the down-regulated in adenoma (DRA) gene correlates with colon tumor progression. *Clin. Cancer Res.* **4,** 1857–1863.

Bai, L., and Pajor, A. M. (1997). Expression cloning of NaDC-2, an intestinal $Na^{(+)}$- or $Li^{(+)}$-dependent dicarboxylate transporter. *Am. J. Physiol.* **273,** G267–G274.

Bieberdorf, F. A., Gordon, P., and Fordtran, J. S. (1972). Pathogenesis of congenital alkalosis with diarrhea: Implications for the physiology of normal ileal electrolyte absorption and secretion. *J. Clin. Invest.* **51,** 1958–1968.

Binder, H. J., Foster, E. S., Budinger, M. E., and Hayslett, J. P. (1987). Mechanism of electroneutral sodium chloride absorption in distal colon of the rat. *Gastroenterology* **93,** 449–455.

Binder, H. J., and Mehta, P. (1989). Short-chain fatty acids stimulate active sodium and chloride absorption *in vitro* in the rat distal colon. *Gastroenterology* **96,** 989–996.

Binder, H. J., and Mehta, P. (1990). Characterization of the butyrate-dependent electroneutral Na-Cl absorption in the rat distal colon. *Pflugers Arch.* **417,** 365–369.

Binder, H. J., and Sandle, G. I. (1994). Electrolyte transport in the mammalian colon. "Physiology In of the Gastrointestinal tract" (L. Johnson, ed.) New York, Raven Press: 2133–2171.

Bond, J. H., Currier, B. E., Buchwald, H., and Levitt, M. D. (1980). Colonic conservation of malabsorbed carbohydrate. *Gastroenterology* **78,** 444–447.

Bookstein, C., DePaoli, A. M., Xie, Y., Niu, P., Musch, M. W., Rao, M. C., and Chang, E. B. (1994). Na^+/H^+ exchangers, NHE-1 and NHE-3 of rat intestine: Expression and localization. *J. Clin. Invest.* **93,** 106–113.

Booth, I. W., Stange, G., Murer, H., Fenton, T. R., and Milla, P. J. (1985). Defective jejunal brush-border Na^+/H^+ exchange: A cause of congenital secretory diarrhoea. *Lancet* **1,** 1066–1069.

Breuer, R. I., Buto, S. K., Christ, M. L., Bean, J., Vernia, P., Paoluzi, P., DiPaulo, M. C., and Caprilli, R. (1991). Rectal irrigation with short-chain fatty acids for distal ulcerative colitis. *Dig. Dis. Sci.* **36,** 185–187.

Browne, J. L., Sanford, P. A., and Smyth, D. H. (1978). Transfer and metabolism of citrate, succinate, alpha-ketoglutarate and pyruvate by hamster small intestine. *Proc. R. Soc. Lond. B. Biol. Sci.* **200,** 117–135.

Burgaut, M. (1987). Occurrence, absorption and metabolism of short chain fatty acids in the digestive tract of mammals. *Comp. Biochem. Physiol.* **86B,** 439–472.

Busch, A. E., Waldegger, S., Herzer, T., Biber, J., Markovich, D., Murer, H., and Lang, F. (1994). Electrogenic cotransport of Na^+ and sulfate in *Xenopus* oocytes expressing the cloned $Na^+SO4(2^-)$ transport protein NaSi-1. *J. Biol. Chem.* **269,** 12407–12409.

Byeon, M. K., Frankel, A., Papas, T. S., Henderson, K. W., and Schweinfest, C. W. (1998). Human DRA functions as a sulfate transporter in Sf9 insect cells. *Protein Exp Purif.* **12,** 67–74.

Byeon, M. K., Westerman, M. A., Maroulakou, I. G., Henderson, K. W., Suster, S., Zhang, X. K., Papas, T. S., Vesely, J., Willingham, M. C., Green, J. E., and Schweinfes, C. W. (1996). The down-regulated in adenoma (DRA) gene encodes an intestine-specific membrane glycoprotein. *Oncogene* **12,** 387–396.

Caverzasio, J., Danisi, G., Straub, R. W., Murer, H., and Bonjour, J. P. (1987). Adaptation of phosphate transport to low phosphate diet in renal and intestinal brush border membrane vesicles: influence of sodium and pH. *Pfluegers Arch* **409**, 333–336.

Chang, H., Tashiro, K., Hirai, M., Ikeda, K., Kurokawa, K., and Fujita, T. (1996). Identification of a cDNA encoding a thiazide-sensitive sodium-chloride cotransporter from the human and its mRNA expression in various tissues. *Biochem. Biophys. Res. Comm.* **223**, 324–328.

Charney, A. N., Micic, L., and Egnor, R. W. (1998). Nonionic diffusion of short-chain fatty acids across rat colon. *Am. J. Physiol.* **274**, G518–G524.

Cheeseman, C. I., Shariff, S., and O'Neill, D. (1994). Evidence for a lactate-anion exchanger in the rat jejunal basolateral membrane. *Gastroenterology* **106**, 559–566.

Chow, A., Dobbins, J. W., Aronson, P. S., and Igarashi, P. (1992). cDNA cloning and localization of a band 3-related protein from ileum. *Am. J. Physiol.* **263**, G345–G352.

Chu, S., and Montrose, M. H. (1995). An Na^+-independent short-chain fatty acid transporter contributes to intracellular pH regulation in murine colonocytes. *J. Gen. Physiol.* **105**, 589–615.

Chu, S., and Montrose, M. H. (1996). Non-ionic diffusion and carrier-mediated transport drive extracellular pH regulation of mouse colonic crypts. *J. Gen. Physiol.* **494**, 783–793.

Dagher, P. C., Behm, T., Taglietta-Kohlbrecher, A., Egnor, R. W., and Charney, A. N. (1996). Dissociation of colonic apical Na/H exchange activity from bulk cytoplasmic pH. *Am. J. Physiol.* **270**, C1799–C1806.

Darrow, D. C. (1945). Congenital alkalosis with diarrhea. *J. Pediatr.* **26**, 519–532.

de la Chapelle, A. (1993). Disease gene mapping in isolated human populations: The example of Finland. *J. Med. Genet.* **30**, 857–865.

DeSoignie, R., and Sellin, J. H. (1994). Propionate-initiated changes in intracellular pH in rabbit colonocytes. *Gastroenterology* **107**, 347–356.

Dohgen, M., Hayahshi, H., Yajima, T., and Suzuki, Y. (1994). Stimulation of bicarbonate secretion by luminal short-chain fatty acid in the rat and human colon *in vitro*. *Japanese J. Physiol.* **44**, 519–531.

Dudeja, P. K., Rao, D. D., Syed, I., Joshi, V., Dahdal, R. Y., Gardner, C., Risk, M. C., Schmidt, L., Bavishi, D., Kim, K. E., Harig, J. M., Goldstein, J. L., Layden, T. J., and Ramaswamy, K. (1996). Intestinal distribution of human Na^+/H^+ exchanger isoforms NHE-1, NHE-2, and NHE-3 mRNA. *Am. J. Physiol.* **271**, G483–G493.

Duffey, M. E., Turnheim, K., Frizzell, R. A., and Schultz, S. G. (1978). Intracellular chloride activities in rabbit gallbladder: Direct evidence for the role of the sodium-gradient in energizing "uphill" chloride transport. *J. Membr. Biol.* **42**, 229–245.

Edmonds, C. J., and Mackenzie, J. (1984). Amiloride sensitive and insensitive sodium pathways and the cellular sodium transport pool of colonic epithelium in rats. *J. Physiol. (Lond.)* **346**, 61–71.

Engelhardt, W. V., Gros, G., Burmester, M., Hansen, K., Becker, G., and Rechkemmer, G. (1994). Functional role of bicarbonate in propionate transport across guinea-pig isolated caecum and proximal colon. *J. Physiol. (London)* **477**, 365–371.

Everett, L. A., Glaser, B., Beck, J. C., Idol, J. R., Buchs, A., Heyman, M., Adawi, F., Hazani, E., Nassi, R. E., Baxevanis, A. D., Sheffield, V. C., and Green, E. D. (1997). Pendred syndrome is caused by mutations in a putative sulfate transporter gene PDS. *Nature Genet.* **17**, 411–422.

Feild, J. A., Zhang, L., Brun, K. A., Brooks, D. P., and Edwards, R. M. (1999). Cloning and functional characterization of a sodium-dependent phosphate transporter expressed in human lung and small intestine. *Biochem. Biophys. Res. Commun.* **258**, 578–582.

Foster, E. S., Budinger, M. E., Hayslett, J. P., and Binder, H. J. (1986). Ion transport in proximal colon of the rat: Sodium depletion stimulates neutral sodium chloride absorption. *J. Clin. Invest.* **77**, 228–235.

Foster, E. S., Dudeja, P. K., and Brasitus, T. A. (1990). Contribution of Cl^-/OH-exchange to electroneutral NaCl absorption in rat distal colon. *Am. J. Physiol.* **258**, G261–G267.

Freel, R. W., Hatch, M., and Vaziri, N. D. (1997). cAMP-dependent sulfate secretion by the rabbit distal colon: A comparison with electrogenic chloride secretion. *Am. J. Physiol.* **273,** C148–C160.

Friedrich, M., Murer, H., and Berger, E. G. (1991). Transport of L-leucine hydroxy analogue and L-lactate in rabbit small-intestinal brush-border membrane vesicles. *Pflugers Arch.* 418, 393–399.

Friedrich, M., Murer, H., Sterchi, E., and Berger, E. G. (1992). Transport of L-leucine hydroxy analogue and L-lactate in human small intestinal brush border membrane vesicles. *Eur. J. Clin. Invest.* 22, 73–78.

Frizzell, R. A., Dugas, M. C., and Schultz, S. G. (1975). Sodium chloride transport by rabbit gallbladder: Direct evidence for a coupled NaCl influx process. *J. Gen. Physiol.* **65,** 769–795.

Frizzell, R. A., Field, M., and Schultz, S. G. (1979). Sodium-coupled chloride transport by epithelial tissues. *Am. J. Physiol.* **236,** F1–F8.

Gamble, J. L., Fahey, K. R., Appleton, J., and MacLachlan, E. (1945). Congenital alkalosis with diarrhea. *J. Pediatr.* **26,** 509–518.

Garcia, C. K., Goldstein, J. L., Pathak, R. K., Anderson, R. G. W., and Brown, M. S. (1994). Molecular characterization of a membrane transporter for lactate, pyruvate, and other monocarboxylates: Implications for the Cori cycle. *Cell* **76,** 865–873.

Genz, A. K., Engelhardt, W., and Busche, R. (1999). Maintenance and regulation of the pH microclimate at the luminal surface of the distal colon of guinea-pig. *J. Physiol. (Lond.)* **517,** 507–519.

Gillen, C. M., Brill, S., Payne, J. A., and Forbush III, B. (1996). Molecular cloning and functional expression of the K-Cl cotransporter from rabbit, rat, and human. *J. Biol. Chem.* **271,** 16237–16244.

Gonda, T., Maouyo, D., Rees, S., and Montrose, M. (1999). Regulation of intracellular pH gradients by a short-chain fatty acid and identified Na/H exchanger isoforms. *Am. J. Physiol.* **276,** G259–G270.

Haila, S., Saarialho-Kere, U., Karjalainen-Lindsberg, M.-L., Lohi, H., Airola, K., Holmberg, C., Hästbacka, J., Kere, J., and Höglund, P. The congenital chloride diarrhea gene (CLD) is expressed in inflammatory colon epithelium, seminal vesicle, sweat gland, and in some malignant colon cells. *Histochem. Cell Biol.* (in press).

Halestrap, A. P., and Price, N. T. (1999). The proton-linked monocarboxylate transporter (MCT) family: Structure, function and regulation. *Biochem. J.* **343 Pt 2,** 281–299.

Harig, J. M., Ng, E. K., Dudeja, P. K., Brasitus, T. A., and Ramaswamy, K. (1996). Transport of n-butyrate into human colonic luminal membrane vesicles. *Am. J. Physiol.* **271,** G415–G422.

Harig, J. M., Soergel, K. H., Barry, J. A., and Ramaswamy, K. (1991). Transport of propionate by human ileal brush-border membrane vesicles. *Am. J. Physiol.* **260,** G776–G782.

Hästbacka, J., de la Chapelle, A., Mahtani, M. M., Clines, G., Reeve-Daly, M. P., Daly, M., Hamilton, B. A., Kusumi, K., Trivedi, B., Weaver, A., Coloma, A., Lovett, M., Buckler, A., Kaitila, I., and Lander, E. S. (1994). The diastrophic dysplasia gene encodes a novel sulfate transporter positional cloning by fine-structure linkage disequilibrium mapping. *Cell* **78,** 1073-1087.

Hästbacka, J., Superti-Furga, A., Wilcox, W. R., Rimoin, D. L., Cohn, D. H., and Lander, E. S. (1996). Atelosteogenesis type II is caused by mutations in the diastrophic dysplasia sulfate-transporter gene (DTDST): Evidence for a phenotypic series involving three chondrodysplasias. *Am. J. Hum. Genet.* **58,** 255–262.

Hattenhauer, O., Traebert, M., Murer, H., and Biber, J. (1999). Regulation of small intestinal Na-P(i) type IIb cotransporter by dietary phosphate intake. *Am. J. Physiol.* **277,** G756–G762.

He, H., Ganapathy, V., Isales, C. M., and Whitford, G. M. (1998). pH-dependent fluoride transport in intestinal brush border membrane vesicles. *Biochim. Biophys. Acta* **1372,** 244–254.

Hemminki, A., Höglund, P., Pukkala, E., Salovaara, R., Järvinen, H., Norio, R., and Aaltonen, L. A. (1998). Intestinal cancer in patients with a germline mutation in the down-regulated in adenoma (*DRA*) gene. *Oncogene* **16,** 681–684.

Hiki, K., D'Andrea, R. J., Furze, J., Crawford, J., Woollatt, E., Sutherland, G. R., Vadas, M. A., and Gamble, J. R. (1999). Cloning, characterization, and chromosomal location of a novel human K$^+$-Cl$^-$ cotransporter. *J. Biol. Chem.* **274**, 10661–10667.

Hildmann, B., Storelli, C., Danisi, G., and Murer, H. (1982). Regulation of Na$^+$-Pi cotransport by 1,25–dihydroxyvitamin D$_3$ in rabbit duodenal brush-border membrane. *Am. J. Physiol.* **242**, G533–G539.

Hilfiker, H., Hattenhauer, O., Traebert, M., Forster, I., Murer, H., and Biber, J. (1998). Characterization of a murine type II sodium-phosphate cotransporter expressed in mammalian small intestine. *Proc. Natl. Acad. Sci. USA* **95**, 14564–14569.

Höglund, P., Haila, S., Scherer, S. W., Tsui, L.-C., Green, E. D., Weissenbach, J., Holmberg, C., de la Chapelle, A., and Kere, J. (1996). Positional candidate genes for congenital chloride diarrhea suggested by high-resolution physical mapping in chromosome region 7q31. *Genome Res.* **6**, 202–210.

Höglund, P., Haila, S., Socha, J., Tomaszewski, L., Saarialho-Kere, U., Karjalainen-Lindsberg, M. L., Airola, K., Holmberg, C., de la Chapelle, A., and Kere, J. (1996). Mutations of the down-regulated in adenoma (*DRA*) gene cause congenital chloride diarrhoea. *Nature Genet.* **14**, 316–319.

Höglund, P., Sistonen, P., Norio, R., Holmberg, C., Dimberg, A., Gustavson, K.-H., de la Chapelle, A., and Kere, J. (1995). Fine mapping of the congenital chloride diarrhea gene by linkage disequilibrium. *Am. J. Hum. Genet.* **57**, 95–102.

Höglund, P., et al. (1998a). Genetic background of congenital chloride diarrhea in high incidence populations: Finland, Poland, Saudia Arabia and Kuwait. *Am. J. Hum. Genet.* **63**, 760–768.

Höglund, P., et al. (1998b). Clustering of private mutations in the congenital chloride/diarrhea/down-regulated in adenoma gene. *Hum. Mutat.* **11**, 321–327.

Holmberg, C. (1986). Congenital chloride diarrhea. *Clin. Gastroenterology* **3**, 583–602.

Holmberg, C., and Perheentupa, J. (1985). Congenital Na$^+$ diarrhea: A new type of secretory diarrhea. *J. Pediatr.* **106**, 56–61.

Holmberg, C., Perheentupa, J., and Launiala, K. (1975). Colonic electrolyte transport in health and in congenital chloride diarrhea. *J. Clin. Invest.* **56**, 302–310.

Holtug, K., McEwan, G. T. A., and Skadhauge, E. (1992). Effects of propionate on the acid microclimate of hen (*Gallus domesticus*) colonic mucosa. *Comp. Biochem. Physiol.* **103A**, 649–652.

Holtzman, E. J., Kumar, S., Faaland, C. A., Warner, F., Logue, P. J., Erickson, S. J., Ricken, G., Waldman, J., and Dunham, P. B. (1998). Cloning, characterization, and gene organization of K-Cl cotransporter from pig and human kidney and *C. elegans*. *Am. J. Physiol.* **275**, F550–F564.

Hoogerwerf, W. A., Tsao, W. C., Devuyst, O., Levine, S. A., Yun, C. H. C., Yip, J. W., Cohen, M. E., Wilson, P. D., Lazenby, A. J., Tse, C. M., and Donowitz, M. (1996). NHE2 and NHE3 are human and rabbit intestinal brush-border proteins. *Am. J. Physiol.* **270**, G29–G41.

Hubel, K. A., Renquist, K., and Shirazi, S. (1987). Ion transport in human cecum, transverse colon, and sigmoid colon *in vitro:* Baseline and response to electrical stimulation of intrinsic nerves. *Gastroenterology* **92**, 501.

Ilundain, A., Larralde, J., and Toval, M. (1987). Iodide transport in rat small intestine: Dependence on calcium. *J. Physiol.* **393**, 19–27.

Katai, K., Miyamoto, K., Kishida, S., Segawa, H., Nii, T., Tanaka, H., Tani, Y., Arai, H., Tatsumi, S., Morita, K., Taketani, Y., and Takeda, E. (1999). Regulation of intestinal Na$^+$-dependent phosphate co-transporters by a low-phosphate diet and 1,25–dihydroxyvitamin D$_3$. *Biochem. J.* **343 Pt 3**, 705–712.

Kavanaugh, M. P., and Kabat, D. (1996). Identification and characterization of a widely expressed phosphate transporter/retrovirus receptor family. *Kidney Int.* **49**, 959–963.

Kere, J., Lohi, H., and Höglund, P. (1999). Genetic disorders of membrane transport III. Congenital chloride diarrhea. *Am. J. Physiol.* **276**, G7–G13.

Kere, J., Sistonen, P., Holmberg, C., and de la Chapelle, A. (1993). The gene for congenital chloride diarrhea maps close to but is distinct from the gene for cystic fibrosis transmembrane conductance regulator. *Proc. Natl. Acad. Sci. USA* **90**, 10686–10689.

Khatri, I. A., Kovacs, S. V., and Forstner, J. F. (1996). Cloning of the cDNA for a rat intestinal Na^+/dicarboxylate cotransporter reveals partial sequence homology with a rat intestinal mucin. *Biochim. Biophys. Acta* **1309**, 58–62.

Kimmich, G. A., Randles, J., and Bennett, E. (1991). Sodium-dependent succinate transport by isolated chick intestinal cells. *Am. J. Physiol.* **260**, C1151–C1157.

Knickelbein, R., Aronson, P. S., Atherton, W., and Dobbins, J. W. (1983). Sodium and chloride transport across rabbit ileal brush border: I. Evidence for Na-H exchange. *Am. J. Physiol.* **245**, G504–G510.

Knickelbein, R., Aronson, P. S., Schron, C. M., Seifter, J. and Dobbins, J. W. (1985). Sodium and chloride transport across rabbit ileal brush border: II. Evidence for Cl^-HCO_3 exchange and mechanism of coupling. *Am. J. Physiol.* **249**, G236–G245.

Knickelbein, R. G., Aronson, P. S., and Dobbins, J. W. (1988). Membrane distribution of sodium-hydrogen and chloride-bicarbonate exchangers in crypt and villus cell membranes from rabbit illium. *J. Clin. Invest.* **82**, 2158–2163.

Knickelbein, R. G., and Dobbins, J. W. (1990). Sulfate and oxalate exchange for bicarbonate across the basolateral membrane of rabbit ileum. *Am. J. Physiol.* **259**, G807–G813.

Langridge-Smith, J. E., and Field, M. (1981). Sulfate transport in rabbit ileum: Characterization of the serosal border anion exchange process. *J. Membr. Biol.* **63**, 207–214.

Langridge-Smith, J. E., Sellin, J. H., and Field, M. (1983). Sulfate influx across the rabbit ileal brush border membrane: Sodium and proton dependence, and substrate specificities. *J. Membr. Biol.* **72**, 131–139.

Lee, D. B., Walling, M. W., and Brautbar, N. (1986). Intestinal phosphate absorption: Influence of vitamin D and non-vitamin D factors. *Am. J. Physiol.* **250**, G369–373.

Liedtke, C. M., and Hopfer, U. (1982). Mechanism of Cl^- translocation across intestinal brush border membrane: II. Demonstration of Cl^--OH-exchange and Cl^- conductance. *Am. J. Physiol.* **242**, G272–G280.

Lohi, H., Kujala, M., Kestilä, M., and Kere, J. Identification of five putative anion transporter genes in human and characterization of PAT1, a candidate gene for pancreatic anion exchanger. (submitted).

Lubcke, R., Haag, K., Berger, E., Knauf, H., and Gerok, W. (1986). Ion transport in rat proximal colon *in vivo*. *Am. J. Physiol.* **251**, G132–G139.

Lucke, H., Stange, G., and Murer, H. (1981). Sulfate-sodium cotransport by brush-border membrane vesicles isolated from rat ileum. *Gastroenterology* **80**, 22–30.

Maher, M. M., Gontarek, J. D., Bess, R. S., Donowitz, M., and Yeo, C. J. (1997). The Na/H exchange isoform NHE3 regulates basal canine ileal Na^+ absorption *in vivo*. *Gastroenterology* **112**, 174–183.

Markovich, D., Forgo, J., Stange, G., Biber, J., and Murer, H. (1993). Expression cloning of rat renal $Na^+/SO4(2^-)$ cotransport. *Proc. Natl. Acad. Sci. USA* **90**, 8073–8077.

Mascolo, N., Rajendran, V. M., and Binder, H. J. (1991). Mechanism of short-chain fatty acid uptake by apical membrane vesicles of rat distal colon. *Gastroenterology* **101**, 331–338.

Melvin, J. E., Park, K., Richardson, L., Schultheis, P. J., and Shull, G. E. (1999). Mouse down-regulated in adenoma (DRA) is an intestinal Cl/HCO_3 exchanger and is up-regulated in colon of mice lacking the NHE3 Na/H exchanger. *J. Biol. Chem.* **274**, 22855–22861.

Moe, A. J., Mallet, R. T., Jackson, M. J., Hollywood, J. A., and Kelleher, J. K. (1988). Effect of Na^+ on intestinal succinate transport and metabolism *in vitro*. *Am. J. Physiol.* **255**, C95–C101.

Montrose, M. H., and Kimmich, G. A. (1986). Quantitative use of weak bases for estimation of cellular pH gradients. *Am. J. Physiol.* **250**, C418–C422.

Moseley, R. H., Höglund, P., Wu, G. D., Silberg, D. G., Haila, S., de la Chapelle, A., Holmberg, C., and Kere, J. (1999). Downregulated in adenoma gene encodes a chloride transporter defective in congenital chloride diarrhea. *Am. J. Physiol.* **276,** G185–G192.

Moseley, R. H., Wu, G. D., Silberg, D. G., Huang, N., Zugger, L., Höglund, P., Haila, S., and Kere, J. (1997). The down regulated in adenoma (*DRA*) gene encodes an anion exchanger that is defective in congenital chloridorrhea. *Gastroenterology* **112,** A387.

Mount, D. B., Mercado, A., Song, L., Xu, J., George, A. L., Jr., Delpire, E., and Gamba, G. (1999). Cloning and characterization of KCC3 and KCC4, new members of the cation-chloride cotransporter gene family. *J. Biol. Chem.* **274,** 16355–16362.

Murer, H., Markovich, D., and Biber, J. (1994). Renal and small intestinal sodium-dependent symporters of phosphate and sulphate. *J. Exp. Biol.* **196,** 167–181.

Musch, M. W., Orellana, S. A., Kimberg, L., Field, M., Halm, D. R., Krasny, E. J., and Frizzell, R. A. (1982). Na⁺-K⁺-Cl⁻ cotransporter in the intestine of a marine teleost. *Nature* **300,** 351–353.

Nader, M., Lamprecht, G., Classen, M., and Seidler, U. (1994). Different regulation by pHᵢ and osmolarity of the rabbit ileum brush-border and parietal cell basolateral anion exchanger. *J. Physiol. (London)* **481,** 605–615.

Nellans, H. N., Frizzell, R. A., and Schultz, S. G. (1973). Coupled sodium-chloride influx across the brush border of rabbit ileum. *Am. J. Physiol.* **225,** 467–475.

Norbis, F., Perego, C., Markovich, D., Stange, G., Verri, T., and Murer, H. (1994). cDNA cloning of a rat small-intestinal Na⁺/SO₄(2⁻) cotransporter. *Pflugers Arch.* **428,** 217–223.

Norio, R., Perheentupa, J., Launiala, K., and Hallman, N. (1971). Congenital chloride diarrhea, an autosomal recessive disease. Genetic study of 14 Finnish and 12 other families. *Clin. Genet.* **2,** 182–192.

Orsenigo, M. N., Tosco, M., Laforenza, U., and Faelli, A. (1997). Facilitated transport of lactate by rat jejunal enterocyte. *J. Membr. Biol.* **158,** 257–264.

Pajor, A. M. (1995). Sequence and functional characterization of a renal sodium/dicarboxylate cotransporter. *J. Biol. Chem.* **270,** 5779–5785.

Pajor, A. M. (1999). Sodium-coupled transporters for Krebs cycle intermediates. *Ann. Rev. Physiol.* **61,** 663–682.

Parkins, F. M. (1971). Active F- transport: Species and age effects with rodent intestine, *in vitro.* *Biochim. Biophys. Acta.* **241,** 507–512.

Payne, J. A., and Forbush III, B. (1995). Molecular characterization of the epithelial Na-K-Cl cotransporter isoforms. *Curr. Opinion Cell Biol.* **7,** 493–503.

Payne, J. A., Stevenson, T. J., and Donaldson, L. F. (1996). Molecular characterization of a putative K-Cl cotransporter in rat brain. A neuronal-specific isoform. *J. Biol. Chem.* **271,** 16245–16252.

Quamme, G. A. (1985). Phosphate transport in intestinal brush-border membrane vesicles: Effect of pH and dietary phosphate. *Am. J. Physiol.* **249,** G168–G176.

Race, J. E., Makhlouf, F. N., Logue, P. J., Wilson, F. H., Dunham, P. B., and Holtzman, E. J. (1999). Molecular cloning and functional characterization of KCC3, a new K-Cl cotransporter. *Am. J. Physiol.* **277,** C1210–C1219.

Rajendran, V. M., and Binder, H. J. (1993). Cl-HCO₃ and Cl-OH exchanges mediate Cl uptake in apical membrane vesicles of rat distal colon. *Am. J. Physiol.* **264,** G874–G879.

Rajendran, V. M., and Binder, H. J. (1994). Apical membrane Cl-butyrate exchange: Mechanism of short chain fatty acid stimulation of active chloride absorption in rat distal colon. *J. Membr. Biol.* **141(1),** 51–58.

Rajendran, V. M., and Binder, H. J. (1999). Distribution and regulation of apical Cl/anion exchanges in surface and crypt cells of rat distal colon. *Am. J. Physiol.* **276,** G132–G137.

Rajendran, V. M., Sangan, P., Black, J., and Binder, H. J. (1997). Anion exchange (AE)1 isoform encodes Cl-HCO₃ exchange in surface cells of rat distal colon. *Gastroenterology* **112,** A396.

Rechkemmer, G., and Engelhardt, W. v. (1988). Concentration- and pH-dependence of short-chain fatty acid absorption in the proximal and distal colon of guinea pig (*Cavia porcellus*). *Comp. Biochem. Physiol.* **19A,** 659–663.

Reuss, L. (1983). Basolateral KCl co-transport in a NaCl-absorbing epithelium. *Nature* **305,** 723–726.

Reynolds, D. A., Rajendran, V. M., and Binder, H. J. (1993). Bicarbonate-stimulated [14C]butyrate uptake in basolateral membrane vesicles of rat distal colon. *Gastroenterology* **105,** 725–732.

Ritzhaupt, A., Ellis, A., Hosie, K. B., and Shirazi-Beechey, S. P. (1998a). The characterization of butyrate transport across pig and human colonic luminal membrane. *J. Physiol. (Lond.)* **507,** 819–830.

Ritzhaupt, A., Wood, I. S., Ellis, A., Hosie, K. B., and Shirazi-Beechey, S. P. (1998b). Identification of a monocarboxylate transporter isoform type 1 (MCT1) on the luminal membrane of human and pig colon. *Biochem. Soc. Trans.* **26,** S120.

Roediger, W. E. W. (1982). Utilization of nutrients by isolated epithelial cells of the rat colon. *Gastroenterology* **83,** 424–429.

Roos, A., and Boron, W. F. (1981). Intracellular pH. *Physiol. Rev.* **61,** 296–434.

Sandle, G. I. (1989). Segmental heterogeneity of basal and aldosterone-induced electrogenic Na transport in human colon. *Pflugers Arch.* **414,** 706–712.

Scheppach, W., Sommer, H., Kirchner, T., Paganelli, G. M., Bartram, P., Christl, S., Richter, F., Dusel, G., and Kasper, H. (1992). Effect of butyrate enemas on the colonic mucosa in distal ulcerative colitis. *Gastroenterology* **103,** 51–56.

Schron, C. M., Knickelbein, R. G., Aronson, P. S., Della Puca, J., and Dobbins, J. W. (1985). pH gradient-stimulated sulfate transport by rabbit ileal brush-border membrane vesicles: Evidence for SO_4-OH exchange. *Am. J. Physiol.* **249,** G607–G613.

Schweinfest, C. W., Henderson, K. W., Suster, S., Kondoh, N., and Papas, T. S. (1993). Identification of a colon mucosa gene that is down-regulated in colon adenomas and adenocarcinomas. *Proc. Natl. Acad. Sci. USA* **90,** 4166–4170.

Scott, D. A., Wang, R., Kreman, T. M., Sheffield, V. C., and Karniski, L. P. (1999). The Pendred syndrome gene encodes a chloride-iodide transport protein. *Nature Genet.* **21,** 440–443.

Sellin, J. H., and DeSoignie, R. (1998). Short-chain fatty acids have polarized effects on sodium transport and intracellular pH in rabbit proximal colon. *Gastroenterology* **114,** 737–747.

Sellin, J. H., DeSoignie, R., and Burlingame, S. (1993). Segmental differences in short-chain fatty acid transport in rabbit colon: Effect of pH and Na. *J. Membrane Biol.* **136,**.147–158.

Silberg, D. G., Wang, W., Moseley, R. H., and Traber, P. G. (1995). The down regulated in adenoma (DRA) gene encodes an intestine-specific membrane sulfate transport protein. *J. Biol. Chem.* **270,** 11897–11902.

Smith, P. L., Orellana, S. A., and Field, M. (1981). Active sulfate absorption in rabbit ileum: Dependence on sodium and chloride and effects of agents that alter chloride transport. *J. Membr. Biol.* **63,** 199–206.

Stein, J., Schroeder, O., Milovic, V., and Caspary, W. F. (1995). Mercaptopropionate inhibits butyrate uptake in isolated apical membrane vesicles of the rat distal colon. *Gastroenterology* **108,** 673–679.

Storelli, C., Corcelli, A., Cassano, G., Hildmann, B., Murer, H., and Lippe, C. (1980). Polar distribution of sodium-dependent and sodium-independent transport system for L-lactate in the plasma membrane of rat enterocytes. *Pflugers Arch.* **388,** 11–16.

Superti-Furga, A., Hästbacka, J., Wilcox, W. R., Cohn, D. H., van der Harten, H. J., Rossi, A., Blau, N., Rimoin, D. L., Steinmann, B., Lander, E. S., and Gitzelmann, R. (1996). Achondrogenesis type IB is caused by mutations in the diastrophic dysplasia sulfate transporter gene. *Nature Genet.* **12,** 100–102.

Tamai, I., Sai, Y., Ono, A., Kido, Y., Yabuuchi, H., Takanaga, H., Satoh, E., Ogihara, T., Amano, O., Izeki, S., and Tsuji, A. (1999). Immunohistochemical and functional characterization of pH-dependent intestinal absorption of weak organic acids by the monocarboxylic acid transporter MCT1. *J. Pharm. Pharmacol.* **51,** 1113–1121.

Tiruppathi, C., Balkovetz, D. F., Ganapathy, V., Miyamoto, Y., and Leibach, F. H. (1988). A proton gradient, not a sodium gradient, is the driving force for active transport of lactate in rabbit intestinal brush-border membrane vesicles. *Biochem. J.* **256,** 219–223.

Turnberg, L. A. (1971). Abnormalities in intestinal electrolyte·transport in congenital chloridorrhoea. *Gut* **12,** 544–551.

Umesaki, Y., Yajima, T., Yokokura, T., and Mutai, M. (1979). Effect of organic acid absorption on bicarbonate transport in rat colon. *Pfluegers Arch.* **379,** 43–47.

Vaandrager, A. B., and deJonge, H. R. (1988). A sensitive technique for the determination of anion exchange activities in brush-border membrane vesicles. Evidence for two exchangers with different affinities for HCO_3^- and SITS in rat intestinal epithelium. *Biochim. Biophys. Acta* **939,** 305–314.

Von Engelhardt, W., and Rechkemmer, G. (1992). Segmental differences of short-chain fatty acid transport across guinea-pig large intestine. *Exp. Physiol.* **77,** 491–499.

Walling, M. W. (1977). Intestinal Ca and phosphate transport: Differential responses to vitamin D_3 metabolites. *Am. J. Physiol.* **233,** E488–E494.

Wolffram, S., Badertscher, M., and Scharrer, E. (1994a). Carrier-mediated transport is involved in mucosal succinate uptake by rat large intestine. *Exp. Physiol.* **79,** 215–226.

Wolffram, S., Bisang, B., Grenacher, B., and Scharrer, E. (1990). Transport of tri- and dicarboxylic acids across the intestinal brush border membrane of calves. *J. Nutr.* **120,** 767–774.

Wolffram, S., Hagemann, C., Grenacher, B., and Scharrer, E. (1992). Characterization of the transport of tri- and dicarboxylates by pig intestinal brush-border membrane vesicles. *Comp. Biochem. Physiol. Comp. Physiol.* **101,** 759–767.

Wolffram, S., Unternahrer, R., Grenacher, B., and Scharrer, E. (1994b). Transport of citrate across the brush border and basolateral membrane of rat small intestine. *Comp. Biochem. Physiol. Physiol.* **109,** 39–52.

Yang, H., Jiang, W., Furth, E. E., Wen, X., Katz, J. P., Sellon, R. K., Silberg, D. G., Antalis, T. M., Schweinfest, C. W., and Wu, G. D. (1998). Intestinal inflammation reduces expression of DRA, a transporter responsible for congenital chloride diarrhea. *Am. J. Physiol.* **275,** G1445–G1453.

Zhao, H., Star, R. A., and Muallem, S. (1994). Membrane localization of H^+ and HCO_3^- transporters in the rat pancreatic duct. *J. Gen. Physiol.* **104,** 57–85.

CHAPTER 9

Molecular Mechanisms of Nucleoside and Nucleoside Drug Transport

James D. Young,* Christopher I. Cheeseman,* John R. Mackey,†** Carol E. Cass,† and Stephen A. Baldwin‡

Departments of *Physiology and †Oncology, Membrane Transport Research Group, University of Alberta, Edmonton, Canada; the Departments of †Experimental Oncology and **Medicine, Cross Cancer Institute, Edmonton, Canada; and the ‡School of Biochemistry and Molecular Biology, University of Leeds, Leeds, United Kingdom

Current Topics in Membranes, Volume 50

329

I. INTRODUCTION

Natural nucleosides and synthetic nucleoside analogs have important physiological and pharmacological activities in humans. Pyrimidine and purine nucleosides derived from dietary or endogenous sources are central "salvage" metabolites and, as precursors of nucleotides in enterocytes of the small intestine and other cell types, play an essential role in intermediary metabolism and biosynthesis. Nucleoside drugs, some of which are administered orally, are widely used in the treatment of cancer and viral diseases. Specialized nucleoside transporter (NT) proteins are required for translocation across cell membranes, and transportability is a critical determinant of intestinal absorption, intracellular metabolism and, for nucleoside drugs, pharmacological actions. NTs also regulate adenosine concentrations in the vicinity of their cell surface receptors and have profound effects on vascular tone, ion secretion, neuromodulation, and other processes.

Two mechanisms of nucleoside transport have been identified from studies of fluxes of radiolabeled nucleosides in various intestinal preparations and other mammalian cells and tissues: the concentrative, inwardly directed sodium/nucleoside cotransporters and the equilibrative bidirectional processes or facilitators.

Great progress has been made in our understanding of nucleoside transport processes since 1994, with the molecular cloning and functional expression of cDNAs encoding the membrane proteins responsible for each of the major nucleoside transport processes of mammalian cells. These proteins, all of which are found in intestine, are members of two previously unrecognized families of membrane proteins: the concentrative nucleoside transporter (CNT) and the equilibrative nucleoside transporter (ENT) proteins. The purpose of this chapter is to describe the molecular properties and functional characteristics of these two families of proteins. This information is presented in the context of what is presently known about the roles of NT proteins in gastrointestinal nucleoside and nucleoside drug absorption, metabolism, and other functions.

II. NUCLEOSIDE TRANSPORT PROCESSES OF MAMMALIAN CELLS

A. Functional Classifications

In human and other mammalian cells, seven nucleoside transport processes that differ in cation dependence, permeant selectivities, and inhibitor sensitivities have been observed (for reviews, see Cass, 1995; Griffith and Jarvis, 1996; Buolamwini, 1997; Wang et al., 1997a; Baldwin et al., 1999; Cass et al., 1999a; 1999b). The major (cit, cif) and minor (cib, csg, cs) concentrative transport processes are Na^+ dependent and have been demonstrated functionally in specialized epithelia such as intestine, kidney, liver, and choroid plexus, in other regions of the brain, and in splenocytes, macrophages, and leukemic cells. Additional tissues where CNT transcripts have been found include heart, skeletal muscle, placenta, pancreas, and lung (section V). The two principal concentrative subtypes (cit, cif) share the ability to transport uridine and, to a lesser extent, adenosine, but are otherwise selective for pyrimidine nucleosides (cit) and purine nucleosides (cif). The most widely distributed of the minor concentrative processes (cib) has broad selectivity for both pyrimidine and purine nucleosides. The equilibrative transport processes (es, ei) are found in most (possibly all) cell types, and transport both pyrimidine and purine nucleosides. Equilibrative processes of the es subtype are inhibited by nanomolar concentrations of the reversible tight-binding inhibitor nitrobenzylthioinosine (NBMPR), whereas the ei subtype is unaffected by NBMPR, or inhibited only by high (micromolar) concentrations. The abbreviations used in transporter acronyms are: c, concentrative; e, equilibrative; s and i, sensitive and insensitive (respectively) to inhibition by NBMPR; f, formycin B (a nonmetabolized purine nucleoside); t, thymidine; g, guanosine; b, broad permeant selectivity.

B. Nucleoside Drug Transport

Therapeutic antiviral and anticancer nucleosides, once activated by cellular kinases, act by means of incorporation into DNA and RNA, DNA strand breakage, perturbation of intracellular nucleotide pools, and inhibition of viral polymerases (Cheson, 1997; Peter and Gambertoglio, 1998). Because the pharmacological targets are intracellular, permeation through the plasma membrane is an obligatory first step in their activity (Cass, 1995). Since nucleoside analogs are generally hydrophilic, diffusion through the plasma membrane is often slow, and transporter-mediated uptake via NTs is the major route of drug influx. There is increasing evidence that membrane nucleoside transport processes are intimately linked to the cellular selectivity, therapeutic efficacy, and systemic toxicities of nucleoside drugs (Mackey *et al.*, 1998a) and to their absorption across the small intestine.

III. NUCLEOSIDES AND THE SMALL INTESTINE

A. Metabolism and Actions of Nucleosides

1. Dietary and Endogenous Nucleosides as Precursors of Cellular Nucleotides and Nucleic Acids

The nucleosides handled by the small intestine come primarily from two sources, the lumen and the bloodstream. The luminal content is made up of nucleotides from cells in the diet and from the cells discarded from the mucosal epithelium itself. The enterocyte population of the mucosa, which represents 80% of the cells of the epithelium, is renewed every 3 to 4 days, and parts of the old cells are lost into the intestinal lumen and their nucleic acids released (Creamer, 1967; Schmidt *et al.*, 1985). The nucleoproteins from both sources are broken down by proteases in the lumen to nucleic acids, which in turn are degraded by pancreatic nucleases to free nucleotides. The uptake of nucleotides by enterocytes is apparently of relatively little significance because the majority are dephosphorylated by alkaline phosphatase and nucleotidases to produce nucleosides. These are then taken up extensively by the mucosa, although further degradation to purines and pyrimidines also occurs.

2. Intestinal Metabolism of Nucleosides

Most cells can synthesize their own nucleotides for RNA and DNA production and other biosynthetic and metabolic requirements, although salvage of pyrimidines and purines is energetically favored over *de novo* synthesis and is likely to be particularly important in rapidly dividing cells such as enterocytes. The available evidence indicates that there is a high rate of metabolism of nucleosides within the enterocytes that releases the end product uric acid into the portal blood (Salati *et al.*, 1984; Stow and Bronk, 1992). The major evidence for the complete breakdown comes from the work of J. R. Bronk, who used an *in vitro* technique in which isolated loops of small intestine were bathed in paraffin and the lumen was perfused with solutions containing substrates. This technique allows for the collection of uncontaminated solution from the serosal surface, but in the absence of vascular perfusion, the rate of absorption is very slow and gives catabolic enzymes much longer to act than is likely to occur *in vivo*.

Thus, although breakdown undoubtedly occurs, it is not clear to what extent this normally takes place.

Experiments designed to measure transport of nucleosides using isolated enterocytes have suggested a relatively low rate of metabolism (Schwenk et al., 1984; Vijayalakshimi and Belt, 1988). Also, the demonstration that under a variety of conditions the nucleotide content of the diet can affect the growth rate and digestive enzyme production of the small intestine shows that before breakdown to uric acid some of these substrates must enter other metabolic pathways (Uay et al., 1990). Indeed, there is evidence of significant salvaging of purines and pyrimidines by the small intestine, with about 2–5% of dietary nucleotides being incorporated into tissue pools (Saviano et al., 1978). During fasting, catabolism of nucleotides is reduced and scavenging is increased, which suggests that the intestine has a limited ability to synthesize its own nucleotides. Currently, the evidence of *de novo* synthesis is confusing, with studies supporting significant rates, limited rates, and no synthesis at all (MacKinnon et al., 1973; Saviano et al., 1981; LeLeiko et al., 1983; Bissonnette 1992; Zaharevitz et al., 1990). These apparent contradictions may result from the different techniques used to study this tissue and its metabolism and the difficulty of maintaining conditions *in vitro* that closely mimic those found *in vivo*. Thus, the pathways for handling nucleotides and nucleosides within the enterocyte are not well defined, but the current consensus appears to be that intestinal epithelial cells rely heavily on exogenous nucleosides.

More recently, cell culture lines have been employed to examine the metabolism and incorporation of nucleosides within the intestinal epithelium, although caution must be used when extending these data to normal tissue. Two such cell lines that have been used are Caco-2, a colonic neoplastic cell line that under appropriate conditions matures into epithelial cells; and IEC-6 cells, which are nonmalignant and crypt-like. Both cell types take up nucleosides rapidly and metabolize them extensively (Sanderson and He, 1994; He et al., 1994). About 80% of the total taken up was incorporated into nucleic acids within 30 minutes, and the remaining 20% was found in the form of nucleotides, free nucleosides, and nucleobases. Among Caco-2 cells, there was no difference in metabolism between well-differentiated and poorly differentiated cells. When fluxes of nucleosides across monolayers of Caco-2 cells from the apical to basolateral surface were examined, purine nucleosides (adenosine, guanosine, inosine) appeared predominantly as nucleobases. Uric acid made up about 30%, with hypoxanthine, xanthine, and guanidine (in varying proportions) accounting for the remainder. Exogenous pyrimidine nucleosides (cytidine, uridine, thymidine) appeared up to 40% as uracil, with the remainder as intact cytidine, uridine, or thymidine. Similar patterns were found when the substrates were presented initially to the basolateral surface, although for the purine nucleosides there was less uric acid production.

We have been studying transepithelial fluxes and metabolism of nucleosides by using vascularly perfused rat jejunum. In this preparation (Hirsh *et al.*, 1996; Hanson and Parsons, 1976), there is independent access to both the luminal and vascular compartments; the vasculature of the tissue is preserved, and there is no delay in sampling the solution perfusing the basolateral membrane. Some uridine crosses the epithelium intact, but a substantial amount appears in the vasculature as uracil (unpublished observation).

B. Transport of Nucleosides

The ability of the intestinal epithelium to rapidly metabolize nucleosides must mean that they can be readily taken up and possibly released by enterocytes by NT-mediated processes.

1. Functional Studies of Intestinal Nucleoside Transport

Early studies of intestinal transport clearly showed that nucleosides could be taken up via Na^+-dependent transporters, which suggests a coupling to the electrochemical energy gradient for Na^+ entry to drive the organic substrate uphill into the enterocyte. If this was the case, then any nucleoside that escaped intracellular metabolism likely would leave the enterocytes via a facilitated mechanism paralleling the model for glucose absorption, in which Na^+-dependent uptake is mediated by SGLT1 and the facilitated exit into the blood by GLUT2. This early work in isolated guinea pig enterocytes used uridine as a substrate and found most of the nucleoside, which accumulated 13-fold, still intact inside the cells even after 1 hour (Schwenk *et al.*, 1984). The transport was high affinity with an apparent K_m of 46 µM, but specificity was not fully defined. A second study that used isolated mouse enterocytes also found Na^+-dependent high-affinity nucleoside uptake, which was NBMPR insensitive (Vijayalakshmi and Belt, 1988). Formycin B was used as the substrate, and the apparent K_m was found to be 45 µM. Effective inhibitors of this uptake included the purine nucleosides adenosine, inosine, guanosine, and deoxyadenosine. Thymidine and cytidine were poor inhibitors, even though thymidine uptake itself was Na^+ dependent. These observations led to the proposal for the existence of two types of Na^+-dependent NTs: pyrimidine-selective *cit* and purine-selective *cif* (see section II.A).

These early experiments used isolated enterocytes, a preparation that loses the polarization of the epithelium. Thus, it could only be assumed that the Na^+-dependent uptake would reside in the brush-border membrane and therefore represent a mechanism for uptake from the intestinal lumen. Membrane vesicle studies of isolated rabbit intestinal and rat renal brush-border preparations showed that Na^+-dependent uridine uptake was located in the apical surface of both intestinal and renal epithelial cells (Lee *et al.*, 1988; Jarvis, 1989). The transport activity in intestinal brush-border vesicles had an apparent K_m of approximately 6 μM. Isolated membranes have a much smaller unstirred water layer compared with intact tissue or isolated cells, and so this lower K_m is to be expected. In contrast, adenosine uptake by rabbit ileal basolateral membrane vesicles was Na^+ independent, but NBMPR sensitivity of the flux (to differentiate between *es*- and *ei*-type mechanisms) was not investigated (Betcher *et al.*, 1990). NBMPR-sensitive (*es*) NT activity has, however, been reported in rabbit kidney basolateral membrane vesicles (Williams *et al.*, 1989). The hexose model for nucleoside movement across epithelia was further supported by the finding that uptake of formycin B across the apical surface of sheets of rabbit jejunum was Na^+ dependent and that exit across the basolateral membrane was Na^+ independent (Roden *et al.*, 1991). The steady-state formycin B content of the epithelium in Na^+ medium increased significantly in the presence of NBMPR, which suggests that basolateral transport of formycin B is at least partly NBMPR sensitive.

More recent studies with rat and human jejunal brush-border membrane vesicles have confirmed the presence of both *cit* and *cif* activity with a Na^+-to-substrate coupling ratio (stoichiometry) of 1:1 (Iseki *et al.*, 1995; Patil and Unadkat, 1997). No NBMPR-sensitive transport was found in the brush border, which suggests that equilibrative NTs are expressed in the basolateral membrane. Intestinal NTs also have been studied in *Xenopus* oocytes injected with mRNA from human, rabbit, and rat small intestine (Jarvis and Griffith, 1991; Terasaki *et al.*, 1993; Huang *et al.*, 1993; Chandrasena *et al.*, 1997). The experiments with human jejunal mRNA identified four different activities: the *cit* and *cif* concentrative NTs and the *es* (NBMPR-sensitive) and *ei* (NBMPR-insensitive) equilibrative NTs. However, for rabbit and rat mRNA, only *cit* (rabbit, rat) and *cif* (rat) concentrative NT activities were detected. An additional broad-specificity Na^+-dependent NT activity (*cib*), whose identity is discussed in section XI, was characterized in oocytes injected with rat jejunal mRNA.

Thus, there appear to be at least two Na^+-dependent NTs (*cit* and *cif*) expressed in the intestinal brush-border membrane, and there is evidence of Na^+-independent transport activity in the basolateral membrane. However, the relative contributions of *es* and *ei* to the latter have not been determined. Moreover, although there is some evidence that the major transport capacity is in the jejunum, with some activity in the duodenum and ileum and none in the colon (Schwenk *et al.*, 1984; Roden *et al.*, 1991; McCloud *et al.*, 1994), there is no information on the distribution of transport along the crypt-villus axis. This information is required if a firm determination is to be made as to the role of NTs in the physiology of the intestine. During their migration up the villus from the crypt, enterocytes express progressively more transport proteins and digestive enzymes. This is related to their absorptive functions. However, if nucleoside uptake by enterocytes is required predominantly for their own metabolic needs, then a very different distribution might be expected. In the crypts, where cell division predominates, additional nucleoside uptake might be needed to supply the nucleoprotein synthetic machinery.

The use of cell lines to look at transport of nucleosides appears to confirm some of the above conclusions. Caco-2 cells express Na^+-dependent and Na^+-independent nucleoside transport activity in their apical membrane. Hu (1992) observed that thymidine uptake across the apical surface of confluent Caco-2 cell monolayers was partially Na^+ dependent (about 30%). However, there was also substantial inhibition by NBMPR. In this study there was no additional characterization of the NTs in the apical surface except for the observation that thymidine did not inhibit uptake of the antiviral nucleoside zidovudine (which was interpreted to enter the cells by passive diffusion). The relative lack of Na^+-dependent nucleoside transport and the high level of Na^+-independent, NBMPR-sensitive uptake in the apical surface suggest that this cell line does not fully represent the mature intestinal epithelial cell phenotype. Other studies in this cell line (Belt *et al.*, 1993) found the Na^+-dependent NT acitivity to have *cib*-type characteristics.

Two cell lines have been used as models of intestinal crypts: IEC-6 and T84. An early study of IEC-6 cells (Jakobs *et al.*, 1990) again showed the expression of both Na^+-dependent and Na^+-independent (and NBMPR-sensitive) adenosine and formycin B transport. The Na^+-dependent component did not become significant until the cells started to reach confluence and mature. Assuming that the incubation medium could not access the basolateral surface of the monolayers in these experiments, it would appear that these cells, unlike enterocytes, express both concentrative and equilibrative NTs in their apical membranes.

Transport of adenosine across both the apical and basolateral surface has been studied in T84 cells (Mun *et al.*, 1998). Uptake across the apical surface was very slow and nonsaturable, whereas transport across the basolateral surface was much faster and conformed to Michaelis-Menten kinetics. Further characterization indicated that this process was Na^+ independent, inhibited by NBMPR, and had an apparent K_m of 114 μM. These results suggest that in the intestinal crypt, nucleoside uptake occurs from the bloodstream, not the lumen, and is mediated predominantly by an NBMPR-sensitive mechanism.

2. Ontogeny of Intestinal Nucleoside Transport

Currently there is very little information on the ontogeny of NT expression in the small intestine. McCloud *et al.*, (1994), using everted sacs of jejunum and ileum, performed a study in rats in which it was shown that the transport capacity for uridine declined after birth. Apparently this was the result of a reduced maximal rate of transport. The uptake route was Na^+ dependent and resembled a *cif*-type system. The V_{max} in the jejunum of suckling, weanling, and adult rats was reported to be 1,768 ± 270, 449 ± 46, and 199 ± 30 pmol/g tissue wet weight 30 s^{-1}, respectively. The data for the ileum showed a very similar pattern. This decline could be, in part, a function of increased muscle mass contributing to the overall tissue weight, which was used as the baseline to express their transport data. Competition experiments indicated that thymidine and cytidine had no effect on uridine uptake. However, the study did not measure thymidine transport, so it remains to be established whether both *cit* and *cif* activities are present in the suckling rat, and whether their relative expression changes during development.

There is evidence that milk contains significant quantities of nucleotides that would be metabolized to nucleosides within the intestinal lumen (see section III.A), and this could explain the high transport capacity observed during suckling. Human milk has been shown to contain 82–402 μM total available nucleosides (Leach *et al.*, 1995), a much higher concentration than had been estimated originally. It has been proposed that these nutrients are important in the development and maturation of the neonatal gut, and thus expression of NTs could well be important for these processes. In addition to possible direct effects on the gut itself, there is evidence that the high-nucleotide content of milk might reduce the incidence of diarrhea (Brunser *et al.*, 1994) and may increase lymphocyte killer cell activity in the neonate (Carver *et al.*, 1991).

C. Nucleosides and Cell Proliferation and Differentiation

There is substantial evidence that both luminal and possibly vascular nucleosides have significant physiological effects on the small intestine. The epithelial cells covering the intestinal villi are produced continuously in the crypts surrounding the base of the villi. Stem cells in the crypts divide, and their daughter cells migrate up onto the villus (Ghishan and Wilson, 1985; Schmidt *et al.*, 1985; Ferraris *et al.*, 1992). As they progress up the villi, these enterocytes mature and gain transport and digestive activity. Currently, it is not well understood what controls the rate of proliferation, the migration rate, or the ultimate cell death at the villus tip. Cell death appears to be a very rapid form of apoptosis, but even the control of this process in the intestine has not been defined. Clearly, a balance between proliferation, migration rate, and death determines the height of the villi, which are shorter in the ileum than in the jejunum. This suggests that there is a relationship between the amount of luminal nutrients the epithelium comes into contact with and some of these processes. A large part of luminal nutrients is absorbed in the jejunum and thus less material will normally reach the ileum.

These processes are difficult to study *in vivo,* and cell culture has offered a good model to start to define the mechanisms involved. Experiments using the Caco-2 (colonic cells) and IEC-6 (crypt-like cells) cell lines have shown that exogenous nucleosides (supplied as nucleotides) affect both proliferation and differentiation rates (Sanderson *et al.*, 1994). In Caco-2 cells, nucleotides had no effect on the rates of cell proliferation or maturation unless the cells were first stressed by the omission of glutamine and essential amino acids from the growth medium. In contrast, the proliferation of IEC-6 cells was stimulated by the addition of nucleotides, even when the cells had a full nutrient content in the growth medium. Normally, IEC-6 cells do not differentiate into enterocyte-like cells unless grown on a substrata such as Matrigel. They then start to express the brush-border enzymes alkaline phosphatase and sucrase. The addition of nucleotides to the growth medium of IEC-6 cells plated onto Matrigel significantly increased the expression of both enzymes, which suggests a role for nucleotides (and therefore nucleosides) in promoting differentiation.

A more recent study, by Tanaka *et al.*, (1996), on cell proliferation in primary cultures of fetal human small intestine showed that the addition of adenosine monophosphate (AMP) to culture medium significantly suppressed crypt cell proliferation, while promoting differentiation and apoptosis. This led to hypotheses regarding a possible role for breast milk nucleotides (and therefore nucleosides) in small intestinal development in the neonate. Supplementation of the diets of weanling rats with nucleotides results in increased mucosal protein, DNA, villus height, and enzyme activity (Uay *et al.*, 1990), and AMP alone shows similar effects on gut growth. Therefore, it is possible that supplementation of the diet may help ameliorate some clinical conditions in which the intestinal epithelium is compromised.

D. Basolateral Nucleoside Transporters and Purinergic Receptor Activity

Intestinal crypts not only are the site of cell proliferation for mucosal cells, but also are responsible for producing watery secretions that keep the mucous layer hydrated and help to flush away pathogenic organisms. These secretions result from the opening of Cl^- channels in the apical membranes of the crypt cells. Na^+ then moves into the lumen via paracellular channels and water follows. One mechanism by which these cells are activated to open Cl^- channels is via adenosine receptors expressed on their basolateral membrane. IEC-6 cells express A_{2b} receptors on their basolateral membrane, and addition of adenosine at concentrations greater than 10^{-6} M promote cAMP generation within the cells and an increase in short-circuit current (I_{sc}) (Strohmeier *et al.*, 1995). cAMP is known to be a second messenger involved in the promotion of Cl^- channel opening in the intestine (Field, 1971), and I_{sc} is a good measure of Cl^- secretion. When *es*-type NT activity in these cells was inhibited with NBMPR, the secretory response to adenosine increased 10- to 15-fold. Dobbins *et al.* (1984) also showed that dipyridamole, another *es* inhibitor, significantly increased Cl^- secretion in response to exogenous adenosine in the rabbit ileum. This led Tally *et al.* (1996) and Mun *et al.* (1998) to postulate that regulation of NT activity in the crypt basolateral membrane may be an important component for controlling intestinal secretion and that drug design needs to take into account this mechanism to avoid secretory complications leading to diarrhea. It should be noted, however, that Strohmeier *et al.* (1995) found A_{2b} receptors only in the human colon, esophagus, and gastric antrum. They did not find receptor mRNA in the ileum, so the response in the small intestine may involve another receptor subtype. Another action of adenosine in the gut potentially linked to purinergic receptors is an increase in blood flow in rat jejunum following intravenous administration of either adenosine or AMP (Obata *et al.*, 1996).

IV. NUCLEOSIDE DRUGS AND THE SMALL INTESTINE

A. Nucleoside Drugs for Anticancer and Antiviral Therapy

Nucleoside analogs are powerful antimetabolites with proven activities against cancer and viral diseases. Nucleoside drugs are routinely used clinically to treat hematologic malignancies (cladribine, cytarabine, fludarabine), certain solid tumors (capecitabine, gemcitabine), and viral diseases including AIDS (zidovudine, zalcitabine, didanosine, stavudine, lamivudine), hepatitis B (lamivudine), herpes simplex and herpes zoster (acyclovir), cytomegalovirus (ganciclovir), and other viral infections (ribavirin). Recent successes in AIDS with combinations of nucleoside drugs, in hepatitis B with lamivudine, in leukemias and lymphomas with cladribine and fludarabine, and in metastatic breast, head and neck, bladder, lung, and pancreatic tumors with gemcitabine have sparked new interest in nucleoside therapeutics, and a number of promising new agents are under development. A number of these drugs are administered orally (section IV.B).

B. Orally Administerd Nucleoside Drugs

Oral drug administration is generally the route of choice for anticancer and antiviral therapy. Drug administration by mouth is preferable to intravenous administration because it avoids the expense and discomfort of intermittent intravenous access, and allows prolonged or continuous drug exposure without the complications of long-term venous catheterization. Patients prefer oral therapy when given a choice (Liu *et al.*, 1997), and it has been proposed that parenteral therapy reduces quality of life (Liu *et al.*, 1997). Table I lists eight antiviral and two anticancer nucleoside drugs that are routinely administered by mouth. Below we summarize the current understanding of intestinal absorption, bioavailability,

TABLE I

Oral Antiviral and Anticancer Nucleoside Analogs

Drug	Oral bioavailability in adults	Indications for oral therapy	Accepted as a permeant by
Acyclovir	15–30%	Patients with primary or recurrent genital herpes simplex	NBT[a]
		Herpes zoster	
		Varicella zoster	
Ganciclovir	6–9%	Maintenance therapy of cytomegalovirus retinitis	NBT[b] hENT1[b]
Ribavirin	40%	Lassa fever	hCNT2[c]
		Kong-Crimean hemorrhagic fever	hENT1[d]
Zidovudine (ZDV)	Nearly complete absorption, but first-pass metabolism reduces bioavailability to 65%	HIV infection, including perinatal administration to reduce HIV transmission	hCNT1[e] rCNT1[f,g] hENT2[h] hfCNT[i] NupC[j] CeENT1[h]
Zalcitabine (ddC)	> 80%	In combination with ZDV in HIV patients who do not respond to ZDV alone	hCNT1[k] rCNT1[f,g] hENT1[h,l] hENT2[h] hfCNT[i] NupC[j] CeENT1[h]
Didanosine (ddI)	20–40%	HIV-infected patients intolerant or failing to respond to ZDV	NBT[l] hCNT2[m] hENT1[h,l] hENT2[h] hfCNT[i] CeENT1[h]
Stavudine (d4T)	78–86%	Patients with advanced HIV infection who are intolerant of, or failing to respond to, standard therapy	ND
Lamivudine (3TC)	80–88%	In combination with ZDV or d4T in patients with progressive HIV infection	ND
		Therapy of chronic hepatitis	

continued

Drug	Oral bioavailability in adults	Indications for oral therapy	Accepted as a permeant by
Cladribine	40–50%	Therapy of low-grade lymphoma, chronic lymphocytic leukemia, and hairy cell leukemia	rCNT2[n] hENT1[o] hENT2[o]
Capecitabine		Anthracycline and paclitaxel-refractory metastatic breast cancer	ND

Included are only those major NT processes for which there has been direct determination of transportability of drug, by assay of transporter-mediated passage of radiolabeled drug across plasma membranes.

Abbreviations: NBT, nucleobase transporter; ND, not determined.

References to drug transport: [a]Mahony *et al.,* (1988); [b]Mahony *et al.,* (1991); [c]Patil *et al.,* (1998); [d]Jarvis *et al.,* (1988); [e]Ritzel *et al.,* (1997); [f]Huang *et al.,* (1994); [g]Yao *et al.,* (1996b); [h]Griffiths *et al.,* (2000); [i]Yao *et al.,* (1998); [j]Loewen *et al.,* (1998); [k]unpublished observation; [l]Domin *et al.,* (1993); [m]Ritzel *et al.,* (1998); [n]Schaner *et al.,* (1997); [o]King and Cass (1994).

and interactions with plasma membrane NTs for these compounds. Transport studies involving recombinant NTs are discussed in more detail in section VI.

1. Acyclovir and Ganciclovir

Acyclovir (9-(2-hydroxyethoxymethyl)guanine; ACV) is an acyclic purine nucleoside analog with activity against herpes simplex, herpes zoster, and varicella viruses. Its oral bioavailability is variable and ~22% on average (Morse *et al.,* 1993), which limits the achievable serum concentrations to below those required to treat severe infections, particularly in the immunocompromised patient. Acyclovir uptake in human erythrocytes is mediated solely by a nucleobase-transporter mechanism, and not by the erythrocyte *es* NT (Mahoney *et al.,* 1998). It is not known whether it is a substrate for other NT processes.

Ganciclovir (9-[(1,3-dihydroxy-2-propoxy)methyl]guanine) is the most effective agent for the treatment of cytomegalovirus infection. Oral bioavailability of ganciclovir is only 6–9% at maximally tolerated doses, however, and oral therapy is better tolerated but less effective than intravenous ganciclovir for the maintenance therapy of cytomegalovirus retinitis in immunocompromised patients (Griffy, 1996). Gancyclovir erythrocyte uptake is primarily nucleobase transporter mediated, but at high-substrate concentrations, a minority is transported by the *es* NT (Mahoney *et al.,* 1991). Although ganciclovir uptake was believed to be an NT-mediated process in cultured hepatocytes (Haberkorn *et al.,* 1997), uptake in MCF-7 breast cancer cells was much slower (Haberkorn, 1998). A nucleobase transporter appears to be the major mediator of placental transfer

(Henderson *et al.*, 1993; Gillstrap *et al.*, 1994). It is not known whether ganciclovir is a substrate for other NT processes.

2. Ribavirin

Ribavirin (1-β-D-ribofuranosyl-1,2,4-triazole-3-carboxamide), a broad-spectrum antiviral agent structurally related to guanosine, is less than 50% bioavailable after oral administration (Morse *et al.*, 1993). Intestinal absorption of ribavirin has been studied in brush-border membrane vesicles prepared from human jejunum. Ribavirin was ~15-fold more potent in inhibiting inosine uptake than in inhibiting thymidine uptake, and inosine was ~400-fold more potent than thymidine in inhibiting ribavirin uptake; these findings support the conclusion that ribavirin uptake is preferentially mediated by the *cif* concentrative NT (Patil *et al.*, 1998). The apparent K_m value of ribavirin transport was 19 μM. The human erythrocyte *es* NT transports ribavirin with a K_m of 440 μM, and there is no evidence for nucleobase transporter–mediated uptake (Jarvis *et al.*, 1998). Because a standard 600 mg dose of ribavirin would be expected to achieve intestinal lumen concentrations of ~1500 μM, saturation of intestinal NT processes may partially account for the limited bioavailability of this drug.

3. Zidovudine, Zalcitabine, Didanosine, Lamivudine, and Stavudine

Zidovudine (3′-azido-2′,3′-dideoxythymidine; AZT, ZDV), used in human immunodeficiency virus (HIV) therapy, exhibits nearly complete intestinal absorption, but first-pass hepatic metabolism reduces systemic bioavailability to 65%. AZT is sufficiently hydrophobic that substantial permeation into target cells can occur by passive diffusion. In human erythrocytes, zidovudine was not an *es* NT substrate (Zimmerman *et al.*, 1987). Using a recirculating in situ perfusion technique, it was found that disappearance of zidovudine from the lumen of different regions of rat intestine followed nonsaturable kinetics over a concentration range of 10 to 1000 μM, leading to the conclusion that mediated transport was not involved in zidovudine absorption (Park and Mitra, 1992). As described in section III.B, the same conclusion was reached in studies of zidovudine uptake by Caco-2 cells (Hu, 1992). However, our experiments with the recombinant *cit* transporter rCNT1 (section VI) suggest that higher concentrations of zidovudine would be required to show evidence of saturation (the apparent K_m for zidovudine influx is 550 μM). In Figure 1, we show an experiment in rat jejunum in which uptake of [^{3}H]zidovudine (10 μM) was measured during single-pass perfusion of Na^+-containing and Na^+-free Krebs-bicarbonate salines. Uptake was Na^+ dependent and inhibited by uridine, consistent with *cit*-mediated uptake (unpublished observation). Studies with recombinant NT proteins demonstrate transport of zidovudine by both the *cit* and *ei* NTs.

Zalcitabine (2′,3′-dideoxycytidine; ddC) is a pyrimidine nucleoside analog used in combination antiretroviral therapy for HIV patients. Its oral bioavail-

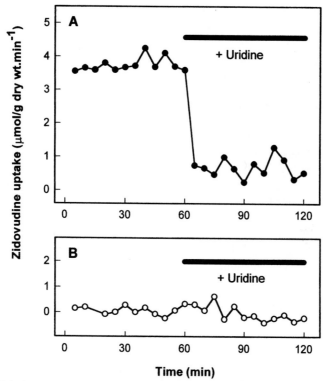

FIGURE 1 Uptake of zidovudine by the isolated rat jejunum. [^{3}H]Zidovudine uptake (10 μM, 37°C) was measured during single-pass perfusion of Na$^+$-containing and Na$^+$-free Krebs-bicarbonate salines with or without 10 mM mannitol or 10 mM uridine (bars indicate the duration of perfusion in the presence of uridine). The Na$^+$ replacement was choline.

ability is high (>80%) (Shelton *et al.*, 1993). Zalcitabine, which exhibits low rates of passive diffusion, is taken up during single-pass perfusion of rat jejunum at a rate several-fold higher than that of zidovudine. Similar to zidovudine (Fig. 1), we have found that uptake is both Na$^+$ dependent and inhibited by uridine, findings consistent with *cit*-mediated uptake (unpublished observation). Zalcitabine is a low-affinity substrate for hENT1 (apparent K_m of 23 mM) (Domin *et al.*, 1993) and has been shown to be transported by recombinant *cit*, *es*, and *ei* NTs (rCNT1 apparent K_m of 500 μM).

Didanosine (2′,3′-dideoxyinosine; ddI), used as an alternative to AZT in HIV therapy, has low and variable bioavailability, ranging from 20–40% (Shelton *et al.*, 1992), and intestinal absorption in an *ex vivo* canine model was higher in the duodenum than in the ileum (Sinko *et al.*, 1997). Nonmediated paracellular diffusion has been proposed to be the major route of intestinal absorption (Sinko *et al.*, 1995). However, in erythrocytes, didanosine cellular entry has been shown to be mediated by both a nucleobase transporter (apparent K_m of 850 μM) and the *es* NT (apparent K_m of 7.4 mM) (Domin *et al.*, 1993). Passive diffusion was low. In further support of the role of mediated transport processes in intestinal transfer, once-daily didanosine was less bioavailable (27%) than twice-daily didanosine (41%), which is consistent with saturation of the absorption process (Drusano *et al.*, 1992). Didanosine, a purine nucleoside analog, is transported by recombinant *cif, es,* and *ei* NTs.

Stavudine (2′,3′-didehydro-3′-deoxythymidine; d4T) is a pyrimidine nucleoside analog reverse-transcriptase inhibitor. Its oral bioavailability is nearly 100% (Rana and Dudley, 1997), and it has been reported in single-pass perfusion studies in rat small intestine that depletion of stavudine from the intestinal perfusate was partly inhibited by thymidine, which suggests involvement of transporter-mediated processes (Waclawski and Sinko, 1996). In brush-border membrane vesicles, uptake was shown to be partially Na^+ dependent. Stavudine interactions with individual NTs have not been studied.

Lamivudine (3′-thiacytidine; 3TC) is a pyrimidine nucleoside analog with antiretroviral activity. Lamivudine therapy reduces serum hepatitis B DNA levels (Lai *et al.*, 1997), and when given for 1 year, produces substantial histological improvement in liver biopsy results for patients with chronic hepatitis B infection (Lai *et al.*, 1998). Triple combination therapy with zidovudine and protease inhibition appears a particularly promising means of treating HIV (Pialoux *et al.*, 1998). The absolute bioavailability of lamivudine is 82% in adults (Yuen *et al.*, 1995). The mechanisms of lamivudine absorption and its interactions with NTs have not been studied.

4. Cladribine and Capecitabine

Cladribine (2-chloro-2′-deoxyadenosine; Cl-dAdo) is a purine nucleoside analog routinely used in the treatment of chronic lymphocytic leukemia, follicular lymphoma, and hairy cell leukemia (Cheson, 1997). It is most commonly administered by intravenous infusion, but there is increasing interest in its use as an oral agent. Although it is acid labile, administration in physiologically buffered saline after omeprazole therapy has achieved 50% bioavailability (Carson *et al.*, 1992). Bioavailability in a larger study was about 40%, and oral cladribine appeared to be as effective as intravenous therapy (Juliusson *et al.*, 1996). Cladribine is known to be an *es* and *ei* NT permeant (King and Cass, 1994), whereas its transportability by recombinant *cif* is species dependent (rat > human) (Schaner *et al.*, 1997; Lang *et al.*, 1999).

Capecitabine (*N*-[1-(5-deoxy-β-D-ribofuranosyl)-5-fluoro-1,2-dihydro-2-oxo-4-pyrimidyl]-*n*-penyl carbamate; Xeloda) is a fluoropyrimidine carbamate nucleoside analog, and was designed as an orally active drug that would selectively be metabolized to the active agent 5′-fluorouracil within tumor tissue. Capecitabine has important clinical activity in anthracycline and paclitaxel-refractory breast cancer (Blum *et al.,* 1999) and is a promising drug for the treatment of advanced colorectal carcinoma (Cassidy *et al.,* 1998). Pharmacological studies show rapid gastrointestinal absorption of capecitabine, and extensive hepatic metabolism of the parent compound, with a maximum plasma concentration 1 hour after ingestion (Budman *et al.,* 1998; Mackean, 1998). The interaction of capecitabine with NTs is under investigation in our laboratories.

In summary, it is clear that some nucleoside antiviral and anticancer drugs are efficiently absorbed. For others, low and variable oral bioavailability is an efficacy-limiting factor, mandating intravenous administration to achieve adequate systemic drug levels. It appears that intestinal NTs may play a significant role in the intestinal absorption of some nucleoside drugs, and may be a major determinant of nucleoside drug bioavailability.

V. IDENTIFICATION OF NT PROTEINS: THE *CNT* AND *ENT* GENE FAMILIES

A. Molecular Cloning of CNT and ENT Proteins

The recent isolation of cDNAs encoding the rat and human proteins of the major NT processes (*cit, cif, es, ei*) operative in plasma membranes of mammalian cells has established that they are mediated by members of two previously unrecognized families of integral membrane proteins (Huang *et al.,* 1994; Che *et al.,* 1995; Yao *et al.,* 1996a; 1997; Griffiths 1997a; 1997b; Ritzel *et al.,* 1997; 1998; Wang *et al.,* 1997b; Crawford *et al.,* 1998). These families, which we have designated as CNT (concentrative nucleoside transporter) and ENT (equilibrative nucleoside transporter) are structurally unrelated to each other and have very different predicted architectural designs. The relationships of the NT proteins identified by molecular cloning to the transport processes previously defined by functional studies are as follows: CNT1, *cit;* CNT2, *cif;* ENT1, *es;* ENT2, *ei.*

The cDNAs encoding rat CNT1 (rCNT1) and human ENT1 (hENT1), the first identified mammalian concentrative and equilibrative NTs, were isolated in our laboratories from rat jejunal epithelium (Huang *et al.,* 1994) and human placenta (Griffiths *et al.,* 1997a), respectively. The rCNT1 cDNA was obtained by expression selection in *Xenopus* oocytes, whereas that for hENT1 was cloned by a PCR strategy based on the N terminus of the human erythrocytic NBMPR-sensitive transporter. The tissue/cell sources of cDNAs encoding the other rat and human CNT and ENT family members were: small intestine (rCNT1, rCNT2, hCNT2, rENT1, rENT2) (Yao *et al.,* 1996a; 1997; Ritzel *et al.,* 1998), kidney (hCNT1, hCNT2) (Ritzel *et al.,* 1997; Wang *et al.,* 1997b), liver (rCNT2) (Che *et al.,* 1995), placenta (hENT2) (Griffith *et al.,* 1997b), and HeLa cells (hENT2)

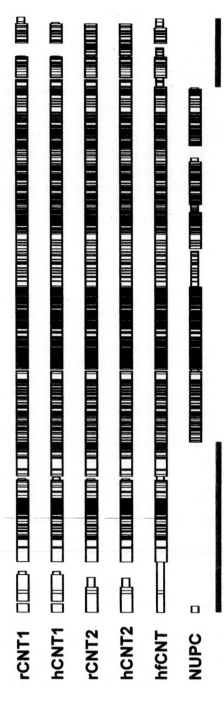

FIGURE 2 A multiple sequence alignment of mammalian, hagfish, and *E. coli* CNT proteins. Analysis of predicted amino acid sequences (Fig. 3) was performed using the MACAW (NCBI) program. Blocks indicate regions of similarity; vertical bars indicate residues conserved in ≥50% of sequences. The right-hand horizontal bar corresponds to the N-terminal tail and first three TMs of the mammalian and hagfish proteins (absent in NupC). The left-hand horizontal bar corresponds to the C-terminal tail of mammalian and hagfish CNTs (also absent from NupC).

rCNT1

hCNT1

rCNT2

hCNT2

hfCNT

NUPC

(Crawford *et al.*, 1998). A CNT1 cDNA has subsequently been isolated from porcine kidney (pCNT1) (Pajor, 1998).

The cDNA cloning of rCNT1 (*cit*), rCNT2 (*cif*), hCNT2 (*cif*), rENT1 (*es*), and rENT2 (*ei*) from human and rat small intestine is consistent with the functional studies described in section III.B, and suggests that small intestine contains all four transporters. The transcript for CNT1 (small intestine, kidney, liver, brain) has a narrower tissue distribution than that for CNT2 (intestine, kidney, liver, brain, heart, skeletal muscle, pancreas, placenta, lung) (Huang *et al.*, 1994; Che *et al.*, 1995; Anderson *et al.*, 1996; Yao *et al.*, 1996a; Ritzel *et al.*, 1997; 1998; Wang *et al.*, 1997b). Those for ENT1 and ENT2 appear to be ubiquitous. The hENT2 transcript, for example, has been shown in tissue blots to be present in small intestine, colon, kidney, liver, brain, heart, skeletal muscle, pancreas, placenta, lung, spleen, thymus, prostate, ovary, and peripheral blood lymphocytes, but to different extents (Crawford *et al.*, 1998). The genes for hCNT1 and hCNT2 have been mapped to chromosome 15q25–25 and 15q15 (possibly 15q13–14), respectively (Ritzel *et al.*, 1997; 1998; Wang *et al.*, 1997b). The genes for hENT1 and hENT2 are present on chromosomes 6p21.1–21.2 and 11q13 (Coe *et al.*, 1997; Williams *et al.*, 1997). The gene for hCNT1 contains 18 exons separated by 17 introns, and all exon-intron junctions have standard acceptor and donor sites (GenBank accession numbers AF187967–78). The translation start and termination codons are in exons 4 and 18, respectively, but the major transcription start site is in exon 1. The gene for hENT1 contains 12 exons (GenBank accession number AF190884).

The NT proteins responsible for the minor mammalian concentrative processes (*cib, cs, csg*) remain to be identified, although, as discussed in later sections of this review, we have recently cloned a cDNA encoding a CNT protein (hfCNT) with *cib*-like transport activity from the ancient marine vertebrate Pacific hagfish (*Eptatretus stouti*) (Loewen *et al.*, 1999). The CNT family also includes the *Escherichia coli* proton/nucleoside cotransporter NupC (Craig *et al.*, 1994; Yao *et al.*, 1999a). Other members of the ENT family include CeENT1, a *Caenorhabditis elegans* NT (Griffiths *et al.*, 2000), FUN26, a *Saccharomyces cerevisiae* NT (Barton and Kaback, 1994; Vickers *et al.*, 2000), and protozoan NTs from *Leishmania donovani* (LdNT1.1, LdNT1.2) (Vasudevan *et al.*, 1998) and *Trypanosoma brucei* (TbNT1, TbNT2) (Maser *et al.*, 1999; Sanchez *et al.*, 1999). Trypanosomes harbouring a defective variant of the TbNT1 protein (also designated TbAT1) are resistant to melaminophenyl arsenicals.

Figures 2 and 3 show a multiple sequence alignment of h/rCNT1, h/rCNT2, hfCNT, and NupC and their phylogenetic relationships with other CNT family members. Figures 4 and 5 present a corresponding analysis of h/rENT1, h/rENT1, CeENT1, and FUN26 in the ENT family. The functional characteristics of these CNT and ENT transporters, determined by production of the recombinant proteins in *Xenopus* oocytes and other heterologous expression systems, are described in sections VI and VII. Current models for CNT and ENT membrane topology are presented in Figs. 6 and 7.

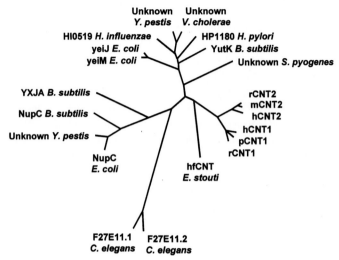

FIGURE 3 Phylogeny of CNT family members. Predicted amino acid sequences of these 21 members of the CNT transporter family were analyzed using the PHYLIP Phylogeny Inference Package (Version 3.5c [1993], University of Washington): rCNT1 (rat CNT1, GenBank accession no. U10279); hCNT1 (human CNT1, GenBank accession no. U62968); pCNT1 (pig kidney CNT1, GenBank accession no. AF009673); rCNT2 (rat CNT2, GenBank accession no. U25055); mCNT2 (mouse CNT2, GenBank accession no. AF079853); hCNT2 (human CNT2, GenBank accession no. AF036109); hfCNT (hagfish *cib* transporter, GenBank accession no. AF132298); F27E11.1 (*Caenorhabditis elegans,* GenBank accession no. AF016413); F27E11.2 (*Caenorhabditis elegans,* GenBank accession no. AF016413); HI0519_HAEIN (*Haemophilus influenzae,* GenBank accession no. U32734); HP1180_HELPY (*Helicobacter pylori,* GenBank accession no. AE000623); YEIM_ECOLI (*Escherichia coli,* Swissprot accession no. P33024); YEIJ_ECOLI (*Escherichia coli,* Swissprot accession no. P33021); YXJA_BACSU (*Bacillus subtilis,* Swissprot accession no. P42312); NUPC_ECOLI (*Escherichia coli,* Swissprot accession no. P33031); NUPC_BACSU (*Bacillus subtilis,* Swissprot accession no. P39141); UNKNOWN_STREP (*Streptococcus pyogenes,* open reading frame present in contig216 from the *S. pyogenes* genome sequencing project, Oklahoma University); YUTK_BACSU (*Bacillus subtilis,* GenBank accession no. Z99120); UNKNOWN_VIBRO (*Vibro cholerae,* open reading frame from TIGR Blast searches of the *V. cholerae* genome sequencing project); UNKNOWN_YERSI (*Yersinia pestis,* two open reading frames present in contig772 and contig770, respectively, from the *Y. pestis* genome sequencing project at the Sanger Center, Cambridge, U.K.).

B. Immunolocalization of rCNT1

We have identified rCNT1 protein in immunoblots of total membranes prepared from rat jejunum and kidney proximal tubule and liver, but not in skeletal muscle, spleen, or testes (Hamilton *et al.,* 2000). As shown in Figure 8 for rat jejunum, immunocytochemical analysis demonstrated expression of rCNT1 in enterocytes at the brush-border membrane but not at the basolateral membrane, a finding we have confirmed by immunoblots of purified brush-border and ba-

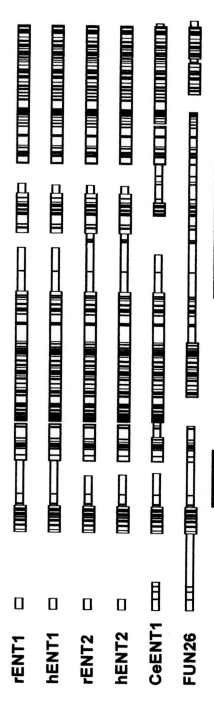

FIGURE 4 A multiple-sequence alignment of mammalian, *C. elegans*, and yeast ENT proteins. Analysis of predicted amino acid sequences (Fig. 5) was performed using the MACAW (NCBI) program. Blocks indicate regions of similarity; vertical bars indicate residues conserved in ≥50% of sequences. The two horizontal bars correspond to the large variable loop regions linking TMs 1–2 (extracellular) and 6–7 (intracellular).

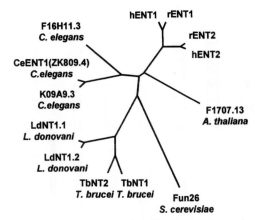

FIGURE 5 Phylogeny of ENT family members. Predicted amino acid sequences of these 11 members of the ENT transporter family were analyzed using the PHYLIP Phylogeny Inference Package (Version 3.5c [1993], University of Washington): hENT1 (human ENT1, GenBank accession no. AF079117); rENT1 (rat ENT1, GenBank accession no. AF015304); hENT2 (human ENT2, GenBank accession no. AF034102); rENT2 (rat ENT2, GenBank accesssion number AF015305); LdNT1.1 (*Leishmania donovani*, GenBank accession no. AF065311); LdNT1.2 (*Leishmania donovani*, GenBank accesion number AF041473); F17O7.13 (*Arabidopsis thaliana*, GenBank accession no. F17O7); FUN26 (*Saccharomyces cerevisiae*, GenBank accession no. L05146); ZK809.4 (*Caenorhabditis elegans*, GenBank accession no. Z68303); K09A9.3 (*Caenorhabditis elegans*, GenBank accession no. Z79601); F16H11.3 (*Caenorhabditis elegans*, GenBank accession no. U55376); TbNT1 (TbAT1) (*Trypanosoma brucei*, GenBank accession no. AF152369); TbNT2 (*Trypanosoma brucei*, GenBank accession no. AF153409).

solateral membrane vesicles. A brush-border location also was identified in kidney proximal tubule. In liver parenchymal cells, rCNT1 was abundant in bile canalicular membranes but largely excluded from sinusoidal membranes. Corresponding immunocytochemical studies of the other NTs are in progress. Although coexpression of rCNT1 and rCNT2 at the brush border is anticipated in the intestine and kidney, it has been reported that isolated rat liver sinusoidal membranes, which we found by immunocytochemistry to lack rCNT1, are enriched in rCNT2 (Felipe *et al.*, 1998).

VI. FUNCTIONAL STUDIES OF MEMBERS OF THE CNT PROTEIN FAMILY

A. CNT1 and CNT2 of Rat and Human Cells

1. Radioisotope Flux Studies: Physiological Nucleosides

hCNT1 and rCNT1 produced in *Xenopus* oocytes mediate saturable Na^+-dependent transport of [^{14}C]uridine (K_m of 40 μM), with a Na^+/uridine coupling

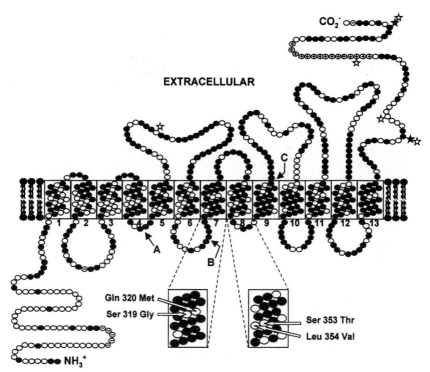

FIGURE 6 Topographical model of hCNT1 and hCNT2. Potential TMs are numbered, and putative N-linked glycosylation sites in hCNT1 and hCNT2 are indicated by solid stars and open stars, respectively. Residues identical in the two proteins are shown as solid ovals. Residues corresponding to insertions in the sequence of hCNT1 or hCNT2 are indicated by ovals containing "+" and "−" signs, respectively. A, B, and C correspond to splice sites used to engineer hCNT1/hCNT2 chimeras.

stoichiometry of 1:1. An apparent K_m of 21 μM was found for rCNT1 transiently expressed in COS-1 cells (Fang *et al.*, 1996). Transport is inhibited by pyrimidine nucleosides (thymidine, cytidine) and adenosine, but not by guanosine or inosine (Huang *et al.*, 1994; Ritzel *et al.*, 1997). Because inhibitors of transport are not necessarily permeants, we determined the transportability of adenosine by direct radioisotope flux measurements. These studies (Yao *et al.*, 1996a) found that adenosine is transported by rCNT1 with a K_m (25 μM) similar to that of uridine, but with a much reduced V_{max}. This new finding raises the possibility that adenosine may function as a physiological inhibitor of CNT1. The nucleoside specificity of hCNT2 and rCNT2, determined by measuring fluxes of a panel of radiolabeled purine and pyrimidine nucleosides, was opposite to that of h/rCNT1, showing a preference for adenosine, other purine nucleosides, and uridine (Yao *et al.*, 1996a; Ritzel *et al.*, 1998. Although the hCNT2 K_m (40 μM)

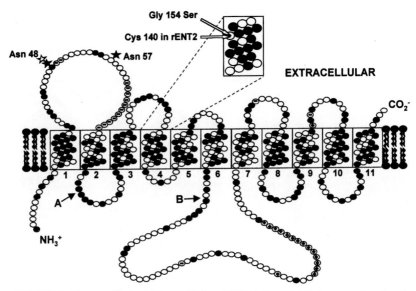

FIGURE 7 Topographical model of hENT1 and hENT2. Potential TMs are numbered, and the single N-linked glycosylation site in hENT1 is indicated by an open star. Putative N-linked glycosylation sites in hENT2 are indicated by solid stars. Residues identical in the two proteins are shown as solid ovals. Residues corresponding to insertions in the sequence of hENT1 or hENT2 are indicated by ovals containing "+" and "−" signs, respectively. A and B correspond to splice sites in hENT1 used to engineer hENT1/rENT1 chimeras.

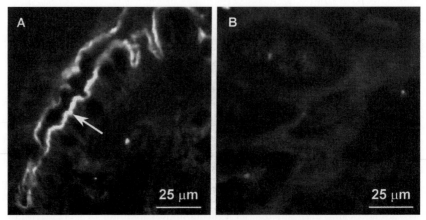

FIGURE 8 Immunolocalization of rCNT1 in rat jejunum. Frozen sections of jejunum were probed with affinity purified antipeptide antibodies raised against residues 618–637 of rCNT1 (1:100 dilution) in the (A) absence or (B) presence (control) of peptide. Immunofluorescence indicated specific detection of rCNT1 at the brush-border membrane of the absorptive epithelial cells (arrow), which was not observed on the basolateral membrane.

was higher for uridine than for adenosine (8 μM), V_{max} values for the two nucleosides were similar. Deoxyadenosine, which undergoes net renal secretion in humans, is less readily transported by hCNT1 and hCNT2 than adenosine (Ritzel et al., 1997; Ritzel et al., 1998). hCNT2 has been shown to have a lower affinity for deoxyadenosine than for adenosine (Wang et al., 1997b).

2. Radioisotope Flux Studies (Nucleoside Drugs)

The difference in substrate specificity between CNT1 and CNT2 has important implications with respect to physiological functions of the two NTs and is reflected in their complementary roles in the transport of different anticancer and antiviral nucleoside drugs. For example, we have used Xenopus expression to establish, for the first time, that mammalian CNT proteins transport antiviral dideoxynucleosides: h/rCNT1 transports zidovudine (3'-azido-3'-deoxythymidine) and zalcitabine (2',3'-dideoxycytidine) but not didanosine (2',3'-dideoxyinosine), whereas hCNT2 transports only didanosine (Huang et al., 1994; Yao et al., 1996b; Ritzel et al., 1997, 1998). Transport was demonstrated by direct measurements of [^{3}H]drug influx and, in the case of zidovudine and zalcitabine, by trans-stimulation of [^{3}H]uridine efflux. As discussed in section IV.B, the apparent K_m values for zidovudine and zalcitabine uptake by rCNT1 were 550 and 500 μM, respectively. The pyrimidine nucleoside drug gemcitabine (2',2'-difluorodeoxycytidine), a deoxycytidine analog, is a good h/rCNT1 permeant (hCNT1 apparent K_m of 24 μM), but is not transported by h/rCNT2 either in oocytes (Mackey et al., 1999) or in transiently transfected HeLa cells (Mackey et al., 1998b). Studies in a stably transfected cell line (L1210/DU5) have demonstrated that floxidine (5-fluoro-2'-deoxyuridine) and 5-fluorouridine also are relatively good rCNT1 permeants (Crawford et al., 1999). Cytarabine (arabinosylcytosine), however, was poorly transported. Consistent with the interaction of adenosine with h/rCNT1 (Yao et al., 1996a; Ritzel et al., 1997), cladribine (2-chloro-2'-deoxyadenosine) functioned as a potent inhibitor of recombinant rCNT1, but also was poorly transported (Crawford et al., 1999). Transient transfection studies in HeLa cells (rCNT2) and stable transfection experiments in CEM-ARAC-8C T-lymphoblastoid cells (hCNT2) showed that CNT2-mediated transport of cladribine is species dependent (rat > human) (Schaner et al., 1997; Lang et al., 2000).

3. Electrophysiological Studies

Electrophysiology provides an important and versatile tool to investigate aspects of electrogenic transporter function not readily accessible by other techniques. We have used the Xenopus oocyte system in conjunction with the 2-microelectrode voltage clamp to (1) verify that hCNT1 mediates inward nucleoside-dependent Na^+ currents, (2) establish the order of Na^+ and nucleoside binding (Na^+ binds first, increasing NT's affinity for nucleoside), and (3) independently confirm that zidovudine and gemcitabine are hCNT1 permeants by demonstrating that hCNT1 produced in oocytes mediates inward Na^+ currents in the presence of both drugs (Smith *et al.*, 1998; Mackey *et al.*, 1999; Yao *et al.*, 2000a). An advantage of electrophysiology is that it can be used to test the transportability of potential new anticancer and antiviral drugs not readily available in radioisotope form. Compounds currently under investigation include capecitabine, an orally active prodrug of 5-fluorouracil that is absorbed intact across the intestinal epithelium and is effective against cancer of the breast (section IV.B).

B. Hagfish CNT (hfCNT)

Hagfish are the most ancient class of extant vertebrates and diverged from the main line of vertebrate evolution approximately 550 million years ago. We have cloned a full-length cDNA encoding hagfish CNT (hfCNT) from hagfish intestinal epithelium (see Fig. 2) (Yao *et al.*, 1998; Loewen *et al.*, 1999). Expression in oocytes identified hfCNT as a Na^+-dependent NT with broad permeant selectivity for both pyrimidine and purine nucleosides (thymidine and inosine apparent K_m values of 45 and 35 μM, respectively). hfCNT also transported zidovudine, zalcitabine, and didanosine. The hfCNT K_{50} value for Na^+ activation was $\geq$ 100 mM, compared with 12 mM for rCNT1, consistent with the high concentration of Na^+ in hagfish plasma (400 mM). As described in section X.A, we have used sequence comparisons between hfCNT and human and rat CNT1/2 in site-directed mutagenesis experiments to identify residues responsible for hCNT1/2 substrate specificity (Loewen *et al.*, 1999). Using a similar rationale, it should also be possible to exploit differences in Na^+ binding between hfCNT and hCNT1/2 to identify corresponding transporter domains and residues involved in Na^+ binding and translocation. The substrate selectivity of hfCNT is reminiscent of mammalian *cib,* which suggests that this functional activity also may be mediated by a CNT protein (section XI.A).

C. Proton-Dependent E. coli NupC

Two concentrative NTs (NupC and NupG) have been identified in *E. coli* (Mygind and Munch-Petersen, 1975; Neuhard and Nygaard, 1987; Craig *et al.*, 1994; Seeger *et al.*, 1995). One (NupC) shows sequence similarity to mammalian CNT1/2 (see Fig. 2) (Craig *et al.*, 1994; Huang *et al.*, 1994). We have expressed *E. coli* NupC in oocytes using the enhanced *Xenopus* expression vector pSP64T, in which the protein coding region of the NupC gene is inserted between flanking 5'- and 3'-untranslated regions of a *Xenopus* β-globin gene (Yao *et al.*, 1999a). This resulted in nucleoside transport activity that was 20-fold higher than that obtained with the standard expression vector pGEM-3Z, allowing us to undertake functional characterization of recombinant NupC in the same heterologous expression system used in our CNT1/2 studies (Loewen *et al.,* 1998; Yao *et al.,* 2000a). Recombinant NupC closely resembles h/rCNT1 in specificity, except for increased adenosine transport and higher substrate affinities (uridine apparent K_m of 3 μM), and also transports zidovudine and zalcitabine. Various anticancer nucleosides, including cladribine, inhibited NupC-mediated uridine influx; this suggests that they too may be NupC permeants. Unlike CNT1/2, NupC did not require Na^+ for activity and was H^+ dependent (Yao *et al.,* 1999a). These experiments (1) establish the utility of the *Xenopus* oocyte expression system for structure/function comparisons between bacterial and mammalian CNTs, and (2) illustrate a potential role for NupC in bacterial cytotoxic drug accumulation. In the clinic, zidovudine treatment has been documented to reduce secondary infection rates caused by bacterial pathogens (Elwel *et al.,* 1987; Shepard *et al.,* 1992; Monno *et al.,* 1997). Similar antibacterial effects have been noted for several anticancer nucleoside drugs (Van Hoff *et al.,* 1976; Peterson *et al.,* 1992), and these may be beneficial in immunocompromised patients. We anticipate that structure/function studies of proton-dependent *E. coli* NupC will assist the identification of transporter domains and residues involved in cation binding and translocation. A number of other bacterial proteins belonging to the CNT family have been identified and require functional characterization (see Fig. 3).

VII. FUNCTIONAL STUDIES OF MEMBERS OF THE ENT PROTEIN FAMILY

A. ENT1 and ENT2 of Rat and Human Cells

1. Physiological Nucleosides and Nucleoside Drugs
Consistent with the well-described characteristics of native *es*-type processes in human cells, hENT1 produced in oocytes (Griffiths *et al.,* 1997a; 1997b) mediates saturable (Na^+-independent) transport of $[^{14}C]$uridine (apparent K_m of 0.2 mM) that is inhibited by purine nucleosides (adenosine, guanosine, inosine), pyrimidine nucleosides (thymidine, cytidine), various anticancer nucleosides (cladribine, cytarabine, fludarabine, gemcitabine), NBMPR (apparent K_i of 2 nM), and nonnucleoside vasodilator drugs (dipyridamole, dilazep, draflazine). rENT1 behaves similarly, except that it is unaffected by micromolar concentrations of NBMPR, dipyridamole, and dilazep (Yao *et al.,* 1997). hENT2 and rENT2 mediate saturable NBMPR-insensitive transport of uridine with apparent K_m values similar to those of hENT1 and rENT1 (Griffiths *et al.,* 1997b; Yao *et al.,* 1997). hENT2 is intermediate between hENT1 and rENT1/2 in sensitivity to inhibition by vasoactive compounds (Griffiths *et al.,* 1997a; 1997b; Yao *et al.,* 1997).

Unlike other pyrimidine and purine nucleosides, cytidine is only a weak inhibitor of recombinant h/rENT2 produced in oocytes (Griffiths *et al.*, 1997b; Yao *et al.*, 1997), a finding also reported for hENT2 stably expressed in the nucleoside transport-deficient cell line CEM/C19 (Crawford *et al.*, 1998). Kinetic characterization of cytidine fluxes in oocytes has shown a marked difference in cytidine affinity between hENT1 and hENT2, with apparent K_m values of 0.56 and >5 mM, respectively (manuscript in preparation). This unexpected functional difference between the ENT1 and ENT2 subtypes also is reflected in their transport of the deoxycytidine anticancer nucleoside analog gemcitabine (hENT1 and hENT2 apparent K_m values of 0.16 and 0.74 mM, respectively) (Mackey *et al.*, 1999). Adenosine has been confirmed to be an hENT1 permeant, and transport is inhibited by NBMPR and vasoactive drugs (Griffiths *et al.*, 1997a).

A common structural feature of the antiviral drugs zidovudine, zalcitabine, and didanosine is the absence of the ribose 3'-hydroxyl group, which greatly reduces their ability to be transported by *es*-type NTs (Gati *et al.*, 1984). We have confirmed a lack of transportability for zidovudine and relatively weak transport of zalcitabine and didanosine in experiments with recombinant hENT1 produced in *Xenopus* oocytes (Griffiths *et al.*, 2000). For hENT2, however, significant transport was observed for all three antiviral nucleosides (Griffiths *et al.*, 2000). When tested at a concentration of 20 μM, hENT2-mediated uptake of zidovudine and didanosine was greater than for zalcitabine, which, like gemcitabine, is an analog of deoxycytidine. As described in section VII.B, the *C. elegans* family member CeENT1 exibits an even greater capacity for transport of these compounds.

2. Nucleobase Transport

Although there are isolated reports in the literature that *ei*-type transporters also transport nucleobases (Plagemann *et al.*, 1988; Griffith and Jarvis, 1996), the presence of multiple pathways for nucleoside and nucleobase entry in most cell types has made functional dissection of these pathways difficult. We now have direct evidence from studies of hypoxanthine uptake by oocytes that recombinant h/rENT2 proteins, but not h/rENT1 proteins, are capable of transporting hypoxanthine (Griffiths *et al.*, 2000). We hypothesize that the physiological function of ENT2 may include transport of nucleobases as well as nucleosides. Such a possibility would explain why most cell types and tissues, including the intestine, coexpress *ei*- and *es*-type NTs. Separate nucleobase-preferring transporters also are operative in mammalian cells (Griffith and Jarvis, 1996).

B. CeENT1, a C. elegans ENT Homolog

There are six ENT homologs of unknown function in the genomic database for *C. elegans,* three of which are identified in Figure 5. The cDNA encoding one of these (ZK809.4) has been cloned by us and tested for transport activity (Griffiths *et al.*, 2000). The encoded protein, which we have designated CeENT1 (see Fig. 4), mediates large fluxes of uridine when expressed in oocytes, confirming its identity as a nucleoside transporter. A remarkable feature of CeENT1 is its ability to transport dideoxynucleosides and derivatives (zidovudine, zalcitabine, and didanosine) as well as physiological nucleosides. A second unexpected observation is that CeENT1, although resistant to inhibition by NBMPR, dilazep and draflazine, is moderately sensitive to inhibition by dipyridamole (IC_{50} 300 nM). These findings suggest that there are multiple structural determinants of vasoactive drug binding.

C. FUN 26, a S. cerevisiae ENT Homolog

FUN26 is a "function unknown" yeast protein (Barton and Kaback, 1994) that was identified as a homolog of mammalian ENTs in the yeast genomic database (Griffiths *et al.*, 1997a). Using an enhanced *Xenopus* expression vector (pGEMHE) similar to pSP64T (Liman *et al.*, 1992), we have established that FUN26 has nucleoside transport activity (Vickers *et al.*, 2000). In *S. cerevisiae*, FUN26 may be localized in intracellular membranes (Vickers *et al.*, 2000).

VIII. STRUCTURE AND TOPOLOGY OF THE CNT PROTEIN FAMILY

A. The Mammalian CNT1 (cit) and CNT2 (cif) Subfamilies

rCNT1 (648 residues, 71 kDa) (Huang *et al.*, 1994) and hCNT1 (650 residues) (Ritzel *et al.*, 1997) are 83% identical in sequence and contain three possible N-glycosylation sites ($Asn_{543/4}$, $Asn_{605/6}$, $Asn_{643/4}$), one of which ($Asn_{543/4}$) is located in TM 12 and is therefore unlikely to be utilized. The other two are in a large (83 residue) exofacial tail at the C terminus of the protein. It now appears that the protein contains 13 predicted TMs (one less than in earlier models) (Fig. 6) (Hamilton *et al.*, 2000). Antipeptide antibodies directed against N- and C-terminal regions of rCNT1 have been used to verify their respective cytosolic and extramembranous locations (Hamilton *et al.*, 2000). Site-directed mutagenesis in combination with oocyte expression and endoglycosidase-F digestion has confirmed Asn_{605} and Asn_{643} as the only two sites of glycosylation in rCNT1 (Hamilton *et al.*, 2000). Aglyco-rCNT1 (Asn_{605} and Asn_{643} converted to Gln) is functional.

hCNT2 (658 residues) (Wang *et al.*, 1997b; Ritzel *et al.*, 1998) is 83% identical to rCNT2 (659 residues) (Che *et al.*, 1995; Yao *et al.*, 1996a) and 72% identical to hCNT1. Hydropathy analyses of h/rCNT2 are consistent with the topol-

ogy model proposed for h/rCNT1, one of the most noticeable differences between the two subfamilies being the greater number of possible N-glycosylation
sites in hCNT2 (six) and rCNT2 (five). As shown in Figured 6, four of the six
possible glycosylation sites in hCNT2 are located at the C terminus (Asn_{600},
Asn_{605}, Asn_{624}, Asn_{653}) and one is the external loop between TMs 5 and 6.

h/rCNT1 and h/rCNT2 contain multiple consensus sites for protein-kinase A
and C phosphorylation, which suggests that both CNTs may be regulated by
protein-kinase dependent mechanisms. It has been reported that *cit* transport activity in human B-lymphocyte cell lines is stimulated by protein kinase C activation (Soler *et al.*, 1998).

B. hfCNT and NupC

hfCNT (683 residues) is 46–48% identical to h/rCNT1 and 44–47% identical
to h/rCNT2 (Loewen *et al.*, 1999). *E. coli* NupC (401 residues, 27% identity to
h/rCNT1) is smaller than its vertebrate counterparts and has three fewer TMs at
the N terminus (see Fig. 2). TMs 1–3 of rCNT1 can be removed (i.e., to correspond to the membrane arrangement of NupC) without loss of function (unpublished results).

IX. STRUCTURE AND TOPOLOGY OF THE ENT PROTEIN FAMILY

A. The Mammalian ENT1(es) and ENT2 (ei) Subfamilies

hENT1 (456 residues, 50kDa) (Griffiths *et al.*, 1997a) and rENT1 (455
residues) (Yao *et al.*, 1997) are 78% identical in amino acid sequence and are
predicted to contain 11 TMs connected by short (≤ 16 residue) hydrophilic regions, with the exceptions of larger loops connecting TMs 1–2 and 6–7 which,
in hENT1, contain 41 and 66 residues, respectively, and three potential sites of
N-linked glycosylation (TMs 1–2: Asn_{48}; TMs 6–7, Asn_{277} and Asn_{288}). Our previous observation (Kwong *et al.*, 1993) that the human erythrocyte NT protein
is glycosylated very close to one end suggests that only the first of the three sites
is glycosylated, and leads to the putative topology illustrated in Figure 7.
Hydropathy analyses of other members of the ENT family are consistent with
this model, the TM 1–2 and 6–7 loop regions being most variable between family members (see Fig. 6). rENT1, which our earlier studies (Kwong *et al.*, 1992)
indicated was more extensively glycosylated than hENT1, has three potential
sites of glycosylation in the loop between TMs 1 and 2, one of which is in the
same location as hENT1 Asn_{48} (Yao *et al.*, 1997). We have used site-directed mutagenesis in combination with oocyte expression and endoglycosidase-F digestion
to confirm the identity of Asn_{48} as the only site of glycosylation in hENT1. Aglyco

hENT1 (Asn$_{48}$ converted to Gln) is functional, and equilibrium binding of [^{3}H]NBMPR to yeast with recombinant hENT1 or hENT1/N48Q has shown that the glycosylation-defective mutant has reduced apparent affinity for NBMPR and increased apparent affinities for dipyridamole and dilazep (Vickers et al., 1999). The reduced apparent affinity of hENT1/N48Q is attributable to an increased rate of dissociation (k$_{off}$) and a decreased rate of association (k$_{on}$) of specifically bound [^{3}H]NBMPR.

hENT2 (456 residues) (Griffiths et al., 1997b; Crawford et al., 1998) is 46% identical in amino acid sequence with hENT1, and contains 13 fewer residues in the loop between TMs 1 and 2 and 14 additional residues in the loop linking TMs 6 and 7 (see Fig. 7). The former loop contains two potential sites of N-linked glycosylation, one of which (Asn$_{48}$) is conserved in h/rENT1. rENT2 (456 residues) (Yao et al., 1997) is 49% identical to rENT1 and 50% identical to hENT1 and, like hENT2, contains two potential sites of N-linked glycosylation in the predicted extracellular loop between TMs 1 and 2. In this case, the glycosylation site of Asn$_{48}$ is conserved as Asn$_{47}$.

B. CeENT and FUN26

CeENT1 (461 residues) (Griffiths et al., 2000) is between 25% and 28% identical in amino acid sequence to the mammalian nucleoside transporters h/rENT1 and h/rENT2 and has a predicted topology similar to that of hENT1, although the loops linking TMs 1–2 and TMs 6–7 are somewhat shorter (30 and 43 residues, respectively). Interestingly, the predicted extracellular loop between TMs 7 and 8 is considerably larger in CeENT1 (34 residues) than the equivalent loop in hENT1 (14 residues). Such variation supports the identification of the region between TMs 7 and 8 as an extramembranous loop. FUN26 (517 residues) (Barton and Kaback, 1994) is 20% identical to hENT1 and has been shown by targeted gene disruption to be nonessential for growth. FUN26 is predicted to have 11 TMs, with a long hydrophilic N terminus (see Fig. 6), a characteristic of yeast permeases (Horak, 1997), and 19- and 78-residue loops, respectively, between TMs 1–2 and 6–7.

X. RELATIONSHIP OF STRUCTURE TO FUNCTION IN THE CNT AND ENT PROTEIN FAMILIES

A. Identification of Residues Responsible for the Pyrimidine and Purine Specificities of hCNT1 and hCNT2

hCNT1 and hCNT2 belong to two different CNT subfamilies and exhibit the strongest residue similarity within TMs of the C-terminal halves of the proteins (see Fig. 6). Functionally, hCNT1 and hCNT2 display *cit-* and *cif*-type transport activities. Therefore, although both hCNT1 and hCNT2 transport uridine, they are otherwise selective for pyrimidine (hCNT1) and purine (hCNT2) nucleosides (except for modest transport of adenosine by hCNT1). We have conducted chimeric and site-directed mutagenesis experiments to identify hCNT1/2 domains and amino acid residues responsible for the marked differences in permeant selectivity between these two transporters (Loewen *et al.*, 1999).

1. hCNT1/hCNT2 Chimeras

Splice sites between hCNT1 and hCNT2 (see *A–C* in Fig. 6) at hCNT1 residues 202 (start of TM 4), 302 (start of TM 7), and 387 (end of TM 9) were engineered to minimize disruption of native TMs and loops and to divide the proteins into four unequal quarters ranging from 85 to 361 residues, each containing two to four TMs (Loewen *et al.*, 1999). RNA transcripts encoding the chimeric proteins were expressed in *Xenopus* oocytes and assayed for (1) functional activity (using uridine as a universal hCNT1/2 permeant) and (2) substrate selectivity (using thymidine and inosine as diagnostic hCNT1 and hCNT2 permeants, respectively). As shown in Figures 9A and B, a domain swap within the C-terminal half of hCNT1 between the second and third splice sites to produce the chimera C1121 (where 1 and 2 denote hCNT1 and hCNT2, respectively) was sufficient to cause the shift from pyrimidine selective (*cit*) to purine selective (*cif*), identifying residues 303–387 (incorporating TMs 7–9) as the determinant of substrate recognition (see Wang and Giacomini, 1997, for chimeric studies between rCNT1 and rCNT2). The substrate specificity of C1121, measured with a full panel of radiolabeled pyrimidine and purine nucleosides, is compared with hCNT1 and hCNT2 in Figure 10.

2. Site-Directed Mutagenesis of hCNT1

Comparisons of sequences of h/rCNT1 and h/rCNT2 in TMs 7, 8, and 9 (see Fig. 9C) identified nine residues that were conserved in the CNT1 and CNT2 transporter subtypes, respectively, but differed between subtypes and might therefore contribute to permeant selectivity. Some were common to h/rCNT1 and hfCNT, the native broad-specificity *cib*-type transporter. Others were common to hfCNT and h/rCNT2. In subsequent experiments, these nine residues in hCNT1 were mutated singly and in combination to the corresponding residues in hCNT2. Three of the mutations were in TM 7 (M1–3), five were in TM 8 (M4–8), and one was in TM9 (M9) (see Fig. 9C). These mutagenesis studies identified two sets of adjacent residues in TMs 7 and 8 of hCNT1 (Ser_{319}/Gln_{320} and Ser_{353}/Leu_{354}) that, when converted to the corresponding residues in hCNT2 (Gly_{313}/Met_{314} [M2/3] and Thr_{327}/Val_{348} [M6/7]), changed the specificity of the transporter from *cit* to *cif* (see Fig. 6) (Loewen *et al.*, 1999). Mutation of Ser_{319} in TM 7 of hCNT1 to Gly (M2) was sufficient to enable transport of purine nu-

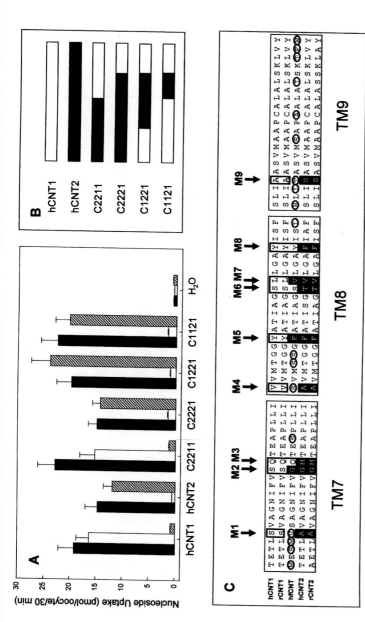

FIGURE 9 Analysis of hCNT1/hCNT2 chimeras. (A) Uptake of ^{14}C-labeled nucleosides by recombinant hCNT1, hCNT2, and chimeras C1122, C1112, C1221, and C1121 expressed in *Xenopus* oocytes. Uptake of uridine (solid bars), thymidine (open bars), and inosine (hatched bars) (20 μM, 20°C, 30 min) in oocytes injected with RNA transcript or water alone were measured in transport buffer containing NaCl. Values are means ± SE of 10–12 oocytes. (B) Schematic representation of hCNT1/hCNT2 chimeric molecules (see Fig. 6 for splice sites). (C) Alignment of predicted amino acid sequences of h/rCNT1, h/rCNT2, and hfCNT in TMs 7, 8, and 9. Positions of conserved residues that differ between hCNT1/rCNT1 and hCNT2/rCNT2 and were selected for mutation in hCNT1 are indicated by arrows (M1–M9). At each of these positions, amino acids in h/rCNT1 and h/rCNT2 in common with hfCNT are shown as open boxes and solid boxes, respectively. Ovals indicate amino acids in hfCNT that differ from conserved residues in either h/rCNT1 or h/rCNT2. Adapted from Loewen *et al.* (1999).

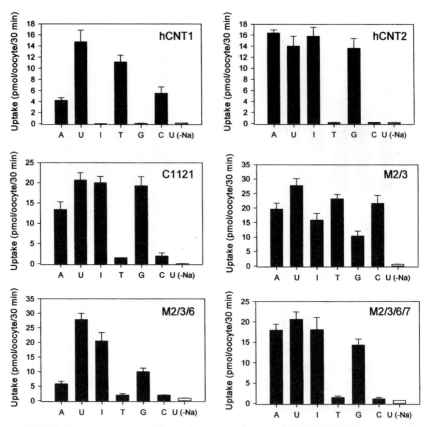

FIGURE 10 Nucleoside specificity of hCNT1, hCNT2, chimera C1121, and mutants M2/3, M2/3/6, and M2/3/6/7. Transporter-mediated nucleoside uptake (A, adenosine; U, uridine; I, inosine; T, thymidine; G, guanosine; C, cytidine) (20 μM, 20°C, 30 min) was measured in *Xenopus* oocytes in transport buffer containing 100 mM NaCl (solid bars) or 100 mM choline chloride (open bars). Mediated transport was calculated as uptake in RNA-injected oocytes minus uptake in oocytes injected with water alone. Values are means ± SE of 10–12 oocytes. Adapted from Loewen *et al.* (1999).

cleosides, while concurrent mutation of the adjacent residue Gln_{320} to Met (M3) (which had no effect on its own) augmented this transport. The resulting substrate selectivity of the combination M2/3 mutant (hCNT1/Ser319Gly/ Gln320Met) is shown in Figure 10. The additional mutation of the TM 8 residue Ser_{353} to Thr (M6) converted the broad-specificity *cib*-type characteristics of mutant M2/3 into a protein (mutant M2/3/6: hCNT1/Ser319Gly/Gln320Met/ Ser353Thr) with *cif*-type characteristics but with relatively low adenosine transport activity (Fig. 10). Concurrent mutation of the adjacent TM8 residue Leu_{354} to Val (M7) (which had no effect on its own) increased the adenosine transport ability of mutant M2/3/6, producing, as shown in Figure 10, a full *cif*-type transport phenotype (mutant M2/3/6/7: hCNT1/Ser319Gly/Gln320Met/ Ser353Thr/ Leu354Val). Apparent K_m values (μM) for nucleoside uptake by M2/3/6/7 were: adenosine, 18; uridine, 29; inosine, 20; and thymidine, 167. On its own, the Ser353 Thr mutation (M6) converted hCNT1 into a transporter with novel uridine-selective transport properties. Mutation of rCNT1 Ser_{318} (the rat counterpart of hCNT1 Ser_{319}) has been shown to lead to an increase in inosine transport similar to that observed with hCNT1/Ser319Gly (Wang and Giacomini, 1999).

3. Helix Modeling of hCNT1

Examination of the aligned sequences of putative TMs 7, 8, and 9 in the CNT family of transporters showed the presence of a number of positions where residue variability is very restricted. Conservation of these characteristic residues suggests that they are involved either in maintaining the structure of the transporters or in binding nucleoside substrates, these being features of the family members that are held in common. Thus, they are likely to face either the putative substrate translocation channel or another helix. The positions of the conserved residues are fairly symmetrically distributed around the circumference of TMs 7, 8, and 9, although TM 8 shows a slightly more asymmetric distribution (Fig. 11A). The distributions of positions that can accommodate polar residues in one or more of the transporters, or at which no polar residue is found, are likewise fairly symmetrical for TM 9. In contrast, TMs 7 and 8 exhibit a more amphipathic character, with predominantly polar residues clustered on one face of the helix and predominantly hydrophobic residues clustered on the other. These distributions of conserved residues, and the existence of conserved residues in the apolar faces of the TMs, suggest that all three TMs are largely sequestered from contact with membrane lipids, presumably by interactions with other transmembrane segments of the protein. The nature of these TMs can be contrasted with, for example, TM 4, which has a much more asymmetric distribution of both conserved and polar residues and which is likely to occupy a position in the transporter structure that is much more exposed to the membrane lipids (Fig. 11A).

Because the loops connecting putative TMs 7, 8, and 9 in the transporter are predicted to be very short (5 and 13 residues), it is likely that these three putative helices are adjacent in the tertiary structure of the protein. The pattern of conserved polar residues within the helices, together with the results of site-directed mutagenesis, allows a model to be proposed for their arrangement in the transporter structure, which is shown in Fig. 11B (Loewen *et al.*, 1999). Although tentative, this model aids interpretation of the experimental results and, more importantly, may be used to make predictions that can be tested by future site-directed mutagenesis experiments. In this model, Ser_{319} of TM7 (corresponding to mutant M2) is placed within the putative substrate translocation channel

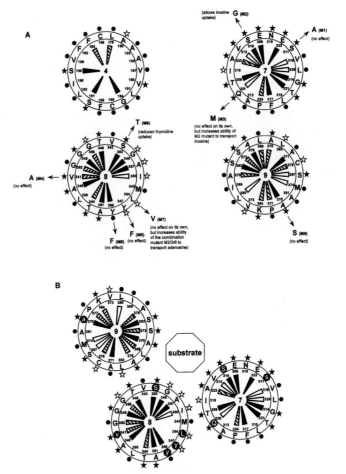

FIGURE 11 Structural features of TMs in the CNT family and the proposed arrangement of TMs 7, 8, and 9 in hCNT1. (A) Helical wheel plots for TMs 4, 7, 8, and 9 of the CNT family of nucleoside transporters, viewed from the extracellular side of the membrane, indicating the degree of residue conservation and the locations of positions accommodating polar residues or where only nonpolar residues are found. The residue identities and sequence positions shown are those of hCNT1. The distribution of residue types is indicated by symbols on the periphery of the plots: (★) a position always occupied by a nonpolar residue, (☆) a position occupied by a polar residue in <15% of the sequences, and (●) a position occupied by a polar residue in >15% of the aligned sequences. The extent of residue variability at each position is indicated by the nature of the central spokes: solid spokes indicate the presence of the same residue or members of a closely related group of residue types (E/Q/D/N; R/K; Y/F/W; A/G/C/S/T/P) in ≥85% of the aligned sequences; hatched spokes indicate the same type of residue conservation in 75–85% of the sequences; open spokes indicate the same type of residue conservation in 65–75% of the sequences, or the presence of members of a less closely related pair of residue types (e.g., I, L) in ≥85% of the sequences. The nature and effects of mutations M1 to M8 also are indicated. (B) Putative arrangement of TMs 7, 8, and 9 in hCNT1 surrounding a substrate translocation pathway. The symbols used are the same as those in A. Residues mutated in the present study are indicated in white on a black background: those that affected the transport activity of the protein are in uppercase letters; those that did not affect transport activity appear in lowercase. Adapted from Loewen *et al.* (1999).

where it is postulated to sterically hinder purine nucleoside transport. Gln_{320} (TM 7) and Leu_{354} (TM 8), in contrast, are predicted to interface with adjacent TMs, and mutations of these residues (M3 and M7, respectively) may exert their effects through altered helix packing. Ser_{353} (TM 8), corresponding to mutant M6, is predicted to face the translocation channel and may participate directly in substrate recognition via hydrogen bonding.

B. Identification of ENT Residues Interacting with NBMPR, Vasoactive Drugs, and PCMBS

hENT1 and rENT1 belong to the same ENT subfamily and are close structural homologs. Functionally, they display *es*-type nucleoside transport activity with similar apparent K_m values for uridine influx and similar IC_{50} values for NBMPR inhibition. In contrast, the two transporters differ markedly in their sensitivity to inhibition by vasoactive drugs. We have exploited this difference to identify regions of hENT1 responsible for vasoactive drug binding. This section also describes site-directed mutagenesis experiments that identify a residue involved in NBMPR binding (hENT1) and PCMBS (p-chloromercuriphenyl sulphonate) inhibition (rENT2).

1. hENT1/rENT1 Chimeras

We have produced reciprocal chimeras between hENT1 and rENT1 by using splice sites at the ends of TM 2 (hENT1 residue 99: position A in Fig. 7) and TM 6 (hENT1 residue 231: position B in Fig. 7) to identify structural domains of hENT1 responsible for transport inhibition by the vasoactive compounds dipyridamole and dilazep (Sundaram *et al.*, 1998). As shown in Figure 12, transplanting the N-terminal half of hENT1 into rENT1 converted rENT1 into a dipyridamole-sensitive transporter, whereas the N-terminal half of rENT1 rendered hENT1 dipyridamole-insensitive (similar results were obtained for dilazep). Domain swaps within the N-terminal halves of hENT1 and rENT1 identified residues 100–231 (incorporating TMs 3–6) as the major site of vasodilator interaction, with secondary contribution from residues 1–99. Because these drugs function as competitive inhibitors of nucleoside transport and NBMPR binding, it is likely that TMs 3–6 also form part of the NBMPR/substrate binding site. Corresponding chimeras engineered between rENT1 (NBMPR sensitive) and rENT2 (NBMPR insensitive) and site-directed mutagenesis of hENT1 and rENT2 (see following) support the involvement of TMs 3–6 in NBMPR and substrate binding.

2. Site-Directed Mutagenesis of rENT2 and hENT1

In an extension of previous experiments in erythrocytes (Jarvis and Young, 1982; 1986; Tse *et al.*, 1985) and HeLa cells (Dahlig-Harley *et al.*, 1981), we

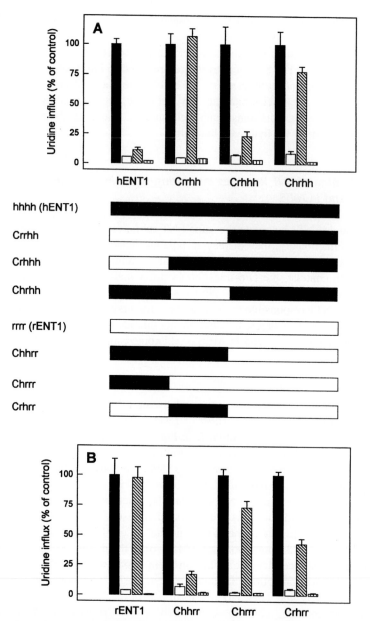

FIGURE 12 Analysis of hENT1/rENT1 chimeras. Initial rates of uridine uptake (10 μM, 20°C) by (A) hENT1 and (B) rENT1 chimeras were measured in *Xenopus* oocytes in the absence (solid bars) or the presence of 1 μM NBMPR (open bars) or dipyridamole (hatched bars). Incubations were 1 min (30 min for Chrhh). Values (means ± SE of 10–12 oocytes) were not corrected for the small contribution of endogenous nucleoside transport activity measured in oocytes injected with water alone (vertical bars) and are expressed as a percentage of the uninhibited flux for each transporter. (See Fig. 7 for splice sites used to construct the hENT1/rENT1 chimeric molecules.) Adapted from Sundaram *et al.* (1998).

have shown that rENT2 in oocytes is inhibited by extracellular PCMBS in a substrate-protectable manner under conditions in which rENT1, hENT1, and hENT2 are unaffected. Sequence alignments between the four ENT proteins showed that the differential sensitivity of rENT2 to extracellular PCMBS correlated with the presence in rENT2 of a unique cysteine residue (Cys_{140}) located in the outer half of TM 4 (see Fig. 7). Mutation of Cys_{140} to Ser rendered rENT2 PCMBS-insensitive, identifying Cys_{140} as an externally accessible residue within, or close to, the translocation channel (Yao *et al.*, 2000b). Mutation of this residue in hENT1 (Gly_{154} in Fig. 7) to the corresponding residue in hENT2 (Ser) converted the transporter from being NBMPR sensitive into one that was NBMPR insensitive (manuscript in preparation). In the *C. elegans* protein CeENT1 (which also is NBMPR insensitive), this residue is Asn.

XI. OTHER NUCLEOSIDE TRANSPORTERS

A. *Mammalian* cib

A broad-specificity Na^+-dependent NT activity (designated *cib*) has been described in *Xenopus* oocytes injected with rat jejunal mRNA (Huang *et al.*, 1993), in human colon (Caco) and myeloid cell lines (HL-60 and U937) (Lee *et al.*, 1991; Belt *et al.*, 1992; 1993), and in rabbit choroid plexus (Wu *et al.*, 1992). A candidate *cib*-type transporter SNST1 that is related to SGLT1 was identified in 1992 in rabbit kidney (Pajor and Wright, 1992). SNST1 encodes a predicted protein of 672 amino acids that is 61% identical to SGLT1. Its identity was established by functional expression in *Xenopus* oocytes. Although the recombinant protein stimulated low levels of Na^+-dependent uptake of uridine, which was inhibited by pyrimidine and purine nucleosides (i.e. *cib*-type pattern), the function of SNST1 remains unclear because (1) the rate of uridine transport observed when SNST1 cDNA was expressed in oocytes was only twofold above endogenous (background) levels, whereas a > ~500-fold stimulation was observed with r/hCNT1 cDNA (Huang *et al.*, 1994; Ritzel *et al.*, 1997), and (2) a *cib*-type transport activity has not been observed in the tissues (kidney, heart) in which the SNST1 message was reported to occur (Conant and Jarvis, 1994; Griffith and Jarvis, 1996). It is possible that the physiological substrate of SNST1 has not yet been identified and that it may be some other low-molecular-weight metabolite for which there is overlapping permeant recognition with nucleosides. Based on our studies of hfCNT and the hCNT1 mutant S319G/Q320M (Loewen *et al.*, 1999), we hypothesise that mammalian *cib* is a member of the CNT protein family.

At present, there is no functional evidence of the existence of *cib*-type transport activity in native intestinal preparations (section III.B). The molecular origin of the *cib*-like transport activity in oocytes injected with rat jejunal mRNA (Huang *et al.*, 1993) remains to be determined but, surprisingly, may result from the presence in intestine of transcript for the amino acid transport–related protein BAT. Unlike conventional transport proteins, BAT has one or perhaps four TMs, and functions as a transport activator within larger, multimeric amino acid transport assemblies (Palacin *et al.*, 1998). Mutations in human BAT cause type-I cystinuria. When produced in oocytes, BAT induces system $b^{0,+}$ amino acid transport activity by activation of cryptic endogenous *Xenopus* transporters. We have found that BAT also induces low but significant Na^+-dependent fluxes of pyruvate and uridine, the latter having *cib*-like characteristics (Yao *et al.*, 1999). Thus, BAT may activate a number of different nutrient transporters native to *Xenopus* oocytes, which raises the possibility that the transport-regulating functions of BAT may not be restricted only to amino acids.

B. Intracellular Nucleoside Transporters

Two organelles, lysosomes and mitochondria, are known to have nucleoside transport activities, (for a recent review, see Cass et al., 1999a). Candidate proteins (ND4 and MTP) that may serve as nucleoside transporters in these organelles have been identified (Hogue et al., 1996; 1997). Moreover, we have established by analysis of transport activity in membrane preparations and proteoliposomes of cultured human choriocarcinoma (BeWo) cells that hENT1 and hENT2 are located in intracellular as well as plasma membranes (Mani *et al.*, 1998). The yeast ENT FUN26 appears to be located exclusively in intracellular membranes (Vickers *et al.*, 2000).

XII. CONCLUDING REMARKS

Nucleosides are important precursors of nucleic acids and energy-rich cellular metabolites, and one (adenosine) has functions as a local hormone in a variety of tissues, including the gastrointestinal system. Cells obtain nucleosides from breakdown of dietary and endogenous nucleotides. The former are important human nutrients and are absorbed as nucleosides by enterocytes of the intestinal mucosa. Enterocytes appear to have a limited capacity for *de novo* nucleotide synthesis and require both dietary and endogenous nucleosides for their own metabolism and differentiation. Nucleoside analogs are important antiviral and anticancer agents and a number are administered orally.

Nucleosides and most nucleoside drugs enter and leave cells by transporter-mediated processes located in the plasma membrane. Intestinal transport of nucleosides involves both inwardly directed, concentrative (Na^+-dependent) and bidirectional, equilibrative (Na^+-independent) mechanisms. The former are present in brush-border membranes, and it is likely, but not yet firmly established, that the latter have a complementary distribution in basolateral membranes. As-yet-undefined nucleoside transport processes also are present in intestinal crypts, which are the site of epithelial Cl^- secretion and mucosal cell proliferation. There is evidence that Cl^- secretion is regulated by adenosine availability to surface purinergic receptors, and that crypt-cell proliferation requires exogenous nucleosides. Both crypt functions, therefore, will be dependent on nucleoside transport activity.

Corresponding nucleoside transport processes are present in other cell types and tissues, but until recently the proteins (NTs) responsible for this transport had not been identified. The efforts of several laboratories, including our own, have now achieved the cDNA cloning and functional characterization of the proteins responsible for each of the major nucleoside transport processes of human and rat cells. These NTs, several of which were first identified in the small intestine, belong to two different new protein families (CNT and ENT). Human and rat CNT1 (hCNT1, rCNT1) are selective for pyrimidine nucleosides (former system *cit*), whereas h/rCNT2 is purine nucleoside selective (former system *cif*). Both h/rENT1 and hENT2 have broad permeant selectivity for pyrimidine and purine nucleosides, but are either sensitive (h/ENT1, former system *es*) or insensitive to inhibition by NBMPR (h/rENT2, former system *ei*).

Previously it has been difficult to dissect and functionally characterize the multiple nucleoside transport pathways that usually are coexpressed in the same cell type or tissue. With the recent availability of cDNAs encoding each of the major NTs found in mammalian cells, it has been possible to express each recombinant transporter individually in an appropriate NT-deficient background such as the *Xenopus* oocyte. These heterologous expression studies have shown a number of important new insights into CNT and ENT function, including the ability of human and rat CNT1, CNT2, and ENT2 to transport several key antiviral nucleoside drugs that were previously believed not to be NT permeants, and the demonstration that ENT2 functions as a nucleobase transporter. Both these features likely are central to the role of NTs in drug absorption and transepithelial nucleoside fluxes, the latter being associated with at least partial intracellular conversion of nucleosides to nucleobases. Progress also has been made in studies of CNT and ENT membrane topology and in the identification of regions of the proteins and individual amino acid residues responsible for substrate specificity (CNTs) and inhibitor binding (ENTs).

As exemplified by the immunolocalization of rCNT1 to the jejunal brush-border membrane, by the demonstration of Na^+-dependent, uridine-inhibitable uptake of zidovudine during perfusion of rat jejunum, and by the characterization of zidovudine transport by recombinant rCNT1 expressed in oocytes, we are now in a position to combine whole-tissue, cellular, subcellular, and molecular approaches to address the many fundamental and unresolved questions concerning the roles of NTs in gastrointestinal function and development.

Available data suggest that intestinal NTs contribute to absorption of some of the orally administered therapeutic nucleosides, and that the bioavailability of such drugs depends, at least in part, on intestinal NTs accepting such compounds as substrates. Studies to define the ability of NTs to accept new anticancer and antiviral nucleoside analogs should be incorporated in drug design and preclinical testing because these may determine the feasibility of oral administration of such compounds. A goal of NT gene-therapy strategies is to develop engineered CNT and ENT proteins tailored to transport particular nucleoside analogs with high selectivity and activity.

Acknowledgments

The authors thank Ms. C.A. Sheldon for performing the zidovudine uptake experiment shown in Figure 1; Drs. A.J. Hirsh and W.P. Gati for undertaking vascular perfusion of rat jejunum; and Dr. S.R. Hamilton for performing the immunolocalization of rCNT1 shown in Fig. 8. Work described in this article was supported in Canada by grants from the Canadian Medical Research Council, the National Cancer Institute of Canada, the Natural Sciences and Research Council of Canada, the Canadian Foundation for AIDS Research, and the Alberta Cancer Board. Research in the United Kingdom was supported by the U.K. Medical Research Council, the Wellcome Trust, and the Biotechnology and Biological Sciences Research Council and Royal Society. J.D.Y. is a Heritage Medical Scientist of the Alberta Heritage Foundation for Medical Research, and C.E.C. is a Terry Fox Cancer Research Scientist of the National Cancer Institute of Canada.

References

Anderson, C. M., Xiong, W., Young, J. D., Cass, C. E., and Parkinson, F. E. (1996). Demonstration of the existence of mRNAs encoding N1/cif and N2/cit sodium/nucleoside cotransporters in rat brain. *Mol. Brain Res.* **42**, 358–361.

Baldwin, S. A., Mackey, J. R., Cass, C. E., and Young, J. D. (1999). Nucleoside transporters: Molecular biology and implications for therapeutic development. *Mol. Med. Today* **5**, 216–224.

Barton, A. B., and Kaback, D. B. (1994). Molecular cloning of chromosome 1 DNA from *Saccharomyces cerevisiae:* Analysis of the genes in FUN38-MAK16-SPO7 region. *J. Bacteriol.* **176**, 1872–1880.

Belt, J. A., Harper, E., and Byl, J. (1992). Na^+-dependent nucleoside transport in human myeloid leukemic cell lines and freshly isolated myeloblasts. *Proc. Am. Assoc. Cancer Res.* **33**, 20.

Belt, J. A., Marina, N. M., Phelps, D. A., and Crawford, C. R. (1993). Nucleoside transport in normal and neoplastic cells. *Adv. Enzyme Regul.* **33**, 235–252.

Betcher, S. L., Forrest, J. N., Jr., Knickelbein, R. G., and Dobbins, J. W. (1990). Sodium-adenosine cotransport in brush-border membranes from rabbit ileum. *Am. J. Physiol.* **259**, G504–G510.

Bissonnette, R. (1992). The *de novo* and salvage pathways for the synthesis and pyrimidine residues of RNA predominate in different locations within the mouse duodenal epithelium. *Cell Tissue Res.* **267**, 131–137.

Blum, J. L., Jones, S. E., Buzdar, A. U., LoRusso, P. M., Kuter, I., Vogel, C., Osterwalder, B., Burger, H. U., Brown, C. S., and Griffin, T. (1999). Multicenter phase II study of capecitabine in paclitaxel-refractory metastatic breast cancer. *J. Clin. Oncol.* **17**, 485–493.

Brunser, O., Espinoza, J., Araya, M., Cruchet, S., and Gil, A. (1994). Effect of dietary nucleotide supplementation on diarrheal disease in infants. *Acta Paediatr.* **83**, 188–191.

Budman, D. R., Meropol, N. J., Reigner, B., Creaven, P. J., Lichtman, S. M., Berghorn, E., Behr, J., Gordon, R. J., Osterwalder, B., and Griffin, T. (1998). Preliminary studies of a novel oral fluoropyrimidine carbamate: Capecitabine. *J. Clin. Oncol.* **16,** 1795–1802.

Buolamwini, J. K. (1997). Nucleoside transport inhibitors: Structure-activity relationships and potential therapeutic applications. *Curr. Med. Chem.* **4,** 35–66.

Carver, J. D., Pimentel, B., Cox, W. I., and Barness, L. A. (1991). Dietary nucleotide effects upon immune function in infants. *Pediatrics* **88,** 359–363.

Casari, G., Andrade, M. A., Bork, P., Boyle, J., Daruvar, A., Ouzounis, C., Schneider, R., Tamames, J., Valencia, A., and Sander, C. (1995). Challenging times for bioinformatics. *Nature* **376,** 647–648.

Cass, C. E. (1995). Nucleoside transport. *In* "Drug Transport in Antimicrobial Therapy and Anticancer Therapy" (N. H. Georgopapadakou, ed.), pp. 403–451. Dekker, New York.

Cass, C. E., Young, J. D., and Baldwin, S. A. (1999a). Recent advances in the molecular biology of nucleoside transporters of mammalian cells. *Biochem. Cell Biol.* **76,** 761–770.

Cass, C. E., Young, J. D., Baldwin, S. A., Cabrita, M. A., Graham, K. A., Griffiths, M., Jennings, L. L., Mackey, J. R., Ng, A. M. L., Ritzel, M. W. L., Vickers, M. F., and Yao, S. Y. M. (1999b). Nucleoside transporters of mammalian cells. *In* "Membrane Transporters as Drug Targets" (G. Amidon and W. Sadee, eds.). Plenum, New York pp 313–352.

Cassidy, J., Dirix, L., Bissett, D., Reigner, B., Griffin, T., Allman, D., Osterwalder, B., and Van Oosterom, A. T. (1998). A phase I study of capecitabine in combination with oral leucovorin in patients with intractable solid tumors. *Clin. Cancer Res.* **4,** 2755–2761.

Chandrasena, G., Giltay, R., Patil, S. D., Bakken, A., and Unadkat, J. D. (1997). Functional expression of human intestinal Na^+-dependent and Na^+-independent nucleoside transporters in *Xenopus laevis* oocytes. *Biochem. Pharmacol.* **53,** 1909–1918.

Che, M., Ortiz, D. F., and Arias, I. M. (1995). Primary structure and functional expression of a cDNA encoding the bile canalicular, purine-specific Na^+-nucleoside cotransporter. *J. Biol. Chem.* **270,** 13596–13599.

Cheson, B. D. (1997). Miscellaneous chemotherapeutic agents. *In* "Cancer: Principles and Practice of Oncology" (V. T. Devita, ed.), Vol. 1, 5th ed., pp. 490–498, Lippincott-Raven, Philadelphia.

Coe, I. R., Griffiths, M., Young, J. D., Baldwin, S. A., and Cass, C. E. (1997). Assignment of the human equilibrative nucleoside transporter (hENT1) to 6p21.1–p21.2. *Genomics* **45,** 459–460.

Conant, A. R., and Jarvis, S. M. (1994). Nucleoside influx and efflux in guinea-pig ventricular myocytes. Inhibition by analogues of lidoflazine. *Biochem. Pharmacol.* **48,** 873–880.

Craig, J. E., Zhang, Y., and Gallagher, M. P. (1994). Cloning of the *nupC* gene of *Escherichia coli* encoding a nucleoside transport system, and identification of an adjacent insertion element, IS 186. *Mol. Microbiol.* **11,** 1159–1168.

Crawford, C. R., Cass, C. E., Young, J. D., and Belt, J. A. (1999). Stable expression of a recombinant sodium-dependent, pyrimidine-selective nucleoside transporter (rCNT1) in a transport-deficient mouse leukemia cell line. *Biochem. Cell Biol.* **76,** 843–851.

Crawford, C. R., Patel, D. H., Naeve, C., and Belt, J. A. (1998). Cloning of the human equilibrative, nitrobenzylmercaptopurine riboside (NBMPR)–insensitive nucleoside transporter *ei* by functional expression in a transport-deficient cell line. *J. Biol. Chem.* **273,** 5288–5293.

Creamer, B. (1967). The turnover of the epithelium of the small intestine. *Br. Med. Bull.* **23,** 226–230.

Dahlig-Harley, E., Eilam, Y., Paterson, A. R. P., and Cass, C. E. (1981). Binding of nitrobenzylthioinosine to high-affinity sites on the nucleoside-transport mechanism of HeLa cells. *Biochem. J.* **200,** 295–305.

Dobbins, J. W., Laurenson, J. P., and Forrest, J. N. (1984). Adenosine and adenosine analogues stimulate adenosine cyclic 3′,5′-monophosphate-dependent chloride secretion in the mammalian ileum. *J. Clin. Invest.* **3,** 929–935.

Domin, B. A., Mahony, W. B., and Zimmerman, T. P. (1993). Membrane permeation mechanisms of 2',3'-dideoxynucleosides. *Biochem. Pharmacol.* **46,** 725–729.

Drusano, G. L., Yuen, G. J., Morse, G., Cooley, T. P., Seidlin, M., Lambert, J. S., Liebman, H. A., Valentine, F. T., and Dolin, R. (1992). Impact of bioavailability on determination of the maximal tolerated dose of 2',3'-dideoxyinosine in phase I trials. *Antimicrob. Agents Chemother.* **36,** 1280–1253.

Elwel, L. P., Ferone, R., Freeman, A. O., and Fyfe, J. A. (1987). Antibacterial activity and mechanism of action of 3'-azido-3'-deoxythumidine. *Antimicob. Agents Chemother.* **31,** 274–280.

Fang, X., Parkinson, F. E., Mowles, D. A., Young, J. D., and Cass, C. E. (1996). Functional characterization of a recombinant sodium-dependent nucleoside transporter with selectivity for pyrimidine nucleosides (cNT1rat) by transient expression in cultured mammalian cells. *Biochem. J.* **317,** 457–465.

Felipe, A., Valdes, R., de Santo, B., Lloberas, J., Casado, J., and Pastor-Anglada, M. (1998). Na$^+$-dependent nucleoside transport in liver: Two different isoforms from the same gene family are expressed in liver cells. *Biochem. J.* **330,** 997–1001.

Ferraris, R. P., Villenas, S. A., and Diamond, J. M. (1992). Regulation of brush-border enzyme activities and enterocyte migration rates in mouse small intestine. *Am. J. Physiol.* **262,** G1047–G1059.

Field, M. (1971). Ion transport in rabbit ileal mucosa. II. Effects of cyclic 3',5'-AMP. *Am. J. Physiol.* **221,** 992–997.

Gati, W. P., Misra, H. K., Knaus, E. E., and Wiebe, L. I. (1984). Structural modifications at the 2'- and 3'-positions of some pyrimidine nucleosides as determinants of their interaction with the mouse erythrocyte nucleoside transporter. *Biochem. Pharmacol.* **33,** 3325–3331.

Ghishan, F. K., and Wilson, F. A. (1985). Developmental maturation of D-glucose transport by rat jejunal brush-border membrane vesicles. *Am. J. Physiol.* **248,** G87–G92.

Gilstrap, L. C., Bawdon, R. E., Roberts, S. W., and Sobhi, S. (1994). The transfer of the nucleoside analog ganciclovir across the perfused human placenta. *Am. J. Obstet. Gynecol.* **170,** 967–972; discussion, 972–973.

Griffith, D. A., and Jarvis, S. M. (1996). Nucleoside and nucleobase transport systems of mammalian cells. *Biochem. Biophys. Acta* **1286,** 153–181.

Griffiths, M., Baldwin, S. A., Yao, S. Y. M., Chomey, E. G., MacGregor, D., Isaac, R. E., Cass, C. E., and Young, J. D. (2000). Identification and molecular characterization of an unusual equilibrative nucleoside transporter from *Caenorhabditis elegans* which mediates uptake of the antiviral nucleoside analogs 3'-azido-3'-deoxythymidine, 2',3'-dideoxycytidine and 2',3'-dideoxyinosine. *J. Biol. Chem.* (submitted).

Griffiths, M., Beaumont, N., Yao, S. Y., Sundaram, M., Boumah, C. E., Davies, A., Kwong, F. Y., Coe, I., Cass, C. E., Young, J. D., and Baldwin, S. A. (1997a). Cloning of a human nucleoside transporter implicated in the cellular uptake of adenosine and chemotherapeutic drugs. *Nature Med.* **3,** 89–93.

Griffiths, M., Yao, S. Y. M., Abidi, F., Phillips, S. E. V., Cass, C. E., Young, J. D., and Baldwin, S. A. (1997b). Molecular cloning and characterization of a nitrobenzylthioinosine-insensitive (*ei*) equilibrative nucleoside transporter from human placenta. *Biochem. J.* **328,** 739–743.

Griffy, K. G. (1996). Pharmacokinetics of oral ganciclovir capsules in HIV-infected persons. *AIDS* **10,** S3–S6 (Suppl. 4).

Haberkorn, U., Altmann, A., Morr, I., Knopf, K. W., Germann, C., Haeckel, R., Oberdorfer, F., and van Kaick, G. (1997). Monitoring gene therapy with herpes simplex virus thymidine kinase in hepatoma cells: Uptake of specific substrates. *J. Nucl. Med.* **38,** 287–294.

Haberkorn, U., Khazaie, K., Morr, I., Altmann, A., Muller, M., and van Kaick, G. (1998). Ganciclovir uptake in human mammary carcinoma cells expressing herpes simplex virus thymidine kinase. *Nucl. Med. Biol.* **25,** 367–373.

Hamilton, S. R., Yao, S. Y. M., Ingram, J. C., Gallagher, M. P., Henderson, P. J. F., Cass, C. E., Young, J. D., and Baldwin, S. A. (2000). Membrane topology and subcellular distribution of the rat concentrative sodium-nucleoside co-transporter rCNT1. (manuscript in preparation).

Hanson, P. J., and Parsons, D. S. (1976). The utilization of glucose and the production of lactate by *in vitro* preparations of rat small intestine: Effect of vascular perfusion. *J. Physiol.* **255**, 775–795.

He Y., Sanderson I. R., and Walker W. A. (1994). Uptake, transport, and metabolism of exogenous nucleosides in intestinal epithelial cell cultures. *J. Nutr.* **124**, 1942–1949.

Henderson, G. I., Hu, Z. Q., Yang, Y., Perez, T. B., Devi, B. G., Frosto, T. A., and Schenker, S. (1993). Ganciclovir transfer by human placenta and its effects on rat fetal cells. *Am. J. Med. Sci.* **306**, 151–156.

Hirsh, A. J., Tsang, R., Kammila, S., and Cheeseman, C. I. (1996). The effect of cholecystokinin and related peptides on jejunal transepithelial hexose transport in the Sprague-Dawley rat. *Am. J. Physiol.* **271**, G755–G761.

Hogue, D. L., Ellison, M. J., Vickers, M. F., and Cass, C. E. (1997). Functional complementation of a membrane transport deficiency in *Saccharomyces cerevisiae* by recombinant ND4 fusion protein. *Biochem. Biophys. Res. Commun.* **238**, 811–816.

Hogue, D. L., Ellison, M. J., Young, J. D., and Cass, C. E. (1996). Identification of a novel membrane transporter associated with intracellular membranes by phenotypic complementation in the yeast *Saccharomyces cerevisiae*. *J. Biol. Chem.* **271**, 9801–9808.

Horak, J. (1997). Yeast nutrient transporters. *Biochim. Biophys. Acta* **1331**, 41–79.

Hu, M. (1992). Comparison of uptake characteristics of thymidine and zidovudine in a human intestinal epithelial model. *J. Pharmaceut. Sci.* **82**, 829–833.

Huang, Q. Q., Yao, S. Y. M., Ritzel, M. W., Paterson, A. R., Cass, C. E., and Young, J. D. (1994). Cloning and functional expression of a complementary DNA encoding a mammalian nucleoside transport protein. *J. Biol. Chem.* **269**, 17757–17760.

Iseki, K., Sugawara, M., Fujiwara, T., Naasani, I., Kobayashi, M., and Miyazaki, K. (1996). Transport mechanisms of nucleosides and the derivative, 6-mercaptopurine riboside across rat intestinal brush-border membranes. *Biochim. Biophys. Acta* **1278**, 105–110.

Jakobs, E. S., Van Os-Corby, D. J., and Paterson, A. R. P. (1990). Expression of sodium-linked nucleoside transport activity in monolayer cultures of IEC-6 intestinal epithelial cells. *J. Biol. Chem.* **265**, 22210–22216.

Jarvis, S. M. (1988). Characterization of sodium-dependent nucleoside transport in rabbit intestinal brush-border membrane vesicles. *Biochim. Biophys. Acta* **979**, 132–138.

Jarvis, S. M., and Griffith, D. A. (1991). Expression of the rabbit intestinal N2 Na$^+$/nucleoside transporter in *Xenopus laevis* oocytes. *Biochem. J.* **278**, 605–607.

Jarvis, S. M., Thorn, J. A., and Glue, P. (1998). Ribavirin uptake by human erythrocytes and the involvement of nitrobenzylthioinosine-sensitive (*es*)-nucleoside transporters. *Br. J. Pharmacol.* **123**, 1587–1592.

Jarvis, S. M., and Young, J. D. (1982). Nucleoside translocation in sheep reticulocytes and fetal erythrocytes: A proposed model for the nucleoside transporter. *J. Physiol.* **324**, 47–66.

Jarvis, S. M., and Young, J. D. (1986). Nucleoside transport in rat erythrocytes: two components with differences in sensitivity to inhibition by nitrobenzylthioinosine and p-chloromercuriphenyl sulphonate. *J. Membr. Biol.* **93**, 1–10.

King, K. M., and Cass, C. E. (1994). Membrane transport of 2-chloro-2′-deoxyadenosine and 2-chloro-2′-arabinofluoro-2′-deoxyadenosine is required for cytotoxicity. *Proc. Annu. Meet. Am. Assoc. Cancer Res.* **35**, A3436.

Kwong, F. Y., Fincham, H. E., Davies, A., Beaumont, N., Henderson, P. J., Young, J. D., and Baldwin, S. A. (1992). Mammalian nitrobenzylthioinosine-sensitive nucleoside transport proteins. Immunological evidence that transporters differing in size and inhibitor specificity share sequence homology. *J. Biol. Chem.* **267**, 21954–21960.

Kwong, F. Y., Wu, J. S., Shi, M. M., Fincham, H. E., Davis, A., Henderson, P. J., Baldwin, S. A., and Young, J. D. (1993). Enzymic cleavage as a probe of the molecular structures of mammalian equilibrative nucleoside transporters. *J. Biol. Chem.* **268**, 22127–22134.

Lai, C. L., Chien, R. N., Leung, N. W., Chang, T. T., Guan, R., Tai, D. I., Ng, K. Y., Wu, P. C., Dent, J. C., Barber, J., Stephenson, S. L., and Gray, D. F. (1998). A one-year trial of lamivudine for chronic hepatitis B. Asia hepatitis lamivudine study group. *N. Engl. J. Med.* **33**, 61–68.

Lai, C. L., Ching, C. K., Tung, A. K., Li, E., Young, J., Hill, A., Wong, B. C., Dent, J., and Wu, P. C. (1997). Lamivudine is effective in suppressing hepatitis B virus DNA in Chinese hepatitis B surface antigen carriers: A placebo-controlled trial. *Hepatology* **25**, 241–244.

Lang, T. T., Young, J. D., and Cass, C. E. (2000). Selective acquisition of sensitivity to cytotoxic nucleosides in a nucleoside-transport deficient T-lymphoblastoid cell line by gene transfer: Functional expression of hCNT2. *Cancer Res.* (submitted).

Leach, J. L., Baxter, J. H., Molitor, B. E., Ramstack, M. B., and Masor, M. L. (1995). Total potentially available nucleosides of human milk by stage of lactation. *Am. J. Clin. Nutr.* **61**, 1224–1230.

Lee, C. W., Cheeseman, C. I., and Jarvis, S. M. (1988). Na^+- and K+-dependent uridine transport in rat renal brush-border membrane vesicles. *Biochim. Biophys. Acta* **942**, 139–149.

Lee, C. W., Sokoloski, J. A., Sartorelli, A. C., and Handschumacher, R. E. (1991). Induction of the differentiation of HL-60 cells by phorbol 12-myristate 13-acetate activates a Na^+-dependent uridine transport system. *Biochem. J.* **274**, 85–90.

LeLeiko N. S., Bronstein A. D., Baliga B. S., and Munro H. N. (1983). *De novo* purine nucleotide synthesis in the rat small and large intestine: Effect of dietary protein and purines. *J. Pediatr. Gastroenterol. Nutr.* **2**, 313–319.

Liman, E. R., Tygat, J., and Hess, P. (1992). Subunit stoichiometry of a mammalian K^+ channel determined by construction of multimeric cDNAs. *Neuron* **9**, 861–871.

Liu, G., Franssen, E., Fitch, M. I., and Warner, E. (1997). Patient preferences for oral versus intravenous palliative chemotherapy. *J. Clin. Oncol.* **15**, 110–115.

Loewen, S. K., Yao, S. Y. M., Cass, C. E., Baldwin, S. A., and Young, J. D. (1999). Identification of amino acid residues responsible for the pyrimidine and purine specificities of human concentrative Na^+-nucleoside cotransporters hCNT1 and hCNT2. *J. Biol. Chem.* **274**, 24475–24484.

Loewen, S. K., Yao, S. Y. M., Turner, R. J., Weiner, J. H., Henderson, P. J. F., Cass, C. E., Baldwin, S. A., Gallagher, M. P., and Young, J. D. (1998). Transport of nucleosides and nucleoside drugs by a recombinant bacterial nucleoside transporter (NupC) expressed in oocytes of *Xenopus laevis*. 8th Fisher Winternational Symposium on Cellular and Molecular Biology. Banff, Alberta, Canada.

Mackean, M., Planting, A., Twelves, C., Schellens, J., Allman, D., Osterwalder, B., Reigner, B., Griffin, T., Kaye, S., and Verweij, J. (1998). Phase I and pharmacologic study of intermittent twice-daily oral therapy with capecitabine in patients with advanced and/or metastatic cancer. *J. Clin. Oncol.* **16**, 2977–2985.

Mackey, J. R., Baldwin, S. A., Young, J. D., and Cass, C. E. (1998a). Nucleoside transport and its significance for anticancer drug resistance. *Drug Resistance Updates* **1**, 310–324.

Mackey, J. R., Mani, R. S., Selner, M., Mowles, D., Young, J. D., Belt, J. A., Crawford, C. R., and Cass, C. E. (1998b). Functional nucleoside transporters are required for gemcitabine and cytotoxicity in cancer cell lines. *Cancer Res.* **58**, 4349–4357.

Mackey, J. R., Yao, S. Y. M., Smith, K. M., Karpinski, E., Baldwin, S. A., Cass, C. E., and Young, J. D. (1999). Gemcitabine transport mediated by recombinant plasma membrane mammalian nucleoside transporters expressed in *Xenopus* oocytes. *J. Natl. Cancer Inst. USA* **91**, 1876–1881.

MacKinnon, A. M., and Deller, D. J. (1973). Purine nucleoside biosynthesis in gastrointestinal mucosa. *Biochim. Biophys. Acta.* **319**, 1–4.

Mahony, W. B., Domin, B. A., McConnell, R. T., and Zimmerman, T. P. (1998). Acyclovir transport into human erythrocytes. *J. Biol. Chem.* **263**, 9285–9291.

Mahony, W. B., Domin, B. A., and Zimmerman, T. P. (1991). Ganciclovir permeation of the human erythrocyte membrane. *Biochem. Pharmacol.* **41,** 263–271.

Mani, R. S., Hammond, J. R., Marjan, J. M. J., Graham, K. A., Young, J. D., Baldwin, S. A., and Cass, C. E. (1998). Demonstration of equilibrative nucleoside transporters (hENT1, hENT2) in intracellular membranes of cultured human choriocarcinoma (BeWo) cells by functional reconstitution in proteoliposomes. *J. Biol. Chem.* **273,** 30818–30825.

Maser, P., Sutterlin, C., Kralli, A., and Kaminsky, R. (1999). A nucleoside transporter from *Trypanosoma brucei* involved in drug resistance. *Science* **285,** 242–244.

McCloud, E., Mathis, R. K., Grant, K. E., and Said, H. M. (1994). Intestinal uptake of uridine in suckling rats: Mechanism and ontogeny. *Prog. Exp. Biol. Med.* **206,** 425–430.

Monno, R., Marcuccio, L., Valenza, M. A., Leone, E., Bitetto, C., Larocca, A., Maggi, P., and Quarto, M. (1997). *In vitro* antimicrobial properties of azidothymidine (AZT). *Acta Microbiol. Immunol. Hung.* **44,** 165–171.

Morse, G. D., Shelton, M. J., and O'Donnell, A. M. (1993). Comparative pharmacokinetics of antiviral nucleoside analogues. *Clin. Pharmacokinet.* **24,** 101–123.

Mun, E. C., Tally, K. J., and Matthews, J. B. (1998). Characterization and regulation of adenosine transport in T84 intestinal epithelial cells. *Am. J. Physiol.* **274,** G261–G269.

Mygind, B., and Munch-Petersen, A. (1975). Transport of pyrimidine nucleosides in cells of *Escherichia coli. J. Biol. Chem.* **254,** 3703–3737.

Neuhard, J., and Nygaard, P. (1987). Pyrimidines and purines. *In* "*Escherichia coli* and *Salmonella typhimurium,*" (F.C. Neidhardt, ed.) pp. 445–473, ASM International, Materials Park, OH.

Obata, T., Yamanaka, Y., Hirata, T., and Uchida, Y. (1996). Role of adenosine in blood flow in rat intestine. *Res. Commun. Mol. Pathol. Pharmacol.* **92,** 369–372.

Pajor, A. M. (1998). Sequence of a pyrimidine-selective Na$^+$/nucleoside cotransporter from pig kidney, pkCNT1. *Biochim. Biophys. Acta* **1415,** 266–269.

Pajor, A. M., and Wright, E. M. (1992). Cloning and functional expression of a mammalian Na$^+$/nucleoside cotransporter. *J. Biol. Chem.* **267,** 3557–3560.

Palacin, M. (1998). Cystinuria calls for heteromultimeric amino acid transporters. *Curr. Opin. Cell Biol.* **10,** 455–461.

Park, G. B., and Mitra, A. K. (1992). Mechanism and site dependency of intestinal mucosal transport and metabolism of thymidine analogs. *Pharmaceut. Res.* **9,** 326–331.

Patil, S. D., Ngo, L. Y., Gue, P., and Unadkat, J. D. (1998). Intestinal absorption of ribavirin is preferentially mediated by the Na$^+$-dependent nucleoside purine (N1) transporter. *Pharmaceut. Res.* **15,** 950–952.

Patil, S. D., and Unadkat, J. D. (1997). Sodium-dependent nucleoside transport in the human intestinal brush-border membrane. *Am. J. Physiol.* **272,** G1314–G1320.

Perry, C. M., and Faulds, D. (1997). Lamivudine. A review of its antiviral activity, pharmacokinetic properties, and therapeutic efficacy in the management of HIV infection. *Drugs* **53,** 657–680.

Peter, K., and Gambertoglio, J. G. (1998). Intracellular phosphorylation of zidovudine (ZDV) and other nucleoside reverse transcriptase inhibitors (RTI) used for human immunodeficiency virus (HIV) infection. *Pharm. Res.* **15,** 819–825.

Peterson, E. M., Brownell, J., and Vince, R. (1992). Synthesis and biological evaluation of 5'-sulfamoylated purinyl carbocyclic nucleosides. *J. Med. Chem.* **35,** 3991–4000.

Pialoux, G., Raffi, F., Brun-Vezinet, F., Meiffredy, V., Flandre, P., Gastaut, J. A., Dellamonica, P., Yeni, P., Delfraissy, J. F., and Aboulker, J. P. (1998). A randomized trial of three maintenance regimens given after three months of induction therapy with zidovudine, lamivudine, and indinavir in previously untreated HIV-1–infected patients. Trilege (Agence Nationale de Recherches sur le SIDA 072) Study Team. *N. Engl. J. Med.* **339,** 1269–1276.

Plagemann, P. G., Wohlhueter, R. M., and Woffendin, C. (1988). Nucleoside and nucleobase transport in animal cells. *Biochim. Biophys. Acta* **947,** 405–443.

Rana, K. Z., and Dudley, M. N. (1997). Clinical pharmacokinetics of stavudine. *Clin. Pharmacokinet.* **33,** 276–284.

Ritzel, M. W. L., Yao, S. Y. M., Huang, M. Y., Elliott, J. F., Cass, C. E., and Young, J. D. (1997). Molecular cloning and functional expression of cDNAs encoding a human Na^+-nucleoside cotransporter (hCNT1). *Am. J. Physiol.* **272,** C707–C714.

Ritzel, M. W. L., Yao, S. Y. M., Ng, A. M. L., Mackey, J. R., Cass, C. E., and Young, J. D. (1998). Molecular cloning, functional expression, and chromosomal localization of a cDNA encoding a human Na^+/nucleoside cotransporter (hCNT2) selective for purine nucleosides and uridine. *Mol. Membr. Biol.* **15,** 203–211.

Roden, M., Paterson, A. R., and Turnheim, K. (1991). Sodium-dependent nucleoside transport in rabbit intestinal epithelium. *Gastroenterology* **100,** 1553–1562.

Salati, L. M., Gross, C. J., Henderson, L. M., and Saviano, D. A. (1984). Absorption and metabolism of adenine, adenosine-5'-mono-phosphate, adenosine, and hypoxanthine by the vascularly perfused rat small intestine. *J. Nutr.* **114,** 753–760.

Sanchez, M. A., Ullman, B., Landfear, S. M., and Carter, N. S. (1999). Cloning and functional expression of a gene encoding a P1 type nucleoside transporter from *Trypanosoma brucei*. *J. Biol. Chem.* **274,** 30244–30249.

Sanderson, I. R., and He, Y. (1994). Nucleotide uptake and metabolism by intestinal epithelial cells. *J. Nutr.* **124,** 131S–137S.

Saviano, D. A., and Clifford, A. J. (1978). Absorption, tissue incorporation, and excretion of free purine bases in the rat. *Nutr. Rep. Int.* **17,** 551–556.

Saviano, D. A., and Clifford, A. J. (1981). Adenine, the precursor of nucleic acids in intestinal cells unable to synthesize purines *de novo*. *J. Nutr.* **111,** 1816–1822.

Schaner, M. E., Wang, J., Zevin, S., Gersin, K. M., and Giacomini, M. (1997). Transient expression of a purine-selective nucleoside transporter ($SPNT_{int}$) in a human cell line (HeLa). *Pharmaceut. Res.* **10,** 1316–1321.

Schmidt, G. H., Wilkinson, M. M., and Ponder, B. A. J. (1985). Cell migration pathway in the intestinal epithelium: An *in situ* marker system using mouse aggregation chimeras. *Cell* **40,** 425–429.

Schwenk, M., Hegazy, E., and Del Pino, V. L. (1984). Uridine uptake by isolated intestinal epithelial cells of guinea pig. *Biochim. Biophys. Acta* **805,** 370–374.

Seeger, C., Poulsen, C., and Dandanall, G. (1995). Identification and characterization of genes involved in xanthosine catabolism in *Escherichia coli*. *J. Bacteriol.* **177,** 5506–5516.

Shelton, M. J., O'Donnell, A. M., and Morse, G. D. (1992). Didanosine. *Ann. Pharmacother.* **26,** 660–670.

Shelton, M. J., O'Donnell, A. M., and Morse, G. D. (1993). Zalcitabine. *Ann. Pharmacother.* **27,** 480–489.

Shepard, A. J., Barr, P., and Lewin, C. S. (1992). The effect of thymidine on the antibacterial and antiviral activity of zidovudine. *J. Pharm. Pharmacol.* **44,** 704–706.

Sinko, P. J., Hu, P., Waclawski, A. P., and Patel, N. R. (1995). Oral absorption of anti-AIDS nucleoside analogues. 1. Intestinal transport of didanosine in rat and rabbit preparations. *J. Pharm. Sci.* **84,** 959–965.

Sinko, P. J., Sutyak, J. P., Leesman, G. D., Hu, P., Makhey, V. D., Yu, H., and Smith, C. L. (1997). Oral absorption of anti-AIDS nucleoside analogues: 3. Regional absorption and *in vivo* permeability of 2',3'-dideoxyinosine in an intestinal-vascular access port (IVAP) dog model. *Biopharm. Drug Dispos.* **18,** 697–710.

Smith, K. M., Cass, C. E., Karpinski, E., and Young, J. D. (1998). Electrophysiological studies of the human Na^+-dependent nucleoside transporter hCNT1. 8th Fisher Winternational Symposium on Cellular and Molecular Biology. Banff, Alberta, Canada.

Soler, C., Felipe, A., Mata, J. F., Casado, J., Celada, A., and Pastor-Anglada, M. (1998). Regulation of nucleoside transport by lipopolysaccharide, phorbol ester, and tumor necrosis factor in human B-lymphocytes. *J. Biol. Chem.* **273,** 26939–26945.

Stow, R. A., and Bronk, J. R. (1992). Nucleoside transport and metabolism in isolated rat jejunum. *J. Physiol.* **468,** 311–324.

Strohmeier, G. R., Reppert S. M., Lencer, W. I., and Madara, J. L. (1995). The A_{2b} adenosine receptor mediates cAMP response to adenosine receptor agonists in human intestinal epithelia. *J. Biol. Chem.* **270,** 2387–2394.

Sundaram, M., Yao, S. Y. M., Cass, C. E., Baldwin, S. A., and Young, J. D. (1998). Chimaeric constructs between human and rat equilibrative nucleoside transporters (hENT1 and rENT1) reveal hENT1 structural domains interacting with coronary vasoactive drugs. *J. Biol. Chem.* **273,** 21519–21525.

Tally, K. J., Hrnjez, B. J., Smith, J. A., Mun, E. C., and Matthews, J. B. (1996). Adenosine scavenging: A novel mechanism of chloride secretory control in intestinal epithelial cells. *Surgery* **120,** 248–254.

Tanaka, M., Lee, K., Martinez-Augustin, O., He, Y., Sanderson, I. R., and Walker, A. (1996). Exogenous nucleotides alter proliferation, differentiation, and apoptosis of human small intestinal epithelium. *J. Nutr.* **126,** 424–433.

Terasaki, T., Kadowaki, A., Higashida, H., Nakayama, K., Tamai, I., and Tsuji, A. (1993). Expression of the Na^+-dependent uridine transport system of rabbit small intestine: Studies with mRNA-injected *Xenopus laevis* oocytes. *Biol. Pharm. Bull.* **16,** 493–496.

Tse, C. M., Wu, J. S., and Young, J. D. (1985). Evidence for the asymmetrical binding of p-chloromercuriphenyl sulphonate to the human erythrocyte nucleoside transporter. *Biochim. Biophys. Acta* **818,** 316–324.

Uay, R., Stringel, G., Thomas, R., and Quan, R. (1990). Effect of dietary nucleosides on growth and maturation of the developing gut in the rat. *J. Paediatr. Gastroenterol. Nutr.* **10,** 497–503.

Vasudevan, G., Carter, N. S., Drew, M. E., Beverley, S. M., Sanches, M. A., Seyfang, A., Ullman, B., and Landfear, S. M. (1998). Cloning of *Leishmania* nucleoside transporter genes by rescue of a transport-deficient mutant. *Proc. Natl. Acad. Sci.* **95,** 9873–9878.

Vickers, M. F., Mani, R. S., Sundaram, M., Hogue, D. L., Young, J. D., Baldwin, S. A., and Cass, C. E. (1999). Functional production and reconstitution of the human equilibrative nucleoside transporter (hENT1) and a glycosylation-defective derivative (hENT1/N48Q). *Biochem. J.* **339,** 21–32.

Vickers, M. F., Yao, S. Y. M., Young, J. D., Baldwin, S. A., and Cass, C. E. (2000). Nucleoside transporter proteins of *S. cerevisiae:* Demonstration of a transporter (FUI1) with high uridine selectivity in plasma membranes and a transporter (FUN26) with broad nucleoside selectivity in intracellular membranes. *J. Biol. Chem.* (in press).

Vijayalakshimi, D., and Belt, J. A. (1988). Sodium-dependent nucleoside transport in mouse intestinal epithelial cells. Two transport systems with differing substrate specificities. *J. Biol. Chem.* **263,** 19419–19423.

Von Hoff, D. D., Slavik, M., and Muggia, F. M. (1976). 5-azacytidine. A new anticancer drug with effectiveness in acute myelogenous leukemia. *Ann. Intern. Med.* **85,** 237–245.

Waclawski, A. P., and Sinko, P. J. (1995). Oral absorption of anti–acquired immune deficiency syndrome nucleoside analogs. 2. Carrier-mediated intestinal transport of stavudine in rat and rabbit preparations. *J. Pharmaceut. Sci.* **84,** 959–965.

Wang, J., and Giacomini, K. M. (1997). Molecular determinants of substrate selectivity in Na^+-dependent nucleoside transporters. *J. Biol. Chem.* **272,** 28845–28848.

Wang, J., and Giacomini, K. M. (1999). Serine 318 is essential for the pyrimidine selectivity of the N2 Na^+-nucleoside transporter. *J. Biol. Chem.* **274,** 2298–2302.

Wang, J., Schaner, M. E., Thomassen, S., Su, S. F., Piquette-Miller, M., and Giacomini, K. M. (1997a). Functional and molecular characteristics of Na$^+$-dependent nucleoside transporters. *Pharm. Res.* **14,** 1524–1532.

Wang, J., Su, S.-F., Dresser, M. J., Schnaner, M. E., Washington, C. B., and Giacomini, K. M. (1997b). Na$^+$-dependent purine nucleoside transporter from human kidney: Cloning and functional characterization. *Am. J. Physiol.* **273,** F1058–F1065.

Williams, J. B., Rexer, B., Sirripurapu, S., John, S., Goldstein, R., Phillips, J. A., Haley, L. L., Sait, S. N., Shows, T. B., Smith, C. M., and Gerhard, D. S. (1997). The human HNP36 gene is localized to chromosome 11q13 and produces alternative transcripts that are not mutated in multiple endocrine neoplasia, type I (MEN I) syndrome. *Genomics* **42,** 325–330.

Williams, T. C., Doherty, A. J., Griffith, D. A., and Jarvis, S. M. (1989). Characterization of sodium-dependent and sodium-independent nucleoside transport systems in rabbit brush-border and basolateral plasma-membrane vesicles from the renal outer cortex. *Biochem. J.* **264,** 223–231.

Wu, X., Yuan, G., Brett, C. M., Hui, A. C. and Giacomini, K. M. (1992). Sodium-dependent nucleoside transport in choroid plexus from rabbit. Evidence for a single transporter for purine and pyrimidine nucleosides. *J. Biol. Chem.* **267,** 8813–8818.

Yao, S. Y., Cass, C. E., and Young, J. D. (1996b). Transport of the antiviral nucleoside analogs 3′-azido-3′-deoxythymidine and 2′,3′-dideoxycytidine by a recombinant nucleoside transporter (rCNT) expressed in *Xenopus laevis* oocytes. *Mol. Pharmacol.* **50,** 388–393.

Yao, S. Y. M., Cass, C. E., and Young, J. D. (2000a). The *Xenopus* oocyte expression system for the expression cloning and characterization of plasma membrane transport proteins. *In* "Membrane Transport: A Practical Approach" (S. A. Baldwin, ed.). Oxford University Press, Oxford (in press).

Yao, S. Y. M., Muzyka, W. R., Cass, C. E., Cheeseman, C. I., and Young, J. D. (1999). Evidence that the transport-related proteins BAT and 4F2hc are not specific for amino acids. *Biochem. Cell Biol.* **76,** 859–865.

Yao, S. Y. M., Ng, A. M. L., Cass, C. E., Baldwin, S. A., and Young, J. D. (1998). Molecular cloning and functional expression of a broad specificity Na$^+$-dependent nucleoside transporter (hfCNT) from an ancient vertebrate, the Pacific hagfish (*Eptatretus stouti*). 8th Fisher Winternational Symposium on Cellular and Molecular Biology. Banff, Alberta, Canada.

Yao, S. Y., Ng, A. M., Muzyka, W. R., Griffiths, M., Cass, C. E., Baldwin, S. A., and Young, J. D. (1997). Molecular cloning and functional characterization of nitrobenzylthioinosine (NBMPR)-sensitive (*es*) and NBMPR-insensitive (*ei*) equilibrative nucleoside transporter proteins (rENT1 and rENT2) from rat tissues. *J. Biol. Chem.* **272,** 28423–28430.

Yao, S. Y., Ng, A. M., Ritzel, M. W., Gati, W. P., Cass, C. E., and Young, J. D. (1996a). Transport of adenosine by recombinant purine- and pyrimidine-selective sodium/nucleoside cotransporters from rat jejunum expressed in *Xenopus laevis* oocytes. *Mol. Pharmacol.* **50,** 1529–1535.

Yao, S. Y. M., Sundaram, M., Chomey, E. G., Cass, C. E., Baldwin, S. A., and Young, J. D. (2000b). Identification of the exofacial cysteine residue (Cys140) responsible for the PCMBS inhibition of rat equilibrative nitrobenzylthioinosine-insensitive nucleoside transporter rENT2. *Biochem. J.* (submitted).

Yuen, G. J., Morris, D. M., Mydlow, P. K., Haidar, S., Hall, S. T., and Hussey, E. K. (1995). Pharmacokinetics, absolute bioavailability, and absorption characteristics of lamivudine. *J. Clin. Pharmacol.* **35,** 1174–1180 [erratum appears in *J. Clin. Pharmacol.* (1996) **36,** 373].

Zaharavitz, D. W., Anderson, L. W., Strong, J. M., Hyman, R., and Cysyk, R. L. (1990). *De novo* synthesis of uracil nucleotides in mouse liver and intestine studied using [^{15}N]alanine. *Eur. J. Biochem.* **187,** 437–440.

Zimmerman, T. P., Mahony, W. B., and Prus, K. L. (1987). 3′-azido-3′-deoxythymidine. An unusual nucleoside analogue that permeates the membrane of human erythrocytes and lymphocytes by nonfacilitated diffusion. *J. Biol. Chem.* **262,** 5748–5754.

CHAPTER 10

Intestinal Transport of Peptides and Amino Acids

Vadivel Ganapathy,* Malliga E. Ganapathy,† and Frederick H. Leibach*
*Department of Biochemistry and Molecular Biology and †Department of Medicine, Medical College of Georgia, Augusta, Georgia 39012.

I. INTRODUCTION

Digestion and absorption of dietary nutrients constitute the principal function of the gastrointestinal tract. This digestive process involves various enzymes secreted by the stomach and pancreas as well as enzymes associated with the intestinal brush-border membrane and cytoplasm. The absorptive process is mediated by specific transport systems present in the intestinal brush-border and basolateral membranes. Digestion of dietary proteins is initiated in the stomach by pepsin and is continued in the intestine by pancreatic proteases (trypsin, chymotrypsin, and elastase) and peptidases (aminopeptidases and carboxypeptidases) and by peptidases present in the brush-border membrane of the enterocytes. The end result of this digestive process is the generation of free amino acids and small peptides. Thus, the breakdown of dietary proteins by the digestive enzymes in the lumen of the gastrointestinal tract does not go to completion prior to absorption, as is the case with dietary carbohydrates. The intestinal brush-border membrane possesses a variety of transport systems that are responsible for the absorption of amino acids and peptides resulting from luminal digestion. Peptidases present in the cytoplasm of the enterocytes carry out the end stage of protein digestion in which the absorbed peptides are hydrolyzed to free amino acids. The transport systems present in the basolateral membrane of the enterocytes mediate the exit of free amino acids from the cells into the portal circulation. Quantitatively, entry of peptides across the intestinal brush-border membrane is the predominant mode of absorption of protein digestion products. However, as a result of participation of cytosolic peptidases in the hydrolysis of absorbed peptides, protein digestion products enter the portal circulation across the intestinal basolateral membrane, predominantly in the form of free amino acids. Small peptides that are resistant to hydrolysis by cytosolic peptidases do enter the blood, but these constitute a relatively minor component. Available evidence suggests that most amino acids are absorbed in the intestine more efficiently when presented to the lumen in the form of small peptides rather than in the free form. Detailed reviews are available in the literature on the topic of intestinal digestion and absorption of dietary proteins (Ganapathy and Leibach, 1998; Ganapathy *et al.*, 1994; Leibach and Ganapathy, 1996). Although there has been little increase in recent years in our current understanding of the process of protein digestion, the field of transport systems that are responsible for the absorptive process is fast expanding. This is attributable to the successful cloning of various transporter proteins that mediate the absorption of peptides and amino acids. As a result, there has been a tremendous gain in our current knowledge of the structural, functional, and regulatory aspects of these transporters in the past few years. The purpose of this review is to focus on the recent developments in the field of peptide and amino acid transporters.

II. PEPTIDE TRANSPORT

A. Functional Characteristics

Initial studies of intestinal peptide transport were prompted by intriguing observations in patients with the amino acid transport defects Hartnup disease and cystinuria; the affected amino acids were found to have normal absorption from the intestine when they were presented in the form of dipeptides (Hellier *et al.,* 1972; Silk *et al.,* 1975; Leonard *et al.,* 1976). These observations suggest that the intestine is capable of absorbing intact peptides and that this process is distinct from the amino acid absorptive process. Subsequently, the existence of a peptide transport system in the intestine was established unequivocally using a variety of intestinal tissue preparations (see review in Matthews, 1975; 1991; Matthews and Adibi, 1976). Investigations of intestinal peptide transport using isolated brush-border membrane vesicles launched a new era in the field of peptide transport (Ganapathy and Leibach, 1983; Ganapathy *et al.,* 1981; 1984; 1985) because this experimental approach made it possible to delineate the functional and energetic aspects of the peptide transport process in the absence of potentially complicating factors such as intracellular hydrolysis and metabolism. One distinguishing feature of intestinal peptide transport is its energization by a transmembrane H^+ gradient rather than a transmembrane Na^+ gradient (Ganapathy and Leibach, 1985, 1991; Ganapathy *et al.,* 1987; Hoshi, 1985). When the transported peptide is zwitterionic, the H^+-dependent transport is electrogenic, associated with the transfer of a net positive charge in the direction of the peptide movement (Ganapathy and Leibach, 1983; Ganapathy *et al.,* 1984; 1985). This suggests that the intestinal peptide transport process involves H^+/peptide cotransport. These findings indicated for the first time the functional relevance of the microclimate acid pH, which has been known to exist on the luminal side of the enterocyte brush-border membrane, to the intestinal absorptive process (Lucas *et al.,* 1975; 1978). This acid pH microclimate is generated by the Na^+-H^+ exchanger in the intestinal brush-border membrane that converts a transmembrane Na^+ gradient to a transmembrane H^+ gradient (Hoshi, 1985; Ganapathy and Leibach, 1991; 1998; Ganapathy *et al.,* 1987; 1994). Because the H^+/peptide cotransport is electrogenic, the inside-negative membrane potential that exists across the enterocyte plasma membrane also serves as a driving force for the transport process. The Na^+-K^+-ATPase localized in the basolateral membrane plays a major role in maintaining the electrochemical H^+ gradient across the brush-border membrane. Chemical modification studies have indicated that specific histidyl residues might be involved as a H^+ acceptor/donor in the intestinal peptide transport mechanism involving H^+ cotransport (Kato *et al.,* 1989; Kramer *et al.,* 1990; Brandsch *et al.,* 1997).

   The intestinal peptide transport system accepts a wide variety of chemically and structurally diverse dipeptides and tripeptides as substrates. Free amino acids are not recognized as substrates by this system. Based on kinetic analysis of the transport process, a single transport system is apparently responsible for the absorption of peptides in the intestine. Thus, the intestinal peptide transport system exhibits very broad substrate specificity. However, the transport system is stereoselective (Asatoor *et al.,* 1973; Das and Radhakrishnan, 1975; Lister *et al.,* 1995). Peptides containing L amino acids are recognized by the transport system with higher affinity than those containing D amino acids. The intestinal peptide transport system interacts not only with physiologically occurring peptides but also with a variety of peptidomimetics. Many of these peptidomimetics are pharmacologically active. Therefore, the intestinal peptide transport system plays an important role in enhancing the oral bioavailability of these peptidomimetic drugs. Examples of pharmacologically relevant drugs that are absorbed in the intestine via the peptide transport system include β-lactam antibiotics (see review in Tsuji, 1995; Tsuji and Tamai, 1996; Amidon and Lee, 1994), bestatin (Inui *et al.*, 1992), ACE inhibitors (Hu and Amidon, 1988; Swaan *et al.*, 1995), and prodrugs such as peptidyl derivatives of α-methyldopa (Tsuji *et al.*, 1990; Hu *et al.*, 1989) and valyl ester of acyclovir (Ganapathy *et al.*, 1998; De Vrueh *et al.*, 1998). Thus, the intestinal peptide transport system has enormous potential in drug delivery via the oral route because drugs that are not absorbed appreciably in the intestine may be modified by amino acid substitutions via peptide or ester linkages such that these modified drugs are recognized by the intestinal peptide transport system. Once inside the enterocyte, cellular peptidases and esterases can hydrolyze these modified drugs to release the pharmacologically active component into the system.

B. Primary Structure of the Intestinal Peptide Transporter

Even though the presence of a specific peptide transport system in the intestine has been known for more than three decades from functional studies, it was not until recently that the protein responsible for the transport function was successfully cloned and its primary structure elucidated. The peptide transporter was first cloned from rabbit small intestine by an expression cloning approach that used the *Xenopus laevis* oocyte expression system (Fei *et al.*, 1994). Subsequently, the transporter was cloned from human and rat intestinal cDNA libraries by homology screening (Liang *et al.*, 1995; Saito *et al.*, 1995). The structural aspects of the cloned transporter, designated PEPT1, have been reviewed (Meredith and Boyd, 1995; Daniel, 1996; Fei *et al.*, 1998). The primary structures of PEPT1s from different animal species are highly homologous. The human PEPT1 consists of 708 amino acids with a predicted molecular mass of ~78 kDa. There are 12 putative transmembrane domains in the protein. Several sites for *N*-linked glycosylation also are present in presumptive extracellular loops. The human PEPT1 mRNA is 3.1 kb in size and is expressed principally in the intestine and, to a small extent, in the kidney. The *PEPT1* gene is located on human chromosome 13q24–q33.

There is evidence of a peptide transport system in mammalian kidney whose functional characteristics are distinct from those of the intestinal peptide transport system (Ganapathy and Leibach, 1996). Even though the renal system is specific for dipeptides and tripeptides and is driven by an electrochemical H^+

gradient, as is the intestinal system, the former exhibits much higher affinity for its substrates than the latter. Unequivocal evidence of the distinct nature of the renal peptide transport system came from the successful cloning of the renal peptide transporter PEPT2 (Liu *et al.*, 1995; Saito *et al.*, 1996; Boll *et al.*, 1996). PEPT2 is not expressed in the intestine. The *PEPT2* gene is located on human chromosome 3q13.3–q21 (Ramamoorthy *et al.*, 1995). The structural and functional characteristics of the renal peptide transporter have been reviewed (Ganapathy and Leibach, 1996; Daniel and Herget, 1997).

C. Functional Characteristics of the Cloned Intestinal Peptide Transporter PEPT1

The characteristics of cloned PEPT1 have been studied by expressing the transporter heterologously in either *X. laevis* oocytes or mammalian cells (Fei *et al.*, 1994; Liang *et al.*, 1995; Saito *et al.*, 1995; Boll *et al.*, 1994). The transporter is energized by a transmembrane electrochemical H^+ gradient, accepts a variety of small peptides as substrates, and exhibits low affinity for its substrates. These characteristics of the cloned intestinal peptide transporter are similar to those of the peptide transport system studied in intestinal brush-border membrane vesicles. Interaction of peptidomimetic drugs also is demonstrable with PEPT1 (Ganapathy *et al.*, 1995; 1997; 1998; Boll *et al.*, 1994; Saito *et al.*, 1995; Tamai *et al.*, 1997; 1998). More recent studies have suggested that PEPT1 is capable of recognizing a more diverse group of compounds as substrates. Examples of these compounds include δ-aminolevulinic acid (Doring *et al.*, 1998a), amino acid arylamides (Borner *et al.*, 1998), p-aminophenylacetic acid (Temple *et al.*, 1998), and ω-amino fatty acids of certain specific carbon-chain length (Doring *et al.*, 1998b). These studies suggest that the presence of a peptide bond is not obligatory for a compound to be recognized as a substrate by PEPT1.

Protein-derived peptides are the physiological substrates for the intestinal peptide transporter. PEPT1 recognizes dipeptides and tripeptides containing neutral, anionic, and cationic amino acids (Mackenzie *et al.*, 1996a; Amasheh *et al.*, 1997; Steel *et al.*, 1997). Interestingly, the transport of peptides by PEPT1 is activated by a H^+ gradient in a similar manner, irrespective of the net charge associated with the peptide substrate. The H^+ activation of transport is saturable and hyperbolic, which suggests that the H^+:peptide stoichiometry remains the same (i.e., 1:1) for differently charged peptides. It appears that only the zwitterionic forms of the peptides are preferentially recognized by PEPT1 as transportable substrates. This seems to be true not only with physiologically occurring peptides but also with β-lactam antibiotics (Wenzel *et al.*, 1996). PEPT1 also seems to recognize preferentially peptides with the peptide bond in the transconfiguration (Brandsch *et al.*, 1998).

PEPT1 transports a broad spectrum of dipeptides and tripeptides consisting of structurally diverse amino acids. Thus, the potential physiological substrates for this transporter include 400 different dipeptides and 8000 different tripeptides. If peptidomimetic xenobiotics are considered as potential substrates, the number of PEPT1 substrates is likely to be greater than 10,000. Doring *et al* (1998b) have identified the minimal structural determinants of substrates to be recognized by PEPT1. Recognition and transport by PEPT1 requires, as a minimum, only two ionized groups: an amino group and a carboxyl group. These two groups need to be separated by a spacer of 4 carbon units with an optimal distance of 500–635 pm. There are several thousand di- and tripeptides and xenobiotics that meet these minimal criteria of structural features necessary for recognition and transport by PEPT1. This explains the extraordinary breadth of structurally diverse substrates that are transported by this transporter.

The operational mechanism of the intestinal peptide transporter has been studied in detail with human and rabbit PEPT1s (Mackenzie *et al.,* 1996b; Nussberger *et al.,* 1997). These studies were carried out in *X. laevis* oocytes expressing the cloned peptide transporter heterologously, and the transporter function was assessed by peptide- or H^+-evoked currents using the two-microelectrode voltage-clamp technique. The transport of peptides via PEPT1 is electrogenic and is coupled to an inwardly directed H^+ gradient. The H^+/peptide cotransport process obeys Michaelis-Menten kinetics. The transporter-associated current is voltage dependent. Under voltage-clamped conditions, the peptide-evoked current is hyperbolically related to extracellular H^+ concentration, with a Hill coefficient of approximately 1. Interestingly, the affinity for the peptide is markedly influenced by extracellular H^+ concentration as well as by membrane potential. The affinity for H^+ also is influenced by membrane potential. In the absence of peptide substrates, PEPT1 exhibits H^+-dependent pre-steady-state transient currents. These currents are likely attributable to reorientation of the unloaded transporter in the membrane following H^+ binding/dissociation to and from the transporter. The binding of H^+ and the translocation of the unloaded carrier are voltage dependent, but it is the reorientation of the unloaded carrier that accounts for most of the charge movements observed. These transient currents are affected by extracellular as well as intracellular H^+ concentration. Thus, the magnitude and direction of the transient charge movements depend on the magnitude and direction of the transmembrane electrochemical H^+ gradient. The transporter appears to possess a single H^+ binding site that is symmetrically accessible from both sides of the membrane. The magnitude and direction of the electrochemical H^+ gradient determine the side at which the site becomes preferentially accessible to H^+ binding. Analysis of H^+- and peptide-induced currents has led to a hypothetical, ordered, simultaneous transport model in which H^+ binds first.

Histidyl residues are highly suitable for H^+ binding/dissociation in H^+-coupled transporters. These residues exhibit a pKa value of 6.8, which is optimal for H^+ binding/dissociation under physiological conditions. Site-directed mutagenesis studies have identified the histidine residues that might participate in H^+ interaction in PEPT1 (Fei *et al.*, 1997; Terada *et al.*, 1996). The His-57 in PEPT1 is the most likely residue involved in H^+ binding/dissociation because mutation of this residue to glutamine or asparagine completely abolishes the transport function. However, the synthesis and stability of the mutant transporter and its insertion into the plasma membrane are not affected. Another study has demonstrated that Tyr-167 in human PEPT1 is essential for transport function (Yeung *et al.*, 1998), but the exact role of this amino acid residue in the transport function of PEPT1 is not known. Recently, an alternative splice variant of human PEPT1 has been identified (Saito *et al.*, 1997). This splice variant consists of 208 amino acids and does not possess peptide transport function but has the ability to modulate the H^+ sensitivity of the wild-type PEPT1 when coexpressed. At the amino acid sequence level, the splice variant differs significantly from the wild-type PEPT1, but it arises from the same gene as the wild type PEPT1. The amino acid residues 18–195 are identical to the amino acid residues 8–185 in wild-type PEPT1, whereas the amino acid residues 1–17 and 196–208 are unique to the splice variant. The localization and physiological relevance of this splice variant have not been investigated in detail.

D. Distribution of the Peptide Transporter in the Intestinal Tract

In situ hybridization studies in the rabbit intestine have shown that PEPT1-specific mRNA transcripts are present all along the small intestine (Freeman *et al.*, 1995). The colon expresses very little PEPT1-specific mRNA. Expression of PEPT1 is undetectable in the stomach. In the small intestine, PEPT1 expression is restricted to enterocytes. Along the crypt-villus axis, the expression is not detectable in the crypt, becomes detectable at the crypt-villus junction, and increases rapidly toward the villus tip. PEPT1 mRNA is most abundant in the duodenum and jejunum and is present at comparatively lower levels in the ileum. Immunolocalization studies have indicated that the PEPT1 protein is expressed exclusively in the brush-border membrane (Sai *et al.*, 1996). However, there is evidence of the presence of a carrier-mediated peptide transport system in the basolateral membrane of the small intestine (Dyer *et al.*, 1990). This transporter is likely distinct from PEPT1 of the brush-border membrane based on the functional characteristics studied in polarized Caco-2 cells (Saito and Inui, 1993; Thwaites *et al.*, 1993). The physiological function of the peptide transporter in the basolateral membrane is to mediate the exit of peptides that are resistant to hydrolysis in the cytoplasm of the enterocytes from the cells into the portal circulation.

E. Regulation of Intestinal Peptide Transporter

A number of studies have shown that the intestinal peptide absorptive process is subject to developmental, dietary, and hormonal regulation (Ganapathy *et al.*, 1994). In laboratory animals, it has been shown that peptide absorptive capacity is maximal at birth and then decreases with age to adult levels (Guandalini and Rubino, 1982; Himukai *et al.*, 1980). A high-protein diet, as well as metabolic changes such as starvation, influences intestinal peptide transport (Ferraris *et al.*, 1988; Lis *et al.*, 1972; Vazquez *et al.*, 1985). With the current availability of probes to analyze the expression of PEPT1 at mRNA and protein levels, recent studies are focusing on the cellular and molecular mechanisms that underlie the regulation of intestinal peptide transport. Erickson *et al.* (1995) have shown that when rats are switched from a low-protein diet to a high-protein diet, the expression of PEPT1 increases significantly as assessed from steady-state mRNA levels.

There is increasing evidence that the intestinal peptide transport function mediated by PEPT1 is regulated by hormones and second-messenger signaling pathways. Activators of protein kinase C inhibit peptide transport in Caco-2 cells of human intestinal origin (Brandsch *et al.*, 1994). These cells express PEPT1. The possibility that protein kinase C may modulate PEPT1 function by phosphorylation of the protein is supported by findings that human PEPT1 possess consensus sites for protein kinase C–mediated protein phosphorylation. Kinetic studies have shown that the inhibition of peptide transport in Caco-2 cells associated with protein kinase C activation is primarily caused by a decrease in maximal velocity with no appreciable change in substrate affinity. However, it is not known whether the observed decrease in maximal velocity is directly related to a reduction in the transport function or to a decrease in the peptide transporter density in the plasma membrane. Protein kinase C–mediated phosphorylation of PEPT1 may affect its transport activity or its turnover. Peptide transport in Caco-2 cells also is inhibited by agents that increase intracellular cAMP levels (Muller *et al.*, 1996). These agents include cholera toxin and *Escherichia coli* heat-labile enterotoxin. Interestingly, human PEPT1 does not possess consensus sites for protein phosphorylation mediated by cAMP-dependent protein kinase. It is possible that the inhibitory action of cAMP on PEPT1 is mediated by protein kinase C resulting from cross-talk between signaling pathways. Alternatively, an accessory protein that might be a substrate for cAMP-dependent protein kinase may be involved in the regulation of the transport function of PEPT1. This alternative mechanism is analogous to the well-characterized regulation of the Na^+/H^+ exchangers by the proteins of the regulatory factor gene family (Donowitz *et al.*, 1998). More recent studies have demonstrated that insulin regulates peptide transport in Caco-2 cells (Thamotharan *et al.*, 1999). Treatment of the cells with insulin at physiological concentrations for a short period significantly activates peptide transport. This activation is associated with an increase in maximal velocity with no detectable change in substrate affinity. The PEPT1 protein density in the plasma membrane is increased in parallel, corresponding to the increase in transport activity. It appears that insulin, through activation of the insulin receptor–associated tyrosine kinase, increases the insertion of PEPT1 into the plasma membrane from a preformed cytoplasmic pool. This mechanism is supported by findings that the effect of insulin is not associated with any noticeable change in steady-state PEPT1 mRNA levels and is blocked by the tyrosine-kinase inhibitor genistein and the microtubule function inhibitor colchicine.

A recent study by Shiraga et al. (1999) has investigated the molecular mechanisms involved in the dietary regulation of the intestinal peptide transporter in the rat. This investigation was performed in intact animals and by using reporter constructs to assess the function of the PEPT1 gene promoter. Feeding the rat a high-protein diet increases peptide transport activity in intestinal brush-border membrane vesicles and, in parallel, increases steady-state levels of PEPT1 mRNA. Interestingly, feeding the rat a diet containing excess levels of the dipeptide glycylphenylalanine or the amino acid phenylalanine also elicits a similar stimulatory effect on PEPT1 expression. However, a diet containing excess levels of the amino acid glycine does not have any effect on PEPT1 expression. Transient transfection analysis of rat PEPT1 gene promoter using luciferase reporter constructs has indicated that several dipeptides and amino acids are capable of activating this promoter, thus resulting in the stimulation of gene expression. This effect is not nonspecific because some amino acids such as glycine, glutamine, valine, and aspartate as well as D-glucose and D amino acids do not have any effect on PEPT1 gene promoter activity. The amino acid/peptide-responsive element is located between -295 and -277 nucleotides relative to the transcription start site in the rat PEPT1 gene. Thus, a high-protein diet or a diet supplemented with excess levels of specific dipeptides or amino acids is capable of enhancing intestinal peptide transporter activity by activating the transcription of the PEPT1 gene.

Recently, Fujita et al (1999) have shown that PEPT1 in Caco-2 cells is regulated by the σ1 receptor. Treatment of the cells with the σ1-specific ligand (+)-pentazocine leads to an increase in steady-state levels of PEPT1 mRNA and in the maximal velocity of H^+-coupled dipeptide transport activity. It appears that these effects might involve up-regulation of the PEPT1 gene. These findings are potentially very important for several reasons. The σ1 receptor is expressed abundantly in the intestinal tract and in cell lines of intestinal origin (Roman et al., 1989; Kekuda et al., 1996). Although the endogenous ligands for the receptor have not been identified unequivocally, progesterone is believed to be the most likely ligand (Su et al., 1988; Ramamoorthy et al., 1995). The σ1 receptor interacts with progesterone with a K_d value of ~90 nM (Ganapathy et al., 1999).

In women, plasma concentrations of this hormone vary markedly under different physiological conditions such as menstrual cycle and pregnancy. The levels are known to be ~500 nM in late pregnancy. It is possible that progesterone upregulates the intestinal absorption of peptides during pregnancy by interacting with the σ1 receptor, and thus plays a crucial role in the maintenance of optimal nutritional status in the pregnant mother and the developing fetus.

There is evidence that neural pathways are involved in the regulation of intestinal peptide transporter (Harcouet *et al.*, 1997). Nifedipine, a calcium channel blocker, enhances the bioavailability of β-lactam antibiotics by increasing intestinal absorption via the peptide transport system. This effect is seen only *in vivo*. It appears that the noradrenergic receptors alpha-1 and alpha-2 are involved in this process (Berlioz *et al.*, 1999). The second-messenger signaling pathways responsible for the regulation via these receptors have not been identified.

An interesting feature of intestinal peptide transport is that its activity is not affected significantly in conditions such as starvation, protein-calorie malnutrition, vitamin deficiency, and intestinal diseases (Matthews, 1975; Ganapathy *et al.*, 1994). Under similar conditions there is significant impairment in amino acid transport activity. These conditions often are associated with mucosal injury, and the molecular mechanisms of the resistance of peptide transport function to mucosal injury have not been studied in detail. A recent study by Tanaka *et al.* (1998) investigated peptide transport function and PEPT1 expression in the rat intestine in response to 5-fluorouracil–induced mucosal injury. Their study showed that although the density of the sodium-dependent glucose transporter protein in the intestinal brush-border membrane decreases significantly after 5-fluorouracil treatment, the density of the PEPT1 protein remains unaffected. Similarly, the levels of amino acid, glucose, and phosphate transporter mRNAs decrease in response to 5-fluorouracil–mediated mucosal injury, but the level of PEPT1 mRNA increases. These findings demonstrate that the resistance of peptide transport function to mucosal injury is attributable to up-regulation of *PEPT1* gene expression. However, the signaling events involved in this process are not known.

III. AMINO ACID TRANSPORT

A. Classification of Amino Acid Transport Systems in the Intestine

The brush-border membrane of the intestinal epithelial cells expresses several amino acid transport systems with significant overlap in their substrate specificity. Based on functional characteristics, these transport systems have been classified as follows: B^0 as a Na^+-dependent transport system for a variety of neutral amino acids; X_{AG}^- as a Na^+-dependent transport system for anionic amino acids; y^+ as a Na^+-independent transport system for cationic amino acids; $b^{0,+}$

as a Na^+-independent transport system for neutral and cationic amino acids as well as for cystine, L as a Na^+-independent transport system for neutral amino acids; IMINO as a Na^+- and Cl^-- dependent transport system for imino acids; and BETA as a Na^+- and Cl^--dependent transport system for β amino acids. The basolateral membrane of the intestinal epithelial cells also expresses several Na^+-dependent and Na^+-independent amino acid transport systems, classified as follows: A as a Na^+-dependent transport system for neutral amino acids and imino acids; ASC as a Na^+-dependent transport system for alanine, serine, and cysteine; and y^+L as a Na^+-independent transport system for cationic amino acids as well as a Na^+-dependent transport system for neutral amino acids. Whereas the amino acid transport systems in the brush-border membrane function in the absorption of amino acids from the lumen into the enterocytes, the amino acid transport systems in the basolateral membrane perform a dual function. They mediate the exit of amino acids from the enterocytes into the blood during the absorptive process and also mediate the entry of amino acids from the blood into the enterocytes under conditions such as starvation or in between meals, in which there is no nutrient supply from the lumen. Several of these transport systems have been cloned and functionally characterized. The structural and functional aspects of amino acid transport systems have been reviewed in great detail (Malandro and Kilberg, 1996; Kanai, 1997; Deves and Boyd, 1998; Palacin et al., 1998). Successful cloning of these transport systems has expanded our current knowledge of the diversity of these transporters. Many of the amino acid transport systems previously classified based on functional properties as independent single systems have since been shown to consist of several subtypes. The tissue distribution of the cloned amino acid transporters varies considerably. Here we focus on the amino acid transporters expressed in the intestine.

B. System ATB^0/ASCT2

The presence of a Na^+-dependent amino acid transport system in the intestinal brush-border membrane with broad specificity for neutral amino acids was identified in the rabbit intestine by Stevens et al. (1982). This system, named B^0, subsequently has been described in the intestine of different animal species, the human intestine, and the human intestinal cell line Caco-2 (Stevens, 1992). An amino acid transport system with characteristics similar to those of B^0 has been cloned from Caco-2 cells and the rabbit intestine (Kekuda et al., 1997). This transporter, designated ATB^0 for amino acid transporter B^0, mediates Na^+-dependent transport of several neutral amino acids including alanine, threonine, glutamine, asparagine, methionine, leucine, isoleucine, tryptophan, and phenylalanine. This transporter is expressed not only in the intestine but also in the human kidney (Kekuda et al., 1997) and the human placenta (Kekuda et al., 1996).

Based on the amino acid sequence, ATB^0 is likely a species homolog of ASCT2 cloned from mouse tissues (Liao and Lane, 1995; Utsunomiya-Tate *et al.*, 1996). Even though ATB^0/ASCT2 transports zwitterionic amino acids in the presence of Na$^+$, controversy exists with regard to whether the transport process is electrogenic (Kekuda *et al.*, 1996; 1997; Utsunomiya-Tate *et al.*, 1996). A more recent study, by Torres-Zamorano *et al.* (1998), has shown that ATB^0/ASCT2 is a Na$^+$-dependent homo- and heteroexchanger of neutral amino acids. However, it is not known whether the exchange of intracellular amino acids is obligatory for the influx of extracellular amino acids via the transporter. If the exchange is not obligatory, the ATB^0/ASCT2-mediated amino acid transport is expected to be electrogenic in the absence of amino acids on the trans-side. Stevens *et al.* (1982) studied the function of the amino acid transport system B^0 in isolated intestinal brush-border membrane vesicles in the absence of amino acids in the intravesicular medium. This would suggest that the transport system is not an obligatory amino acid exchanger. Predictably, the transport process mediated by B^0 in this experimental system is electrogenic (Schell *et al.*, 1983). Even though it is known that cloned ATB^0/ASCT2 is expressed in the intestine, immunolocalization studies to show the presence of this transporter in the brush-border membrane of enterocytes have not been conducted.

Glutamine is one of the excellent substrates for ATB^0/ASCT2. Because the physiological and clinical relevance of this amino acid to the normal function of the intestine is being increasingly recognized (van der Hulst *et al.*, 1996; Buchman, 1996), the role of ATB^0/ASCT2 in providing this amino acid to the enterocytes needs further attention. The gene for ATB^0/ASCT2 maps to human chromosome 19q13.3 (Kekuda *et al.*, 1996). It has been suggested that this gene is a potential candidate for the genetic disorder Hartnup disease, which involves a defect in the intestinal absorption of neutral amino acids (Kekuda *et al.*, 1997).

The function and expression of ATB^0/ASCT2 may be modulated by certain hormones. In cultured human placental trophoblast cells, which constitutively express ATB^0/ASCT2, epidermal growth factor enhances the transport function of ATB^0/ASCT2 and this effect is associated with an increase in the steady-state levels of the transporter mRNA (Torres-Zamorano *et al.*, 1997). In the intestine, growth hormone induces system B^0 transport activity in short bowel syndrome (Iannoli *et al.*, 1997). In Caco-2 cells, the expression of system B^0 is upregulated by protein kinase C (Pan and Stevens, 1995).

The human ATB^0/ASCT2 consists of 541 amino acids with 10 putative transmembrane domains. Amino acid sequence homology predicts this transporter to be a member of a superfamily consisting of the neutral amino acid transporters and the anionic amino acid transporters. There are two potential N-glycosylation sites in the putative extracellular domain between the transmembrane domains 3 and 4, and two potential sites for protein-kinase C-dependent phosphorylation in putative intracellular domains. mRNA transcripts for this transporter are detectable in placenta, lung, skeletal muscle, kidney, intestine, pancreas, and the human intestinal cell lines Caco-2 and HT-29. Rasko *et al.* (1999) have made a very interesting and important observation indicating that $ATB^0/ASCT2$ also serves as a receptor for the RD114/simian type-D viruses. The N-linked glycosylation sites of $ATB^0/ASCT2$ appear to play a critical role in the binding of the virus. The transport function of $ATB^0/ASCT2$ is significantly impaired in cells infected with the RD114/type-D virus. Because this transporter mediates the cellular uptake of a large number of neutral amino acids (essential as well as nonessential), the impairment of the transporter function as a consequence of the virus binding to the transporter and the resultant infection of the cells might account for the pathogenesis of diseases associated with these viruses.

C. System ASCT1

The intestine expresses a Na^+-dependent transport system specific for the short-chain neutral amino acids alanine, serine, and cysteine. This system, called ASC, does not interact with bulkier neutral amino acids such as glutamine or leucine. A Na^+-dependent amino acid transporter has been cloned from human brain that exhibits transport characteristics similar to those of the ASC system (Arriza *et al.*, 1993; Shafqat *et al.*, 1993). The cloned transporter, designated ASCT1, has a widespread expression pattern in peripheral tissues and is likely to be expressed in the intestine. ASCT1 consists of 532 amino acids and is closely related structurally to $ATB^0/ASCT2$ and anionic amino acid transporters. The gene coding for ASCT1 is located on human chromosome 2p13–p15. ASCT1 mediates the Na^+-dependent transport of alanine, serine, and cysteine but not of glutamine or leucine. Thus, the substrate specificity of ASCT1 is different from that of $ATB^0/ASCT2$. However, similar to $ATB^0/ASCT2$, ASCT1 acts as a Na^+-dependent amino acid exchanger (Zerangue and Kavanaugh, 1996). The exchange process is electroneutral when studied in *X. laevis* oocytes, but whether the transporter is an obligatory exchanger is not known. Furthermore, despite the electroneutral Na^+-dependent amino acid exchange, the transport via ASCT1 is associated with Cl^--dependent currents. It appears that these currents are not directly associated with amino acid transport but rather are attributable to Cl^- channel-like activity of the transporter (Zerangue and Kavanaugh, 1996a). In this respect, ASCT1 is similar to $ATB^0/ASCT2$ because the latter also is a Na^+-dependent electroneutral amino acid exchanger and, at the same time, is associated with currents (Kekuda *et al.*, 1996; 1997).

D. System EAAT3

X_{AG}^- refers to a Na^+-dependent amino acid transport process specific for the anionic amino acids aspartate and glutamate. Recent cloning studies, however, have shown that X_{AG}^- consists of at least five subtypes with differential tissue distribution pattern (Kanai, 1997; Palacin *et al.,* 1998). These anionic amino acid transporters have been renamed as excitatory amino acid transporters (EAATs) because of the role of glutamate as an excitatory amino acid in the brain. Among the five subtypes of EAATs (EAAT1–5) known thus far, only EAAT3 is expressed in the intestine. EAAT3 was originally cloned from rabbit intestine (Kanai and Hediger, 1992) and was later shown to be expressed also in neurons in humans and in laboratory animals (Kanai *et al.,* 1994; Arriza *et al.,* 1994; Shashidharan *et al.,* 1994). The human EAAT3 consists of 525 amino acids and its gene maps to chromosome 9p24 (Smith *et al.,* 1994). Based on amino acid sequence homology, ASCT1, ATB^0/ASCT2, and EAAT3 belong to a superfamily of Na$^+$-coupled amino acid transporters.

Detailed kinetic studies have been carried out with EAAT3 in *X. laevis* oocytes expressing the transporter heterologously (Kanai *et al.,* 1994; 1995; Zerangue and Kavanaugh, 1996b). EAAT3 mediates the Na$^+$-coupled transport of glutamate and aspartate. The transporter function is enhanced by K$^+$ on the trans-side. The affinity of EAAT3 for its substrate is high, with K_t values in micromolar range. Interestingly, EAAT3 is stereospecific for glutamate, preferring the L isomer, but does not differentiate between the L-aspartate and D-aspartate. Even though all five EAAT isoforms mediate Na$^+$-coupled transport of glutamate and aspartate, they can be differentiated based on their pharmacological profiles (Palacin *et al.,* 1998). Transmembrane H$^+$ gradient also plays a role in the energetics of the transporter. Stoichiometry of the cotransported amino acid and ions is such that either one or two positive charges are transported along with one glutamate molecule in the direction of glutamate movement. EAAT3 mediates the cotransport of one glutamate, two or three Na$^+$, and one H$^+$, and the countertransport of one K$^+$. In addition to this complex stoichiometry, EAAT3 is associated with ligand (Na$^+$/glutamate)-gated Cl$^-$-channel activity, but the currents resulting from this channel activity are not thermodynamically coupled to glutamate transport. The Cl$^-$-channel property is not unique to EAAT3 because the other members of the gene family, including ASCT1 and ATB^0/ASCT2, also appear to possess this characteristic (Kekuda *et al.,* 1996; 1997; Sonders and Amara, 1996). There is evidence to suggest that EAAT3 functions as a homomultimer *in vivo,* with individual oligomers held together noncovalently (Haugeto *et al.,* 1996).

E. System y$^+$/CAT1

y$^+$ refers to a Na$^+$-independent transport system for the cationic amino acids lysine, arginine, and ornithine. This system also is capable of transporting the neutral amino acids homoserine and cysteine, but only in the presence of Na$^+$.

Cloning studies have indicated that system y^+ consists of at least four subtypes, designated CAT1–4 for cationic amino acid transporters 1–4. In addition, CAT2 also is expressed as the alternative splice variant CAT2a. Among these various subtypes, only CAT1 is expressed in the intestine (MacLeod and Kakuda, 1996).

CAT1 originally was cloned from mouse as the ecotropic murine leukemia virus receptor (Albritton *et al.*, 1989) and later was shown to be a cationic amino acid transporter (Kim *et al.*, 1991; Wang *et al.*, 1991). The human CAT1 consists of 629 amino acids, and its gene maps to chromosome 13q12–q14 (Yoshimoto *et al.*, 1991; Albritton *et al.*, 1992). CAT1 has been expressed heterologously in *X. laevis* oocytes and its transport function studied (Kim *et al.*, 1991; Wang *et al.*, 1991; Kakuda *et al.*, 1993; Kavanaugh *et al.*, 1994). These studies have shown that CAT1 is a Na^+-independent high-affinity transporter for lysine, arginine, and ornithine, with K_t values in the range of 70–200 μM. The transport process is electrogenic, and membrane potential influences the kinetic parameters (substrate affinity and maximal velocity) of the transport process. The transporter exhibits stereoselectivity, preferring L isomers over D isomers. Homoserine and cysteine are transported by CAT1, but only in the presence of Na^+.

The function as a receptor for murine ecotropic leukemia virus is unique to mouse and rat CAT1. Human CAT1 does not exhibit this characteristic. A specific amino acid sequence (Asn-Val-Lys-Tyr-Gly-Glu) in mouse CAT1 has been shown to be the recognition site for the virus (Yoshimoto *et al.*, 1993; Albritton *et al.*, 1993). This sequence also is present in rat CAT1 but not in human CAT1, which demonstrates a molecular basis for the observed species-specific differences in the function of CAT1 as the receptor for the virus. Interaction of mouse CAT1 with the virus causes a decrease in its transport activity (Wang *et al.*, 1992). The virus envelope glycoprotein gp70 is responsible for the binding of the virus to mouse CAT1, and this binding causes a steric hindrance that results in inhibition of the transport function.

F. System $b^{0,+}$

$b^{0,+}$ refers to a Na^+-independent amino acid transport system specific for neutral and cationic amino acids and also for cystine. Transport studies using intestinal brush-border membrane vesicles have provided evidence of the presence of system $b^{0,+}$ in this membrane (Stevens *et al.*, 1982). cDNA of a transporter-associated protein was cloned independently by three groups of investigators; when expressed in *X. laevis* oocytes, the cDNA induces $b^{0,+}$-like amino acid transport activity (Bertran *et al.*, 1992a; Tate *et al.*, 1992; Wells and Hediger, 1992). These characteristics include Na^+-independent transport of cystine, the cationic amino acids arginine and lysine, and neutral amino acids such as leucine and histidine. Interestingly, the cDNA codes for a protein that is not highly hydrophobic and that possesses one or four transmembrane domains uncharacteristic of typical transporters. This membrane topology may not allow formation of a suitable pore in the membrane for the translocation of the polar amino acids. Most transporters possess 10–12 membrane-spanning domains. Therefore it is generally believed that the cloned protein is a subunit of the $b^{0,+}$ transport system (Palacin, 1994). When expressed heterologously in *X. laevis* oocytes, the cloned protein apparently associates with another protein in the oocyte membrane that is expressed endogenously to constitute the complete functional transporter complex with characteristics of the $b^{0,+}$ system. Accordingly, the cloned protein is currently known as rBAT for "related to $b^{0,+}$ amino acid transporter." rBAT is expressed in the intestine and the kidney. The protein localizes to the brush-border membrane of the absorptive cells in these tissues (Furriols *et al.*, 1993; Pickel *et al.*, 1993). Human rBAT consists of 685 amino acids, and the gene coding for the protein maps to chromosome 2p16.3–p21 (Bertran *et al.*, 1993; Lee *et al.*, 1993).

The $b^{0,+}$ transport system induced by rBAT in *X. laevis* oocytes behaves as an obligatory amino acid exchanger involving $b^{0,+}$-specific amino acids in the influx and efflux processes (Busch *et al.*, 1994; Ahmed *et al.*, 1995; Chillaron *et al.*, 1996). Because $b^{0,+}$ recognizes neutral as well as basic amino acids, and because the transport process does not involve inorganic ions, the ionic nature of the amino acids participating in the influx and efflux processes determines the electrical nature of the overall transport process. If $b^{0,+}$ mediates homoexchanger (i.e., neutral amino acid / neutral amino acid, or basic amino acid / basic amino acid), the transport process is electroneutral. If $b^{0,+}$ mediates heteroexchange (i.e., neutral amino acid / basic amino acid), the transport process is electrogenic, with the direction of the current being determined by the ionic nature of the amino acids involved in the influx and efflux processes. Chillaron *et al.* (1996) have suggested that $b^{0,+}$ in the brush-border membrane of the intestinal and renal absorptive cells is capable of active absorption of cationic amino acids and cystine by functioning in concert with other amino acid transport systems in the brush-border and basolateral membranes of these cells.

The predicted membrane topology of rBAT precludes the possibility that this protein by itself constitutes the active $b^{0,+}$ amino acid transporter. It is likely that rBAT forms an active transporter complex by interacting with another protein. Thus, $b^{0,+}$ may be a heterodimeric complex of which rBAT is a subunit. The molecular mass of rBAT detected under reduced conditions in the intestine is 90 kDa (Furriols *et al.*, 1993; Mosckovitz *et al.*, 1993). However, protein crosslinking studies have provided evidence of the presence of rBAT as a heterodimeric complex of a much higher molecular mass of ~125 kDa (Wang and Tate, 1995). Thus, the functional $b^{0,+}$ transporter complex likely is made up of the 90-kDa rBAT (heavy chain) and another protein of ~40 kDa in size (light chain). Successful cloning of this light chain, called $b^{0,+}$AT for $b^{0,+}$ amino acid transporter, has been reported from our laboratory (Rajan et $al.,$ 1999) as well as by Feliubadalo et $al.$ (1999). We have cloned $b^{0,+}AT$ from rabbit small intestine. Interestingly, when the transport of a neutral amino acid is measured, the rabbit $b^{0,+}AT/rBAT$ complex has no detectable transport activity. However, $b^{0,+}AT$ is able to interact with the 4F2 heavy chain, a protein structurally related to rBAT (see below) to induce the transport of neutral amino acids. When the transport of arginine is measured, both $b^{0,+}AT/rBAT$ and $b^{0,+}AT/4F2$ heavy-chain complexes exhibit transport activity. More recent studies from our laboratory have shown that both complexes exhibit $b^{0,+}$-like transport activity, but with significant differences in their affinities for neutral amino acids (Rajan et $al.,$ 2000a). The affinity of the $b^{0,+}AT/4F2$ heavy chain complex for neutral amino acids is much higher than that of the $b^{0,+}AT/rBAT$ complex. These findings provide evidence that there are two different subtypes of the $b^{0,+}$ amino acid transport system: $b^{0,+}1$, which consists of the $rBAT/b^{0,+}AT$ heteromeric complex, and $b^{0,+}2$, which consists of the $4F2hc/b^{0,+}AT$ heteromeric complex. Northern blot analysis has shown that $b^{0,+}AT$-specific mRNA is expressed predominantly in the small intestine and kidney (Rajan et $al.,$ 1999; Feliubadalo et $al.,$ 1999). $b^{0,+}AT$ consists of 487 amino acids and has 12 putative transmembrane domains. The $b^{0,+}AT$ gene maps to human chromosome 19q13.1.

G. System y^+L

y^+L is an amino acid transporter that mediates the Na^+-independent transport of cationic amino acids and the Na^+-dependent transport of neutral amino acids. This transport system, originally observed in erythrocytes (Deves et $al.,$ 1992), is expressed in the intestine.

When rBAT was cloned and identified as an inducer of $b^{0,+}$ amino acid transport activity, it was recognized that rBAT displayed a significant structural similarity to the large subunit of human lymphocyte activation antigen 4F2, also called 4F2hc (for heavy chain of the 4F2 antigen) or CD98 (Lumadue et $al.,$ 1987; Quackenbush et $al.,$ 1987; Teixeira et $al.,$ 1987). This led to investigations of the possible involvement of 4F2hc in amino acid transport. These investigations have shown that 4F2hc, when expressed heterologously in $X.$ $laevis$ oocytes, induces an amino acid transport activity with characteristics similar to those of y^+L (Bertran et $al.,$ 1992b; Wells et $al.,$ 1992; Fei et $al.,$ 1995). Interestingly, similar to rBAT, 4F2hc is not highly hydrophobic and possesses a single putative transmembrane domain. This structural property makes it unlikely that this protein would constitute the functional y^+L transport system by itself. It is possible that 4F2hc also forms a heterodimeric complex with another protein in the membrane to produce the functional y^+L transporter. Recently, the other subunit of the y^+L transporter has been successfully cloned (Torrents et $al.,$ 1998;

Pfeiffer *et al.*, 1999). This protein, called y^+LAT1 for y^+L amino acid transporter-1, associates with 4F2hc to form a heterodimeric complex via disulfide linkage with functional y^+L transport activity. When coexpressed in *X. laevis* oocytes, 4F2hc and y^+LAT1 together induce the Na^+-independent transport of cationic amino acids and the Na^+-dependent transport of neutral amino acids. The two subunits are linked by a disulfide bond in the heterodimeric complex, but the disulfide linkage may not be obligatory for the transport function of the complex. Unlike 4F2hc, which contains a single putative transmembrane domain, y^+LAT1 contains 12 putative transmembrane domains.

4F2hc is expressed ubiquitously in animal tissues, including the intestine and the kidney, and its localization in the absorptive cells of these tissues is limited to the basolateral membrane (Quackenbush *et al.*, 1986). Human 4F2hc consists of 529 amino acids, and the gene encoding this protein is located on chromosome 11q12–q13 (Gottesdiener *et al.*, 1988). y^+LAT1 also is expressed in several tissues including the intestine and the kidney. Human y^+LAT1 consists of 511 amino acids, and the gene coding for this protein maps to chromosome 14q11.2 (Torrents *et al.*, 1998; Pfeiffer *et al.*, 1999).

y^+L also behaves as an amino acid exchanger (Chillaron *et al.*, 1996). However, unlike $b^{0,+}$, the amino acid exchange via y^+L is asymmetric. The preferred mode of exchange is the Na^+-dependent influx of neutral amino acids coupled to the Na^+-independent efflux of basic amino acids. Thus, the y^+L transporter located in the basolateral membrane of the intestinal and renal epithelial cells is likely to mediate the efflux of basic amino acids from the cells into the blood, in exchange for the Na^+-dependent entry of neutral amino acids from the blood into the cells.

H. System L

System L is a Na^+-independent amino acid transporter specific for neutral amino acids. Interaction with the amino acid 2-amino-2-carboxybicyclo-[2,2,1]heptane (BCH) is a unique characteristic of this transport system. Attempts to clone this transporter have led to the discovery that 4F2hc is at least a component of system L (Broer *et al.*, 1995; 1997). These findings came as a surprise because earlier studies showed that 4F2hc also is a component of the amino acid transport system y^+L. This suggested that 4F2hc may be a common subunit that constitutes either system L or system y^+L, depending on the nature of the second subunit that heterodimerizes with 4F2hc. Kanai *et al.* (1998) have cloned a protein from rat glioma cells that induces system L activity in *X. laevis* oocytes when coexpressed with 4F2hc. This protein, designated LAT1 for L amino acid transporter-1, is not capable of amino acid transport on its own. However, as a heterodimer with 4F2hc, LAT1 is capable of amino acid transport with the known characteristics of system L, including interaction with the system L amino acid prototype BCH. Human LAT1 has been cloned and characterized

(Mastroberardino *et al.*, 1998; Prasad *et al.*, 1999). Human LAT1 consists of 507 amino acids with 12 putative transmembrane domains (Prasad *et al.*, 1999). When coexpressed with 4F2hc in mammalian cells, human LAT1 induces Na^+-independent amino acid transport activity specific for neutral amino acids including leucine and BCH. However, mRNA transcripts for LAT1 are not detectable in the intestine (Kanai *et al.*, 1998; Prasad *et al.*, 1999). This is interesting because functional studies have clearly indicated the presence of system L in the intestine (Stevens *et al.*, 1982). 4F2hc is expressed in the intestine, however. This apparent discrepancy was resolved with the successful cloning of LAT2, which is expressed in the intestine (Segawa *et al.*, 1999; Pineda *et al.*, 1999; Rajan *et al.*, 2000b). LAT2 consists of 535 amino acids with a predicted molecular mass of 58 kDa, and has 12 putative transmembrane domains. LAT2 interacts with 4F2hc to induce system L-like amino acid transport activity. However, the 4F2hc/LAT2-mediated transport process differs from the 4F2hc/LAT1-mediated transport process with respect to substrate specificity, substrate affinity, tissue distribution, interaction with D amino acids, and pH dependence. The 4F2hc/LAT2-associated transport process has broad specificity toward neutral amino acids, with K_t values in the range of 100–1000 μM, does not interact with D amino acids to any significant extent, and is stimulated by acidic pH. In contrast, the 4F2hc/LAT1-associated transport process has narrower specificity toward neutral amino acids but with comparatively higher affinity (K_t values in the range of 10–20 μM). It interacts with some D amino acids with high affinity and is not influenced by pH. LAT2 is expressed primarily in the small intestine and kidney, whereas LAT1 exhibits much broader tissue distribution. The gene for LAT2 maps to human chromosome 14q11.2–13 (Pineda *et al.*, 1999).

I. System ATB$^{0,+}$

The amino acid transport system $B^{0,+}$ mediates the Na^+-dependent transport of both neutral and cationic amino acids (Van Winkle *et al.*, 1985). This system is expressed in the intestine (Chen *et al.*, 1994; Munck and Munck, 1995). Successful cloning of the $B^{0,+}$ transport system ATB$^{0,+}$ has been reported by Sloan and Mager (1999). Human ATB$^{0,+}$ consists of 642 amino acids and belongs to the family of Na^+/Cl^--dependent neurotransmitter transporters. The gene for ATB$^{0,+}$ is located on human chromosome Xq22.1–24.

J. System BETA/TAUT

The amino acid transport system BETA mediates the Na^+- and Cl^--coupled transport of β amino acids such as β-alanine, taurine, and hypotaurine. The most physiologically relevant substrate for this transporter is taurine. Although this amino acid is not found in proteins, it is the most abundant free amino acid in many tissues. The nutritional needs for taurine are met primarily by intestinal absorption from dietary sources. The BETA system is expressed in the brush-border membrane of the intestinal absorptive cells (Miyamoto *et al.*, 1989; 1990). The human colon carcinoma cells also express this transporter (Tiruppathi *et al.*, 1992; Brandsch *et al.*, 1993). The transporter in these cells is subject to regulation by protein kinase C (Brandsch *et al.*, 1993) and protein kinase A (Brandsch *et al.*, 1995).

The BETA system has been cloned and functionally characterized (Ramamoorthy *et al.*, 1994). Because taurine is the physiologically relevant substrate, this system also is called TAUT (for taurine transporter). Even though TAUT originally was cloned from human placenta, the same transporter is expressed in the intestine. Human TAUT consists of 620 amino acids with 12 putative transmembrane domains, and the gene coding for the transporter maps to chromosome 3p24–p26.

K. Other Amino Acid Transport Systems

The intestine expresses a Na^+-coupled transport system for proline called IMINO, a Na^+-coupled transport system preferring phenylalanine called PHE, and the ubiquitously expressed system A, which is a Na^+-dependent transport system for neutral amino acids including glutamine. Recent molecular cloning studies have shown that system A consists of at least two subtypes; ATA1 (amino acid transporter A1) that is expressed predominantly in the brain and ATA2 (amino acid transporter A2) that is expressed ubiquitously including small intestine (Varoqui *et al.*, 2000; Sugawara *et al.*, 2000). Rat ATA2 consists of 504 amino acids and mediates Na^+-dependent electrogenic transport of neutral short-chain amino acids. Human ATA2, cloned recently from our laboratory, consists of 506 amino acids. The gene for ATA2 is located on human chromosome 12q (unpublished data). The transport systems IMINO and PHE have been characterized solely by functional studies. There is no information available on the molecular and structural aspects of these transporters.

Table I describes the structural and functional characteristics of the amino acid transporters that have been cloned thus far and shown to be expressed in the intestine.

IV. GENETIC DEFECTS IN INTESTINAL PEPTIDE AND AMINO ACID TRANSPORT

There are no reports in the literature on genetic disorders of intestinal peptide transport. However, a number of genetic disorders associated with intestinal amino acid transporters have been reported. Notable among these disorders are Hartnup

TABLE I

Intestinal Amino Acid Transporters

Amino acid transporter	Substrates	Ion dependence	Subunit composition	Human gene locus	Genetic defect
ATA2	Short-chain neutral amino acids	Na^+		12q	
ASCT1	Ala, Ser, Cys	Na^+		2p13–15	
ATB[0]/ASCT2	Ala, Ser, Cys, Thr, Gln, Asp	Na^+		19q13.3	Hartnup disorder (?)
EAAT3	Glu, Asp	Na^+, K^+, H^+		9p24	Dicarboxylic aminoaciduria (?)
y[+]/CAT1	Arg, Lys, Orn	—		13q12–14	
ATB[0,+]	Cationic and neutral amino acids	Na^+, Cl^-		Xq22.1–24	
BETA/TAUT	Taurine, β-alanine	Na^+, Cl^-		3p24–26	
b[0,+]1	Cationic and neutral amino acids	—	rBAT	2p16.3–21	Type I cystinuria
			b[0,+]AT	19q13.1	Non-type-I cystinuria
b[0,+]2	Cationic and neutral amino acids	—	4F2hc	11q12–13	Non-type-I cystinuria
			b[0,+]AT	19q13.1	
y[+]L1	Cationic and neutral amino acids	Na^+ (only for neutral amino acids)	4F2hc	11q12–13	Lysinuric protein intolerance
			y[+]LAT1	14q11.2	
L2	Neutral amino acids	H^+	4F2hc	11q12–13	
			LAT2	14q11.2–13	

disorder, cystinuria, and lysinuric protein intolerance (Ganapathy and Leibach, 1998). Interestingly, many of these amino acid transport defects affect amino acid absorption not only in the intestine but also in the kidney. Obviously, the same gene products related to these transport systems are expressed in these two absorptive tissues. Consequently, these disorders result in urinary hyperexcretion of the substrates of the affected transport systems. It is intriguing to note that many of these amino acid transport disorders generally are not associated with protein malabsorption. This suggests that intestinal absorption of free amino acids may be quantitatively less important than absorption of peptides in the assimilation of dietary proteins.

Hartnup disorder is an autosomal-recessive disease associated with an impairment of neutral amino acid absorption in the intestine and the kidney (Levy, 1995). Even though urinary excretion of the affected amino acids is elevated in individuals with the disorder, the clinical consequences of this disorder generally are mild. Most of the clinical symptoms arise from the deficiency of niacin. Normally, niacin is synthesized endogenously to a significant extent from the neutral amino acid tryptophan. Defective absorption of this amino acid in the intestine and increased urinary elimination in the kidney are likely the contributing factors for niacin deficiency in these patients. The amino acid transporter system B^0 has been suggested as a candidate for the transport defect seen in Hartnup disorder, but no studies have been reported on the possible involvement of the $ATB^0/ASCT2$ gene in this disease.

Cystinuria is a disorder associated with defective intestinal and renal absorption of basic amino acids and cystine (Segal and Thier, 1995). The major clinical symptom of cystinuria is nephropathy resulting from kidney stones. Cystine is only sparingly soluble in water, and therefore defective renal reabsorption of this amino acid increases its concentration in the tubular lumen precipitating crystallization and stone formation. Therefore, therapeutic strategies are targeted toward increasing urine volume, alkalinizing urine to facilitate the solubility of cystine, and chemically modifying cystine to enhance its solubility. The cystinuria gene maps to chromosome 2 (Pras *et al.*, 1994), a location coinciding with the locus of the *rBAT* gene. Because *rBAT* is associated with $b^{0,+}$ amino acid transport activity with ability to transport cystine, these findings and the matching chromosomal loci of the cystinuria gene and the *rBAT* gene prompted investigations of the possible involvement of *rBAT* in the disorder (Palacin *et al.*, 1998). These investigations have demonstrated that mutations in the *rBAT* gene are responsible for several cases of cystinuria. There are three types of cystinuria based on the clinical phenotype. Type I and type II are associated with a defective transport system for cystine in the intestine as well as the kidney, whereas type III is associated with a defective transport system for cystine in the kidney but not in the intestine. Mutations in the *rBAT* gene are linked to type-I cystinuria. Recent studies have shown that mutations in the $b^{0,+}$AT gene are responsible for non–type-I cystinuria (Feliubadalo *et al.*, 1999).

Lysinuric protein intolerance is an autosomal-recessive disorder caused by defective absorption of basic amino acids (arginine, lysine, and ornithine) in the intestine and the kidney (Simell, 1995). Patients with this disease have low plasma levels of these three amino acids. This causes impairment in protein synthesis and urea cycle, which provide the molecular basis of the clinical symptoms, namely growth retardation, neurological complications, hyperammonemia, and protein intolerance. The absorption of cystine in the intestine and the kidney is not affected in lysinuric protein intolerance, which distinguishes this disorder from cystinuria. The transport defect in lysinuric protein intolerance is restricted to the basolateral membrane of the intestinal and renal absorptive cells (Rajante *et al.*, 1980). The gene for the disease maps to chromosome 14q (Lauteala *et al.*, 1997). With the recent findings that the gene coding for y^+LAT1, a subunit of the y^+L amino acid transport system, is located on chromosome 14q and that the activity of this transport system is restricted to the basolateral membrane of the intestine and the kidney, it appears that the y^+LAT1 gene is a prime candidate for the disorder (Torrents *et al.*, 1998). Although 4F2hc is a component of the y^+L system, it is ubiquitously expressed, and forms a component of the amino acid transport system L. Therefore, it is unlikely to be the site of defect in lysinuric protein intolerance. Furthermore, the gene locus for 4F2hc (11q12–q13) is different from that for lysinuric protein intolerance. The transport characteristics of system y^+L coincide with the functional defects observed in the disease. This transport system functions in the Na^+-independent efflux of basic amino acids from the intestinal and renal absorptive cells across the basolateral membrane coupled to the Na^+-dependent influx of neutral amino acids. A genetic defect associated with an impairment of this transport system is expected to interfere with the intestinal and renal absorption of basic amino acids. Consistent with these reasonings is the fact that mutations in the y^+LAT1 gene have been reported for patients with lysinuric protein intolerance (Torrents *et al.*, 1999; Borsani *et al.*, 1999).

Dicarboxylic aminoaciduria is a genetic defect with impaired absorption of acidic amino acids (glutamate and aspartate) in the intestine and the kidney. Neither the gene locus nor the molecular defect linked to the disease is known. Based on the expression pattern and functional characteristics, the acidic amino acid transporter EAAT3 is a prime candidate for the disease. This is supported by the observations that EAAT3-knockout mice develop dicarboxylic aminoaciduria (Peghini *et al.*, 1997).

V. CONCLUDING REMARKS

The multitude of transporters for peptides and amino acids expressed in the intestinal absorptive cells play a crucial role in the maintenance of protein nutrition. Even though peptide and amino acid transport mechanisms exist in the brush-border membrane and the basolateral membrane of these cells, the individual transporters expressed in these two membranes are different in several ways. This results in quantitative as well as qualitative differences in the transport of peptides and amino acids across the brush-border membrane and the basolateral membrane. These differences pertain to substrate specificity and driving forces. Differences in the regulation of these transporters also may exist in the two membranes. This is obligatory for the function of the intestinal absorptive cell in the vectorial transfer of protein digestion products from the lumen into the blood. Successful cloning of several of these transporters in recent years has provided insight into the structural, functional, and regulatory aspects of individual transport systems. Physiologically, the transport of peptides is much more important than the transport of amino acids across the brush-border membrane, whereas the transport of amino acids is much more important than the transport of peptides across the basolateral membranes. Available information on various genetic disorders of amino acid transport attest to this difference in the relative importance of the two processes. Defects in amino acid transport systems that are expressed primarily in the brush-border membrane do not generally result in protein malnutrition (e.g., Hartnup disorder and cystinuria). In contrast, defects in amino acid transport systems that are expressed primarily in the basolateral membrane do result in impairment of protein nutrition (e.g., lysinuric protein intolerance). It is likely that the opposite is true in the case of peptide transport systems. Defects in PEPT1, which is responsible for peptide transport across the brush-border membrane, are expected to have far more detrimental consequences on protein nutrition than defects in peptide transport systems, which function in the transport of peptides across the basolateral membrane. Whether this is true remains to be seen because there are no known genetic defects in peptide transport systems.

Acknowledgments

This work was supported by National Institutes of Health grants DK 28389 (F.H.L.), GM 54122 (M.E.G.), and HD 24451 (V.G.). The authors thank Ida O. Walker for excellent secretarial assistance.

References

Ahmed, A., Peter, G. J., Taylor, P. M., Harper, A. A., and Rennie, M. J. (1995). Sodium-independent currents of opposite polarity evoked by neutral and cationic amino acids in neutral and basic amino acid transporter cRNA-injected oocytes. *J. Biol. Chem.* **270,** 8482–8486.

Albritton, L. M., Bowcock, A. M., Eddy, R. L., Morton, C. C., Tseng, L., Farrer, L. A., Cavalli-Sforza, L. L., Shows, T. B., and Cunningham, J. M. (1992). The human cationic amino acid transporter (ATRC1): Physical and genetic mapping to 13q12-q14. *Genomics* **12,** 430–434.

Albritton, L. M., Kim, J. W., Tseng, L., and Cunningham, J. M. (1993). Envelope binding domain in the cationic amino acid transporter determines the host range of ecotropic murine retroviruses. *J. Virol.* **67,** 2091–2096.

Albritton, L. M., Tseng, L., Scadden, D., and Cunningham, J. M. (1989). A putative murine ecotropic retrovirus receptor gene encodes a multiple membrane-spanning protein and confers susceptibility to virus infection. *Cell* **57,** 659–666.

Amasheh, S., Wenzel, U. Boll, M., Dorn, D., Weber, W., Clauss, W., and Daniel, H. (1997). Transport of charged dipeptides by the intestinal H^+/peptide symporter PepT1 expressed in *Xenopus laevis* oocytes. *J. Membr. Biol.* **155**, 247–256.

Amidon, G. L., and Lee, H. J. (1994). Absorption of peptides and peptidomimetic drugs. *Ann. Rev. Pharmacol. Toxicol.* **34**, 321–341.

Arriza, J. L., Fairman, W. A., Wadiche, J. I., Murdoch, G. H., Kavanaugh, M. P., and Amara, S. G. (1994). Functional comparisons of three glutamate transporter subtypes cloned from human motor cortex. *J. Neurosci.* **14**, 5559–5569.

Arriza, J. L., Kavanaugh, M. P., Fairman, W. A., Wu, Y. N., Murdoch, G. H., North, R. A., and Amara, S. G. (1993). Cloning and expression of a human neutral amino acid transporter with structural similarity to the glutamate transporter gene family. *J. Biol. Chem.* **268**, 15329–15332.

Asatoor, A. M., Chadha, A., Milne, M. D., and Prosser, D. I. (1973). Intestinal absorption of stereoisomers of dipeptides in the rat. *Clin. Sci. Mol. Med.* **45**, 199–212.

Berlioz, F., Julien, S., Tsocas, A., Chariot, J., Carbon, C., Farinotti, R., and Roze, C. (1999). Neural modulation of cephalexin intestinal absorption through the di- and tripeptide brush border transporter of rat jejunum *in vivo*. *J. Pharmacol. Exp. Ther.* **288**, 1037–1044.

Bertran, J., Magagnin, S., Werner, A., Markovich, D., Biber, J., Testar, X., Zorzano, A., Kuhn, L. C., Palacin, M., and Murer, H. (1992b). Stimulation of y^+-like amino acid transport by the heavy chain of human 4F2 surface antigen in *Xenopus laevis* oocytes. *Proc. Natl. Acad. Sci. USA* **89**, 5606–5610.

Bertran, J., Werner, A., Chillaron, J., Nunes, V., Biber, J., Testar, X., Zorzano, A., Estivill, X., Murer, H., and Palacin, M. (1993). Expression cloning of a human renal cDNA that induces high affinity transport of L-cystine shared with dibasic amino acids in *Xenopus* oocytes. *J. Biol. Chem.* **268**, 14842–14849.

Bertran, J., Werner, A., Moore, M. L., Stange, G., Markovich, D., Biber, J., Testar, X., Zorzano, A., Palacin, M., and Murer, H. (1992a). Expression cloning of a cDNA from rabbit kidney cortex that induces a single transport system for cystine and dibasic and neutral amino acids. *Proc. Natl. Acad. Sci. USA* **89**, 5601–5605.

Boll, M., Herget, M., Wagener, M., Weber, W. M., Markovich, D., Biber, J., Clauss, W., Murer, H., and Daniel, H. (1996). Expression cloning and functional characterization of the kidney cortex high-affinity proton-coupled peptide transporter. *Proc. Natl. Acad. Sci. USA* **93**, 284–289.

Boll, M., Markovich, D., Weber, W. M., Korte, H., Daniel, H., and Murer, H. (1994). Expression cloning of a cDNA from rabbit small intestine related to proton coupled transport of peptides, β-lactam antibiotics, and ACE inhibitors. *Pflugers Arch.* **429**, 146–149.

Borner, V., Fei, Y. J., Hartrodt, B., Ganapathy, V., Leibach, F. H., Neubert, K., and Brandsch, M. (1998). Transport of amino acid aryl amides by the intestinal H^+/peptide cotransport system, PEPT1. *Eur. J. Biochem.* **255**, 698–702.

Borsani, G., Bassi, M. T., Sperandeo, M. P., Grandi, A. D., Buoninconti, A., Riboni, M., Manzoni, M., Incerti, B., Pepe, A., Andria, G., Ballabio, A., and Sebastio, G. (1999). SLC7A7, encoding a putative permease-related protein, is mutated in patients with lysinuric protein intolerance. *Nature Genet.* **21**, 297–301.

Brandsch, M., Brandsch, C., Ganapathy, M. E., Chew, C. S., Ganapathy, V., and Leibach, F. H. (1997). Influence of proton and essential histidyl residues on the transport kinetics of the H^+/peptide cotransport systems in intestine (PEPT1) and kidney (PEPT2). *Biochim. Biophys. Acta* **1324**, 251–262.

Brandsch, M., Miyamoto, Y., Ganapathy, V., and Leibach, F. H. (1993). Regulation of taurine transport in human colon carcinoma cell lines (HT-29 and Caco-2) by protein kinase C. *Am. J. Physiol.* **264**, G939–G946.

Brandsch, M., Miyamoto, Y., Ganapathy, V., and Leibach, F. H. (1994). Expression and protein kinase C-dependent regulation of peptide/H^+ cotransport system in the Caco-2 human colon carcinoma cell line. *Biochem. J.* **299**, 253–260.

Brandsch, M., Ramamoorthy, S., Marczin, N., Catravas, J. D., Leibach, J. W., Ganapathy, V., and Leibach, F. H. (1995). Regulation of taurine transport by *Escherichia coli* heat-stable entero-toxin and guanylin in human intestinal cell lines. *J. Clin. Invest.* **96**, 361–369.

Brandsch, M., Thunecke, F., Kullertz, G., Schutkowski, M., Fischer, G., and Neubert, K. (1998). Evidence for the absolute conformational specificity of the intestinal H^+/peptide symporter PEPT1. *J. Biol. Chem.* **273**, 3861–3864.

Broer, S., Broer, A., and Hamprecht, B. (1995). The 4F2 surface antigen is necessary for system L-like transport activity in C6-BU-1 glioma cells: Evidence by expression studies in *Xenopus laevis* oocytes. *Biochem. J.* **312**, 863–870.

Broer, S., Broer, A., and Hamprecht, B. (1997). Expression of the surface antigen 4F2hc affects system L-like neutral amino acid transport activity in mammalian cells. *Biochem. J.* **324**, 535–541.

Buchman, A. L. (1996). Glutamine: Is it a conditionally required nutrient of the human gastro-intestinal system? *J. Am. Coll. Nutr.* **15**, 199–205.

Busch, A., Herzer, T., Waldegger, S., Schmidt, F., Palacin, M., Biber, J., Markovich, D., Murer, H., and Lang, F. (1994). Opposite directed currents induced by the transport of dibasic and neutral amino acids *Xenopus* oocytes expressing the protein rBAT. *J. Biol. Chem.* **269**, 25581–25586.

Chen, J., Zhu, Y., and Hu, M. (1994). Mechanisms and kinetics of uptake and efflux of L-methionine in an intestinal epithelial model (Caco-2). *J. Nutr.* **124**, 1907–1916.

Chillaron, J., Estevez, R., Mora, C., Wagner, C. A., Suessbrich, H., Lang, F., Gelpi, J. L., Testar, X., Busch, A., Zorzano, A., and Palacin, M. (1996). Obligatory amino acid exchange via systems $b^{0,+}$-like and y^+L-like. A tertiary active transport mechanism for renal reabsorption of cystine and dibasic amino acids. *J. Biol. Chem.* **271**, 17761–17770.

Daniel, H. (1996). Function and molecular structure of brush border membrane peptide/H^+ sym-porters. *J. Membr. Biol.* **154**, 197–203.

Daniel, H., and Herget, M., (1997). Cellular and molecular mechanisms of renal peptide transport. *Am. J. Physiol.* **273**, F1–F8.

Das, M., and Radhakrishnan, A. N. (1975). Studies on a wide-spectrum intestinal dipeptide uptake system in the monkey and in the human. *Biochem. J.* **146**, 133–139.

Deves, R., and Boyd, C. A. R. (1998). Transporters for cationic amino acids in animal cells: Discovery, structure and function. *Physiol. Rev.* **78**, 487–545.

Deves, R., Chavez, P., and Boyd, C. A. R. (1992). Identification of a new transport system (y^+L) in human erythrocytes that recognizes lysine and leucine with high affinity. *J. Physiol.* **454**, 491–501.

De Vrueh, R. L. A., Smith, P. L., and Lee, C. P. (1998). Transport of L-valine-acyclovir via the oligopeptide transporter in the human intestinal cell line, Caco-2. *J. Pharmacol. Exp. Ther.* **286**, 1166–1170.

Donowitz, M., Khurana, S., Tse, C. M., and Yun, C. H. (1998). G protein-coupled receptors in gas-trointestinal physiology. III. Asymmetry in plasma membrane signal transduction: Lessons from brush-border Na^+/H^+ exchangers. *Am. J. Physiol.* **274**, G971–G977.

Doring, F., Walter, J., Will, J., Focking, M., Boll, M., Amasheh, S., Clauss, W., and Daniel, H. (1998a). Delta-aminolevulinic acid transport by intestinal and renal peptide transporters and its physiological and clinical implications. *J. Clin. Invest.* **101**, 2761–2767.

Doring, F., Will, J., Amasheh, S., Clauss, W., Ahlbrecht, H., and Daniel, H. (1998b). Minimal mol-ecular determinants of substrates for recognition by the intestinal peptide transporter. *J. Biol. Chem.* **273**, 23211–23218.

Dyer, J., Beechey, R. B., Gorvel, J. P., Smith, R. T., Wootton, R., and Shirazi-Beechey, S. P. (1990). Glycyl-L-proline transport in rabbit enterocyte basolateral membrane vesicles. *Biochem. J.* **269**, 565–571.

Erickson, R. H., Gum, J. R., Jr., Lindstrom, M. M., McKean, D., and Kim, Y. S. (1995). Regional expression and dietary regulation of rat small intestinal peptide and amino acid transporter mRNAs. *Biochem. Biophys. Res. Commun.* **216**, 249–257.

Fei, Y. J., Ganapathy, V., and Leibach, F. H. (1998). Molecular and structural features of the proton-coupled oligopeptide transporter superfamily. *Prog. Nucleic Acid Res. Mol. Biol.* **58,** 239–261.

Fei, Y. J., Kanai, Y., Nussberger, S., Ganapathy, V., Leibach, F. H., Romero, M. F., Singh, S. K., Boron, W. F., and Hediger, M. A. (1994). Expression cloning of a mammalian proton-coupled oligopeptide transporter. *Nature* **368,** 563–566.

Fei, Y. J., Liu, W., Prasad, P. D., Kekuda, R., Oblak, T. G., Ganapathy, V., and Leibach, F. H. (1997). Identification of the histidyl residue obligatory for the catalytic activity of the human H^+/peptide cotransporters PEPT1 and PEPT2. *Biochemistry* **36,** 452–460.

Fei, Y. J., Prasad, P. D., Leibach, F. H., and Ganapathy, V. (1995). The amino acid transport system y^+L induced in *Xenopus laevis* oocytes by human choriocarcinoma cell (JAR) mRNA is functionally related to the heavy chain of the 4F2 cell surface antigen. *Biochemistry* **34,** 8744–8751.

Feliubadalo, L., Font, M., Purroy, J., Rousaud, F., Estivill, X., Nunes, V., *et al.* (1999). Non-type I cystinuria caused by mutations in SLC7A9, encoding a subunit ($b^{0,+}AT$) of rBAT. *Nature Genet.* **23,** 52–57.

Ferraris, R. P., Diamond, J. M., and Kwan, W. W. (1988). Dietary regulation of intestinal transport of the dipeptide carnosine. *Am. J. Physiol.* **255,** G143–G150.

Freeman, T. C., Bentsen, B. S., Thwaites, D. T., and Simmons, N. L. (1995). H^+/di-tripeptide transporter (PepT1) expression in the rabbit intestine. *Pflugers Arch.* **430,** 394–400.

Fujita, T., Majikawa, Y., Umehisa, S., Okada, N., Yamamoto, A., Ganapathy, V., and Leibach, F. H. (1999). σ Receptor ligand-induced up-regulation of the H^+/peptide transporter PEPT1 in the human intestinal cell line Caco-2. *Biochem. Biophys. Res. Commun.* **261,** 242–246.

Furriols, M., Chillaron, J., Mora, C., Castello, A., Bertran, J., Camps, M., Testar, X., Vilaro, S., Zorzano, A., and Palacin, M. (1993). rBAT, related to L-cystine transport, is localized to the microvilli of proximal straight tubules, and its expression is regulated in kidney by development. *J. Biol. Chem.* **268,** 27060–27068.

Ganapathy, M. E., Brandsch, M., Prasad, P. D., Ganapathy, V., and Leibach, F. H. (1995). Differential recognition of β-lactam antibiotics by intestinal and renal peptide transporters, PEPT1 and PEPT2. *J. Biol. Chem.* **270,** 25672–25677.

Ganapathy, M. E., Huang, W., Wang, H., Ganapathy, V., and Leibach, F. H. (1998). Valacyclovir: A substrate for the intestinal and renal peptide transporters PEPT1 and PEPT2. *Biochem. Biophys. Res. Commun.* **246,** 470–475.

Ganapathy, M. E., Prasad, P. D., Huang, W., Seth, P., Leibach, F. H., and Ganapathy, V. (1999). Molecular and ligand-binding characterization of the σ-receptor in the Jurkat human T lymphocyte cell line. *J. Pharmacol. Exp. Ther.* **289,** 251–260.

Ganapathy, V., Brandsch, M., and Leibach, F. H. (1994). Intestinal transport of amino acids and peptides. *In* "Physiology of the Gastrointestinal Tract" (L.R. Johnson, ed.), 3rd ed., pp. 1773–1794. Raven Press, New York.

Ganapathy, V., Burckhardt, G., and Leibach, F. H. (1984). Characteristics of glycylsarcosine transport in rabbit intestinal brush border membrane vesicles. *J. Biol. Chem.* **259,** 8954–8959.

Ganapathy, V., Burckhardt, G., and Leibach, F. H. (1985). Peptide transport in rabbit intestinal brush border membrane vesicles. Studies with a potential-sensitive dye. *Biochim. Biophys. Acta* **816,** 234–240.

Ganapathy, V., and Leibach, F. H. (1983). Role of pH gradient and membrane potential in dipeptide transport in intestinal and renal brush border membrane vesicles from the rabbit. Studies with L-carnosine and glycyl-L-proline. *J. Biol. Chem.* **258,** 14189–14192.

Ganapathy, V., and Leibach, F. H. (1985). Is intestinal peptide transport energized by a proton gradient? *Am. J. Physiol.* **249,** G153–G160.

Ganapathy, V., and Leibach, F. H. (1986). Carrier-mediated reabsorption of small peptides in renal proximal tubule. *Am. J. Physiol.* **251,** F945–F953.

Ganapathy, V., and Leibach, F. H. (1991). Proton-coupled solute transport in the animal cell plasma membrane. *Curr. Opin. Cell Biol.* **3,** 695–701.

Ganapathy, V., and Leibach, F. H. (1996). Peptide transporters. *Curr. Opin. Nephrol. Hypertens.* **5,** 395–400.

Ganapathy, V., and Leibach, F. H. (1998). Protein digestion and assimilation. *In* "Textbook of Gastroenterology" (T. Yamada, ed.), 3rd ed., pp. 456–467. Lippincott Williams & Wilkins, Philadelphia.

Ganapathy, V., Mendicino, J. F., and Leibach, F. H. (1981). Transport of glycyl-L-proline into intestinal and renal brush border vesicles from rabbit. *J. Biol. Chem.* **256,** 118–124.

Ganapathy, V., Miyamoto, Y., and Leibach, F. H. (1987). Driving force for peptide transport in mammalian intestine and kidney. *Contrib. Infusion Ther. Clin. Nutr.* **17,** 54–68.

Gottesdiener, K. M., Karpinski, B. A., Lindstein, T., Strominger, J. L., Jones, N. H., Thompson, C. B., and Leiden, J. M. (1988). Isolation and structural characterization of the human 4F2 heavy chain gene, an inducible gene involved in T-lymphocyte activation. *Mol. Cell. Biol.* **8,** 3809–3819.

Guandalini, S., and Rubino, A. (1992). Development of dipeptide transport in the intestinal mucosa of rabbits. *Pediatr. Res.* **16,** 99–103.

Harcouet, L., Lebrec, D., Roze, C., Carbon, C., and Farinotti, R. (1997). Increased intestinal absorption of cefixime by nifedipine in the rat intestinal perfusion model: Evidence for a neural regulation. *J. Pharmacol. Exp. Ther.* **281,** 738–745.

Haugeto, O., Ullensvang, K., Levy, L. M., Chaudhry, F. A., Honore, T., Nielsen, M., Lehre, K. P., and Danbolt, N. C. (1996). Brain glutamate transporter proteins form homomultimers. *J. Biol. Chem.* **271,** 27715–27722.

Hellier, M. D., Holdsworth, C. D., Perrett, D., and Thirumalai, C. (1972). Intestinal dipeptide transport in normal and cystinuric subjects. *Clin. Sci.* **43,** 659–668.

Himukai, M., Konno, T., and Hoshi, T. (1980). Age-dependent change in the intestinal absorption of dipeptides and their constituent amino acids in the guinea pig. *Pediatr. Res.* **14,** 1272–1275.

Hoshi, T. (1985). Proton-coupled transport of organic solutes in animal cell membranes and its relation to Na^+ transport. *Jpn. J. Physiol.* **35,** 179–191.

Hu, M., and Amidon, G. L. (1988). Passive and carrier-mediated intestinal absorption components of captopril. *J. Pharm. Sci.* **77,** 1007–1011.

Hu, M., Subramanian, P., Mosberg, H. I., and Amidon, G. L. (1989). Use of the peptide carrier system to improve the intestinal absorption of L-alpha-methyldopa: Carrier kinetics, intestinal permeabilities, and *in vitro* hydrolysis of dipeptidyl derivatives of L-alpha-methyldopa. *Pharm. Res.* **6,** 66–70.

Iannoli, P., Miller, J. H., Ryan, C. K., Gu, L. H., Ziegler, T. R., and Sax, H. C. (1997). Human growth hormone induces system B transport in short bowel syndrome. *J. Surg. Res.* **69,** 150–158.

Inui, K. I., Tomita, Y., Katsura, T., Okano, T., Takano, M., and Hori, R. (1992). H^+-coupled active transport of bestatin via the dipeptide transport system in rabbit intestinal brush border membranes. *J. Pharmacol. Exp. Ther.* **260,** 482–486.

Kakuda, D. K., Finley, K. D., Dionne, V. E., and MacLeod, C. L. (1993). Two distinct gene products mediate y^+ type cationic amino acid transport in *Xenopus* oocytes and show different tissue expression patterns. *Transgene* **1,** 91–101.

Kanai, Y. (1997). Family of neutral and acidic amino acid transporters: Molecular biology, physiology and medical implications. *Curr. Opin. Cell Biol.* **9,** 565–572.

Kanai, Y., and Hediger, M. A. (1992). Primary structure and functional characterization of a high-affinity glutamate transporter. *Nature* **360,** 467–471.

Kanai, Y., Nussberger, S., Romero, M. F., Boron, W. F., Hebert, S. C., and Hediger, M. A. (1995). Electrogenic properties of the epithelial and neuronal high affinity glutamate transporter. *J. Biol. Chem.* **270,** 16561–16568.

Kanai, Y., Segawa, H., Miyamoto, K. I., Uchino, H., Takeda, E., and Endou, H. (1998). Expression cloning and characterization of a transporter for large neutral amino acids activated by the heavy chain of 4F2 antigen (CD98). *J. Biol. Chem.* **273**, 23629–23632.

Kanai, Y., Stelzner, M., Nussberger, S., Khawaja, S., Hebert, S. C., Smith, C. P., and Hediger, M. A. (1994). The neuronal and epithelial human high affinity glutamate transporter. Insights into structure and mechanism of transport. *J. Biol. Chem.* **269**, 20599–20606.

Kato, M., Maegawa, H., Okano, T., Inui, K. I., and Hori, T. (1989). Effect of various chemical modifiers on H^+-coupled transport of cepharadine via dipeptide carriers in rabbit intestinal brush border membranes: Role of histidine residues. *J. Pharmacol. Exp. Ther.* **251**, 745–749.

Kavanaugh, M. P., Wang, H., Zhang, Z., Zhang, W., Wu, Y. N., Dechant, E., North, R., and Kabat, D. (1994). Control of cationic amino acid transport and retroviral receptor functions in a membrane protein family. *J. Biol. Chem.* **269**, 15445–15450.

Kekuda, R., Prasad, P. D., Fei, Y. J., Leibach, F. H., and Ganapathy, V. (1996). Cloning and functional expression of the human type 1 sigma receptor (hSigma R1). *Biochem. Biophys. Res. Commun.* **229**, 553–558.

Kekuda, R., Prasad, P. D., Fei, Y. J, Torres-Zamorano, V., Sinha, S., Yang-Feng, T., Leibach, F. H., and Ganapathy, V. (1996). Cloning of the sodium-dependent, broad-scope, neutral amino acid transporter B^0 from a human placental choriocarcinoma cell line. *J. Biol. Chem.* **271**, 18657–18661.

Kekuda, R., Torres-Zamorano, V., Fei, Y. J., Prasad, P. D., Li, H. W., Mader, L. D., Leibach, F. H., and Ganapathy, V. (1997). Molecular and functional characterization of intestinal Na^+-dependent neutral amino acid transporter B^0. *Am. J. Physiol.* **272**, G1463–G1472.

Kim, J. W., Closs, E. I., Albritton, L. M., and Cunningham, J. M. (1991). Transport of cationic amino acids by the mouse ecotropic retrovirus receptor. *Nature* **352**, 725–728.

Kramer, W., Girbig, F., Petzoldt, E., and Leipe, I. (1990). Inactivation of the intestinal uptake system for β-lactam antibiotics by diethylpyrocarbonate. *Biochim. Biophys. Acta* **943**, 288–296.

Lauteala, T., Sistonen, P., Savontaus, M. L., Mykkanen, J., Simell, J., Lukkarinen, M., Simell, O., and Aula, P. (1997). Lysinuric protein intolerance (LPI) gene maps to the long arm of chromosome 14. *Am. J. Hum. Genet.* **60**, 1479–1486.

Lee, W. S., Wells, R. G., Sabbag, R. V., Mohandas, T. K., and Hediger, M. A. (1993). Cloning and chromosomal localization of a human kidney cDNA involved in cystine, dibasic, and neutral amino acid transport. *J. Clin. Invest.* **91**, 1959–1963.

Leibach, F. H., and Ganapathy, V. (1996). Peptide transporters in the intestine and the kidney. *Ann. Rev. Nutr.* **16**, 99–119.

Leonard, J. V., Marrs, T. C., Addison, J. M., Burston, D., Clegg, K. M., Lloyd, J. K., Matthews, D. M., and Seakins, J. W. (1976). Intestinal absorption of amino acids and peptides in Hartnup disorder. *Pediatr. Res.* **10**, 246–249.

Levy, H. L. (1995). Hartnup disorder. *In* "The Metabolic and Molecular Bases of Inherited Disease" (C. R. Scriver, A. L. Beaudet, W. S. Sly, and D. Valle, eds.), pp. 3629–3642. McGraw-Hill, New York.

Liang, R., Fei, Y. J., Prasad, P. D., Ramamoorthy, S., Han, H., Yang-Feng, T. L., Hediger, M. A., Ganapathy, V., and Leibach, F. H. (1995). Human intestinal H^+-peptide cotransporter. Cloning, functional expression, and chromosomal localization. *J. Biol. Chem.* **270**, 6456–6463.

Liao, K., and Lane, M. D. (1995). Expression of a novel insulin-activated amino acid transporter gene during differentiation of 3T3-L1 preadipocytes into adipocytes. *Biochem. Biophys. Res. Commun.* **208**, 1008–1015.

Lis, M. T., Matthews, D. M., and Crampton, R. F. (1972). Effects of dietary restriction and protein deprivation on intestinal absorption of protein digestion products in the rat. *Br. J. Nutr.* **28**, 443–446.

Lister, N., Sykes, A. P., Bailey, P. D., Boyd, C. A. R., and Bronk, J. R. (1995). Dipeptide transport and hydrolysis in isolated loops of rat small intestine: Effects of stereospecificity. *J. Physiol.* **484**, 173–182.

Liu, W., Liang, R., Ramamoorthy, S., Fei, Y. J., Ganapathy, M. E., Hediger, M. A., Ganapathy, V., and Leibach, F. H. (1995). Molecular cloning of PEPT2, a new member of the H^+/peptide cotransporter family, from human kidney. *Biochim. Biophys. Acta* **1235**, 461–466.

Lucas, M. L., Cooper, B. T., Lei, F. H., Johnson, I. T., Holmes, G. K. T., Blair, J. A., and Cooke, W. T. (1978). Acid microclimate in coeliac and Crohn's disease: A model for folate malabsorption. *Gut* **19**, 735–742.

Lucas, M. L., Schneider, W., Haberich, F. J., and Blair, J. A. (1975). Direct measurement by pH-microelectrode of the pH microclimate in rat proximal jejunum. *Proc. R. Soc. Lond.* **192**, 39–48.

Lumadue, J. A., Glick, A. B., and Ruddle, F. H. (1987). Cloning, sequence analysis, and expression of the large subunit of the human lymphocyte activation antigen 4F2. *Proc. Natl. Acad. Sci. USA* **84**, 9204–9208.

Mackenzie, B., Fei, Y. J., Ganapathy, V., and Leibach, F. H. (1996a). The human intestinal H^+/oligopeptide cotransporter hPEPT1 transports differently charged dipeptides with identical electrogenic properties. *Biochim. Biophys. Acta* **1284**, 125–128.

Mackenzie, B., Loo, D. D. F., Fei, Y. J., Liu, W., Ganapathy, V., Leibach, F. H., and Wright, E. M. (1996b). Mechanisms of the human intestinal H^+-coupled oligopeptide transporter hPEPT1. *J. Biol. Chem.* **271**, 5430–5437.

MacLeod, C. L., and Kakuda, D. K. (1996). Regulation of CAT: Cationic amino acid transporter gene expression. *Amino Acids* **11**, 171–191.

Malandro, M. S., and Kilberg, M. S. (1996). Molecular biology of mammalian amino acid transporters. *Ann. Rev. Biochem.* **65**, 305–336.

Mastroberardino, L., Spindler, B., Pfeiffer, R., Skelly, P. J., Loffing, J., Shoemaker, C. B., and Verrey, F. (1998). Amino acid transport by heterodimers of 4F2hc/CD98 and members of a permease family. *Nature* **395**, 288–291.

Matthews, D. M. (1975). Intestinal absorption of peptides. *Physiol. Rev.* **55**, 537–608.

Matthews, D. M. (1991). "Protein Absorption. Development and Present State of the Subject." Wiley–Liss, New York.

Matthews, D. M., and Adibi, S. A. (1976). Peptide absorption. *Gastroenterology* **71**, 151–161.

Meredith, D., and Boyd, C. A. R. (1995). Oligopeptide transport by epithelial cells. *J. Membr. Biol.* **145**, 1–12.

Miyamoto, Y., Nakamura, H., Hoshi, T., Ganapathy, V., and Leibach, F. H. (1990). Uphill transport of β-alanine in intestinal brush border membrane vesicles. *Am. J. Physiol.* **259**, G372–G379.

Miyamoto, Y., Tiruppathi, C., Ganapathy, V., and Leibach, F. H. (1989). Active transport of taurine in rabbit jejunal brush border membrane vesicles. *Am. J. Physiol.* **257**, G65–G72.

Mosckovitz, R., Yan, N., Heimer, E., Felix, A., Tate, S. S., and Udenfriend, S. (1993). Characterization of the rat neutral and basic amino acid transporter utilizing antipeptide antibodies. *Proc. Natl. Acad. Sci. USA* **90**, 4022–4026.

Muller, U., Brandsch, M., Prasad, P. D., Fei, Y. J., Ganapathy, V., and Leibach, F. H. (1996). Inhibition of the H^+/peptide cotransporter in the human intestinal cell line Caco-2 by cyclic AMP. *Biochem. Biophys. Res. Commun.* **218**, 461–465.

Munck, L. K., and Munck, B. G. (1995). Transport of glycine and lysine on the chloride-dependent beta-alanine ($B^{0,+}$) carrier in rabbit small intestine. *Biochim. Biophys. Acta* **1235**, 93–99.

Nussberger, S., Steel, A., Trotti, D., Romero, M. F., Boron, W. F., and Hediger, M. A. (1997). Symmetry of H^+ binding to the intra- and extracellular side of the H^+-coupled oligopeptide cotransporter PepT1. *J. Biol. Chem.* **272**, 7777–7785.

Palacin, M. (1994). A new family of proteins (rBAT and 4F2hc) involved in cationic and zwitterionic amino acid transport: A tale of two proteins in search of a transport function. *J. Exp. Biol.* **196**, 123–137.

Palacin, M., Estevez, R., Bertran, J., and Zorzano, A. (1998). Molecular biology of mammalian plasma membrane amino acid transporters. *Physiol. Rev.* **78**, 969–1054.

Pan, M., and Stevens, B. R. (1995). Differentiation and protein kinase C-dependent regulation of alanine transport via system B. *J. Biol. Chem.* **270**, 3582–3587.

Peghini, P., Jansen, J., and Stoffel, W. (1997). Glutamate transporter EAAC-1-deficient mice develop dicarboxylic aminoaciduria and behavioral abnormalities but not neurodegeneration. *EMBO J.* **16**, 3822–3832.

Pfeiffer, R., Rossier, G., Spindler, B., Meier, C., Kuhn, L., and Verrey, F. (1999). Amino acid transport of y$^+$L-type by heterodimers of 4F2hc/CD98 and members of the glycoprotein-associated amino acid transporter family. *EMBO J.* **18**, 49–57.

Pickel, V. M., Nirenberg, M. J., Chan, J., Moscovitz, R., Udenfriend, S., and Tate, S. S. (1993). Ultrastructural localization of a neutral and basic amino acid transporter in rat kidney and intestine. *Proc. Natl. Acad. Sci. USA* **90**, 7779–7783.

Pineda, M., Fernandez, E., Torrents, D., Estevez, R., Lopez, C., Camps, M., Lloberas, J., Zorzano, A., and Palacin, M. (1999). Identification of a membrane protein, LAT-2, that co-expresses with 4F2 heavy chain, an L-type amino acid transport activity with broad specificity for small and large zwitterionic amino acids. *J. Biol. Chem.* **274**, 19738-19744.

Pras, E., Arber, N., Aksentijevich, I., Katz, G., Shapiro, J. M., Prosen, L., Gruberg, L., Harel, D., Liberman, U., Weissenbach, J., Pras, M., and Kastner, D. L. (1994). Localization of a gene causing cystinuria to chromosome 2P. *Nature Genet.* **6**, 415–419.

Prasad, P. D., Wang, H., Huang, W., Kekuda, R., Rajan, D. P., Leibach, F. H., and Ganapathy, V. (1999). Human LAT1, a subunit of system L amino acid transporter: Molecular cloning and transport function. *Biochem. Biophys. Res. Commun.* **255**, 283–288.

Quackenbush, E., Clabby, M., Gottesdiener, K. M., Barbosa, J., Jones, N. H., Strominger, J. L., Speck, S., and Leiden, J. M. (1987). Molecular cloning of complementary DNAs encoding the heavy chain of the human 4F2 cell-surface antigen: A type II membrane glycoprotein involved in normal and neoplastic cell growth. *Proc. Natl. Acad. Sci. USA* **84**, 6526–6530.

Quackenbush, E. J., Gougos, A., Baumal, R., and Letarte, M. (1986). Differential localization within human kidney of five membrane proteins expressed on acute lymphoblastic leukemia cells. *J. Immunol.* **136**, 118–124.

Rajan, D. P., Huang, W., Kekuda, R., George, R. L., Wang, J., Conway, S. J., Devoe, L. D., Leibach, F. H., Prasad, P. D., and Ganapathy, V. (2000a). Differential influence of the 4F2 heavy chain and the protein related to b$^{0,+}$ amino acid transport on substrate affinity of the heteromeric b$^{0,+}$ amino acid transporter. *J. Biol. Chem.* **275**, 14331–14335.

Rajan, D. P., Kekuda, R., Huang, W., Devoe, L. D., Leibach, F. H., Prasad, P. D., and Ganapathy, V. (2000b). Cloning and functional characterization of a Na$^+$-independent, broad-specific neutral amino acid transporter from mammalian intestine. *Biochim. Biophys. Acta* **1463**, 6–14.

Rajan, D. P., Kekuda, R., Huang, W., Wang, H., Devoe, L. D., Leibach, F. H., Prasad, P. D., and Ganapathy, V. (1999). Cloning and expression of a b$^{0,+}$-like amino acid transporter functioning as a heterodimer with 4F2hc instead of rBAT. A new candidate gene for cystinuria. *J. Biol. Chem.* **274**, 29005–29010.

Rajante, J., Simell, O., and Perheentupa, J. (1980). Basolateral membrane transport defect for lysine in lysinuric protein intolerance. *Lancet* **1**, 1219–1221.

Ramamoorthy, J. D., Ramamoorthy, S., Mahesh, V. B., Leibach, F. H., and Ganapathy, V. (1995). Cocaine-sensitive σ-receptor and its interaction with steroid hormones in the human placental syncytiotrophoblast and in choriocarcinoma cells. *Endocrinology* **136**, 924–932.

Ramamoorthy, S., Leibach, F. H., Mahesh, V. B., Han, T., Yang-Feng, T., Blakely, R. D., and Ganapathy, V. (1994). Functional characterization and chromosomal localization of a cloned taurine transporter from human placenta. *Biochem. J.* **300**, 893–900.

Ramamoorthy, S., Liu, W., Ma, Y. Y., Yang-Feng, T. L., Ganapathy, V., and Leibach, F. H. (1995). Proton/peptide cotransporter (PEPT2) from human kidney: Functional characterization and chromosomal localization. *Biochim. Biophys. Acta* **1240**, 1–4.

Rasko, J. E. J., Battini, J. L., Gottschalk, R. J., Mazo, I., and Miller, A. D. (1999). The RD114/simian type D retrovirus receptor is a neutral amino acid transporter. *Proc. Natl. Acad. Sci. USA* **96**, 2129–2134.

Roman, F., Pascaud, X., Chomette, G., Bueno, L., and Junien, J. L. (1989). Autoradiographic localization of sigma opioid receptors in the gastrointestinal tract of the guinea pig. *Gastroenterology* **97**, 76–82.

Sai, Y., Tamai, I., Sumikawa, H., Hayashi, K., Nakanishi, T., Amano, O., Numata, M., Iseki, S., and Tsuji, A. (1996). Immunolocalization and pharmacological relevance of oligopeptide transporter PepT1 in intestinal absorption of β-lactam antibiotics. *FEBS Lett.* **392**, 25–29.

Saito, H., and Inui, K. I. (1993). Dipeptide transporters in apical and basolateral membranes of the human intestinal cell line Caco-2. *Am. J. Physiol.* **265**, G289–G294.

Saito, H., Motohashi, H., Mukai, M., and Inui, K. I. (1997). Cloning and characterization of a pH-sensing regulatory factor that modulates transport activity of the human H^+/peptide cotransporter, PEPT1. *Biochem. Biophys. Res. Commun.* **237**, 577–582.

Saito, H. Okuda, M., Terada, T., Sasaki, S., and Inui, K. I. (1995). Cloning and characterization of a rat H^+/peptide cotransporter mediating absorption of beta-lactam antibiotics in the intestine and kidney. *J. Pharmacol. Exp. Ther.* **275**, 1631–1637.

Saito, H., Terada, T., Okuda, M., Sasaki, S., and Inui, K. I. (1996). Molecular cloning and tissue distribution of rat peptide transporter PEPT2. *Biochim. Biophys. Acta* **1280**, 173–177.

Schell, R. E., Stevens, B. R., and Wright, E. M. (1983). Kinetics of sodium-dependent solute transport by rabbit renal and jejunal brush border vesicles using a fluorescent dye. *J. Physiol.* **335**, 307–318.

Segal, S., and Thier, O. (1995). Cystinuria. *In* "The Metabolic and Molecular Bases of Inherited Disease" (C. R. Scriver, A. L. Beaudet, W. S. Sly, and D. Valle, eds.), pp. 3581–3601. McGraw–Hill, New York.

Segawa, H., Fukasawa, Y., Miyamoto, K. I., Takeda, E., Endou, H., and Kanai, Y. (1999). Identification and functional characterization of a Na^+-independent neutral amino acid transporter with broad substrate selectivity. *J. Biol. Chem.* **274**, 19745–19751.

Shafqat, S., Tamarappoo, B. K., Kilberg, M. S., Puranam, R. S., McNamara, J. O., Guadano-Ferraz, A., and Fremeau, R. T., Jr. (1993). Cloning and expression of a novel Na^+-dependent neutral amino acid transporter structurally related to mammalian Na^+/glutamate cotransporters. *J. Biol. Chem.* **268**, 15351–15355.

Shashidharan, P., Huntley, G. W., Meyer, T., Morrison, J. H., and Plaitakis, A. (1994). Neuron-specific human glutamate transporter: Molecular cloning, characterization and expression in human brain. *Brain Res.* **662**, 245–250.

Shiraga, T., Miyamoto, K. I., Tanaka, H., Yamamoto, H., Taketani, Y., Morita, K., Tamai, I., Tsuji, A., and Takeda, E. (1999). Cellular and molecular mechanisms of dietary regulation of rat intestinal H^+/peptide transporter PepT1. *Gastroenterology* **116**, 354–362.

Silk, D. B. A., Perrett, D., and Clark, M. L. (1975). Jejunal and ileal absorption of dibasic amino acids and an arginine-containing dipeptide in cystinuria. *Gastroenterology* **68**, 1426–1432.

Simell, O. (1995). Lysinuric protein intolerance and other cationic aminoacidurias. *In* "The Metabolic and Molecular Bases of Inherited Disease" (C. R. Scriver, A. L. Beaudet, W. S. Sly, and D. Valle, eds.), pp. 3603–3628. McGraw–Hill, New York.

Sloan, J. L., and Mager, S. (1999). Cloning and functional expression of a human Na^+- and Cl^--dependent neutral and cationic amino acid transporter $B^{0,+}$. *J. Biol. Chem.* **274**, 23740–23745.

Smith, C. P., Weremowicz, S., Kanai, Y., Stelzner, M., Morton, C. C., and Hediger, M. A. (1994). Assignment of the gene coding for the human high-affinity glutamate transporter EAAC1 to 9p24: Potential role in dicarboxylic aminoaciduria and neurodegenerative disorders. *Genomics* **20**, 335–336.

Sonders, M. S., and Amara, S. G. (1996). Channels in transporters. *Curr. Opin. Neurobiol.* **6**, 294–302.

Steel, A., Nussberger, S., Romero, M. F., Boron, W. F., Boyd, C. A. R., and Hediger, M. A. (1997). Stoichiometry and pH dependence of the rabbit proton-dependent oligopeptide transporter PepT1. *J. Physiol.* **498**, 563–569.

Stevens, B. R. (1992). Amino acid transport in intestine. *In* "Mammalian Amino Acid Transport: Mechanisms and Control" (M. Kilberg and D. Haussinger, eds.), pp. 149–163. Plenum, New York.

Stevens, B. R., Ross, H. J., and Wright, E. M. (1982). Multiple transport pathways for neutral amino acids in jejunal brush border vesicles. *J. Membr. Biol.* **66**, 213–225.

Su, T. P., London, E. D., and Jaffe, J. H. (1988). Steroid binding at σ receptors suggests a link between endocrine, nervous, and immune systems. *Science* **240,** 219–221.

Sugawara, M., Nakanishi, T., Fei, Y. J., Huang, W., Ganapathy, M. E., Leibach, F. H., and Ganapathy, V. (2000). Cloning of an amino acid transporter with functional characteristics and tissue expression pattern identical to that of system A. *J. Biol. Chem.* 10.1074/jbc.C000205200.

Swaan, P. W., Stehouwer, M. C., and Tukker, J. J. (1995). Molecular mechanism for the relative binding affinity to the intestinal peptide carrier. Comparison of three ACE inhibitors: Enalapril, enalaprilat, and lisinopril. *Biochim. Biophys. Acta* **1236,** 31–38.

Tamai, I., Nakanishi, T., Hayashi, K., Terao, T., Sai, Y., Shiraga, T., Miyamoto, K., Takeda, E., Higashida, H., and Tsuji, A. (1997). The predominant contribution of oligopeptide transporter PepT1 to intestinal absorption of β-lactam antibiotics in the rat small intestine. *J. Pharm. Pharmacol.* **49,** 796–801.

Tamai, I., Nakanishi, T., Nakahara, H., Sai, Y., Ganapathy, V., Leibach, F. H., and Tsuji, A. (1998). Improvement of L-dopa absorption by dipeptidyl derivation, utilizing peptide transporter PepT1. *J. Pharm. Sci.* **87,** 1542–1546.

Tanaka, H., Miyamoto, K. I., Morita, K., Haga, H., Segawa, H., Shiraga, T., Fujioka, A., Kouda, T., Taketani, Y., Hisano, S., Fukui, Y., Kitagawa, K., and Takeda, E. (1998). Regulation of the PepT1 transporter in the rat small intestine in response to 5-fluorouracil-induced injury. *Gastroenterology* **114,** 714–723.

Tate, S. S., Yan, N., and Udenfriend, S. (1992). Expression cloning of a Na^+-independent neutral amino acid transporter from rat kidney. *Proc. Natl. Acad. Sci. USA* **89,** 1–5.

Teixeira, S., Di Grandi, S., and Kuhn, L. C. (1987). Primary structure of the human 4F2 antigen heavy chain predicts a transmembrane protein with a cytoplasmic NH_2 terminus. *J. Biol. Chem.* **262,** 9574–9580.

Temple, C. S., Stewart, A. K., Meredith, D., Lister, N. A., Morgan, K. M., Collier, I. D., Vaughn-Jones, R. D., Boyd, C. A. R., Bailey, P. D., and Bronk, J. R. (1998). Peptide mimics as substrates for the intestinal peptide transporter. *J. Biol. Chem.* **273,** 20–22.

Terada, T., Saito, H., Mukai, M., and Inui, K. I. (1996). Identification of the histidine residues involved in substrate recognition by a rat H^+/peptide cotransporter, PEPT1. *FEBS Lett.* **394,** 196–200.

Thamotharan, M., Bawani, S. Z., Zhou, X., and Adibi, S. A. (1999). Hormonal regulation of oligopeptide transporter (Pept-1) in a human intestinal cell line. *Am. J. Physiol.* **276,** C821–C826.

Thwaites, D. T., Brown, D. A., Hirst, B. H., and Simmons, N. L. (1993). H^+-coupled dipeptide (glycylsarcosine) transport across apical and basal borders of human intestinal Caco-2 cell monolayers displays distinctive characteristics. *Biochim. Biophys. Acta* **1151,** 237–245.

Tiruppathi, C., Brandsch, M., Miyamoto, Y., Ganapathy, V., and Leibach, F. H. (1992). Constitutive expression of the taurine transporter in a human colon carcinoma cell line. *Am. J. Physiol.* **263,** G625–G631.

Torrents, D., Estevez, R., Pineda, M., Fernandez, E., Lloberas, J., Shi, Y. B., Zorzano, A., and Palacin, M. (1998). Identification and characterization of a membrane protein (y^+L amino acid transporter-1) that associates with 4F2hc to encode the amino acid transport activity y^+L. A candidate gene for lysinuric protein intolerance. *J. Biol. Chem.* **273,** 32437–32445.

Torrents, D., Mykkanen, J., Pineda, M., Feliubadalo, L., Estevez, R., Cid, R. D., Sanjurjo, P., Zorzano, A., Nunes, V., Huoponen, K., Reinikainen, A., Simell, O., Savontaus, M. L., Aula, P., and Palacin, M. (1999). Identification of SLC7A7, encoding y^+LAT1, as the lysinuric protein intolerance gene. *Nature Genet.* **21,** 293–296.

Torres-Zamorano, V., Kekuda, R., Leibach, F. H., and Ganapathy, V. (1997). Tyrosine phosphorylation—and epidermal growth factor–dependent regulation of the sodium-coupled amino acid transporter B^0 in the human placental choriocarcinoma cell line JAR. *Biochim. Biophys. Acta* **1356,** 258–270.

Torres-Zamorano, V., Leibach, F. H., and Ganapathy, V. (1998). Sodium-dependent homo- and hetero-exchange of neutral amino acids mediated by the amino acid transporter ATB^0. *Biochem. Biophys. Res. Commun.* **245,** 824–829.

Tsuji, A. (1995). Intestinal absorption of β-lactam antibiotics. *In* "Peptide-Based Drug Design" (M. D. Taylor and G. L. Amidon, eds.), pp. 101–134. Am. Chem. Soc., Washington, DC.

Tsuji, A., and Tamai, I. (1996). Carrier-mediated intestinal transport of drugs. *Pharm. Res.* **13,** 963–977.

Tsuji, A., Tamai, I., Nakanishi, M., and Amidon, G. L. (1990). Mechanism of the absorption of the dipeptide α-methyldopa-phe in intestinal brush border membrane vesicles. *Pharm. Res.* **7,** 308–309.

Utsunomiya-Tate, N., Endou, H., and Kanai, Y. (1996). Cloning and functional characterization of a system ASC-like Na^+-dependent neutral amino acid transporter. *J. Biol. Chem.* **271,** 14883–14890.

van der Hulst, R. R., von Meyenfeldt, M. F., and Soeters, P. B. (1996). Glutamine: An essential amino acid for the gut. *Nutrition* **12,** S78–S81.

Van Winkle, L. J., Christensen, H. N., and Campione, A. L. (1985). Na^+-dependent transport of basic, zwitterionic, and bicyclic amino acids by a broad-scope system in mouse blastocysts. *J. Biol. Chem.* **260,** 12118–12123.

Varoqui, H., Zhu, H., Yao, D., Ming, H., and Erickson, J. D. (2000). Cloning and functional identification of a neuronal glutamine transporter. *J. Biol. Chem.* **275,** 4049–4054.

Vazquez, J. A., Morse, E. L., and Adibi, S. A. (1985). Effect of starvation on amino acid and peptide transport and peptide hydrolysis in humans. *Am. J. Physiol.* **249,** G563–G566.

Wang, H., Dechant, E., Kavanaugh, M. P., North, R. A., and Kabat, D. (1992). Effects of ecotropic murine retroviruses on the dual function cell surface receptor/basic amino acid transporter. *J. Biol. Chem.* **267,** 23617–23624.

Wang, H., Kavanaugh, M. P., North, R. A., and Kabat, D. (1991). Cell-surface receptor for ecotropic murine retroviruses is a basic amino acid transporter. *Nature* **352,** 729–731.

Wang, Y., and Tate, S. S. (1995). Oligomeric structure of a renal cystine transporter: Implications in cystinuria. *FEBS Lett.* **368,** 389–392.

Wells, R. G., and Hediger, M. A. (1992). Cloning of a rat kidney cDNA that stimulates dibasic and neutral amino acid transport and has sequence similarity to glucosidases. *Proc. Natl. Acad. Sci. USA* **89,** 5596–5600.

Wells, R. G., Lee, W., Kanai, Y., Leiden, J. M., and Hediger, M. A. (1992). The 4F2 antigen heavy chain induces uptake of neutral and dibasic amino acids in *Xenopus* oocytes. *J. Biol. Chem.* **267,** 15285–15288.

Wenzel, U., Gebert, I., Weintraut, H., Weber, W. M., Clauss, W., and Daniel, H. (1996). Transport characteristics of differently charged cephalosporin antibiotics in oocytes expressing the cloned intestinal peptide transporter PepT1 and in human intestinal Caco-2 cells. *J. Pharmacol. Exp. Ther.* **277,** 831–839.

Yeung, A. K., Basu, S. K., Wu, S. K., Chu, C., Okamoto, C. T., Hamm-Alvarez, S. F., von Grafenstein, H., Shen, W. C., Kim, K. J., Bolger, M. B., Haworth, I. S., Ann, D. K., and Lee, V. H. L. (1998). Molecular identification of role for tyrosine 167 in the function of the human intestinal proton-coupled dipeptide transporter (hPepT1). *Biochem. Biophys. Res. Commun.* **250,** 103–107.

Yoshimoto, T., Yoshimoto, E., and Meruelo, D. (1991). Molecular cloning and characterization of a novel human gene homologous to the murine ecotropic retroviral receptor. *Virology* **185,** 10–17.

Yoshimoto, T., Yoshimoto, E., and Meruelo, D. (1993). Identification of amino acid residues critical for infection with ecotropic murine leukemia retrovirus. *J. Virol.* **67,** 1310–1314.

Zerangue, N., and Kavanaugh, M. P. (1996a). ASCT-1 is a neutral amino acid exchanger with chloride channel activity. *J. Biol. Chem.* **271,** 27991–27994.

Zerangue, N., and Kavanaugh, M. P. (1996b). Flux coupling in a neuronal glutamate transporter. *Nature* **383,** 634–637.

CHAPTER 11

Electrogenic Transepithelial Na$^+$ Transport in the Colon

Jean-Daniel Horisberger

Institute of Pharmacology and Toxicology, University of Lausanne, Bugnon 27, CH-10005 Lausanne, Switzerland

I. INTRODUCTION

The balance of the most abundant ions (Na, K, Cl, HCO$_3^-$) is regulated by tight control of the ionic composition of the fluids leaving the body. For instance, to prevent Na loss, the concentration of Na in excreta must be decreased to values much lower than that present in the extracellular fluid. This control is carried out by the epithelia lining the distal portion of the excretory organs, epithelia that are responsible for the final ionic composition of the excreta. To prevent ionic back-leak, these epithelia must provide an effective barrier to ion movement. This barrier is related to the unique nature of the tight junctions in these epithelia, which have very low ionic permeability, resulting in low conductance of the paracellular pathway. These epithelia therefore are designated as "tight epithelia" by contrast with the "leaky epithelia" in which the ionic permeability of the paracellular pathway is one or two orders of magnitude higher. This high resistance of the tight junctions is dependent on the integrity of the epithelium and is altered in pathological states, for instance early in the course of colon cancer development (Soler *et al.,* 1999). The mechanism used to create a large transepithelial Na^+ gradient is common to all tight epithelia; it is the electrogenic Na transport mediated by the amiloride-sensitive Na channel (ENaC). The principle of this mechanism is illustrated in Figure 1. Together with the low permeability of the paracellular pathway, this mechanism allows the generation and maintenance of a >100-fold transepithelial Na gradient and is therefore very efficient in promoting Na conservation. However, its transport capacity is limited and cannot prevent Na losses when the tight epithelia are exposed to a large flow of fluid with high Na concentrations.

Under normal circumstances, urine is the largest volume of excreta, and the regulation of Na balance is provided mostly by the kidney, in particular by controlled Na reabsorption in the distal part of the nephron, the collecting tubule. However, large volumes of fluid also can be lost in other ways, for instance through sweat or through feces in the case of diarrhea. Therefore, it is not surprising that the distal portion of the sweat glands (sweat ducts) and the distal colon are lined with tight epithelia capable of electrogenic Na transport. The colonic epithelium also performs electroneutral NaCl absorption mediated by other transport systems (Sandle, 1998); the present review will focus on the ENaC-mediated electrogenic Na transport.

Because of a variable salt intake and the possibility of large variations in Na loss through the sweat glands or the digestive tract, the reabsorption of Na must be tightly regulated. This regulation must be rapid and efficient because a sizable portion of the total body Na may be ingested or lost over a short time. The colon has a major homeostatic role under such conditions (Rubens and Lambert, 1972).

Even if the other main active element of transepithelial Na transport, namely the basolateral membrane Na,K-pump, is also regulated (Verrey *et al.,* 1996), the rate-limiting step is the entry of Na into the cell through the apical membrane, and therefore the rate of transepithelial Na transport is essentially controlled by the number of open Na channels in the apical membrane. In this review, present knowledge about the structure, function, and distribution of ENaC will be summarized as will the complex (and not yet fully understood) regulation of ENaC.

The presence of the Na channel in the apical membrane of tight epithelia is useful not just for the reabsorption of Na ions; the electrical gradient created by Na entry through the apical Na channel also can be used to drive K secretion through K channels located in the same membrane. This phenomenon was first described in the kidney cortical collecting tubule (Koeppen and Giebisch, 1985), but it also operates in the distal colon and will be discussed herein.

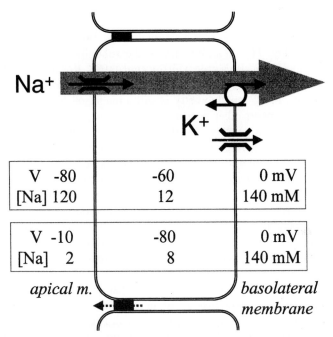

FIGURE 1 Mechanism of electrogenic Na transport mediated by the epithelial Na channel. Na ions enter the cell from the luminal fluid through ENaC following their electrochemical gradient, which is the sum of the electrical potential across the apical membrane and the chemical potential owing to the difference in Na concentrations across the same membrane. The low intracellular Na concentration is maintained by the basolateral Na,K-pump. The Na,K-pump has a steep dependence on intracellular Na (with a Hill coefficient between 2 and 3) and normally works well below its maximal capacity; therefore it is able to maintain intracellular Na concentration within a narrow range in spite of large variations of the rate of Na entry. The tight junctions (indicated by the small shaded rectangles between the cells) have low permeability to Na and other ions, and prevent the back-leak of sodium toward the lumen (dotted arrow). The lower part of the scheme indicates values of Na concentrations, and membrane and transepithelial potential for two characteristic situations with a high (top rectangle) or a low (bottom rectangle) luminal Na concentration. At high luminal Na concentrations, the apical membrane is strongly depolarized [apical membrane potential (V_a) = + 20 mV] with a membrane potential tending toward the Na equilibrium potential; the electrochemical gradient (V_a + E_{Na}) driving Na entry may be rather small: +20 (V_a) + (−60) (E_{Na}) = −40 mV. At low luminal Na concentrations, the apical membrane potential is mainly determined by the whole-cell K conductance and is close to the K equilibrium potential ($\approx$−80 mV), and an electrochemical gradient driving Na entry is maintained in spite of the reverse chemical gradient: −70 (V_a) + 36 (E_{Na}) = −34 mV, even if the luminal Na concentration is much lower than the 5- to 10-mM intracellular Na concentration (see more complete description of this phenomenon in Schultz *et al.*, 1977).

II. ENaC STRUCTURE

The epithelial Na channel is a heteromultimeric protein made of three subunits: α, β, and γ. The stoichiometry of the subunits is probably $2\alpha/1\beta/1\gamma$ tetramer, arranged in a $\alpha\beta\alpha\gamma$ order (Firsov *et al.*, 1998; Kosari *et al.*, 1998), similar to other members of the ENaC/degenerin protein family (Coscoy *et al.*, 1998); however, ENaC stoichiometry is still a matter of debate because different sets of data support other possible arrangements, including a total of nine (Snyder *et al.*, 1998; Cheng *et al.*, 1998) or eight (Eskandari *et al.*, 1999) subunits. The primary structure of the three subunits is homologous and predicts a similar membrane topology with two transmembrane segments (M1 and M2), intracellular N- and C-terminal domains, and a large extracellular loop containing highly conserved cysteine-rich sequences (Canessa *et al.*, 1994a; Canessa *et al.*, 1994b). According to a number of recent publications (Kellenberger *et al.*, 1999a; Kellenberger *et al.*, 1999b; Zhang *et al.*, 1999), the selective part of the channel pore is formed by the segment immediately preceding M2 (Fig. 2), and structural models of this region have been proposed (Kellenberger *et al.*, 1999b).

Following the discovery of mutations and deletions leading to a gain-of-function and salt-sensitive hypertension (Shimkets *et al.*, 1994), the C-terminal domain was shown to contain a "PY" motif, which is essential for the regulation of plasma-membrane channel density (see section entitled Control of Channel Density). The discovery of a mutation in the N-terminal intracellular domain of the β subunit and the systematic study of a segment of this domain in the α subunit have shown that this region is implicated in the gating of the channel (Gründer *et al.*, 1997; Gründer *et al.*, 1999). More detailed information concerning the structure of ENaC can be found in recent reviews (Horisberger, 1998; Fyfe *et al.*, 1998; Benos and Stanton, 1999).

III. FUNCTIONAL PROPERTIES

The channel resulting from heterologous expression of the α, β, and γ subunits reproduces reliably the physiological and pharmacological properties of the highly Na-selective channel naturally expressed in the apical membrane of tight epithelia (Canessa *et al.*, 1994a; Horisberger, 1998). These properties have been described in detail elsewhere (Garty and Palmer, 1997) and are summarized briefly below:

1. ionic selectivity: A very high selectivity for Na over K, with a Li conductance 1.5- to 2-fold larger than the Na conductance
2. a single-channel conductance of approximately 5 pS for Na and 8 pS for Li

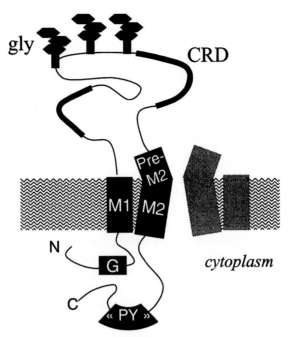

FIGURE 2   Molecular structure of ENaC. The scheme shows the membrane topology of an ENaC subunit; only one subunit is shown, and the transmembrane part of a second subunit is outlined. The complete channel is expected to be made of the assembly of four homologous subunits. The two transmembrane domains (M1 and M2 putative α helices) are represented as shaded cylinders. The N-terminal (N) and C-terminal (C) domains are located inside the cell. The large extracytoplasmic loop bears several glycosylation sites (gly; only three are shown, but the number of potential glycosylation sites vary from 5 to 11 in the three subunits) and two cysteine-rich domains (CRD). The pore of the channel is made of the M2 transmembrane segment and the segment immediately preceding M2 (pre-M2). The N-terminal domain contains a segment involved in the gating of the channel, and the C-terminal domain includes a P-P-P-x-Y motif ("PY"), which is known to interact with the "WW" domain of regulatory proteins (see text).

3. highly variable gating kinetics remarkable for the presence of long (seconds) open and closed states
4. high affinity for amiloride (K_I 0.1–0.5 μM) or benzamil (K_I 10–50 nM)

Amiloride-sensitive channels with slightly or markedly different properties have been described in epithelia other than "tight epithelia," in particular in airway epithelia (Palmer, 1992; Ruckes *et al.*, 1997; Marunaka, 1996). Although subunits of ENaC may be involved, the exact nature of the proteins responsible for these channels and, in particular, their subunit composition, are still a matter of debate. Expression of subunit combinations other than α + β + γ subunits may result in channels with different properties (Fyfe and Canessa, 1998; Zhang *et al.*, 1999; Bonny *et al.*, 1999). Reconstitution in lipid bilayers of ENaC subunits obtained either by *in vitro* transcription or by artificial expression have yielded single-channel currents that were amiloride-sensitive but differed in most physiological characteristics (selectivity, single-channel conductance, gating kinetics) (Ismailov *et al.*, 1995; Ismailov *et al.*, 1996a; Berdiev *et al.*, 1998). The relationship between these reconstituted channels and the channels observed in cell plasma membrane (after natural or artificial expression) is not yet clear.

IV. TISSUE DISTRIBUTION

Amiloride-sensitive electrogenic Na transport has been extensively studied in amphibian tight epithelia (frog skin, toad urinary bladder) as a model of Na transport by the distal part of the nephron. However, electrogenic Na transport and ENaC subunit expression also have been observed in other organs: the distal colon as described below, the excretory ducts of the sweat gland (Roudier-Pujol *et al.*, 1996) and salivary glands (Dinudom *et al.*, 1993), and airway epithelia (Voilley *et al.*, 1994; Farman *et al.*, 1997). ENaC also is expressed in keratinocytes (Roudier-Pujol *et al.*, 1996), but its role in skin is not yet understood because it appears that the skin epithelium does not perform vectorial electrogenic Na transport (Brouard *et al.*, 1999).

Measurement of transepithelial amiloride-sensitive current in the distal colon indicated that the properties of the apical Na channel were very similar to those described for urinary tight epithelia (Edmonds, 1981; Wills *et al.*, 1982; Zeiske *et al.*, 1982). Indeed, ENaC initially was cloned from a rat colon cDNA library because this tissue proved to be the most abundant source of mRNA expressing amiloride-sensitive Na channel function (Canessa *et al.*, 1993; 1995; Lingueglia *et al.*, 1993). The three subunits of ENaC are expressed in the distal colon, especially after hormonal stimulation (see section V.). Although electrogenic Na transport (and ENaC expression) is essentially restricted to the distal part of the colon under normal conditions, a low-salt diet or chronic salt wasting can stimulate the expression of ENaC, not only along the whole colon but also in the ileum (Koyama *et al.*, 1999); these observations have demonstrated that the expression of ENaC (and probably the whole differentiation program of tight epithelia) is flexible and can extend beyond the anatomic limits of the colon.

Within the distal colon epithelium, ENaC is not evenly distributed; *in situ* hybridization and immunocytochemistry studies showed the presence of ENaC subunits mostly in surface cells (Smith *et al.*, 1993; Duc *et al.*, 1994; Watanabe *et al.*, 1999). This observation was confirmed by physiological studies showing the presence of a sizable aldosterone-induced Na conductance only in surface cells (Kockerling *et al.*, 1996; Lomax *et al.*, 1996), whereas cells found at the bottom of the crypts showed mostly a cAMP-inducible chloride conductance (Ecke *et al.*, 1996).

The expression of ENaC is regulated in the colon during development; the abundance of mRNA for the three subunits increases gradually from fetal day 19 to day 30 of life (Watanabe *et al.*, 1998); this developmental profile is specific to the colon and differs from that observed either in the kidney (Watanabe *et al.*, 1999) or in the lung (Watanabe *et al.*, 1998).

V. ENaC REGULATION

Sodium deprivation, or any other cause of decreased extracellular fluid volume, results in increased adrenal production of aldosterone, and this mineralocorticoid hormone is a well-known stimulator of electrogenic Na transport (Horisberger and Rossier, 1992). The mechanism of this stimulation is complex and not yet fully understood, but significant advances have been made recently by the observation of the transcriptional or posttranscriptional effects on the ENaC subunits themselves and by the identification of aldosterone-induced proteins (AIP) that may be responsible for a part of the effect of aldosterone on electrogenic Na transport. These questions have been well summarized in a recent review (Verrey, 1999), and the following section will focus mainly on the differences between colon and the other aldosterone target epithelia.

A. Transcriptional and Posttranscriptional Control of ENaC Synthesis

Because it provides the rate-limiting step of Na transport, ENaC was one of the first candidates for the transcriptional effect of aldosterone, a hormone that is known to increase the density of active Na channels in the apical membrane (Palmer *et al.*, 1982; Clauss *et al.*, 1993). The effect of aldosterone, or Na deprivation, on the Na transport rate by the distal colon epithelium of several vertebrate species was found qualitatively similar to that observed in urinary tight epithelia (Frizzell and Schultz, 1978; Will *et al.*, 1980; Clauss *et al.*, 1985; Epple *et al.*, 1995). This effect is a high-affinity ($\sim$1 nM) response (Fromm *et al.*, 1993). Sodium deprivation, which induces the release of aldosterone, has similar effects (Turnheim *et al.*, 1986; Sandle *et al.*, 1984; Wang *et al.*, 2000). In many studies, stimulation of Na transport in the colon also could be obtained by glucocorticoid treatment (Bastl *et al.*, 1980; Clauss *et al.*, 1985; Charney *et al.*, 1981; Halevy *et al.*, 1988). The mechanism of the response to glucocorticoids does not appear to be attributable to the occupancy of the mineralocorticoid receptor because this response is not inhibited by spironolactone (Halevy *et al.*, 1988; Charney *et al.*, 1981) and because the mineralocorticoid receptor of the colonic epithelium is "protected" by the presence of an active 11β-OH-steroid dehydrogenase (Pacha and Miksik, 1994). Some kind of cooperativity between mineralo- and glucocorticoid receptors may be essential for a full Na transport response (Fromm *et al.*, 1990; Grotjohann *et al.*, 1999). The essential role of the mineralocorticoid receptor for the control of the activity of colonic ENaC also is demonstrated by the marked reduction in amiloride-sensitive transepithelial potential and short-circuit current in knockout mice lacking the mineralocorticoid receptor (Bleich *et al.*, 1999). There are also circadian variations in the activity of ENaC in the colon, as estimated by the amplitude of the amiloride-sensitive potential difference measured across the distal colonic epithelium *in vivo* (Clauss *et al.*, 1988; Wang *et al.*, 2000), and these variations probably reflect the circadian cycle of steroid hormones.

In spite of physiological effects similar to those observed in urinary tight epithelia, the effects of steroid hormones on ENaC subunit mRNA have shown a pattern of response in the colon that is clearly different from that observed in the kidney or lung. In airway epithelia, the mRNAs of the three subunits are induced by glucocorticoids, whereas mineralocorticoids or salt deprivation have no effects (Lingueglia *et al.*, 1994; Stokes and Sigmund, 1998); physiologically this effect of glucocorticoids appears to be most important in the maturation of the lung and airways at the time of birth (Tchepichev *et al.*, 1995). In kidney, the transcriptional effects of mineralocorticoids on ENaC subunits are somewhat controversial: no large effect of either mineralo- or glucocorticoids could be detected initially (Renard *et al.*, 1995), but other studies demonstrated a slow and modest increase in the mRNA abundance of mostly the α subunit following aldosterone treatment (Asher *et al.*, 1996; May *et al.*, 1997). In contrast, in the distal colon a number of studies have demonstrated that (1) there is constitutive expression of the α subunit of ENaC, (2) this expression of the α subunit is not influenced by steroid hormones (3) the baseline expression of the β and γ subunits is very low, *but* (4) it may be strongly stimulated by either gluco- or mineralocorticoid hormones (Asher *et al.*, 1996; Stokes and Sigmund, 1998; Renard *et al.*, 1995).

In urinary epithelia, the main early effect on the rate of ENaC protein synthesis appears to be posttranscriptional, mostly through an effect on the translation of the α subunit; that is, a clear increase in protein synthesis takes place before any detectable change in mRNA abundance (May *et al.*, 1997). It is not yet known whether a similar posttranscriptional mechanism is at work in the colonic epithelium.

VI. CONTROL OF CHANNEL DENSITY

For a given rate of protein synthesis, the density of channels at the apical cell surface depends on two main processes: the distribution of channels between intracellular and surface pools and the rate of recovery and/or protein degradation from the surface pool.

The control of the movement of ENaC from an intracellular pool to the surface is not well understood. In the well-studied model of the *Xenopus* oocyte expressing ENaC, only a small proportion of the synthesized ENaC protein appears as functional channels at the oocyte membrane (Valentijn *et al.*, 1998). Thus, this model may be only partially useful to understand the control of apical membrane expression in tight epithelia that express ENaC naturally.

Vasopressin is known to increase Na reabsorption in the collecting tubule (Schafer and Hawk, 1992) and the amphibian model tight epithelia (Verrey, 1994). This effect occurs through an increase in the number of active Na channels in the apical membrane (Li *et al.*, 1982; Garty and Edelman, 1983; Marunaka and Eaton, 1991). Recently it was shown that expression of ENaC in *Xenopus* oocytes using the guinea pig α subunit produced channels that were strongly activated by cAMP stimulation, which suggests that protein kinase A has a direct effect on these channels (Schnizler *et al.*, 2000). A similar stimulatory effect of vasopressin has been observed in amphibian colon (Krattenmacher and Clauss, 1988), but in mammalian colon, vasopressin does not increase the electrogenic Na transport rate, and it may decrease active Cl$^-$ secretion (Bridges *et al.*, 1983; 1984). The precise mechanism of the transfer of new Na channels to the apical membrane is not yet known, but an increased rate of new channels (Snyder, 2000) and membrane insertion (Erlij *et al.*, 1999) suggests an exocytotic process, and phosphorylation of ENaC also appears to be involved (Shimkets *et al.*, 1998).

A somewhat better understanding of the regulation of ENaC density by control of the half-life of this protein in the apical membrane has resulted from the detailed study of channel mutants responsible for Liddle's syndrome, a hereditary form of salt-sensitive hypertension. The clinical study of patients with Liddle's syndrome suggests an inappropriate excessive Na reabsorption in the kidney (Liddle *et al.*, 1963; Botero-Velez *et al.*, 1994). Expression of the identified mutants, first in *Xenopus* oocytes and later in transgenic mice, showed a gain of function that could be attributable to both an increase in channel density and open probability (Hansson *et al.*, 1995; Schild *et al.*, 1996; Pradervand *et al.*, 1999). The results of several different studies indicated an increased half-life of Liddle's mutant ENaC at the plasma membrane of cells in which ENaC was expressed (Shimkets *et al.*, 1997; Kellenberger *et al.*, 1998; Snyder, 2000). Studies of Liddle's mutations of ENaC β and γ subunits have shown that the "PY" motif present in the intracellular C-terminal domain of these subunits (see Fig. 2) is essential for the interaction of ENaC with the "WW" domain of the ubiquitin ligase Nedd4 (a ubiquitin ligase is an enzyme responsible for the attachment of ubiquitin to proteins [Staub *et al.*, 1996; 1997]). Because ubiquitination is a well-known signal that tags proteins for degradation, it was initially proposed that the increased membrane density of Liddle's mutants was attributable to a reduced protein degradation rate. However, results from several laboratories have indicated involvement of the endocytic retrieval of ENaC from the apical membrane (endocytosis that may or may not be followed by further protein degradation). The short half-life of wild-type ENaC at the plasma membrane may be explained by an active endocytic process (Chalfant *et al.*, 1999a; Shimkets *et al.*, 1997; Staub *et al.*, 1997). The expression of a dominant negative dynamin mutant (Shimkets *et al.*, 1997) or the injection of dynamin antibodies (Hopf *et al.*, 1999) enhanced the expression of amiloride-sensitive current; these observations suggest that the lifetime of the normal ENaC in the plasma membrane is limited by clathrin-mediated endocytosis.

Both the N-terminal and the C-terminal intracellular domains appear to be involved in the endocytic retrieval of ENaC. First, concerning the C-terminal domain, endocytosis of several membrane proteins has been shown to be promoted by ubiquitination of the protein (Hicke, 1999). Thus, the binding of Nedd4 to ENaC and the following ubiquitination reaction are expected to result in stimulation of ENaC endocytosis. The loss of or the alteration of the "PY" motif results in a deficient association with Nedd4, reduced ubiquitination, and thereby decreased endocytosis and a longer half-life of Liddle's mutant channels (Abriel *et al.*, 1999). Another interpretation of the same effect has been proposed: the tyrosine residue in the "PY" motif of the β and γ subunits might be part of internalization signal sequences ("PPXY" and "YXXL") such as those encountered in other membrane proteins known to be internalized via clathrin-coated pits (Shimkets *et al.*, 1997).

Second, concerning the N-terminal domain, this segment, in each subunit, contains several lysine residues that are potential sites for the attachment of ubiquitin by the ubiquitin ligase. Multiple mutations of several of these lysine residues led to both a decrease in ENaC ubiquitination and an increase of the channel density at the oocyte surface (Staub *et al.*, 1997). Another mechanism was proposed following the observation that the truncation of about 50 residues from the α subunit N terminus resulted in a longer half-life and increased density of the channel in the membrane: this segment contains a "KGDK" motif similar to a known endocytic motif of other membrane proteins (Chalfant *et al.*, 1999a).

By what mechanism is the endocytic retrieval rate of ENaC regulated? Presently this question has no complete answer. A recent publication indicates that the increased density of ENaC following vasopressin treatment is *not* due to a decreased retrieval rate (Snyder, 2000). The best-characterized influence on ENaC half-life is the effect of intracellular Na concentration. An increase in intracellular Na concentration ($[Na]_i$) is known to reduce apical membrane conductance, a phenomenon described as *feedback inhibition* (Garty and Palmer, 1997); a similar phenomenon has been well described in the colon (Nzegwu and

Levin, 1992; Frizzell and Schultz, 1978; Kirk and Dawson, 1985). Although this regulation also could involve a change in the open probability of the channel (see below), recently published observations (Kellenberger *et al.*, 1998) suggest that the rate of channel retrieval from the surface may be mainly responsible. Although a stable amiloride-sensitive current was observed when a stable [Na]$_i$ was maintained, oocytes expressing wild-type ENaC showed a decrease in the amiloride-sensitive current as [Na]$_i$ was allowed to increase; in contrast, this down-regulation was not observed in channels bearing Liddle's mutation.

Although all the studies showing the effect of Liddle's mutations on channel half-life have been carried out with *in vitro* artificial expression systems, the pathophysiology of the patients suffering from Liddle's syndrome and the alterations of the salt balance in the animal model of this syndrome strongly suggest that the same modifications of ENaC biology are present in the epithelia of these organisms. A more direct indication was provided by measurements of the amiloride-sensitive transepithelial potential in the distal colon; mice bearing Liddle's mutation on the β subunit of ENaC showed an increased Na transport rate compared with normal mice (Pradervand *et al.*, 1999). Therefore, the available evidence supports the hypothesis of an increase in Na reabsorption in the colon as well as in urinary epithelia in Liddle's syndrome.

VII. CONTROL OF CHANNEL ACTIVITY

The study of the regulation of ENaC activity has been rendered difficult because of large spontaneous variations of the open probability of these channels (Palmer and Frindt, 1988; 1996). A number of intracellular and extracellular factors have been suggested as ENaC regulators, but several observations are not easily reproduced in different experimental models and the role of most of these factors is still a matter of debate.

A number of intracellular factors are known to influence ENaC activity. The channel is sensitive to *intracellular pH* (pH$_i$) (Palmer, 1985; Palmer and Frindt, 1987; Chalfant *et al.*, 1999b), an effect that is fast and reversible and can be observed in isolated membrane preparations (Abriel and Horisberger, 1999). It is not clear whether the pH$_i$ sensitivity of ENaC is sufficient to be responsible for a physiological regulatory role; a decrease in the pH$_i$ to 6.5 is needed to decrease the Na current by about half.

An increase in *intracellular Ca* also has been shown to result in a decrease in ENaC activity (Palmer and Frindt, 1987; Garty *et al.*, 1987; Silver *et al.*, 1993). The mechanism of this inhibition, however, remains controversial; some studies are compatible with a direct effect of intracellular calcium activity on ENaC activity (Ishikawa *et al.*, 1998), whereas others do not show evidence of a direct effect (Palmer and Frindt, 1987; Abriel and Horisberger, 1999).

The phenomenon of *feedback inhibition,* mentioned earlier in the control of channel density, also may be involved in the regulation of ENaC activity. Studies of ENaC expressed in MDCK cells suggest a direct inhibitory effect of intracellular Na (Ishikawa *et al.,* 1998), but studies with ENaC expressed the *Xenopus* oocyte do not show evidence of a direct control by intracellular Na on ENaC activity (Abriel and Horisberger, 1999). The direct or indirect effect of calcium on ENaC activity has been proposed as part of the mechanism responsible for feedback inhibition; that is, the increase in intracellular Na would lead to an increase in calcium activity because of the presence of the basolateral Na/Ca exchanger (Grinstein and Erlij, 1978). The feedback regulation of amiloride-sensitive Na conductance has been well described in distal colonic epithelium (Turnheim *et al.,* 1983).

The single-channel inward current carried by ENaC is not a linear function of the extracellular Na concentration; rather it shows saturation with a half concentration around 20 mM (Palmer and Frindt, 1988). In addition, the macroscopic Na current is known to decrease rapidly after a sudden increase in extracellular Na concentration, a phenomenon described as Na "self-inhibition" (Garty and Palmer, 1997; Garty and Benos, 1988). This regulation appears to depend directly on the luminal Na concentration and to be distinct from the secondary increase in intracellular Na (Palmer and Frindt, 1996), but its precise mechanism is not known.

When expressed in *Xenopus* oocytes, the activity of ENaC can be strongly stimulated by treatment with extracellular proteases (Chraïbi *et al.,* 1998) and in frog epithelial cells Na transport is inhibited by protease inhibitors, an effect that can be reversed by exposure to trypsin (Vallet *et al.,* 1997). The discovery of the expression of a serine protease in the same cells (Vallet *et al.,* 1997) strongly suggests that a posttranslational proteolytic modification of ENaC, of the α, β, or γ subunits, or of an associated protein, may play a regulatory role in the expression of active channels at the cell surface. The observation of a shift in the molecular weight of the γ subunit of ENaC after aldosterone treatment suggests a regulatory proteolytic cleavage of this subunit (Masilamani *et al.,* 1999). The full mechanism of this regulation and the precise substrate of the protease still need to be determined.

The apical membrane Na conductance shows an inward rectification that is more marked than that observed for single-channel currents (Palmer *et al.,* 1998). This suggests that the open probability of the channel is voltage dependent, in the sense that it is increased by membrane hyperpolarization (in contrast with the voltage-gated Na channels of excitable cells). Single-channel measurements show only a slight voltage dependence under steady-state conditions (Palmer and Frindt, 1988), but large increases in open probability also have been observed after membrane hyperpolarization (Palmer and Frindt, 1996). The ENaC activation caused by a low luminal Na concentration and the activation owing to membrane hyperpolarization are not additive, and thus may share a common mechanism (Palmer *et al.*, 1998).

All these mechanisms, saturability, control by extracellular ("self-inhibition") and intracellular Na ("feedback inhibition"), and the effect of membrane potential, work in the same direction and tend to enhance the efficiency of Na reabsorption when luminal Na concentration is low and to limit Na entry at high Na concentration.

An interaction between ENaC and the cystic fibrosis transmembrane conductance regulator (CFTR) was first indicated by the increased amiloride-sensitive transepithelial potential observed in the airway epithelia of patients with cystic fibrosis (Boucher *et al.*, 1986), and similar stimulation was observed in animal models of CFTR (Grubb and Boucher, 1997). An inhibitory effect of the presence or activity of CFTR on the amiloride-sensitive current has been observed in several preparations naturally expressing these two membrane proteins, such as the colon (Ecke *et al.*, 1996; Mall *et al.*, 1999) and artificial expression systems (Stutts *et al.*, 1995; Mall *et al.*, 1996; Briel *et al.*, 1998). Although a direct interaction of an intracellular domain of CFTR with ENaC has been observed *in vitro* (Ismailov *et al.*, 1996b), the exact mechanism of this interaction is not fully understood. The relevance of this effect might be important in distal colonic epithelium because both CFTR and ENaC may be highly expressed in this tissue; however, its significance is not yet clear because of the different cellular distribution of ENaC and CFTR. As stated earlier, ENaC is expressed mostly in surface cells, whereas CFTR is found in crypt cells (Smith *et al.*, 1993; Duc *et al.*, 1994; Watanabe *et al.*, 1999; Kockerling *et al.*, 1996; Lomax *et al.*, 1996; Ecke *et al.*, 1996). Inhibition of ENaC by CFTR could play a role in the balance between fluid absorption or secretion in the cells along the crypts, particularly cells of the mid-crypt zone in which both proteins are expressed.

The negative interaction between ENaC and CFTR appears to be specific to the airway and colonic epithelia. It is not observed in urinary epithelia, in which antidiuretic hormone induces a simultaneous activation of the apical membrane Na and Cl conductances (Verrey, 1994), or in sweat glands, where the activity of CFTR is required for ENaC function (Reddy *et al.*, 1999).

VIII. ENaC AND K SECRETION

The distal colon epithelium is capable of both K reabsorption and K secretion, similar to the collecting tubule of the kidney (Bastl *et al.*, 1978; Fisher *et al.*, 1976; McCabe *et al.*, 1982; Halm and Dawson, 1984a; 1984b). This ability is essential for the preservation of K balance, especially when kidney function is impaired (as in renal failure) (Bastl *et al.*, 1977) or perturbed through the effects of diuretic drugs. Available evidence indicates that K reabsorption is performed by an apical membrane K-exchange ATPase (Meneton *et al.,* 1998), the so-called colonic H,K-ATPase (Crowson and Shull, 1992) even though this P-type ATPase is expressed not only in the colon but also in kidney, and it transports not only H but also Na in exchange for K (Jaisser and Beggah, 1999). Available evidence also suggests that K secretion occurs in the colon along the same principle that is known to drive K secretion in the cortical collecting tubule of the kidney (Koeppen and Giebisch, 1985); K is accumulated into the cell by the basolateral Na,K-ATPase, and the presence of K channels in the apical membrane allows a flow of potassium from the cell to the lumen according to its electrochemical gradient (Fig. 3). In the cortical collecting tubule there is good evidence that ROMK is the apical K channel responsible for K secretion (Palmer *et al.,* 1997).

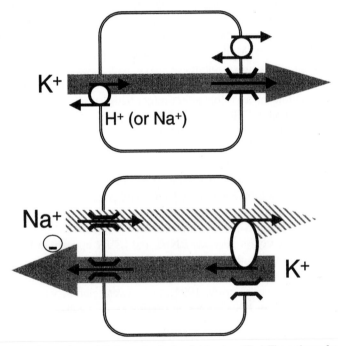

FIGURE 3 K secretion and K absorption in the distal colon. (Top) The pathway for K absorption by a distal colonic cell with the "colonic H,K-ATPase" in the apical membrane, which exchanges luminal K for cellular H or Na, and basolateral K channels; the molecular nature of these channels is not yet determined. (Bottom) The principle of the K secretion mechanism. The basolateral Na,K-ATPase maintains a low intracellular Na concentration and imports K from the basolateral side of the cell. In the presence of K channels in the apical membrane, K is driven from the cell to the lumen according to its electrochemical gradient. Electrogenic Na entry through ENaC depolarizes the apical membrane potential and provides a favorable electrochemical gradient for K secretion. Under mineralocorticoid stimulation, with a high Na,K-ATPase and high ENaC activity, the system works as an efficient exchanger of luminal Na for serosal K.

In colon, the apical K channel associated with K secretion has been characterized as barium- and tetraethylammonium-sensitive (Wilkinson *et al.,* 1993), but its molecular nature is not yet known. As illustrated in Figure 3, the activity of ENaC in the same membrane is important for two reasons in this process: (1) it provides Na, which is the rate-limiting substrate to the Na,K-ATPase promoting the basolateral import of K, and (2) it depolarizes the apical membrane and creates an electrochemical gradient favorable for K secretion toward the lumen (Halm and Dawson, 1984b). Thus, the K-secreting capacity of the colon is strongly linked to the Na-reabsorbing rate (Skadhauge and Thomas, 1979), and the activity of ENaC is essential not only for Na absorption but also for efficient K secretion. Accordingly, the permissive effect of corticosteroid hormones on K secretion can be largely explained by both the stimulation of ENaC and the increased Na,K-ATPase activity (Hayslett *et al.,* 1980; Sandle *et al.,* 1984; Martin *et al.,* 1986), but a direct effect of steroids on the abundance and/or activity of K channels also appears to be present (Sandle *et al.,* 1985; Edmonds and Willis, 1988; Rechkemmer and Halm, 1989; Halm and Halm, 1994).

IX. CONCLUDING REMARKS

The distal colon may play an important role in Na and K balance by the ability of colonic epithelium to perform Na absorption and K secretion; this ability is most important in cases of impaired renal function or when large volumes of intestinal fluid are being lost. The amiloride-sensitive Na channel present in the apical membrane of surface colonic cells is essential for both Na reabsorption and K secretion. Available evidence indicates that the structural, pharmacological, and biophysical properties of ENaC in colon are similar to those observed in urinary epithelia, but the mechanism of its hormonal regulation shows significant differences when compared with the kidney or airway epithelial sodium channels.

Acknowledgments

The author is grateful to Bernard Rossier and Olivier Bonny for their careful reading of the manuscript and helpful suggestions. The work on the epithelial sodium channel is supported by the Human Frontier Science Program (Grant No. RG–464/96).

References

Abriel, H., and Horisberger, J.-D. (1999). Feedback inhibition of amiloride-sensitive epithelial sodium channel in *Xenopus leavis* oocyte. *J. Physiol. (Lond.)* **516,** 31–43.

Abriel, H., Loffing, J., Rebhun, J. F., Pratt, J. H., Schild, L., Horisberger, J.-D., Rotin, D., and Staub, O. (1999). Defective regulation of the epithelial Na$^+$ channel by Nedd4 in Liddle's syndrome. *J. Clin. Invest.* **103,** 667–673.

Asher, C., Wald, H., Rossier, B. C., and Garty, H. (1996). Aldosterone-induced increase in the abundance of Na$^+$ channel subunits. *Am. J. Physiol. Cell Physiol.* **40,** C605–C611.

Bastl, C., Hayslett, J. P., and Binder, H. J. (1977). Increased large intestinal secretion of potassium in renal insufficiency. *Kidney Int.* **12,** 9–16.

Bastl, C., Kliger, A. S., Binder, H. J., and Hayslett, J. P. (1978). Characteristics of potassium secretion in the mammalian colon. *Am. J. Physiol.* **234,** F48–F53.

Bastl, C. P., Binder, H. J., and Hayslett, J. P. (1980). Role of glucocorticoids and aldosterone in maintenance of colonic cation transport. *Am. J. Physiol.* **238,** F181–F186.

Benos, D. J., and Stanton, B. A. (1999). Functional domains within the degenerin/epithelial sodium channel (Deg/ENaC) superfamily of ion channels. *J. Physiol. (Lond.)* **520,** 631–644.

Berdiev, B. K., Karlson, K. H., Jovov, B., Ripoll, P. J., Morris, R., Loffingcueni, D., Halpin, P., Stanton, B. A., Kleyman, T. R., and Ismailov, I. I. (1998). Subunit stoichiometry of a core conduction element in a cloned epithelial amiloride-sensitive Na^+ channel. *Biophys. J.* **75,** 2292–2301.

Bleich, M., Warth, R., Schmidt-Hieber, M., Schulz-Baldes, A., Hasselblatt, P., Fisch, D., Berger, S., Kunzelmann, K., Kriz, W., Schutz, G., and Greger, R. (1999). Rescue of the mineralocorticoid receptor knock-out mouse. *Pflugers Arch.* **438,** 245–254.

Bonny, O., Chraibi, A., Loffing, J., Jaeger, N. F., Gründer, S., Horisberger, J. D., and Rossier, B. C. (1999). Functional expression of a pseudohypoaldosteronism type I mutated epithelial Na^+ channel lacking the pore-forming region of its alpha subunit. *J. Clin. Invest.* **104,** 967–974.

Botero-Velez, M., Curtis, J. J., and Warnock, D. G. (1994). Brief report: Liddle's syndrome revisited—A disorder of sodium reabsorption in the distal tubule. *N. Engl. J. Med.* **330,** 178–181.

Boucher, R. C., Stutts, M. J., Knowles, M. R., Cantley, L., and Gatzy, J. T. (1986). Na^+ transport in cystic fibrosis respiratory epithelia. *J. Clin. Invest.* **78,** 1245–1252.

Bridges, R. J., Nell, G., and Rummel, W. (1983). Influence of vasopressin and calcium on electrolyte transport across isolated colonic mucosa of the rat. *J. Physiol. (Lond.)* **338,** 463–475.

Bridges, R. J., Rummel, W., and Wollenberg, P. (1984). Effects of vasopressin on electrolyte transport across isolated colon from normal and dexamethasone-treated rats. *J. Physiol. (Lond.)* **355,** 11–23.

Briel, M., Greger, R., and Kunzelmann, K. (1998). Cl^- transport by cystic fibrosis transmembrane conductance regulator (CFTR) contributes to the inhibition of epithelial Na^+ channels (ENaCs) in xenopus oocytes co-expressing CFTR and ENaC. *J. Physiol. (Lond.)* **508,** 825–836.

Brouard, M., Casado, M., Djelidi, S., Barrandon, Y., and Farman, N. (1999). Epithelial sodium channel in human epidermal keratinocytes: Expression of its subunits and relation to sodium transport and differentiation. *J. Cell Sci.* **112,** 3343–3352.

Canessa, C. M., Horisberger, J.-D., and Rossier, B. C. (1993). Functional cloning of the epithelial sodium channel: Relation with genes involved in neurodegeneration. *Nature* **361,** 467–470.

Canessa, C. M., Horisberger, J.-D., Schild, L., and Rossier, B. C. (1995). Expression cloning of the epithelial sodium channel. *Kidney Int.* **48,** 950–955.

Canessa, C. M., Merillat, A.-M., and Rossier, B. C. (1994b). Membrane topology of the epithelial sodium channel in intact cells. *Am. J. Physiol. Cell Physiol.* **267,** C1682–C1690.

Canessa, C. M., Schild, L., Buell, G., Thorens, B., Gautshi, Y., Horisberger, J.-D., and Rossier, B. C. (1994a). The amiloride-sensitive epithelial sodium channel is made of three homologous subunits. *Nature* **367,** 463–467.

Chalfant, M. L., Denton, J. S., Berdiev, B. K., Ismailov, I. I., Benos, D. J., and Stanton, B. A. (1999b). Intracellular H^+ regulates the alpha-subunit of ENaC, the epithelial Na^+ channel. *Am. J. Physiol. Cell Physiol.* **45,** C477–C486.

Chalfant, M. L., Denton, J. S., Langloh, A. L., Karlson, K. H., Loffing, J, Benos, D. J., and Stanton, B. A. (1999a). The NH2 terminus of the epithelial sodium channel contains an endocytic motif. *J. Biol. Chem.* **274,** 32889–32896.

Charney, A. N., Wallach, J., Ceccarelli, S., Donowitz, M., and Costenbader, C. L. (1981). Effects of spironolactone and amiloride on corticosteroid-induced changes in colonic function. *Am. J. Physiol.* **241,** G300–G305.

Cheng, C., Prince, L. S., Snyder, P. M., and Welsh, M. J. (1998). Assembly of the epithelial Na$^+$ channel evaluated using sucrose gradient sedimentation analysis. *J. Biol. Chem.* **273,** 22693–22700.

Chraïbi, A., Vallet, V., Firsov, D., Kharoubi-Hess, S., and Horisberger, J.-D. (1998). Protease modulation of the activity of the epithelial sodium channel expressed in *Xenopus* oocyte. *J. Gen. Physiol.* **111,** 1–12.

Clauss, W., Durr, J., Skadhauge, E., and Hornicke, H. (1985). Effects of aldosterone and dexamethasone on apical membrane properties and Na-transport of rabbit distal colon in vitro. *Pflugers Arch.* **403,** 186–192.

Clauss, W., Durr, J. E., Krattenmacher, R., Hornicke, H., and Van Driessche, W. (1988). Circadian rhythm of apical Na-channels and Na-transport in rabbit distal colon. *Experientia* **44,** 608–610.

Clauss, W., Hoffmann, B., Krattenmacher, R., and Van Driessche, W. (1993). Current-noise analysis of Na absorption in the embryonic coprodeum: Stimulation by aldosterone and thyroxine. *Am. J. Physiol. Regul. Integr. Comp. Physiol.* **265,** R1100–R1108.

Coscoy, S., Lingueglia, E., Lazdunski, M., and Barbry, P. (1998). The FMRFamide activated sodium channel is a tetramer. *J. Biol. Chem.* **273,** 8317–8322.

Crowson, M. S., and Shull, G. E. (1992). Isolation and characterization of a cDNA encoding the putative distal colon H$^+$,K$^+$-ATPase. Similarity of deduced amino acid sequence to gastric H$^+$,K$^+$-ATPase and Na$^+$,K$^+$-ATPase and mRNA expression in distal colon, kidney, and uterus. *J. Biol. Chem.* **267,** 13740–13748.

Dinudom, A., Young, J. A., and Cook, D. I. (1993). Amiloride-sensitive Na$^+$ current in the granular duct cells of mouse mandibular glands. *Pflugers Arch.* **423,** 164–166.

Duc, C., Farman, N., Canessa, C. M., Bonvalet, J.-P., and Rossier, B. C. (1994). Cell-specific expression of epithelial sodium channel α, β, and gamma subunits in aldosterone-responsive epithelia from the rat: Localization by *in situ* hybridization and immunocytochemistry. *J. Cell Biol.* **127,** 1907–1921.

Ecke, D., Bleich, M., and Greger, R. (1996). The amiloride inhibitable Na$^+$ conductance of rat colonic crypt cells is suppressed by forskolin. *Pflugers Arch.* **431,** 984–986.

Edmonds, C. J. (1981). Amiloride sensitivity of the transepithelial electrical potential and of sodium and potassium transport in rat distal colon in vivo. *J. Physiol. (Lond.)* **313,** 547–559.

Edmonds, C. J., and Willis, C. L. (1988). Potassium secretion by rat distal colon during acute potassium loading: Effect of sodium, potassium intake, and aldosterone. *J. Physiol. (Lond.)* **401,** 39–51.

Epple, H. J., Schulzke, J. D., Schmitz, H., and Fromm, M. (1995). Enzyme- and mineralocorticoid receptor-controlled electrogenic Na$^+$ absorption in human rectum in vitro. *Am. J. Physiol. Gastrointest. Liver Physiol.* **32,** G42–G48.

Erlij, D., De Smet, P., Van Mesotten, D., and Driessche, W. (1999). Forskolin increases apical sodium conductance in cultured toad kidney cells (A6) by stimulating membrane insertion. *Pflugers Arch.* **438,** 195–204.

Eskandari, S., Snyder, P. M., Kreman, M., Zampighi, G. A., Welsh, M. J., and Wright, E. M. (1999). Number of subunits comprising the epithelial sodium channel. *J. Biol. Chem.* **274,** 27281–27286.

Farman, N., Talbot, C. R., Boucher, R., Fay, M., Canessa, C., Rossier, B., and Bonvalet, J. P. (1997). Noncoordinated expression of alpha-, beta-, and gamma-subunit mRNAs of epithelial Na$^+$ channel along rat respiratory tract. *Am. J. Physiol. Cell Physiol.* **41,** C131–C141.

Firsov, D., Gautschi, I., Merillat, A. M., Rossier, B. C., and Schild, L. (1998). The heterotetrameric architecture of the epithelial sodium channel (ENaC). *EMBO J.* **17,** 344–352.

Fisher, K. A., Binder, H. J., and Hayslett, J. P. (1976). Potassium secretion by colonic mucosal cells after potassium adaptation. *Am. J. Physiol.* **231,** 987–994.

Frizzell, R. A., and Schultz, S. G. (1978). Effect of aldosterone on ion transport by rabbit colon in vitro. *J. Membr. Biol.* **39,** 1–26.

Fromm, M., Schulzke, J. D., and Hegel, U. (1990). Aldosterone low-dose, short-term action in adrenalectomized glucocorticoid-substituted rats: Na, K, Cl, HCO_3, osmolyte, and water transport in proximal and rectal colon. *Pflugers Arch.* **416,** 573–579.

Fromm, M., Schulzke, J. D., and Hegel, U. (1993). Control of electrogenic Na^+ absorption in rat late distal colon by nanomolar aldosterone added *in vitro. Am. J. Physiol. Endocrinol. Metab.* **264,** E68–E73.

Fyfe, G. K., and Canessa, C. M. (1998). Subunit composition determines the single channel kinetics of the epithelial sodium channel. *J. Gen. Physiol.* **112,** 423–432.

Fyfe, G. K., Quinn, A.-M., and Canessa, C. M. (1998). Structure and function of the Mec-ENaC family of ion channels. *Semin. Nephrol.* **18,** 138–151.

Garty, H., Asher, C., and Yeger, O. (1987). Direct inhibition of epithelial Na^+ channels by a pH-dependent interaction with calcium, and by other divalent ions. *J. Membr. Biol.* **95,** 151–162.

Garty, H., and Benos, D. J. (1988). Characteristics and regulatory mechanisms of the amiloride-blockable Na^+ channel. *Physiol. Rev.* **68,** 309–372.

Garty, H., and Edelman, I. S. (1983). Amiloride-sensitive trypsinization of apical sodium channels. Analysis of hormonal regulation of sodium transport in toad bladder. *J. Gen. Physiol.* **81,** 785–803.

Garty, H., and Palmer, L. G. (1997). Epithelial sodium channels—Function, structure, and regulation. *Physiol. Rev.* **77,** 359–396.

Grinstein, S., and Erlij, D. (1978). Intracellular calcium and the regulation of sodium transport in the frog skin. *Proc. R. Soc. Lond., Ser. B,* **202,** 353–360.

Grotjohann, I., Schulzke, J. D., and Fromm, M. (1999). Electrogenic Na^+ transport in rat late distal colon by natural and synthetic glucocorticosteroids. *Am. J. Physiol. Gastrointest. Liver Physiol.* **39,** G491–G498.

Grubb, B. R., and Boucher, R. C. (1997). Enhanced colonic Na^+ absorption in cystic fibrosis mice versus normal mice. *Am. J. Physiol. Gastrointest. Liver Physiol.* **35,** G393–G400.

Gründer, S., Firsov, D., Chang, S. S., Jaeger, N. F., Gautschi, I., Schild, L., Lifton, R. P., and Rossier, B. C. (1997). A mutation causing pseudohypoaldosteronism type 1 identifies a conserved glycine that is involved in the gating of the epithelial sodium channel. *EMBO J.* **16,** 899–907.

Gründer, S., Jaeger, N. F., Gautschi, I., Schild, L., and Rossier, B. C. (1999). Identification of a highly conserved sequence at the N-terminus of the epithelial Na^+ channel alpha subunit involved in gating. *Pflugers Arch.* **438,** 709–715.

Halevy, J., Boulpaep, E. L., Budinger, M. E., Binder, H. J., and Hayslett, J. P. (1988). Glucocorticoids have a different action than aldosterone on target tissue. *Am. J. Physiol.* **254,** F153–F158.

Halm, D. R., and Dawson, D. C. (1984a). Potassium transport by turtle colon: Active secretion and active absorption. *Am. J. Physiol.* **246,** C315–C322.

Halm, D. R., and Dawson, D. C. (1984b). Control of potassium transport by turtle colon: Role of membrane potential. *Am. J. Physiol.* **247,** C26–C32.

Halm, D. R., and Halm, S. T. (1994). Aldosterone stimulates K secretion prior to onset of Na absorption in guinea pig distal colon. *Am. J. Physiol. Cell Physiol.* **266,** C552–C558.

Hansson, J. H., Nelson-Williams, C., Suzuki, H., Schild, L., Shimkets, R. A., Lu, Y., Canessa, C. M., Iwasaki, T., Rossier, B. C., and Lifton, R. P. (1995). Hypertension caused by mutation in the gamma subunit of the epithelial sodium channel: Genetic heterogeneity of Liddle's syndrome. *Nature Genet.* **11,** 76–82.

Hayslett, J. P., Myketey, N., Binder, H. J., and Aronson, P. S. (1980). Mechanism of increased potassium secretion in potassium loading and sodium deprivation. *Am. J. Physiol.* **239,** F378–F382.

Hicke, L. (1999). Gettin' down with ubiquitin: Turning off cell-surface receptors, transporters and channels. *Trends Cell Biol.* **9,** 107–112.

Hopf, A., Schreiber, R., Mall, M., Greger, R., and Kunzelmann, K. (1999). Cystic fibrosis transmembrane conductance regulator inhibits epithelial Na^+ channels carrying Liddle's syndrome mutations. *J. Biol. Chem.* **274,** 13894–13899.

Horisberger, J.-D. (1998). Amiloride-sensitive Na channels. *Curr. Opin. Cell Biol.* **10**, 443–449.

Horisberger, J.-D., and Rossier, B. C. (1992). Aldosterone regulation of gene transcription leading to control of ion transport. *Hypertension* **19**, 221–227.

Ishikawa, T., Marunaka, Y., and Rotin, D. (1998). Electrophysiological characterization of the rat epithelial Na$^+$ channel (rENaC) expressed in MDCK cells—Effects of Na$^+$ and Ca^{2+}. *J. Gen. Physiol.* **111**, 825–846.

Ismailov, I. I., Awayda, M. S., Berdiev, B. K., Bubien, J. K., Lucas, J. E., Fuller, C. M., and Benos, D. J. (1996a). Triple-barrel organization of ENaC, a cloned epithelial Na$^+$ channel. *J. Biol. Chem.* **271**, 807–816.

Ismailov, I. I., Awayda, M. S., Jovov, B., Berdiev, B. K., Fuller, C. M., Dedman, J. R., Kaetzel, M. A., and Benos, D. J. (1996b). Regulation of epithelial sodium channels by the cystic fibrosis transmembrane conductance regulator. *J. Biol. Chem.* **271**, 4725–4732.

Ismailov, I. I., Berdiev, B. K., and Benos, D. J. (1995). Regulation by Na$^+$ and Ca^{2+} of renal epithelial Na$^+$ channels reconstituted into planar lipid bilayers. *J. Gen. Physiol.* **106**, 445–466.

Jaisser, F., and Beggah, A. T. (1999). The nongastric H$^+$-K$^+$-ATPases: Molecular and functional properties. *Am. J. Physiol. Renal Fluid Electrolyte Physiol.* **45**, F812–F824.

Kellenberger, S., Gautschi, I., Rossier, B. C., and Schild, L. (1998). Mutations causing Liddle syndrome reduce sodium-dependent downregulation of the epithelial sodium channel in the *Xenopus* oocyte expression system. *J. Clin. Invest.* **101**, 2741–2750.

Kellenberger, S., Gautschi, I., and Schild, L. (1999b). A single point mutation in the pore region of the epithelial Na$^+$ channel changes ion selectivity by modifying molecular sieving. *Proc. Natl. Acad. Sci. USA* **96**, 4170–4175.

Kellenberger, S., Hoffmann-Pochon, N., Gautschi, I., Schneeberger, E., and Schild (1999a). On the molecular basis of ion permeation in the epithelial Na$^+$ channel. *J. Gen. Physiol.* **114**, 13–30.

Kirk, K. L., and Dawson, D. C. (1985). Passive cation permeability of turtle colon: Evidence for a negative interaction between intracellular sodium and apical sodium permeability. *Pflugers Arch.* **403**, 82–89.

Kockerling, A., Sorgenfrei, D., and Fromm, M. (1996). Electrogenic Na$^+$ absorption of rat distal colon is confined to surface epithelium: A voltage-scanning study. *Am. J. Physiol. Cell Physiol.* **264**, C1285–C1293.

Koeppen, B. M., and Giebisch, G. (1985). Cellular electrophysiology of potassium transport in the mammalian cortical collecting tubule. *Pflugers Arch.* **405**, 143–146.

Kosari, F., Sheng, S. H., Li, J. Q., Mak, D. D., Foskett, J. K., and Kleyman, T. R. (1998). Subunit stoichiometry of the epithelial sodium channel. *J. Biol. Chem.* **273**, 13469–13474.

Koyama, K., Sasaki, I., Naito, H., Funayama, Y., Fukushima, K., Unno, M., Matsuno, S., Hayashi, H., and Suzuki, Y. (1999). Induction of epithelial Na$^+$ channel in rat ileum after proctocolectomy. *Am. J. Physiol. Gastrointest. Liver Physiol.* **39**, G975–G984.

Krattenmacher, R., and Clauss, W. (1988). Electrophysiological analysis of sodium-transport in the colon of the frog (Rana esculenta). *Pflugers Arch.* **411**, 606–612.

Li, J. H. Y., Palmer, L. G., Edelman, I. S., and Lindemann, B. (1982). The role of sodium-channel density in the natriferic response of the toad urinary bladder to an antidiuretic hormone. *J. Membr. Biol.* **64**, 77–89.

Liddle, G. W., Bledsoe, T., and Coppage, W. S., Jr. (1963). A familial renal disorder simulating primary aldosteronism but with negligible aldosterone secretion. *Trans. Assoc. Am. Physicians* **76**, 199–213.

Lingueglia, E., Renard, S., Waldmann, R., Voilley, N., Champigny, G., Plass, H., Lazdunski, M., and Barbry, P. (1994). Different homologous subunits of the amiloride-sensitive Na$^+$ channel are differently regulated by aldosterone. *J. Biol. Chem.* **269**, 13736–13739.

Lingueglia, E., Voilley, N., Waldmann, R., Lazdunski, M., and Barbry, P. (1993). Expression cloning of an epithelial amiloride-sensitive Na+ channel: A new channel type with homologies to *Caenorhabditis elegans* degenerins. *FEBS Lett.* **318**, 95–99.

Lomax, R. B., McNicholas, C. M., Lombes, M., and Sandle, G. I. (1996). Aldosterone-induced apical Na^+ and K^+ conductances are located predominantly in surface cells in rat distal colon. *Am. J. Physiol. Gastrointest. Liver Physiol.* **266**, G71–G82.

Mall, M., Bleich, M., Kuehr, J., Brandis, M., Greger, R., and Kunzelmann, K. (1999). CFTR-mediated inhibition of epithelial Na^+ conductance in human colon is defective in cystic fibrosis. *Am. J. Physiol. Gastrointest. Liver Physiol.* **40**, G709–G716.

Mall, M., Hipper, A., Greger, R., and Kunzelmann, K. (1996). Wild type but not delta-F508 CFTR inhibits Na^+ conductance when coexpressed in *Xenopus* oocytes. *FEBS Lett.* **381**, 47–52.

Martin, R. S., Oszi, P., Brocca, S., Arrizurieta, E., and Hayslett, J. P. (1986). Failure of potassium adaptation *in vivo* in the colon of aldosterone-deficient rats. *J. Lab. Clin. Med.* **108**, 241–245.

Marunaka, Y. (1996). Amiloride blockable Ca^{2+}-activated Na^+-permeant channels in the fetal distal lung epithelium. *Pflugers Arch.* **431**, 748–756.

Marunaka, Y., and Eaton, D. C. (1991). Effects of vasopressin and cAMP on single amiloride-blockable Na channels. *Am. J. Physiol.* **260**, C1071–C1084.

Masilamani, S., Kim, G. H., Mitchell, C., Wade, J. B., and Knepper, M. A. (1999). Aldosterone-mediated regulation of ENaC alpha, beta, and gamma subunit proteins in rat kidney. *J. Clin. Invest.* **104**, R19–R23.

May, A., Puoti, A., Gaeggeler, H.-P., Horisberger, J.-D., and Rossier, B. C. (1997). Early effect of aldosterone on the rate of synthesis of the epithelial sodium channel alpha subunit in A6 renal cells. *J. Am. Soc. Nephrol.* **8**, 1813–1822.

McCabe, R., Cooke, H. J., and Sullivan, L. P. (1982). Potassium transport by rabbit descending colon. *Am. J. Physiol.* **242**, C81–C86.

Meneton, P., Schultheis, P. J., Greeb, J., Nieman, M. L., Liu, L. H., Clarke, L. L., Duffy, J. J., Doetschman, T., Lorenz, J. M., and Shull, G. E. (1998). Increased sensitivity to K^+ deprivation in colonic H,K-ATPase-deficient mice. *J. Clin. Invest.* **101**, 536–542.

Nzegwu, H. C., and Levin, R. J. (1992). Dietary restriction sensitizes the rat distal colon to aldosterone. *J. Physiol. (Lond.)* **447**, 501–512.

Pacha, J., and Miksik, I. (1994). Distribution of 11 beta-hydroxysteroid dehydrogenase along the rat intestine. *Life Sciences* **54**, 745–749.

Palmer, L. G. (1985). Modulation of apical Na permeability of the toad urinary bladder by intracellular Na, Ca, and H. *J. Membr. Biol.* **83**, 57–69.

Palmer, L. G. (1992). Epithelial Na channels, function and diversity. *Ann. Rev. Physiol.* **54**, 51–66.

Palmer, L. G., Choe, H., and Frindt, G. (1997). Is the secretory K channel in the rat CCT ROMK. *Am. J. Physiol. Renal Fluid Electrolyte Physiol* **42**, F404–F410.

Palmer, L. G., and Frindt, G. (1987). Effects of cell Ca and pH on Na channels from rat cortical collecting tubule. *Am. J. Physiol.* **253**, F333–F339.

Palmer, L. G., and Frindt, G. (1988). Conductance and gating of epithelial Na channels from rat cortical collecting tubule. *J. Gen. Physiol.* **92**, 121–138.

Palmer, L. G., and Frindt, G. (1996). Gating of Na channels in the rat cortical collecting tubule—Effects of voltage and membrane stretch. *J. Gen. Physiol.* **107**, 35–45.

Palmer, L. G., Li, J. H. Y., Lindemann, B., and Edelman, I. S. (1982). Aldosterone control the density of sodium channels in the toad urinary bladder. *J. Membr. Biol.* **64**, 91–102.

Palmer, L. G., Sackin, H., and Frindt, G. (1998). Regulation of Na^+ channels by luminal Na^+ in rat cortical collecting tubule. *J. Physiol. (Lond.)* **509**, 151–162.

Pradervand, S., Wang, Q., Burnier, M., Beermann, F., Horisberger, J. D., Hummler, E., and Rossier, B. C. (1999). A mouse model for Liddle's syndrome. *J. Am. Soc. Nephrol.* **10**, 2527–2533.

Rechkemmer, G., and Halm, D. R. (1989). Aldosterone stimulates K secretion across mammalian colon independent of Na absorption. *Proc. Natl. Acad. Sci. USA* **86**, 397–401.

Reddy, M. M., Light, M. J., and Quinton, P. M. (1999). Activation of the epithelial Na^+ channel (ENaC) requires CFTR Cl^- channel function. *Nature* **402**, 301–304.

Renard, S., Voilley, N., Bassilana, F., Lazdunski, M., and Barbry, P. (1995). Localization and regulation by steroids of the alpha, beta, and gamma subunits of the amiloride sensitive Na$^+$ channel in colon, lung and kidney. *Pflugers Arch.* **430**, 299–307.

Roudier-Pujol, C., Rochat, A., Escoubet, B., Eugene, E., Barrandon, Y., Bonvalet, J. P., and Farman, N. (1996). Differential expression of epithelial sodium channel subunit mRNAs in rat skin. *J. Cell Sci.* **109**, 379–385.

Rubens, R. D., and Lambert, H. P. (1972). The homeostatic function of the colon in acute gastroenteritis. *Gut* **13**, 915–919.

Ruckes, C., Blank, U., Moller, K., Rieboldt, J., Lindemann, H., Munker, G., Clauss, W., and Weber, W. M. (1997). Amiloride-sensitive Na$^+$ channels in human nasal epithelium are different from classical epithelial Na$^+$ channels. *Biochem. Biophys. Res. Commun.* **237**, 488–491.

Sandle, G. I. (1998). Salt and water absorption in the human colon—A modern appraisal. *Gut* **43**, 294–299.

Sandle, G. I., Foster, E. S., Lewis, S. A., Binder, H. J., and Hayslett, J. P. (1985). The electrical basis for enhanced potassium secretion in rat distal colon during dietary potassium loading. *Pflugers Arch.* **403**, 433–439.

Sandle, G. I., Hayslett, J. P., and Binder, H. J. (1984). Effect of chronic hyperaldosteronism on the electrophysiology of rat distal colon. *Pflugers Arch.* **401**, 22–26.

Schafer, J. A., and Hawk, C. T. (1992). Regulation of Na$^+$ channels in the cortical collecting duct by AVP and mineralocorticoids. *Kidney Int.* **41**, 255–268.

Schild, L., Lu, Y., Gautschi, I., Schneeberger, E., Lifton, R. P., and Rossier, B. C. (1996). Identification of a PY motif in the epithelial Na channel subunits as a target sequence for mutations causing channel activation found in Liddle syndrome. *EMBO J.* **15**, 2381–2387.

Schnizler, M., Mastroberardino, L., Reifarth, F. W., Weber, W.-M., Verrey, F., and Clauss, W. (2000). cAMP sensitivity conferred to the epithelial Na$^+$ channel by α-subunit cloned from guinea-pig colon. *Pflugers Arch.* **439**, 579–587.

Schultz, S. G., Frizzell, R. A., and Nellans, H. N. (1977). Active sodium transport and the electrophysiology of rabbit colon. *J. Membr. Biol.* **33**, 351–384.

Shimkets, R. A., Lifton, R., and Canessa, C. M. (1998). *In vivo* phosphorylation of the epithelial sodium channel. *Proc. Natl. Acad. Sci. USA* **95**, 3301–3305.

Shimkets, R. A., Lifton, R. P., and Canessa, C. M. (1997). The activity of the epithelial sodium channel is regulated by clathrin-mediated endocytosis. *J. Biol. Chem.* **272**, 25537–25541.

Shimkets, R. A., Warnock, D. G., Bositis, C. M., Nelson-Williams, C., Hansson, J. H., Schambelan, M., Gill, J. R., Jr., Ulick, S., Milora, R. V., Findling, J. W., Canessa, C. M., Rossier, B. C., and Lifton, R. P. (1994). Liddle's syndrome: Heritable human hypertension caused by mutations in the beta subunit of the epithelial sodium channel. *Cell* **79**, 407–414.

Silver, R. B., Frindt, G., Windhager, E. E., and Palmer, L. G. (1993). Feedback regulation of Na channels in rat CCT. I. Effects of inhibition of Na pump. *Am. J. Physiol. Renal Fluid Electrolyte Physiol.* **264**, F557–F564.

Skadhauge, E., and Thomas, D. H. (1979). Transepithelial transport of K$^+$, NH4$^+$, inorganic phosphate, and water by hen (*Gallus domesticus*) lower intestine (colon and coprodeum) perfused luminally *in vivo*. *Pflugers Arch.* **379**, 237–243.

Smith, P. R., Bradford, A. L., Dantzer, V., Benos, D. J., and Skadhauge, E. (1993). Immunocytochemical localization of amiloride-sensitive sodium channels in the lower intestine of the hen. *Cell Tissue Res.* **272**, 129–136.

Snyder, P. M. (2000). Liddle's syndrome mutations disrupt cAMP-mediated translocation of the epithelial Na$^+$ channel to the cell surface. *J. Clin. Invest.* **105**, 45–53.

Snyder, P. M., Cheng, C., Prince, L. S., Rogers, J. C., and Welsh, M. J. (1998). Electrophysiological and biochemical evidence that deg/ENaC cation channels are composed of nine subunits. *J. Biol. Chem.* **273**, 681–684.

Soler, A. P., Miller, R. D., Laughlin, K. V., Carp, N. Z., Klurfeld, D. M., and Mullin, J. M. (1999). Increased tight junctional permeability is associated with the development of colon cancer. *Carcinogenesis* **20,** 1425–1431.

Staub, O., Dho, S., Henry, P. C., Correa, J., Ishikawa, T., Mcglade, J., and Rotin, D. (1996). WW domains of Nedd4 bind to the proline-rich PY motifs in the epithelial Na$^+$ channel deleted in Liddle's syndrome. *EMBO J.* **15,** 2371–2380.

Staub, O., Gautschi, I., Ishikawa, T., Breitschopf, K., Ciechanover, A., Schild, L., and Rotin, D. (1997). Regulation of stability and function of the epithelial Na channel (ENaC) by ubiquitination. *EMBO J.* **16,** 6325–6336.

Stokes, J. B., and Sigmund, R. D. (1998). Regulation of rENaC mRNA by dietary NaCl and steroids: Organ, tissue, and steroid heterogeneity. *Am. J. Physiol.* **274,** C1699–C1707.

Stutts, M. J., Canessa, C. M., Olsen, J. C., Hamrick, M., Cohn, J. A., Rossier, B. C., and Boucher, R. C. (1995). CFTR as a cAMP-dependent regulator of sodium channel. *Science* **269,** 847–850.

Tchepichev, S., Ueda, J., Canessa, C., Rossier, B. C., and Obrodovich, H. (1995). Lung epithelial Na channel subunits are differentially regulated during development and by steroids. *Am. J. Physiol. Cell Physiol.* **38,** C805–C812.

Turnheim, K., Plass, H., Grasl, M., Krivanek, P., and Wiener, H. (1986). Sodium absorption and potassium secretion in rabbit colon during sodium deficiency. *Am. J. Physiol.* **250,** F235–F245.

Turnheim, K., Thompson, S. M., and Schultz, S. G. (1983). Relation between intracellular sodium and active sodium transport in rabbit colon: Current-voltage relations of the apical sodium entry mechanism in the presence of varying luminal sodium concentrations. *J. Membr. Biol.* **76,** 299–309.

Valentijn, J. A., Fyfe, G. K., and Canessa, C. M. (1998). Biosynthesis and processing of epithelial sodium channels in *Xenopus* oocytes. *J. Biol. Chem.* **273,** 30344–30351.

Vallet, V., Chraïbi, A., Gaeggeler, H.-P., Horisberger, J.-D., and Rossier, B. C. (1997). An epithelial serine protease activates the amiloride-sensitive sodium channel. *Nature* **389,** 607–610.

Verrey, F. (1994). Antidiuretic hormone action in A6 cells: Effect on apical Cl and Na conductances and synergism with aldosterone for NaCl reabsorption. *J. Membr. Biol.* **138,** 65–76.

Verrey, F. (1999). Early aldosterone action: Toward filling the gap between transcription and transport. *Am. J. Physiol. Renal Fluid Electrolyte Physiol.* **46,** F319–F327.

Verrey, F., Beron, J., and Spindler, B. (1996). Corticosteroid regulation of renal Na,K-ATPase. *Miner. Electrolyte Metab.* **22,** 279–292.

Voilley, N., Lingueglia, E., Champigny, G., Mattéi, M.-G., Waldmann, R., Lazdunski, M., and Barbry, P. (1994). The lung amiloride-sensitive Na$^+$ channel: Biophysical properties, pharmacology, ontogenesis, and molecular cloning. *Proc. Natl. Acad. Sci. USA* **91,** 247–251.

Wang, Q., Horisberger, J.-D., Maillard, M., Brunner, H. R., Rossier, B. C., and Burnier, M. (2000). Salt and angiotensin II-dependent variations in amiloride-sensitive rectal PD in mice. *Clin. Exp. Pharmacol. Physiol.* **27,** 60–66.

Watanabe, S., Matsushita, K., McCray, P. B., and Stokes, J. B. (1999). Developmental expression of the epithelial Na$^+$ channel in kidney and uroepithelia. *Am. J. Physiol. Renal Fluid Electrolyte Physiol* **45,** F304–F314.

Watanabe, S., Matsushita, K., Stokes, J. B., and McCray, P. B. (1998). Developmental regulation of epithelial sodium channel subunit mRNA expression in rat colon and lung. *Am. J. Physiol. Gastrointest. Liver Physiol.* **38,** G1227–G1235.

Wilkinson, D. J., Kushman, N. L., and Dawson, D. C. (1993). Tetraethylammonium-sensitive apical K$^+$ channels mediating K$^+$ secretion by turtle colon. *J. Physiol. (Lond.)* **462,** 697–714.

Will, P. C., Lebowitz, J. L., and Hopfer, U. (1980). Induction of amiloride-sensitive sodium transport in the rat colon by mineralocorticoids. *Am. J. Physiol.* **238,** F261–F268.

Wills, N. K., Zeiske, W., and Van Driessche, W. (1982). Noise analysis reveals channel conductance fluctuations in the apical membrane of rabbit colon. *J. Membr. Biol.* **69,** 187–197.

Zeiske, W., Wills, N. K., and Van Driessche, W. (1982). Na$^+$ channels and amiloride-induced noise in the mammalian colon epithelium. *Biochim. Biophys. Acta* **688,** 201–210.

Zhang, P., Fyfe, G. K., Grichtchenko, I. I., and Canessa, C. M. (1999). Inhibition of $\alpha\beta$ epithelial sodium channels by external protons indicates that the second hydrophobic domain contains structural elements for closing the pore. *Biophys. J.* **77,** 3043–3051.

CHAPTER 12

Molecular Physiology of Mammalian Epithelial Na^+/H^+ Exchangers NHE2 and NHE3

Mark Donowitz and Ming Tse
Departments of Medicine and Physiology, GI Division, Johns Hopkins University School of Medicine, Baltimore, Maryland

I. OVERVIEW

This review will deal primarily with recent advances in understanding of the physiological function and molecular regulation of the mammalian *epithelial-specific* Na^+/H^+ exchangers (NHE) NHE2 and NHE3. These two exchangers contribute to small intestinal, colonic, gallbladder, renal, and salivary gland Na^+ and HCO_3^- absorption, and along with ENaC and Na^+-linked substrate cotransporters (including Na^+ amino acid and the Na^+ glucose cotransporters), carry out most of the renal and intestinal Na^+ absorption. Topics covered include (1) an overview of NHEs including the identified members of the Na^+/H^+ exchanger gene family and structure/function relationships of the NHE N terminus; (2) cellular and subcellular locations; (3) physiological functions; (4) regulation of transport activity; and (5) diseases in which the epithelial NHEs play a pathogenic role. Recent reviews have been published of molecular studies of all the mammalian NHEs (Wakabayashi *et al.*, 1997a; Noel and Pouyssegur, 1995; Orlowski and Grinstein, 1997; Yun *et al.*, 1995a; Tse *et al.*, 1993a).

A. Six Molecularly Identified NHE Isoforms

Since our initial recognition that there was a gene family of mammalian Na^+/H^+ exchangers in 1991 (Tse *et al.*, 1991), the number of cloned isoforms has increased to six (Sardet *et al.*, 1989; Tse *et al.*, 1992; 1993b; Orlowski *et al.*,

TABLE I

Cloned Mammalian Na^+/H^+ Exchangers

(Number of amino acids derived from DNA sequence/Estimated size kDa)

NHE1	815	Human	90,773	Sardet *et al.*, 1989
	817	Bovine	91,028	
	816	Rabbit	90,727	Tse *et al.*, 1991
	820	Rat	91,657	
	820	Mouse	91,478	
	818	Pig	91,023	
	822	Chinese hamster	92,014	
NHE2	812	Human	91,530	Malakooti *et al.*, 1999
	809	Rabbit	90,787	Tse *et al.*, 1993b
	813	Rat	91,413	Orlowski *et al.*, 1992
NHE3	834	Human	92,906	Brant *et al.*, 1995
	832	Rabbit	92,997	Tse *et al.*, 1992
	831	Rat	92,747	Orlowski *et al.*, 1992
	839	Opossum	94,776	Amemiya *et al.*, 1995
NHE4	717	Rat	81,531	Orlowski *et al.*, 1992
NHE5	896	Human	99,023	Klanke *et al.*, 1995; Baird *et al.*, 1999
	898	Rat	99,063	Attaphitaya *et al.*, 1999
NHE6	669	Human	74,170	Numata *et al.*, 1998

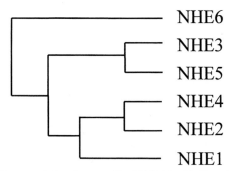

FIGURE 1 Dendrogram of cloned mammalian NHEs identified in epithelial cells. Human sequences were used except for NHE4 (rat). Similarities were compared at the amino acid level with PC gene. Note that NHE2 and NHE4, and NHE3 and NHE5 are the most related isoforms.

1992; Wang *et al.,* 1993; Klanke *et al.,* 1995; Baird *et al.,* 1999; Attaphitaya *et al.,* 1999; Numata *et al.,* 1998) (Table I and Fig. 1). All the cloned isoforms occur in epithelial cells except NHE5. This review will deal primarily with two isoforms that are found predominantly, but not entirely, in the apical membranes of some epithelial cells, and are the only two isoforms that contribute to transcellular Na^+ absorption as part of their function. The other cloned NHEs will be commented on only for comparison.

NHE1 and NHE4 are considered the "housekeeping isoforms," which take part in regulation of intracellular pH and intracellular volume, and perhaps affect cell division by increasing intracellular pH; at least NHE1 appears to be involved in organizing the cytoskeleton (Hooley *et al.,* 1996; Tominaga and Barber, 1998; Tominaga *et al.,* 1998). NHE1 is present in almost all mammalian cells, whereas NHE4 is present primarily in epithelial cells, specifically in rat kidney and rabbit intestine and kidney. Although NHE2 and NHE3 also can alter intracellular pH and intracellular volume, these have not been shown to be among their major physiological functions (see chapter by MacLeod in this volume).

Only one NHE isoform, NHE6, has been shown to be exclusively an intracellular isoform (Numata *et al.,* 1998). NHE6 was cloned from yeast and a mammalian homologue identified (Numata *et al.,* 1998; Nass *et al.,* 1997). In the yeast, it is in the prevacuolar compartment (Nass *et al.,* 1997). This corresponds to a lysosomal compartment in mammalian cells. However, in mammalian cells, it has been suggested that NHE6 is present in the mitochondrial inner membrane (Numata *et al.,* 1998), although its location is still under investigation. Interestingly, a homologue in plants has been shown to be present in the tonoplast and golgi/ER fractions, and functions to allow adaptation to growth in high-saline conditions (Apse *et al.,* 1999).

The number of members of the Na^+/H^+ exchanger gene family that have not been molecularly identified is unknown, but physiological studies indicate that at least several more isoforms may exist. These include (1) a mammalian NHE isoform related to the yeast prevascuolar NHE (Numata et al., 1998; Nass et al., 1997), (2) an amiloride-insensitive hippocampal isoform (Raley-Susman et al., 1991), and (3) the renal proximal tubule S3 segment apical isoform. There are three other possible NHE isoforms. One appears to be a Cl^--dependent isoform in the rat colon crypt (Rajendran et al., 1995; 1999; Binder et al., 1997). However, the Cl^- dependence appears to relate to indirect linking to a Cl^- channel. Consequently, it is likely but not certain that a unique NHE isoform is involved, especially given the preliminary results of Grinstein that NHE3 can be affected by Cl^- (Grinstein, personal communication). The second isoform is potentially jejunal because message is low for NHE2 and NHE3 in rabbit jejunum compared with ileum, whereas functional Na^+/H^+ exchange is comparable in the jejunum and ileum (Tse et al., 1992; 1993b). However, protein levels of NHE2 and NHE3 have not been quantitatively compared, and until this is accomplished, rationale for another jejunal isoform is inconclusive. The third type is potential alternatively spliced isoforms of NHE2, based on the presence of multiple message sizes (Ghishan et al., 1995). Several different-sized messages for NHE2 have been identified (Tse et al., 1993b), but the only splice variant claimed for NHE2 is a clone that lacks the N-terminal 150 amino acids. However, it has not been established whether this is a splice variant or a result of cloning artifact. Our analysis of the sequence of the human clone reported by Ghishan is that it has an almost identical C-terminal sequence to that of rat NHE2 (Tse, unpublished observation). Recently, Malakooti et al. (1999) reported a sequence of human NHE2 (Apse et al., 1999) and raised the possibility of cloning artifacts in the previously reported sequence of human NHE2.

The cloned NHE isoforms are all separate genes and, with one exception, are present on different chromosomes (Table II). Note that only the related NHE2 and NHE4 are on the same chromosome. Pseudogenes have been identified only for NHE3 (Kokke et al., 1996). No definite NHE splice variants have been identified.

B. Membrane Topology

Based on similarity in hydrophobicity predicted by multiple algorithms, it has been assumed that all members of the NHE gene family have similar membrane topology. However, experimental studies to determine NHE topology have been limited until recently and still are incomplete. Our current model for the topology of NHE3 is shown in Figure 2. It consists of 11 membrane-spanning domains

TABLE II

Chromosomal Location of Human and Rat *NHE* Gene Family

NHE isoform	Human chromosome location	Rat chromosome location	References
NHE1	1p	5	Szpirer *et al.,* 1994
NHE2	2q	9	Ghishan *et al.,* 1995; Brant *et al.,* 1995
NHE3	5p[a]	1	Kokke *et al.,* 1996; Szpirer *et al.,* 1994; Brant *et al.,* 1993
NHE4	2	9	Kokke *et al.,* 1996
NHE5	16q	—	Klanke *et al.,* 1995
NHE6	—	—	

[a]Pseudogene on human chromosome 10 (Kokke *et al.,* 1996).

(MSDs), with the N terminus being outside the cell and an intracellular C terminus. The reasons for concluding that the N terminus is extracellular include (1) ability to identify NHE3 immunocytochemically when it is epitope tagged at amino acid 30 without permeabilization (Noel *et al.,*, 1996; Shrode *et al.,* 1998), and (2) ability of limited proteolysis with chymotrypsin to digest only this part of NHE1 (Zizak *et al.,* 2000). The C terminus is intracellular based on the ability to visualize a C-terminal epitope only with permeabilization (true for NHE1, 2, and 3) (Hoogerwerf *et al.,* 1996a). However, studies with monoclonal antibodies showed that the C-terminal 131 amino acids of rabbit NHE3 could be recognized in renal brush borders without permeabilization, which suggests access of the antibodies to the epitope without permeabilization (Biemesderfer *et al.,* 1998). This appeared true of two separate epitopes, one between aa 702 and 756 and the other between aa 756 and 832. We believe that the C-terminal 372 aa of NHE3 are intracellular. However, the explanation for some accessibility without permeabilization is not known, and the observation suggests that the last 130 aa may be part of some sort of tertiary structure that allows access of the antibody from the extracellular surface. In fact, preliminary studies indicate that the NHE1 C terminus has at least secondary structure. A fusion protein of the C terminus of NHE1 was purified, and circular dichroism spectroscopy used to analyze the types of structures that were present. The C terminus appeared to be 35% alpha-helical, 17% beta-turn, and 48% random coil (Gebreselassie *et al.,* 1998).

All models of NHE topology suggest that it is made up of two domains, an approximately 500 aa N terminus and an approximately 300 aa C terminus, with

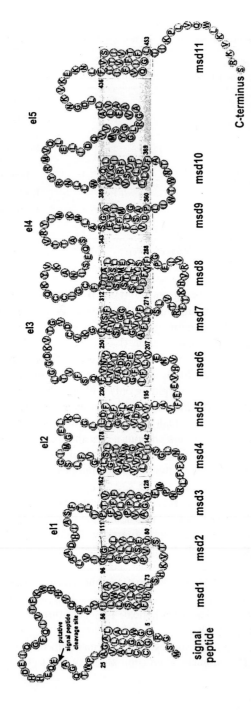

FIGURE 2 NHE topology. This current model of NHE3 topology shows (1) a cleaved signal peptide, (2) extracellular N terminus, (3) 11 α-helical membrane-spanning domains, (4) intracellular C terminus, and (5) large 5th extracellular loop that is postulated to serve as "P loop." M indicates the membrane-spanning domain; el denotes the extracellular loop. This model is discussed in Zizak et al., 2000.

the transition possibly being around Pro 460 of NHE3, based on hydropathy analysis. The NHE N-terminus is made up of multiple hydrophobic alpha-helical structures, which presumably are multiple membrane-spanning domains. When NHE1, 2, or 3 is truncated to include only this domain (approximately the entire N-terminal domain plus the N-terminal ~15 amino acids of the C-terminal domain), the protein is still capable of carrying out Na^+/H^+ exchange, but at a much slower rate than when the C terminus of any one of these isoforms is included (Yun *et al.*, 1995b). This suggests that at least part of the NHE C terminus is needed to allow normal function of the transport domain, probably by interaction with the N-terminal domain.

Application of multiple analyses of hydrophobicity suggests that NHE3 has between 10 and 14 MSDs (Zizak *et al.*, 2000; Wakabayashi *et al.*, 2000). However, only three types of analyses have experimentally tested NHE topology. The first study was of NHE1 and used chymotryptic proteolysis, plus cell surface biotinylation, to support a 12-MSD model (Shrode *et al.*, 1998). While the identity of specific MSDs was not undertaken, several conclusions were reached: (1) the first extracellular loop is freely accessible to chymotryptic cleavage and thus lies extracellularly; (2) no other parts of NHE1 are sensitive to chymotrypsin, consistent with other parts of the protein not being freely exposed externally; and (3) the 4th and 5th putative extracellular loops are not accessible to chymotryptic cleavage in spite of having cleavable amide linkages. Thus, they are not free externally but rather may be folded into the membrane or lie next to the membrane such that they are sterically protected (Zizak *et al.*, 2000).

Another method was used to assess the topology of NHE3 based on *in vitro* transcription/translation of putative MSDs singly or in pairs using two expression vectors, only one of which contained a known MSD for orientation (Zizak *et al.*, 2000). This approach (1) identified a probable cleaved signal peptide, and (2) was consistent with 11 MSDs, the boundaries of at least pairs of which were defined; these form the basis for the MSD boundaries shown in Figure 2. The only contradictory data by this analysis related to putative extracellular loop 5, which was found to have some membrane-spanning properties but did not appear to behave consistently like MSDs. Thus, by two separate techniques, the putative 5 extracellular loop is postulated as associating with the plasma membrane although not in ways that cross the membrane like a typical α-helical MSDs. Possibilities include that it is a "P loop," which partially enters the membrane but then exits from the same side it entered, or are located next to the membrane. We suggest that this area may be involved in forming the transport domain or in determining the selectivity of the transporter, as has been shown in the past for "P loops" in other transporters. Of importance is that this part of several transport proteins (near the C terminus of transport domain) has been involved in either carrying out transport or determining selectivity of transport

(K channel, NaPi cotransporter)[1] (Yellen *et al.*, 1991; Murer, privileged communication).

C. Structure/Function Studies—The NHE N-Terminal Transport Domain

Truncating the NHE3 C terminus to aa 475 leads to a protein that carries out Na^+/H^+ exchange with 1:1 stoichiometry of Na^+ and H^+, forms dimers, has a H^+ modifier site, and exhibits amiloride sensitivity (similar results occurred when NHE1 and NHE2 were truncated at the comparable amino acid) (Levine *et al.*, 1995; Fafournoux *et al.*, 1994). In addition, this is the part of the protein that is glycosylated, based on which isoform is considered (Sardet *et al.*, 1990; Counillon *et al.*, 1994; Tse *et al.*, 1994; Haworth *et al.*, 1993). Although the NHE N-termini appear highly homologous, attempts to make chimeras of the NHEs in their N termini greatly reduced Na^+/H^+ exchange function or totally abolished it. Orlowski and Kandasamy (1996) made chimeras of rat NHE1 and rat NHE3 and found that replacing the 7th MSD (terminology according to Fig. 2) of NHE3 with the comparable domain of NHE1 decreased activity only moderately, and that replacing the 8th MSD of NHE1 with that of NHE3 maintained Na^+/H^+ exchange function. All other substitutions led to large decreases in function or no measurable function at all. We have carried out similar studies with rabbit NHE3 and NHE1 and found that a series of chimeras that substituted MSDs 1–3, 4–7, and 8–11 all failed to transport (Cavet *et al.*, 1999). Part of the reason for this failure was that some chimeras did not leave the ER or golgi, but even when the chimeras reached the plasma membrane, they did not function in Na^+/H^+ exchange. This indicates that there are major differences in the tertiary structure of the N-termini of NHE1 and NHE3 that are critical for function.

---

[1]After this manuscript was submitted, a third approach, cysteine mutagenesis, was reported to evaluate the membrane topology of NHE1 (Wakabayashi *et al.*, 2000). The following conclusions were reached: (1) deleting all 9 cysteines produced a protein that performed Na^+/H^+ exchange; (2) there was no cleaved signal peptide; (3) the amino acids in intracellular loops 2 and 4 after MSD 3 and MSD 7, respectively (terminology as in Fig. 2) seem to dip into the membrane; (4) extracellular loop 5 is a MSD and supplies an intracellular and extracellular loop, whereas MSD 9 from Fig. 1 may be a "P loop." These results for NHE1 further support the observation that extracellular loop 5 from Fig. 1 is a critical functional domain, and its exact topology and function need clarification. The fact that such differences in the N termini of NHE1 and NHE3 exist (cleaved signal peptide vs intact MSD) suggests that this part of the NHE family, which is highly divergent among the isoforms, may have a very different function among the isoforms. Please note virtual identity of MSDs by the Wakabayashi and Zizak models except for (1) presence of cleaved signal peptide; and (2) area between MSD 8 and 11 (Zizak nomenclature, Fig. 2). This area contains 2 MSDs and a "P loop" in both models, but they are in reverse order.

1. Na$^+$ Transport

The NHEs have K_m^{Na} of approximately 10–20 mM and follow Michaelis–Menten kinetics, indicating involvement of one Na$^+$ per exchanger per cycle of transport (Levine *et al.*, 1993; Orlowski, 1993a; 1993b; Yu *et al.*, 1993). The main exception is rat NHE2, which has a lower affinity for Na$^+$, with a K_m^{Na} of approximately 50 mM (Yu *et al.*, 1993). There is also a species difference between rabbit and rat NHE3, with rabbit NHE3 having a K_m^{Na} close to that of the other isoforms (approximately 15–20 mM) while rat NHE3 has a K_m^{Na} of ~5 mM (Levine *et al.*, 1993; Bookstein *et al.*, 1997). For all NHEs studied (NHE1, 2, 3, 4), external cations have the affinity H$^+$>Li$^+$>Na$^+$>K$^+$ (Orlowski, 1993a; Yu *et al.*, 1993). Whether one or two sites are used for Na$^+$ and H$^+$ transport has not been established in dimers.

The Na$^+$ transport domain has not been identified. It has been assumed that the most highly conserved part of NHE is involved in Na$^+$ and/or H$^+$ transport—this is MSD 5–6 in Figure 2, which is 84–87% identical among NHE1–3, including the connecting intracellular loop. However, there is no experimental data to support that this domain functions in Na$^+$ or H$^+$ exchange, other than the finding that a mutation in this domain eliminated at least 90% of Na$^+$/H$^+$ exchange activity (Fafournoux *et al.*, 1994) (NHE1, E2611). Domains involved in amiloride sensitivity (MSDs 3 and 8 in Fig. 2) also are candidates for Na$^+$ transport.

Exchange of Na$^+$ for H$^+$ occurs with a stoichiometry of 1:1, with no change induced by any mutagenesis thus far. As recently demonstrated and reviewed, a study that was thought to show the association of transporter activation with a transepithelial current was not related to a change in stoichiometry but rather to the presence of a separate [H$^+$] conductance that was sensitive to change in extracellular [H+] concentrations (Demaurex *et al.*, 1995).

2. H$^+$ Modifier Site

All NHEs have an intracellular H$^+$ modifier site (Sardet *et al.*, 1989; Orlowski *et al.*, 1992; Levine *et al.*, 1993; Orlowski, 1993a; Yu *et al.*, 1993; Tse *et al.*, 1993c) that causes the relationship of Na$^+$/H$^+$ exchange and intracellular [H$^+$] to follow a sigmoidal relationship, which is characterized by a Hill coefficient of 2–3. This is present in all NHEs and is one of their signatures. This relationship is similar among NHE isoforms under basal conditions. Truncation of the NHE C terminus does not remove the cooperative relationship between rate of Na$^+$/H$^+$ exchange and intracellular [H$^+$], which indicates that the intracellular H$^+$ modifier is within the N terminus plus ~15 aa of the cytoplasmic domain (Yun *et al.*, 1995b; Levine *et al.*, 1995; Fafournoux *et al.*, 1994; Levine *et al.*, 1993). Which amino acids are involved in the H$^+$ modifier is not known. In a bacterial Na$^+$/H$^+$ antiporter, the modifier has been shown to contain a functionally important His at the C terminus of MSD7 (Gerchman *et al.*, 1993; Padan *et al.*, 1999). Mutation of this His (His 226) changed the pH dependence of the exchanger, causing an acidic shift in the resting pH$_i$ in the mutant His 226 Arg, an alkaline shift in mutant His 226 Asp, and no change in mutants His 226 Cys or His 226 Ser. Whether His is involved in creating the H^+ modifier of mammalian NHEs is not known. Extensive mutagenesis of His in NHE1 failed to significantly alter the function of the exchanger or the H^+ modifier site (Wang *et al.*, 1995). It has been postulated by Wakabayashi and coworkers that the first ~90 aa of the NHE1 cytoplasmic domain are part of the H^+ modifier or interact with the H^+ modifier and are necessary for the H^+ modifier to act (see later) (Wakabayashi, *et al.*, 1992; Ikeda *et al.*, 1997).

3. Amiloride Sensitivity

A characteristic of all NHEs is inhibition by amiloride and related compounds (Harris and Fliegel, 1996b). The sole exception is a still uncloned hippocampal Na^+/H^+ exchanger (Raley-Susman *et al.*, 1991). It has been assumed that identification of the part of the NHE responsible for amiloride sensitivity would provide insight into the Na^+ transport domain because in many (but not all) studies of NHE3, amiloride is a competitive inhibitor of NHE Na^+ transport. However, studies with NHEs in intact tissue have suggested a noncompetitive interaction of Na^+ and amiloride, which suggests nonidentity of amiloride and Na^+ binding sites. That a more complex picture of the amiloride binding site may emerge could have been predicted. Until now, two domains in the NHE N-terminus have been shown to partially determine amiloride sensitivity: (1) NHE1 MSD 3, specifically amino acids Phe 161, Leu 163, and Gly 174 (Counillon *et al.*, 1993a; 1993b; 1997; Yun *et al.*, 1993a), and (2) NHE MSD 8 (Orlowski and Kandasamy, 1996). MSD 3 possesses 3 Pro residues, which disrupt the helical structure. These are conserved among the NHE isoforms in which amiloride sensitivity has been determined. The presence of multiple sites that are involved in determining amiloride sensitivity suggests that NHE MSDs are folded to allow amiloride to bind in a plane to both MSDs involved and suggests that NHEs fold in a way that MSD 3 and 8 are in close opposition. Moreover, the Na^+ binding site must be present in a closely placed but probably distinct site.

Multiple classes of drugs inhibit the NHEs (Harris and Fliegel, 1999a). These include amiloride and its 5′-amino substituted analogues and HOE694 and analogues of the latter and cimetidine and harmaline (Orlowski *et al.*, 1992; Wang *et al.*, 1993). All have sensitivity NHE1>NHE2>NHE3, although NHE1 and NHE2 have sensitivity similar to amiloride itself (Levine *et al.*, 1993; Orlowski, 1993a). Recently two drugs have been found to have greater effects on NHE3 than on NHE2 or NHE1. HOECHST S3226, an analogue of HOE 694, is marked by a large difference in IC50 for NHE3 based on species (Schwark *et al.*, 1998). A separate type of inhibition of NHE3 was found for the drug squalamine (Akhter *et al.*, 1999). This compound has a steroid nucleus and a spermidine side chain and is made endogenously by the shark liver (Fig. 3). It inhibits NHE3 but not NHE2 or NHE1. It acts over approximately 1 hour rather than having the immediate effect of the other drugs. This indicates a different type of action.

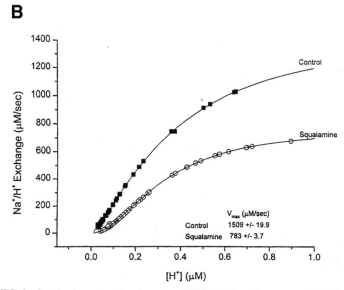

FIGURE 3 Squalamine, a novel endogenous NHE3 inhibitor. Structure and kinetic effect on NHE3 stably expressed in PS120 fibroblasts. The hour delay in inhibition demonstrates a unique inhibitory mechanism. Reprinted with permission from Akhter *et al.*, 1999.

Perhaps it acts on some aspect of signal transduction that has an impact on NHE3. The presence of an endogenous shark inhibitor that comes in contact with the shark gut and perhaps the kidney, via renal excretion, suggests that even in mammalian cells there might be endogenous inhibitors of NHE3 that might control Na^+ homeostatasis.

4. Dimerization

NHEs exist as a dimer, which is visible on Western blotting using SDS-PAGE, even in the absence of any cross-linking agents (Wakabayashi *et al.*, 1997a; Noel and Pouyssegur, 1995; Sardet *et al.*, 1989; 1990; Tse *et al.*, 1992;

Fafournoux *et al.*, 1994). That the membrane domain is involved in forming the dimer has been suggested based on truncation studies with NHE1 containing the N terminus plus ~15 aa in the C terminus (Fafournoux *et al.*, 1994). While NHE exists at least as a dimer, no functional consequences have been suggested. In fact, when part of the dimer is made up of an almost-dead mutant, no significant effect on Na^+/H^+ exchange is detected (Fafournoux *et al.*, 1994). The lack of a dominant-negative effect of the poorly functioning mutant, while not definite, raised the possibility that the NHEs might exist as a dimer but function as a monomer. It appears that heterodimers do not form when multiple NHEs are present in the same cell. For instance, when NHE2 and NHE3 are expressed in the same fibroblast, there is no evidence that they heterodimerize or that isoforms with intermediate amiloride sensitivities arise (Nath *et al.*, 1999).

5. Glycosylation

NHE1 is both N- and O-glycosylated on the first extracellular loop of the N-terminal domain (see Fig. 2) (Counillon *et al.*, 1994). Deglycosylating NHE1 did not have any effect on plasma membrane location, amiloride sensitivity, or transport function (Counillon *et al.*, 1994), although function in a polarized cell has not been determined. NHE2 is not N-glycosylated but is O-glycosylated, whereas NHE3 is neither N- nor O-glycosylated (Counillon *et al.*, 1994; Tse *et al.*, 1994). There is a suggestion that in renal epithelial cells of the rabbit, NHE3 might be glycosylated. However, rat renal cortex NHE3 was not glycosylated (Bizal *et al.*, 1996; Soleimani *et al.*, 1996). Thus, NHE2 and NHE3 are the first transport proteins present in the plasma membrane that are neither N-glycosylated nor known to be associated with N-glycosylated proteins. These results appear to indicate that N-glycosylation is not necessary for the plasma membrane location or function of all transport proteins. In an intact tissue, deglycosylation of NHE3 decreased V_{max} by 80%, but it was never determined whether this was a specific effect of deglycosylation of NHE3 or if the effect indicated deglycosylation of another protein involved in the regulation of NHE3 (Yusufi *et al.*, 1988).

II. CELLULAR AND SUBCELLULAR LOCATION OF NHE2 AND NHE3

A. Cellular Localization

The cellular and subcellular localizations of NHE2 and NHE3 in the intestine, kidney, and salivary glands have been successfully defined; several groups developed anti-NHE polyclonal, and in one case monoclonal, antibodies. There is general agreement on the localization of NHE2 and NHE3 in these organs; however, a few points of controversy remain, which may be related to differences in sensitivities and nonspecific staining of the available antibodies.

1. Stomach

Species variation seems important in defining which NHE isoforms are present in gastric epithelial cells. In most species NHE1, 2, and 4 are thought to be on the basolateral membrane of parietal cells (Bachmann *et al.,* 1998; Joutsi *et al.,* 1996; Kaneko *et al.,* 1992; Schultheis *et al.,* 1998a). Among rat and rabbit gastric mucosa, NHE3 is present only in rat gastric parietal cells (Seidler *et al.,* 1997).

2. Small Intestine/Colon

All groups agree that both NHE2 and NHE3 are present in the apical membrane of villus cells of small intestine and surface cells of the colon (Hoogerwerf *et al.,* 1996; Bookstein *et al.,* 1994; Bookstein *et al.,* 1997). The only exception is that NHE2, but not NHE3, is present in the apical membrane of rabbit descending colon (Hoogerwerf *et al.,* 1996).

NHE3 appears to be present in a larger amount in the ileum than in the colon of several species (Hoogerwerf *et al.,* 1996). This is also true in human intestine at the message level (ileum>jejunum>proximal=distal colon) (Dudeja *et al.,* 1996). Conversely, at the message level NHE2 is present in largest amount in the human distal colon>small intestine>proximal colon (Dudeja *et al.,* 1996). Arguing that the ileum/right colon function acts as a unit, NHE2 at least in humans is predominantly expressed in the distal colon, whereas NHE3 is found more proximally (Malakooti *et al.,* 1999; Hoogerwerf *et al.,* 1996). However, no careful quantitation of the amount of NHE3 or NHE2 along the horizontal intestinal axis has been reported.

There is some disagreement as to how far down the villus/crypt axis NHE2 and NHE3 occur. In rabbit ileum, Hoogerwerf *et al.* (1996) demonstrated both NHE2 and NHE3 in the brush border of the entire villus of small intestine, in surface cells of colon, and also in the apical membrane of the approximately upper half of the crypt. Given that this distribution encompasses the enterocytes that take part in Na^+ absorption, we suggest that apical NHE2 and NHE3 are responsible, and in fact, define the Na^+ absorptive cells. However, other groups found that message for NHE3 was more in the small intestinal and colonic surface than in crypt cells, whereas NHE2 was equally present in surface and crypt cells (Malakooti *et al.,* 1999; Bookstein *et al.,* 1994; Dudeja *et al.,* 1996). Another group reported that in rat small intestine NHE3 was present only in the upper 50% of the villus, whereas the upper crypt contained NHE2 in addition to the surface and villus enterocytes (Ikuma *et al.,* 1999).

3. Gallbladder

NHE3 is present in the surface cells of the gallbladder. Only preliminary studies have been reported suggesting that NHE2 also is present in the apical surface of gallbladder epithelial cells (Silviani *et al.,* 1996; Cremaschi *et al.,* 1992).

4. Kidney

With respect to the kidney, there also is more consensus than controversy (Bookstein *et al.*, 1997; Biemesderfer *et al.*, 1993; Soleimani *et al.*, 1994a; 1994b; Amemiya *et al.*, 1995a; Grishan *et al.*, 1995; Sun *et al.*, 1997; 1998; Chambrey *et al.*, 1998; Mrkic *et al.*, 1993). NHE3 is present in the brush border of the rat and rabbit proximal tubule (except for the rat proximal tubule S3 segment) and in large amounts in the apical membrane of the thick ascending limb of Henle and perhaps small amounts in the thin descending limb of Henle. NHE2 is present in the apical membrane of the thick ascending limb along with NHE3, and in the apical membrane of the distal convoluted tubules and connecting tubules and in a subsection of the thin medullary limbs, but not in collecting ducts. In addition, NHE2 but not NHE3 appears to be present in the rat and rabbit macula densa cell apical membrane, whereas NHE4 (and not NHE1) is present on the basolateral membranes of these cells (Peti-Peterdi *et al.*, 1998; Chambrey, privileged communication). There appears to be a species specificity of NHEs present in renal tissues. In the rabbit, NHE2 is present in the kidney, but data are contradictory with regard to the rat kidney. Chang and coworkers (Bookstein *et al.*, 1997) reported that NHE2 is not found in rat kidney message or protein. In contrast, immunostaining has shown that NHE2 is present in rat cortical thick ascending limb, distal convoluted tubule, collecting duct, and medullary thick ascending loop of Henle (Chambrey *et al.*, 1998).

It is uncertain whether NHE2 is present in renal proximal tubules; most groups have not found any by immunocytochemistry or functionally by amiloride analogue sensitivity (Bookstein *et al.*, 1997; Chambrey *et al.*,1998; Schultheis *et al.*, 1998a). One group presented preliminary results showing that a small amount was present on the apical membrane (Yip *et al.*, 1995). Furthermore, NHE2 activity is absent in the proximal tubule of NHE3 knockout mice (Schultheis *et al.*, 1998b). However, NHE2 message was identified in an SV40 transfected rabbit proximal tubule cell line, the RKPC-2 cell (Mrkic *et al.*, 1993).

Some controversy exists concerning the epithelial NHEs present in the inner medullary collecting duct. This cell functionally lacks an apical exchanger, but the basolateral exchanger physiologically demonstrates an intermediate amiloride analogue sensitivity that resembles that of NHE2. This segment contains message for NHE2 (Soleimani *et al.*, 1994a). However, immunocytochemical demonstration of basolateral NHE2 in this segment has not been established experimentally.

5. Salivary Glands

Salivary glands also possess NHE2 and NHE3. In rat submandibular glands, NHE2 and NHE3 are present in the apical membrane of duct cells (He *et al.*, 1997; 1998; Park *et al.*, 1999; Lee *et al.*, 1998). Based on amiloride analogue sensitivity, NHE2 was the functionally significant apical NHE, with no clear functional contribution from NHE3 (Lee *et al.,* 1998). Controversy exists concerning the localization of NHE2 and NHE3 in the submandibular gland acinar cells; one group believes that the acinus apical membrane has NHE2 and NHE3, and another did not find either in the acinus (He *et al.,* 1997; Lee *et al.,* 1998). A cell culture model of a salivary gland epithelial cell containing apical NHE2 has been described (He *et al.,* 1998).

B. Subcellular Localization

When overexpressed in AP-1 fibroblasts and PS120 fibroblasts, NHE3 is present in two pools, in the plasma membrane and in a juxtanuclear location (Fig. 4). Grinstein and colleagues used a technique based on study of chimeras of a fluorescent intracellular probe and organellar targeting signals to colocalize intracellular NHE3 with markers of the recycling endosome (both transferrin receptor and cellubrevin) but not with the golgi (D'Souza *et al.,* 1998; Kim *et al.,* 1998; Demaurex *et al.,* 1998; Lukacs *et al.,* 1997). This supports the view that intracellular NHE3 predominantly is in an endosomal pool. However, others using similar techniques found that cellubrevin and the transferrin receptor were not in an identical pool (Machen, privileged communication). Quantitation of the amount of plasma membrane vs intracellular NHE3 indicated that only ~15% of NHE3 but ~90% of NHE1 is on the plasma membrane under basal conditions

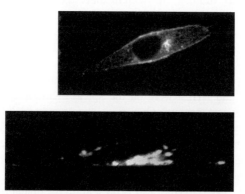

FIGURE 4   Juxtanuclear location of NHE3 in stably transfected PS120/NHE3-GFP cells. PS120 cells were stably transfected with NHE3 EGFP (green fluorescent protein). Multiple xy sections of living cells were obtained by confocal microscopy. Shown is a single xy section (left) and the reconstructed xz distribution. NHE3-EGFP is present in the plasma membrane and in a juxtanuclear compartment. Reprinted and modified with permission from Janecki *et al.,* 1999b.

when expressed in PS120 cells and AP1 (Akhter *et al.,* 1999; Kurashima *et al.,* 1998). A similar predominantly intracellular distribution of NHE3 was found in OK cells, containing endogenous or transfected NHE3 (see book cover, Akhter, Lee, Kovbosnjiak, Tse, Donowitz, unpublished observations). However, not all cells have the majority of NHE3 in an intracellular pool; for instance, although not quantitated to the extent of these cell culture models, rabbit renal proximal tubules and ileal villus cells have the majority of NHE3 in a pool that colocalizes with brush border markers (Zhang *et al.,* 1998; 1999; Yip *et al.,* 1998; Hensley *et al.,* 1989). Also, in Caco-2 cells, NHE3 is present on the apical membrane, with ~80% in the brush border and ~20% in a subapical compartment (Fig. 5) (Janecki *et al.,* 1998; 1999a). The latter is suggested to represent the recycling endosome in these cells. The differences in this distribution has important consequences for the regulation of NHEs by endocytosis/exocytosis. If there is a large subapical pool, stimulation by addition to the plasma membrane is likely to be an effective mechanism of stimulatory regulation and a less effective mechanism of downregulation, and vice versa. No general conclusion can be reached about the differences in the types of cells that express a higher vs lower percentage of NHE3 on the plasma membrane. Note that differences in plasma membrane distribution do not appear to be related to cell culture models vs intact tissue or to NHE overexpression after transfection.

The NHE3 in the rat proximal tubule has been localized to the microvillus membrane by electron microscopy and to a subapical membrane population, consistent with an endosomal site (Biemesderfer *et al.,* 1997). However, an inactive storage pool of NHE3 has been postulated to exist in intermicrovillus clefts in proximal tubules, in which there is association with megalin (Biemesderfer *et al.,* 1999).

In Caco-2 cells, based on which clone was studied, there have been differences in apical membrane expression of NHE2 and NHE3 (Jarecki *et al.,* 1999a; McSwine *et al.,* 1998). Three clones were compared, including relatively early passage and late passage (Janecki *et al.,* 1999a). In the early postconfluency period, both NHE1 and NHE3 were on the apical membrane, although little endogenous NHE3 was expressed. Over time, the amount of NHE3 on the apical membrane increased and the amount of NHE1 decreased. NHE3 was never present on the basolateral membrane. In other clones of Caco-2 cells, NHE2 is said to have been present on the apical surface (McSwine *et al.,* 1998). In HT-29 cells, another colon cancer cell line, NHE2 and not NHE3 is present on the apical surface (Gonda *et al.,* 1999). NHE3 also is present on the apical membrane of some renal cell lines. OK cells, a model proximal tubule cell line from the opossum kidney, has endogenous apical NHE3 but no NHE2 (Amemiya *et al.,* 1995b). LLC-PK1 cells, another proximal tubule cell line, generally do not express apical NHE3 or NHE2, although a clone that expressed significant amounts of NHE3

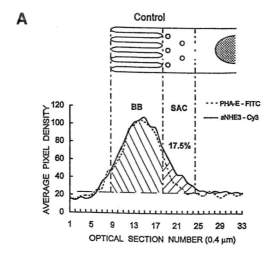

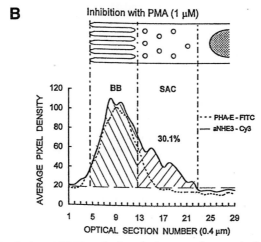

FIGURE 5 Distribution of NHE3 on the Caco-2 plasma membrane and subapical compartment as determined by confocal analysis. Caco-2 cells (clone PF-11) were grown on filters and studied 17–22 days postconfluency, at which time they expressed endogenous NHE3 on the apical membrane. (A) The location of endogenous NHE3 is shown in comparison with the brush border as marked from outside the cell with FITC-labeled lectin PHA-E. In this single cell, 82.5% of NHE3 coincided with the brush border, and 17.5% was in the subapical compartment. In a series of cells, 81.3% of NHE3 was in the Caco-2 brush border and 18.7% was in the subapical compartment. (B) The effect of 30-min exposure to PMA (1 μM) decreased the percentage of total cell NHE3 in the brush border by approximately 15% and increased the percentage of NHE3 in the subapical compartment by 50%, along with inhibiting brush-border Na^+/H^+ exchange by 28%. The protein-kinase C inhibition of NHE3 in Caco-2 cells occurs both by decreasing the amount of NHE3 in the brush border and by decreasing the turnover number of each exchanger. Reprinted with permission from Janecki *et al.,* 1998.

has been described (Haggerty *et al.*, 1988). MDCK cells have neither NHE2 nor NHE3 on the apical membrane endogenously (Helmle-Kolb *et al.*, 1997).

III. PHYSIOLOGICAL FUNCTIONS OF APICAL MEMBRANE NHES AND THOSE ATTRIBUTED TO NHE2 AND/OR NHE3

A. Intact Tissue Studies

1. GI Tract

a. Stomach. The role for Na^+/H^+ exchange in the stomach is not well understood, but it appears to be in volume control as part of H^+ secretion (Bachmann *et al.*, 1998; Joutsi *et al.*, 1996; Kaneko *et al.*, 1992; Lamprecht *et al.*, 1993) or in dealing with entry into the parietal cell of secreted acid (Kaneko *et al.*, 1992).

b. Small Intestine. In the small intestine, apical NHEs take part in neutral NaCl absorption, HCO_3 absorption, and H^+ secretion (Donowitz and Welsh, 1987). In the neutral NaCl absorptive process, one molecule of Na^+ is absorbed along with one molecule of Cl^-. This occurs not by a single transport protein carrying both Na^+ and Cl^-, but rather by the linking of brush border Na^+/H^+ exchange and Cl^-/HCO_3^- exchange (Donowitz and Welsh, 1987; Knickelbein *et al.*, 1983; 1990; Rajendran and Binder, 1990). It has been suggested that this linkage is by small changes in intracellular pH, catalyzed by carbonic anhydrase (Knickelbein *et al.*, 1983). This transport process is not part of the Na and Cl cotransporter gene family (Na:K:2 Cl transporter, NaCl cotransporter, or KCl cotransporters). The apical membrane Cl^-/HCO_3^- exchanger has not been definitely identified molecularly, but has been suggested to be the gene product of the *dra* gene (for *downregulated in adenoma*). Based on the species, the jejunum and ileum have approximately equal neutral NaCl absorption (rabbit, rat) or the jejunum has much less than the ileum (human) (reviewed in Donowitz and Welsh, 1987). The jejunum transports the majority of water and accompanying Na^+ in the GI tract, but does so with low efficiency. The ileum absorbs the second largest volume per 24 hours and does so with a higher efficiency than the jejunum but a lower efficiency than the colon. The colon lowers stool Na^+ concentration to <10 mM. Neutral NaCl absorption has several functions. (1) This appears to be the major way Na^+ is absorbed in the intestine in the period between meals. (2) In the ileum, it is the major way Na^+ is absorbed in the postprandial state (Maher *et al.*, 1996; 1997). There is a large ileal postprandial increase in absorption of water and electrolytes, and this appears to be mediated by neurohumoral stimulation of NHE3 (Maher *et al.*, 1996; 1997). The absorption of the digested food end-products glucose, and amino acids in the jejunum is so efficient that there is relatively little of these products in the ileal lumen to stimulate absorption. (3) Neutral NaCl absorption is inhibited in most significant diarrheal diseases (Donowitz and Welsh, 1997). (4) It has been suggested that the increased intestinal absorption associated with cystic fibrosis might be partially caused by stimulation of neutral NaCl absorption, which might contribute to the meconium ileus, which is the earliest manifestation of cystic fibrosis (Berschneider *et al.,* 1988).

The neutral NaCl absorptive process is variable in rate and based on the status of digestion. Under basal (fasting) conditions, neutral NaCl absorption occurs, as does brush-border Na^+/H^+ exchange. It has been assumed, but not shown, that in the immediate postprandial state there is transient inhibition of neutral NaCl absorption in the intestine, accompanied by intestinal secretion of water and electrolytes. In the later postprandial state, there is a marked increase in Na^+ absorption in the small intestine (what occurs in the colon has not been defined). The cause of the increased Na^+ absorption in the small intestine appears to be different in jejunum and ileum. In the jejunum, the increased Na^+ absorption appears to be caused by stimulation by end products of digestion (mediated by Na^+/D-glucose/D-galactose and Na^+/L amino acid transporters). However, in the ileum, there are limited luminal products of digestion to stimulate water and Na^+ absorption. Rather, neurohumoral stimulation of NHE3 is triggered in response to digestion (Maher *et al.,* 1996).

The apical membrane Na^+/H^+ exchanger has been studied by vesicle technology in the small intestine and colon. Of note was lack of ability to demonstrate the presence of an intracellular H^+ modifier site in either ileal or colonic apical membrane exchanger (Knickelbein *et al.,* 1990; Rajendran and Binder, 1990). However, the intracellular modifier site also could not be demonstrated in the basolateral exchanger, suggesting that the technique used could not detect its presence. Separation of rabbit ileal cells along the villus/crypt axis revealed far more Na^+/H^+ exchange in the villus compartment, and decline throughout the crypt, with no Na^+/H^+ exchange on the apical membrane at least in the lower crypt (Knickelbein *et al.,* 1988). However, where in the crypt apical Na^+/H^+ exchange stopped was not clarified.

c. Colon. The proximal colon behaves similarly to the ileum in terms of Na^+ absorption and probably acts as a functional unit with the ileum, with neutral NaCl absorption serving the same absorptive function as in the ileum. The distal colon is involved in concentrating fluid, with the epithelial Na^+ channel having a more major role than neutral NaCl absorption. An unusual aspect of Cl^- dependence of Na^+/H^+ exchange was identified in rat distal colonic crypt cell apical membranes. This Cl^- dependence was inhibited by a Cl^- channel blocker, DIDS, and this was believed to indicate dependence of Na^+/H^+ exchange on a linked Cl channel, perhaps CFTR (Rajendran *et al.,* 1995; 1999; Binder *et al.,* 1997). Cloning of this NHE has not been reported.

d. Gallbladder.   In the gallbladder, neutral NaCl absorption is critical to Na^+ reabsorption, which concentrates the bile (Cremaschi *et al.,* 1992). The Na^+ absorptive process is inhibited by protein-kinase C and probably by cAMP characteristics of NHE3 but not NHE2.

In addition, it has been suggested that NHE3 may be present in pancreatic zymogen granules, where its function can only be speculated on as relating to acid/base homeostasis in this storage/secretory compartment (Thevenod, 1996; Anderie *et al.,* 1998).

The relative functional roles of NHE2 and NHE3 in the intestinal tract have not been studied adequately enough to determine their contributions to the physiological functions of most of the tissues in which they occur. In ileum and colon, the role of NHE3 vs NHE2 in neutral NaCl absorption appears to vary, based on the species and whether basal or stimulated conditions are studied. In the canine ileum, both basal- and meal-stimulated Na^+ and water absorption are entirely attributable to NHE3 (Maher *et al.,* 1997; 1996). In the rabbit ileum and avian ileum, basal Na^+ absorption is provided 50% each by NHE2 and NHE3 (Wormmeester *et al.,* 1998; Donowitz *et al.,* 1998). In the rabbit proximal colon, both NHE2 and NHE3 contribute approximately equally to neutral NaCl absorption. In rabbit distal colon there is no NHE3 and only a small amount of apical Na^+/H^+ exchange, which is explained entirely by NHE2 (Hoogerwerf *et al.,* 1996). Conversely, in rat proximal and distal colon, NHE3 accounts for most basal brush-border Na^+/H^+ exchange (~85% in both) (Ikuma *et al.,* 1999). In the avian colon, approximately 85% of basal Na^+/H^+ exchange is attributable to NHE2 (Donowitz *et al.,* 1998). Human colon has not been studied in detail, but message for NHE2 is highest in the distal colon, where there is minimal NHE3 message (Malakooti *et al.,* 1999; Dudeja *et al.,* 1996).

2. Kidney

The functions of the different segments of kidney in terms of Na^+ and HCO_3^- absorption and H^+ extrusion include the proximal tubule carrying out the largest amount of Na^+ absorption, but with lower efficiency than other parts of the nephron, and with the majority of this transport being $Na\,HCO_3$ absorption. Approximately 80% of filtered $Na\,HCO_3$ is absorbed proximally, with ~60% being attributable to NHE3 (Alpern, 1990; Preisig *et al.,* 1997). Na^+ concentration occurs in the distal tubule/collecting ducts. HCO_3^- is absorbed in the proximal tubule and in the thick ascending limb of Henle (TAL), owing to Na^+/H^+ exchange. Na^+ is absorbed in the proximal tubule mostly by BB Na^+/H^+ exchange, whereas in the TAL, apical Na^+/H^+ exchange explains ~10% of Na^+ absorption.

3. Salivary Glands

The saliva has its volume regulated by the resorptive capacity of the salivary ducts. Apical NHEs take part in the resorption, although their quantitative contribution has not been determined (Lee *et al.,* 1998).

B. Lessons from Studies of NHE2 and NHE3 Knockout Mice

1. NHE2

The only phenotype of NHE2 knockout mice relates to gastric pathology (Wormmeester *et al.,* 1998). In NHE2 knockout mice, the parietal cells form, but during weaning and later they are damaged and disappear, as do (to a lesser extent) the chief cells. Because the parietal cells appear to be the stem cells for the chief cells, these results are consistent with NHE2 being involved in protecting the parietal cells from damage related to acid secretion. The specific function NHE2 is subserving has not been determined, although it is speculated that NHE2 is involved with compensation for changes in intracellular volume associated with acid secretion. Another potential function of NHE2 is to remove accumulation of intracellular H^+ (Kaneko *et al.,* 1992). Confirmation of a protective rule of NHE2 in gastric cells was reported recently. Isolated gut gastric cells were damaged by exposure to acid. This damage was decreased by EGF by a process that involved NHE2 and was PI 3-kinase dependent (Furukawa *et al.,* 1999). The mechanism of this action is not known, nor is it known why the other BLM NHEs (NHE1, NHE4) do not compensate for the absence of NHE2. This is especially surprising because the highest amount of NHE4 message is in parietal cells, suggesting an important function in these cells (Orlowski *et al.,* 1992). There is no recognized consequence of NHE3 knockout in the mouse stomach (Schultheis *et al.,* 1998b).

2. NHE3

That NHE3 is critical for intestinal and renal Na^+ absorption was confirmed by studies of the NHE3 knockout mouse (Schultheis *et al.,* 1998b). These mice have modest diarrhea, borderline dehydration, marked increased sensitivity to Na^+ deprivation, and increased fluid in the cecum/colon, with alkaline intestinal contents and an enlarged small intestine and colon. Interestingly, there was no evidence that intestinal NHE2 compensated for the lack of NHE3 under basal conditions; that is, there did not appear to be an increase in the amount of NHE2 in the small intestine or colon in these mice. These mice had both intestinal and renal compensatory mechanisms. The intestinal compensatory mechanisms involved increased intestinal Na^+ absorption, including hypertrophy of the small intestine and colon and an increase in colonic apical ENaC and apical H^+-K^+-ATPase (Schultheis *et al.,* 1998b).

Renal adaptation was induced owing to chronic dehydration and decreased proximal tubule Na absorption in the NHE3 knockout mouse (Wang *et al.,* 1999; Nakamura *et al.,* 1999). There was adaptive stimulation of HCO_3 absorption in the cortical collecting duct and outer medullary collecting duct, that is, locations downstream from where the defect in HCO_3^- absorption occurred (proximal tubule and TAL), with an increase in H-ATPase and H-K-ATPase as well as basolateral expression of the Cl^-/HCO_3^- exchanger AE-1. Of interest, there was no measurable increase in serum aldosterone or renal renin mRNA, suggesting that renal compensatory mechanisms can make up for decreased proximal tubule Na^+ absorption.

What are the relative contributions of NHE2 and NHE3 to epithelial Na^+ absorption, and what is the unique contribution of each? In circumstances in which only one of the two isoforms is present, this question is answerable. For instance, in renal proximal tubule, NHE3 is responsible for all basal Na^+/H^+ exchange. However, in conditions in which both are present, although tools are available to separate their relative contributions, only few studies with this approach have been reported. Consequently the physiological advantage of having both NHE2 and NHE3 contribute has not been clarified. It does appear, however, that they do not compensate for each other. Although both NHE2 and NHE3 contribute to mammalian Na^+ absorption in all intestinal segments (except rabbit distal colon), in the NHE3 knockout mouse, these animals are at least borderline dehydrated basally and have significant mortality when placed on a low-Na^+ diet, yet there appears to be no compensation by increased expression of NHE2 under normal Na intake conditions (Schultheis *et al.,* 1998b; Nakamura *et al.,* 1999). Similarly, in the NHE2 knockout mouse, damage to parietal cells owing to the absence of NHE2 does not lead to overexpression of the other NHEs present in the same cell and apparently in the same membrane (basolateral membrane NHE1 and NHE4) (Schultheis *et al.,* 1998a).

Moreover, as described above, the relative contributions of NHE2 and NHE3 to basal Na^+/H^+ exchanger have been defined in the small intestine and colon of multiple species and shown to differ widely. However, no insight has been provided concerning the advantages of having NHE2 vs NHE3 as the apical Na^+/H^+ exchanger.

Similarly, the role of NHE2 and NHE3 in renal segments that contain both is not understood. However, unlike in the intestine, most renal segments contain NHE2 or NHE3 but not both. For instance, the proximal tubule contains only NHE3 (see above discussion of uncertainty concerning NHE2 presence). NHE2 is the distal convoluted tubule brush-border NHE. NHE2 (but not NHE3) also appears in IMCD cells, even though all NHE is on the basolateral membrane of these cells. Moreover, only NHE2 occurs apically in macula densa cells. In the TAL, both NHE2 and NHE3 are present in the brush border; however, their relative contributions to Na^+ absorption in this segment have not been defined.

Differences in the relative roles of these two functionally distinct epithelial isoforms do not appear to relate to differences in their developmental expression. For both NHE2 and NHE3, the lowest intestinal message/protein expression is at 2 weeks of age, with increases thereafter (postweaning) such that similar levels are present at weeks 3 and 6 and the adult stage for both NHE2 and NHE3 (Collins *et al.,* 1997; 1998). Perhaps their relative contributions differ based on the way they respond to regulation transcriptionally or by responses over short periods to growth factors and protein kinases.

IV. REGULATION

A. Short Term

1. Relevance to Digestive Physiology

Neutral NaCl absorption and presumably Na^+/H^+ exchange are up- and down-regulated in the postprandial state as part of neurohumoral regulation of the digestive process. Immediately after eating, these processes appear to be inhibited, and then shortly after digestion is initiated, they are greatly stimulated (Donowitz and Welsh, 1987). *In vivo* and *in vitro* studies of neurohumoral regulation of intestinal NaCl absorption have mimicked this up- and downregulation that occurs postprandially. Understanding how NHE2 and NHE3 respond to these stimuli is important for understanding normal digestive physiology. Similarly, protein-kinase regulation of renal brush-border Na^+/H^+ exchanger function has been demonstrated, which leads to changes in water/electrolyte handling and in concentrating function of the kidney. However, the relevance of these rapid changes in brush-border Na^+/H^+ exchange to renal physiology and pathophysiology still is unclear. Extensive studies of short-term regulation of NHE2 and NHE3 have been performed. The models studied have included intact intestine and kidney, cell culture models of epithelial cells often transfected with NHE2 or NHE3 (predominantly the intestinal HT-29 and Caco-2 cells, renal lines OK, LLC-PK1, and MDCK cells) and fibroblasts that lack endogenous Na^+/H^+ exchangers. Surprisingly, models of regulation, including intracellular mechanisms, are similar among the models, although some aspects of NHE regulation have been exquisitely cell or isoform specific (including differences between the same NHE isoform cloned from different species).

2. Kinetics

Extensive kinetic characterization of the transport function of cloned NHEs has been carried out, generally when expressed in simple cells such as fibroblasts. The kinetic studies have been somewhat limited by technical problems with the methods most commonly used. For instance, with $^{22}Na^+$ uptake, the driving force contributed by intracellular H^+ ions is not always considered quantitatively, whereas measurements of amiloride-sensitive Na^+-dependent alkalinization in intracellular pH, which generally take intracellular buffering capacity into consideration, are limited by the rapidity of the Na^+-dependent changes in intracellular pH, such that an underestimation of V_{max} occurs. In addition, some acidification conditions used may damage cells and are not physiological.

Nonetheless, general patterns in regulation have emerged. The general kinetic response of NHE1 has been stimulation by multiple growth factors, protein kinases, hyperosmolarity, and others by changes in $K'(H^+)_i$ (generally an increase in sensitivity to H^+) (Fafournoux *et al.*, 1994; Sardet *et al.*, 1990; 1991; Levine *et al.*, 1993; Kapus *et al.*, 1994; Wakabayashi *et al.*, 1994a; 1992; Grinstein *et al.*, 1985). The only exception is the response to a decrease in intracellular ATP, which decreases both V_{max} and $K'(H^+)_i$ (Kapus *et al.*, 1994).

In contrast, the general pattern of response of NHE2 and NHE3 to changes in growth factors, kinases, and hyperosmolarity has been via changes in V_{max} for NHE3 and NHE2 in response to both stimulatory and inhibitory regulation (Yun *et al.*, 1995b; Levine *et al.*, 1993). Several exceptions to this rule have been identified. These include the effect of removal of ATP, which decreases the affinity for intracellular H^+ ions, as well as decreases the V_{max} (Levine *et al.*, 1993), and the inhibitory response of NHE3 to squalamine (Akhter *et al.*, 1999). This endogenous substance made by the shark liver inhibits NHE3, with an effect that takes 1 hour to become maximum and is not seen at all with the initial exposure (see earlier comments and Fig. 3). Squalamine lowers both the NHE3 V_{max} and the affinity for intracellular H^+ (Akhter *et al.*, 1999). Another exception is the inhibition of NHE3 by cAMP, which also inhibits both V_{max} and $K'(H^+)_i$ (Lamprecht *et al.*, 1998 and Nudarajah, Cha, Lee, Tse, Donowitz, unpublished). Changes in V_{max} is the type of kinetic regulation that most efficiently would allow NHE2 and NHE3 to carry out their major role of transcellular Na^+ absorption. In contrast, regulation of NHE1, which is involved in housekeeping functions, including pH_i homeostatis, would occur best by responding to small changes in intracellular pH by a $K'(H^+)_i$ mechanism.

3. Structure/Function Studies of the NHE C-Terminus Regulatory Domain

This ~300–400 aa domain is hydrophilic, is located intracellularly, and is the regulatory portion of the exchanger. When NHE1, 2, and 3 were truncated to include ~15 aa of the C terminus (after the proline, which appears to mark the end of the transporting domain), all exchangers could transport only very slowly (Yun *et al.*, 1995b). Most importantly, the truncated NHE transporters could not be regulated by growth factors and protein kinases (Yun *et al.*, 1995a; 1995b; Tse *et al.*, 1993a; Levine *et al.*, 1995; Cabado *et al.*, 1996). Thus the C terminus of NHE is called the regulatory domain.

The N- and C-termini of the NHEs interact with each other in a very specific manner (Yun *et al.*, 1995b; Ikeda *et al.*, 1997). Wakabayashi and coworkers (Ikeda *et al.*, 1997) presented preliminary studies of the C-terminus of NHE1, in which it was divided into four functional domains. Domain I, the approximately 90 aa closest to the N-terminus, has a pH_i-maintenance function, appears necessary to maintain the intracellular H^+ modifier site, and is involved in the ATP sensitivity of the exchanger. Domain II, the next C-terminal 50 aa, is silent in terms of pH_i sensitivity, but along with domain I is critical for all regulation of NHE1 by kinases and growth factors. Domain III, the next C-terminal 25 aa is an inhibitory domain that can be overcome by elevating Ca^{2+} and allowing calmodulin to bind to this part of the protein. This domain may interact with cytoplasmic domain I or with the $[H^+]_i$ modifier site. Domain IV is the part of the NHE1 protein that is phosphorylated basally and increases its phosphorylation with protein-kinase regulation, accounting for 50% of growth factor/kinase stimulation of NHE1.

Some N- plus C-termini interactions seem specific and some general. Any NHE N-terminus plus any C-terminus allows Na^+/H^+ exchange activity to occur at a rate approaching wild type (Yun *et al.*, 1995b). However, regulation by growth factors and protein kinases is more restricted. Chimeras made from the entire N-terminus of one NHE isoform plus the entire C-terminus of another NHE have provided some insights into which part of the NHE accounts for the kinetics of regulation. The kinetics of regulation are determined by the N terminus. For instance, serum stimulated the chimera N1C3 (N terminus of NHE1 and C terminus of NHE3) by a change in $K'(H^+)_i$, whereas serum stimulated N3C1 by a change in V_{max} (Yun *et al.*, 1995b). Regulation by other agonists was more complicated; specifically, regulation by EGF/FGF and protein-kinase C occurred only when the epithelial isoforms NHE2 or NHE3 were on both N-terminal and C-terminal domains, and no response occurred when NHE1 made up either domain (Yun *et al.*, 1995b). When chimeras consisted of NHE2 and NHE3, the C-terminal isoform dictated whether the response to protein kinase C was stimulatory (NHE2 on C-terminus) or inhibitory (NHE3 on C-terminus) (Yun *et al.*, 1995b). Adding complexity, cAMP was able to regulate a chimera of N3C1 (N-terminus NHE3, C-terminus NHE1) by an effect on the chimera $K'(H^+)_i$, but had no effect on N1C3 (Cabado *et al.*, 1996). This demonstrated that the NHE1 C terminus can interact with the NHE3 H^+ modifier site in such a way to transmit regulation by cAMP. An interpretation of these results is that the epithelial isoforms NHE2 and NHE3 resemble each other at the tertiary structure level sufficiently to allow interactions of their N- and C-termini, whereas NHE1 differs enough not to be able to allow adequate interaction for regulation by growth factors and protein kinase C, although it is similar enough to allow an effect of protein kinase A.

The NHE C-terminal regulatory domain consists of multiple subdomains (Yun *et al.*, 1995b; Levine *et al.*, 1995; Levine *et al.*, 1993; Nath *et al.*, 1999; Cabado *et al.*, 1996). C-terminal truncations of NHE1, 2, and 3 demonstrated that all protein kinase, growth factor, and hyperosmolar regulation required the NHE C-terminus. Thus, the C-terminus is involved in most regulation. Of note, serum, hyperosmolarity, and ATP depletion change regulation of NHE3 with only ~15 aa of the C-terminus intact (Yun *et al.*, 1995b; Levine *et al.*, 1995; Wakabayashi *et al.*, 1994a). Thus, either these agents act on this part of the C-terminus or they directly affect the transport domain, including the intracellular loops. For NHE1, primarily through the studies of Pouyssegur and coworkers and Wakabayashi *et al.*, it was shown that the C-terminus responds to stimuli by a critical central C-terminal domain (aa 585–636) plus an area immediately C terminal that has two sequential calmodulin (CaM) binding domains, a high- and a low-affinity binding site for calmodulin (Cabado *et al.*, 1996; Wakabayashi *et al.*, 1997b; 1994b; Bertrand *et al.*, 1994). The high-affinity site is inhibitory under basal Ca^{2+} conditions and is relieved with an elevation of intracellular Ca^{2+}. The C-terminus takes part in phosphorylation-dependent and phosphorylation-independent regulation, with phosphorylation occurring in the C-terminal ~130 aa and accounting for approximately 50% of NHE1 regulation (Bianchini and Pouyssegur, 1996; Wakabayashi *et al.*, 1994a).

In contrast to this single major regulatory domain in the C-terminus of NHE1, the organization of the C-terminus of NHE2 and NHE3 appears much more complex. Based on studies of C-terminal truncation mutants, it was found that the C-terminus of both NHE2 and NHE3 consists of a stimulatory domain and an inhibitory domain, which appear to act independently (Fig. 6) (Levine *et al.*, 1995; Nath *et al.*, 1999). The stimulatory domain is closer to the N terminus, and the inhibitory domain is closer to the C-terminus. Both stimulatory and inhibitory domains are made up of multiple subdomains, most of which appear to act independently because they have additive effects on Na^+/H^+ exchange (Levine *et al.*, 1995). The stimulatory domains of both NHE2 and NHE3 include ele-

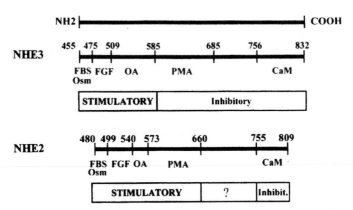

FIGURE 6 Similarity in organization of C-terminal regulatory domains of NHE2 and NHE3. The C-terminal regulatory domains of NHE2 and NHE3 are made up of a stimulatory and an inhibitory domain, both of which consist of multiple subdomains. Shown in this figure is the striking similarity of organization of these regulatory domains in NHE2 and NHE3. This suggests that, in spite of low homology at the amino acid level, the C-terminal cytoplasmic domains of NHE2 and NHE3 are probably highly homologous in terms of tertiary structure. Reprinted with permission from Nath *et al.*, 1999.

ments that respond to serum, EGF and FGF, and okadaic acid. The inhibitory domain of NHE3 responds to protein kinase A and C and to calmodulin (Levine *et al.,* 1995). The inhibitory domain of NHE2 responds to calmodulin (Nath *et al.,* 1999). This organization of the NHE2 and NHE3 C-terminus is remarkably similar, the major difference being that protein kinase A and C stimulate NHE2 but inhibit NHE3, although they act on domains in similar locations in their respective C-termini (Yun *et al.,* 1995b; Nath *et al.,* 1999; Cabado *et al.,* 1996).

The C-terminus of NHE3 is involved in basal as well as stimulated regulation. Under basal conditions, NHE2 and NHE3 are stimulated by okadaic acid, indicating that basal ser/thr phosphorylation is inhibiting them (Yun *et al.,* 1995b; Nath *et al.,* 1999) are inhibited by genistein, indicating basal tyrosine phosphorylation is stimulating them (Nath *et al.,* 1996a); and are stimulated by the calmodulin inhibitor KN-62, indicating that basal CaM kinase is inhibiting them (Nath *et al.,* 1996a; Donowitz *et al.,* 1997). It is important to note that these responses do not indicate whether NHE2 or NHE3 are the immediate substrate of these kinases or are affected indirectly. In addition, the C-terminal 140 aa of NHE3 act as an inhibitory domain by a mechanism involving endocytosis (see page 473). When removed, the rate of the truncated NHE3 increases to 5–6 times that of wild-type NHE1 (Akhter *et al.,* 2000). This inhibitory domain of NHE3 has two subdomains: the C-terminal 76 aa and the immediately N-terminal 64 aa. The agonists that act on the C-terminal 76 aa (aa 757–832) include CaM kinase II, tyrosine kinases, and squalamine; the agonists that act on aa 690–755 have not been identified (Yun *et al.,* 1995b; Nath *et al.,* 1996a).

4. The NHE C-Termini Associate with Multiple Regulatory Proteins

This is true for NHE1, 2, and 3. The NHE C-termini binding proteins have been identified by coprecipitation, overlay studies, pull-down assays, and yeast two-hybrid approaches. Identification of NHE C-terminal binding proteins has just begun, and there are almost certainly many yet-to-be-identified proteins. The function of this binding is just being understood, but it appears there are several types: cytoskeletal binding and an association with regulatory proteins (Table III).

a. Proline-Rich Domain of NHE2. The C-terminal domain of NHE2 has two proline-rich areas (Chow *et al.,* 1999) that are conserved among species: proline-rich area I (aa 743–750 in rat NHE2) and area II (aa 786–792). Using an overlay approach, proline-rich area I was shown to bind promiscuously to SH3 domains of multiple proteins (including c-Src, ras-gap, p85 subunit of PI 3-kinase, α-spectrin, and neural-Src). Proline-rich domain II bound to some SH3 domains but with more specificity, binding most avidly to the SH3 domain of p85 and c-Src>n-Src, Abl and not to that of ras-gap and α-spectrin. No clear functional role for these proline-rich areas was determined, and their mutation did not pre-

TABLE III

Identified Associated Regulatory Proteins Binding to the NHE C Terminus

NHE1:	calmodulin (Bertrand *et al.*, 1994; Wakabayashi *et al.*, 1994b); calcineurin homologous protein (CHP) (Lin and Barber, 1996); dnaK, (bacterial equivalent of mammalian 70-kDa heat-shock protein (Silva *et al.*, 1995); ezrin; 24-kDa unidentified protein (Goss *et al.*, 1996)
NHE2:	calmodulin (Nath *et al.*, 1999); 32-, 34-kDa unidentified proteins (Tse, unpublished observations); see discussions of binding to Pro-rich areas above
NHE3:	calmodulin (Donowitz *et al.*, 1997); CHP (Tse, Lee, Barber and Donowitz, unpublished observations); NHERF and E3KARP (Yun *et al.*, 1997; 1998); megalin (Biemesderfer *et al.*, 1999)

vent hyperosmolarity and ATP depletion from altering rat NHE2. Truncations of the NHE2 C-terminus to include the part of the C-terminus containing these pro-line-rich domains and expression in the polarized renal proximal tubule cell line LLC-PK1 revealed that truncation to aa 731 but not to aa 777 decreased the percentage of NHE2 targeted to the plasma membrane, and that which reached the plasma membrane was more on the basolateral membrane and less on the apical membrane than would occur normally. Thus, it was concluded that NHE2 aas 731–776 were involved in membrane targeting and apical membrane location of NHE2. Whether proline-rich domain I is involved in this targeting has not been established.

b. Association with Cytoskeleton. NHE3 is associated with the apical membrane cytoskeleton. This association is very different from that shown for NHE1. It is not known whether NHE2 is associated with cytoskeleton. NHE1 colocalizes in the lamillopodia or leading edge of fibroblasts, where it colocalizes with ezrin, vinculin, talin, and actin (Grinstein *et al.*, 1993). Recent studies by Barber suggested that NHE1 may be involved in forming or stabilizing the cytoskeleton. She showed that NHE1 binds to ezrin and, in addition, that deletion of the NHE1 domain that binds to cytoskeleton leads not only to disorganization of the cytoskeleton but also to a lesser association of ezrin with the plasma membrane (Barber, privileged communication).

One way NHE3 associates with the brush-border cytoskeleton is by being indirectly anchored to ezrin (Lamprecht *et al.*, 1998; Yun *et al.*, 1998). The association with ezrin appears to be via a gene family called the regulatory factor (RF) gene family, the recognized members of which contain two highly homologous PDZ domains and serve as linking proteins. These proteins are NHERF (Na^+/H^+ exchanger regulatory factor, also called EBP50 for *ezrin binding proton 50*) and E3KARP (NHE3 kinase A regulatory protein, also called TKA-1) (Fig. 7).

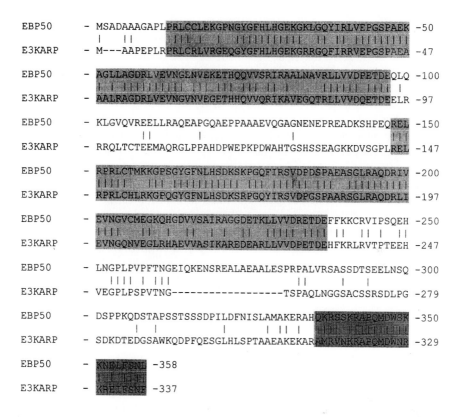

```
EBP50    -  MSADAAAGAPLPRLCCLEKGPNGYGFHLHGEKGKLGQYIRLVEPGSPAEK  -50
            |  ||        |||| |   |  ||||||||||||  || ||  ||||||||
E3KARP   -  M---AAPEPLRPRLCRLVRGEQGYGFHLHGEKGRRGQFIRRVEPGSPAEA  -47

EBP50    -  AGLLAGDRLVEVNGENVEKETHQQVVSRIAALNAVRLLVVDPPETDEQLQ  -100
            |   |||||||||||| |    ||| |||  ||  || |||||||  ||||   |
E3KARP   -  AALRAGDRLVEVNGVNVEGETHHQVVQRIKAVEGQTRLLVVDQETDEELR  -97

EBP50    -  KLGVQVREELLRAQEAPGQAEPPAAAEVQGAGNENEPREADKSHPEQREL  -150
                      ||        |            |                    |||
E3KARP   -  RRQLTCTEEMAQRGLPPAHDPWEPKPDWAHTGSHSSEAGKKDVSGPLREL  -147

EBP50    -  RPRLCTMKKGPSGYGFNLHSDKSKPGQFIRSVDPDSPAEASGLRAQDRIV  -200
            |||||   |||  |||||||||||| |||  ||||  ||| ||||||||||
E3KARP   -  RPRLCHLRKGPQGYGFNLHSDKSRPGQYIRSVDPGSPAARSGLRAQDRLI  -197

EBP50    -  EVNGVCMEGKQHGDVVSAIRAGGDETKLLVVDRETDEFFKKCRVIPSQEH  -250
            ||||   ||   ||  ||  | ||     |||||||||  ||   ||  |  ||
E3KARP   -  EVNGQNVEGLRHAEVVASIKAREDEARLLVVDPETDEHFKRLRVTPTEEH  -247

EBP50    -  LNGPLPVPFTNGEIQKENSREALAEAALESPRPALVRSASSDTSEELNSQ  -300
               |||| | |||                        ||     |     |
E3KARP   -  VEGPLPSPVTNG-----------------TSPAQLNGGSACSSRSDLPG  -279

EBP50    -  DSPPKQDSTAPSSTSSSDPILDFNISLAMAKERAHQKRSSKRAPQMDWSK  -350
               |   |         |  |||      |||| |||||||||
E3KARP   -  SDKDTEDGSAWKQDPFQESGLHLSPTAAEAKEKARAMRVNKRAPQMDWNR  -329

EBP50    -  KNELFSNL  -358

E3KARP   -  KRELFSNF  -337
```

Identity : 176 (52.2%)

FIGURE 7 Amino acid composition and amino acid alignment of NHERF and E3KARP. The deduced amino acid structures of human E3KARP and rabbit NHERF are compared with the two similar PDZ domains (lightly shaded areas) and the ezrin binding domain (darkly shaded areas). Note that these proteins, while ~50% identical, are only highly homologous in these areas. Modified with permission from Yun *et al.*, 1997.

NHERF was cloned based on obtaining a partial sequence via microsequencing of a protein that could reconstitute cAMP inhibition of renal proximal tubule brush-border Na^+/H^+ exchange (Weinman *et al.*, 1993; 1995). E3KARP was cloned by a yeast two-hybrid approach using a strategy of selecting clones to study which bound to the NHE3 C-terminus but not to the NHE1 C-terminus (Yun *et al.*, 1997). Both NHERF and E3KARP are found in almost all tissues, and NHERF is present in most epithelial cell culture models, whereas E3KARP is somewhat less expressed in some of these (Cha *et al.*, 1999). Exceptions include the PS120 fibroblast used for many studies of cloned NHEs in which there is no E3KARP and no or minimal NHERF expression; Caco-2 cells and T84 cells, which have a small amount of NHERF and E3KARP compared with other epithelial cell lines; and OK, the any MDCK renal epithelial cell lines, which have little or no E3KARP (Cha *et al.,* 1999). What explains the differences in expression of these RF gene products is not known. As an extreme example, PS120 cells are derived from Chinese hamster lung fibroblasts. Another Chinese hamster–derived fibroblast line (ovary), AP-1, has been used for NHE expression studies and has one of the highest expression levels of NHERF.

Multiple interactions of NHERF and/or E3KARP with NHE3 have been described. (1) They co-precipitate (Yun *et al.,* 1997; 1998). (2) The part of NHERF or E3KARP that interacts with NHE3 is the second PDZ domain plus the beginning of the C terminus (Yun *et al.,* 1998). It is unusual for a PDZ-domain interacting protein to interact with a domain in addition to the PDZ domain. Crystallization of the PDZ domain of PSD95 with its binding partner has shown that the binding pocket with substrate in place is entirely in the PDZ domain (Cabral *et al.,* 1996; Doyle *et al.,* 1996). The predicted structure of the E3KARP PDZ domain II is very highly homologous to the crystallized PDZ domain of PSD95 (Donowitz, Cha, Yun, Amsel, unpublished observations). Thus, the E3KARP or NHERF PDZ-domain II interaction with NHE3 is almost certainly unique. (3) The part of NHE3 that binds E3KARP or NHERF is between aa 586 and 711 (Yun *et al.,* 1998). This domain also has been shown to be necessary for PKA and PKC regulation (Yun *et al.,* 1995b; Cabado *et al.,* 1996). (4) NHERF or E3KARP binds to the N terminus of ezrin (Yun *et al.,* 1998; Reczek *et al.,* 1997). It is not known if this binding of ezrin to NHE3 via E3KARP or NHERF anchors ezrin to the plasma membrane, that is, if a given ezrin molecule that binds NHE3 also is able to bind to another protein in the plasma membrane. The parts of NHERF and E3KARP that bind ezrin are the C-terminal ~30 aa, which are highly homologous to each other and are rich in basic amino acids (Yun *et al.,* 1998; Reczek *et al.,* 1997), as are other ezrin binding proteins (see Fig. 7). (5) The interactions of NHERF and E3KARP are strong enough to co-precipitate both NHE3 and ezrin simultaneously, although the amount of ezrin co-precipitated is small (Yun *et al.,* 1998). (6) NHERF colocalizes with NHE3 and ezrin based on immunocytochemistry; in PS120 cells, both NHERF and ezrin are in the plasma membrane, and to a lesser extent in the cytoplasm (Yun *et al.,* 1998). They do not colocalize in the juxtanuclear region, where only NHE3 is found (Yun *et al.,* 1998). Thus, NHERF and/or E3KARP link NHE3 to ezrin and the brush-border cytoskeleton. The first PDZ domain of these regulatory factor proteins binds other proteins. Those identified to date include CFTR, β2 adrenergic receptors, and P2Y1 purinergic receptors (Wang *et al.,* 1998; Hall *et al.,* 1998a; 1998b). In addition, NHERF dimerizes and thus could help form a multiprotein complex. Whether NHE3 takes part in such a complex is not known and is just one of many unanswered questions relating to the role of this recently recognized gene family in regulation of Na^+/H^+ exchange. Also, it is not yet known what percentage of NHE3 in the plasma membrane associates with ezrin and NHERF and/or E3KARP; whether NHE3 binds directly to ezrin in addition to this indirect binding; whether intracellular NHE3 associates with these proteins; and whether NHE3 moves on and off the plasma membrane accompanied by the regulatory factor proteins. Of note is that NHERF association with the β2 adrenergic receptor does involve movement of NHERF on and off the plasma membrane (Hall *et al.*, 1988a). The functional significance of the cytoskeletal attachment of NHE3 identified to date relates to cAMP regulation and will be discussed under that topic.

NHE3 is regulated by mechanisms involving phosphorylation and vesicle trafficking. NHERF and E3KARP appear to be involved in regulation by phosphorylation.

5. Regulation Involving Phosphorylation of NHE3 by cAMP

cAMP and cGMP inhibit intestinal NaCl absorption, including brush border Na^+/H^+ exchange throughout most of the intestinal tract including in the gallbladder, and inhibit Na^+/H^+ exchange in the proximal tubule. An exception is that cGMP does not have effects on the duodenum and distal colon of the rabbit (Field *et al.*, 1978). When cloned NHE1, 2, and 3 were studied in AP1 fibroblasts, cAMP stimulated NHE1 and NHE2 and inhibited NHE3, whereas cGMP had no effect (Kapus *et al.*, 1994; Cabado *et al.*, 1996). However, when studied in PS120 cells, cAMP failed to affect NHE1, NHE2, or NHE3 (Levine *et al.*, 1993), although it did stimulate trout βNHE (Borgese *et al.*, 1994; Malapert *et al.*, 1997), indicating that at least some of the cAMP-dependent machinery to affect Na^+/H^+ exchange was intact.

Orlowski, Grinstein, and coworkers (Cabado *et al.*, 1996) characterized the domains in the C-terminus of NHE3 that are involved in cAMP inhibition of NHE3 by studying c-terminal truncations. A region between aa 579 and 684 was shown to be necessary for cAMP inhibition (see Fig. 6). In addition, the chimera consisting of NHE3 on the N-terminus and NHE1 on the C-terminus (N3C1) was inhibited by cAMP, whereas there was no effect of cAMP on N1C3 (Cabado *et al.*, 1996).

The mechanism by which cAMP regulates NHE3 has been shown to involve phosphorylation by protein kinase AII of NHE3. Various groups of investigators have contributed to understanding this complex regulation (Yip *et al.*, 1997; Moe *et al.*, 1995; Zhao *et al.*, 1999; Kurashima *et al.*, 1997; Zizak *et al.*, 1999). NHE3 is phosphorylated in the basal state (Yip *et al.*, 1997; Moe *et al.*, 1995; Kurashima *et al.*, 1997). There is no evidence that the basal phosphorylation of NHE3 is involved in cAMP inhibition. In contrast, cAMP stimulation of NHE3 phosphorylation is required for the inhibition of NHE3 (Kurashima *et al.*, 1997). cAMP stimulates phosphorylation of NHE3. S605 is involved, and its increase in phosphorylation by cAMP-dependent protein kinase II is necessary for ~50% of cAMP inhibition of NHE3 function (Kurashima *et al.*, 1997). Also, NHE3 S634 was shown to be necessary for cAMP inhibition of NHE3, and when both NHE3 S605 and S634 were mutated, cAMP had minimal effect on NHE3 activity (Kurashima *et al.*, 1997). However, S634 was not phosphorylated by cAMP-dependent protein kinase, and the mechanism of its role in regulation is still not clear (Kurashima *et al.*, 1997). However, Moe and coworkers (Zhao *et al.*, 1999) observed that another serine, S552, also was phosphorylated in response to cAMP, and mutating it almost totally prevented cAMP inhibition of NHE3 (larger inhibitory effect than mutation of S605) in the same cell model used by Grinstein and Orlowski. Because this lies outside the domain shown to be necessary for cAMP inhibition of NHE3, its role in cAMP regulation is not yet clear.

The mechanism of cAMP inhibition of NHE3 and its dependence on the regulatory factor gene family members NHERF and E3KARP were studied by Yun, Lamprecht, Weinman, Zizak, Tse, Donowitz and associates (Lamprecht *et al.*, 1998; Yun *et al.*, 1998; Zizak *et al.*, 1999). They found that (1) it was cAMP-dependent protein-kinase AII that was responsible for inhibition (Lamprecht *et al.*, 1998); (2) the regulatory factor gene family members NHERF and E3KARP and NHE3 did not act as AKAPs (kinase A anchoring proteins (Lamprecht *et al.*, 1998); and (3) kinase A did not change the phosphorylation of either NHERF or E3KARP in PS120 cells. Although NHERF was phosphorylated under basal conditions (Lamprecht *et al.*, 1998), mutating three series in NHERF led to it not being a phosphoprotein under basal conditions in addition to it not being phosphorylated by cAMP, yet still allowing the cAMP-induced inhibition of NHE3 as well as the phosphorylation of NHE3 (Zizak *et al.*, 1999).

The mechanism by which RF members allow protein kinase AII to inhibit NHE3 include (1) linking NHE3 to ezrin—the last ~30 aa of these RF proteins are required for ezrin binding (Cha *et al.*, 1999; Reczek *et al.*, 1997) (NHERF truncated by C-terminal 30 aa does not bind to ezrin, and this cDNA does not reconstitute cAMP inhibition of NHE3; Cha *et al.*, 1999, Weinman *et al.*, 2000), and (2) allowing protein kinase AII to phosphorylate NHE3—cAMP phosphorylates NHE3 in PS120 transfected with either NHERF or E3KARP on two phosphopeptides, and this does not occur in the absence of these RF (Fig. 8) (Zizak *et al.*, 1999). Ezrin is an AKAP, although with a binding affinity lower than that for other AKAPs that have been shown to allow protein kinase A effects (Drensfield *et al.*, 1997). The current model is that RF bind NHE3 at their second PDZ domain and simultaneously bind ezrin, thus allowing protein kinase AII to be in the correct location to phosphorylate NHE3 (Fig. 9).

In PS120 cells transfected with NHERF, cAMP exposure did not change the percentage of NHE3 on the plasma membrane (Cavet, Akhter, Tse, Donowitz, unpublished observations). Thus, there is no evidence that cAMP inhibition of NHE3 in fibroblasts is associated with a decrease in the amount of plasma membrane on NHE3. In addition, there is no evidence that NHERF/E3KARP are in-

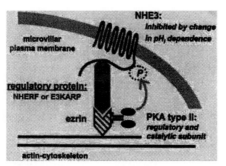

FIGURE 8 Model of NHE3 attachment to the cytoskeletal protein ezrin by means of the linking function of the regulatory factor gene family members NHERF and E3KARP. In addition, the proposed mechanism is shown by which NHERF/E3KARP allows ezrin, acting as an AKAP, to present protein kinase A II to phosphorylate NHE3. Reprinted with permission from Lamprecht *et al.*, 1998.

volved in the 50% inhibition of NHE3 by cAMP, which involves S634. However, as described below, cAMP inhibition of NHE3 in epithelial cells may involve a contribution of vesicle trafficking.

Studies with NHE2 have demonstrated that, like NHE1 and NHE3, it is phosphorylated in the basal state (Tse, unpublished observation). However, changes in its phosphorylation have not been reported. There are multiple unanswered questions about cAMP regulation of NHE2 and NHE3. For NHE2: What is the associated protein or other cofactor missing in PS120 cells, because NHE2 does not appear to bind NHERF or E3KARP and is stimulated by cAMP in AP-1 cells but not affected in PS120 cells? For NHE3: How does the protein kinase AII induced phosphorylation lead to inhibition of Na^+/H^+ exchange activity? What accounts for cAMP inhibition of NHE3 by both V_{max} and $K'(H^+)_i$ unlike most other NHE3 regulation having an effect only on V_{max}? What is the role of S634 and perhaps S552 in the effect of cAMP on NHE3? What binds the RF PDZ1 domain when NHERF binds NHE3 and is this relevant to regulation of NHE3? How is specificity achieved in the interaction of the regulatory factor gene family and NHE3, and is the interaction constitutive or does it change as part of signal transduction? (This is relevant because it has been suggested that basal NHE3 phosphorylation is needed for protein kinase C inhibition.) (Wiederkehr *et al.*, 1999).

For cGMP, the situation is still not well defined. In neither AP-1 cells nor PS120 cells does cGMP affect NHE1, NHE2, or NHE3 (Levine *et al.*, 1993; Wakabayashi *et al.*, 1994a). This is not unexpected because, in most cells in culture, cGMP-dependent protein-kinase II, the isoform involved in regulation of Na^+ absorption, is downregulated (Pfeifer *et al.*, 1996). Whether this is the explanation for the lack of regulation of NHE3 by cGMP must be tested experimentally.

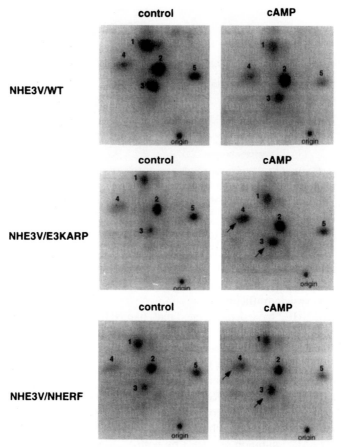

FIGURE 9 Members of the regulatory factor gene family are necessary for cAMP to phosphorylate NHE3. When expressed in PS120 fibroblasts, cAMP only phosphorylated NHE3, based on two-dimensional phosphopeptide analysis, in the presence of either NHERF or E3KARP. Reprinted with permission from Zizak *et al.*, 1999.

6. Regulation Involving Membrane Trafficking—Regulation by Protein-Kinase C and EGF/FGF

There are several agonists that regulate NHE3 but have not been shown to change its phosphorylation; if changes occur, they do not correlate with the changes in transport (Yip *et al.*, 1997; Wiederkehr *et al.*, 1999). These include the growth factors EGF/FGF and phorbol esters acting via protein kinase C. Evidence of a role for membrane trafficking in regulation of NHE3 comes from intact tissue studies and studies in cell culture models of intestine (Caco-2), kidney (OK), and PS120 fibroblasts.

Two models in which movement of NHE3 from the plasma membrane occurs as part of signal transduction have been studied in intact renal tissues: parathyroid hormone inhibition of renal brush border Na^+/H^+ exchange and acute hypertension (Zhang et al., 1998; 1999; Yip et al., 1998; Hensley et al., 1989). Parathyroid hormone exposure acutely inhibits rat proximal tubule brush border Na^+/H^+ exchange, shortens the brush border, and redistributes Na^+/H^+ exchange activity to intracellular vesicles that are enriched in acid phosphatase and galactosyltransferase (Zhang et al., 1999; Hensley et al., 1989). These studies used sorbitol density gradients and were extended to examine the location of NHE3 specifically. Comparable percentages of NHE3 and NaPi2 were redistributed by PTH exposure from apical membranes to fractions containing intracellular membrane markers, although the response of removal of NaPi2 occurred more rapidly than that of NHE3 (McSwine et al., 1998). The PTH analogues that caused these effects are coupled to adenylate cyclase, implicating cAMP in this redistribution.

Induction of acute hypertension in the rat reversibly redistributed NHE3 from the proximal tubule brush border to intracellular vesicles. A similar study using confocal microscopy demonstrated NHE3 redistributing from the brush border to the base of microvilli (Yip et al., 1998). This location of NHE3 also was found in kidneys of spontaneously hypertensive rats, indicating that hypertension acutely and chronically is associated with NHE3 being in a brush border location different than conditions of normal blood pressure.

The debate over what constitutes the storage pool of NHE3 is not yet resolved. NHE3 colocalizes with megalin in the intermicrovillus cleft region in the renal proximal tubule, which is the area in which clathrin-coated pits form (Biemesderfer et al., 1999). Megalin appears to be involved in forming a complex for NHE3 in trafficking it to and from the rest of the brush border into what has been suggested as being a storage pool in which NHE3 has decreased activity. In addition, some density gradient separations strongly support involvement of an intravesicular compartment in trafficking as well. The concept of an intracellular vesicular pool of NHE3 is supported by studies using electron microscopy. NHE3 in proximal tubule cells was shown to be in a pool in subapical membrane vesicles, the nature of which was not further identified (Cremaschi et al., 1992).

Further evidence for trafficking of NHE3 in short-term regulation comes from studies of NHE3 in the human colon cancer cell line Caco-2 cells. These cells undergo differentiation after confluency. Part of this differentiation is the appearance of NHE3 in the apical membrane over time (Noel and Pouyssegur, 1995; Janecki et al., 1998). Janecki and coworkers (1998; 1999a) used a confocal microscope approach to demonstrate the NHE3 moved on and off the brush border as part of kinase/growth factor regulation of NHE3 (see Fig. 5). They defined the brush border by use of the lectin PHA-E from the outside and actin from the inside of the cell, and defined the percentage of NHE3 that colocalized with the brush border vs that which was in the area below the brush border but above the nucleus, called the subapical compartment. This is an area believed to contain the recycling endosomes in Caco-2 cells. Under basal conditions, approximately 80% of NHE3 was present in the brush border and 20% in the subapical compartment. Phorbol ester, acting by protein kinase C, inhibited NHE3 by ~30% (Fig. 5). It also decreased the percentage of NHE3 in the brush border by ~15% and increased the percentage in the subapical compartment by ~50%. Similar results were obtained by reversible cell-surface biotinylation by studying the apical surface of these cells (Janecki et al., 1998). In 20 minutes, 13% of NHE3 was internalized from the brush border under control conditions, and this increased to 28% with phorbol ester exposure. The fact that the effect on transport exceeded the change in amount of NHE3 on the brush border suggests that protein kinase C also decreases the NHE3 turnover number. This C kinase effect was not associated with a change in size of the brush border. Conversely, EGF stimulated the NHE3 transport rate by 20% in Caco-2 cells, an effect associated with movement of NHE3 from the subapical compartment into the brush border, such that no distinguishable subapical compartment was present after EGF (Janecki et al., 1999a). However, EGF also caused a comparable 20% increase in size of the brush border, and thus it could not be determined from these studies whether the EGF stimulation of NHE3 was specific or just occurred as part of the increase in size of the brush border, a phenomenon previously described in intact rabbit small intestine by Gall and coworkers (Opleta-Madsen et al., 1999).

Most detailed studies and evidence of plasma membrane trafficking/recycling of NHE3 have been carried out in AP-1 and PS120 fibroblasts. Studies compared NHE1 and NHE3 in these cells. NHE3 trafficks continually under basal conditions, whereas NHE1 does not appear to cycle (D'Souza et al., 1998; Akhter et al., 2000; Kurashima et al., 1998). In these cells, NHE3 was present both in a juxtanuclear compartment and in the plasma membrane, and NHE1 was almost exclusively on the plasma membrane. NHE3 colocalized with transferrin receptor and cellubrevin but not with markers of lysosomes and golgi (Kurashima et al., 1998). Grinstein and Orlowski concluded that the intracellular NHE-containing compartment was the recycling endosome. The juxtanuclear location of NHE3 was disrupted by disrupting microtubules with colchicine, but this did not affect basal NHE3 function (Kurashima et al., 1998). Brefeldin A compacted the juxtanuclear compartment, increased the percent of NHE3 on the plasma membrane, and stimulated NHE3 transport activity, consistent with an increase in the plasma membrane amount of NHE3 (Kurashima et al., 1998). Basal Na^+/H^+ exchange was dependent on PI 3-kinase (Kurashima et al., 1998). Inhibiting PI 3-kinase with wortmannin or LY294002 almost completely inhibited NHE3 transport activity, inhibited the amount of NHE3 on the plasma membrane, and increased the size of the NHE3-containing juxtanuclear compartment (Kurashima *et al.,* 1998). Thus, NHE3 cycles on and off the plasma membrane under basal conditions. The exocytosis of NHE3 under basal conditions is dependent on PI 3-kinase activity, but endocytosis of NHE3 is at most minimally dependent on PI 3-kinase. In contrast, PI 3-kinase inhibitors do not affect NHE1, consistent with NHE1 being present on the plasma membrane and not trafficking under basal conditions (Ma *et al.,* 1994).

The separation of endo- and exocytosis of NHE3 was most clearly achieved by studying NHE3 truncation mutants (Akhter *et al.,* 2000). These showed that there are two domains in the NHE3 C terminus that are involved in endocytosis of the exchanger. Truncating NHE3 at aa 756 increased the rate of transport 2–3 times and to aa 690 5–6 times, both of which caused comparable increases in the percentage of NHE3 on the plasma membrane (E3/756 increased to 30% from wild-type 14%, and E3/690 increased to 60%). These truncation mutants failed to internalize over 3 hours based on reversible cell-surface biotinylation, indicating marked inhibition of endocytosis. These mutants were not affected by inhibiting PI 3-kinase, supporting that under steady-state conditions they were not trafficking, and that with the marked slowing of endocytosis there was a similar limitation in exocytosis. That endo- and exocytosis of NHE3 could be controlled separately was demonstrated by the observation that serum had a similar stimulatory effect on these truncation mutants and wild-type NHE3 (Akhter *et al.,* 2000).

Extension of the studies of growth factor stimulation of NHE3 in PS120 cells was achieved by study of chimeras of NHE3 transfected with red shifted GFP (EGFP) on the C terminus (to allow study in living cells), and using FM dyes to mark the plasma membrane (Janecki *et al.,* 2000). This allowed quantitation of size of the plasma membrane and distribution of NHE3-EGFP between the plasma membrane and intracellular pool based on confocal microscopy. This chimera functioned similarly to wild-type NHE3 based on percentage of NHE3 on the plasma membrane, stimulation by FGF, and inhibition by PI 3-kinase. This chimera was inhibited by 50% in basal transport rate and percentage on the plasma membrane by inhibitors of PI 3-kinase. This supports the view that plasma-membrane NHE3 trafficks under basal conditions. Unknown is the nature of the noncycling pool of NHE3, especially whether the cytoskeletal-anchored (ezrin/NHERF/E3KARP) NHE3 also recycles or represents this noncycling pool. FGF stimulated NHE3-EGFP, with stimulation of transport rate exceeding the increase in the amount of NHE3-EGFP on the plasma membrane, indicating that FGF increased both the amount of plasma-membrane NHE3 as well as the turnover number of the exchanger. Inhibiting PI 3-kinase blocked approximately half the FGF-stimulated NHE3 transport and amount (Janecki *et al.,* 2000). Thus PI 3-kinase is involved in stimulation of both basal and stimulated NHE3. The nature of the PI 3-kinase–independent component of FGF stimulation is not known. Of note, PI 3-kinase might be regulating NHEs at a step in addition to trafficking, because the PI 3-kinase inhibitor wortmannin also inhibited PDGF stimulation of NHE1 (Ma *et al.,* 1994).

As reviewed, the data available suggest that regulation of NHE3 is either by changes in its phosphorylation or by changes in membrane trafficking. However, recently Moe and coworkers (Fan *et al.,* 1999) showed that both mechanisms may be involved in response to some agonists. They described a model of PTH inhibition of rat renal proximal tubule brush border NHE3 by a PTH analogue that acts by elevation of cAMP, in which both mechanisms appeared to occur in response to cAMP, although with different time courses. They showed that PTH inhibited NHE3 and increased its phosphorylation without changing the amount of plasma membrane NHE3 for the initial 30–60 minutes of PTH exposure. However, between 4 and 12 hours after PTH exposure, the amount of plasma membrane NHE3 also decreased. In addition, Alpern, Moe and coworkers (Peng *et al.,* 1999) showed that in OK cells, a renal proximal tubule model, endothelin acutely stimulates NHE3 by acting on the endothelin B receptors. The NHE3 stimulation was associated with an increase in NHE3 phosphorylation on both Ser and Thr and was associated with an increase in the amount of NHE3 on the apical membrane. This study implicated both phosphorylation and trafficking in stimulation of NHE3. Inhibition of the actin cytoskeleton with cytochalasin D did not alter endothelin stimulation of NHE3 activity, phosphorylation, or apical membrane amount of NHE3. However, inhibition of tyrosine kinases plus prevention of elevation of intracellular Ca^{2+} with BAPTAM prevented the endothelin stimulation of NHE3 activity, but not amount or phosphorylation, implicating another step in activation of NHE3 after movement of NHE3 to the apical membrane and its phosphorylation. Thus, careful studies of growth factor/kinase regulation of NHE3 must be studied over time to determine whether both phosphorylation and regulation of amount of plasma-membrane NHE3 are critical for acute regulation and to define the interrelationships of these two mechanisms.

7. Other

There are many other well-studied examples of acute regulation of NHE2/NHE3 in which the mechanisms have not been thoroughly understood and specifically have not been shown to involve change in NHE phosphorylation or trafficking. Some of these will be described.

a. Hyper-Hypo-osmolarity. NHEs are involved in the reswelling of cells that occurs after cells shrink upon exposure to hyperosmolarity. NHEs involved include NHE1 (see chapter by MacLeod in this volume). Whether NHE2 and NHE3 have a role in responding to osmotic stimuli in renal epithelial cells is not known, although in the intestine they do not appear to have a major role. The potential of their involvement in responding to changes in osmolarity led to evaluation of changes in NHE activity owing to both hypertonicity and hypotonicity. All NHEs are affected by hyperosmolarity, with the response being induced by the resulting cell shrinkage. NHE1 is stimulated and NHE3 is inhibited by hyperosmolarity (Kapus *et al.,* 1994; Shrode *et al.,* 1996; Nath *et al.,* 1996b; Bianchini *et al.,* 1995; Soleimani *et al.,* 1994c; Singh *et al.,* 1996). NHE2 appeared to be regulated inconsistently by hyperosmolarity. Rat NHE2 expressed in AP-1 fibroblasts or in the epithelial proximal tubule model LLC-PK1 is stimulated, whereas rabbit NHE2 expressed in PS120 fibroblasts and in the intestinal cell model Caco-2 is inhibited by hyperosmolarity (Levine *et al.,* 1993; Nath *et al.,* 1999; 1996b; Kapus *et al.,* 1994; Singh *et al.,* 1996; and Donowitz, Levine, Tse, unpublished observations). That the cell type of expression rather than subtle difference in the species specificity of NHE2 (rat vs rabbit) determined the differential response to hyperosmolarity was suggested by Wakabayashi by expression of rat and rabbit NHE2 in PS120 cells; he demonstrated inhibition of both by hyperosmolarity (Wakabayashi, privileged communication). The change in NHE activity by cell shrinkage is via a change in V_{max} for NHE2 and NHE3 and a change in $K'(H^+)_i$ for NHE1, and is rapid in onset and reversible. Conversely, hypotonicity inhibits rat NHE1 and NHE2 but does not affect NHE3 (Kapus *et al.,* 1994).

The mechanism of the change in NHE activity with hyperosmolarity has not been defined, but does not involve elevation in intracellular Ca^{2+} and is not mediated by protein kinases C or A, or calmodulin (Kapus *et al.,* 1994; Nath *et al.,* 1996b). Data are contradictory about the involvement of tyrosine phosphorylation (Szaszi *et al.,* 1997; Krump *et al.,* 1997; Kapus *et al.,* 1999; Good, 1995). Shrinkage does induce tyrosine phosphorylation of multiple proteins, including cortactin, but does not lead to a change in NHE tyrosine phosphorylation (Nath *et al.,* 1996b; Szaszi *et al.,* 1997; Krump *et al.,* 1997; Kapus *et al.,* 1999). In PS120 cells, inhibiting tyrosine kinases with genistein did not affect hyperosmolar regulation of NHE1, 2, or 3, making it unlikely that changes in tyrosine phosphorylation play a role in the response to hyperosmolarity (Nath *et al.,* 1996b). However, in the rat renal medullary thick ascending limb, tyrosine phosphorylation may be involved in hyperosmolar inhibition of apical Na^+/H^+ exchange, as well as basal apical Na^+/H^+ exchange (Good, 1995). Specifically, both basal Na^+/H^+ exchange and the inhibition caused by hyperosmolarity were altered by inhibiting tyrosine phosphorylation. Basal apical NHE3 was stimulated by inhibiting tyrosine kinases (implicating tyrosine kinases in the inhibition of NHE3 under basal conditions). Also, inhibiting tyrosine kinases prevented the hyperosmolar inhibition of apical Na^+/H^+ exchange.

b. ATP Depletion. NHE1, NHE2, and NHE3 respond similarly to acute depletion of ATP with inhibition of V_{max} and $K'(H^+)_i$, with the magnitude of the effect greatest on NHE3 (Levine *et al.,* 1993; Counillon *et al.,* 1993b; Kapus *et al.,* 1994; Goss *et al.,* 1994). This ATP effect is the only regulatory effect that inhibits Na^+/H^+ exchange and involves both V_{max} and $K'(H^+)_{li}$ for NHE1–3. In addition, ATP depletion leads to reduction of the Hill coefficient to approximately 1. The last fact strongly implies that the H^+ modifier site function requires ATP, although which ATP-dependent process is involved is not known. The changes in NHE function caused by ATP depletion do not involve changes in NHE (at least NHE1) phosphorylation (Demaurex *et al.,* 1997; Aharonovitz *et al.,* 1999). In fact, recent studies of NHE1 suggest involvement of an intermediate protein (Aharonovitz *et al.,* 1999).

   c. Elevation in Intracellular Ca^{2+}.   Regulation of NHE3 by elevating intracellular Ca^{2+} occurs but is not a direct effect. In rabbit ileum, ionomycin and thapsigargin inhibit neutral NaCl absorption, but this is mediated by protein-kinase C and not by Ca^{2+} directly or by Ca^{2+}/calmodulin (Cohen *et al.,* 1991; Donowitz *et al.,* 1986; 1989). The ileal effect of thapsigargin is similar to the effect of carbachol, which also appears to be mediated by protein kinase C. However, under basal conditions, ileal NaCl absorption is stimulated by inhibition of calmodulin and calmodulin kinase II (Rood *et al.,* 1988; 1989; Cohen *et al.,* 1990). This suggests that under basal conditions, Ca^{2+}/calmodulin, probably acting through calmodulin kinase II, inhibits NaCl absorption. The stimulus that activates Ca^{2+}/calmodulin under basal conditions is not known. When NHE3 is expressed in PS120 cells, elevating intracellular Ca^{2+} with thapsigargin or ionomycin does not acutely alter Na^+/H^+ exchange (Levine *et al.,* 1993). It has been assumed, but not shown, that protein kinase C is not activated by elevation of intracellular Ca^{2+} alone in these cells. Similar to results in intact tissue, inhibition of calmodulin and calmodulin-kinase II with KN-62 stimulates basal NHE3 activity (the effect of KN-62 exceeds the effect of calmodulin inhibitors). In this area, recent results from our laboratory are different from those we reported previously (Levine *et al.,* 1995). This effect requires the C-terminal 76 amino acids of NHE3 (Levine *et al.,* 1995). Although not studied in as much detail as NHE3, elevating intracellular Ca^{2+} with ionomycin or thapsigargin also did not alter NHE2 (Nath *et al.,* 1999). NHE2 is stimulated by protein kinase C (Nath *et al.,* 1999). However, Ca^{2+}/calmodulin is a major regulator of NHE2 expressed in PS120 fibroblasts under basal conditions (Nath *et al.,* 1999). Inhibiting calmodulin with W13 stimulated the NHE2 V_{max}. The effect of inhibiting calmodulin was exerted in the NHE2 C-terminal 54 amino acids. Calmodulin bound to the NHE2 C terminus in a similar area (C-terminal 80 amino acids) with an intermediate affinity (Kd 300 nM) (Fig. 10). Thus, NHE2 and NHE3 are regulated by calmodulin actins via similar C-terminal domains (Fig. 6); however the effect appears to be via Ca^{2+}/calmodulin for NHE2 but via calmodulin kinase II for NHE3.

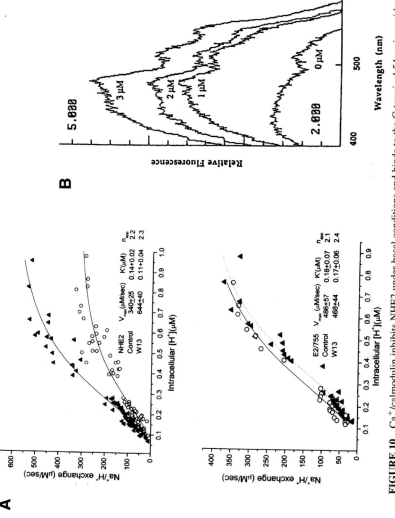

FIGURE 10 $Ca^+/$calmodulin inhibits NHE2 under basal conditions and binds to the C-terminal 54 amino acids. (A) The calmodulin inhibitor W13 stimulated NHE2 V_{max}. The effect of inhibiting calmodulin was exerted in the C-terminal 54 amino acids based on C-terminal truncation studies. (There was no effect of W13 on NHE2/755 NHE2 with truncation of the C-terminal 76 amino acids of NHE3.) (B) CaM binds to NHE2 in the C-terminal 80 amino acids. The binding of dansylated calmodulin was to a fusion protein of the C-terminal 80 amino acids of NHE2. Reprinted with permission from Nath et al., 1999.

Wakabayashi's studies extended the understanding of Ca^{2+} regulation of NHE3 using N-terminal chimeras of NHE3 and NHE1 (Wakabayashi et al., 1995). In contrast to NHE3, NHE1 is stimulated by elevating intracellular Ca^{2+} (Wakabayashi et al., 1994b; 1997b; Bertrand et al., 1994). Wakabayashi and coworkers showed that under basal conditions of $[Ca^{2+}]$, NHE1 is inhibited by a domain in the middle of the C terminus (aa 637–656) that has a high affinity for calmodulin binding (Kd 30 nM) (Wakabayashi et al., 1994b; 1997b; Bertrand et al., 1994). Even though NHE3 was not affected by elevation of intracellular Ca^{2+}, Wakabayashi et al. (1995) also determined that NHE3 bound calmodulin. Moreover, the N3C1 chimera, but not the N1C3 chimera, was stimulated by elevating intracellular Ca^{2+} similarly to NHE1, an effect that required the high-affinity calmodulin binding domain of the NHE1 C terminus. The Ca^{2+} stimulation of N3C1 was by a change in $K'(H^+)_i$ and not in V_{max}. Thus the C terminus (of NHE1) determined the kinetics of regulation of Na^+/H^+ exchange. This is different from other studies of N3C1 and N1C3 chimeras in which, in response to serum, the N terminus determined the kinetics of regulation (Yun et al., 1995b). These studies established that although NHE3 is not affected by elevation in intracellular Ca^{2+}, the autoinhibitory domain of NHE1, which is affected by Ca^{2+}, can interact with and regulate the H^+ modifier site of NHE3.

    d. Other Agonists. (1) *Adenosine:* In the A6 distal nephron cell line expressing transfected NHE3 on the apical membrane, stimulation of an apical membrane adenosine A1 receptor inhibited both apical NHE3 and basolateral endogenous (xenopus) NHE (Di Sole et al., 1999). This effect appeared to be mediated by protein kinase C. Stimulation of a basolateral adenosine A2 receptor inhibited apical NHE3 but stimulated basolateral NHE, effects mimicked by cAMP and prevented by a PKA inhibitor. Thus, this cell type responds to adenosine with protein kinase C activation in response to apical A1 adenosine receptors, and with protein kinase A in response to basolateral A2 adenosine receptors, both of which inhibit apical NHE3. (2) *Short-chain fatty acids:* Short-chain fatty acids are major regulators of colonic Na^+ absorption. A human colon cancer cell line, HT29-Cl, has NHE2 on the apical membrane and NHE1 on the basolateral membrane (Gonda et al., 1999). Exposure to the short-chain fatty acid proprionate from either the apical or basolateral surface stimulated NHE2 (Gonda et al., 1999). The apical Na^+/H^+ exchanger was activated more than the basolateral exchanger. This suggests differential activation of apical vs basolateral NHEs in response to intracellular acidification, although differences in buffering in the different cell domains could be contributing as well. (3) *G proteins:* The role of guanine nucleotide binding proteins has been explored primarily regarding NHE1 by the studies of Barber et al. (Hooley et al., 1996; Tominaga and Barber, 1998; Tominaga et al., 1998; Lin et al., 1996; Voyno-Yasenetskaya et al., 1994; Barber and Ganz, 1992). Their results implicate both heterotrimeric and small G proteins in regulation of NHE1, with members of the Rho/Rac family acting to link cytoskeleton to NHE1. Still unclear is whether there are any direct effects of G proteins in the regulation of NHE1. G protein effects on NHE3 and NHE2 have been studied in much less detail. Preliminary results suggesting that G proteins directly regulate NHE3 are weakened by the facts that receptor-related regulation of NHE3 and NHE2 occurs very frequently and that there are apical and basolateral membrane G protein–coupled receptors in epithelial cells that affect NHE3 and NHE2.

8. Effect of Basolateral Membrane NHE1 on Regulation of Apical Na$^+$/H$^+$ Exchange

In studies of NHE2 and NHE3 in epithelial cells, it has been assumed that regulation by growth factors and protein kinases is directly on the apical exchanger. However, that this may be a oversimplification was suggested by Good and coworkers, who used isolated perfused medullary thick ascending limb of the rat (Good *et al.*, 1995; Watts *et al.*, 1999). Their data suggested that basolateral Na$^+$/H$^+$ exchange was necessary and partially responsible for basal apical membrane Na$^+$/H$^+$ exchange activity. In addition, they showed that nerve growth factor inhibited both apical and basolateral Na$^+$/H$^+$ exchange in these cells, and that the effect on apical Na$^+$/H$^+$ exchange was dependent on the inhibition of basolateral Na$^+$/H$^+$ exchange rather than via an effect directly on apical Na$^+$/H$^+$ exchange. However, the mechanism of this cross-talk was not identified. Although these studies were made more difficult to interpret by the fact that these cells may have more than one apical NHE, they indicate that studies of regulation in epithelial cells must attempt to separate direct effects of regulation of apical exchangers from effects that occur indirectly by changes in basolateral NHE1.

B. Long Term

In addition to the short-term regulation by growth factors and protein kinases that acutely regulate NHE2 and NHE3 over minutes to hours, these exchangers are also regulated over longer periods (hours to days). The specific mechanisms are not well understood because the genetic organization of NHE2 is totally unknown and there has been only one reported study of NHE3 (Kandasamy and Orlowski, 1996). These long-term regulatory mechanisms represent a form of memory in terms of the response to agonists and include effects of chronic acidosis, glucocorticoids and mineralocorticoids, protein kinases C and A, thyroid hormone, and hyperosmolarity. The genomic organization of the rat NHE3 gene has been reported (Kandasamy and Orlowski, 1996). This gene spans 40 kilobases, contains 17 exons, contains an atypical TATA-box and CCAAT-box, and several putative cis-acting elements including glucocorticoid and thyroid receptor recognition sites, as well as AP-1, AP-2, C/EBP, NF-I, OCT-1, PEA3, and SP1 response elements. Glucocorticoid stimulation of gene expression has been documented.

1. Response to Chronic Acid-Base Changes

These studies have largely been carried out in renal tissues (Alpern *et al.*, 1993; 1995; Wu *et al.*, 1996; Ambuhl *et al.*, 1996; Laghmani *et al.*, 1997; 1999; Soleimani *et al.*, 1992; 1995; Yamaji *et al.*, 1994; 1997; Amemiya *et al.*, 1995b; Yamaji *et al.*, 1995). Chronic metabolic acidosis increases the capacity of the proximal tubule and the cortical collecting duct to secrete H^+, whereas metabolic alkalosis decreases proximal tubule brush border Na^+/H^+ exchange (Alpern *et al.*, 1993; 1995; Wu *et al.*, 1996). The changes in the proximal tubule relate to changes in the Na^+/H^+ exchanger, with an increase in V_{max} of brush border Na^+/H^+ exchange and an increase in basolateral Na^+/HCO_3^- cotransport. The increase in proximal tubule Na^+/H^+ exchange activity requires protein synthesis and represents an increase in the amount of NHE3 (approximately $1.5\times$ increase after 5 days of chronic metabolic acidosis). In rats, the response to chronic metabolic acidosis developed slowly, with a 90% increase in the NHE3 protein level at 14 days of acidosis, compared with a 28% increase at 3 days and a 59% increase at 7 days. In these rats, NHE3 increased in the S1 and S2 segments of the proximal tubule and in the medullary thick ascending limb, the normal location of NHE3. This NHE3 effect was specific in that similar acid exposure to fibroblasts decreased message and protein for NHE1. Of note, however, chronic metabolic acidosis also increased NHE1 in basolateral membranes of rat renal proximal tubules, cultured mouse proximal tubules, and cultured inner medullary collecting duct cells. Chronic acidosis also increased the amounts of NHE3 mRNA and protein, and the magnitude of the increase in NHE3 mRNA protein and transport activity was similar (Laghmani *et al.*, 1997). The stimulation of NHE3 by chronic acidosis was duplicated in OK cells, with functional changes detectable 12 hours after acid exposure, with an increase in mRNA at 24 hours (Amemiya *et al.*, 1995b). Similar stimulation of NHE3 with chronic acidosis occurred in a clone of LLC-PK1 cells. Effects of chronic acidosis on renal NHE2 have not been reported in detail. Inner medullary collecting ducts cultured in chronic acid media decreased NHE1 mRNA. Conversely, chronic metabolic alkalosis decreased the NHE3 transport activity and NHE3 mRNA and protein in rat renal MTAL cells (Laghmani *et al.*, 1999). The chronic acidosis stimulation of message for NHE3 and NHE1 was associated with changes in multiple classes of genes. Chronic acidosis led to increased transcription of the immediate early genes *c-fos*, *c-jun*, *junB*, and *EGR1*, and an increase in nuclear AP-1 activity (Yamaji *et al.*, 1994). There was some cell specificity in that the described changes in immediate early genes occurred in the SV40-transformed mouse proximal tubule MCT cells, but no c-Jun activation occurred with similar acid exposure to 3T3 fibroblasts. The mechanism of activation appears to involve activation of the Src family of tyrosine kinases (Yamaji *et al.*, 1997; Yamaji *et al.*, 1995). In MCT cells and OK cells, chronic exposure to acid increased c-Src activity twofold, an effect that occurred rapidly but persisted and correlated with induction of intracellular acidosis (Yamaji *et al.*, 1997). Pharmacological inhibitors of tyrosine kinases prevent the acidosis-induced increase in NHE3, and overexpression of the endogenous Src kinase inhibitor Csk prevented acidosis stimulation of NHE3 protein and mRNA amount (Yamaji *et al.*, 1995). These effects were specific in that Csk overexpression and tyrosine kinase inhibitors did not alter dexamethasone-induced stimulation of NHE3 (see later). Thus, c-Src is involved at least in mediating the chronic acidosis stimulation of NHE3 mRNA.

2. Long-Term Activation of Protein Kinases C and A

Long-term activation of protein kinase C and A: Application of phorbol esters for 2 hours increased Na^+/H^+ exchange for more than 24 hours in proximal tubule cells by a mechanism involving transcription/translation, and was associated with an increase in NHE1 message/protein (Horie *et al.*, 1992). These studies were carried out in primary cultures of renal proximal tubules in which C kinase acutely stimulated Na^+/H^+ exchange. A long-term increase in cAMP in OK cells with application of 8-Br-cAMP for 24 hours increased Na^+/H^+ exchange activity, an effect requiring protein synthesis (Cano *et al.*, 1993). This effect was unusual in that the direction of the long-term effect (stimulation) was opposite that of the acute effect (inhibition).

3. Mineralo- and Glucocorticoid Effects on NHE2 and NHE3

Glucocorticoids were used initially to suggest that NHE3 was the intestinal apical membrane Na^+/H^+ exchanger involved in Na^+ absorption (Yun *et al.*, 1993b). This was based on the observation that glucocorticoid treatment for 18–72 hours was associated with stimulation of intestinal jejunal, ileal, and colonic Na^+ absorption. In these studies, glucocorticoids stimulated NHE3 message and protein, but not message or protein of NHE1 or NHE2 (Yun *et al.*, 1993b). The regional effects of glucocorticoids were specific in that in the rat, dexamethasone increased NHE3 message in the ileum and proximal colon but not in the jejunum or distal colon, and not in the whole kidney (Cho *et al.*, 1994). Conversely, adrenalectomy lowers NHE3 message in the ileum and proximal colon but not in the jejunum (Cho *et al.*, 1994).

Glucocorticoids also stimulated renal proximal tubule Na^+/H^+ exchange (Cho *et al.*, 1994; Loffing *et al.*, 1998; Baum *et al.*, 1993; 1994; 1996; Ambuhl *et al.*, 1999; Soleimani *et al.*, 1990). Dexamethasone increased renal proximal tubule brush-border Na^+/H^+ exchange, and increased the protein and message for NHE3 but not for NHE1; the magnitude of the increase of NHE3 message was greater after 2 days than after 1 day of glucocorticoid treatment (Loffing *et al.*,

1998). In addition, changes in NHE3 transport activity occurred by 3–4 hours after glucocorticoid treatment, a time at which there was no detectable change in NHE3 mRNA. Thus, glucocorticoids stimulate NHE3 activity and message in renal proximal tubules, but it remains unclear whether the effects at 3–4 hours indicate a separate glucocorticoid nontranscriptional effect or whether these differences reflect differences in sensitivity of measurements of transport and mRNA (Baum et al., 1994). Of interest, kidney glucocorticoids act synergistically with acidosis to stimulate NHE3. This occurs by effects on both transcription and trafficking (Ambuhl et al., 1999; Baum et al., 1996).

Mineralocorticoid (aldosterone) levels have been increased chronically by several different methods and effects on NHEs studied. These methods include chronic Na^+ depletion and chronic K^+ depletion. Na^+ depletion led to conflicting results in studies of rat colon (Ikuma et al., 1999; Cho et al., 1998). Chronic aldosterone elevation stimulates NaCl absorption in the rat proximal colon and inhibits it in the rat distal colon. Seven days of a low-Na^+ diet affected brush-border Na^+/H^+ exchange and NHE2 and NHE3 differently in the proximal and distal colon (Ikuma et al., 1999). In the rat proximal colon, based on HOE694 sensitivity, NHE3 made up approximately 85% of basal Na^+/H^+ exchange. Seven days of a low-Na^+ diet caused approximately a doubling of brush border Na^+/H^+ exchange, which was caused by approximately equal increases in NHE2 and NHE3 activity and no change in basolateral Na^+/H^+ exchange activity. This was associated with comparable increases in amounts of NHE2 and NHE3 and in the amount of mRNA for both isoforms. In rat distal colon, under basal conditions, NHE2 accounts for ~25% of basal brush border Na^+/H^+ exchange and NHE3 for ~75%; following the low-Na^+ diet, there was reduction of both NHE2 and NHE3 activity to values approaching zero (Ikuma et al., 1999). This was associated with reduction of NHE3 and NHE2 mRNA and protein, although the reduction of NHE2 mRNA and protein was modest, suggesting an inhibitory effect on NHE2 function in addition to the reduction of protein. In separate contradictory studies, aldosterone administration of IP for 3 days stimulated NHE3 but not NHE2 in the proximal colon and had no effect on Na^+/H^+ exchange in the distal colon (Cho et al., 1998). However, the model used in the latter study appeared to employ submaximal conditions, which failed to alter neutral NaCl absorption in the rat colon and produced a serum aldosterone level lower than that observed with longer dietary Na^+ depletion.

Additionally, some species show differences with respect to whether the response to increased mineralocorticoids is via NHE2 or NHE3. This was made clear by studies of avian intestine in response to a low-Na^+ diet. Two weeks of a low-Na^+ diet caused approximately doubling of brush border Na^+/H^+ exchange in avian ileum and colon. This was almost entirely attributable to an increase in the amount of NHE2 with minimal or no increase in NHE3 (Donowitz et al., 1998).

Chronic hypokalemia also has been used to increase aldosterone levels. However, consideration that the mechanism of the effect of chronic hypokalemia is entirely by a transcriptional mechanism must be reinterpreted given recent results in which OK cells exposed to low-K^+ medium for 8 hours increased NHE3 activity, with changes in NHE3 mRNA and protein levels becoming detectable only at 24 hours of exposure (Amemiya *et al.*, 1999).

4. Long-Term Exposure to Hyperosmolarity

Exposing LLC-PK1 cells to hyperosmolarity for 48 hours inhibited NHE3 transport activity but did not alter the amount of NHE3 mRNA or protein levels (Soleimani *et al.*, 1998). This indicates a posttranslational effect. Whether the mechanism of this inhibition is different from that of the initial inhibition of NHE3 by hyperosmolarity was not clarified. In contrast, similar hyperosmolar exposure stimulated rat NHE2 expressed in LLC-PK1 cells and IMCD cells. Because rat NHE2 is stimulated by hyperosmolarity, the issue of whether this is a prolongation of what has been studied over shorter times in response to hyperosmolarity or an additional or separate mechanism was not clarified (Singh *et al.*, 1996).

Thus, for multiple models, although chronic regulation of NHE3 has been documented, including evidence that transcriptional regulation occurs, the mechanisms of regulation have not been well studied and the relative roles of changes in transcription, mRNA stability, and protein turnover have not been compared.

5. Thyroid Hormones

In intact animals, hyperthyroidism increases and hypothyroidism decreases renal proximal tubule NaCl and Na/HCO_3 absorption and Na^+/H^+ exchange V_{max} (Yonemura *et al.*, 1990; Kinsella *et al.*, 1986). In rat proximal tubules, message for NHE2 and NHE3 increased when hypothyroid and hyperthyroid animals were compared, but no differences in the amount of NHE3 protein were found (Azuma *et al.*, 1996). The stimulatory effect on NHE3 of the hyperthyroid state was duplicated in OK cells in culture, in which T3 increased the NHE3 V_{max} (Cano *et al.*, 1999). The mechanism of the effect of T3 in OK cells involves activation of NHE3 gene transcription, which leads to an increase in NHE3 mRNA and protein levels.

V. DISEASES IN WHICH THE EPITHELIAL NHEs PLAY A PATHOGENIC ROLE

Although there have been suggestions of a role for NHEs in hypertension and diabetes, there is no clear experimental evidence of the involvement of NHE2 or NHE3 in causing sequelae of these conditions (Soleimani and Singh, 1995).

There is a single disease in which there is functional absence or decrease of intestinal brush border Na^+/H^+ exchange, congenital Na^+ diarrhea (Holmberg and Perheentupa, 1985; Booth *et al.*, 1985; Keller *et al.*, 1990; Fell *et al.*, 1992).

This is one of several congenital diarrheal diseases, the others being Na-glucose/galactose malabsorption, congenital chloridorrhea, and microvillus inclusion disease. Congenital Na$^+$ diarrhea has several forms, distinguished by severity. The disease has onset *in utero* with polyhydramnios, and acidosis and dehydration develop shortly after birth. The disease can be severe, leading to death in the first days/weeks of life, or moderate, allowing the patient to survive on TPN or with time becoming less severe (and in one case totally resolve). The children who survive grow and develop normally. In two cases, the disease was shown to be associated with decreased or absent jejunal brush border Na$^+$/H$^+$ exchange activity and normal Na$^+$ dependent D-glucose uptake based on vesicle studies of jejunal biopsies (Booth *et al.,* 1985; Keller *et al.,* 1990). Whether colonic Na$^+$ absorption is normal is not clear. One case was suggestive of defective colonic Na$^+$ absorption; however, in another case colonic Na$^+$ transport appeared to be normal.

We have studied the DNA of 5 children believed to have congenital Na$^+$ diarrhea and the small intestinal biopsy from three such children. In all cases, the duodenal/jejunal biopsies were normal, and immunocytochemical staining for NHE2 and NHE3 showed that both exchangers were in a normal location and were present in normal amounts. The DNA for the index case was sequenced, including the ORF and all intron-exon junctions. These were normal. Thus, although abnormal jejunal Na$^+$/H$^+$ exchange appears to be the cause of this disease, to date this is a functional abnormality and does not appear related to abnormalities in the NHE3 sequence. Explanations for the disorder potentially include that (1) this is an example of a disease caused by a quantitative decrease in the amount of NHE3, perhaps attributable to abnormalities in the NHE3 promoter; (2) there is a still-to-be-cloned jejunal NHE that explains jejunal Na$^+$/H$^+$ exchange and contributes to intestinal water homeostatis; and (3) that there is an abnormality not in the NHE3 sequence itself but in the associated regulatory proteins, which prevents it from functioning.

Although molecular studies of the *NHE* gene family have begun providing insight into how epithelial Na$^+$ absorption occurs, it remains unclear what functions are provided by having contributions from both NHE2 and NHE3. In addition, although diseases that result from genetic abnormalities of SGLT1 and ENaC have been identified, genetic diseases of NHE2 and NHE3 have not been identified with certainty or understood at a molecular level. However, it has been only 9 years since the demonstration that there was a mammalian *NHE* gene family, and it is likely that genetic diseases of NHE2 and NHE3 remain to be identified.

References

Aharonovitz, O., Demaurex, N., Woodside, M., and Grinstein, S. (1999). ATP dependence is not an intrinsic property of Na$^+$/H$^+$ exchanger: NHE1: Requirement for an ancillary factor. *Am. J. Physiol.* **276,** C1303–1311.

Akhter, S., Cavet, M., Tse, C. M., and Donowitz, M. (2000). C-terminal domains of Na^+/H^+ exchanger isoform 3 are involved in the basal and serum-stimulated membrane trafficking of the exchanger. *Biochemistry* **39,** 1990–2000, 2000.

Akhter, S., Nath, S. K., Tse, C. M., Williams, J., Zasloff, M., and Donowitz, M. (1999). Squalamine, a novel cationic steroid, specifically inhibits the brush-border Na^+/H^+ exchanger isoform NHE3. *Am. J. Physiol.* **276,** C136–C144.

Alpern, R. J., Yamaji, Y., Cano, A., Horie, S., Miller, R. T., Moe, O. W., and Preisig, P. A. (1993). Chronic regulation of the Na/H antiporter. *J. Lab. Clin. Med.* **122,** 137–140.

Alpern, R. J. (1990). Cell mechanisms of proximal tubule acidification. *Physiol. Rev.* **70,** 79–114.

Alpern, R. J., Moe, O. W., and Preisig, P. A. (1995). Chronic regulation of the proximal tubular Na/H antiporter: From HCO_3 to Src. *Kidney Int.* **48,** 1386–1396.

Amemiya, M., Loffing, J., Lotscher, M., Kaissling, B., Alpern, R. J., and Moe, O. W. (1995a). Expression of NHE-3 in the apical membrane of rat renal proximal tubule and thick ascending limb. *Kidney Int.* **48,** 1206–1215.

Amemiya, M., Tabei, K., Kusano, E., Asano, Y., and Alpern, R. J. (1999). Incubation of OKP cells in low-K^+ media increases NHE3 activity after early decrease in intracellular pH. *Am. J. Physiol.* **276,** C711–C716.

Amemiya, M., Yamaji, Y., Cano, A., Moe, O. W., and Alpern, R. J. (1995b). Acid incubation increases NHE3 mRNA abundance in OKP cells. *Am. J. Physiol.* **269,** C126–C133.

Ambuhl, P. M., Amemiya, M., Danczkay, M., Lotscher, M., Kaissling, B., Moe, O. W., Preisig, P. A., and Alpern, R. J. (1996). Chronic metabolic acidosis increases NHE3 protein abundance in rat kidney. *Am. J. Physiol.* **271,** F917–F925.

Ambuhl, P. M., Yang, X., Peng, Y., Preisig, P. A., Moe, O. W., and Alpern, R. J. (1999). Glucocorticoids enhance acid activation of the Na^+/H^+ exchanger 3 (NHE3). *J. Clin. Invest.* **103,** 429–435.

Anderie, I., Blum, R., Haase, W., Grinstein, S., and Thevenod, F. (1998). Expression of NHE1 in rat pancreatic zymogen granule membranes. *Biochem. Biophys. Res. Commun.* **246,** 330–336.

Apse, M. P., Aharon, G. S., Snedden, W. A., and Blumwald, E. (1999). Salt tolerance conferred by overexpression of a vacuolar Na^+/H^+ antiport in Arabidopsis. *Science* **285,** 1256–1258.

Attaphitaya, S., Park, K., and Melvin, J. E. (1999). Molecular cloning and functional expression of a rat Na^+/H^+ exchanger (NHE5) highly expressed in brain. *J. Biol. Chem.* **274,** 4383–4388.

Azuma, K. K., Balkovetz, D. F., Magyar, C. E., Lescale-Matys, L., Zhang, Y., Chambrey, R., Warnock, D. G., and McDonough, A. A. (1996). Renal Na^+/H^+ exchanger isoforms and their regulation by thyroid hormone. *Am. J. Physiol.* **270,** C585–C592.

Bachmann, O., Sonnentag, T., Siegel, W. K., Lamprecht, G., Weichert, A., Gregor, M., and Seidler, U. (1998). Different acid secretagogues activate different Na^+/H^+ exchanger isoforms in rabbit parietal cells. *Am. J. Physiol.* **275,** G1085–G1093.

Baird, N. R., Orlowski, J., Szabo, E. Z., Zaun, H. C., Schultheis, P. J., Memon, A. G., and Shull, G. E. (1999). Molecular cloning, genomic organization, and functional expression of Na^+/H^+ exchanger isoform 5 (NHE5) from human brain. *J. Biol. Chem.* **274,** 4377–4382.

Barber, D. L., and Ganz, M. B. (1992). Guanine nucleotides regulate beta-adrenergic activation of Na/H exchange independently of receptor coupling to Gs. *J. Biol. Chem.* **267,** 20607–20612.

Baum, M., Amemiya, M., Dwarakanath, V., Alpern, R. J., and Moe, O. W. (1996). Glucocorticoids regulate NHE3 transcription in OKP cells. *Am. J. Physiol.* **270,** F164–F169.

Baum, M., Cano, A., and Alpern, R. J. (1993). Glucocorticoids stimulate Na^+/H^+ antiporter in OKP cells. *Am. J. Physiol.* **264,** F1027–F1031.

Baum, M., Moe, O. W., Gentry, D. L., and Alpern, R. J. (1994). Effect of glucocorticoids on renal cortical NHE3 and NHE1 mRNA. *Am. J. Physiol.* **267,** F437–F442.

Berschneider, H. M., Knowles, M. R., Azizkhan, R. G., Boucher, R. C., Tobey, N. A., Orlando, R. C., and Powell, D. W. (1988). Altered intestinal chloride transport in cystic fibrosis. *FASEB J.* **2,** 2625–2629.

Bertrand, B., Wakabayashi, S., Ikeda, T., Pouyssegur, J., and Shigekawa, J. (1994). The Na$^+$/H$^+$ exchanger isoform 1 (NHE1) is a novel member of the calmodulin-binding proteins. Identification and characterization of calmodulin-binding sites. *J. Biol. Chem.* **269,** 13703–13709.

Bianchini, L., Kapus, A., Lukacs, G., Wasan, S., Wakabayashi, S., Pouyssegur, J., Yu, F. H., Orlowski, J., and Grinstein, S. (1995). Responsiveness of mutants of NHE1 isoform of Na$^+$/H$^+$ isoform of Na$^+$/H$^+$ antiport to osmotic stress. *Am. J. Physiol.* **269,** C998–C1007.

Bianchini, L., and Pouyssegur, J. (1996). Regulation of Na$^+$/H$^+$ exchanger isoform NHE1: Role of phosphorylation. *Kidney Int.* **49,** 1038–1044.

Biemesderfer, D., DeGray, B., and Aronson, P. S. (1998). Membrane topology of NHE3. Epitopes within the carboxyl-terminal hydrophilic domain are exoplasmic. *J. Biol. Chem.* **273,** 12391–12396.

Biemesderfer, D. Nagy, T., DeGray, B., and Aronson, P. S. (1999). Specific association of megalin and the Na$^+$/H$^+$ exchanger isoform NHE3 in the proximal tubule. *J. Biol. Chem.* **274,** 17518–17524.

Biemesderfer, D., Pizzonia, J., Abu-Alfa, A., Exner, M., Reilly, R., Igarashi, P., and Aronson, P. S. (1993). NHE3: A Na$^+$/H$^+$ exchanger isoform of renal brush border. *Am. J. Physiol.* **265,** F736–F742.

Biemesderfer, D., Rutherford, P. A., Nagy, T., Pizzonia, J. H., Abu-Alfa, A. K., and Aronson, P. S. (1997). Monoclonal antibodies for high-resolution localization of NHE3 in adult and neonatal rat kidney. *Am. J. Physiol.* **273,** F289–F299.

Binder, H. J., Singh, S. K., Geiber, J. P., and Rajendran, V. M. (1997). Novel transport properties of colonic crypt cells: Fluid absorption and Cl-dependent Na-H exchange. *Comp. Biochem. Physiol. A* **118,** 265–269.

Bizal, G. L., Howard, R. L., Bookstein, C., Rao, M. C., Chang, E. B., and Soleimani, M. (1996). Glycosylation of the Na$^+$/H$^+$ exchanger isoform NHE-3 is species specific. *J. Lab. Clin. Med.* **128,** 304–312.

Bookstein, C., DePaoli, A. M., Xie, Y., Musch, M. W., Rao, M. C., and Chang, E. B. (1994). Na$^+$/H$^+$ exchangers, NHE-1 and NHE-3, of rat intestine. Expression and localization. *J. Clin Invest* **93,** 106–113.

Bookstein, C., Xie, Y., Rabenau, K., Musch, M. W., McSwine, R. L., Rao, M. C., and Chang, E. B. (1997). Tissue distribution of Na$^+$/H$^+$ exchanger isoforms NHE2 and NHE4 in rat intestine and kidney. *Am. J. Physiol.* **273,** C496–C505.

Booth, I. W., Stange, G., Murer, H., Fenton, T. R., and Milla, P. J. (1985). Defective jejunal brush-border Na$^+$/H$^+$ exchange: A cause of congenital secretory diarrhoea. *Lancet* **8437,** 1066–1069.

Borgese, F., Malapert, M., Fievet, B., Pouyssegur, J., and Motais, R. (1994). The cytoplasmic domain of the Na$^+$/H$^+$ exchangers (NHEs) dictates the nature of the hormonal response: Behavior of a chimeric human NHE1/trout beta NHE antiporter. *Proc. Natl. Acad. Sci. USA* **91,** 5431–5435.

Brant, S. R., Bernstein, M., Wasmuth, J. J., Taylor, E. W., McPherson, J. D., Li, X., Walker, S. A., Pouyssegur, J., Donowitz, M., Tse, C. M., and Jabs, W. W. (1993). Physical and genetic mapping of a human apical epithelial Na$^+$/H$^+$ exchanger (NHE-3) isoform to chromosome 5p15.3. *Genomics* **15,** 668–672.

Brant, S. R., Yun, C. H. C., Donowitz, M., and Tse, C. M. (1995). Cloning, tissue distribution, and functional analysis of the human Na$^+$/H$^+$ exchanger isoform, NHE3. *Am. J. Physiol.* **26,** C198–C206.

Cabado, A. G., Yu, F. H., Kapus, A., Lukacs, G., Grinstein, S., and Orlowski, J. (1996). Distinct structural domains confer cAMP sensitivity and ATP dependence to the Na$^+$/H$^+$ exchanger NHE3 isoform. *J. Biol. Chem.* **271,** 3590–3599.

Cabral, M., Petosa, J. H., Sutcliffe, M. J., Raza, S., Byron, O., Poy, F., Marfatia, S. M., Chishti, A. H., and Liddington, R. C. (1996). Crystal structure of a PDZ domain. *Nature* **382,** 649–652.

Cano, A., Baum, M., and Moe, O. W. (1999). Thyroid hormone stimulates the renal Na/H exchanger NHE3 by transcriptional activation. *Am. J. Physiol.* **276**, C102–C108.

Cano, A., Preisig, P., and Alpern, R. J. (1993). Cyclic adenosine monophosphate acutely inhibits and chronically stimulates Na/H antiporter in OKP cells. *J. Clin. Invest.* **92**, 1632–1638.

Cavet, M., Akhter, S., Sanchez, F., Donowitz, M., and Tse, C. M. (1999). Cloned Na^+/H^+ exchangers (NHE1-3) have similar turnover numbers. *FASEB J.* **13**, A400.

Cha, B., Nadarajah, J., Wang, X.-T., Lamprecht, G., Guggino, S. E., and Yun, C. (1999). Functional role of Na/H exchanger regulatory factor (NHERF) and localization in intestine and kidney. *Gastroenterology* **116**, A3773.

Chambrey, R., Warnock, D. G., Podevin, R.-A., Bruneval, P., Mandet, C., Belair, M.-F., Bariety, J., and Paillard, M. (1998). Immunolocalization of the Na^+/H^+ exchanger isoform NHE2 in rat kidney. *Am. J. Physiol.* **275**, F379–F386.

Cho, J. H., Musch, M. W., Bookstein, C. M., McSwine, R. L., Rabenau, K., and Chang, E. B. (1998). Aldosterone stimulates intestinal Na^+ absorption in rats by increasing NHE3 expression of the proximal colon. *Am. J. Physiol.* **274**, C586–C594.

Cho, J. H., Musch, M. W., DePaoli, A. M., Bookstein, C. M., Xie, Y., Burant, C. F., Mrinalini, C., and Chang, E. B. (1994). Glucocorticoids regulate Na^+/H^+ exchange expression and activity in region–and tissue-specific manner. *Am. J. Physiol.* **267**, C796–C803.

Chow, C. W., Woodside, M., Demaurex, N., Yu, F. H., Plant, P., Rotin, D., Grinstein, S., and Orlowski, J. (1999). Proline-rich motifs of the Na^+/H^+ exchanger 2 isoform. *J. Biol. Chem.* **274**, 10481–10488.

Cohen, M. E., Reinlib, L., Watson, A. J., Gorelick, F., Rys-Sikora, K., Tse, M., Rood, R. P., Czernik, A. J., Sharp, G. W., and Donowitz, M. (1990). Rabbit ileal villus cell brush border Na^+/H^+ exchange is regulated by Ca^{2+}/calmodulin-dependent protein kinase II, a brush border membrane protein. *Proc. Natl. Acad. Sci. USA* **87**, 8990–8994.

Cohen, M. E., Wesolek, J., McCullen, J., Rys-Sikora, K., Pandol, S., Rood, R. P., Sharp, G. W., and Donowitz, M. (1991). Carbachol and elevated Ca^{2+}-induced translocation of functionally active protein kinase to the brush border of rabbit ileal Na^+ absorbing cells. *J. Clin. Invest.* **88**, 855–863.

Collins, J. F., Kiela, P. R., Xu, H., Zeng, J., and Ghishan, F. K. (1998). Increased NHE2 expression in rat intestinal epithelium during ontogeny is transcriptionally regulated. *Am. J. Physiol.* **275**, C1143–C1150.

Collins, J. F., Xu, H., Kiela, P. R., Zeng, J., and Ghishan, F. K. (1997). Functional and molecular characterization of NHE3 expression during ontogeny in rat jejunal epithelium. *Am. J. Physiol.* **273**, C1937–C1946.

Counillon, L., Franchi, A., and Pouyssegur, J. (1993b). A point mutation of the Na^+/H^+ exchange gene (NHE1) and amplification of the mutated allele confer amiloride resistance upon chronic acidosis. *Proc. Natl. Acad. Sci. USA* **90**, 4508–4512.

Counillon, L., Noel, J., Reithmeier, R. A. F., and Pouyssegur, J. (1997). Random mutagenesis reveals a novel site involved in inhibitor interaction within the fourth transmembrane segment of the Na^+/H^+ exchanger-1. *J. Biol. Chem.* **36**, 2951–2959.

Counillon, L., Pouyssegur, J., and Reitmeier, R. A. F. (1994). The Na^+/H^+ exchanger NHE-1 possesses N- and O-linked glycosylation restricted to the first N-terminal extracellular domain. *Biochemistry* **33**, 10463–10469.

Counillon, L., Scholz, W., Lang, H. J., and Pouyssegur, J. (1993a). Pharmacological characterization of stably transfected Na^+/H^+ antiporter isoforms using amiloride analogs and a new inhibitor exhibiting anti-ischemic properties. *Mol. Pharmacol.* **44**, 1041–1045.

Cremaschi, D., Meyer, G., Rossetti, C., Botta, G., and Palestini, P. (1992). The nature of the neutral Na^+-Cl^- coupled entry at the apical membrane of rabbit gallbladder epithelium: I. Na^+/H^+, Cl/HCO_3-double exchange and Na^+-Cl-symport. *J. Membr. Biol.* **129**, 221–235.

Demaurex, N., Furuya, W., D'Souza, S., Bonifacino, J. S., and Grinstein, S. (1998). Mechanism of acidification of the trans-Golgi network (TGN). In situ measurements of pH using retrieval of TGN38 and furin from the cell surface. *J. Biol. Chem.* **273**, 2044–2051.

Demaurex, N., Romanek, R. R., Orlowski, J., and Grinstein, S. (1997). ATP dependence of Na^+/H^+ exchange. Nucleotide specificity and assessment of the role of phospholipids. *J. Gen. Physiol.* **109**, 117–128.

Demaurex, N., Orlowski, J., Brisseau, G., Woodside, M., and Grinstein, S. (1995). The mammalian Na^+/H^+ antiporters NHE-1, NHE-2, and NHE-3 are electroneutral and voltage independent, but can couple to an H^+ conductance. *J. Gen. Physiol.* **106**, 85–111.

Di Sole, F., Casavola, V., Mastroberardino, L., Verrey, F., Moe, O. W., Burckhardt, G., Murer, H., and Helmle-Kolb, C. (1999). Adenosine inhibits the transfected Na^+-H^+ exchanger NHE3 in *Xenopus laevis* renal epithelial cells (A6/C1). *J. Physiol.* **515**, 829–842.

Donowitz, M., Cheng, H. Y., and Sharp, G. W. (1986). Effects of phorbol esters on sodium and chloride transport in rat colon. *Am. J. Physiol.* **251**, G509–G517.

Donowitz, M., Cohen, M. E., Gould, M., and Sharp, G. W. (1989). Elevated intracellular Ca^{2+} acts through protein kinase C to regulate rabbit ileal NaCl absorption. Evidence for sequential control by Ca^{2+}/calmodulin and protein kinase C. *J. Clin. Invest.* **83**, 1953–1962.

Donowitz, M., De La Horra, C., Calonge, M. L., Wood, I. S., Dyer, J., Gribble, S. M., De Medina, F. S., Tse, C. M., Shirazi-Beechey, S. P., and Ilundain, A. A. (1998). In birds, NHE2 is major brush-border Na^+/H^+ exchanger in colon and is increased by a low-NaCl diet. *Am. J. Physiol.* **274**, R1659–R1669.

Donowitz, M., Kambadur, R., Zizak, M., Nath, S., Akhter, S., and Tse, M. (1997). NHE3 is a calmodulin binding protein: Calmodulin inhibits NHE3 by binding to the C-terminal 76 amino acids of NHE3. *Gastroenterology* **112**, A360.

Donowitz, M., and Welsh, M. J. (1987). Regulation of mammalian small intestinal electrolyte secretion. In "Physiology of the Gastrointestinal Tract" (L. R. Johnson, ed.), 2nd Ed. Raven Press, New York.

Doyle, D. A., Lee, A., Lewis, J., Kim, E., Sheng, M., and MacKinnon, R. (1996). Crystal structures of a complexed and peptide-free membrane protein-binding domain: Molecular basis of peptide recognition by PDZ. *Cell* **85**, 1067–1076.

Drensfield, D. T., Bradford, A. J., Smith, J., Martin, M., Roy, C., Mangeat, P. H., and Goldenring, J. R. (1997). Ezrin is a cyclic AMP-dependent protein kinase anchoring protein. *EMBO J.* **16**, 35–43.

D'Souza, S., Garcia-Cabado, A., Yu, F., Teter, K., Lukacs, G., Skorecki, K., Moore, H. P., Orlowski, J., and Grinstein, S. (1998). The epithelial sodium-hydrogen antiporter Na^+/H^+ exchanger 3 accumulates and is functional in recycling endosomes. *J. Biol. Chem.* **273**, 2035–2043.

Dudeja, P. K., Rao, D. D., Syed, I., Joshi, V., Dahdal, R. Y., Gardner, C., Risk, M. C., Schmidt, L., Bayishi, D., Kim, K. E., Harig, J. M., Goldstein, F. L., Layden, T. J., and Ramaswamy, K. (1996). Intestinal distribution of human Na^+/H^+ exchanger isoforms NHE-1, NHE-2, and NHE-3 mRNA. *Am. J. Physiol.* **271**, G483–G493.

Fafournoux, P., Noel, J., and Pouyssegur, J. (1994). Evidence that the Na^+/H^+ exchanger isoforms NHE-1 and NHE-3 exist as stable dimers with a high degree of specificity for homodimers. *J. Biol. Chem.* **268**, 2589–2596.

Fan, L., Wiederkehr, M. R., Collazo, R., Wang, H., Crowder, L. A., and Moe, O. W. (1999). Dual mechanisms of regulation of Na/H exchanger NHE-3 by parathyroid hormone in rat kidney. *J. Biol. Chem.* **274**, 11289–11295.

Fell, J. M., Miller, M. P., Finkel, Y., and Booth, I. W. (1992). Congenital sodium diarrhea with a partial defect in jejunal brush border membrane sodium transport, normal rectal transport, and resolving diarrhea. *J. Pediatr. Gastroenterol. Nutr.* **15**, 112–116.

Field, M., Graf, L. H., Jr., Laird, W. J., and Smith, P. L. (1978). Heat-stable enterotoxin of *Escherichia coli: In vitro* effects on guanylate cyclase activity, cyclic GMP concentration, and ion transport in small intestine. *Proc. Natl. Acad. Sci. USA* **75**, 2800–2804.

Furukawa, O., Matsui, H., Suzuki, N., and Okabe, S. (1999). Epidermal growth factor protects rat epithelial cells against acid-induced damage through the activation of Na^+/H^+ exchangers. *J. Pharmacol. Exp. Ther.* **288**, 620–626.

Gebreselassie, D., Rajarathnam, K., and Fliegel, L. (1998). Expression, purification, and characterization of the carboxyl-terminal region of the Na^+/H^+ exchanger. *Biochem. Cell Biol.* **76**, 837–842.

Gerchman, Y., Olami, Y., Rimon, A., Taglicht, D., Schuldiner, S., and Padan, E. (1993). Histidine-226 is part of the pH sensor of NhaA, a Na^+/H^+ antiporter in *Escherichia coli. Proc. Natl. Acad. Sci. USA* **90**, 1212–1216.

Ghishan, F. K., Knobel, S., Barnard, J. A., and Breyer, M. (1995). Expression of a novel sodium-hydrogen exchanger in the gastrointestinal tract and kidney. *J. Membr. Biol.* **144**, 267–271.

Ghishan, F. K., Knobel, S. M., and Summar, M. (1995). Molecular cloning, sequencing, chromosomal localization, and tissue distribution of the human Na^+/H^+ exchanger (SLC9A2). *Genomics* **30**, 25–30.

Gonda, T., Maouyo, D., Rees, S. E., and Montrose, M. H. (1999). Regulation of intracellular pH gradients by identified Na/H exchanger isoforms and a short-chain fatty acid. *Am. J. Physiol.* **276**, G259–G270.

Good, D. W. (1995). Hyperosmolality inhibits bicarbonate absorption in rat medullary thick ascending limb via a protein-tyrosine kinase-dependent pathway. *J. Biol. Chem.* **270**, 9883–9889.

Good, D. W., George, T., and Watts, B. A., III (1995). Basolateral membrane Na^+/H^+ exchange enhances HCO_3 absorption in rat medullary thick ascending limbs: Evidence for functional coupling between basolateral and apical membrane Na^+/H^+ exchangers. *Proc. Natl. Acad. Sci. USA* **92**, 12525–12529.

Goss, G., Orlowski, J., and Grinstein, S. (1996). Co-immunoprecipitation of a 24-kDa protein with NHE1, the ubiquitous isoform of the Na^+/H^+ exchanger. *Am. Physiol. Soc.* **270**, C1493–C1502.

Goss, G. G., Woodside, M., Wakabayashi, S., Pouyssegur, J., Waddell, T., Downey, G. P., and Grinstein, S. (1994). ATP dependence of NHE-1, the ubiquitous isoform of the Na^+/H^+ antiporter. Analysis of phosphorylation and subcellular localization. *J. Biol. Chem.* **269**, 8741–8748.

Grinstein, S., Cohen, S., Goetz, J. D., Rothstein, A., and Gelfand, E. W. (1985). Characterization of the activation of Na^+/H^+ exchange in lymphocytes by phorbol esters. Change in the cytoplasmic pH-dependence of the antiport. *Proc. Natl. Acad. Sci. USA* **82**, 1429–1433.

Grinstein, S., Woodside, M., Waddell, T. K., Downey, G. P., Orlowski, J., Pouyssegur, J., Wong, D. C., and Foskett, J. K. (1993). Focal localization of the NHE-1 isoform of the Na^+/H^+ antiport: Assessment of effects on intracellular pH. *EMBO J.* **12**, 5209–5218.

Haggerty, J. G., Agarwal, N., Cragoe, E. J., Jr., Adelberg, E. A., and Slayman, C. W. (1988). LLC-PK mutant with increased Na^+/H^+ exchange and decreased sensitivity to amiloride. *Am. J. Physiol.* **255**, C495-C501.

Hall, R. A., Ostedgaard, L. S., Premont, R. T., Blitzer, J. T., Rahman, N., Welsh, M. J., and Lefkowitz, R. J. (1998b). A C-terminal motif found in the beta2-adrenergic receptor, P2Y1 receptor and cystic fibrosis transmembrane conductance regulator determines binding to the Na^+/H^+ exchanger regulatory factor family of PDZ proteins. *Proc. Natl. Acad. Sci. USA* **95**, 8496–8501.

Hall, R. A., Premont, R. T., Chow, C. W., Blitzer, J. T., Pitcher, J. A. Claing, A., Stoffel, R. H., Barak, L. S., Shenolikar, S., Weinman, E. J., Grinstein, S., and Lefkowitz, R. J. (1998a). The beta2-adrenergic receptor interacts with the Na^+/H^+ exchange. *Nature* **392**, 626-630.

Harris, C., and Fliegel, L. (1999a). Amiloride and the Na$^+$/H$^+$ exchanger protein: Mechanism and significance of inhibition of the Na$^+$/H$^+$ exchanger (review). *Int. J. Mol. Med.* **3**, 315–321.

Harris, C., and Fliegel, L. (1999b). Amiloride and the Na$^+$/H$^+$ exchanger protein: Mechanism and significance of inhibition of the Na$^+$/H$^+$ exchanger. *Int. J. Mol. Med.* **3**, 315–321.

Haworth, R. S., Frohlich, O., and Fliegel, L. (1993). Multiple carbohydrate moieties on the Na$^+$/H$^+$ exchanger. *Biochem. J.* **289**, 637–640.

He, X., Kuijpers, G. A. J., Goping, G., Kulakusky, J. A., Zheng, C., Delporte, C., Tse, C. M., Redman, R. S., Donowitz, M., Pollard, H. B., and Baum, B. J. (1998). A polarized salivary cell monolayer useful for studying transepithelial fluid movement *in vitro*. *Pflugers Arch.* **435**, 375–381.

He, X., Tse, C. M., Donowitz, M., Alper, S. L., Gabriel, S. E., and Baum, B. J. (1997). Polarized distribution of key membrane transport proteins in the rat submandibular gland. *Pflugers Arch.* **433**, 260–268.

Helmle-Kolb, C., Di Sole, F., Forgo, J., Hilfiker, H., Tse, C. M., Casavola, V., Donowitz, M., and Murer, H. (1997). Regulation of the transfected Na$^+$/H$^+$-exchanger NHE3 in MDCK cells by vasotocin. *Pflugers Arch.* **434**, 123–131.

Hensley, C. B., Bradley, M. E., and Mircheff, A. K. (1989). Parathyroid hormone-induced translocation of Na-H antiporters in rat proximal tubules. *Am. J. Physiol.* **257**, C637–C645.

Holmberg, C., and Perheentupa, J. (1985). Congenital Na$^+$ diarrhea: A new type of secretory diarrhea. *J. Pediatr.* **106**, 56–61.

Hoogerwerf, W. A., Tsao, S. C., Devuyst, O., Levine, S. A., Yun, C. H., Yip, J. W., Cohen, M. E., Wilson, P. D., Lazenby, A. J., Tse, C. M., and Donowitz, M. (1996a). NHE2 and NHE3 are human and rabbit intestinal brush-border proteins. *Am. J. Physiol.* **270**, G29–G41.

Hooley, R. C., Yu, Y., Symons, M., and Barber, D. L. (1996). G alpha 13 stimulates Na$^+$/H$^+$ exchange through distinct Cdc42-dependent and RhoA-dependent pathways. *J. Biol. Chem.* **271**, 6152–6158.

Horie, S., Moe, O., Miller, R. T., and Alpern, R. J. (1992). Long-term activation of protein kinase C causes chronic Na/H antiporter stimulation in cultured proximal tubule cells. *J. Clin. Invest.* **89**, 365–372.

Ikeda, T., Schmitt, B., Pouyssegur, J., Wakabayashi, S., and Shigekawa, M. (1997). Identification of cytoplasmic subdomains that control pH-sensing of the Na$^+$/H$^+$ exchanger (NHE1): pH-maintenance, ATP-sensitive, and flexible loop domains. *J. Biochem.* (*Tokyo*) **121**, 295–303.

Ikuma, M., Kashgarian, M., Binder, H. J., and Rajendran, V. M. (1999). Differential regulation of NHE isoforms by sodium depletion in proximal and distal segments of rat colon. *Am. J. Physiol* **276**, G539–G549.

Janecki, A., Montrose, M., Tse, C. M., Sanchez de Medina, F., Zweibaum, A., and Donowitz, M. (1999a). Development of an endogenous epithelial Na$^+$/H$^+$ exchanger (NHE3) in three clones of Caco-2 cells. *Am. J. Physiol.* **277**, G292–G305.

Janecki, A. J., Montrose, M. H., Zimniak, P., Zweibaum, A., Tse, C. M., Khurana, S., and Donowitz, M. (1998). Subcellular redistribution is involved in acute regulation of the brush border Na$^+$/H$^+$ exchanger isoform 3 in human colon adenocarcinoma cell line Caco-2. Protein kinase C–mediated inhibition of the exchanger. *J. Biol. Chem.* **273**, 8790-8798.

Janecki, A. J., Janecki, M., Akhter, S., and Donowitz, M. (2000b). Basic fibroblast growth factor stimulates surface expression and activity of Na(+)/H(+) exchanger NHE3 via mechanism involving phosphatidylinositol 3-kinase. *J. Biol. Chem.* **275**, 8133–8142.

Joutsi, T., Paimela, H., Bhowmik, A., Kiviluoto, T., and Kivilaakso, E. (1996). Role of Na$^+$-H$^+$-antiport in restitution of isolated guinea pig gastric epithelium after superficial injury. *Dig. Dis. Sci.* **41**, 2187–2194.

Kandasamy, R. A., and Orlowski, J. (1996). Genomic organization and glucocorticoid transcriptional activation of the rat Na$^+$/H$^+$ exchanger NHE3 gene. *J. Biol. Chem.* **271**, 10551–10559.

Kaneko, K., Guth, P. H., and Kaunitz, J. D. (1992). Na$^+$/H$^+$ exchanger regulates intracellular pH of rat gastric surface cells *in vivo*. *Pflugers Arch.* **421**, 322–328.

Kapus, A., Grinstein, S., Wasan, S., Kandasamy, R., and Orlowski, J. (1994). Functional characterization of three isoforms of the Na$^+$/H$^+$ exchanger stably expressed in Chinese hamster ovary cells. *J. Biol. Chem.* **269**, 23544–23552.

Kapus, A., Szaszi, K., Sun, J., Rizoli, S., and Rotstein, O. D. (1999). Cell shrinkage regulates Src kinases and induces tyrosine phosphorylation of cortactin, independent of the osmotic regulation of Na$^+$/H$^+$ exchangers. *J. Biol. Chem.* **274**, 8093–8102.

Keller, K. M., Wirth, S., Baumann, W., Sule, D., and Booth, I. W. (1990). Defective jejunal brush border membrane sodium/proton exchange in association with lethal familial protracted diarrhoea. *Gut* **31**, 1156–1158.

Kim, J. H., Johannes, L., Goud, B., Antony, C., Lingwood, C. A., Daneman, R., and Grinstein, S. (1998). Noninvasive measurement of the pH of the endoplasmic reticulum at rest and during calcium release. *Proc. Natl. Acad. Sci. USA* **95**, 2997–3002.

Kinsella, J. L., Cujdik, T., and Sacktor, B. Kinetic studies on the stimulation of Na$^+$/H$^+$ exchange activity in renal brush border membranes isolated from thyroid hormone-treated rats. *J. Membr. Biol.* **91**, 183–191.

Klanke, C. A., Su, Y. R., Callen, D. F., Wang, Z., Meneton, P., Baird, N., Kandasamy, R. A., Orlowski, J., Otterud, B. E., Leppert, M., Shull, G. E., and Memon, A. G. (1995). Molecular cloning and physical and genetic mapping of a novel human Na$^+$/H$^+$ exchanger (NHE5/SLC9A5) to chromosome 16q22.1. *Genomics* **25**, 615–622.

Knickelbein, R. G., Aronson, P. S., Atherton, W., and Dobbins, J. W. (1983). Sodium and chloride transport across rabbit ileal brush border. *Am. J. Physiol.* **245**, G504–G510.

Knickelbein, R. G., Aronson, P. S., and Dobbins, J. W. (1988). Membrane distribution of sodium-hydrogen and chloride-bicarbonate exchangers in crypt and villus cell membranes from rabbit ileum. *J. Clin. Invest.* **82**, 2158–2163.

Knickelbein, R. G., Aronson, P. S., and Dobbins, J. W. (1990). Characterization of Na$^+$/H$^+$ exchangers on villus cells in rabbit ileum. *Am. J. Physiol.* **259**, G802–G806.

Kokke, F. T., Elsawy, T., Bengtsson, U., Wasmuth, J. J., Wang, J. E., Tse, C. M., Donowitz, M., and Brant, S. R. (1996). A NHE3-related pseudogene is on human chromosome 10; The functional gene maps to 5p15.3. *Mamm. Genome* **7**, 235–236.

Krump, E., Nikitas, K., and Grinstein, S. (1997). Induction of tyrosine phosphorylation and Na$^+$/H$^+$ exchanger activation during shrinkage of human neutrophils. *J. Biol. Chem.* **272**, 17303–17311.

Kurashima, K., Szabo, E. Z., Lukacs, G., Orlowski, J., and Grinstein, S. (1998). Endosomal recycling of the Na$^+$/H$^+$ exchanger NHE3 isoforms is regulated by the phosphatidylinositol 3-kinase pathway. *J. Biol. Chem.* **273**, 20828–20836.

Kurashima, K., Yu, F. H., Cabado, A. G., Szabo, E. Z., Grinstein, S., and Orlowski, J. (1997). Identification of sites required for down-regulation of Na$^+$/H$^+$ exchanger NHE3 activity by cAMP-dependent protein kinase. *J. Biol. Chem.* **272**, 28672–28678.

Laghmani, K., Borensztein, P., Ambuhl, P., Froissart, M., Bichara, M., Moe, O. W., Alpern, R. J., and Paillard, M. (1997). Chronic metabolic acidosis enhances NHE3 protein abundance and transport activity in the rat thick ascending limb by increasing NHE3 mRNA. *J. Clin. Invest.* **99**, 24–30.

Laghmani, K., Chambrey, R., Froissart, M., Bichara, M., Paillard, M., and Borensztein, P. (1999). Adaption of NHE3 in the rat thick ascending limb: Effects of high sodium intake and metabolic alkalosis. *Am. J. Physiol.* **276**, F18–F26.

Lamprecht, G., Seidler, U., and Classen, M. (1993). Intracellular pH-regulating ion transport mechanisms in parietal cell basolateral membrane vesicles. *Am. J. Physiol.* **265**, G903–G910.

Lamprecht, G., Weinman, E. J., and Yun, C. H. (1998). The role of NHERF and E3KARP in the cAMP-mediated inhibition of NHE3. *J. Biol. Chem.* **273**, 29972–29978.

Lee, M. G., Schultheis, P. J., Yan, M., Shull, G. E., Bookstein, C., Chang, E., Tse, M., Donowitz, M., Park, K., and Muallem, S. (1998). Membrane-limited expression and regulation of Na^+/H^+ exchanger isoforms by P2 receptors in the rat submandibular gland duct. *J. Physiol.* **513,** 341–357.

Levine, S. A., Montrose, M. H., Tse, C. M., and Donowitz, M. (1993). Kinetics and regulation of three cloned mammalian Na^+/H^+ exchangers stably expressed in a fibroblast cell line. *J. Biol. Chem.* **268,** 25527–25535.

Levine, S. A., Nath, S. K., Yun, C. H., Yip, J. W., Donowitz, M., and Tse, C. M. (1995). Separate C-terminal domains of the epithelial specific brush border Na^+/H^+ exchanger isoform NHE3 are involved in stimulation and inhibition by protein kinases/growth factors. *J. Biol. Chem.* **270,** 13716–13725.

Lin, X., and Barber, D. L. (1996). A calcineurin homologous protein inhibits GTPase-stimulated Na^+/H^+ exchange. *Proc. Natl. Acad. Sci. USA* **93,** 12631–12636.

Lin, X., Voyno-Yasenetskaya, T. A., Hooley, R., Lin, C. Y., Orlowski, J., and Barber, D. L. (1996). Galpha12 differentially regulates Na^+/H^+ exchanger isoforms. *J. Biol. Chem.* **271,** 22604–22610.

Loffing, J., Lotscher, M., Kaissling, B., Biber, J., Murer, H., Seikaly, M., Alpern, R. J., Baum, M., and Moe, O. W. (1998). Renal Na/H exchanger NHE3 and Na-PO4 cotransporter NaPi-2 protein expression in glucocorticoid excess and deficient states. *J. Am. Soc. Nephrol.* **9,** 1560–1567.

Lotscher, M., Kaissling, B., Biber, J., Murer, H., and Levi, M. (1997). Role of microtubules in the rapid regulation of renal phosphate transport in response to acute alterations in dietary phosphate content. *J. Clin. Invest.* **99,** 1302–1312.

Lukacs, G. L., Segal, G., Kartner, N., Grinstein, S., and Zhang, F. (1997). Constitutive internalization of cystic fibrosis transmembrane conductance regulator occurs via clathrin-dependent endocytosis and is regulated by protein phosphorylation. *Biochem. J.* **328,** 353–361.

Ma, Y. H., Reusch, H. P., Wilson, E., Escobedo, J. A., Fantl, W. J., Williams, L. T., and Ives, H. E. (1994). Activation of Na^+/H^+ exchange by platelet-derived growth factor involves phosphatidylinositol 3-kinase and phospholipase C gamma. *J. Biol. Chem.* **269,** 30734–30739.

Maher, M. M., Gontarek, J. D., Bess, R. S., Donowitz, M., and Yeo, C. J. (1997). The Na^+/H^+ exchange isoform NHE3 regulates basal canine ileal Na^+ absorption *in vivo. Gastroenterology* **112,** 174–183.

Maher, M. M., Gontarek, J. D., Jimenez, R. E., Donowitz, M., and Yeo, C. J. (1996). Role of brush border Na^+/H^+ exchange in canine ileal absorption. *Dig. Dis. Sci.* **41,** 654–659.

Malakooti, J., Dahdal, R. Y., Schmidt, L., Layden, T. J., Dudeja, P. D., and Ramaswamy, K. (1999). Molecular cloning, tissue distribution, and functional expression of the human Na^+/H^+ exchanger NHE2. *Am. J. Physiol.* **277,** G383–G390.

Malapert, M., Guizouarn, H., Fievet, B., Jahns, R., Garcia-Romeu, F., Motais, R., and Borgese, F. (1997). Regulation of Na^+/H^+ antiporter in trout red blood cells. *J. Exp. Biol.* **200,** 353–360.

McSwine, R. L., Musch, M. W., Bookstein, C., Xie, Y., Rao, M., and Chang, E. B. (1998). Regulation of apical membrane Na^+/H^+ exchangers NHE2 and NHE3 in intestinal epithelial cell line C2/bbe. *Am. J. Physiol.* **275,** C693–C701.

Moe, O. W., Amemiya, M., and Yamaji, Y. (1995). Activation of protein kinase A acutely inhibits and phosphorylates Na^+/H^+ exchanger NHE-3. *J. Clin. Invest.* **96,** 2187–2194.

Mrkic, B., Tse, C. M., Forgo, J., Helmle-Kolb, C., Donowitz, M., and Murer, H. (1993). Identification of PTH-responsive Na/H-exchanger isoforms in a rabbit proximal tubule cell line (RKPC-2). *Pflugers Arch.* **424,** 377–384.

Nakamura, S., Amlal, H., Schultheis, P. J., Galla, J. H., Shull, G. E., and Soleimani, M. (1999). HCO_3^- reabsorption in renal collecting duct of NHE-3 deficient mouse: A compensatory response. *Am. J. Physiol.* **276,** F914–F921.

Nass, R., Cunningham, K. W., and Rao, R. (1997). Intracellular sequestration of sodium by a novel Na^+/H^+ exchanger in yeast is enhanced by mutations in the plasma membrane H^+-ATPase. *J. Biol. Chem.* **272,** 26145–26152.

Nath, S. K., Akhter, S., Levine, S., Tse, C. M., and Donowitz, M. (1996a). Tyrosine kinase is required for calmodulin to inhibit basal activity of the epithelial brush border Na^+/H^+ exchanger, NHE3. *Gastroenterology* **110**, A349.

Nath, S. K., Hang, C. Y., Levine, S. A., Yun, C. H., Montrose, M. H., Donowitz, M., and Tse, C. M. (1996b). Hyperosmolarity inhibits the Na^+/H^+ exchanger isoforms NHE2 and NHE3: An effect opposite to that on NHE1. *Am. J. Physiol.* **270**, G431–G441.

Nath, S. K., Kambadur, R., Yun, C. H., Donowitz, M., and Tse, C. M. (1999). NHE2 contains subdomains in the COOH terminus for growth factor and protein kinase regulation. *Am. J. Physiol.* **276**, C873–C882.

Noel, J., and Pouyssegur, J. (1995). Hormonal regulation, pharmacology, and membrane sorting of vertebrate Na^+/H^+ exchanger isoforms. *Am. J. Physiol.* **268**, C283–C296.

Noel, J., Roux, D., and Pouyssegur, J. (1996). Differential localization of Na^+/H^+ exchanger isoforms (NHE1 and NHE3) in polarized epithelial cell lines. *J. Cell. Sci.* **109**, 929–939.

Numata, M., Petrecca, K., Lake, N., and Orlowski, J. (1998). Identification of a mitochondrial Na^+/H^+ exchanger. *J. Biol. Chem.* **273**, 6951–6959.

Opleta-Madsen, K., Hardin, J., and Gall, D. G. (1999). Epidermal growth factor upregulates intestinal electrolyte and nutrient transport. *Am. J. Physiol.* **260**, G807–G814.

Orlowski, J. (1993a). Heterologous expression and functional properties of amiloride high affinity (NHE-1) and low affinity (NHE-3) isoforms of the rat Na/H exchanger. *J. Biol. Chem.* **268**, 16369–16377.

Orlowski, J., and Grinstein, S. (1997). Na^+/H^+ exchangers of mammalian cells. *J. Biol. Chem.* **272**, 22373–22376.

Orlowski, J., and Kandasamy, R. A. (1996). Delineation of transmembrane domains of the Na^+/H^+ exchanger that confer sensitivity to pharmacological antagonists. *J. Biol. Chem.* **271**, 19922–19927.

Orlowski, J., Kandasamy, R. A., and Shull, G. E. (1992). Molecular cloning of putative members of the Na^+/H^+ exchanger gene family. cDNA cloning, deduced amino acid sequence, and mRNA tissue expression of the rat Na^+/H^+ exchanger NHE-1 and two structurally related proteins. *J. Biol. Chem.* **267**, 9331–9339.

Padan, E., Gerchman, Y., Rimon, A., Rothman, A., Dover, N., and Carmel-Harel, O. (1999). The molecular mechanism of regulation of the NhaA Na^+/H^+ antiporter of *Escherichia coli*, a key transporter in the adaptation to Na^+/H^+. *Novartis Found. Symp.* **221**, 183–196.

Park, K., Olschowka, J. A., Richardson, L. A., Boostein, C., Chang, E. B., and Melvin, J. E. (1999). Expression of multiple Na^+/H^+ exchanger isoforms in rat parotid acinar and ductal cells. *Am. J. Physiol.* **276**, G470–G478.

Peng, Y., Moe, O. W., Chu, T., Preisig, P. A., Yanagisawa, M., and Alpern, R. J. (1999). ETB receptor activation leads to activation and phosphorylation of NHE3. *Am. J. Physiol.* **276**, C938–C945.

Peti-Peterdi, J., Bebok, Z., St. John, P. L., Chambrey, R., Abrahamson, D. R., Warnock, D. G., and Bell, P.D. (1998). Basolateral Na^+/H^+ exchange (NHE4) in cells of the macula densa. *J. Am. Soc. Nephrol.* **9**, A52.

Pfeifer, A., Aszodi, A., Seidler, U., Ruth, P., Hofmann, F., and Fassler, R. (1996). Intestinal secretory defects and dwarfism in mice lacking cGMP-dependent protein kinase II. *Science* **274**, 2082–2086.

Pizzonia, J. H., Biemesderfer, D., Abu-Alfa, A. K., Wu, M. S., Exner, M., Isenring, P., Igarashi, P., and Aronson, P. S. (1998). Immunochemical characterization of Na^+/H^+ exchanger isoform NHE4. *Am. J. Physiol.* **275**, F510–F517.

Preisig, P. A., Ives, H. E., Cragoe, E. J., Jr., Alpern, R. J., and Rector, F. C., Jr. (1987). Role of the Na^+/H^+ antiporter in rat proximal tubule bicarbonate absorption. *J. Clin. Invest.* **80**, 970–978.

Rajendran, V. M., and Binder, H. J. (1990). Characterization of Na$^+$/H$^+$ exchange in apical membrane vesicles of rat colon. *J. Biol. Chem.* **265,** 8408–8414.

Rajendran, V. M., Geibel, J., and Binder, H. J. (1995). Chloride-dependent Na-H exchange. A novel mechanism of sodium transport in colonic crypts. *J. Biol. Chem.* **270,** 11051–11054.

Rajendran, V. M., Geibel, J., and Binder, H. J. (1999). Role of Cl channels in Cl-dependent Na-H exchange. *Am. J. Physiol.* **276,** G73–G78.

Raley-Susman, K. M., Cragoe E. J., Jr., Sapolsky, R. M., and Kopito, R. R. Regulation of intracellular pH in cultured hippocampal neurons by an amiloride-insensitive Na$^+$/H$^+$ exchanger. *J. Biol. Chem.* **266,** 2739–2745.

Reczek, D., Berryman, M., and Bretscher, A. (1997). Identification of EBP50: A PDZ-containing phosphoprotein that associates with members of the ezrin-radixin-moesin family. *J. Cell. Biol.* **139,** 169–179.

Rood, R. P., Emmer, E., Wesolek, J., McCullen, J., Husain, Z., Cohen, M. E., Braithwaite, R. S., Murer, H., Sharp, G. W., and Donowitz, M. (1988). Regulation of the rabbit ileal brush-border Na$^+$/H$^+$ exchanger by an ATP-requiring Ca^{++}/calmodulin-mediated process. *J. Clin. Invest.* **82,** 1091–1097.

Rood, E., Rood, R. P., Wesolek, J. H., Cohen, M. E., Braithwaite, R. S., Sharp, G. W., Murer, H., and Donowitz, M. (1989). Role of calcium and calmodulin in the regulation of the rabbit ileal brush-border membrane Na$^+$/H$^+$ antiporter. *J. Membr. Biol.* **108,** 207–215.

Sardet, C., Counillon, C., Franchi, A., and Pouyssegur, J. (1990). The Na$^+$/H$^+$ antiporter is a glycoprotein of 110 kDa phosphorylated by growth factors in quiescent cells. *Science* **247,** 723–726.

Sardet, C., Counillon, C., Franchi, A., and Pouyssegur, J. (1991). Thrombin, EGF and okadaic acid activate the Na$^+$/H$^+$ exchanger, NHE1, by phosphorylating a set of common sites. *J. Biol. Chem.* **266,** 19166–19171.

Sardet, C., Franchi, A., and Pouyssegur, J. (1989). Molecular cloning, primary structure and expression of the human growth factor-activatable Na$^+$/H$^+$ antiporter. *Cell* **56,** 271–280.

Schultheis, P. J., Clarke, L. L., Meneton, P., Harline, M., Boivin, G. P., Stemmermann, G., Duffy, J. J., Doetschman, T., Miller, M. L., and Shull, G. E. (1998a). Targeted disruption of the Murine Na$^+$/H$^+$ exchanger isoform 2 gene causes reduced viability of gastric parietal cells and loss of net acid secretion. *J. Clin. Invest.* **101,** 1243–1253.

Schultheis, P. J., Clark, L. L., Meneton, P., Miller, M. L., Soleimani, M., Gawenis, L. R., Riddle, T. M., Duffy, J. J., Doetschman, T., Wang, T., Giebisch, G., Aronson, P. S., Lorenz, J. N., and Shull, G. E. (1998b). Renal and intestinal absorptive defects in mice lacking the NHE3 Na$^+$/H$^+$ exchanger. *Nature Genet.* **19,** 282–285.

Schwark, J. R., Jansen, H. W., Lang, J. H., Krick, W., Burckhardt, G., and Hropot, M. (1998). S3226, a novel inhibitor of Na$^+$/H$^+$ exchanger subtype 3 in various cell types. *Pflugers Arch.* **436,** 797–800.

Seidler, B., Rossmann, H., Murray, A., Orlowski, J., Tse, C. M., Donowitz, M., and Shull, G. (1997). Expression of the Na$^+$/H$^+$ exchanger isoform NHE1–4 mRNA in different epithelial cell types of rat and rabbit gastric mucosa. *Gastroenterology* **110,** A285.

Shrode, L. D., Cabado, A. G., Goss, G. G., and Grinstein, S. (1996). Role of the Na$^+$/H$^+$ antiporter isoforms in cell volume regulation. *In* "Na$^+$/H$^+$ Exchanger" (L. Fliegel, ed). Chapman & Hall, NYL, pp 102–122.

Shrode, L. D., Gan, B. S., D'Souza, S. J., Orlowski, J., and Grinstein, S. (1998). Topological analysis of NHE1, the ubiquitous Na$^+$/H$^+$ exchanger using chymotryptic cleavage. *Am. J. Physiol.* **275,** C431–C439.

Siffert, W., and Dusing, R. (1996). Na$^+$/H$^+$ exchange in hypertension and in diabetes—Facts and hypotheses. *Basic Res. Cardiol.* **91,** 179–190.

Singh, G., Orlowski, J., and Soleimani, M. (1996). Transient expression of Na$^+$/H$^+$ exchanger isoform NHE-2 in LLC-PK$_i$ cells: Inhibition of endogenous NHE-3 and regulation by hypertonicity. *J. Membr. Biol.* **151,** 261–268.

Silva, N. L., Haworth, R. S., Singh, D., and Fliegel, L. (1995). The carboxyl-terminal region of the Na$^+$/H$^+$ exchanger interacts with mammalian heat shock protein. *Biochemistry* **34**, 10412–10420.

Silviani, V., Colombani, V., Heyries, L., Gerolami, A., Cartouzou, G., and Marteau, C. (1996). Role of the NHE3 isoform of the Na$^+$/H$^+$ exchanger in sodium absorption by the rabbit gallbladder. *Pflugers Arch.* **432**, 791–796.

Soleimani, M., Bergman, J. A., Hosford, M. A., and McKinney, T. D. (1990). Potassium depletion increases luminal Na$^+$/H$^+$ exchange and basolateral Na$^+$: HCO$_3$ cotransport in rat renal cortex. *J. Clin. Invest.* **86**, 1076–1083.

Soleimani, M., Bizal, G. L., McKinney, T. D., and Hattabaugh, Y. J. (1992). Effect of *in vitro* metabolic acidosis on luminal Na$^+$/H$^+$ exchange and basolateral Na:HCO$_3$ cotransport in rabbit kidney proximal tubules. *J. Clin. Invest.* **90**, 211–218.

Soleimani, M., Bookstein, C., Bizal, G. L., Musch, M. W., Hattabaugh, Y. J., Rao, M. C., and Chang, E. B. (1994b). Localization of the Na$^+$/H$^+$ exchanger isoform NHE-3 in rabbit and canine kidney. *Biochim. Biophys. Acta* **1195**, 89–95.

Soleimani, M., Bookstein, C., McAteer, J. A., Hattabaugh, Y. J., Bizal, G. L., Musch, M. W., Villereal, M., Rao, M. C., Howard, R. L., and Chang, E. B. (1994c). Effect of high osmolality on Na$^+$/H$^+$ exchange in renal proximal tubule cells. *J. Biol. Chem.* **269**, 15613–15618.

Soleimani, M., Bookstein, C., Singh, G., Rao, M. C., Chang, E. B., and Bastani, B. (1995). Differential regulation of Na$^+$/H$^+$ exchange and H$^{(+)}$-ATPase by pH and HCO$_3$ in kidney proximal tubules. *J. Membr. Biol.* **144**, 209–216.

Soleimani, M., and Singh, G. (1995). Physiologic and molecular aspects of the Na$^+$/H$^+$ exchangers in health and disease processes. *J. Invest. Med.* **43**, 419–430.

Soleimani, M., Singh, G., Bizal, G. L., Gullans, S. R., and McAteer, J. A. (1994a). Na$^+$/H$^+$ exchanger isoforms NHE-2 and NHE-1 in inner medullary collecting duct cells: Expression, functional localization, and differential regulation. *J. Biol. Chem.* **269**, 27973–27978.

Soleimani, M., Singh, G., Bookstein, C., Rao, M. C., Chang, E. B., and Dominguez, J. H. (1996). Inhibition of glycosylation decreases Na$^+$/H$^+$ exchange activity, blocks NHE-3 transport to the membrane, and increases NHE-3 mRNA expression in LLC-PK1 cells. *J. Lab. Clin. Med.* **127**, 565–573.

Soleimani, M., Watts, B. A., III, Singh, G., and Good, D. W. (1998). Effect of long-term hyperosmolality on the Na$^+$/H$^+$ exchanger isoform NHE3 in LLC-PK1 cells. *Kidney Int.* **53**, 423–431.

Sun, A. M., Liu, Y., Centracchio, J., and Dworkin, L. D. (1998). Expression of Na$^+$/H$^+$ exchanger isoforms in inner segment of inner medullary collecting duct. *J. Membr. Biol.* **164**, 293–300.

Sun, A. M., Liu, Y., Dworkin, L. D., Tse, C. M., Donowitz, M., and Yip, K. P. (1997). Na$^+$/H$^+$ exchanger isoform 2 (NHE2) is expressed in the apical membrane of the medullary thick ascending limb. *J. Membr. Biol.* **160**, 85–90.

Szaszi, K., Buday, L., and Kapus, A. (1997). Shrinkage-induced protein tyrosine phosphorylation in Chinese hamster ovary cells. *J. Biol. Chem.* **272**, 16670–16678.

Szpirer, C., Szpirer, J., Riviere, M., Levan, G., and Orlowski, J. (1994). Chromosomal assignment of four genes encoding Na/H exchanger isoforms in human and rat. *Mamm. Genome* **5**, 153–159.

Thevenod, A. F. (1996). Evidence for involvement of a zymogen granule Na$^+$/H$^+$ exchanger in enzyme secretion from rat pancreatic acinar cells. *J. Membr. Biol.* **152**, 195–205.

Tse, C. M., Brant, S. R., Walker, S., Pouyssegur, J., and Donowitz, M. (1992). Cloning and sequencing of a rabbit cDNA encoding an intestinal and kidney specific Na$^+$/H$^+$ exchanger isoform (NHE-3). *J. Biol. Chem.* **267**, 9340–9346.

Tse, M., Levine, S., Yun, C., Brant, S., Counillon, L. T., Pouyssegur, J., and Donowitz, M. (1993a). Structure/function studies of the epithelial isoforms of the mammalian Na$^+$/H$^+$ exchanger gene family. *J. Membr. Biol.* **135**, 93–108.

Tse, C. M., Levine, S. A., Yun, C. H. C., Brant, S. R., Pouyssegur, Jr., Montrose, M. H., and Donowitz, M. (1993c). Functional characteristics of a cloned epithelial Na^+/H^+ exchanger (NHE3): Resistance to amiloride and inhibition by protein kinase C. *Proc. Natl. Acad. Sci. USA* **90**, 9110–9114.

Tse, C. M., Levine, S. A., Yun, C. H. C., Khurana, S., and Donowitz, M. (1994). Na^+/H^+ exchanger-2 is an O-linked but not an N-linked sialoglycoprotein. *Biochemistry* **33**, 12954–12961.

Tse, C. M., Levine, S. A., Yun, C. H. C., Montrose, M. H., Little, P. J., Pouyssegur, J., and Donowitz, M. (1993b). Cloning and expression of a rabbit cDNA encoding a serum-activated ethylisopropylamiloride-resistant epithelial Na^+/H^+ exchanger isoform (NHE2). *J. Biol. Chem.* **268**, 11917–11924.

Tse, C. M., Ma, A. I., Yang, V. W., Watson, A. J., Levine, S., Montrose, M. H., Potter, J., Sardet, C., Pouyssegur, J., and Donowitz, M. (1991). Molecular cloning and expression of a cDNA encoding the rabbit ileal villus cell basolateral membrane Na^+/H^+ exchanger. *EMBO. J.* **10**, 1957–1967.

Tominaga, T., and Barber, D. L. (1998). Na-H exchange acts downstream of RhoA to regulate integrin-induced cell adhesion and spreading. *Mol. Biol. Cell.* **9**, 2287–2303.

Tominaga, T., Ishizaki, T., Narumiya, S., and Barber, D. L. (1998). p160Rock mediates RhoA activation of Na-H exchange. *Embo. J.* **17**, 4712–4722.

Voyno-Yasenetskaya, T., Conklin, B. R., Gilbert, R. L., Hooley, R., Bourne, H. R., and Barber, D. L. (1994). G alpha 13 stimulates Na/H exchange. *J. Biol. Chem.* **269**, 4721–4724.

Wakabayashi, S., Bertrand, B., Ikeda, T., Pouyssegur, J., and Shigekawa, M. (1994b). Mutation of calmodulin-binding site renders the Na^+/H^+ exchanger (NHE1) highly H^+-sensitive and $Ca2^+$ regulation-defective. *J. Biol. Chem.* **269**, 13710–13715.

Wakabayashi, S., Bertrand, B., Shigekawa, M., Fafournoux, P., and Pouyssegur, J. (1994a). Growth factor activation and "H^+-sensing" of the Na^+/H^+ exchanger isoform 1 (NHE1). Evidence for an additional mechanism not requiring direct phosphorylation. *J. Biol. Chem.* **269**, 5583–5588.

Wakabayashi, S., Fafournoux, P., Sardet, C., and Pouyssegur, J. (1992). The Na^+/H^+ antiporter cytoplasmic domain mediates growth factor signals and controls "H^+-sensing." *Proc. Natl. Acad. Sci. USA* **89**, 2424–2428.

Wakabayashi, S., Ikeda, T., Iwamoto, T., Pouyssegur, J., and Shigekawa, M. (1997b). Calmodulin-binding autoinhibitory domain controls "pH-sensing" in the Na^+/H^+ exchanger NHE1 through sequence-specific interaction. *Biochemistry* **36**, 12854–12861.

Wakabayashi, S., Ikeda, T., Noel, J., Schmitt, B., Orlowski, J., Pouyssegur, J., and Shigekawa, M. (1995). Cytoplasmic domain of the ubiquitous Na^+/H^+ exchanger NHE1 can confer Ca^{2+} responsiveness to the apical isoform NHE3. *J. Biol. Chem.* **270**, 26460–26465.

Wakabayashi, S., Pang, T., Su, X., and Shigekawa, M. (2000). A novel topology model of the Na^+/H^+ exchanger NHE1. *J. Biol. Chem.* **275**, 7542–7949.

Wakabayashi, S., Shigekawa, M., and Pouyssegur, J. (1997a). Molecular physiology of vertebrate Na^+/H^+ exchangers. *Physiol. Rev.* **77**, 51–74.

Wang, D., Balkovetz, D. F., and Warnock, D. G. (1995). Mutational analysis of transmembrane histidines in the amiloride-sensitive Na^+/H^+ exchanger. *Am. J. Physiol.* **269**, C392–C402.

Wang, S., Raab, P. J., Guggino, W. B., and Li, M. (1998). Peptide binding consensus of the NHE-FR-PDZ1 domain matches the C-terminal sequence of cystic fibrosis transmembrane conductance regulator (CFTR). *FEBS Lett.* **427**, 103–108.

Wang, T., Yang, C. L., Abbiati, T., Schultheis, P. J., Shull, G. E., Giebisch, G., and Aronson, P. S. (1999). Mechanism of proximal tubule bicarbonate absorption in NHE3 null mice. *Am. J. Physiol.* **277**, F298–302.

Wang, Z., Orlowski, J., and Shull, G. E. (1993). Primary structure and functional expression of a novel gastrointestinal isoform of the rat Na^+/H^+ exchanger. *J. Biol. Chem.* **268**, 11925–11928.

Watts B. A., III, Thampi, G., and Good, D. W. (1999). Nerve growth factor inhibits HCO_3 absorption in renal thick ascending limb through inhibition of basolateral membrane Na^+/H^+ exchange. *J. Biol. Chem.* **274**, 7841–7847.

Weinman, E. J., Steplock, D., Donowitz, M., and Shenolikar, S. (2000). NHERF associations with sodium-hydrogen exchanger isoform 3 (NHE3) and ezrin are essential for cAMP-mediated phosphorylation and inhibition of NHE3. *Biochemistry* **39**, 6123–6129.

Weinman, E. J., Steplock, D., and Shenolikar, S. (1993). cAMP-mediated inhibition of the renal brush border membrane Na^+/H^+ exchanger requires a dissociable phosphoprotein cofactor. *J. Clin. Invest.* **92**, 1781–1786.

Weinman, E. J., Steplock, D., and Shenolikar, S. (1995). Characterization of a protein cofactor that mediates protein kinase A regulation of the renal brush border membrane Na^+/H^+ exchanger. *J. Clin. Invest.* **95**, 2143–2149.

Wiederkehr, M. R., Zhao, H., and Moe, O. W. (1999). Acute regulation of Na^+/H^+ exchanger NHE3 activity by protein kinase C: Role of NHE3 phosphorylation. *Am. J. Physiol.* **276**, C1205–C1217.

Wormmeester, L., De Medina, F. S., Kokke, F., Tse, C. M., Khurana, S., Bowswer, J., Cohen, M. E., and Donowitz, M. (1998). Quantitative contribution of NHE2 and NHE3 to rabbit ileal brush-border Na^+/H^+ exchange. *Am. J. Physiol.* **274**, C1261-C1272.

Wu, M.-S., Biemesderfer, D., Giebisch, G., and Aronson, P. S. (1996). Role of NHE3 in mediating renal brush border Na^+/H^+ exchange. *J. Biol. Chem.* **271**, 32749–32752.

Yamaji, Y., Moe, O. W., Miller, R. T., and Alpern, R. J. (1994). Acid activation of immediate early genes in renal epithelial cells. *J. Clin. Invest.* **94**, 1297–1303.

Yamaji, Y., Tsuganezawa, H., Moe, O. W., and Alpern, R. J. (1997). Intracellular acidosis activates c-Src. *Am. J. Physiol.* **272**, C886–C893.

Yamaji, Y., Amemiya, M., Cano, A., Preisig, P. A., Miller, R. T., Moe, O. W., and Alpern, R. J. (1995). Overexpression of Csk inhibits acid-induced activation of NHE3. *Proc. Natl. Acad. Sci. USA* **92**, 6274–6278.

Yellen, G., Jurman, M. E., Abramson, T., and MacKinnon, R. (1991). Mutations affecting internal TEA blockade identify the probable pore-forming region of a K^+ channel. *Science* **251**, 939–942.

Yip, J. W., Ko, W. H., Viberti, G., Huganir, R. L., Donowitz, M., and Tse, C. M. (1997). Regulation of the epithelial brush border Na^+/H^+ exchanger isoform 3 stably expressed in fibroblasts by fibroblast growth factor and phorbol esters is not through changes in phosphorylation of the exchanger. *J. Biol. Chem.* **272**, 18473–18480.

Yip, K. P., Tse, C. M., McDonough, A. A., and Marsh, D. J. (1998). Redistribution of Na^+/H^+ exchanger isoform NHE3 in proximal tubules induced by acute and chronic hypertension. *Am. J. Physiol.* **275**, F565–F575.

Yip, K. P., Tse, C. M., McDonough, A. N., Donowitz, M., and Marsh, D. J. (1995). Differential translocation of Na^+/H^+ exchanger isoforms NHE2 and NHE3 in rat proximal tubule during acute hypertension. *J. Am. Soc. Nephrol.* **6**, 218A.

Yonemura, K., Cheng, L., Sacktor, B., and Kinsella, J. L. (1990). Stimulation by thyroid hormone of Na^+/H^+ exchange activity in cultured opossum kidney cells. *Am. J. Physiol.* **258**, F333–F338.

Yu, F. H., Shull, G. E., and Orlowski, J. (1993). Functional properties of the rat Na/H exchanger NHE-2 isoform expressed in Na/H exchanger-deficient Chinese hamster ovary cells. *J. Biol. Chem.* **268**, 25536–25541.

Yun, C. H., Lamprecht, G., Forster, D. V., and Sidor, A. (1998). NHE3 kinase A regulatory protein E3KARP binds the epithelial brush border Na^+/H^+ exchanger NHE3 and the cytoskeletal protein ezrin. *J. Biol. Chem.* **273**, 25856–25863.

Yun, C. H., Oh, S., Zizak, M., Steplock, D., Tsao, S., Tse, C. M., Weinman, E. J., and Donowitz, M. (1997). cAMP-mediated inhibition of the epithelial brush border Na^+/H^+ exchanger, NHE3, requires an associated regulatory protein. *Proc. Natl. Acad. Sci. USA* **94**, 3010–3015.

Yun, C. H., Tse, C. M., and Donowitz, M. (1995b). Chimeric Na^+/H^+ exchangers: An epithelial membrane-bound N-terminal domain requires an epithelial cytoplasmic C-terminal domain for regulation by protein kinases. *Proc. Natl. Acad. Sci. USA* **92**, 10723–10727.

Yun, C. H. C., Gurubhagavatula, S., Levine, S. A., Montgomery, J. L., Brant, S. R., Cohen, M. E., Cragoe, E., Jr., Pouyssegur, J., Tse, C. M., and Donowitz, M. (1993b). Glucocorticoid stimula-

tion of ileal Na⁺ absorptive cell brush border Na⁺/H⁺ exchange and association with an increase in message for NHE3, an epithelial Na⁺/H⁺ exchanger isoform. *J. Biol. Chem.* **268,**
206–211.

Yun, C. H. C., Tse, C. M., Nath, S. K., Levine, S. A., and Donowitz, M. (1993a). Leu 143 in the
fourth membrane spanning domain is critical for amiloriction of an epithelial Na⁺/H⁺ exchanger isoform (NHE-2). *Biochem. Biophys. Res. Commun.* **193,** 532–539.

Yun, C. H. C., Tse, C. M., Nath, S. K., Levine, S. R., and Donowitz, M. (1995a). Mammalian
Na⁺/H⁺ exchanger gene structure and function studies. *Am. J. Physiol.* **269,** G1–G11.

Yusufi, A. N. K., Szczepanska-Konkel, M., and Dousa, T. P. (1988). Role of N-linked oligosaccharides in the transport activity of the Na⁺/H⁺ antiporter in rat renal brush-border membrane. *J.
Biol. Chem.* **236,** 13683–13691.

Zhang, Y., Magyar, C. E., Norian, J. M., Holstein-Rathlou, N. H., Mircheff, A. K., and McDonough,
A. A. (1998). Reversible effects of acute hypertension on proximal tubule sodium transporters.
Am. J. Physiol. **274,** C1090–C1100.

Zhang, Y., Norian, J. M., Magyar, C. E., Holstein-Rathlou, N. H., Mircheff, A. K., and McDonough,
A. A. (1999). *In vivo* PTH provokes apical NHE3 and NaPi2 redistribution and Na-K-ATPase
inhibition. *Am. J. Physiol.* **276,** F711–F719.

Zhao, H., Wiederkehr, M. R., Fan, L., Collazo, R. L., Crowder, L. A., and Moe, O. W. (1999). Acute
inhibition of Na/H exchanger NHE-3 by cAMP. Role of protein kinase A and NHE-3 phos-
phoserines 552 and 605. *J. Biol. Chem.* **274,** 3978–3987.

Zizak, M., Cavet, M., Bayle, D., Tse, M., Hallen, S., Sachs, G., and Donowitz, M. (XXXX). Na⁺/H⁺
exchanger NHE3 has 11 membrane spanning domains and a cleaved signal peptide. Topology
analysis using *in vitro* transcription/translation. *Biochemistry,* in Press, 2000.

Zizak, M., Lamprecht, G., Steplock, D., Tariq, N., Shenolikar, S., Donowitz, M., Yun, C. H. C., and
Weinman, E. J. (1999). cAMP-induced phosphorylation and inhibition of Na⁺/H⁺ exchanger
(NHE3) is dependent on the presence but not the phosphorylation of NHERF. *J. Biol. Chem.*
274, 24753–24758.

CHAPTER 13

Molecular Aspects of Intestinal Brush-Border Na$^+$/Glucose Transport

Ernest M. Wright

Department of Physiology, UCLA School of Medicine, Los Angeles, California 90095-1751

I. SUMMARY

The intestinal brush-border membrane protein that plays a major part in the absorption of D-glucose from the diet is the Na$^+$/glucose cotransporter (SGLT1). The transporter also plays a major role in salt and water absorption, and provides the basis for oral rehydration therapy used to combat secretory diarrhea. Mutations in the gene coding for SGLT1 cause malabsorption of glucose and galactose, and this results in life-threatening diarrhea. In this chapter I review the status of research on the structure and function of the cotransporter, and highlight our current concept of how Na^+ and sugar transport are coupled by SGLT1 and how water transport is linked to sugar transport.

II. INTRODUCTION

The daily food intake on a Western diet contains about 350 g of carbohydrate, 70 g of protein, 100 g of fat, and a liter of water. Digestion of the carbohydrate yields around 1 mole of D-glucose, which is completely absorbed in the small intestine. Glucose absorption is accompanied by fluid absorption amounting to around 8 liters (1 liter of water ingested and 7 liters of fluid secreted to aid digestion). It is the mature enterocytes on the upper third of the villi that are responsible for intestinal absorption. Figure 1 shows a model for sugar absorption across these cells. Glucose is absorbed by a two-stage process: the first is the "active" accumulation of sugars across the brush-border membrane by the Na^+/glucose cotransporter (SGLT1), and the second is the downhill transport of glucose out of the enterocyte into the blood across the basolateral membrane by a facilitated glucose transporter (GLUT2). The sodium gradient across the brush border is maintained by the basolateral Na^+/K^+-pump (ATPase); that is, the Na^+ that enters the cell across the brush border along with sugar is pumped out across the basolateral membrane. The net result is that glucose and salt are absorbed, and this is the scientific basis for the oral rehydration therapy (Greenough, 1989)

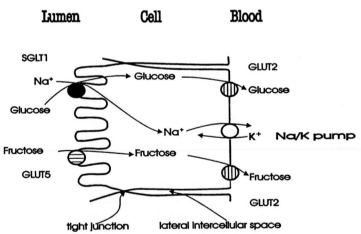

FIGURE 1   A model for glucose and fructose transport across the enterocytes of the small intestine. SGLT1 is the Na^+/glucose cotransporter, GLUT5 the brush-border fructose uniporter, GLUT2 the basolateral hexose uniporter, and Na^+/K^+ pump the basolateral 3 sodium/2 potassium exchange pump.

used so successfully to combat secretory diarrhea. Fructose, an ever increasing component of the diet, is absorbed across the enterocyte by a parallel process. Transport across both the brush-border and basolateral membranes occurs by facilitated diffusion. A private transporter allows fructose through the brush-border membrane (GLUT5), whereas a less selective transporter (GLUT2) allows fructose and glucose to pass through the basolateral membrane. Fructose does not significantly stimulate salt and water absorption across the intestine of most animals. Although it is clear that a paracellular pathway (tight junctions and lateral intercellular spaces) allows the passive flux of ions across the intestinal epithelium, there is little direct evidence that either sugars or water uses this route.

III THE SODIUM/GLUCOSE COTRANSPORTER (SGLT1)

Almost 40 years ago, Bob Crane and his colleagues made a conceptual breakthrough that has had a long-lasting impact on the field of bioenergetics (see Wright *et al.*, 1994). Crane proposed that the active transport of glucose across the intestinal brush border was driven by the sodium gradient across the membrane; in other words, the transporter was a protein that coupled the transport of glucose to the downhill transport of sodium, sodium/glucose cotransport. In the 1960s and 1970s this concept was tested extensively for the transport of sugars and many other substrates in cells throughout the body, and it is now generally accepted that Na or H cotransport is the mechanism by which cells accumulate substrates such as amino acids, neurotransmitters, vitamins, osmolytes, and ions such as iodide and phosphate. Perhaps the most definite evidence of cotransport came from experiments on brush border plasma membrane vesicles, where it was shown that sodium electrochemical gradients alone could drive the accumulation of glucose and amino acids within the vesicle (see Hopfer, 1987).

The rabbit intestinal Na/glucose cotransporter was identified in 1981–1984 as a 70- to 75-kDa protein on SDS-PAGE of brush-border membranes by using photoaffinity labeling, monoclonal antibodies, and fluorescent group-specific reagents (see Wright *et al.*, 1994). The cDNA coding for the transporter was isolated in this laboratory in 1987 using a novel expression cloning method, and the human isoform was isolated in 1989 by homology cloning (see Wright *et al.*, 1994). The open reading frames code for 73-kDa proteins with either 662 (rabbit) or 664 (human) residues, and there is 84% identity (94% similarity) between the two amino acid sequences. The proteins are referred to as rabbit and human SGLT1. There is a single N-linked glycosylation site (N248) that is used in enterocytes to increase the actual mass of the protein to 84kDa. SGLT1 was the first member of a new gene family, and this family now has over 55 members (see Turk and Wright, 1997) including sodium-dependent transporters for proline, panthothenic acid, and iodide found in diverse cells from bacteria to man.

Is the cloned transporter SGLT1 responsible for Na-dependent sugar transport across the intestinal brush border? All studies to date provide compelling evidence that this is the case. (1) Northern blotting indicates that SGLT1 mRNA is abundant in the intestinal mucosa, but not in other tissues (Fig. 1; Pajor and Wright, 1992). SGLT1 mRNA also was detected in the renal cortex and outer renal medulla, but at lower levels of abundance. (2) Antibodies raised against two different peptide sequences of the cloned transporter recognize the same 75-kDa band in Western blots of rabbit brush-border membranes (Hirayama et al., 1991). (3) SGLT1 antibodies specifically recognize an antigen in the brush-border membrane of enterocytes but do not immunoreact with proteins in crypt epithelial cells or goblet cells (Hwang et al., 1991; Takata et al., 1992). (4) The cation and sugar selectivity, and phlorizin sensitivity of the cloned transporter expressed in oocytes agree well with those for the transporter in intestinal brush-border membrane (Ikeda et al., 1989). (5) Mutations in the SGLT1 gene have been shown to account for the defect in sugar absorption in patients with glucose-galactose malabsorption (see Wright, 1998).

IV. STRUCTURE AND FUNCTION OF SGLT1

In our laboratory we have focused on understanding how SGLT1 couples the transport of sodium and glucose across the plasma membrane. So far, most of these studies have involved characterization of the cloned transporter expressed in *Xenopus laevis* oocytes. The advantages of this heterologous expression system for SGLT1 are that native oocytes do not express the cotransporter to any significant level, but these cells efficiently translate SGLT1 cRNA and insert the fully functional protein in the plasma membrane. In cRNA-injected oocytes, the protein is readily detected in membranes using Western blotting, immunocytochemistry, and freeze-fracture electron microscopy (Lostao et al., 1995; Zampighi et al., 1995). The latter technique directly demonstrates that more than 1×10^{11} copies of SGLT1 (>4,000 per μm^2) may be inserted into the oocyte plasma membrane. In one series of experiments 250,000 copies of SGLT1 were inserted into the plasma membrane every second, and after 6 days the maximum rate of glucose uptake was 20 nmol per oocyte per hour (Hirsh et al., 1996). Na/glucose cotransport also can be measured as a sugar-induced inward Na^+ current, and this has been validated by measuring simultaneously the sugar-induced currents and either Na^+ or sugar influxes (Mackenzie et al., 1998). In oocytes expressing SGLT1 the sugar-induced currents range from 100 to 2,000 nA depending on the level of expression, whereas in water-injected oocytes the currents are less than 1 nA. Likewise, sodium-dependent sugar uptake values (50 μM αMDG and 100 mM Na^+) typically are in the range of 300–700 pmoles per oocyte per hour for SGLT1-expressing oocytes and less than 1 pmole per oocyte per hour in control oocytes. With signal-to-noise ratios such as this, it is then straightforward to determine the kinetics of Na$^+$/glucose cotransport with great precision (see Parent *et al.*, 1992a; Panayotova-Heiermann *et al.*, 1995).

V. KINETIC MODEL

We have developed a simple kinetic model to account for the kinetics of Na$^+$/glucose transport by the SGLT1 expressed in oocytes (Parent *et al.*, 1992b). This six-state, ordered kinetic model is reproduced in Figure 2. Sodium is shown to bind first to the outside face of the transporter, and this results in a conformational change that opens up the sugar-binding site. In our model we have assumed that two sodium ions bind simultaneously in order to keep the model simple, with only 14 rate constants and two additional parameters. This approximation is probably justified at high external sodium concentrations, but the data suggest that this may not hold at low-sodium concentrations (Hazama *et al.*, 1997). Once the Na$^+$ and sugar have bound at the external face, there is a transition where the ligands now face the cytoplasm, where the sugar and then the sodium ions dissociate. Finally, the empty carrier completes the reaction cycle

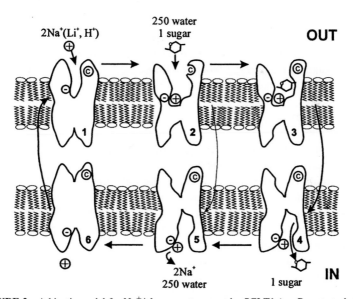

FIGURE 2 A kinetic model for Na$^+$/glucose cotransport by SGLT1 (see Parent *et al.*, 1992b). As indicated, both Li$^+$ and H$^+$ can replace Na$^+$ as the driving cation for sugar transport, and 250 water molecules also are cotransported during each cycle. In the absence of sugar, SGLT1 behaves as a Na$^+$ (Li$^+$ and H$^+$) uniporter, but the turnover number is only 10% of that for cotransport.

by switching the ligand binding sites from facing in to facing out. Thus, in a complete reaction cycle, 2 Na^+ ions and 1 sugar molecule are transported from the external to the internal surface of the membrane. In the absence of sugar, there is transport of Na^+ through the protein (uncoupled Na^+ transport).

There is substantial experimental evidence in favor of this model, including that showing (1) 2 Na^+ ions are transported along with each glucose molecule (Mackenzie *et al.,* 1998); (2) Na^+ is transported in the absence of glucose with similar kinetics to the full cycle ($K_{0.5}$ 3 mM, Hill coefficient 2, activation energy 20 Kcal/mole, phlorizin K_i 1 μM) (Mackenzie *et al.,* 1998, Panayotova-Heiermann *et al.,* 1998; Loo *et al.,* 1999); (3) at saturating membrane potentials (-150 mV) the maximal rate of Na^+/glucose transport is independent of Na^+ concentration, but the maximum rate of Na transport is a function of the glucose concentration (Parent *et al.,* 1992b); this precludes an ordered scheme in which 1 Na^+ binds before glucose, followed by binding of the second Na^+; and (4) voltage jump experiments in which the transporter exhibits pre-steady-state currents. These SGLT1 capacitive currents represent the redistribution of the states of the transporter between C2 and C6 (see Fig. 1).

The human transporter turns over about 60 times per second at 22°C, and depending on the membrane potential the rate-limiting step is either the dissociation of Na^+ from C5 or the rate of isomerization of the protein from states 6 to 5. In the absence of sugar, the turnover of the partial reaction is only about 12 per second owing to the slow isomerization of the protein from C2 to C5.

In general, the quantitative model we have developed accounts well for both the steady-state and pre-steady-state behavior of the transporter (Parent *et al.,* 1992b; Hazama *et al.,* 1997; Loo *et al.,* 1998).

A. Water Transport

In addition to behaving as a Na^+/glucose cotransporter and a Na^+ uniporter, SGLT1 acts as a low-conductance water channel (see Loo *et al.,* 1999). The water permeability is approximately 1–2% of the water channel AQP1. In the presence of Na^+, phlorizin inhibits this passive water permeability with a K_i of 3–5 μM, but this inhibitor has no effect in the absence of Na^+. We interpret these latter observations as showing that the water channel depends on the conformational state of the cotransporter.

Perhaps more surprisingly, cotransporters such as SGLT1 also behave as water cotransporters in the presence of sugar. There is a strict stoichiometry between water flow and sugar transport with a coupling ratio of 2 Na^+, 1 glucose, and 250 water molecules (Loo *et al.,* 1996; Meinild *et al.,* 1998). This cotransport of water is independent of the osmotic gradient, and water transport can even occur against an osmotic gradient. A molecular basis for water cotransport is discussed later.

B. Sugar Selectivity

The substrate selectivity of SGLT1 has been extensively examined using current measurements (Birnir *et al.,* 1990; Lostao *et al.,* 1994; Hager *et al.,* 1995; Diez-Sampedro *et al.,* 2000). Figure 3 illustrates the protocol. In this single oocyte expressing the human SGLT1, the cell was held at a membrane potential of −50 mV and the current was recorded when 1 mM αMDG was added to the Na$^+$ buffer superfusing the cell. This produced an inward current of 870 nA, which returned to baseline after washing out the sugar. Measuring the currents produced by different concentrations of αMDG, and plotting the sugar currents as a function of sugar concentration can then be used to determine the kinetics

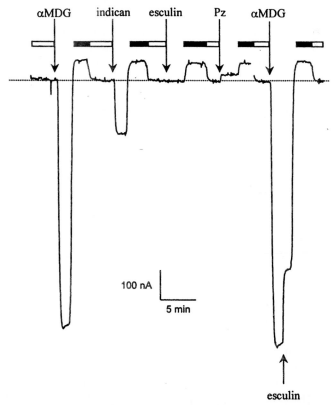

FIGURE 3 Sugar-induced Na$^+$ currents in an oocyte-expressing human SGLT1. The membrane potential was clamped at −50 mV, and the current was recorded as a function of time. The dotted line indicates the baseline current in the absence of sugar. After each sugar test, the oocyte was superfused with a Na$^+$-free buffer (solid rectangles) to wash the sugar from the bath. The sugar concentrations were: αMDG 1 mM; indican 2 mM, and esculin 10 mM. Reproduced with permission from Diez-Sampedro *et al.,* 2000.

of αMDG transport. Nonlinear curve-fitting procedures are then used to estimate the apparent affinity for sugar ($K_{0.5}$) and the maximum rate of transport (I_{max}). The apparent affinities (0.2–0.5 mM) and maximal rates of transport are similar for αMDG, D-glucose, and D-galactose, whereas the affinity for 3-O-methyl-glucose is about an order of magnitude lower ($K_{0.5}$ 5 mM). A wide variety of pyranose sugars are transported, including D- and L-xylose, D-fucose, and even 2-DEOXY-D-glucose, but the apparent affinities may be very low, for example, 2-DEOXYglucose $K_{0.5} = 50$ mM. This contradicts accepted dogma, which states that 2DOG and xylose are not substrates for SGLT1, and this raises questions about the results of experiments in which these sugars are used to monitor passive sugar uptake in the intestine.

Figure 3 also shows that glucosides such as indican are transported by SGLT1. However, kinetic analysis indicates that although SGLT1 has a higher affinity for indican ($K_{0.5}$ 0.06 mM) than αMDG, the maximum rate of transport is only about 10% of that for hexoses. In contrast, esculin is not transported, but it does inhibit αMDG transport with a K_i of 13 mM. In general, glucosides with phenyl or napthylalene rings lying in the same plan as the pyranose ring are transported by SGLT1, indicating that the conformational changes underlying cotransport must be very large (see Fig. 2).

The glucoside phlorizin is the most potent inhibitor of Na/glucose transport, with a K_i of 200 nM for human and 10 nM for rat (Hirayama *et al.*, 1996). This molecule has a second aromatic ring folded back over the pyranose ring in such as way as to present a bulky cross-section to the transporter. This probably accounts for the lack of transport of this molecule in the wild-type protein. In a SGLT1/SGLT3 chimera protein phlorizin is actually transported with high affinity (Panayotova-Heiermann *et al.*, 1996), which suggests that the discrimination between substrate and blocker is quite subtle.

C. Cation Selectivity

In addition to sodium, other cations can drive the cotransport of sugars through SGLT1. These include H^+ and Li^+ (Hirayama *et al.*, 1994; 1997; Panayotova-Heiermann *et al.*, 1998). The SGLT1 H^+ affinity is more than 1,000-fold higher ($K_{0.5}$ 5 μM) than Na^+ (4 mM) or Li^+ (12 mM), but the affinity for sugar is higher in Na^+ ($K_{0.5}$ 2 mM) than in either H^+ or Li^+ (4 and 11 mM, respectively). The maximum rate of transport is higher in H^+ than in Na^+ or Li^+ by a factor of 2. The coupling coefficients are 2 for both Na^+/sugar and H^+/sugar transport (Quick, unpublished results). These results suggest that SGLT1 can operate in two modes, a high-capacity low-affinity transporter when driven by protons, and a lower-capacity high-affinity transporter when driven by sodium. This may be physiologically important as sugars pass from the stomach into the duodenum and jejunum. Kinetic modeling suggests that these kinetic effects are attributable to differences in the rate constants for cation binding and dissociation at the two faces of the protein (see Fig. 2). Other monovalent cations are poor driving cations for this transporter. H$^+$ and Li$^+$ also are transported by SGLT1 in the absence of sugars, and the kinetics are similar to those for coupled transport (Panayotova-Heiermann *et al.*, 1998).

D. Backwards Transport

Cotransporters are expected to be able to operate in the reverse direction (i.e., transport Na$^+$ and sugar out of the cell) when the driving forces are favorable. This is explicit in our kinetic model of Na$^+$/glucose cotransport (Parent *et al.*, 1992b). The first evidence for SGLT1 came from experiments in which oocytes were preloaded with a nonmetabolized glucose analog, αMDG, and transporter currents were recorded when sugar was removed from the external bathing solution (Umbach *et al.*, 1990). Outward currents were observed when the membrane potential was depolarized. Outward currents were analyzed in more detail using excised patch-clamp techniques in which the cytoplasmic surface of the plasma membrane faced the bath solution (Sauer *et al.*, 2000; Eskandari *et al.*, 1999). In both of these studies the pipette solution contained less than 10 mM Na$^+$ and no sugar, and transporter currents were recorded as a function of bath (cytoplasmic) Na$^+$ and sugar concentrations, that is, the exact inverse of experiments in intact oocytes. The results demonstrate that SGLT1 can transport Na$^+$ and sugar out of the cell, but that the affinities for sugar and Na$^+$ are orders of magnitude lower than for inward transport ($K_{0.5}$ for αMDG >7 mM and for Na$^+$ >50 mM). Phlorizin also inhibits outward transport from the cytoplasmic surface, but as expected, with a much lower affinity than from the outside. Thus, SGLT1 can operate in the reverse mode, but there is considerable asymmetry in the ligand-binding constants (see also Parent *et al.*, 1992b). The asymmetry assures that SGLT1 is poised to accumulate sugars in the cell, as required for efficient sugar absorption in the intestine (see Fig. 1).

E. Secondary Structure

The current secondary structure model for SGLT1 is shown in Figure 4. In this model the protein consists of 14 transmembrane α-helices, with the hydrophobic N-terminal domain and the C-terminus of the 14th helix facing the extracellular surface of the membrane. The N-linked glycosylation site is indicated on the 4th extracellular hydrophilic domain, between helices 7 and 8. This model is supported by theoretical analysis using sophisticated computer algorithms

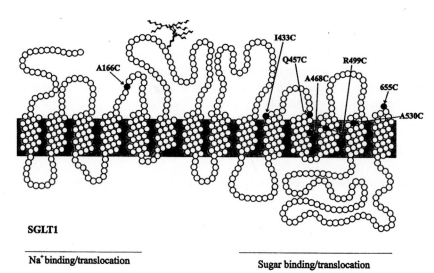

SGLT1

Na⁺ binding/translocation   Sugar binding/translocation

FIGURE 4 A secondary structure model for SGLT1 shows the putative Na^+ and sugar binding/translocation domains. Also shown are several cysteine mutants that have been used in our experimental studies.

(PredictProtein and Memsat; see Turk and Wright, 1997) and by substantial experimental studies on both SGLT1 (see Turk and Wright, 1997) and closely related family members such as the Na^+/proline cotransporter PutP (Jung *et al.*, 1998; Wegner *et al.*, 2000) and the Na^+/iodide transporter NIS (Levy *et al.*, 1998).

F. Tertiary/Quaternary Structure

So far, little information is available about the structure of membrane transporters, apart from electron microscopic evidence that functional SGLTs exist in the plasma membrane as monomers. Freeze-fracture electron microscopic methods have been used to examine the density and size of transporters expressed in oocyte plasma membranes (Zampighi *et al.*, 1995; Eskandari *et al.*, 1998). The density of SGLT1 proteins in the plasma membrane is directly proportional to the density estimated from pre-steady-state charge measurements, and the area of the protein in the membrane is consistent with the area of a 14 transmembrane protein. The area measurements were calibrated by measurements on membrane proteins that had been crystallized: in both the two-dimensional projection maps and freeze fractures, each helix of these proteins occupies an average of 1.40 ± 0.03 nm^2. The SGLT1 protein in the oocyte membrane is elliptical (a/b ratio of 1.2) with an area of 21 ± 3 nm^2. Recent electron microscopic studies with the purified *Vibrio parahaemolyticus* SGLT (vSGLT) reconstituted into proteolipo-somes confirmed that this functional protein also is a monomer (Turk *et al.,* 2000). These results are in direct contrast to those of previous irradiation inacti-vation experiments (Stevens *et al.,* 1990) that suggested that in rabbit brush bor-ders SGLT1 exists as a homotetramer. The explanation for this discrepancy in re-sults is not yet clear, but we note that similar discrepancies exist for other transport proteins such as lactose permease and the renal Na$^+$/phosphate cotransporter.

Biochemical cross-linking studies are in progress to determine the helical packing of the vSGLT1 (Xie *et al.,* 2000). These preliminary results suggest that helices 4 and 5 are in close proximity to helices 10 and 11. It should also be noted that our attempt to locate salt bridges between helices has not been suc-cessful (Panayotova-Heiermann *et al.,* 1998). Although there are several candi-dates for conserved residues involved in salt bridges with lysine 321 in helix 8 (e.g., E225 in helix 6, D294 in helix 7), mutagenesis experiments have been frus-trated by poor expression of the mutants in the plasma membrane. This is a fa-miliar problem with SGLT1 in that only 1 of 23 mutations causing glucose-galactose malabsorption is correctly targeted to the plasma membrane (see Wright, 1998). Even conservative alanine-to-valine mutations (as positions 304, 388, and 468) result in missorting of the protein in the cell, and in one case, A468V, replacing valine with cysteine partially restores normal trafficking to the plasma membrane (see Lam *et al.,* 1999). This indicates that simply increasing the bulk of the side chain at position 468 produces graded perturbations in traf-ficking of the protein to the plasma membrane.

G. Sugar Binding/Translocation

Our current hypothesis is that the C-terminal domain of the protein is involved in sugar binding and translocation, and that the N-terminal domain is involved in sodium binding and translocation. This stems originally from studies on chimeras constructed from the pig SGLT1 and the low-affinity pig SGLT2 (Panayotova-Heiermann *et al.,* 1996). These results suggest that the C-terminal domain of SGLT1 (residues 407–662, see Fig. 4) determines the sugar selectivity and sugar affinity of the transporter. This was further reinforced by the results of experiments in which the truncated protein (C5) was expressed in *Xenopus* oocytes or in *E. coli,* purified and reconstituted into proteoliposomes (Panayotova-Heiermann *et al.,* 1997; 1999). Transport assays (Fig. 5) show that in both sys-tems C5 behaves as a low-affinity sugar uniporter, which is sensitive to phloretin but not to phlorizin. Figure 5 shows that in oocytes, tracer αMDG uptake was 20-fold higher in C5-expressing oocytes than in control oocytes, and that 50 mM αMDG inhibited this tracer uptake by 50%. The C5-mediated αMDG uptake into proteoliposomes was blocked completely by 200 mM cold αMDG. These

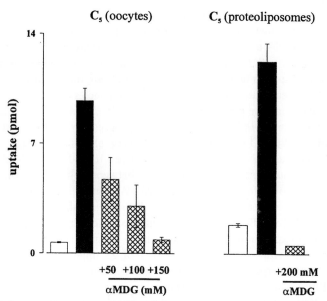

FIGURE 5 Sugar transport by C5 expressed in oocytes and reconstituted into proteoliposomes. C5 cRNA was injected into oocytes, and after 5 days, 50 μM 14C-αMDG uptake was measured in the presence and absence of cold (50–150 mM) αMDG in the absence of sodium. No increase in sugar uptake was observed with the addition of sodium. Shown are the uptake values in pico-moles/oocyte/7 minutes for both control and cRNA-injected oocytes (data from Panayotova-Heiermann *et al.,* 1997). C5 also was expressed in *E. coli* cells, and C5 protein was purified and re-constituted into liposomes. Sugar (5 μM ^{3}H-D-glucose) uptake was measured after 15 minutes. Note the higher uptake into proteolipsomes, and this was blocked by the addition of 200 mM αMDG and 0.5 mM phloretin, but not by 200 mannitol or L-glucose (not shown) (data from Panayotova-Heiermann *et al.,* 1999).

results have led us to speculate that helices 10–13 (and their hydrophilic linkers) form the sugar translocation pathway through SGLT1. Helix 14 is absent in some other members of the SGLT1 family (Turk and Wright, 1997), and sugar transport has been detected after deletion of helix 14 from C5 (Panayotova-Heiermann *et al.,* 1997).

H. Sodium Sensitivity

Although coexpression of C5 and N9 in oocytes failed to restore Na$^+$/glucose cotransport activity, similar experiments with vibrio SGLT N7 and C7 in bacteria resulted in full Na$^+$/glucose transport activity (Xie *et al.,* 2000). On the basis of the transport activity of the full-length proteins, the truncated C5, and the split vSGLT, we suggest that the N-terminal half of the proteins determines the sodium sensitivity of SGLTs. We have not been able to successfully express the N-terminal domains of either SGLT1 or vSGLT in a functional form. Neither N7 nor N9 SGLT1 constructs expressed in oocytes produced detectable changes in the electrical properties (steady-state conductance or presteady currents) of the cell. In the case of the Na/proline cotransporter, mutagenesis studies also suggest that the N-terminal domain is involved in Na$^+$ binding and/or coupling of Na$^+$ transport to proline transport (see Wegener et al., 2000; Quick et al., 1999). In terms of the coupling between sodium and sugar transport through SGLT1, it is now imperative to have a helical packing model of the transporter in which the precise relationships between the N and C termini are established. The only transporter in which the helical packing is known is the H$^+$/lactose transporter (E. coli. lactose permease; Kaback, 1998).

I. Support for the Alternating Access Model

Further progress in our understanding of the mechanism of Na$^+$/glucose co-transport was made possible as a result of the experiments described above on the identification of the sugar binding/translocation domain and the identification of missense mutations responsible for GGM (see Fig. 4). Three GGM mutations were located in the C5 region of the protein, and one of these, Q457R, was inserted normally into the plasma (see Wright, 1998). We further examined the role of Q457 in sugar permeation by replacing the glutamine residue with cysteine (Q457C) and testing the effect of alkylating reagents (Loo et al., 1998). Q457C SGLT1 retained full transport function, apart from an increase in the $K_{0.5}$ for sugar from 0.4 to 6 mM. However, treating the mutant with alkylating reagents (e.g., methanethiosulfonates [MTS]), blocked sugar transport. This was fully reversed by reducing agents such as DTT. Although sugar transport was blocked, sugar still bound with a dissociation constant very similar to the affinity for sugar transport after MTS treatment; this suggests that 457C is not part of the actual sugar binding site.

The inhibition of Q457C by MTS reagents occurred only in the presence of Na$^+$ at negative membrane potentials. The reagents failed to block sugar transport in the absence of Na$^+$, in the presence of Na$^+$ and phlorizin or sugar, and in the presence of Na$^+$ at depolarized membrane potentials. The degree of inhibition by MTS reagents varied with membrane potential in direct proportion to the probability of the protein being in the C2 conformation (see Fig. 2). We further suggest that Q457C is in the sugar translocation pathway because phlorizin can protect against MTS inhibition.

There is also direct evidence that voltage-dependent conformational changes in SGLT1 are closely linked to local environmental changes around residue 457. This was obtained by labeling Q457C with tetramethylrhodamine-6-maleimide

(TMR6M). The fluorophore labeled only the SGLT1 with a cysteine at position 457, and only in the presence of Na^+ at negative potentials. Phlorizin and sugar also blocked labeling. Fluorescence of the labeled protein was measured under voltage-clamp conditions, and it was found that after rapid voltage jumps the fluorescence changes with the same time course and magnitude as the SGLT1 charge movements. Phlorizin blocked both the fluorescence changes and the charge movements. An example is shown in Figure 6. In this experiment the membrane potential was first clamped at -90 mV in the presence of 100 mM external NaCl. Under these conditions, 80% of the transporter is in the C2 conformation (see Fig. 2). After a baseline fluorescence signal was recorded, the membrane potential was jumped rapidly to $+50$ mV for 40 ms, where 95% of the transporter is in the C6 conformation (see Fig. 2). These voltage jumps were accompanied by a transient increase in protein fluorescence. At $+50$ mV the fluorescence increased to a steady-state value with two time constants: a fast one of 100 μs and a slow one of 7 μs; after the potential returned to -90 mV, the fluorescence returned toward baseline with both fast and slow time constants. The magnitude of the steady-state fluorescence change was directly proportional to the SGLT1 pre-steady-state charge transfer. The time constants of the ON and OFF fluorescence changes also followed the time course of the pre-steady-state charge movements. No such changes in fluorescence were recorded with SGLT1 labeled with TMRM at A166C or 665C (see Fig. 4).

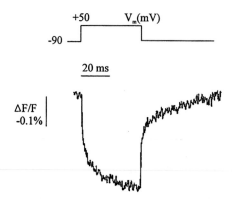

FIGURE 6 Fluorescence changes in rhodamine-labeled Q457C SGLT1 occurring during rapid voltage jumps. The cysteine mutant was expressed in an oocyte, and Q457C was labeled with tetrarhodamine maleimide in 100 mM Na^+ at -100 mV (Loo et al., 1998). The fluorescence and membrane currents (not shown) were recorded as the membrane potential was jumped from -90 to $+50$ mV, as described previously (Loo et al., 1998). Both the change in fluorescence and the transient currents were blocked by the addition of 1 mM phlorizin (not shown). Control experiments with non-injected oocytes, wild-type SGLT1 oocytes, and other cysteine mutants (A166C and 665C) did not show voltage-dependent changes in fluorescence.

These experiments demonstrate that voltage-induced changes in conformation from C2 to C6 (see Fig. 2) are accompanied by local environmental changes at, or close to, residue 457. Altogether, the fluorescence and MTS inhibition studies of Q457C give direct support for the alternating access model of Na$^+$/glucose cotransport. In particular, the conformational dependent labeling of Q457C by fluorescent and nonfluorescent alkylating reagents provides evidence that SGLT1 exists in the C1, C2, C3, and C6 conformations. Additional work in progress further indicates that residue I443 on helix 10, A468 on helix 11, R449 on helix 12, and A530 on helix 13 behave in a similar fashion to Q457 on helix 11. Our working hypothesis is that Na$^+$ causes local movement of helices 10–13, perhaps helical tilting or rotation, to open up a large pocket on the external face of the protein, and this pocket actually may be part of the pathway for sugar translocation. The pocket may be large enough to accommodate large sugars such as glucosides, phlorizin, and 200–300 water molecules, and further sugar-induced movements of helices 10–13 may expose the substrates and water to the cytoplasmic face of the protein where they dissociate to complete the transport process.

J. Future Directions

To gain further information about the mechanism of sodium, sugar, and water transport by SGLT1 and related proteins, it is necessary to obtain structural information about the proteins in their different conformations. An initial approach is to determine the nature of helical packing in the protein, and this may be carried out by methods similar to that used for the H$^+$/lactose symporter, e.g., by cysteine cross-linking experiments with homofunctional cysteine reagents with different length spacers between the functional groups (see Kaback, 1998). This requires a transporter with cysteine-less background, and thus we propose to use the vSGLT because it contains only one cysteine residue and this is not required for function (Xie *et al.*, 2000). Ultimately, two- and three-dimensional crystal structures are required to model the structure of the transporter, and as a starting point we have purified the vibrio SGLT to homogeneity using a poly-His tagged protein and metal chelate chromatography (Turk *et al.*, 2000). This protein is fully functional when reconstituted into proteoliposomes. We anticipate having a wealth of structural information about this class of membrane proteins in the years to come, and this will give insight into the actual transport mechanisms.

Acknowledgments

The studies discussed in this chapter were made possible by the creativity and hard work of a dedicated group of colleagues, students, and assistants in the author's laboratory over the past 15 years, and by the financial support provided by the National Institutes of Health (DK19567, DK44602, and DK44582).

References

Birnir, B., Loo, D. D. F., and Wright, E. M. (1991). Voltage clamp studies of the Na^+/glucose cotransporter cloned from rabbit small intestine. *Pflugers Arch.* **418,** 79–85.

Diez-Sampedro, A., Lostao, M. P., Wright, E. M., and Hirayama, B. A. (2000). Glycoside binding and translocation in Na^+-dependent glucose cotransporters: Comparison of SGLT1 and SGLT3. *J. Membr. Biol.* (in press).

Eskandari, S., Loo, D. D. F., and Wright, E. M. (1999). Functional asymmetry of the sodium/glucose cotransporter. Experimental Biology 99, Washington, D.C. *FASEB J.* 13, A399.

Eskandari, S., Wright, E. M., Kreman, M., Starace, D. M., and Zampighi, G. A. (1998). Structural analysis of cloned membrane proteins by freeze-fracture electron micoscopy. *PNAS* **95,** 11235–11240.

Greenough, W. B. (1989). Oral rehydration therapy: An epithelial transport success story. *Arch. Dis. Child.* **64,** 419–422.

Hager, K., Hazama, A., Kwon, H. M., Loo, D. D. F., Handler, J. S., and Wright, E. M. (1995). Kinetics and specificity of the renal Na^+/*myo*-inositol cotransporter expressed in *Xenopus* oocytes. *J. Membr. Biol.* **143,** 103–113.

Hazama, A., Loo, D. D. F., and Wright, E. M. (1997). Presteady-state currents of the Na^+/glucose cotransporter (SGLT1). *J. Membr. Biol.* **155,** 175–186.

Hirayama, B. A., Lostao, M. P., Panayotova-Heiermann, M., Loo, D. D. F., Turk, E., and Wright, E. M. (1996). Kinetic and specificity differences between rat, human and rabbit Na/glucose cotransporters (SGLT1). *Am J Physiol.* **270,** G919–G926.

Hirayama, B. A., Wong, H. C., Smith, C. D., Hagenbuch, B. A., Hediger, M. A., and Wright, E. M. (1991). Intestinal and renal Na^+/glucose cotransporters share common structures. *Am. J. Physiol.* **261,** C296–C304.

Hirayama, B. A., Loo, D. D. F., and Wright, E. M. (1997). Cation effects on protein conformation and transport in the Na^+/glucose cotransporter. *J. Biol. Chem.* **272,** 2110–2115.

Hirsch, J. R., Loo, D. D. F., and Wright, E. M. (1996). Regulation of Na^+/glucose cotransporter expression by protein kinases in *Xenopus laevis* oocytes. *J. Biol. Chem.* **271,** 14740–14746. Na^+/glucose cotransporter (SGLT1). *J. Membr. Biol.* **155,** 175–186.

Hopfer, U. (1987). Mentrane transport mechanisms fo rhexoses and amino acies in the small intestine. *In* "Physiology of the Gastrointestinal Tract" (L. R. Johnson, ed), 2nd ed, pp. 1499–1526. New York, Raven Press.

Hwang, E.-S., Hirayama, B. A., and Wright, E. M. (1991). Distribution of the SGLT1 Na^+/glucose cotransporter and mRNA along the crypt-villus axis of rabbit small intestine. *Biochem. Biophys. Res. Commun.* **181,** 1208–1217.

Ikeda, T. S., Hwang, E.-S., Coady, M. J., Hirayama, B. A., Hediger, M. A., and Wright, E. M. (1989). Characterization of a Na^+/glucose cotransporter cloned from rabbit small intestine. *J. Membr. Biol.* **110,** 87–95.

Jung, H., Rübenhagen, R., Tebbe, S., Leifker, K., Tholema, N., Quick, M., and Schmid, R. (1998). Topology of the Na^+/proline transporter of *Escherichia coli. J. Biol. Chem.* **273,** 26400–26407.

Kaback, H. R. (1998). Structure/function studies on the lactose permease of *Escherichia coli. Acta Physiol. Scand. Suppl.* **643,** 21–33.

Lam, J. T., Martin, M. G., Turk, E., Bosshard, N. U., Steinmann, B., and Wright, E. M. (1998). Missense mutations in SGLT1 cause glucose-galactose malabsorption by trafficking defects. *Biochim. Biophys. Acta* **1453,** 297–303.

Levy, O., De la Vieja, A., Ginter, C. S., Riedel, C., Dai, G., and Carrasco, N. (1998). N-linked glycosylation of the thyroid Na^+/I^- symporter. *J. Biol. Chem.* **273,** 22657–22663.

Loo, D. D. F., Hirayama, B. A., Gallardo, E. M., Lam, J. T., Turk, E., and Wright, E. M. (1998). Conformational changes couple Na^+ and glucose transport. *Proc. Natl. Acad. Sci..* **95,** 7789–7794.

Loo, D. D. F., Hirayama, B. A., Meinild, A.-K., Chandy, G., Zeuthen, Z., and Wright, E. M. (1999). Passive water and ion transport by cotransporters. *J. Physiol.* **518.1,** 195–202.

Loo, D. D. F., Zeuthen, T., Chandy, G., and Wright, E. M. (1996). Cotransport of water by the Na$^+$/glucose cotransporter. *Proc. Natl. Acad. Sci.* **93**, 13367–13370.

Lostao, M. P., Hirayama, B. A., Loo, D. D. F., and Wright, E. M. (1994). Phenylglucosides and the Na$^+$/glucose cotransporter (SGLT1) expressed in oocytes using tracer uptake and electrophysiological methods. *J. Membr. Biol.* **142**, 162–170.

Lostao, M. P., Hirayama, B. A., Panayotova-Heiermann, M., Samposna, S. L., Bok, D., and Wright, E. M. (1995). Arginine-427 in the Na/glucose cotransporter (SGLT1) is involved in trafficking to the plasma membrane. *FEBS Lett.* **377**, 181–184.

Mackenzie, B., Loo, D. D. F., and Wright, E. M. (1998) Relations between Na$^+$/glucose cotransporter (SGLT1) currents and fluxes. *J. Membr. Biol.* **162**, 101–106.

Meinild, A.-K., Klaerke, D., Loo, D. D. F., Wright, E. M., and Zeuthen, T. (1998). The human Na$^+$/glucose cotransporter is a molecular water pump. *J. Physiol.* **508**, 15–21.

Pajor, A. M., and Wright, E. M. (1992). Sequence[1], tissue distribution and functional expression of a mammalian Na$^+$/nucleoside cotransporter. *J. Biol. Chem.* **267**, 3557–3560.

Panayotova-Heiermann, M., Eskandari, S., Zampighi, G. A., and Wright, E. M. (1997). Five transmembrane helices form the sugar pathway through the Na$^+$/glucose transporter. *J. Biol. Chem.* **272**, 20324–20327.

Panayotova-Heiermann, M., Leung, D. W., Hirayama, B. A., and Wright, E. M. (1999). Purification and functional reconstitution of a truncated human Na$^+$/glucose cotransporter (SGLT1) expressed in *E. coli*. *FEBS Lett.* **459**, 386–390.

Panajotova-Heiermann, M., Loo, D. D. F., Klong, C.-T., Lever, J. E., and Wright, E. M. (1996). Sugar binding to Na$^+$/glucose cotransporters is determined by the C-terminal half of the protein. *J. Biol. Chem.* **271**, 10029–10034.

Panayotova-Heiermann, M., Loo, D. D. F., and Wright, E. M. (1995). Kinetics of steady-state currents and charge movements associated with the rat Na$^+$/glucose cotransporter. *J. Biol. Chem.* **270**, 27099–27105.

Panayotova-Heiermann, M., Loo, D. D. F., and Wright, E. M. (1998). Neutralization of conservative charged transmembrane residues in the Na$^+$/glucose cotransporter SGLT1. *Biochemistry,* **37**, 10522–10528.

Parent, L., Supplisson, S., Loo, D. F., and Wright, E. M. (1992). Electrogenic properties of the cloned Na$^+$/glucose cotransporter. Part I. Voltage-clamp studies. *J. Membr. Biol.* **125**, 49–62.

Parent, L., Supplisson, S., Loo, D. F., and Wright, E. M. (1992) Electrogenic properties of the cloned Na$^+$/glucose cotransporter. Part II. A transport model under non rapid equilibrium conditions. *J. Membr. Biology* **125**, 63–79.

Quick, M., Stolting, S., and Jung, H. (1999). Role of conserved Arg40 and Arg117 in the Na$^+$/proline transporter of *Escherichia coli*. *Biochemistry* **38**, 13523–13529.

Sauer, G. A., Nagel, G., Koepsell, H., Bamberg, E., and Hartung, K. (2000). Voltage and substrate dependence of the inverse transport model of the rabbit Na$^+$/glucose cotransporter (SGLT1). *FEBS Lett.* **469**, 98–100.

Stevens, B. R., Fernandez, A., Hirayama, B., Wright, E. M., and Kempner, E. S. (1990). Intestinal brush border membrane Na$^+$/glucose cotransporter functions *in situ* as a homotetramer. *Proc. Natl. Acad. Sci. USA* **87**, 1456–1460.

Takata, K., Kasahara, M., Oka, Y., and Hirano, H. (1992). Mammalian sugar transporters. Their localization and link to cellular functions. *Acta Histochem. Cytochem.* **26**, 175–178.

Turk, E., Kim, O., leCoutre, J., Whitelegge, J. P., Eskandari, S., Lam, J. T., Kreman, M., Zampighi, G., Faull, K. F., and Wright, E. M. (2000). Molecular characterization of *Vibrio parahaemolyticus* vSGLT: A model for sodium-coupled sugar cotransporters *J. Biol. Chem.* (in press).

Turk, E., and Wright, E. M. (1997). Membrane topological motifs in the SGLT cotransporter family. *J. Membr. Biol.* **159**, 1–20.

Umbach, J. A., Coady, M. J., and Wright, E. M. (1990). The intestinal Na$^+$/glucose cotransporter expressed in *Xenopus* oocytes is electrogenic. *Biophys. J.* **57**, 1217–1224.

Wegener, C., Tebbe, S., Steinhoff, H.-J., and Jung, H. (2000). Spin labeling analysis of structure and dynamics of the Na⁺/proline transporter of *Escherichia coli. Biochemistry* **39**, 4831–4837.

Wright, E. M. (1998). Glucose galactose malabsorption. *Am. J. Physiol.* **275**, G879–G882.

Wright, E. M., Loo, D. D. F., Heierman, M., and Boorer, K. J. (1994). Mechanisms of Na⁺/sugar cotransport. *Biochem. Soc. Trans.* **22**, 646–650.

Xie, Z., Turk, E., and Wright, E. M. (2000). Characterization of the *Vibrio parahaemolyticus* Na⁺/glucose cotransporter: A bacterial member of the SGLT family *J. Biol. Chem.* (in press).

Zampighi, G. A., Kreman, M., Boorer, K. J., Loo, D. D. F., Bezanilla, F., Chando, G., Hall, J. E., and Wright, E. M. (1995). A method for determining the unitary functional capacity of cloned channels and transporters expressed in *Xenopus laevis* oocytes. *J. Membr. Biol.* **148**, 65–78.

Index

ISBN 0-12-153350-6

90065

9 780121 533502